Pharmaceuticals

Volume 3

J. L. McGuire (Editor)

WILEY-VCH

Pharmaceuticals

J. L. McGuire (Editor)

Volume 1 Introduction
Cardiovascular Drugs

Volume 2 Neuropharmaceuticals
Gastrointestinal Drugs
Respiratory Tract

Volume 3 Antiinfectives
Endocrine and Metabolic Drugs

Volume 4 Miscellaneous Drugs
Related Technology
Indexes

Pharmaceuticals

Classes, Therapeutic Agents, Areas of Application

J. L. McGuire (Editor)

Volume 3

Antiinfectives
Endocrine and Metabolic Drugs

Weinheim · New York · Chichester · Brisbane · Singapore · Toronto

Dr. J. L. McGuire (Editor)
Johnson & Johnson
Science & Technology/Business Development
One Johnson & Johnson Plaza
New Brunswick, NJ 08933
USA

> This book was carefully produced. Nevertheless, editor, authors and publisher do not warrant the information contained therein to be free of errors. Readers are advised to keep in mind that statements, data, illustrations, procedural details or other items may inadvertently be inaccurate.

Cover illustration: courtesy of BASF Aktiengesellschaft, Ludwigshafen, Germany.

Library of Congress Card No.: Applied for.

British Library Cataloguing-in-Publication Data: A catalogue record for this book is available from the British Library.

Die Deutsche Bibliothek – CIP Cataloguing-in-Publication-Data:
A catalogue record for this publication is available from Die Deutsche Bibliothek.

ISBN 3-527-29874-6

© WILEY-VCH Verlag GmbH, D-69469 Weinheim (Federal Republic of Germany), 2000

Printed on acid-free paper.

All rights reserved (including those of translation in other languages). No part of this book may be reproduced in any form – by photoprinting, microfilm, or any other means – nor transmitted or translated into a machine language without written permission from the publishers. Registered names, trademarks, etc. used in this book, even when not specifically marked as such, are not to be considered unprotected by law.

Composition: Rombach GmbH, D-79115 Freiburg
Printing: Strauss Offsetdruck GmbH, D-69509 Mörlenbach
Bookbinding: Wilhelm Osswald & Co., D-67433 Neustadt (Weinstraße)

Contents

Antiinfectives ... 949

1 Antibiotics ... 951

1. Introduction ... 952
2. Chemotherapeutic Use of Antibiotics ... 955
3. Classification of Antibiotics ... 959
4. Antibiotics in Current Use ... 1017
5. Fermentation ... 1073
6. Isolation and Purification of Antibiotics; Quality Specifications ... 1084
7. Analytical Measurements and Quality Control ... 1089
8. Economic Aspects ... 1093
9. References ... 1095

2 Synthetic Chemotherapeutic Agents ... 1107

1. Introduction ... 1108
2. Chemotherapy of Bacterial Infections ... 1109
3. Chemotherapy of Protozoan Infections ... 1146
4. Chemotherapy of Viral Infections ... 1172
5. References ... 1189

3 Antimycotics ... 1199

1. Introduction ... 1199
2. Azole Antimycotics ... 1201
3. Polyene Antimycotics ... 1232
4. Flucytosine ... 1238
5. Griseofulvin ... 1241
6. Ciclopirox ... 1244
7. Thiocarbamates ... 1246
8. Allylamines ... 1248
9. Amorolfine ... 1252
10. Unspecific Topical Antimycotics ... 1254
11. Recent Developments and Outlook ... 1256
12. References ... 1256

4 Anthelmintics ... 1265

1. Introduction ... 1265
2. Parasitic Worms ... 1266
3. Anthelmintic Drugs ... 1277
4. References ... 1287

5 HIV and AIDS Therapeutics ... 1291

1. Introduction — HIV and AIDS ... 1292
2. Reverse Transcriptase Inhibitors ... 1294
3. Protease Inhibitors ... 1304
4. Emerging Concepts in AIDS Therapy ... 1315
5. References ... 1315

Endocrine and Metabolic Drugs 1319

6 Steroids 1321

1. Nomenclature 1321
2. Sterols........................ 1322
3. Bile Acids 1328
4. Steroid Hormones 1330
5. Sapogenins 1332
6. Steroid Alkaloids 1333
7. Steroid Lactones.............. 1335
8. References 1336

7 Peptides and Protein Hormones 1339

1. Introduction................... 1340
2. Gonadoliberin, Thyroliberin, Gonadotropins, Thyrotropin, Inhibin, and Related Hormones 1348
3. Parathyroid Hormone and the Calcitonin Family 1362
4. Corticoliberin – Proopiomelanocortin Cascade 1369
5. Blood Pressure Regulating Peptides . 1388
6. Cholecystokinin and Gastrin 1413
7. Secretin Family................ 1421
8. References 1449

8 Chemical Contraception 1493

1. Introduction................... 1493
2. Natural Methods 1493
3. Barrier Methods............... 1495
4. Hormonal Methods 1497
5. Intrauterine Devices (IUDs)...... 1506
6. Sterilization................... 1507
7. Contragestational Drugs 1508
8. New Approaches 1510
9. References 1514

9 Thyrotherapeutic Agents 1519

1. Introduction................... 1520
2. Thyroid Hormones as Therapeutic Agents....................... 1521
3. Thyroid Depressants as Therapeutic Agents....................... 1524
4. References 1531

10 Hormones 1533

1. Introduction................... 1535
2. Amino Acid Hormones 1555
3. Steroid Sex Hormones 1564
4. Adrenal Steroid Hormones 1601
5. Cholecalciferol 1636
6. References 1641

11 Oral Antidiabetic Drugs 1653

1. Introduction................... 1653
2. Stimulators of Insulin Secretion.... 1654
3. Biguanides................... 1660
4. Inhibitors of Intestinal Carbohydrate Digestion.................... 1661
5. Enhancers of Insulin Action....... 1665
6. References 1667

Antiinfectives

Antiinfectives comprise approximately 15 % of the worldwide pharmaceutical market. This category is made up of both antibiotics (produced by living organisms, an example being penicillin) and synthetic or semi-synthetic antiinfective agents. It includes three main classes antibacterial, antiviral and antifungal agents. Antibacterials, comprising the largest class, are made up of many different groups, examples being the macrolide antibiotics, cephalosporins or quinolones. Antivirals as a class are growing in terms of new agents and usage. Much of the growth is linked to patient populations infected by specific viral diseases, such as HIV, and an increase in the general awareness of various viral infections. Improved topical and systemic antifungal agents continue to evolve. Chemotherapy of parasitic infections is not covered as that is a specialized area of pharmaceuticals.

Antibiotics

MASAJI OHNO, Faculty of Pharmaceutical Sciences, University of Tokyo, Tokyo, Japan (Chaps. 1–4)
MASAMI OTSUKA, Faculty of Pharmaceutical Sciences, University of Tokyo, Tokyo, Japan (Chaps. 1–4)
MORIMASA YAGISAWA, Japan Antibiotics Research Association, Tokyo, Japan (Chaps. 1–4)
SHINICHI KONDO, Institute of Microbial Chemistry, Tokyo, Japan (Chaps. 1–4)
HEINZ ÖPPINGER, Hoechst Aktiengesellschaft, Frankfurt, Federal Republic of Germany (Chaps. 5–7)
HINRICH HOFFMANN, Hoechst Aktiengesellschaft, Frankfurt, Federal Republic of Germany (Chaps. 5–7)
DIETER SUKATSCH, Hoechst Aktiengesellschaft, Frankfurt, Federal Republic of Germany (Chaps. 5–7)
LEO HEPNER, L. Hepner and Associates, Ltd., London, United Kingdom (Chap. 8)
CELIA MALE, L. Hepner and Associates, Ltd., London, United Kingdom (Chap. 8)

1.	Introduction	952
1.1.	General Definition	952
1.2.	Historical Development and Classification	952
1.3.	Nomenclature	955
2.	Chemotherapeutic Use of Antibiotics	955
2.1.	Microbial Pathogens	955
2.2.	Tumor Cells	956
2.3.	Enzyme Inhibitors	956
2.4.	Chemotherapeutic Uses	956
2.5.	Use in Agriculture	957
2.6.	Resistance	957
2.7.	Units	958
2.8.	Analysis	959
3.	Classification of Antibiotics	959
3.1.	β-Lactams	959
3.1.1.	Natural Penicillins	959
3.1.2.	Semisynthetic Penicillins	960
3.1.3.	Natural Cephalosporins	963
3.1.4.	Semisynthetic Cephalosporins	964
3.1.5.	Cephamycins	965
3.1.6.	1-Oxacephems	969
3.1.7.	Clavulanic Acids	970
3.1.8.	Penems	971
3.1.9.	Carbapenems	972
3.1.10.	Nocardicins	975
3.1.11.	Monobactams	976
3.2.	Tetracyclines	977
3.2.1.	Structure and Properties	977
3.2.2.	Anhydrotetracyclines	978
3.2.3.	Anthracyclines	979
3.3.	Aminoglycosides	981
3.4.	Nucleosides	988
3.4.1.	N-Nucleosides	989
3.4.2.	C-Nucleosides	993
3.4.3.	Carbocyclic Nucleosides	994
3.4.4.	An Exceptional Nucleoside	995
3.5.	Macrolides	995
3.5.1.	12-Membered Ring Macrolides	996
3.5.2.	14-Membered Ring Macrolides	996
3.5.3.	16-Membered Ring Macrolides	998
3.6.	Ansamycins	1000
3.7.	Peptides	1001
3.7.1.	The Bleomycin Group	1002
3.7.2.	The Gramicidin Group	1004
3.7.3.	The Polymyxins	1005
3.7.4.	The Bacitracins	1006
3.7.5.	Large-Ring Peptide Antibiotics Containing Lactone Linkages	1006
3.7.6.	The Actinomycin Group	1007
3.7.7.	Other Peptide Antibiotics	1008

3.8.	Other Important Antibiotics and Intermediates	1008	6.	Isolation and Purification of Antibiotics; Quality Specifications	1084
4.	Antibiotics in Current Use	1017	6.1.	Isolation	1084
5.	Fermentation	1073	6.2.	Purification Techniques, Sterile End Products, Official Regulations	1087
5.1.	Screening	1073			
5.2.	Selection, Mutation, and Maintenance of Strains	1074	7.	Analytical Measurements and Quality Control	1089
5.3.	Process Development Leading to Large-Scale Production	1075	7.1.	Microbiological Analysis	1089
5.4.	Fermentation Technology	1079	7.2.	Isotopically Labeled Antibiotics	1092
5.4.1.	Maintenance of the Strain and Production of Inoculum	1079	8.	Economic Aspects	1093
5.4.2.	Treatment Before and During Fermentation	1081	9.	References	1095

1. Introduction

1.1. General Definition

In 1942, WAKSMAN defined antibiotics as chemical substances produced by microorganisms and capable of inhibiting the growth of microorganisms [20]. Great effort has been devoted to the worldwide search for new antibiotics, and numerous compounds possessing various biological activities, that is, antibacterial, antiviral, antifungal, antitumor, and enzyme-inhibiting activities, have been discovered. These substances are mostly of microbial origin but are also semisynthetic in some cases. They have a wide variety of structural characteristics. Each is entered in the index of antibiotics. The area defined by the term "antibiotics" is therefore expanding, although WAKSMAN's original definition is still basically valid.

1.2. Historical Development and Classification

In 1877, PASTEUR observed that saprophytic bacteria inhibited the growth of pathogenic anthrax organisms. His was the first scientific description of the antagonism phenomenon. Production of a certain metabolic substance seemed to be responsible for the inhibition. PASTEUR suggested the therapeutic potential of this type of growth repression. VUILLEMIN used the term "antibiosis" to describe the inhibition of the growth of one organism by another. The potential utility of "bacteriotherapy" was recognized and enormous experimental efforts were made to investigate the antagonism phenomenon.

In 1894, METCHNIKOFF reported the repressive effect of *Pseudomonas* on *Vibrio cholerae*. From the culture of a *Penicillium*, GOSIO isolated an antibacterial crystalline substance, mycophenolic acid, in 1896. Other results also demonstrated the ability of various microbes to produce antibacterial substances [21].

It was in 1929 that FLEMING observed that a culture of a *Penicillium* inhibited the growth of bacteria [22]. He demonstrated the production of an antibacterial substance in the culture broth and named it penicillin. Although he suggested the promising therapeutic utility of penicillin, none of the attempts to isolate penicillin were successful and immediate attention was not attracted for the next decade. Instead, synthetic chemotherapeutics, such as sulfonamides, became objects of general interest after the discovery of prontosil by DOMAGK in 1935 [23]. The antibiotics, many of which were known even before the discovery of penicillin, remained without great importance for decades. Only the outbreak of the Second World War in 1939 led to an intense worldwide search for drugs to treat infections and wounds. Toward the end of the 1930s, FLOREY, CHAIN, and co-workers began to investigate penicillin in the course of their systematic study of antibacterial substances. They demonstrated the marked activity and therapeutic value of penicillin in 1940 and the "antibiotic era" began [24]. The production of penicillin had until that time been unsatisfactory, and favorable conditions for the effective formation of the antibiotic were explored in the United States. An active culture of the penicillin-producing organism was sought and submerged fermentation was developed. The use of lactose as a carbon source and the addition of cornsteep liquor to the nutrients were found effective. Irradiation of the culture with X rays or ultraviolet light produced mutant strains. These findings set the stage for the industrial production of penicillin.

After the discovery of penicillin, BROTZU began a search for antibiotic-producing organisms and examined a culture of *Cephalosporium* spp. isolated from the sea near a sewage outlet in Sardinia. It secreted substances active against gram-positive bacteria. In September of 1948, BROTZU sent his organism to ABRAHAM at Oxford for detailed inspection. Several antibiotics were isolated from the culture and named cephalosporin [25]. Particular attention was attracted by cephalosporin C, crystallized from the crude mixture of antibiotics, because of its low toxicity and its resistance to penicillinase.

Penicillins and cephalosporins, both of which possess a β-lactam ring as a structural characteristic, are designated β-lactam antibiotics. Extensive attempts to improve their antibacterial spectra through chemical modifications led to the development of many kinds of semisynthetic penicillins and cephalosporins. Moreover, nontraditional β-lactams have recently been discovered and commercialized [26]–[29].

Other novel antibiotics were discovered among the products of fungi, bacteria, and actinomycetes [30]. As a result of a search for water-soluble and heat-stable substances which would be active against gram-negative bacteria, WAKSMAN isolated actinomycin in 1940, streptothricin in 1942, and streptomycin in 1944 from cultures of actinomycetes. Various other antibiotics were isolated from microbes in France, Germany, Japan, the United Kingdom, the United States, and other countries.

Actinomycin is a peptide antibiotic effective against tumor cells as well as bacteria. Peptide antibiotics, one of the major antibiotics groups, possess diverse activities, i.e., antibacterial, antifungal, and antitumor activities. Their structures are varied and feature complicated modes of connection of often unusual amino acids. Ring peptides, linear peptides, lactonic peptides, and peptides containing hydroxy acids have been isolated. The group includes actinomycins, gramicidins, polymyxins, and colistins.

Streptomycin is used for infections of gram-positive and gram-negative bacteria and as a specific medicine for tuberculosis. Kanamycin, discovered by UMEZAWA in 1957, is especially effective against resistant bacteria. Paromomycin, spectinomycin, and ribostamycin were isolated later. These antibiotics are called aminoglycosides because their structural units are amino sugars, sugars, and amino acids. They are water soluble and basic in nature. The mechanism of resistance to aminoglycosides has been closely investigated and derivatives for use active against the resistant bacteria have been developed.

A yellow substance showing antibacterial activity was found in 1948 among products of *Streptomyces aureofaciens* and was named aureomycin. Terramycin was isolated from the fermentation broth of *Streptomyces rimosus*. The chemical and structural similarities of the two soon became apparent; they each have a linearly fused tetracyclic structure of six-membered rings. This parent skeleton is designated "tetracycline," and aureomycin and terramycin are now called chlortetracycline and oxytetracycline, respectively.

Chloramphenicol was first isolated as a product of *Streptomyces venezuelae;* it showed a broad antimicrobial spectrum. Chloramphenicol is an unusual natural product because it possesses both chloro and nitro groups in its structure. It is the only antibiotic produced commercially by an entirely chemical synthesis.

Macrolides are macrocyclic lactones to which sugars are attached. Various clinically important antibiotics are macrolides. Erythromycin, isolated in 1952, has two sugars connected to different sites on the 14-membered ring aglycone. Dimethylamino sugars are often found in macrolide antibiotics. Macrolides are classified as 12-, 14-, or 16- membered ring macrolides.

Polyene antibiotics have a conjugated olefinic structure in the macrocyclic lactone moiety, as is the case with amphotericin B, and sometimes lack the amino sugar moiety. Polyenes are produced mainly by *Streptomyces* and show antifungal activity. Ansamycin has an aliphatic "ansa" bridge spanning two nonadjacent positions of the aromatic system. Rifamycin is a representative ansamycin possessing a 1,4-naphthoquinone moiety. Benzenoid ansamycins having a hydroquinone moiety are also known.

In 1950, rhodomycin was isolated from the culture broth of *Streptomyces purpurascens* by BROCKMANN and his co-workers. Structural analysis disclosed that rhodomycin is a glycoside, combining an amino sugar and a 7,8,9,10-tetrahydro-5,12-naphthacenequinone moiety. Structurally similar antibiotics have since been discovered, and the generic name "anthracycline" has been assigned. The aglycone of anthracycline is called anthracyclinone. Daunorubicin and doxorubicin are representative anthracycline antitumor antibiotics. Aclacinomycin, recently isolated by Umezawa, possesses three sugars and is of interest because of its low toxicity.

The first nucleoside antibiotic, cordycepin, was isolated in 1950. Various nucleoside antibiotics of unusual structure have subsequently been isolated and are important as antibacterials, antineoplastics, and agricultural chemicals.

In addition to the antibiotics previously described, various unclassified antibiotics have been isolated. Their structures are not always fully known, although they exhibit significant biological activity. The most important antibiotics and derivatives are discussed in this article. Some representative compounds that have recently been introduced in clinics are also described. These include the new semisynthetic penicillins and the cephalosporins. Clearly the field of antibiotics is experiencing dynamic growth.

1.3. Nomenclature

In principle, it is the privilege of the discoverer to name to his or her new antibiotic. Usually antibiotics are named after the producing organisms or some aspect of their chemical and biological nature. In accordance with the suggestions of the Nomenclature Committee of the American Society of Microbiology [31], names of antibiotics should be based on (a) the family to which the antibiotic belongs, (b) the chemical structure of the compound, or (c) some property of the antibiotic. If, for some reason, a name cannot be given to the new antibiotic, a code designation may be given.

2. Chemotherapeutic Use of Antibiotics

2.1. Microbial Pathogens

Microorganisms that can be treated by chemotherapy include bacteria, fungi (→ Antimycotics), viruses, rickettsia, and protozoa (→ Synthetic Chemotherapeutic Agents).

Outside the cytoplasmic membrane, bacteria have a rigid shell called the cell wall that is not seen in mammalian cells. The main constituent of the cell wall is peptidoglycan, a crosslinked structure of long parallel chains of polysaccharides and short peptide chains. Bacteria are separated into two classes based on the results of a Gram's stain. Gram-positive bacteria hold the color of the primary stain and gram-negative ones are decolorized and are stained by the counterstain. The structure and constituents of the cell walls of gram-positive and gram-negative bacteria are slightly, but distinctly, different.

Some antibiotics are effective only against gram-positive bacteria whereas others are active against gram-negative ones. Streptomycin and kanamycin are effective against mycobacteria as well as gram-positive and gram-negative bacteria. Chloramphenicol and tetracyclines are broad-spectrum antibiotics active against not only the usual bacteria but also rickettsia.

2.2. Tumor Cells

Certain antibiotics are effective for clinical treatment of cancer. Cancer cells function abnormally, escaping growth regulation. Antibiotics occupy an important position among the various agents for cancer chemotherapy. At present, doxorubicin, daunorubicin, mitomycin C, bleomycin, actinomycin D, chromomycin A_3, and neocarzinostatin are used clinically. They interact with DNA to inhibit polymerases of DNA and RNA, or to cause DNA strand breakage. Normal cells are also damaged to some extent; the selective toxicity is generally based on the unusually rapid multiplication of the tumor cells. Some of these antibiotics also possess anti-neoplastic activity because they inhibit the synthesis of DNA.

2.3. Enzyme Inhibitors

The action of antibiotics can be interpreted as a direct or indirect inhibition of certain enzyme systems. The activity of an antibiotic can be recognized only when its site of action is critically important to the maintenance of the life of the cell. However, various enzyme inhibitors of microbial origin have been found useful as medicines even though they do not exhibit antibiotic activity [32]. For example, pepstatin, an inhibitor of pepsin isolated by UMEZAWA in 1970 from *Streptomyces testaceus,* shows promise for the treatment of gastric ulcer, but its therapeutic effects are not based on any antibiotic action.

2.4. Chemotherapeutic Uses

Chemical and bacteriological diagnoses are especially important for successful chemotherapy because the choice of drug depends primarily on the sensitivity of the microorganism to the drug. The antimicrobial activity of an antibiotic is expressed as the minimum inhibitory concentration (MIC) measured by the dilution method. Antibiotics are administered by hypodermic, intramuscular, or intravenous injections, or as internal medicines. For external applications, they are given according to the nature of the antibiotic and the characteristics of the disease. Doses large enough to maintain a sufficient drug concentration in the blood and tissues are prescribed. Various undesirable side effects of antibiotics have been reported. One of the serious side effects is the allergic reaction to penicillins. Oto- and nephrotoxicity result from the long-term use of aminoglycosides in quantity. Aplastic anemia caused by chloramphenicol also has been reported. Antitumor and antiviral antibiotics are generally highly toxic.

2.5. Use in Agriculture

Although antibiotics were originally developed for use against microbial diseases in humans, they are also applicable to agriculture. Several antibiotics are used in the treatment of animal and plant diseases. Kasugamycin and blastcidin S are effective against rice blast disease. Polyoxins are selectively effective against certain species of phytopathogenic fungi. Penicillins, tetracyclines, mikamycins, erythromycins, tylosins, spiramycins, and thiopeptins are used in animal feeds to stimulate growth. The mechanism of growth enhancement induced by these antibacterial antibiotics is unknown. The use of antibiotics as an additive in animal fodder has led to considerable improvement in agricultural production. Antibiotics, on the one hand, prevent breeding diseases, afflictions, and mortality in young animals. On the other hand, antibiotics, even in very small quantities (on the order of 5–20 mg per kilogram of fodder mixture), enhance the growth and full utilization of fodder in, e.g., pigs and poultry. For all antibiotics that are used simultaneously as therapeutic agents and fodder additives, there is the danger of developing a resistance. Therefore, worldwide efforts are directed at avoiding the use of the therapeutically important antibiotics, particularly the penicillins and the tetracyclines, as feed additives. Instead, these antibiotics are used only for genuine veterinary treatment in compliance with certain controls and quarantine guarantees.

2.6. Resistance

An organism becomes resistant to an antibiotic if it survives continued contact. Antibiotics repress the growth of the sensitive organism in a culture, resulting in the survival of naturally resistant organisms. Microbial resistance can be acquired through a spontaneous or induced mutation. Various examples of cross-resistance have been observed and several cross-resistant groups of antibiotics are recognized. Combined use of two antibiotics retards the appearance of resistant organisms.

2.7. Units

The production, isolation, and processing of commercial products require careful control for all pharmaceuticals. Because of the extremely high sensitivity and the danger of diminished activity, this consideration has always been particularly important.

The Oxford unit (O. U.) is defined as the amount of penicillin that just prevents the growth of a certain *Staphylococcus aureus* species. Very pure crystalline penicillin salts generally have constant biological activities and the Oxford unit has been replaced by the international unit: 1 mg of pure benzylpenicillin sodium contains 1670 O. U.; the O. U. specific to this salt was declared to be the international unit (I. U., usually abbreviated U). Conversely, 0.6 µg of benzylpenicillin sodium has the activity of 1 I. U. Because the biological activity comes from the penicillin nucleus, the change to another cation leads to a change in activity proportional to the molecular mass. This change can be calculated. The activities of the chief penicillin salts are:

benzylpenicillin sodium	1670 U/mg
benzylpenicillin potassium	1598 U/mg
benzylpenicillin procaine	1011 U/mg
penicillin-2-hydroxyprocaine	1008 U/mg
penicillin-*N,N'*-dibenzyl-ethylenediamine	1213 U/mg
penicillin-*N*-ethylpiperidine	1328 U/mg

The mass of 1 U, for benzylpenicillin sodium, 0.6 µg, is extremely small. The following larger units of mass are used in production and trade:

1 Mega U	$= 1 \times 10^6$ I. U.
	= 600 mg benzylpenicillin sodium
	= ca. 1 g benzylpenicillin procaine
1 Mio Mega U	$= 1 \times 10^{12}$ I. U.
	= 600 kg benzylpenicillin sodium
	= ca. 1 t benzylpenicillin procaine

The activities of some older penicillins are given in I. U.

phenoxymethyl-penicillin:	1 mg free acid	1699 U
phenethicillin:	1 mg D-potassium salt	1476 U
	1 mg L-potassium salt	1470 U
penicillin O:	1 mg potassium salt	1612 U

All other penicillins that are used therapeutically can be made very pure and the preparations are dosed and traded in mass units (µg, mg, g, kg).

2.8. Analysis

The practical determination of active substances in penicillins and other antibiotics can be divided among three types of methods [25]:

1) Microbiological testing (see Chap. 7).
2) Determination of the contents by chemical or enzymatic conversion followed by a physical method, such as colorimetry.
3) Purely physical methods, such as UV or IR absorption.

3. Classification of Antibiotics

3.1. β-Lactams

The β-lactam group includes natural penicillins, semisynthetic penicillins, natural cephalosporins, semisynheric cephalosporins, cephamycins, 1-oxacephems, clavulanic acids, penems, carbapenems, nocardicins, and monobactams.

3.1.1. Natural Penicillins

Penicillin was discovered in 1929 by FLEMING [22]. At first it was obtained as a mixture of several similar compounds, but these were later separated from each other. The β-lactam structure of penicillin was proposed by ABRAHAM and CHAIN and supported by WOODWARD, but it was opposed by those who believed in the alternative thiazolidine-oxazole structure [33]. The β-lactam structure was finally established by an X-ray crystallographic analysis performed by HODGKIN and LOW [34]. Penicillins G, F, K, X, and N, dihydropenicillin F, and isopenicillin N have been isolated from the fermentation broths of *Penicillium notatum* or *P. chrysogenum*. These compounds differ only in the R moiety of structure **1**.

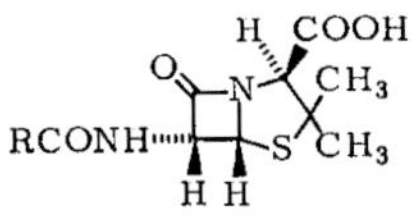

1

Name	R
Penicillin G (Benzylpenicillin) [61-33-6]	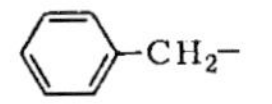 ⟨C₆H₅⟩-CH$_2$-
Penicillin F [118-53-6]	CH$_3$CH$_2$CH=CHCH$_2$-
Dihydropenicillin F [4493-18-9]	CH$_3$(CH$_2$)$_4$-
Penicillin K [525-97-3]	CH$_3$(CH$_2$)$_6$-
Penicillin X [525-91-7]	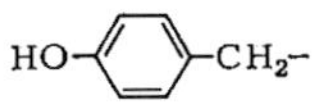 HO-⟨C₆H₄⟩-CH$_2$-
Penicillin N [525-94-0]	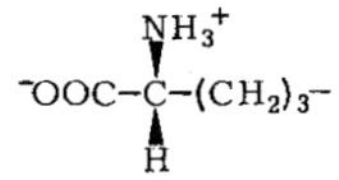 ⁻OOC-C(NH$_3^+$)(H)-(CH$_2$)$_3$-
Isopenicillin [58678-43-6]	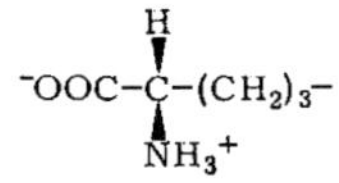 ⁻OOC-C(H)(NH$_3^+$)-(CH$_2$)$_3$-

Of these, penicillin G shows good stability, activity, and rate of production by microorganisms. Total synthesis of penicillin V was achieved by SHEEHAN and HENERY-LOGAN in 1957 [35]. Biogenic syntheses of penicillin–cephalosporin antibiotics also have been reported [36], [37].

3.1.2. Semisynthetic Penicillins

Several limitations have become apparent concerning the antibiotic activity of benzylpenicillin. This drug is not very active against gram-negative bacteria; it is inactivated by penicillinase produced by resistant organisms, and it is not suitable for oral administration because it breaks down under acidic conditions. Penicillins having different side chains have been made by adding appropriate precursors to the fermentation [33]. Various penicillins have been obtained biosynthetically. Among these is phenoxy-methylpenicillin, the first used by oral administration. In contrast, 6-aminopenicillanic acid (6-APA, **2**, R is NH$_2$) can be prepared by either enzymatic or chemical means. Penicillin amidase or penicillin acylase cleaves the side chain of penicillin to produce 6-APA. The amide bond of the side chain is also efficiently cleaved by treatment with phosphorus pentachloride [25, p. 27]. Penicillins with modified side chains (2) have been synthesized from 6-APA via the acyl chloride method, the ethyl or isobutyl chloroformate method, or the dicyclohexylcarbodiimide method in order to improve the antibacterial spectra and increase the stability against penicillinase [33, p. 59]. Ampicillin is active against gram-negative bacteria, and carbenicillin and sulbenicillin are effective against *Pseudomonas*. Ampicillin and amoxicillin are suitable for oral administration. Methicillin, oxacillin, cloxacillin, dicloxacillin, flucloxacillin, and naficillin are resistant to β-lactamase. Mecillinam has an unusual amidino side chain and is relatively stable and effective against gram-negative bacteria.

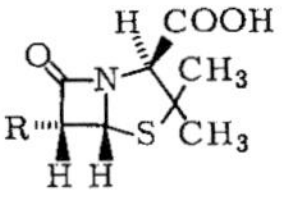

Antibiotics

Structure **2**: penicillin core with R-CONH- substituent

Name	R	Name	R
Penicillin V (Phenoxymethylpenicillin) [87-08-1]	C₆H₅OCH₂CONH– (biosynthetic)	Dicloxacillin [3116-76-5]	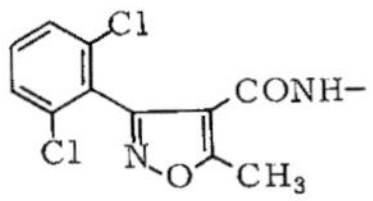 3-(2,6-dichlorophenyl)-5-methyl-isoxazole-4-carboxamide
Penicillin O [87-09-2]	CH₂=CHCH₂SCH₂CONH–		
Phenethicillin [147-55-7]	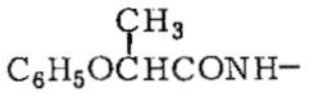 C₆H₅OCH(CH₃)CONH–	Flucloxacillin [5250-39-5]	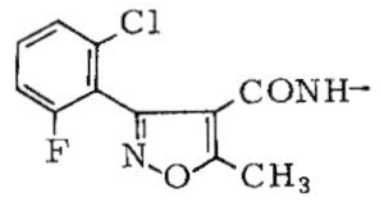 3-(2-chloro-6-fluorophenyl)-5-methyl-isoxazole-4-carboxamide
Propicillin [551-27-9]	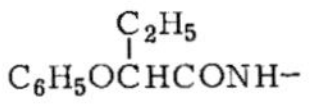 C₆H₅OCH(C₂H₅)CONH–		
Phenbenicillin [1926-48-3]	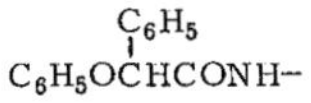 C₆H₅OCH(C₆H₅)CONH–	Ampicillin [69-53-4]	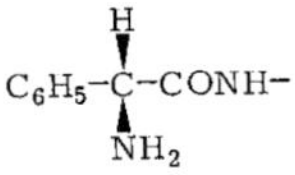 C₆H₅-CH(NH₂)-CONH–
Carbenicillin [4697-36-3]	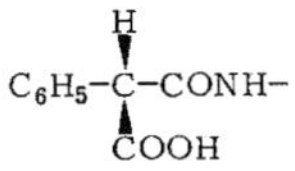 C₆H₅-CH(COOH)-CONH–	Apalcillin [63469-19-2]	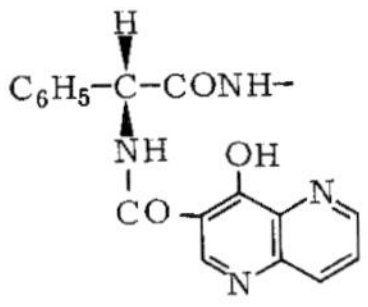 C₆H₅-CH(NH-CO-(4-hydroxy-1,5-naphthyridin-3-yl))-CONH–
Sulbenicillin [41744-40-5]	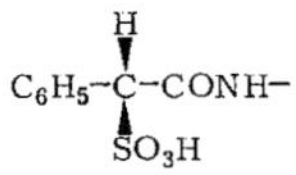 C₆H₅-CH(SO₃H)-CONH–		
Ticarcillin [34787-01-4]	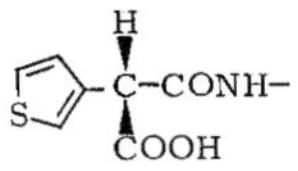 3-thienyl-CH(COOH)-CONH–	Mezlocillin [51481-65-3]	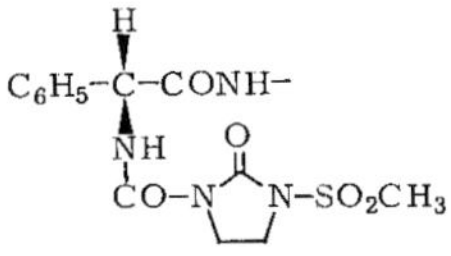 C₆H₅-CH(NH-CO-N(SO₂CH₃)-imidazolidinon-1-yl)-CONH–
Methicillin [61-32-5]	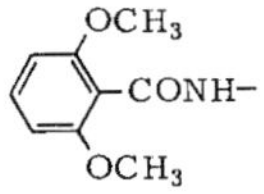 2,6-dimethoxybenzamido (2,6-(OCH₃)₂-C₆H₃-CONH–)	Piperacillin [61477-96-1]	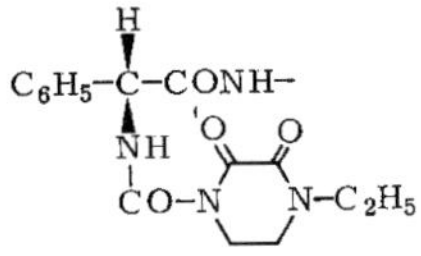 C₆H₅-CH(NH-CO-N(4-ethyl-2,3-dioxopiperazin-1-yl))-CONH–
Nafcillin [147-52-4]	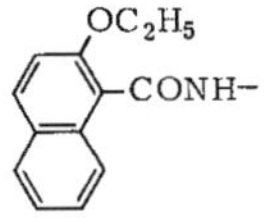 2-ethoxy-1-naphthamido	Amoxicillin [26787-78-0]	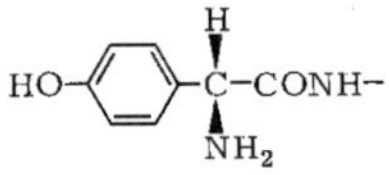 HO-C₆H₄-CH(NH₂)-CONH–
		Cyclacillin [3485-14-1]	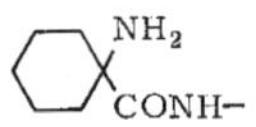 1-aminocyclohexyl-CONH–
Oxacillin [66-79-5]	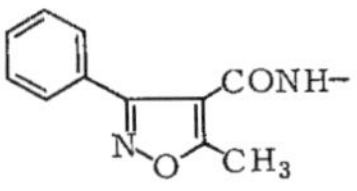 3-phenyl-5-methylisoxazole-4-carboxamide	Hetacillin [3511-16-8]	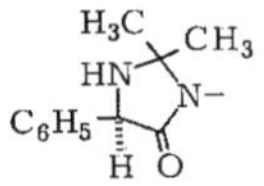 C₆H₅-CH(2,2-dimethyl-imidazolidinon)-CONH–
Cloxacillin [61-72-3]	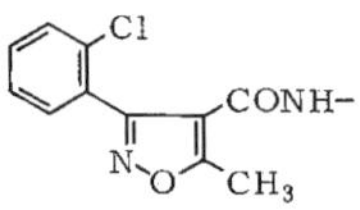 3-(2-chlorophenyl)-5-methylisoxazole-4-carboxamide	Mecillinam [32887-01-7]	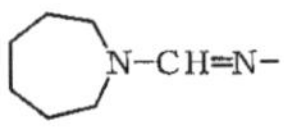 hexamethyleneimino-CH=N–

Modification of the carboxyl group has been found to be effective for the purpose of oral administration, and penicillin esters (**3**) have been developed [33, p. 59]. These are absorbed and hydrolyzed by the small intestine to release free acids of the parent penicillins.

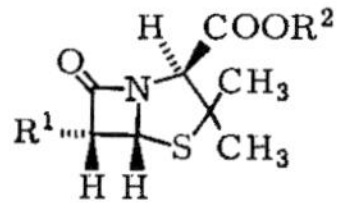

3

Name	R^1	R^2
Talampicillin [47747-56-8]	$C_6H_5-\overset{H}{\underset{NH_2}{C}}-CONH-$	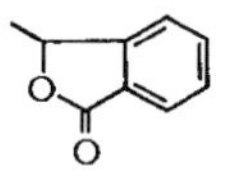 (phthalidyl)
Pivampicillin [33817-20-8]	$C_6H_5-\overset{H}{\underset{NH_2}{C}}-CONH-$	$-CH_2OCO-\underset{CH_3}{\overset{CH_3}{C}}-CH_3$
Bacampicillin [50972-17-3]	$C_6H_5-\overset{H}{\underset{NH_2}{C}}-CONH-$	$-\underset{OCOOC_2H_5}{CH}-CH_3$
Pivmecillinam [32886-97-8]	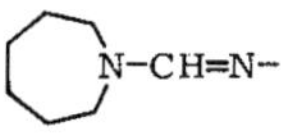 (azepan-1-yl)N-CH=N-	$-CH_2OCO-\underset{CH_3}{\overset{CH_3}{C}}-CH_3$

3.1.3. Natural Cephalosporins

The fermentation broth of *Cephalosporium* spp. isolated by BROTZU contained several antibiotics: cephalosporin P, penicillin N, and cephalosporin C. Cephalosporin P was shown to be an acidic steroidal substance. Cephalosporin C was active against gram-negative bacteria, resistant to β-lactamase, and much less toxic than penicillin. The chemical structure of cephalosporin C was determined by ABRAHAM and NEWTON [25]. It consists of 7-aminocephalosporanic acid (7-ACA) and D-α-aminoadipic acid (see **4**). Treatment of cephalosporin C with acetyl esterase yields deacetylcephalosporin C, which exhibits about 20% of the antibacterial activity of cephalosporin C. The allylic acetoxy group of cephalosporin C can be hydrogenated in the presence of palladium–charcoal catalyst to yield deacetoxycephalosporin C, which shows 10% of the activity of the parent cephalosporin C [25]. WOODWARD and his co-workers synthesized cephalosporin C in a fully stereospecific manner [38].

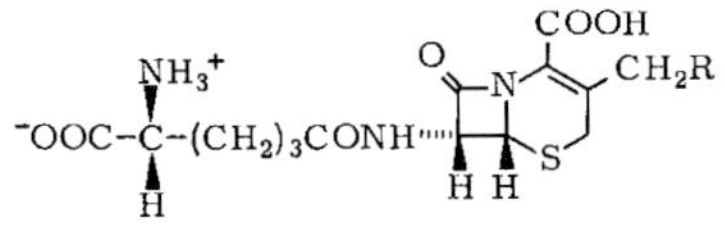

$^-OOC-\underset{H}{\overset{NH_3^+}{C}}-(CH_2)_3CONH\cdots\underset{H\ H}{\overset{\overset{\overset{O}{\|}}{}}{\underset{}{\square}}}\underset{S}{\overset{COOH}{\underset{}{N}}}CH_2R$

4

Name	R
Cephalosporin C [61-24-5]	OCOCH$_3$
Deacetylcephalosporin C [1476-46-6]	OH
Deacetoxycephalosporin C [26924-74-3]	H

3.1.4. Semisynthetic Cephalosporins

Because the antibacterial activity of cephalosporin C itself is relatively low, the development of a more active derivative is desirable. The phosphorus pentachloride method has been applied to the cephalosporin system to produce 7-aminocephalosporanic acid (7-ACA) in high yield [25, p. 27]. The 3'-acetoxy group of cephalosporin is easily replaced by various nucleophiles [25, p. 134]. Modification of the 7-amino group and the 3' group make possible the various cephalosporin derivatives **4a** [33, p. 59]. Cephaloridine, cefazolin, and cefamandole are active against gram-negative bacteria. Cefuroxime, cefotaxime, and ceftizoxime have a methoxyimino group and a 2-aminothiazole ring in common and are resistant to β-lactamase. Cefoperazone is particularly active against *Pseudomonas*. All of these are used by injection. On the other hand, several cephalosporins are used only by oral administration. These include cephalexin, cephaloglycin, cefradine, cefadroxil, cefaclor, cefroxadine, and cefatrizine.

Cephalosporins are also obtainable via the ring expansion reaction of penicillin sulfoxide first devised by MORIN [25, p. 183]. Thus, cephalexin is produced by chemical conversion of phenoxymethylpenicillin or benzylpenicillin.

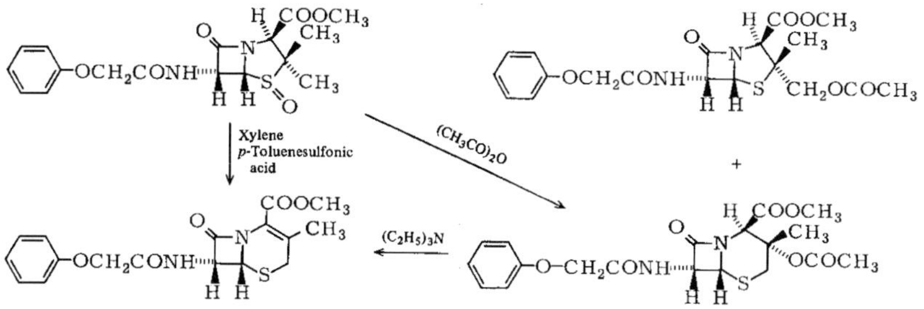

3.1.5. Cephamycins

Substances similar to cephalosporin C were found among the products of various streptomycetes and were characterized by the presence of a 7α-methoxy group. They are named cephamycins after their cephem skeleton (see **5**) and their production by streptomycetes [39], [29, vol. 1, p. 199]. They are strongly resistant to β-lactamase and effective against gram-negative bacteria and bacteria that have acquired resistance to penicillins and cephalosporins. Semisynthetic cephamycins (**5a**) with improved activities are obtained by chemical transformations.

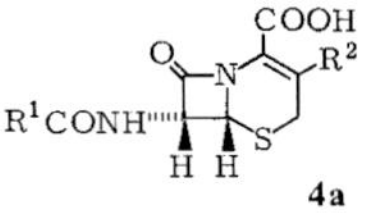

Name	R^1	R^2	Name	R^1	R^2
Cephalexin [15686-71-2]	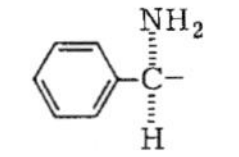 phenyl-CH(NH$_2$)-	-CH$_3$			
Cefaclor [53994-73-3]	"	-Cl	Cefoperazone [62893-19-0]	4-HO-C$_6$H$_4$-CH(NH-CO-N(piperazine-2,3-dione-N-C$_2$H$_5$))-	-CH$_2$S-(1-methyltetrazol-5-yl)
Cephaloglycin [3577-01-3]	"	-CH$_2$OCOCH$_3$			
Cephradine [38821-53-3]	cyclohexa-1,4-dienyl-CH(NH$_2$)-	-CH$_3$	Cefamandole [34444-01-4]	phenyl-CH(OH)-	-CH$_2$S-(1-methyltetrazol-5-yl)
Cefroxadine [51762-05-1]	"	-OCH$_3$	Cefotiam [61622-34-2]	(2-aminothiazol-4-yl)-CH$_2$-	-CH$_2$S-(1-(2-dimethylaminoethyl)tetrazol-5-yl)
Cefadroxil [50370-12-2]	4-HO-C$_6$H$_4$-CH(NH$_2$)-	-CH$_3$			
Cephapirin [21593-23-7]	(pyridin-4-yl)-S-CH$_2$-	-CH$_2$OCOCH$_3$	Ceftezole [26973-24-0]	(tetrazol-1-yl)-CH$_2$-	-CH$_2$S-(1,3,4-thiadiazol-2-yl)
Cephalothin [153-61-7]	(thien-2-yl)-CH$_2$-	-CH$_2$OCOCH$_3$	Cefazolin [25953-19-9]	(tetrazol-1-yl)-CH$_2$-	-CH$_2$S-(5-methyl-1,3,4-thiadiazol-2-yl)
Cephacetrile [10206-21-0]	N≡C-CH$_2$-	-CH$_2$OCOCH$_3$	Cefmenoxime [65085-01-0]	(2-aminothiazol-4-yl)-C(=N-OCH$_3$)-	-CH$_2$S-(1-methyltetrazol-5-yl)
Cefsulodin [62587-73-9]	phenyl-CH(SO$_3^-$)-	-CH$_2$-N$^+$(pyridinium-4-CONH$_2$)	Ceftizoxime [68401-81-0]	"	-H
Cephaloridine [50-59-9]	(thien-2-yl)-CH$_2$-	-CH$_2$-N$^+$(pyridinium)	Cefotaxime [63527-52-6]	"	-CH$_2$OCOCH$_3$
Cefatrizine [51627-14-6]	4-HO-C$_6$H$_4$-CH(NH$_2$)-	-CH$_2$S-(1H-1,2,3-triazol-4-yl)	Cefuroxime [55268-75-2]	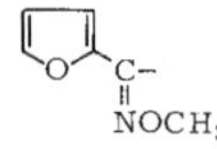 (fur-2-yl)-C(=N-OCH$_3$)-	-CH$_2$OCONH$_2$

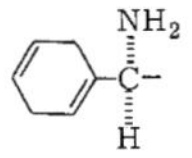

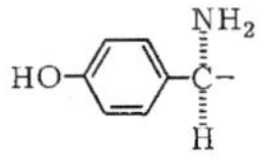

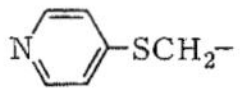

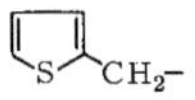

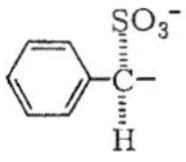

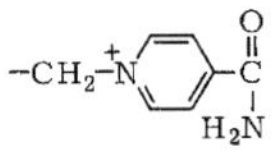

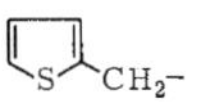

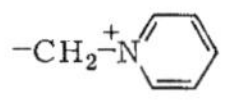

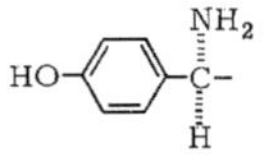

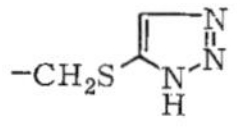

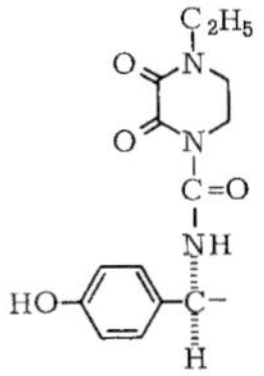

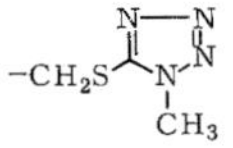

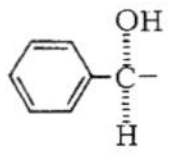

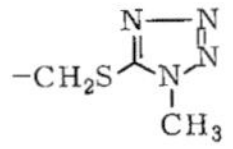

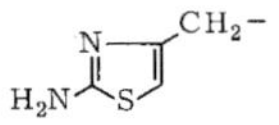

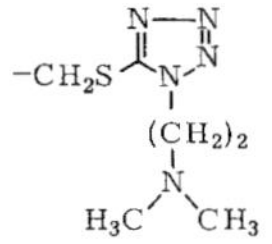

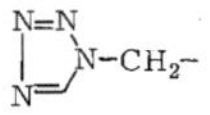

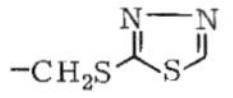

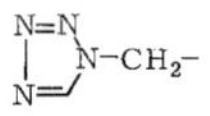

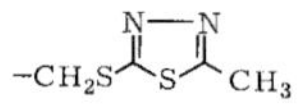

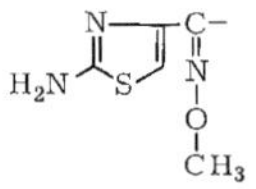

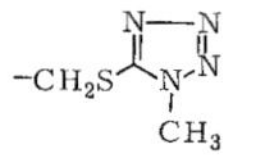

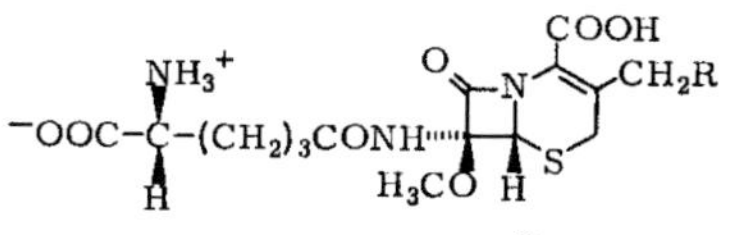

$$^-OOC-\underset{H}{\overset{NH_3^+}{C}}-(CH_2)_3CONH\cdots\underset{H_3CO\ H}{\overset{O}{\underset{\|}{C}}}\overset{COOH}{\underset{S}{N}}CH_2R$$

5

Name	R	Producing organism
7 α-Methoxy-cephalosporin C [32178-82-8]	—OCOCH$_3$	*Streptomyces lipmannii* *St. lactamdurans*
Cephamycin C [34279-51-1]	—OCONH$_2$	*St. clavuligerus* *St. lactamdurans* *St. jumonjiensis*
Cephamycin A [34279-78-2]	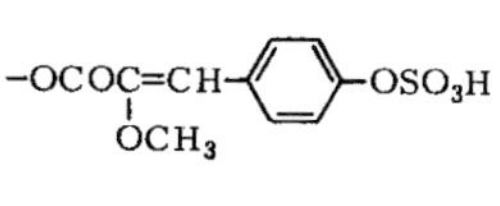 —OCOC(OCH$_3$)=CH—C$_6$H$_4$—OSO$_3$H	*St. griseus* *St. chartreusis* *St. cinnamonensis* *St. fimdriatus*
Cephamycin B [34279-77-1]	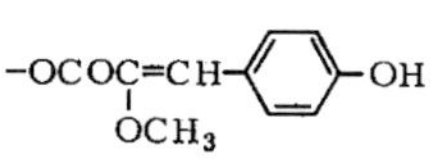 —OCOC(OCH$_3$)=CH—C$_6$H$_4$—OH	*St. halstedii* *St. rochei* *St. viridochromogenes*
C-2801X [62851-50-7]	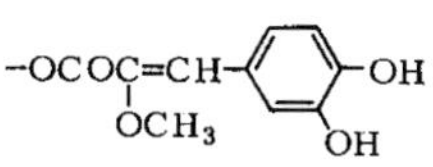 —OCOC(OCH$_3$)=CH—C$_6$H$_3$(OH)—OH	*St. heteromorphous* *St. panagensis*

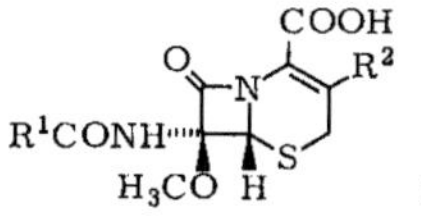

$$R^1CONH\cdots\underset{H_3CO\ H}{\overset{O}{\underset{\|}{C}}}\overset{COOH}{\underset{S}{N}}R^2$$

5a

Name	R^1	R^2
Cefoxitin [35607-66-0]	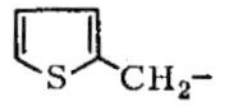 2-thienyl-CH$_2$—	—CH$_2$OCONH$_2$
Cefmetazole [56796-20-4]	N≡CCH$_2$SCH$_2$—	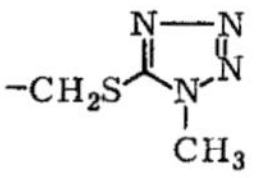 —CH$_2$S-(1-methyl-tetrazol-5-yl)
Cefotetan [69712-56-7]	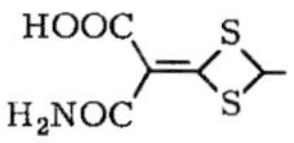 (HOOC)(H$_2$NOC)C=C(S-CH$_2$-S)	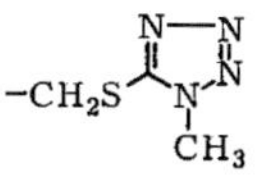 —CH$_2$S-(1-methyl-tetrazol-5-yl)

Cefoxitin, synthesized from cephamycin C, is particularly active against various gram-negative bacteria and stable to β-lactamase. Cefmetazole, produced from 7-ACA, has almost the same antibacterial spectrum as cefoxitin and maintains its high concentration in blood. Cefotetan, recently under development, is reported to be more active than cefoxitin against gram-negative bacteria (for structures see p. 1034).

Chemical modification of cephamycins requires special devices for the following reasons. 7-Aminocephamycinoic acid (7-ACMA; **6**), which corresponds to the 7-ACA of cephalosporins, is not easily isolated because of its instability.

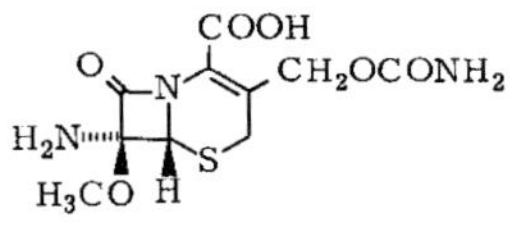

6

7-Aminocephamycinoic acid
(7-ACMA)
[62041-14-9]

It has methoxy and amino groups on the same carbon atom of the β-lactam ring and the elimination of the protonated amino group is quite facile, because of the electron-donating nature of the methoxy group. Moreover, the usual phosphorus pentachloride method cannot be applied to the side chain cleavage of cephamycin C because a strong N–P bond is formed by the reaction of the carbamate moiety of cephamycin C with phosphorus pentachloride [40]. Instead, exchange of the α-aminoadipoyl side chain for another acyl group is achieved by treating the fully protected cephamycin C with the appropriate acyl chloride in the presence of a neutral acid scavenger. This is followed by the simultaneous removal of the amino protective group and the α-aminoadipoyl group [41]. The side chain transformation is also effected using an acyl chloride and partially hydrated molecular sieves [42].

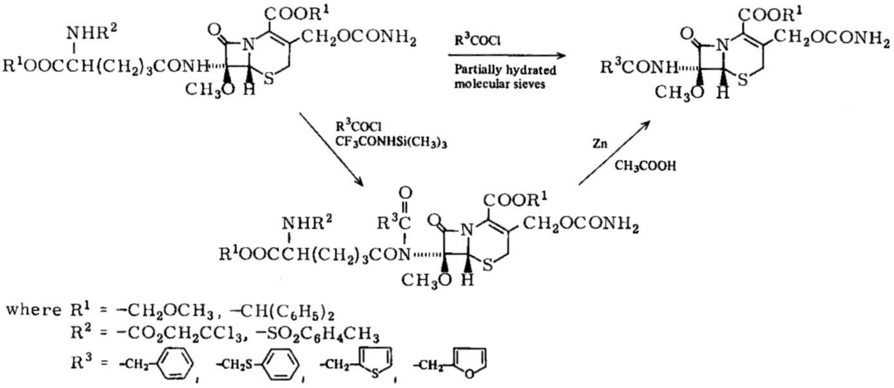

where $R^1 = -CH_2OCH_3, -CH(C_6H_5)_2$
$R^2 = -CO_2CH_2CCl_3, -SO_2C_6H_4CH_3$
$R^3 = $ -CH$_2$-⟨phenyl⟩, -CH$_2$S-⟨phenyl⟩, -CH$_2$-⟨thienyl⟩, -CH$_2$-⟨furyl⟩

Chemical conversion of cephamycin into 7-ACMA ester has been reported [43].

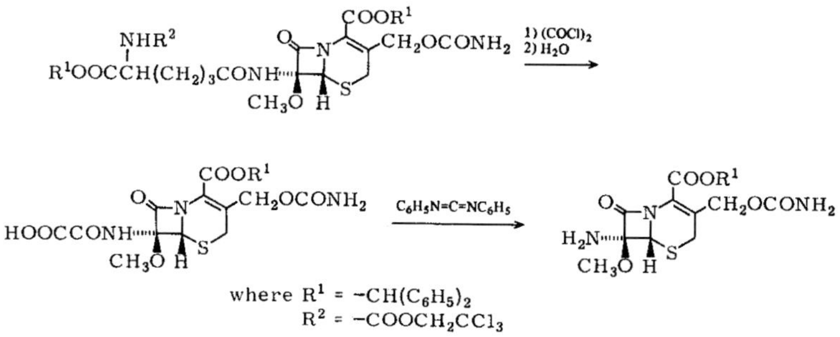

3.1.6. 1-Oxacephems

In addition to the modification of the side chains of natural β-lactam antibiotics, totally or partially synthetic nuclear analogs of penicillins and cephalosporins have been explored extensively [28], [33, p. 59]. In 1974 WOLFE reported the first 1-oxacephem (**7**) derived from penicillin, but its antibacterial activity remains unknown because the amino and carboxy protective groups have not been removed. Racemic 1-oxacephalothin (**8**), synthesized by CHRISTENSEN and his co-workers in the same year, was found to be antibacterially active, suggesting that the sulfur atom is not always necessary for the expression of antibiotic activity. Racemic 1-oxacefamandole is twice as active as cefamandole and the activity of optically active 1-oxacephalothin (**8**) is four to eight times as high as that of cephalothin. NAGATA and his co-workers discovered latamoxef (**9**) (moxalactam, 6059-S), which exhibits strong activity against pathogenic anaerobes, such as *Bacteroides fragilis,* as well as gram-negative bacteria, including *Pseudomonas* [29, vol. 2, p. 1]. It is completely stable against various β-lactamases and has low toxicity. A high plasma-peak level and long duration are maintained. Latamoxef is a nuclear analog of cephamycin that has a 2-(4-hydroxyphenyl)malonylamino side chain and a 1-methyltetrazolylthio moiety; it is produced on an industrial scale by a totally chemical process starting with *epi*-penicillin S-oxide [44], [45].

Name	Structure
Methyl (6S,7S)-3-methyl-8-oxo-7-phthalimido-5-oxa-1-azabicyclo[4.2.0]-oct-2-ene-2-carboxylate [54997-17-0]	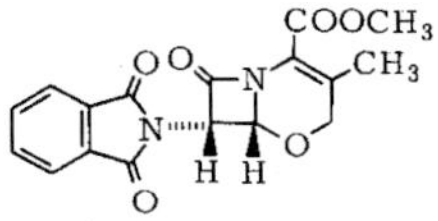 **7**
1-Oxacephalothin [54214-83-4]	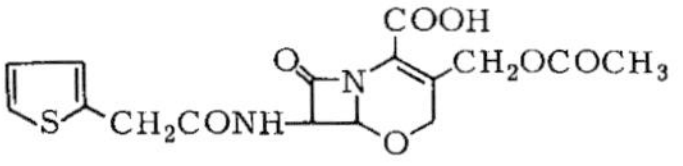 **8**
Latamoxef (Moxalactam, 6059-S) [64952-97-2]	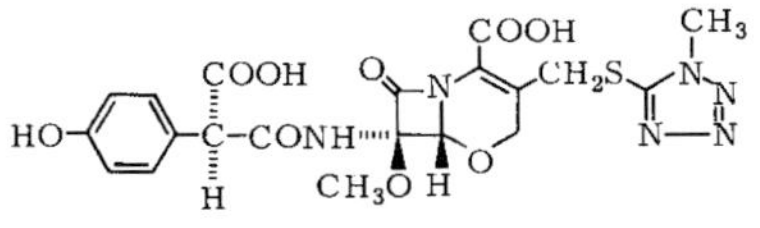 **9**

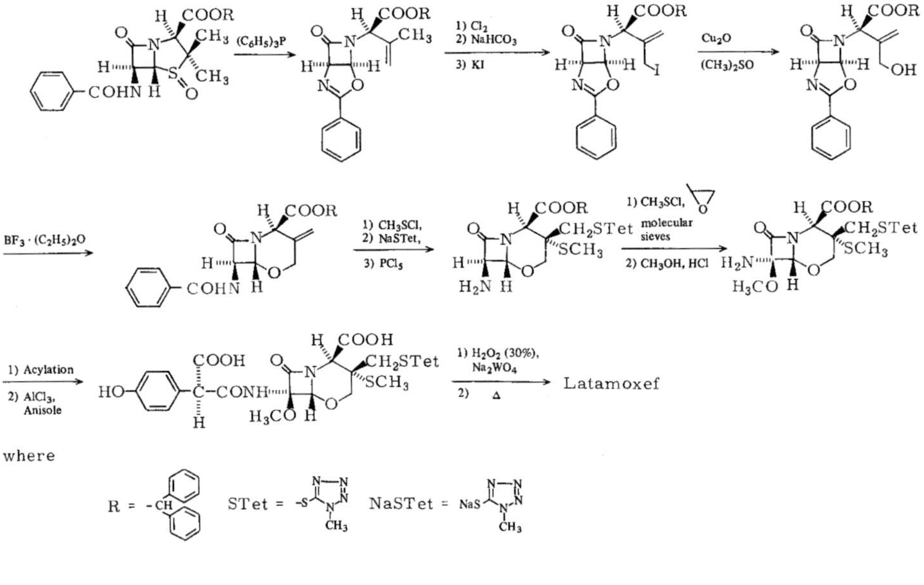

3.1.7. Clavulanic Acids

In the course of screening the substances inhibiting β-lactamase, which is responsible for bacterial resistance to penicillins and cephalosporins, a potent β-lactamase inhibitor, clavulanic acid, was isolated from *Streptomyces clavuligerus* [46]. The antibiotic activity of clavulanic acid is not strong, but it has a broad antibacterial spectrum. Instead, it is effective synergistically when used with β-lactamase-sensitive penicillins and cephalosporins against β-lactamase-producing organisms. Clavulanic acid is characterized by its 1-oxadethiapenam ring system and lack of the side chain at position 6.

Several congeners of clavulanic acid have been isolated [27], [29, vol. 2, p. 361], [47]. Clavam-2-carboxylic acid, 2-hydroxymethylclavam, and 2-formyloxymethylclavam exhibit antifungal activity [48].

Name	Structure
Clavulanic acid [58001-44-8]	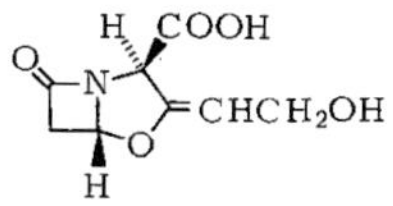 (β-lactam fused ring with H, COOH, =CHCH$_2$OH)
ß-Hydroxypropionyl clavulanic acid [64675-12-3]	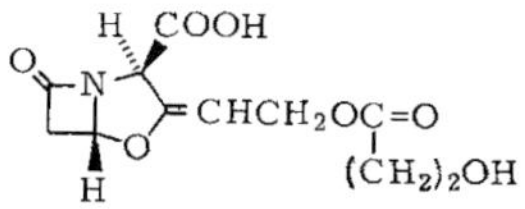 (β-lactam fused ring with H, COOH, =CHCH$_2$OC=O, (CH$_2$)$_2$OH)
Clavam-2-carboxylic acid [71657-61-9]	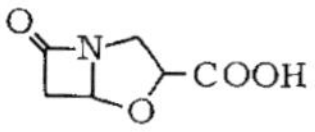 (β-lactam fused ring with COOH)
2-Hydroxymethyl-clavam [66036-39-3]	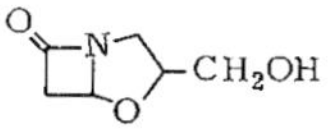 (β-lactam fused ring with CH$_2$OH)
2-Formyloxymethyl-clavam [66036-40-6]	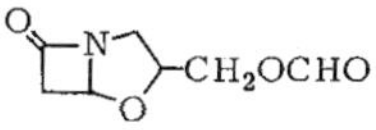 (β-lactam fused ring with CH$_2$OCHO)

3.1.8. Penems

The penem ring system has not been found in nature; it has been designed artificially by WOODWARD [26, p. 167], [28], [29, vol. 2, p. 315]. That the antibacterial activity of β-lactam antibiotics is based on their ability to acylate enzymes is widely accepted. In penicillins, the rigid, nonplanar bicyclic system enhances the reactivity of the β-lactam ring by diminishing the delocalization of the unshared electron pair of the amide nitrogen onto the adjacent carbonyl group. On the other hand, in cephalosporins, where the β-lactam nitrogen is bonded almost planar, the double bond of the six-membered ring interacts with the unshared electrons of the β-lactam nitrogen, diminishing the delocalization to the amide carbonyl. Therefore, the β-lactam ring of cephalosporins is cleaved easily. Penems combine the two structural elements, the five-membered ring and the double bond. 6-Acylaminopenem-3-carboxylic acids (**10**), 6-unsubstituted penem-3-carboxylic acids (**11**), and 6-alkylpenem-3-carboxylic acids (**12**) have been synthesized. The β-lactam moiety with two asymmetric centers derives from penicillin, and the five-membered ring fused with it is formed by the intramolecular Wittig reaction [29, vol. 2, p. 315]. Various 2-substituents, such as H, CH$_3$, C$_2$H$_5$, CH$_2$C$_6$H$_5$, and SCH$_2$CH$_2$NHCOCH$_3$, have been introduced. Although the activity of compound **10** with R = CH$_3$ was disappointing, presumably because of its low stability, 6-unsubstituted penems (**11**) exhibit powerful antibiotic activity [49]. 6-Monoalkylpenems (**12**) show interesting activity in general. The penems having bulky substituents at C-6 are biologically inactive because of the low reactivity of their β-lactam rings.

Name	Structure
6-Acylaminopenem-3-carboxylic acid	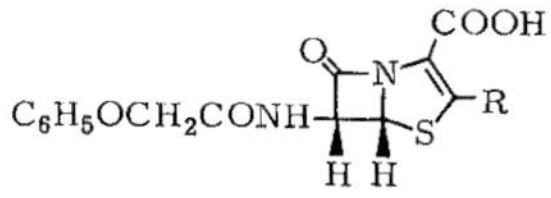 C$_6$H$_5$OCH$_2$CONH— ... —R **10**
Penem-3-carboxylic acid	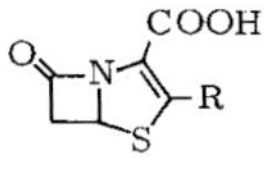 **11**
6-Alkylpenem-3-carboxylic acid	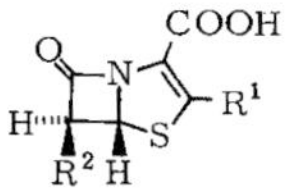 **12**

3.1.9. Carbapenems

Cabapenems are a family of antibiotics having the 1-azabicyclo[3.2.0]hept-2-ene system [27], [29, vol. 2, p. 227]. The first carbapenem antibiotic, thienamycin, was discovered at Merck in 1976 among the fermentation products of *Streptomyces cattleya* [50]. Antibiotics of this type have been isolated one after another in the search for inhibitors of bacterial cell wall synthesis and β-lactamase. From the fermentation broth of *Streptomyces olivaceus*, the Beecham group isolated olivanic acids MM4550, MM13902, MM17880, MM22380, MM22381, MM22382, and MM22383 [51]–[54]. Epithienamycin A, B, C, D, E, and F were found by the Merck group [55]–[58]. Some of olivanic acids and epithienamycins are identical. Olivanic acid MM4550 is identical to MC696-SY2-A found by UMEZAWA as a product of *Streptomyces fulvoviridis* [59]. The antibiotics designated PS-5, –6, and –7 were isolated by Sanraku-Ocean in collaboration with Panlabs from *Streptomyces cremeus*, subsp. *auratilis* A271 [60]–[63]. Carpetimycin A and B, reported by the Kowa Company, are products of *Streptomyces* spp. [64], [65]. Asparenomycin A was isolated by the Shionogi research group from *Streptomyces tokunonensis* and *Streptomyces argenteolus* [66].

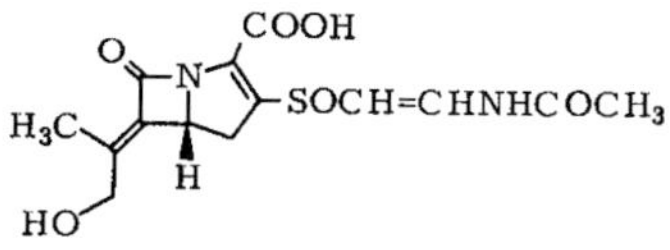

Asparenomycin
[76466-24-5]

Carbapenems are classified into three classes according to the mode of substitution on the β-lactam ring, that is, *trans*-carbapenems (thienamycin, epithienamycins C and D, PS-5, F-6, and F-7), *cis*-carbapenems (epithienamycin A, B, E, and F, MC696-SY2-A, carpetimycin A and B), and ene-carbapenems (asparenomycin A, B, and C). The 5R configuration seems significant for biological activity. The instability of the carbapenems and low broth titer cause difficulties in the determination of the structure.

Structure	Name	R^1	R^2
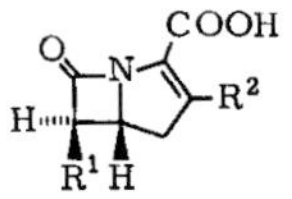 [structure with COOH, R², R¹, H]	Thienamycin [59995-64-1]	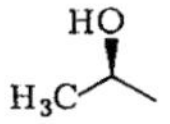 H_3C-CH(OH)-	$-SCH_2CH_2NH_2$
	Epithienamycin C Olivanic acid MM 22381 [63599-16-6]	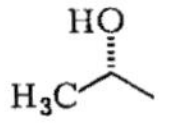 H_3C-CH(OH)-	$-SCH_2CH_2NHCOCH_3$
	Epithienamycin D Olivanic acid MM 22383 [65322-98-7]	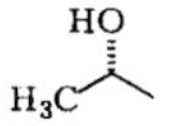 H_3C-CH(OH)-	$-SCH=CHNHCOCH_3$
	PS-5 [78856-77-6]	CH_3CH_2-	$-SCH_2CH_2NHCOCH_3$
	PS-6 [72615-19-1]	$(CH_3)_2CH-$	$-SCH_2CH_2NHCOCH_3$
	PS-7 [72615-18-0]	CH_3CH_2-	$-SCH=CHNHCOCH_3$
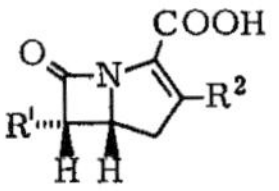 [structure with COOH, R², R¹, H H]	Epithienamycin A Olivanic acid MM 22380 [63582-78-5]	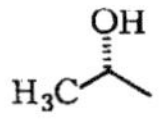 H_3C-CH(OH)-	$-SCH_2CH_2NHCOCH_3$
	Epithienamycin B Olivanic acid MM 22382 [65376-20-7]	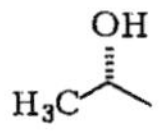 H_3C-CH(OH)-	$-SCH=CHNHCOCH_3$
	Epithienamycin E Olivanic acid MM 13902 [79057-46-8]	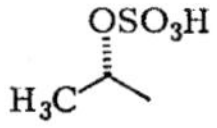 H_3C-CH(OSO_3H)-	$-SCH=CHNHCOCH_3$
	Epithienamycin F Olivanic acid MM 17880 [79057-45-7]	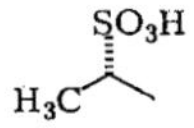 H_3C-CH(SO_3H)-	$-SCH_2CH_2NHCOCH_3$
	Olivanic acid MM 4550 MC696-SY2-A [76985-32-5]	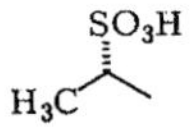 H_3C-CH(SO_3H)-	$-S(=O)CH=CHNHCOCH_3$
	Carpetimycin A [76025-73-5]	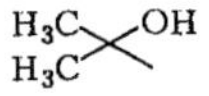 $(H_3C)_2C(OH)-$	$-S(=O)CH=CHNHCOCH_3$
	Carpetimycin B [76094-36-5]	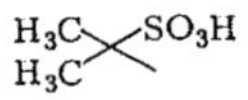 $(H_3C)_2C(SO_3H)-$	$-S(=O)CH=CHNHCOCH_3$

Thienamycin is active against a wide range of gram-positive and gram-negative bacteria, including the ones resistant to conventional β-lactam antibiotics. Carpetimycins and asparenomycin are also effective against resistant bacteria.

Chemical modifications and a great deal of synthetic study of carbapenems have been undertaken to improve the stability of the carbapenem skeleton and compensate for the low productivity of the microbes. This is one of the most important fields of antibiotics [67]–[80], [69, p. 1142].

Naturally occurring carbapenems have several functional groups that have been subjected to chemical modifications to improve their stability and antibacterial potency [54]. The aminoethylthio side chain plays an important role in extending the antibiotic activity, especially the antipseudomonal activity, and is also thought to be a cause of the instability of carbapenems, presumably by intramolecular aminolysis of the β-lactam ring. Derivatives of the aminoethylthio group, carboxyl group, and hydroxyethyl side chain are being sought.

3.1.10. Nocardicins

A mutant strain of *Escherichia coli* showing specific supersensitivity to β-lactam antibiotics has been developed at the Fujisawa Research Laboratories and used to isolate nocardicins from *Nocardia uniformis* by a screening procedure [81]. The nocardicin structure has been elucidated by spectroscopic analysis and chemical degradation. The noncardicins are monocyclic β-lactam antibiotics [27], [29, vol. 2, p. 165], [47, p. 281]. Several congeners differing in the side chains have been isolated. Among them, significant activity is exhibited only by nocardicin A, which is active against gram-negative bacteria, especially *Pseudomonas aeruginosa*, *Proteus*, and *Neisseria*, but inactive against gram-positive bacteria.

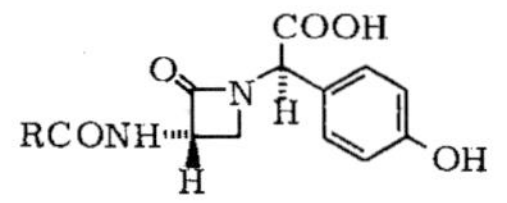

Name	R
Nocardicin A [39391-39-4]	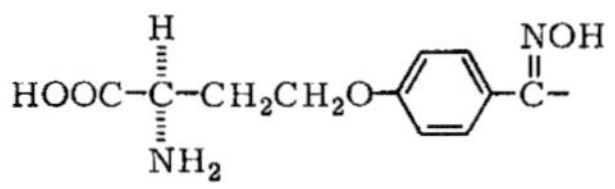 HOOC-CH(NH₂)-CH₂CH₂O-C₆H₄-C(=NOH)-
Nocardicin B [60134-71-6]	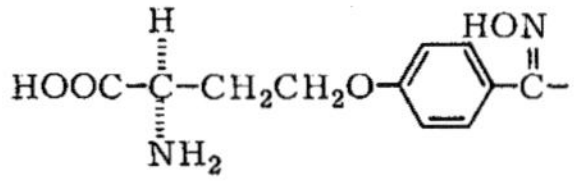 HOOC-CH(NH₂)-CH₂CH₂O-C₆H₄-C(=NOH)- (isomer)
Nocardicin C [59511-12-5]	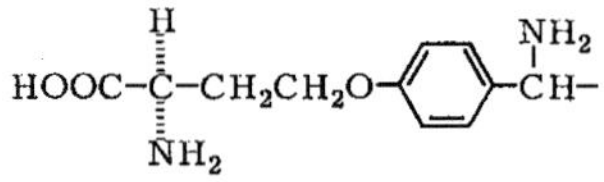 HOOC-CH(NH₂)-CH₂CH₂O-C₆H₄-CH(NH₂)-
Nocardicin D [61425-17-0]	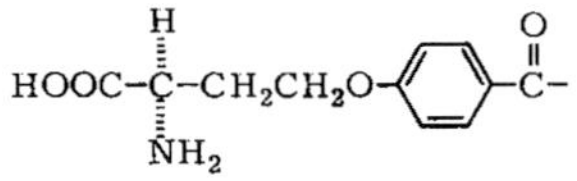 HOOC-CH(NH₂)-CH₂CH₂O-C₆H₄-C(=O)-
Nocardicin E [63555-59-9]	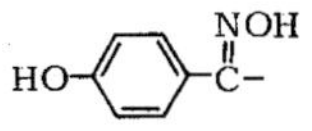 HO-C₆H₄-C(=NOH)-
Nocardicin F [63598-46-9]	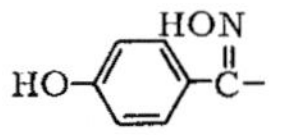 HO-C₆H₄-C(=NOH)- (isomer)
Nocardicin G [65309-11-7]	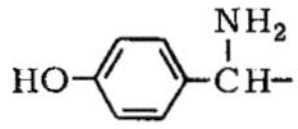 HO-C₆H₄-CH(NH₂)-

3.1.11. Monobactams

Sulfazecin was isolated in 1981 by the Takeda group as a product of *Pseudomonas acidophila* by screening using organisms highly sensitive to β-lactams [82]. The structure was shown to be a monocyclic β-lactam. *Isosulfazecin*, a diastereomer of sulfazecin, also was isolated by the same group. In the same year, the Squibb group reported on a group of monocyclic β-lactams produced by *Agrobacterium, Chromobacterium,* and *Gluconobacter* [83]. A compound (SQ 26 445) identical to sulfazecin was included. SYKES proposed the name "monobactam" for compounds characterized by the 3-acylamino-2-oxoazetidine-1-sulfonic acid group. In monobactams, the β-lactam ring presumably is activated by the electronic effect of the sulfonate moiety alone, in contrast to the case of penicillins and cephalosporins. Because the antibacterial activity of sulfazecin is not satisfactory, many derivatives have been synthesized chemically [29, vol. 3, p. 339]. Among them *aztreonam* (SQ 26 776), synthesized from threonine, has been found highly effective [84].

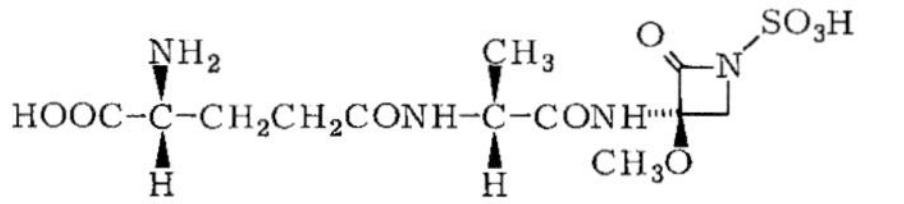

Sulfazecin [*77912-79-9*]

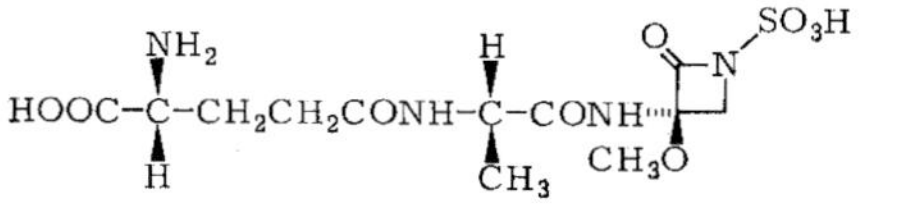

Isosulfazecin [*77900-75-5*]

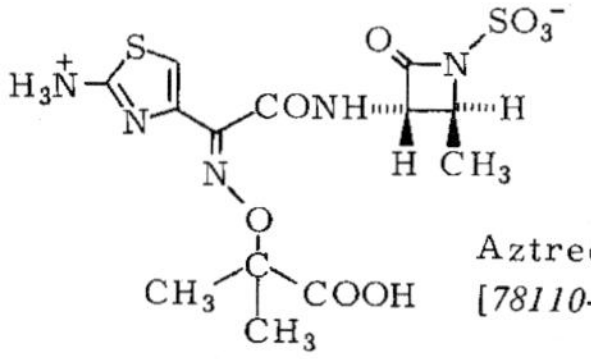

Aztreonam [*78110-38-0*]

3.2. Tetracyclines

The discovery of the tetracyclines, the first being aureomycin (7-chlorotetracycline), was one of the great successes of the worldwide screenings, i.e., testing of media samples and other materials, for the presence of antibiotic-producing microorganisms. This search began during the early 1940s and continues today.

The first patents and publications of Lederle Laboratories [85], marked the beginning of an extensive stream of publications and patents that reflect the medical, industrial, and economic importance of the tetracyclines [86], [87].

3.2.1. Structure and Properties

The linear four-ring-system skeleton is characteristic of the tetracyclines (**13**) and has given the whole group its name. The strongly conjugated system of keto and enol groups is of particular significance for the biological activity. The structures of the first tetracyclines were elucidated and proved by synthetic work, e.g., that of MUXFELDT et al., shortly after their discovery and parallel to their clinical testing and industrial development. The chief tetracyclines are listed under structure **13**.

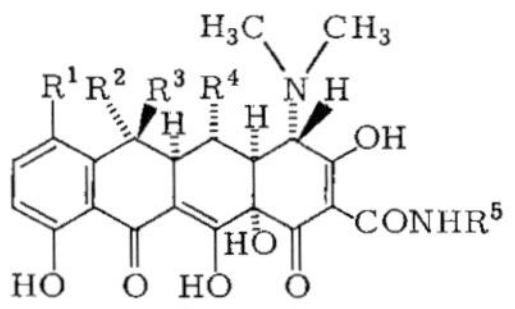

$$\text{13}$$

Structure 13: tetracycline core with substituents R^1, R^2, R^3, R^4 on ring A/B, $N(CH_3)_2$, OH, and $CONHR^5$ groups.

Name	Production	R^1	R^2	R^3	R^4	R^5
Tetracycline [6416-04-2]	*Streptomyces aureofaciens*	H	CH_3	OH	H	H
Chlortetracycline (aureomycin) [57-62-5]	*Streptomyces aureofaciens*	Cl	CH_3	OH	H	H
Demethylchlortetracycline (demeclocycline) [127-33-3]	*S. aureofaciens* *S. viridifaciens*	Cl	H	OH	H	H
Oxytetracycline [79-57-2]	*S. rimosus*	H	CH_3	OH	OH	H
Methacycline [914-00-1]	Semisynthetic	H	$=CH_2$		OH	H
Doxycycline [564-25-0]	Semisynthetic	H	CH_3	H	OH	H
Rolitetracycline [751-97-3]	Semisynthetic	H	CH_3	OH	H	$CH_2N\!\bigcirc$
Minocycline [10118-90-8]	Synthetic	$N(CH_3)_2$	H	H	H	H

The tetracyclines are bright yellow compounds, amphoteric, and with the exception of rolitetracycline and similarly constructed derivates insoluble in water at the isoelectric point. Their salts, e.g., hydrochlorides, are soluble in water and can be administered either parenterally or orally, although the low pH of the solution causes some problems in the latter instance.

3.2.2. Anhydrotetracyclines

Tetracyclines are aromatized in ring C by dehydration with concentrated acids, e.g., aqueous hydrochloric acid or anhydrous hydrogen chloride in acetone, to form anhydrotetracyclines (**14**, wherein R^1-R^5 are similar to those of **13**) [88]–[90]. These compounds have less biological activity than the starting compounds. Their formation, like that of the *epi*-anhydrotetracyclines, must be avoided because they are toxic to the kidneys. An interesting fact is that antibiotics with the anhydrotetracycline structure also are formed in nature and can be isolated from cultures of microorganisms, e.g., chelocardin from *Nocardia sulfurea* [91].

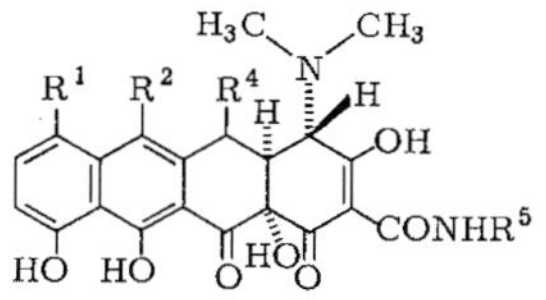

14

3.2.3. Anthracyclines

Structurally these antibiotics belong to the tetracyclines. They are characterized by the *p*-quinone structure of ring C in addition to the aromatic nuclei B and D (see **15**). Here $R^1 - R^5$ are simple substituents, such as H, OH, or CH_3. Only R^6 is a sugar or similarly complex group. Although anthracyclines show antibacterial activity, they have not been used as antibiotics because of their relatively high toxicity and strong side effects. The antitumor activity of rhodomycin was discovered by ARCAMONE et al. in 1961 [92], and various antitumor anthracyclines were subsequently isolated [93]. The most important anthracyclines are listed under structure **15**. Daunorubicin and doxorubicin are representative anthracyclines. Aclarubicin was found by UMEZAWA during a search for anthracyclines that might have lower cardiac toxicity than doxorubicin. Anthracyclines exert their effect by interacting with DNA, the primary cellular receptor [94].

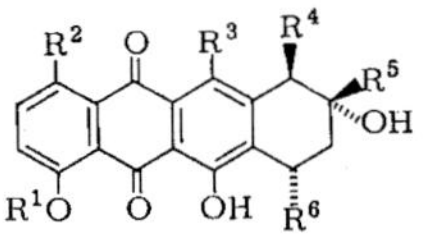

15

Name	R^1	R^2	R^3	R^4	R^5	R^6
Daunorubicin [20830-81-3]	CH_3	H	OH	H	$COCH_3$	A
Doxorubicin [23214-92-8]	CH_3	H	OH	H	$COCH_2OH$	A
Carminomycin I [50935-04-1]	H	H	OH	H	$COCH_3$	A
Baumycin A_2 [64253-71-0]	CH_3	H	OH	H	$COCH_3$	B (R = CH_2OH)
Baumycin B_2 [64312-53-4]	CH_3	H	OH	H	$COCH_3$	B (R = COOH)
Rhodomycin A [1404-50-8]	H	H	OH	C	CH_2CH_3	C
Rhodomycin B [1404-52-0]	H	H	OH	OH	CH_2CH_3	C
Aklavin [60504-57-6]	H	H	H	$COOCH_3$	CH_2CH_3	C
Cinerubin [34044-10-5]	H	OH	H	$COOCH_3$	CH_2CH_3	D
Aclarubicin [57576-44-0]	H	H	H	$COOCH_3$	CH_2CH_3	D

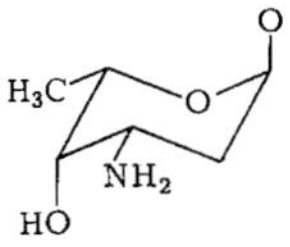

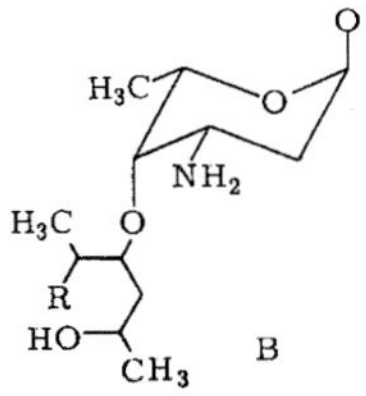

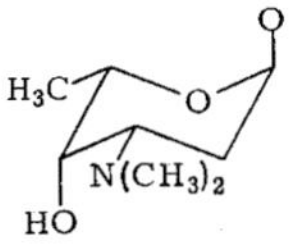

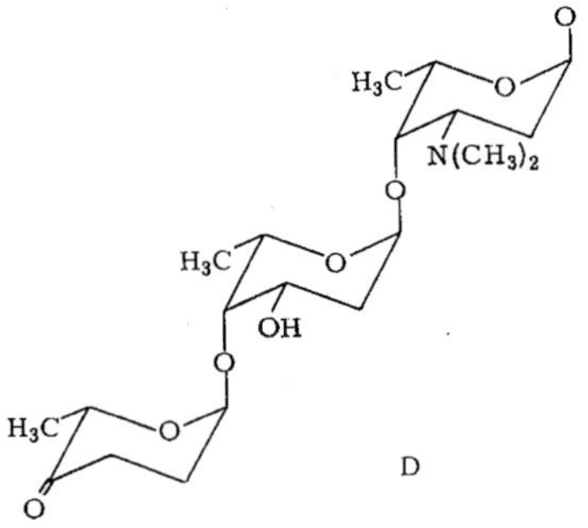

3.3. Aminoglycosides

WAKSMAN initiated the screening of antibiotics and, after finding actinomycin and streptothricin, he discovered the first useful aminoglycoside, streptomycin, in 1944 [95]–[99]. After wide use of penicillin, streptomycin, chloramphenicol, and tetracycline, resistant organisms appeared in hospital patients. In 1957, staphylococci and gram-negative organisms resistant to all the known antibiotic drugs caused serious infections; kanamycin was discovered at that time by UMEZAWA and was introduced clinically. However, in 1965, kanamycin-resistant strains appeared. In 1967, the enzymatic mechanism of resistance to aminoglycoside antibiotics was elucidated. UMEZAWA suggested that 3′-phosphotransferase and 6′-acetyltransferase, which transferred the terminal phosphate of ATP to the 3′-hydroxyl group of kanamycin, neomycin, and paromomycin or the acetyl group of acetyl-CoA to the 6′-amino group, were involved in the mechanism of resistance [100], [101]. In order to prove this enzymatic mechanism of resistance conclusively, 3′-deoxykanamycin A and 3′,4′-dideoxykanamycin B were synthesized and used to demonstrate the inhibition of the growth of resistant strains [102]. This conclusively demonstrated the enzymatic mechanism of resistance. Effective derivatives were obtained not only by deoxygenation but also by modification of the 1-amino group, which was involved in binding to the enzymes.

More than 150 naturally occurring aminoglycosides have been isolated from culture filtrates of *Streptomyces, Streptoverticillium, Nocardia, Micromonospora, Streptoalloteichus, Dactylosporangium, Saccharopolyspora,* and other bacterial strains. They can be devided into:

1) **Noncyclitol aminoglycosides**
    a) Monosaccharide derivatives
    3-amino-3-deoxy-D-glucose, nojirimycin, *N*-carbamoyl-D-glucosamine, streptozotocin, prumycin
    b) Disaccharides (trehalosamines)
    trehalosamine, mannosyl glucosaminide, 4-amino-4-deoxytrehalose
    c) Diaminosorbitol aminoglycosides (sorbistins)
    sorbistins A_1, A_2, B, D
    d) Glycocinnamoylspermidines
    LL-BM123β, γ_1, γ_2, glysperins A, B, C
2) **Aminoglycosides containing neutral cyclitols and monoaminocyclitols**
    kasugamycin, myomycins A, B, C, LL-BM782α_1, α_{1a}, α_2, minosaminomycin, LL-BM123α, validamycins A, B, C, D, E, F
3) **Aminoglycosides containing streptamine and related aminocyclitols**
    a) Streptidine aminoglycosides (streptomycins)
    streptomycin, mannosidostreptomycin, dihydrostreptomycin, hydroxystreptomycin (reticulin), *N*-demethylstreptomycin, mannosidohydroxystreptomycin, glebomycin (bluensomycin)

b) Actinamine aminoglycosides (spectinomycins)

spectinomycin (actinospectacin), dihydrospectinomycin

c) 4-Substituted deoxystreptamine aminoglycosides (neamines)

neamine, paromamine, nebramine (nebramycin 8), lividamine, NK-1003, seldomycin factor 2, gentamines C_1, $C_{1\alpha}$, C_2, apramycin (nebramycin 2), oxyapramycin (nebramycin 7)

d) 5-Substituted deoxystreptamine aminoglycosides (destomycins)

hygromycin B, destomycins A, B, C, A-396–1, SS-56-C, A16316-C

e) 4,5-Disubstituted deoxystreptamine aminoglycosides (neomycins)

neomycins B, C, LP-B, LP-C, paromomycins I, II, lividomycins A, B, mannosyl paromomycin, ribostamycin, xylostasin, ribosyl paromamine (LL-BM408α), butirosins A, B, BU-1709E_1, E_2, BU-1975C_1, C_2

f) 4,6-Disubstituted deoxystreptamine aminoglycosides (kanamycins)

kanamycins A, B, C, NK-1001, NK-1012–1, NK-1013–1, NK-1013–2, tobramycin (nebramycin 6), 6″-O-carbamoylkanamycin B (nebramycin 4), 6″-O-carbamoyltobramycin (nebramycin 5′), 2′-N-carbamoyltobramycin (nebramycin 11), 3″-deamino-3″-hydroxytobramycin (nebramycin 12), 6′-N-carbamoyltobramycin (nebramycin 13), gentamicins A, A_1, A_2, A_3, A_4, B, B_1, C_1, C_{1a}, C_2, C_{2a}, C_{2b} (sagamicin), X_2, JI-20A, JI-20B, G-418, I-1, 6′-C-methylgentamicin A (II-2), 6′-C-methylgentamicin A (III-1), VII-1, VII-3, VII-5, seldomycin factors 1, 3, 5, sisomicin, verdamicin, G-52, 66–40B, 66–40C, 66–40D, 3″-N-demethylsisomicin (66–40G)

4) **Aminoglycosides containing 1,4-diaminocyclitols**

a) Fortamine aminoglycosides (fortimicins)

fortimicins A, C, D, KG_3, 3-O-demethylfortimicin A, sporaricin A, 2″-N-carbamoylsporaricin A, 2″-N-formylsporaricin A, istamycins A (sannamycin A), A_1, A_2, B, B_1, C, C_1, dactimicin (SF-2052), 2″-N-formylfortimicin A

b) Non-glycine fortamine aminoglycosides

fortimicins B, E (AE, KH), AH, AI, AK (KI), AL, AM, AO, AP, AQ, AS, KE, KF, KG, KG_1, KG_2, KO_1, KQ, sporaricin B (KA-6606 II), KA-6606 V, VI, sannamycins B (KA-7038 II, istamycin A_0), C(KA-7038 VI), KA-7038 III, IV, V, VII, istamycin B_0, C_0

Kanamycins are produced by *Streptomyces kanamyceticus*, gentamicins by *Micromonospora purpurea*, and butirosins by *Bacillus circulans*. Many compounds analogous to aminoglycoside antibiotics are produced by the same strain. For example, more than 20 compounds structurally analogous to gentamicin have been isolated from a culture filtrate of a *Micromonospora* strain [103], [104].

Most aminoglycoside antibiotics that are important for chemotherapy contain 1,3- or 1,4-diaminocyclitols named actinamine, 2-deoxystreptamine, fortamine, or streptidine. Among these naturally occurring aminoglycosides, dihydrostreptomycin, kanamycin A, kanamycin B, lividomycin A, ribostamycin, sisomicin, spectinomycin, streptomycin, tobramycin, a mixture of gentamicins C_1, C_2, and C_{1a}, a mixture of neomycins B and C, and a mixture of paromomycins I and II are commercially available as chemotherapeutic agents useful in treating infections. Hygromycin B and destomycin A are used as animal anthelmintics. Kasugamycin and validamycin A are used for the prevention of plant diseases. Among resistant bacteria of clinical origin, the most important mechanism of resistance to aminoglycoside antibiotics is the inactivation by O-phosphorylation, O-nucleotidylation, or N-acetylation of specific sites of the antibiotic. The gene for these enzymes is located on a plasmid. Organisms with resistance resulting from permeability barriers to drugs have been isolated, but ribosomal resistance to aminoglycosides is very rare in organisms isolated clinically. Studies of the enzymatic mechanism of resistance to aminoglycosides have been reviewed extensively [105]–[108].

Semisynthetic Aminoglycosides have been made. Based on the enzymatic mechanism of resistance, studies of the chemical synthesis of derivatives that inhibit the growth of resistant strains have been initiated. 3'-Deoxykanamycin A has been synthesized and used to inhibit the growth of resistant strains having aminoglycoside-3'-phosphotransferase enzymes. Dibekacin (3',4'-dideoxykanamycin B), synthesized from kanamycin B, shows a strong activity not only against resistant staphylococci and gram-negative organisms but also against *Pseudomonas* [102]. These results prove the enzymatic mechanism of resistance.

Streptomycin, the first aminoglycoside antibiotic, was discovered by WAKSMAN. This drug is produced by *Streptomyces griseus* and extracted from the culture filtrate by adsorption on a column of Amberlite IRC-50 resin. The hydrogen chloride–calcium chloride (3 HCl · 1/2 CaCl$_2$) complex salt of streptomycin is easily crystallized from an anhydrous methanol solution. Streptomycin is also produced by several other strains: *Streptomyces bikiniensis, Streptomyces olivaceus, Streptomyces poonensis, Streptomyces mashuensis, Streptomyces galbus, Streptomyces rameus,* and *Streptomyces erythrochromogenes* subsp. *narutoensis*.

The early structural studies have been reviewed [109]. The two anomeric configurations were found to be α-L by application of Hudson's rules of isorotation and NMR spectral analysis. The absolute structure of streptomycin has been confirmed by X-ray analysis of its oxime selenate [110]. Streptomycin has been synthesized by oxidation of dihydrostreptomycin [111].

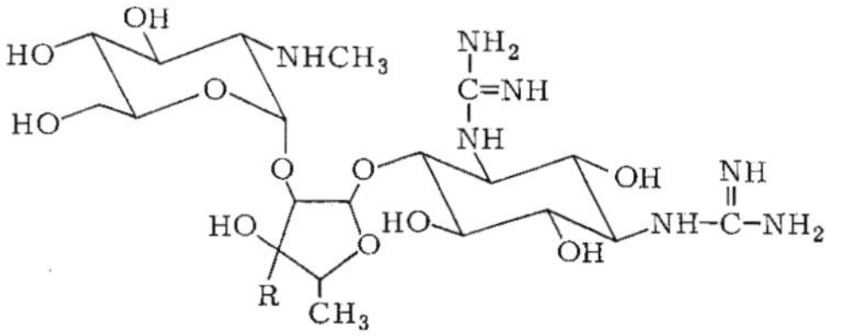

Streptomycin
[*57-92-1*]
R = CHO

Dihydrostreptomycin
[*128-46-1*]
R = CH$_2$OH

Spectinomycin (Actinospectacin, M-141) is produced by *Streptomyces spectabilis* and *Streptomyces flavopersicus*. It is also produced by *Streptomyces hygroscopicus* subsp. *sagamiensis*. Spectinomycin hexahydrate is crystallized from an aqueous acetone solution. This antibiotic is labile, especially in acidic solution.

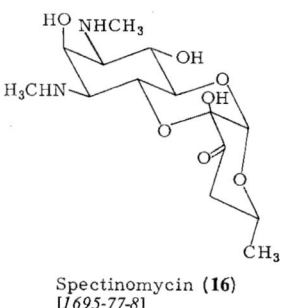

Spectinomycin (**16**)
[*1695-77-8*]

The structure of spectinomycin (**16**) was revealed by chemical studies, and its stereochemistry was determined by X-ray analysis of its dihydrobromide pentahydrate crystal [112]. Total synthesis of spectinomycin has been accomplished [113], [114].

Neomycin (Fradiomycin), a mixture of *Neomycins B and C*, is produced by *Streptomyces fradiae* and by *Streptomyces albogriseolus*. It is marketed as a mixture that contains 85–90% neomycin B [115]. Neomycins B and C are extremely stable in neutral or alkaline aqueous solution.

The final structure and stereochemistry of neomycins B and C were established in 1962. Neomycin C has also been synthesized also [116].

Paromomycin (Catenulin, Aminosidin, Hydroxymycin, Zygomycin A) is produced by *Streptomyces rimosus* subsp. *paromomycinus*; the structures of two isomers, paromomycins I and II, have been proposed. Paromomycin I is the main component of paromomycin preparations.

Catenulin, produced by *Streptomyces catenulae*; amminosidin, produced by *Streptomyces chrestomyceticus*; hydroxymycin, produced by *Streptomyces paucisporogenes*; and zygomycin A, produced by *Streptomyces pulveraceus*, are identical with paromomycin. Zygomycin A_1 is identical with paromomycin I and zygomycin A_2 is identical with paromomycin II.

The absolute configurations of the paromomycins have been determined along with those of other deoxystreptamine-containing aminoglycosides and dihydrostreptomycin [117], [118].

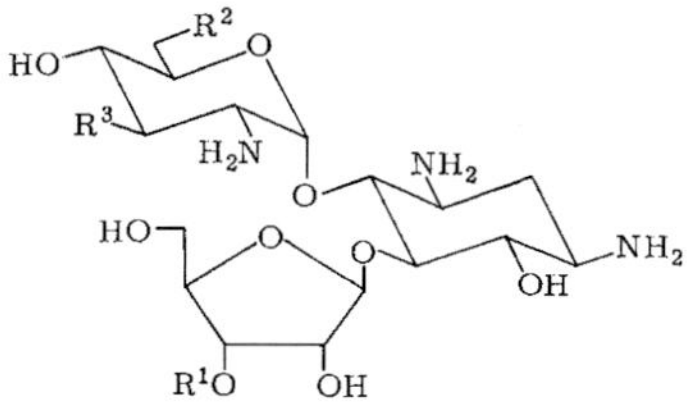

Name	R^1	R^2	R^3
Ribostamycin [25546-65-0]	H	NH$_2$	OH
Neomycin B [119-04-0]	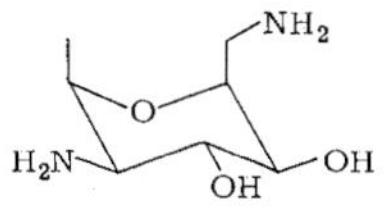 (2,6-diamino-2,6-dideoxy-hexose)	NH$_2$	OH
Paromomycin I [7542-37-2]	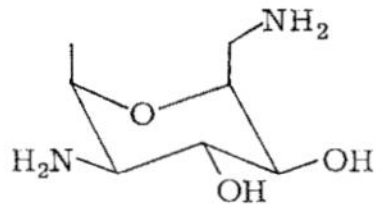 (6-amino-6-deoxy-hexose)	OH	OH
Lividomycin A [36441-41-5]	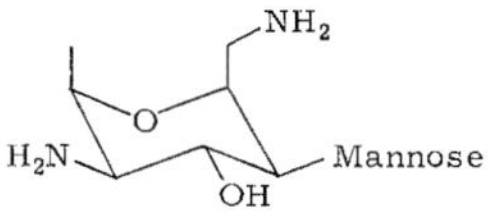 (aminosugar–Mannose)	OH	H

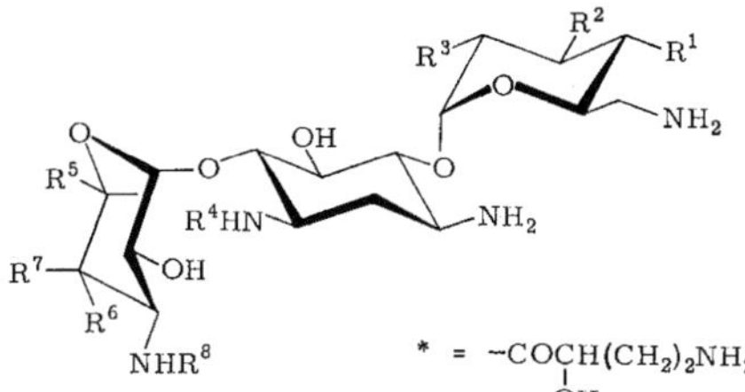

* = –COCH(OH)(CH$_2$)$_2$NH$_2$

Name	R^1	R^2	R^3	R^4	R^5	R^6	R^7	R^8
Kanamycin [59-01-8]	OH	OH	OH	H	CH$_2$OH	OH	H	H
Bekanamycin [4696-76-8]	OH	OH	NH$_2$	H	CH$_2$OH	OH	H	H
Tobramycin [32986-56-4]	OH	H	NH$_2$	H	CH$_2$OH	OH	H	H
Gentamicin C$_{1a}$ [26098-04-4]	H	H	NH$_2$	H	H	CH$_3$	OH	CH$_3$
Dibekacin [34493-98-6]	H	H	NH$_2$	H	CH$_2$OH	OH	H	H
Amikacin [37517-28-5]	OH	OH	OH	*	CH$_2$OH	OH	H	H

Classification of Antibiotics

Ribostamycin (SF-733) is produced by *Streptomyces ribosidificus*. The free base is crystallized from methanol solution. The structure has been determined by chemical methods and total synthesis has been undertaken [119].

Kanamycin A is produced by *Streptomyces kanamyceticus*. The monosulfate monohydrate and the free base are crystallized from aqueous methanol. Kanamycin A is extremely stable in neutral or alkaline aqueous solutions.

The stereochemical structure of kanamycin A has been confirmed by X-ray analysis of its monosulfate monohydrate and monoselenate monohydrate crystals [120]. Total synthesis of kanamycin A has been achieved [121].

Kanamycin B (Bekanamycin, Aminodeoxykanamycin) is one of the two minor components that have been isolated from the culture filtrate of kanamycin-producing *Streptomyces kanamyceticus*. Kanamycin B has a retention factor (R_f) of 0.37, whereas kanamycin A, the major component, shows R_f 0.21–0.26. (R_f is a measure of the relative mobilities of substance and solvent in a chromatographic system.) The free base kanamycin B is crystallized from aqueous *N,N*-dimethylformamide.

Gentamicin (Gentamicin C Complex), a mixture of *Gentamicins C_1, C_{1a}, and C_2*, is the antibiotic complex produced by *Micromonospora purpurea* and *Micromonospora echinospora*. Gentamicins C_1 and C_2 are the principal products, and C_1 is itself a mixture of two major components designated C_1 and C_{1a}. Gentamicin C complex, which consists of the mixture of C_1 and C_{1a} (60–80%) and of C_2 (20–40%), has been used as a chemotherapeutic agent. The structures and the stereochemistry of the gentamicin C components have been reported [122].

Sisomicin (66–40, Rickamicin) is the major antibiotic produced by *Micromonospora inyoensis*. Its structure has been elucidated [123].

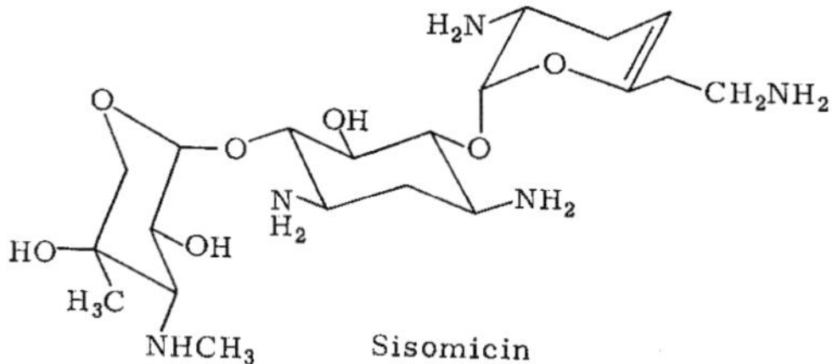

Sisomicin
[32385-11-8]

Dibekacin (3',4'-Dideoxykanamycin B, DKB) was the first drug developed on the basis of the enzymatic mechanism of resistance to aminoglycosides. Dibekacin is synthesized from kanamycin B by the application of the Tipson-Cohen deoxygenation method after selective N- and O-protections [102]. A modified synthetic route via the 3',4'-epoxy compound has given a high yield of more than 40% on an industrial scale [124].

Amikacin (1-N-[(S-)-4-Amino-2-hydroxy-butyryl]kanamycin, BB-K$_8$) is synthesized in 22% yield from the 6'-N-protected kanamycin by selective 1-N-acylation with N-protected (S)-4-amino-2-hydroxybutyric acid using the active ester method [125]. The most important point of this synthesis is the selective protection of all of the amino groups except the 1-amino group. The 3,6'-di-N-protected kanamycin has been obtained in 95% yield by selective N-protection using chelation with Co^{2+}, Ni^{2+}, and Cu^{2+} [126]. A new 3,6',3''-tri-N-protection method using the selective 3''-Ntrifluoroacetylation of the 3,6'-di-N-protected kanamycin with ethyl trifluoroacetate has been developed and the synthesis via 3,6'-di-N(benzyloxycarbonyl)–3''-N-(trifluoroacetyl)kanamycin gives amikacin in a yield of more than 60% [127].

The Fortimicin Group is a new type of deoxyaminoglycoside antibiotics, each consisting of glycine and a pseudodisaccharide, that have been found by screening. Fortimicin is produced by a *Micromonospora* species, sporaricin by *Saccharopolyspora hirsuta* subsp. *kobensis*, istamycin by *Streptomyces tenjimariensis*, sannamycin by *Streptomyces sannanensis*, and dactimicin by *Dactylosporangium matsuzakiense* [128]–[131], [128, p. 1061]. The structures of the fortimicin antibiotics are shown. These antibiotics strongly inhibit the growth of gram-positive and gram-negative bacteria but most *Pseudomonas* strains are resistant to them.

The axial amino group at C-1 in fortimicin A and istamycin A can be acetylated by aminoglycoside acetyltransferase(3)-I, but the equatorial amino group at C-1 in sporaricin A and istamycin B is scarcely acetylated. Other aminoglycoside-modifying enzymes that participate in the resistance to deoxystreptamine-containing aminoglycosides do not inactivate fortimicin-group antibiotics.

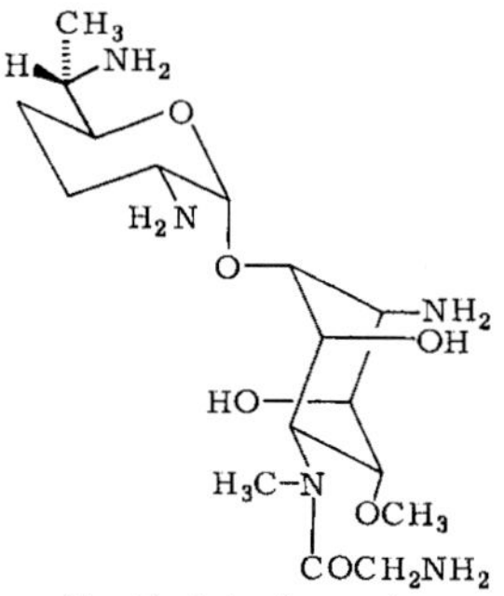

Fortimicin A
[55779-06-1]

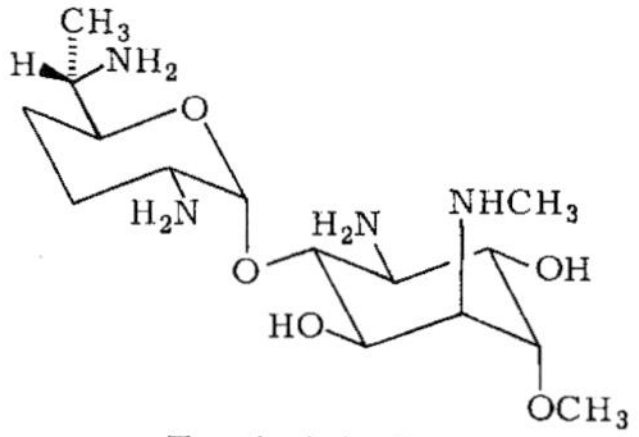

Fortimicin B
[54783-95-8]

3-O-Demethyl derivatives of sporaricin A and istamycin B exhibit good activity not only against gram-positive and gram-negative bacteria but also against most *Pseudomonas* strains. These derivatives will be developed as valuable chemotherapeutic agents in the near future.

3.4. Nucleosides

The biological effects associated with metabolic processes and specific enzyme control mechanisms are diverse in naturally occurring nucleosides and their synthetic analogs. Nucleosides exhibit several biological effects, including antibiotic, anticancer, and antiviral activities. They possess antimitotic and immunosuppressive activities and cardiovascular and other effects [132]–[134]. Moreover, it should be kept in mind that nucleoside analogs can assume other functional roles not as yet recognized, and that further therapeutic applications can be expected in the future. These analogs are obtained predominantly from microbial sources.

The nucleoside antibiotics consist of a heterocyclic base aglycone and a carbohydrate or a carbocyclic ring linked by a carbon–nitrogen (N-nucleoside) or a carbon–carbon bond (C-nucleoside). The nucleoside antibiotics fall somewhat outside the normal field of antibiotics with respect to their activity spectra and hence to their use. They are important for use against fungi, viruses, and certain types of cancer cells. Some typical nucleoside antibiotics are mentioned here.

3.4.1. N-Nucleosides

5-Azacytidine (**17**), a triazine analog of cytidine produced by *Streptoverticillium ladakanus*. It is active against some bacterial strains, Ehrlich ascitic tumor, leukemia L1210, and certain other leukemias. 5-Azacytidine also inhibits the DNA synthesis of bacteriophage T4 [135].

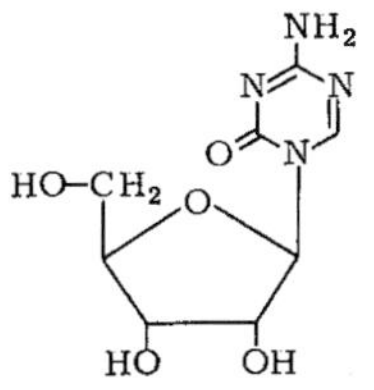

17
5-Azacytidine
[320-67-2]

Bredinin (**18**), produced by *Eupenicillium brefeldianum*, shows marked immunosuppressive activity in mice, interferes with replication of *Vaccinia* virus in vitro, and inhibits leukemia L 5178 cells and *Candida albicans* [136].

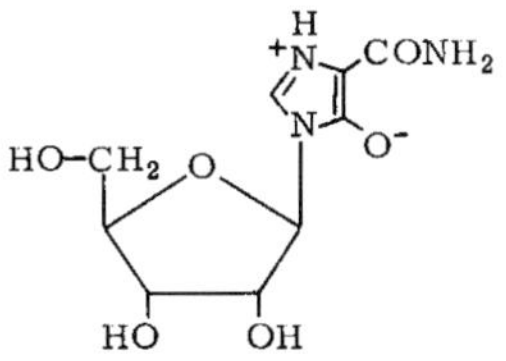

18
Bredinin
[50924-49-7]

Coformycin (**19a**) is isolated from *Nocardia interforma* along with formycin. Coformycin shows a synergistic effect with formycin on Yoshida rat sarcoma cells because of its strong inhibition of adenosine deaminase, which inactivates formycin. Coformycin, having a characteristic seven-membered ring base moiety, is thought to be a typical example of a "transition-state analog" in the adenosine deaminase reaction. 2'-Deoxycoformycin (**19b**) has also been isolated also [137]–[139].

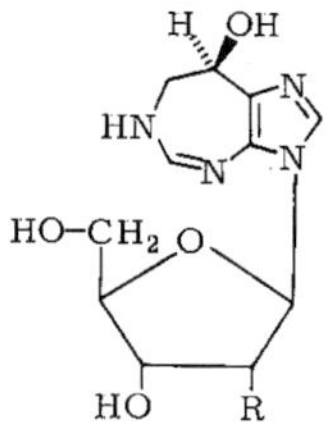

19a: R = OH
Coformycin
[11033-22-0]

19b: R = H
2′-Deoxycoformycin
[53910-25-1]

Cordycepin (20), 3′-deoxyadenosine, was one of the first nucleoside antibiotics isolated from *Cordyceps militaris*. It inhibits *Bacillus subtilis*, *Mycobacterium tuberculosis*, KB cell cultures, and Ehrlich ascites tumor cells.

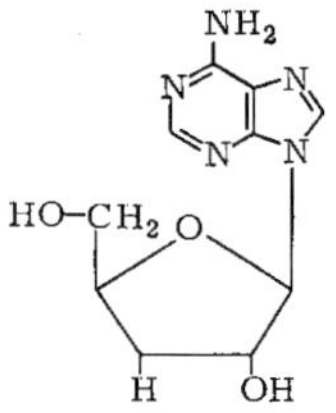

20
Cordycepin
[73-03-0]

Crotonoside (21), isolated from *Croton tiglium* seeds, acts as a vasopressor [140].

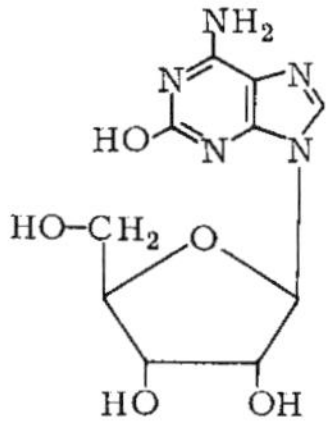

21
Crotonoside
[1818-71-9]

Nebularine (22), produced by the mushroom *Agaricus* (*Clitocybe*) *nebularis*, inhibits the growth of *Mycobacterium tuberculosis* and *Brucella abortus*, and is markedly cytotoxic to mammalian cells, whereas the purine base is relatively nontoxic. Nebularine is toxic to Sarcoma 180 cells [141], [142].

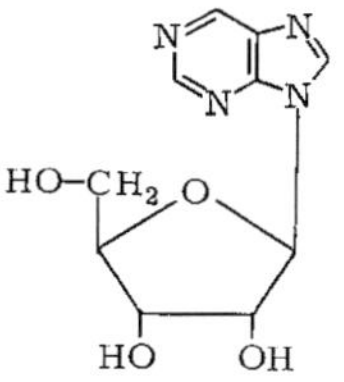

22
Nebularine
[550-33-4]

Toyocamycin (**23a**), produced by *Streptomyces toyocaensis*, strongly inhibits *Candida albicans*, *Trichophyton interdigitale*, and *Mycobacterium tuberculosis* and is also active against NF-sarcoma cells [143]. *Tubercidin* (**23b**) and *sangivamycin* (**23c**) also belong to this class [144].

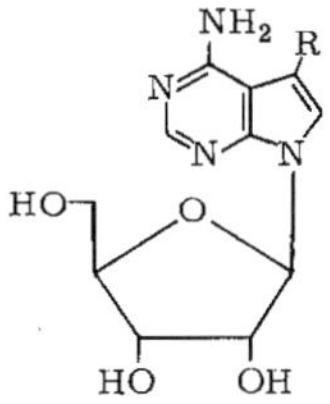

23a: R = CN   Toyocamycin [606-58-6]
23b: R = H   Tubercidin [69-33-0]
23c: R = CONH$_2$ Sangivamycin [18417-89-5]

Polyoxins A–O (**24**, **25**, **26**) are peptide nucleosides produced by *Streptomyces cacaoi*. They possess various heterocycles, e.g., uracil, thymine, 5-hydroxymethyluracil, uracil-5-carboxylic acid, or formylimidazolone. The compounds **24** and **26** are particularly active against sheath-blight in rice, *Pellicularia sasakii*, and are widely used as agricultural drugs [145]–[147].

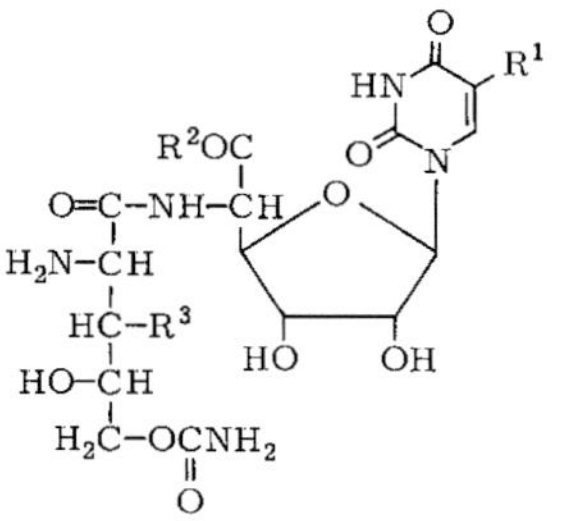

24

Name	R¹	R²	R³
Polyoxin A [19396-03-3]	CH_2OH	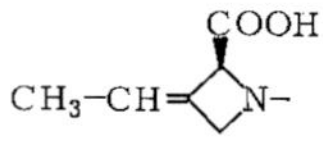 $CH_3-CH=\!\!\!<\!\!\!\begin{array}{c}COOH\\N-\end{array}$	OH
Polyoxin B [19396-06-6]	CH_2OH	HO	OH
Polyoxin D [22976-86-9]	COOH	HO	OH
Polyoxin E [22976-87-0]	COOH	HO	H
Polyoxin F [23116-76-9]	COOH	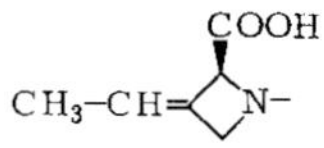 $CH_3-CH=\!\!\!<\!\!\!\begin{array}{c}COOH\\N-\end{array}$	OH
Polyoxin G [22976-88-1]	CH_2OH	HO	H
Polyoxin H [24695-54-3]	CH_3	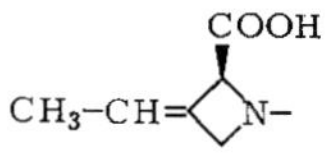 $CH_3-CH=\!\!\!<\!\!\!\begin{array}{c}COOH\\N-\end{array}$	OH
Polyoxin J [22976-89-2]	CH_3	HO	OH
Polyoxin K [22886-46-0]	H	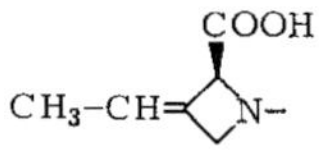 $CH_3-CH=\!\!\!<\!\!\!\begin{array}{c}COOH\\N-\end{array}$	OH
Polyoxin L [22976-90-5]	H	HO	OH
Polyoxin M [34718-88-2]	H	HO	H

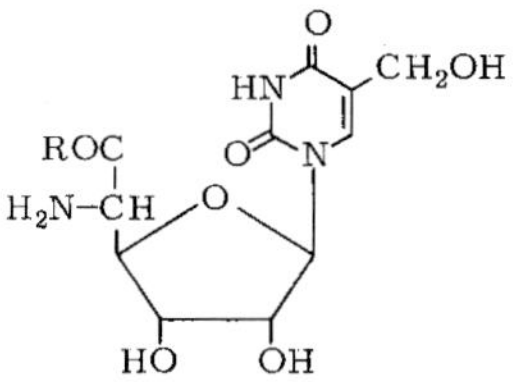

25

Polyoxin C [21027-33-8]   R = OH

Polyoxin I [22886-33-5]   R = (2-carboxymethylidene-azetidinyl group with CH₃)

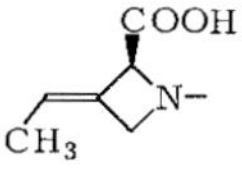

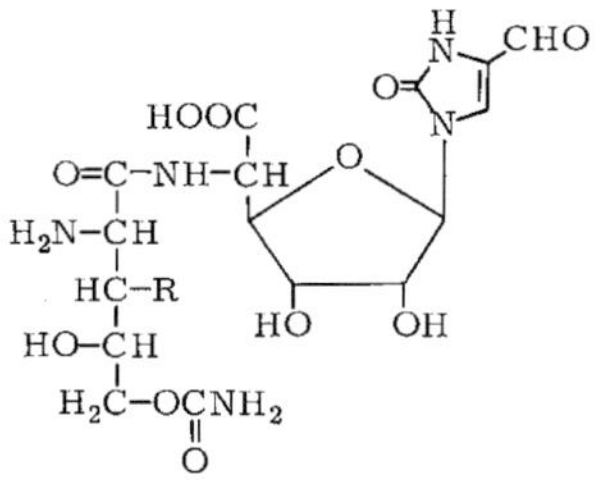

26

Polyoxin N [37362-29-1]   R = OH

Polyoxin O [37362-28-0]   R = H

3.4.2. C-Nucleosides

Formycin (27) (Formycin A) is isolated from *Nocardia interforma* and from *Streptomyces lavendulae* [148], [149]. The antibiotic is effective against *Xanthomonas oryzae* and *Pellicularia filamentosa*. Its activity against Yoshida rat sarcoma cell is enhanced by coformycin. *Formycin B* (**28**) inhibits *Xanthomonas oryzae* and interferes with multiplication of influenza A virus in the cells of chick chorioallantoic membrane [150]. *Oxoformycin B* (**29**) shows no activity against *Xanthomonas oryzae* [151].

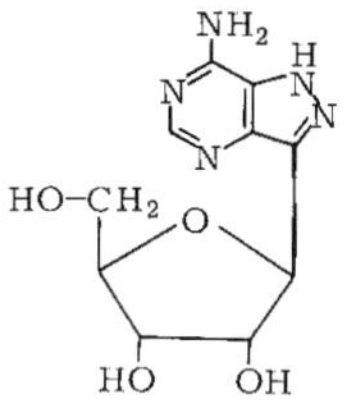

27
Formycin
[6742-12-7]

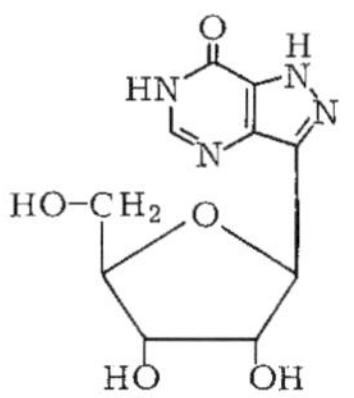

28
Formycin B
[13877-76-4]

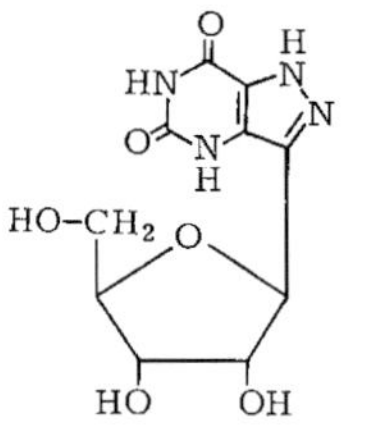

29
Oxoformycin B
[19246-88-9]

Showdomycin (**30**), isolated from *Streptomyces showdoensis*, is very active against *Streptococcus hemolyticus*. It is moderately active against other gram-positive and gram-negative bacteria and also effective against Ehrlich ascites tumor in mice and HeLa cells [152]. *Oxazinomycin* (**31**) belongs to this class of nucleosides [153].

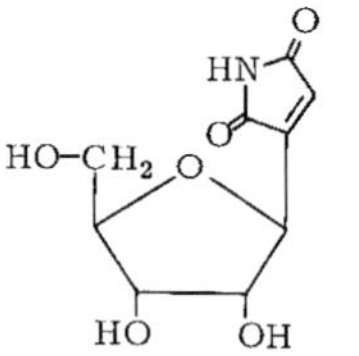

30
Showdomycin
[16755-07-0]

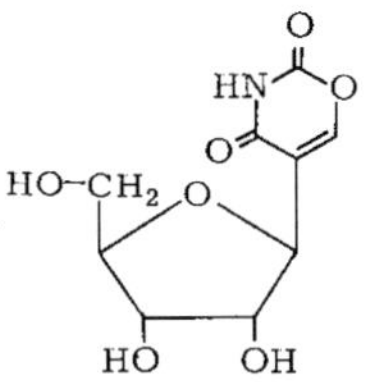

31
Oxazinomycin
[32388-21-9]

3.4.3. Carbocyclic Nucleosides

Since the pioneering synthesis of the racemic carbocyclic analog of adenosine by SHEALY and CLAYTON and the subsequent isolation of *aristeromycin* (**32**) from *Streptomyces citricolor*, the interest in this class of compounds has been renewed by the isolation of a new carbocyclic nucleoside, *neplanocin A* (**33**). The latter exhibits remarkable antitumor activity against L 1210 leukemia in mice, and its synthetic analogs are now being studied extensively [154]–[156].

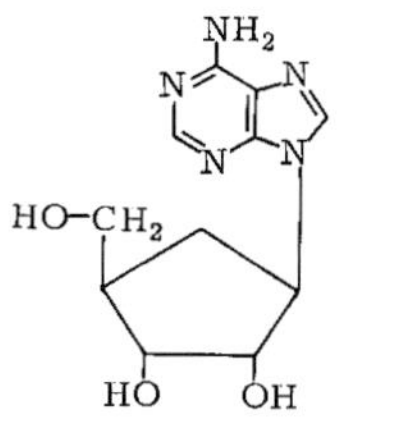

32
Aristeromycin
[19186-33-5]

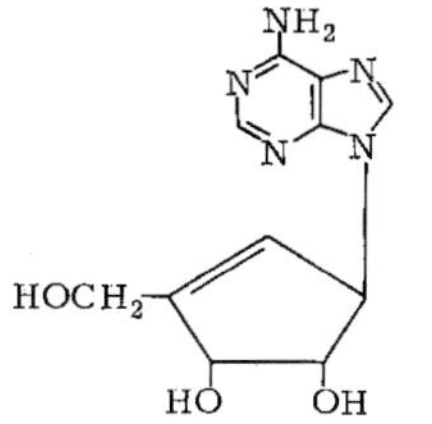

33
Neplanocin A
[72877-50-0]

3.4.4. An Exceptional Nucleoside

Blasticidin S (34) has a pyran ring as the sugar moiety and inhibits various gram-positive and gram-negative bacteria. It is particularly effective against *Pericularia oryzae* and is now used as an agricultural drug [157].

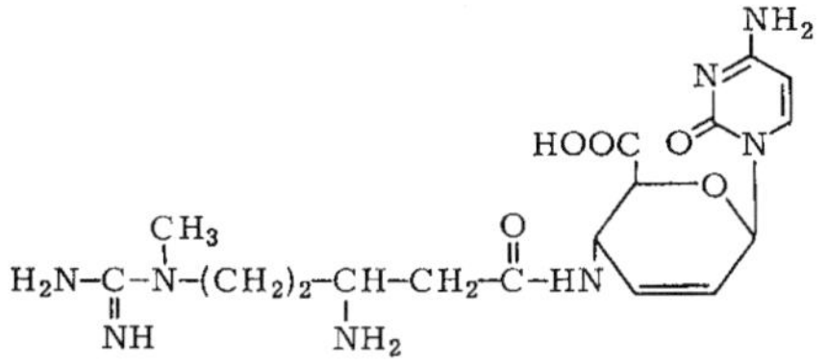

34
Blasticidin S
[2079-00-7]

3.5. Macrolides

This group of antibiotics is characterized by excellent antibacterial activity, particularly against gram-positive bacteria. Macrolides can be defined and distinguished from the other groups of antibiotics by the unique feature of their chemical structure. They are polyfunctional macrocyclic lactones and the majority of them contain at least one amino sugar moiety, which is the cause of the basicity of the molecules. Neutral macrolides containing only a neutral sugar moiety are also known. Recently these antibiotics have become targets in the aldol strategy of organic synthesis to construct their polyhydroxy functions stereoselectively [158], [159]. The antibiotics are classified as either 12-, 14-, or 16-membered ring macrolides according to the ring size of the aglycone.

3.5.1. 12-Membered Ring Macrolides

Methymycin (**35**), produced by *Streptomyces venezuelae*, was first shown to be a 12-membered lactone, comprising the aglycone or methynolide and D-desosamine [160], [161]. *Neomethymycin* (**36**) has an isomeric structure.

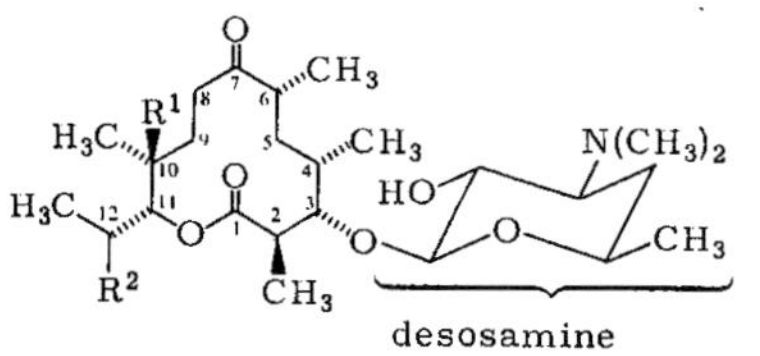

Methymycin (**35**) R^1 = OH, R^2 = H
[*497-72-3*]
Neomethymycin (**36**) R^1 = H, R^2 = OH
[*497-73-4*]

3.5.2. 14-Membered Ring Macrolides

The erythromycins, produced by *Streptomyces erythreus*, are clinically important macrolides and are the most widely investigated 14-membered ring macrolides [162]. Extensive chemical and X-ray crystallographic studies of *erythromycin A* (**37**) have established its structure as well as those of its minor components, *erythromycin B* (**38**), *erythromycin C* (**39**), and *erythromycin D* (**40**). Erythromycin is effective against streptococcal and pneumococcal infections. Derivatives of erythromycin, modified in the cladinose ring, the desosamine ring, and the aglycone moiety (especially at C-9 of the aglycone), have been described. Their characterization contributed greatly to understanding the chemistry and structure–activity relationships of the macrolide antibiotics [163].

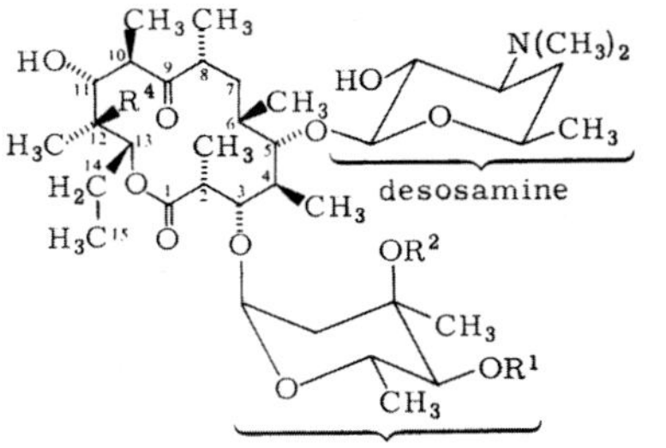

cladinose: R¹ = H, R² = CH₃
mycarose: R¹ = R² = H

Erythromycin A (**37**) [*114-07-8*]
  R¹ = R³ = H, R² = CH₃, R⁴ = OH
Erythromycin B (**38**) [*527-75-3*]
  R¹ = R³ = R⁴ = H, R² = CH₃
Erythromycin C (**39**) [*1675-02-1*]
  R¹ = R² = R³ = H, R⁴ = OH
Erythromycin D (**40**) [*33442-56-7*]
  R¹ = R² = R³ = R⁴ = H

Picromycin (**41**), narbomycin (**42**), and oleandomycin (**43**) belong to the 14-membered lactones.

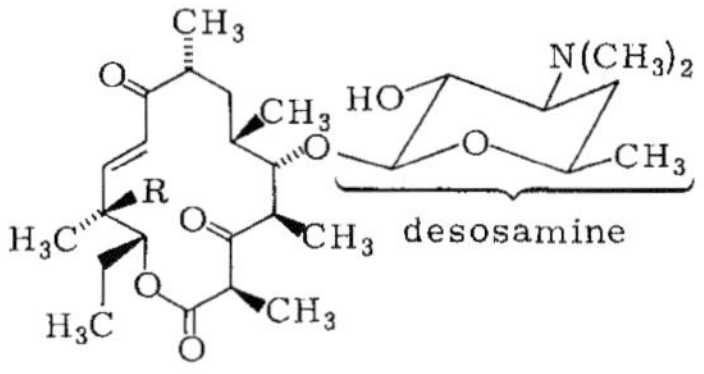

Picromycin (**41**) [*19721-56-3*] R = OH
Narbomycin (**42**) R = H

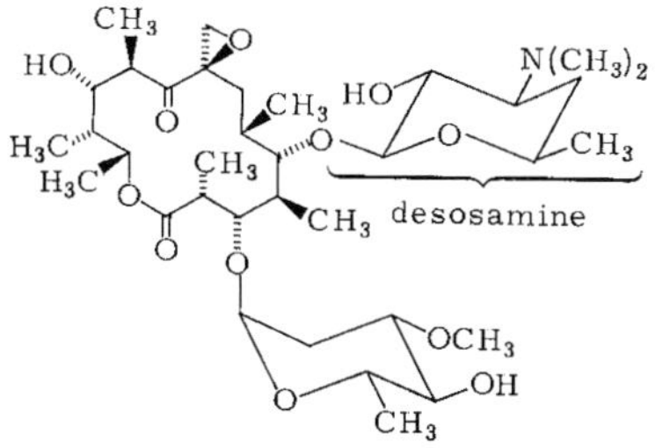

Oleandomycin (**43**) [*3922-90-5*]

3.5.3. 16-Membered Ring Macrolides

Stereochemical structures of carbomycins, leucomycins, spiramycins, and other macrolides have been disclosed by extensive chemical studies [164], [165]. These compounds have a formylmethyl group at C-6 and two conjugated double bonds in the 16-membered lactone ring. They differ only in the nature of the acyl substituents at C-3 and C-4. For instance, the leucomycin complex (kitasamycin), produced by *Streptomyces kitasatoensis* [166] is a mixture of ten similar components. These are *leucomycin A_1* (**44**, *turimycin H_5*), *leucomycin A_3* (**45**, *josamycin*, YL-704A_3 or *platenomycin A_3*), *leucomycin A_4* (**46**), *leucomycin A_5* (**47**), *leucomycin A_6* (**48**, YL-704B_3 or *platenomycin B_3*), *leucomycin A_7* (**49**), *leucomycin A_8* (**50**), *leucomycin A_9* (**51**), *leucomycin U* (**52**), and *leucomycin V* (**53**) [164], [167].

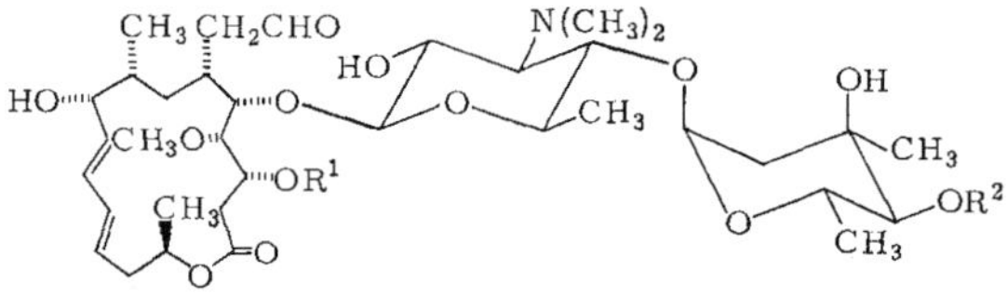

Leucomycin A_1 (Turimycin H_5) (**44**) R^1 = H, R^2 = COCH$_2$CH(CH$_3$)$_2$
[16846-34-7]

Leucomycin A_3 (Josamycin, YL-704 A_3 Turimycin A_5) (**45**) R^1 = COCH$_3$, R^2 = COCH$_2$CH(CH$_3$)$_2$
[16846-24-5]

Leucomycin A_4 (**46**) R^1 = COCH$_3$, R^2 = COCH$_2$CH$_2$CH$_3$
[18361-46-1]

Leucomycin A_5 (**47**) R^1 = H, R^2 = COCH$_2$CH$_2$CH$_3$
[18361-45-0]

Leucomycin A_6 (YL-704 B_3) (**48**) R^1 = COCH$_3$, R^2 = COCH$_2$CH$_3$
[18361-48-3]

Leucomycin A_7 (**49**) R^1 = H, R^2 = COCH$_2$CH$_3$
[18361-47-2]

Leucomycin A_8 (**50**) R^1 = COCH$_3$, R^2 = COCH$_3$
[18361-50-7]

Leucomycin A_9 (**51**) R^1 = H, R^2 = COCH$_3$
[18361-49-4]

Leucomycin U (**52**) R^1 = COCH$_3$, R^2 = H
[31642-61-2]

Leucomycin V (**53**) R^1 = H, R^2 = H
[22875-15-6]

The spiramycins (foromacidins) produced by *Streptomyces ambofaciens* [168] have been separated into three components, namely *spiramycin I* (**54**), *spiramycin II* (**55**), and *spiramycin III* (**56**).

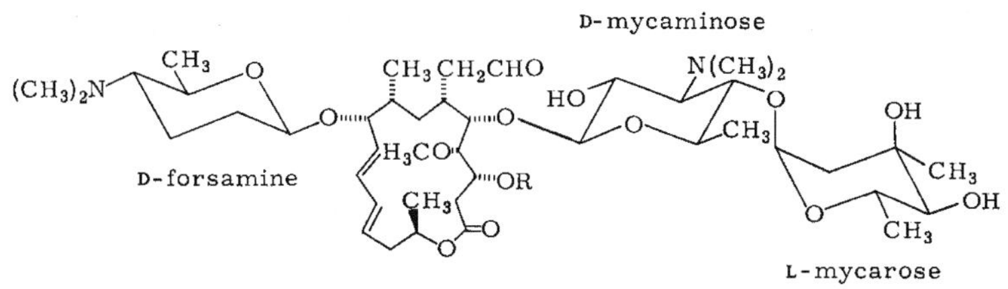

Spiramycin I (Foromacidin A) (**54**): R = H
[24916-50-5]

Spiramycin II (Foromacidin B) (**55**): R = COCH$_3$
[24916-51-6]

Spiramycin III (Foromacidin C) (**56**): R = COCH$_2$CH$_3$
[24916-52-7]

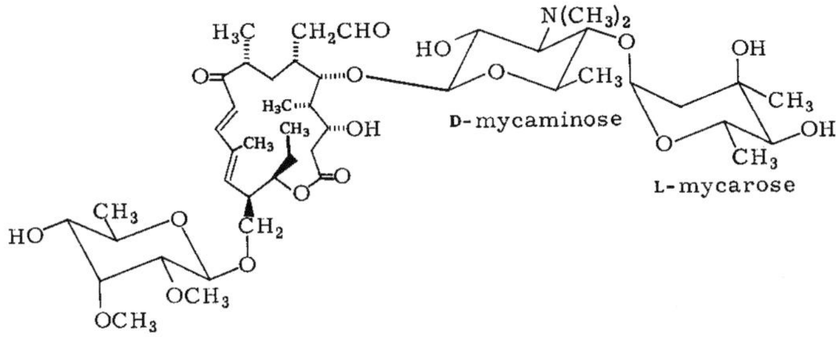

Tylosin (**57**)
[1401-69-0]

Many other macrolides, such as *rosamicin* [169], [170], *cirvamycin A$_1$* [171], [172], *juvenimicin A$_2$*, *juvenimicin A$_4$* [173], [174], *deltamycin* [175], *carbomycin* [176], [177], *angolamycin* [178], *tylosin* (**57**) [179]–[181], and the *mycinamicins* [182], [183], belong to this class.

Since 1950, the structures of more than 90 macrolides have been elucidated, and this knowledge has had a great impact on modern organic synthesis. The classification of the above-mentioned macrolides according to ring size is useful, but very schematic. There is a marked change in the antibiotic activity spectrum corresponding to changes in the ring size, even though many of the substituents and the degree of unsaturation differ considerably.

Macrolides with ca. 10–16 ring members are very strongly antibacterial, as are the smaller lactones. Because there is no cross-resistance among these macrolides, they are mainly used to treat bacterial infections that are resistant to other antibiotics.

Macrolides with ca. 16–40 ring members show very little activity against bacteria, often none at all, but they are highly effective against fungi, yeasts, etc.

The very large macrolides are highly effective against not only fungi, but also viruses and tumors.

3.6. Ansamycins

The ansamycins are a clinically important class of antibiotics with a characteristic structure. They have an aliphatic "ansa" bridge that connects two nonadjacent positions of an aromatic system [184]. The name "ansamycin" is based on the term "ansa compounds" [185]. Ansamycins are classified into two groups based on the nature of the aromatic moiety, i.e., benzoquinoid and naphthoquinoid ansamycins. Geldanamycin and maytansinoids belong to the benzoquinoid ansamycins and are studied as potential antitumor agents. The naphthoquinoid ansamycins include rifamycins, tolypomycins, streptovaricins, halomicins, and naphthomycin. The naphthoquinoid ansamycins are the major group of known ansamycins.

The rifamycins, produced by *Nocardia mediterranei*, have great therapeutic value [186], [188]. Their chemistry is very similar to that of the macrolides. After many attempts to separate, isolate, and purify the naturally occurring rifamycins, *rifamycin B* (**58**), *rifamycin O* (**59**), and *rifamycin S* (**60**) were found among the fermentation products. Rifamycin B was moderately effective against gram-positive bacteria. The oxidation of rifamycin B gave rifamycin O, which can be hydrolyzed to the more active rifamycin S. The latter can be reduced to *rifamycin SV* (**61**) using ascorbic acid. Rifamycin SV is converted, through its formyl derivative (**62**), to the therapeutically important *rifampicin* (**63**). In order to obtain an antibiotic with a broader spectrum and good oral absorption characteristics, thousands of derivatives of rifampicin have been prepared. Rifamycins B, O, and S all are used as starting materials for the modifications. The rifampicins have strong biological activity against gram-positive microorganisms and mycobacteria, particularly *Mycobacterium tuberculosis* [189]–[191]. Most of the ansamycins are weakly active against viruses, and certain derivatives, such as *3-formylrifamycin S* (**62**), have been found to be active against certain tumors. No cross-resistance of the ansamycins to most of the other antibiotics has been observed.

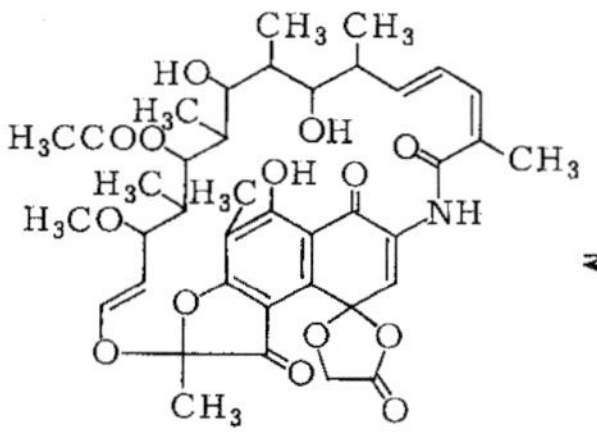

Rifamycin O (**59**)
[*14487-05-9*]

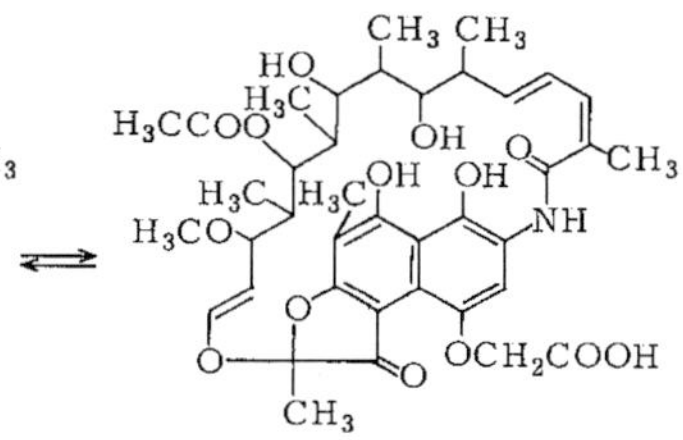

Rifamycin B (**58**)
[*13929-35-6*]

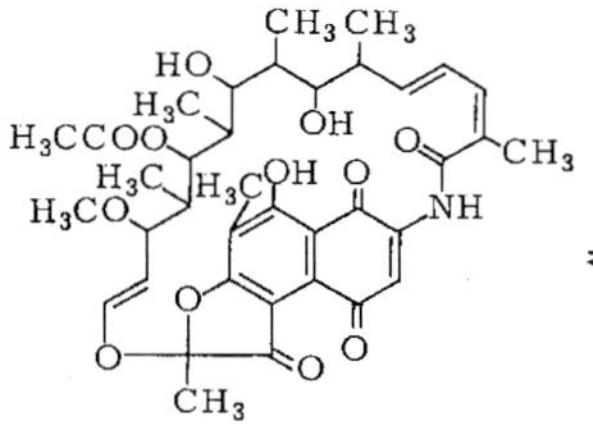

Rifamycin S (**60**)
[*13553-79-2*]

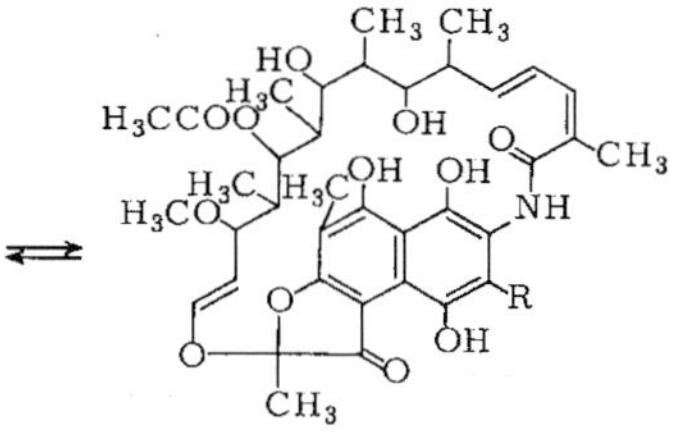

Rifamycin SV (**61**)
[*6998-60-3*]
R = H

3-Formylrifamycin S (**62**)
[*13292-22-3*]
R = CHO

Rifampicin (**63**)
[*13292-46-1*]
R = CH=N-N N-CH$_3$

3.7. Peptides

The various important functions of the living organism are frequently mediated by oligopeptides and proteins, which exist in the most diverse structures. It is therefore not surprising that a large number of low molecular mass peptides, oligopeptides, and protein-like substances are found among the antibiotics of microbial origin. Although peptide antibiotics consist of amino acids linked by peptide bonds, they differ from the proteins and peptides of higher animals and plants in many respects [192]. The following characteristics frequently are found in the peptide antibiotics:

1) Molecular masses of the antibiotics are smaller (in the range of 500–1500) than those of peptide hormones, which are frequently much larger.
2) The antibiotics contain some uncommon amino acids that are not found in proteins and peptide hormones of animal or plant origin. The usual amino acids are infrequently detected or are found in modified forms.

3) Lipids and other moieties not of amino acid character are found in many peptide antibiotics.
4) The peptide antibiotics frequently contain D-amino acid residues, whereas peptides of plant and animal origin consist solely of L-amino acid residues.
5) Virtually all of the peptide antibiotics resist hydrolysis by proteolytic enzymes, which are otherwise effective in hydrolyzing peptides of plant and animal origin.
6) The antibiotics are often cyclic peptides.
7) Families of closely related peptide antibiotics are frequently produced by the same microorganism.

3.7.1. The Bleomycin Group

The bleomycins, a group of glycopeptide antibiotics produced by *Streptomyces verticillus* [193], make up one of the most widely used groups of antitumor antibiotics, effective against squamous cell carcinoma and malignant lymphoma. Extensive degradation studies have shown the main structural features to be a peptide containing unusual amino acid residues and a disaccharide of uncommon sugars. The complete structure has been elucidated by chemical studies and X-ray crystallographic analysis of P-3A, a biosynthetic intermediate structurally related to bleomycin [194], [195]. This structure has been verified by the total synthesis of *bleomycin A_2* (**64**). The naturally occurring bleomycins are obtained in copper-chelated form as a mixture of congeners that differ only in the substituents at the C terminus of *bleomycinic acid* (**65**), the common structural unit. Among these are *bleomycin A_1* (**66**), *demethylbleomycin A_2* (**67**), *bleomycin A_2* (**64**), *bleomycin A_2' ^{a}* (**68**), *bleomycin A_2' ^{b}* (**69**), *bleomycin A_2' ^{c}* (**70**), *bleomycin A_5* (**71**), *bleomycin A_6* (**72**), *bleomycin B_1* (**73**), *bleomycin B_2* (**74**), and *bleomycin B_4* (**75**). Metal-free bleomycin can be prepared by treatment with hydrogen sulfide. A mixture of metal-free bleomycins consisting mainly of A_2 (55–70%) and B_2 (25–32%) has been used for clinical treatment because the mixture has an effect superior to that of A_2 alone on human squamous cell carcinoma. The copper ion in bleomycin is replaced by iron after penetration into cells. A bleomycin–iron complex that exerts antitumor activity is formed. More than 300 bleomycin analogs have been prepared by chemical modifications or fermentations. *Pepleomycin* (**76**), *possessing improved properties, has recently been brought into clinical use* [196] *and the tallysomycins A and B*, **77** and **78**, respectively, are similar glycopeptides [197], [198].

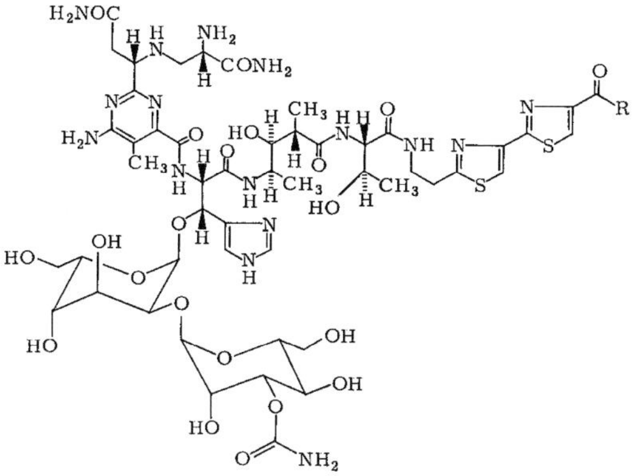

Bleomycinic acid (**64**) [*37364-66-2*]: R = OH

Bleomycin A$_1$ (**65**) [*58995-26-9*]: R = NH(CH$_2$)$_3$S̈CH$_3$ (O above S)

Bleomycin demethyl A$_2$ (**66**) [*41089-03-6*]: R = NH(CH$_2$)$_3$SCH$_3$
Bleomycin A$_2$ (**67**) [*11116-31-7*]: R = NH(CH$_2$)$_3$S$^+$(CH$_3$)$_2$, X$^-$
Bleomycin A$_{2'}$-a (**68**) [*73666-81-6*]: R = NH(CH$_2$)$_4$NH$_2$
Bleomycin A$_{2'}$-b (**69**) [*41138-53-8*]: R = NH(CH$_2$)$_3$NH$_2$

Bleomycin A$_{2'}$-c (**70**) [*62960-69-4*]: R = NH(CH$_2$)$_2$-(imidazole)

Bleomycin A$_5$ (**71**) [*11116-32-8*]: R = NH(CH$_2$)$_3$NH(CH$_2$)$_4$NH$_2$
Bleomycin A$_6$ (**72**) [*37293-17-7*]: R = NH(CH$_2$)$_3$NH(CH$_2$)$_4$NH(CH$_2$)$_3$NH$_2$
Bleomycin B$_1$ (**73**) [*41138-54-9*]: R = NH$_2$

Bleomycin B$_2$ (**74**) [*9060-10-0*]: R = NH(CH$_2$)$_4$NHC(=NH)NH$_2$

Bleomycin B$_4$ (**75**) [*9060-11-1*]: R = NH(CH$_2$)$_4$NHC(=NH)NH(CH$_2$)$_4$NHC(=NH)NH$_2$

Pepleomycin (**76**) [*68247-85-8*]: R = NH(CH$_2$)$_3$NH-CH(CH$_3$)-C$_6$H$_5$

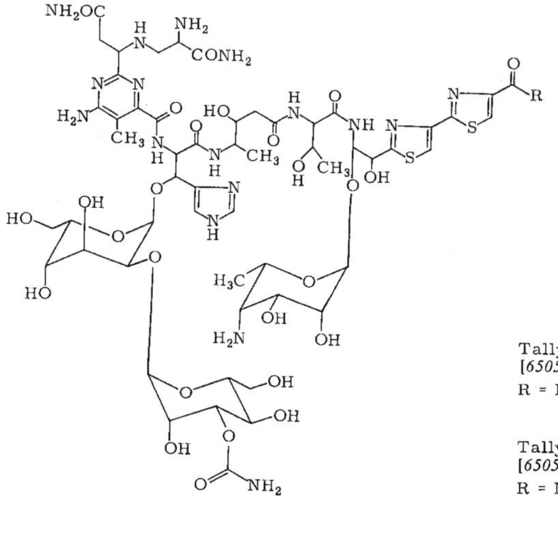

Tallysomycin A (**77**)
[*65057-90-1*]
R = NH(CH$_2$)$_3$CH(NH$_2$)CH$_2$CONH(CH$_2$)$_3$NH(CH$_2$)$_4$NH$_2$

Tallysomycin B (**78**)
[*65057-91-2*]
R = NH(CH$_2$)$_3$NH(CH$_2$)$_4$NH$_2$

3.7.2. The Gramicidin Group

Generally speaking, peptide antibiotics with simple straight-chain structures play a minor role and are mentioned here only for review purposes. However, peptide antibiotics with cyclic structures, which often contain D-amino acids, are of considerable importance. The usual nomenclature is used here for the constituents of the amino acids: Ala = alanine; Asp = asparagine; Cys = cysteine; Dab = α,γ-diaminobutyric acid; Glu = glutamic acid; Gly = glycine; His = histidine; Ileu = isoleucine; Leu = leucine; Lys = lysine; Orn = ornithine; Phe = phenylalanine; Pro = proline; Thr = threonine; Trp = tryptophan; Tyr = tyrosine; Val = valine; Meval = *N*-methylvaline; Sar = sarcosine; Ser = serine; MOA = 6-methyloctanoic acid; IOA = isooctanoic acid. The arrow indicates the direction of the amide bond between the amino acids. Thus, the arrow begins where the carbonyl group attaches and ends where the amino group attaches (– CO – NH →).

The gramicidin group consists of a series of open chain peptides and a series of cyclic peptides. Clinically important *gramicidin S* (**79**) was isolated from a strain of *Bacillus brevis* [199].

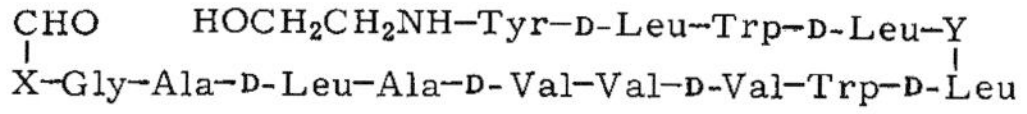

```
CHO      HOCH₂CH₂NH-Tyr-D-Leu-Trp-D-Leu-Y
 |                                      |
 X-Gly-Ala-D-Leu-Ala-D-Val-Val-D-Val-Trp-D-Leu
```

	X	Y
Valine-gramicidin A	Val	Trp
Isoleucine-gramicidin A	Ile	Trp
Valine-gramicidin B	Val	Phe
Isoleucine-gramicidin B	Ile	Phe
Valine-gramicidin C	Val	Tyr
Isoleucine-gramicidin C	Ile	Tyr

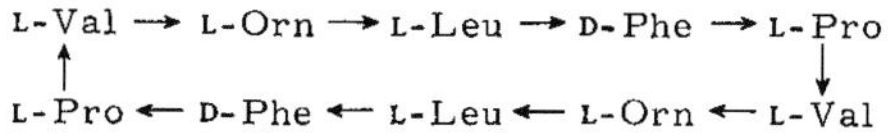

```
L-Val → L-Orn → L-Leu → D-Phe → L-Pro
  ↑                              ↓
L-Pro ← D-Phe ← L-Leu ← L-Orn ← L-Val
```
Gramicidin S (**79**)
[*113-73-5*]

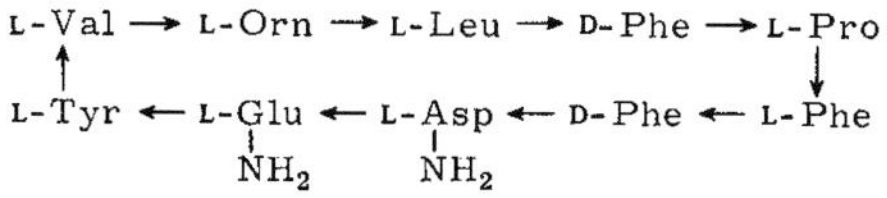

```
L-Val → L-Orn → L-Leu → D-Phe → L-Pro
  ↑                              ↓
L-Tyr ← L-Glu ← L-Asp ← D-Phe ← L-Phe
         |       |
        NH₂     NH₂
```
Tyrocidine A
[*1481-70-5*]

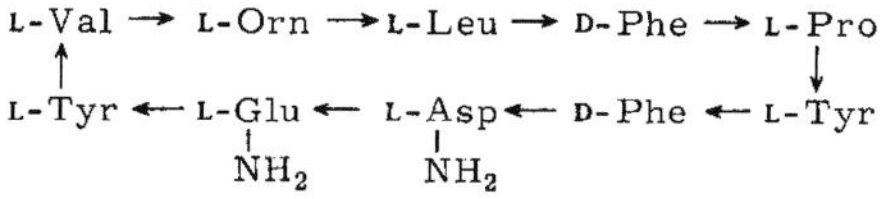

```
L-Val → L-Orn → L-Leu → D-Phe → L-Pro
  ↑                              ↓
L-Tyr ← L-Glu ← L-Asp ← D-Phe ← L-Tyr
         |       |
        NH₂     NH₂
```
Tyrocidine B
[*865-28-1*]

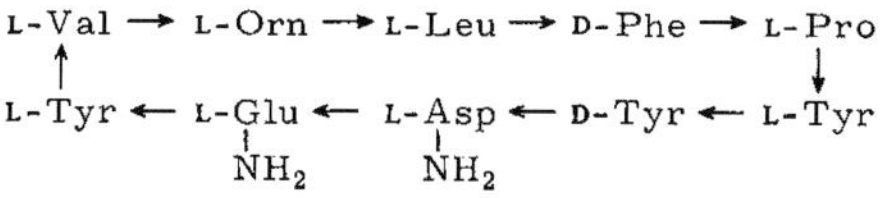

```
L-Val → L-Orn → L-Leu → D-Phe → L-Pro
  ↑                              ↓
L-Tyr ← L-Glu ← L-Asp ← D-Tyr ← L-Tyr
         |       |
        NH₂     NH₂
```
Tyrocidine C
[*3252-29-7*]

3.7.3. The Polymyxins

Among the polymyxins, *polymyxins B, D, and E* are of particular importance. The *colistins A and B* correspond to *polymyxins E_1 and E_2*, respectively, as shown in **80** [200]. These are active only against gram-negative bacteria.

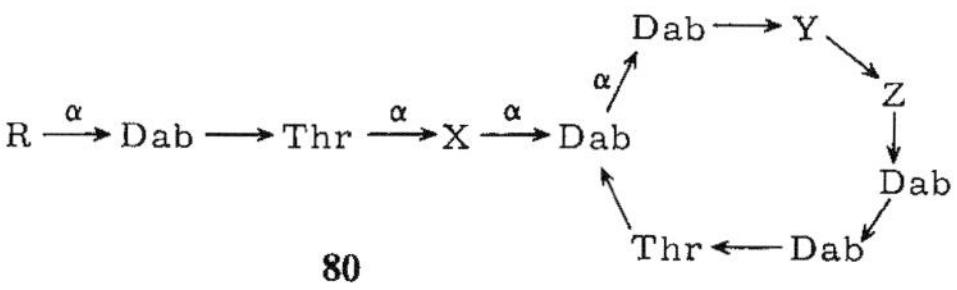

```
                          Dab → Y
                         ↗       ↘
R →α Dab → Thr →α X →α Dab        Z
                         ↑        ↓
                         Thr ← Dab
    80
```

3.7.4. The Bacitracins

Bacitracin, produced by *Bacillus subtilis* [201] and *Bacillus licheniformis* [202], is active against gram-positive bacteria.

	R	X	Y	Z
Polymyxin A₁ (M1) [65454-50-4]	MOA	D-Dab	D-Leu	Thr
Polymyxin A₂ (M2) [65454-51-5]	IOA	D-Dab	D-Leu	Thr
Polymyxin B₁ [4135-11-9]	MOA	Dab	D-Phe	Leu
Polymyxin B₂ [34503-87-2]	IOA	Dab	D-Phe	Leu
Polymyxin D₁ [10072-50-1]	MOA	D-Ser	D-Leu	Thr
Polymyxin D₂ [34167-45-8]	IOA	D-Ser	D-Leu	Thr
Polymyxin E₁ (colistin A) [7722-44-3]	MOA	Dab	D-Leu	Leu
Polymyxin E₂ (colistin B) [7239-48-7]	IOA	Dab	D-Leu	Leu

Dab = α,γ-diaminobutyric acid; MOA = (+)-6-methyloctanoic acid; IOA = isooctanoic acid

After extensive study on the chemistry of bacitracins, a revised structure was proposed for *bacitracin A* (**81**) [203].

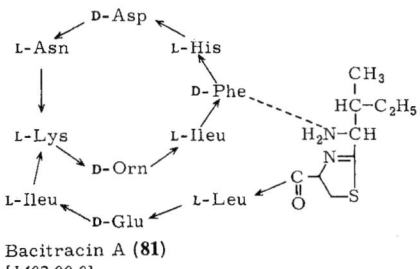

Bacitracin A (**81**)
[1402-99-9]

3.7.5. Large-Ring Peptide Antibiotics Containing Lactone Linkages

This important group can be classified with macrolide antibiotics because it contains a lactone moiety, but it is more properly regarded as a group of cyclic peptides. These peptides generally contain about 16- to 19-membered rings, including many amide bonds and lactone linkages. There are no C–C double bonds in the ring.

3.7.6. The Actinomycin Group

These cyclic peptide antibiotics are also known as chromopeptide antibiotics. The *actinomycins* (**82**), first isolated by WAKSMAN in 1940 from *Streptomyces antibioticus*, are of interest because they have found use in therapy of tumors, particularly Wilm's tumor.

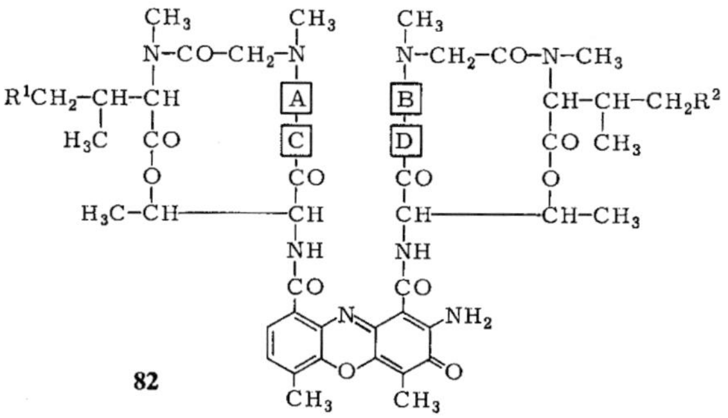

Name	A	B	C	D	R^1	R^2
Actinomycin C_2 [2612-14-8]	L-Pro	L-Pro	D-Val	D-allo-Ile	H	H
Actinomycin C_{2a} (VI) [17914-41-9]	L-Pro	L-Pro	D-allo-Ile	D-Val	H	H
Actinomycin C_3 (VII) [6156-47-4]	L-Pro	L-Pro	D-all-Ile	D-allo-Ile	H	H
Actinomycin D (C_1, IV) [50-76-0]	L-Pro	L-Pro	D-Val	D-Val	H	H
Actinomycin E_1 [1402-41-1]	L-Pro	L-Pro	D-allo-Ile	D-allo-Ile	H	CH_3
Actinomycin E_2 [1402-42-2]	L-Pro	L-Pro	D-allo-Ile	D-allo-Ile	CH_3	CH_3
Actinomycin F_3 [1402-46-6]	Sar	Sar	D-allo-Ile	D-allo-Ile	H	H
Actinomycin F_8 (II, A_{II}) [32934-48-8]	Sar	Sar	D-Val	D-Val	H	H
Actinomycin F_9 (III, A_{III}) [60469-11-8]	L-Pro	Sar	D-Val	D-Val	H	H
Actinomycin $X_{0\beta}$ (I) [1402-60-4]	L-Pro	Pro(4-OH)	D-Val	D-Val	H	H
Actinomycin X_2 (V, A_V, B_V) [1402-61-5]	L-Pro	Pro(4-O)	D-Val	D-Val	H	H

They are also used as a biochemical tool because of their specific inhibition of DNA-primed RNA synthesis. Structurally they have two pentapeptide lactone rings attached to a phenoxazinone chromophore. Several of the actinomycins isolated were found to be identical and the nomenclature is confusing. For example, actinomycin D and actinomycin C_3 are also known as dactinomycin and cactinomycin, respectively [204].

3.7.7. Other Peptide Antibiotics

There are many peptide antibiotics derived from *Streptomyces* species. Among these are amphomycin, capreomycin, distamycin, the enduracidins, mikamycin, neocarzinostatin (antitumor), stendomycin, viomycin, and virginiamycin.

3.8. Other Important Antibiotics and Intermediates

Cycloheximide (83) is produced by *Streptomyces griseus* [205]. It is highly effective against fungi and is therefore used mainly for plant protection. The analogs streptovitacin, naramycin B, and streptimidone have also been isolated, and also are called glutarimide antibiotics because of their common structural moiety.

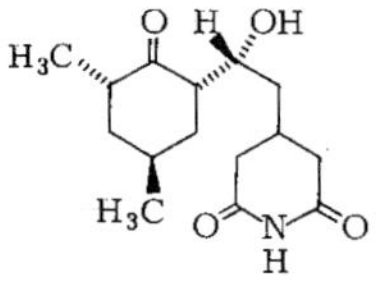

Cycloheximide (**83**)
[*66-81-9*]

Cycloserine (**84**) (*oxymycin, seromycin, orientomycin*) is the simplest antibiotic, D-4-amino-3-isoxazolidine, isolated from many *Streptomyces* species [206]. Cycloserine is now produced only synthetically and used particularly for tuberculosis of the lungs and for leprosy with *p*-aminosalicylic acid (PAS) or isonicotinic acid hydrazide (INH).

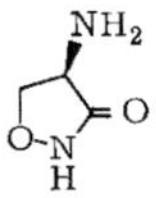

Cycloserine (**84**)
[*68-41-7*]

Variotin (85) (*pecilocin*) is produced by *Paecilomyces varioti* [207]. It is an oily, ester-like substance with an aromatic odor and is particularly active against fungi. Variotin is used against trichophytes.

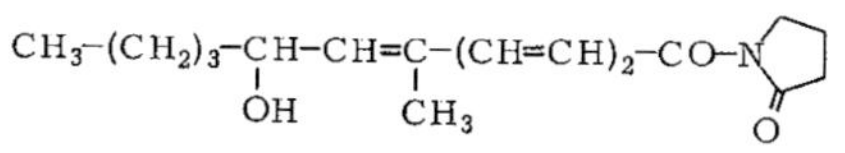

CH₃-(CH₂)₃-CH-CH=C-(CH=CH)₂-CO-N
         | |
         OH   CH₃

Variotin (**85**)
[*19504-77-9*]

Sarkomycin A (**86**) is produced by *Streptomyces erythrochromogenes* [208]. It is active not only as an antibiotic but also as an antitumor agent.

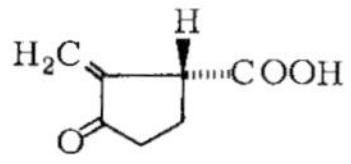

Sarkomycin (**86**)
[*489-21-4*]

Novobiocin, isolated from a culture filtrate of *Streptomyces nieveus,* has the structure **87**, consisting of the aglycone, novobiocic acid, and 3-*O*-carbamoylnoviose. The monosodium or calcium salt is used in therapy and is active mainly against gram-positive bacteria [209].

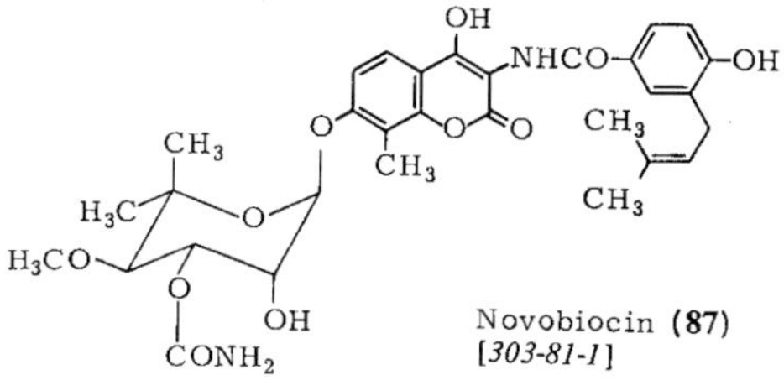

Novobiocin (**87**)
[*303-81-1*]

Griseofulvin (**88**) is produced by *Penicillium griseofulvum, P. janczewskii,* and *Nigrospora oryzae*. It is unique in possessing the spirocarbon moiety [210], [211]. Griseofulvin is very active against fungi, and it is used orally to treat fungal infections of human skin.

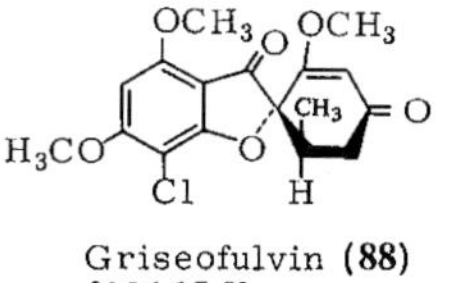

Griseofulvin (**88**)
[*126-07-8*]

Chloramphenicol (**89**), the first of the so-called broad-spectrum medicinal antibiotics, was originally obtained from *Streptomyces venezuelae* in 1947 [212]. It is now manufactured by a chemical process, and the parent compound and its esters are commercially available. Chloramphenicol is active against rickettsia, chlamydiae, and mycoplasmas, as well as a wide range of gram-positive and gram-negative bacteria. However, use is limited by the risk of bone marrow damage or aplastic anemia at too high or too prolonged an application [213].

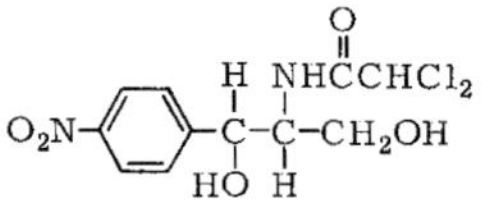

Chloramphenicol (**89**)
[*56-75-7*]

Mitomycins ((**90**)–(**93**)) are a group of unique chemical structures in which three different carcinostatic functions — aziridine, carbamate, and quinone — are arranged around a pyrro[1,2-*a*]indole nucleus [214]. The first mitomycins were discovered in 1956 by HATA in a culture filtrate of *Streptomyces caespitosus*. These compounds, designated mitomycins A and B, show highly potent antibacterial activity and moderate antitumor activity, but they are quite toxic in mice. In 1958, mitomycin C, an extremely valuable antitumor drug, was isolated from *Streptomyces caespitosus* [215], [216]. In 1960, another mitomycin, porfiromycin (**93**), was isolated from *Streptomyces ardus*.

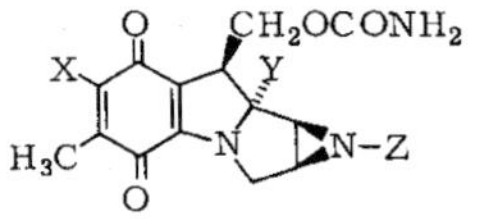

Name	X	Y	Z
Mitomycin A (**90**) [*4055-39-4*]	OCH_3	OCH_3	H
Mitomycin B (**91**) [*4055-40-7*]	OCH_3	OH	CH_3
Mitomycin C (**92**) [*50-07-7*]	NH_2	OCH_3	H
Porfiromycin (**93**) [*801-52-5*]	NH_2	OCH_3	CH_3

Fumagillin (**94**) is a useful polyene antibiotic; feeding it to honeybees with natural *Nosema apis* infections suppressed the disease and led to considerably increased honey production [217].

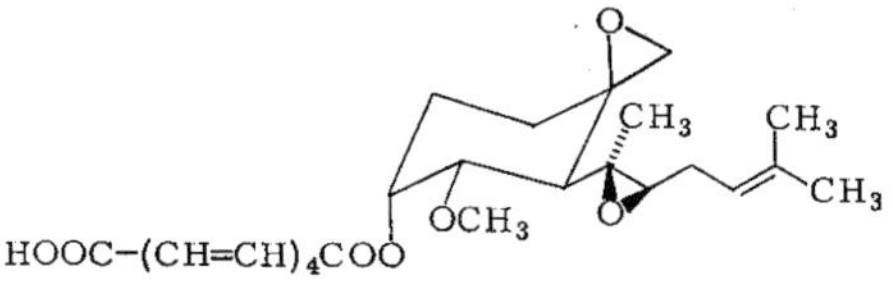

Fumagillin (**94**)
[*23110-15-8*]

Monensins (**95**) are useful polyether antibiotics that control infections of *Coccidia*. They are particularly important to the poultry industry [218].

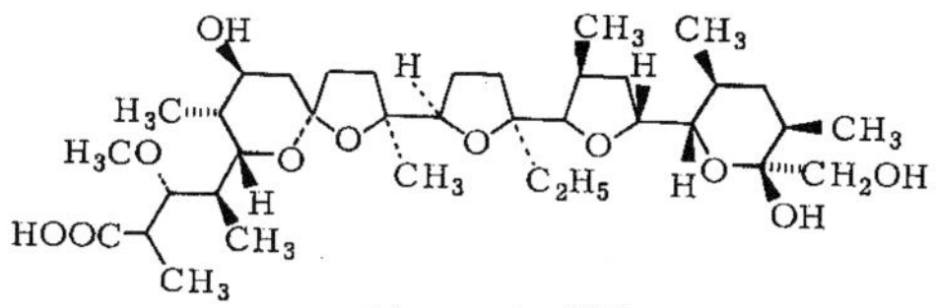

Monensin (**95**)
[*17090-79-8*]

Pyrrolnitrin (**96**), isolated from *Pseudomonas* species, is highly active against fungi, particularly trichophyte species.

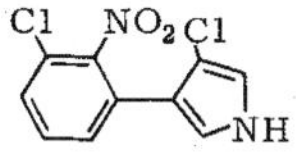

Pyrrolnitrin (**96**)
[*1018-71-9*]

Fosfomycin (**97**) is unique in possessing a simple epoxide ring and has a broad activity spectrum against gram-positive and gram-negative bacteria [219].

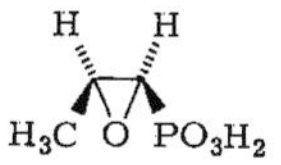

Fosfomycin (**97**)
[*23155-02-4*]

Fusidic acid (**98**) has a steroidal skeleton, but it is markedly different from the usual steroid hormones in biological activity. Fusidic acid has been isolated from *Fusidium coccineum* and is particularly active against *Staphylococcus, Clostridium, Neisserias, Corynebacterium diphtheriae,* and *Mycobacterium tuberculosis* [220].

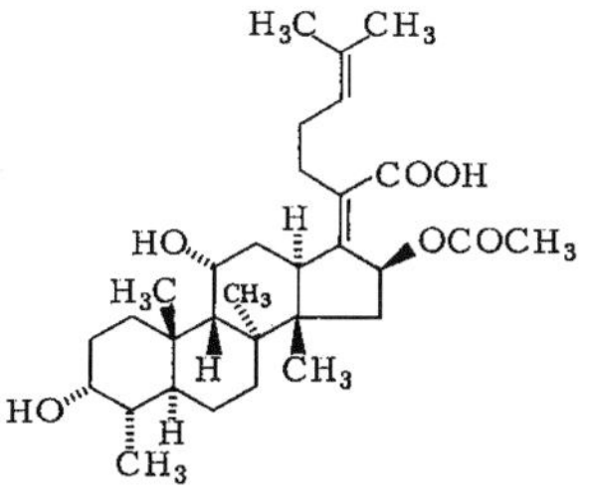

Fusidic acid (**98**)
[*6990-06-3*]

D-(*p*-Hydroxyphenyl)glycine (D-HPG) is widely used in large amounts as an important intermediate for the synthesis of amoxicillin and several other semisynthetic β-lactam antibiotics.

Industrially, D-HPG has been produced by resolving the racemic DL-HPG obtained by the usual nonenzymatic synthesis. The optical resolution by means of fractional crystallization of the corresponding diastereomeric salt or by the predominant crystallization of the corresponding aromatic sulfonic acid salt are typical methods. However, these methods suffer disadvantages. The former requires multistep reactions and an expensive resolving agent.

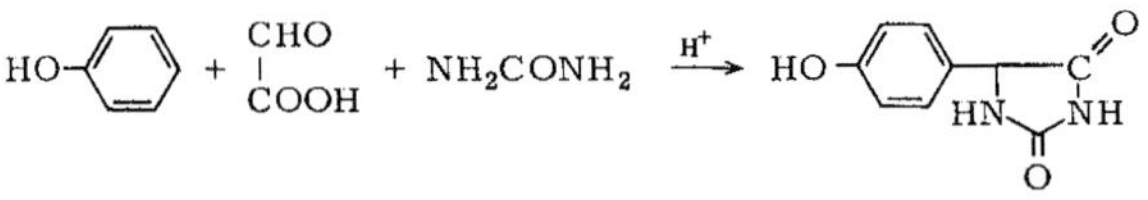

$$HO-\langle\bigcirc\rangle- + \underset{COOH}{\overset{CHO}{|}} + NH_2CONH_2 \xrightarrow{H^+} HO-\langle\bigcirc\rangle-\underset{HN\diagdown_{C}\diagup NH}{\overset{H}{\underset{|}{C}}}-\overset{O}{\underset{\|}{C}}$$

DL-Hydantoin

$$\xrightarrow[\text{Hydantoinase}]{\text{Asymmetric hydrolysis}} HO-\langle\bigcirc\rangle-\underset{NHCONH_2}{\overset{H}{\underset{|}{C}}}-COOH \xrightarrow{HNO_2} HO-\langle\bigcirc\rangle-\underset{NH_2}{\overset{H}{\underset{|}{C}}}-COOH$$

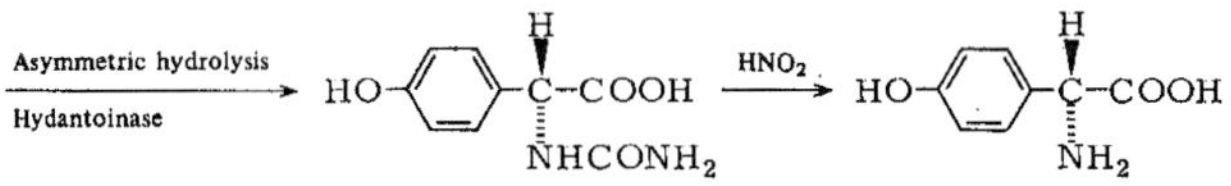

N-Carbamoyl-D-p-hydroxyphenylglycine
D-p-Hydroxyphenylglycine
$[\alpha]_D^{20} = -156°$ (1% solution in NH_2Cl)

The optical purity of the product is not very high. The yield obtained from one cycle of the latter resolution method is very low, and both methods require racemization and recyclization steps in the optical resolution process in order to increase the total yield of D-HPG.

The process developed by YAMADA et al. involves subjecting DL-5-(p-hydroxyphenyl)hydantoin (DL-HPH) to an enzymatic asymmetric hydrolysis utilizing microorganisms [221]. A new method of synthesizing DL-HPH has also been developed [222]. Thus, a high-purity D-HPG is produced industrially as shown in the sequence at the top of this page.

The substrate (DL-HPH) for the enzymatic reaction is synthesized starting with relatively inexpensive basic materials: phenol, glyoxylic acid, and urea. The enzymatic reaction hydrolyzes the hydantoin stereospecifically to afford the acid with the D configuration. The remaining substrate with the L configuration is automatically racemized under enzymatic reaction conditions. Consequently, DL-HPH can be quantitatively transformed into the desired acid, D-HPG. In this manner a very effective optical resolution is conducted kinetically. The D-N-carbamoyl(p-hydroxyphenyl)glycine thus otained is easily converted to D-HPG by treatment with nitrous acid.

As has been mentioned, the process for preparing D-HPG has been greatly simplified and provides D-HPG of very high optical purity when the enzymatic reaction is used.

This novel process has been developed industrially by Kanegafuchi Chemical Industry Company.

D-Phenylglycine is widely used in large quantity as the side chain that makes ampicillin and cephalexin orally acceptable. Industrially, D-phenylglycine has been produced by a conventional resolution method or by kinetic resolution with acylase. Recently, another enzymatic procedure has been developed by YAMADA et al. [221], [222] as shown below.

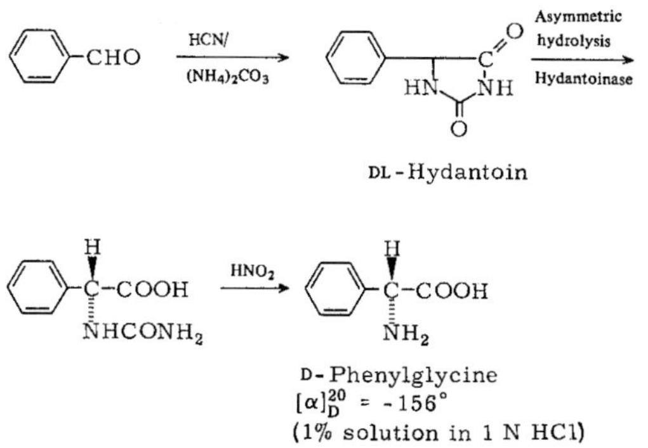

Enediynes. A class of antitumor antibiotics possessing characteristic enediynechromophores have been isolated [223]. These include neocarzinostatin (**98a**) [224], [225], calicheamicins (**98b**) [226], [227], esperamicins (**98c**) [228], [229], dynemicin A (**98d**) [230], C-1027 (**98e**) [231], [232], kedarcidin (**98f**) [233]–[236], and maduropeptin (**98g**) [237].

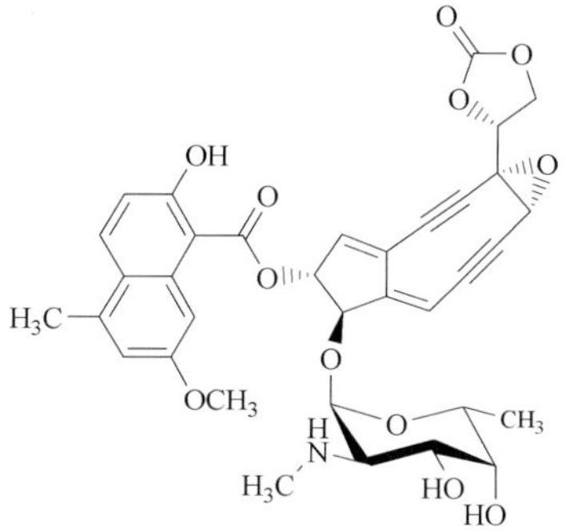

(**98a**) Neocarzinostain chromophore

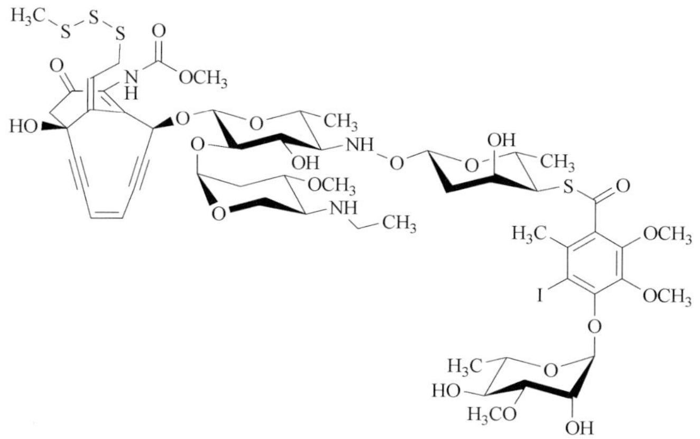

(**98b**) Calicheamicin γ'_1

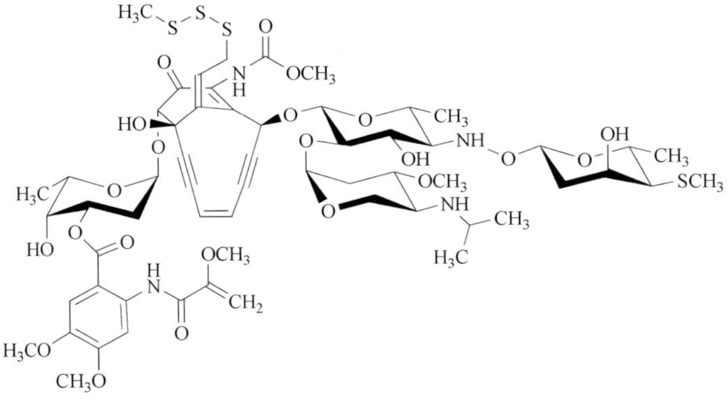

(98c) Esperamicin A₁

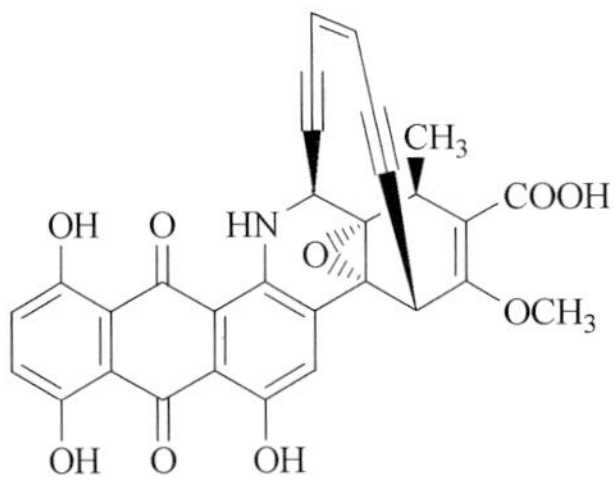

(98d) Dynemicin A

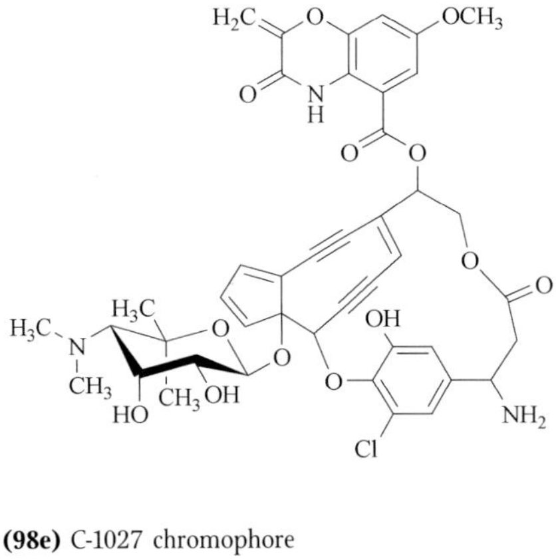

(98e) C-1027 chromophore

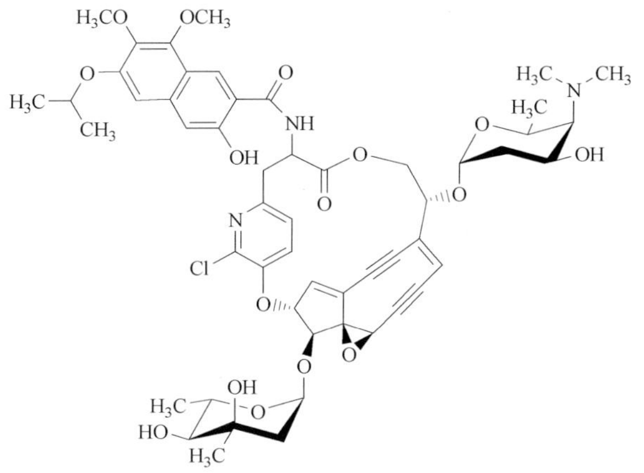

(98f) Kedarcidin chromophore

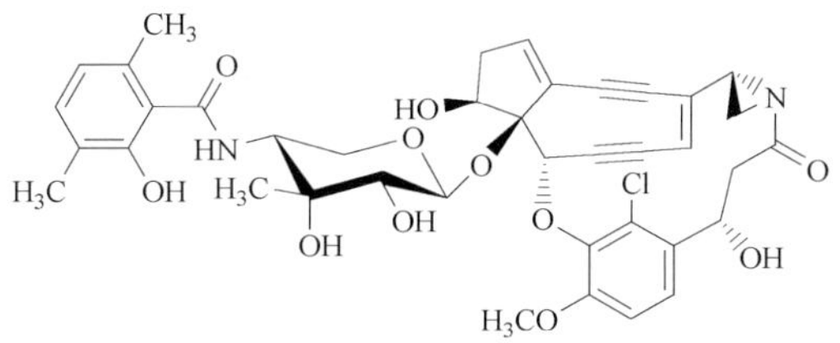

(98g) Maduropeptin chromophore

Neocarzinostatin was isolated from *Streptomyces carzinostaticus* Var. F-41 as a complex of a chromophore and an apoprotein [224], [225]. The role of the apoprotein is to transport and stabilize the chromophore that is responsible for the anticancer activity of neocarzinostatin coming mainly from its capability to cleave DNA. The DNA damage is initiated by nucleophilic attack at C12 of neocarzinostatin chromophore by a sulfur nucleophile leading to the formation of a labile cumulene intermediate that undergoes a facile cycloaromatization [238]. The resulting biradical abstracts hydrogen atom of the deoxyribose of DNA to induce degradation. A polymer conjugated derivative of neocarzinostatin was prepared and administered via the tumor-feeding artery showing increased stability in blood and the immunogenicity was much lower than the parental neocarzinostatin [239]. Similar biradical formation via the Bergman cyclization of the other enediyne class antibiotics has been proposed [223].

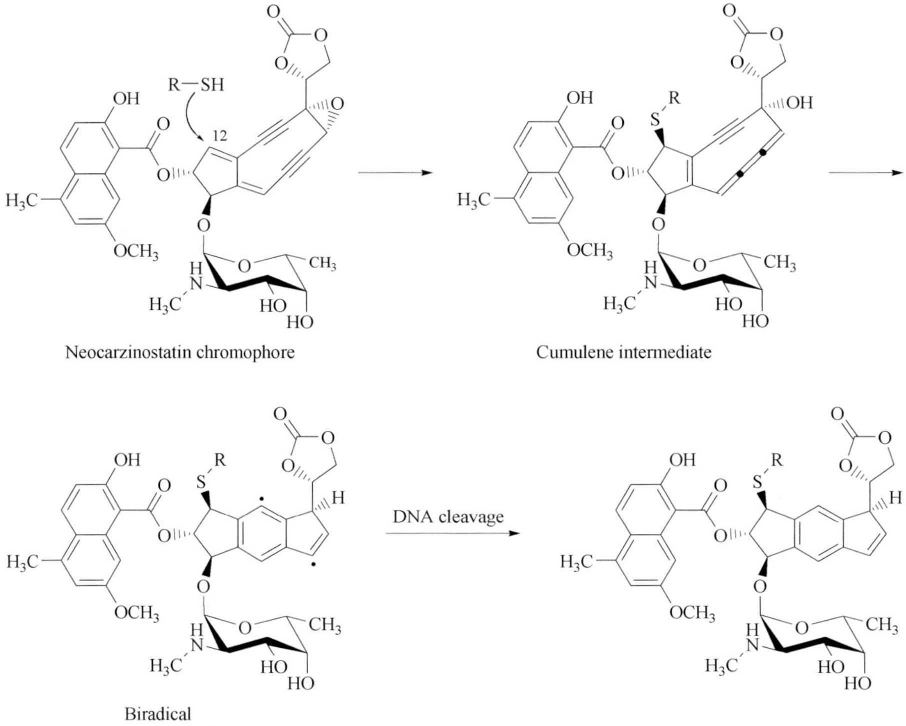

Antibiotics

Neocarzinostatin chromophore → Cumulene intermediate →

Biradical → DNA cleavage

1016

4. Antibiotics in Current Use

Benzylpenicillin potassium [113-98-4] (**99**), penicillin G potassium, $C_{16}H_{17}KN_2O_4S$, M_r 372.49.

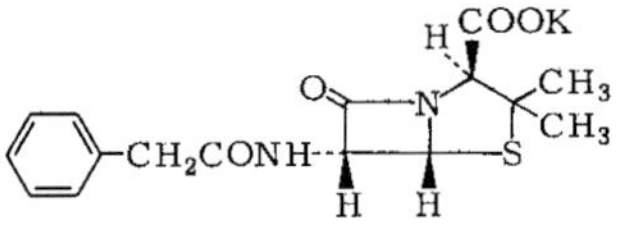

99

Benzylpenicillin potassium was the first crystalline penicillin produced on an industrial scale [22], [24]. It is produced in the pure state by the addition of phenylacetate to a culture of *Penicillium chrysogenum*. The crystalline benzylpenicillin potassium contains 1598 U/mg.

Benzylpenicillin procaine [6130-64-9] (**100**), penicillin G procaine, $C_{16}H_{18}N_2O_4S \cdot C_{13}H_{20}N_2O_2 \cdot H_2O$, M_r 588.73.

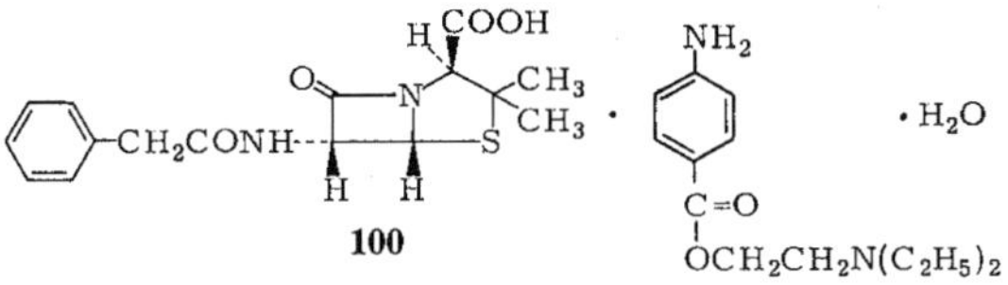

100

Benzylpenicillin procaine was developed as a dilatorily acting benzylpenicillin, only slightly soluble in water. It has been used with peanut oil or carboxymethylcellulose as an oil suspension or an aqueous suspension, respectively [240].

Benzylpenicillin benzathine [41372-02-5] (**101**), penicillin G benzathine, $(C_{16}H_{18}N_2O_4S)_2 \cdot C_{16}H_{20}N_2 \cdot 4\ H_2O$, M_r 981.21.

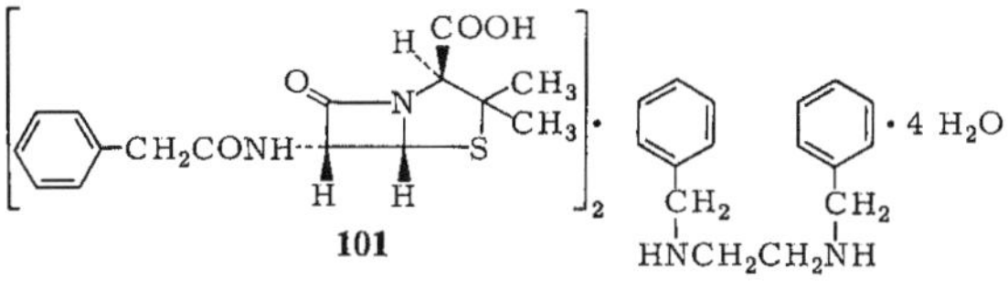

101

Benzylpenicillin benzathine was developed as a dilatorily acting benzylpenicillin that maintains an effective serum concentration for 2 days following a single intramuscular injection of 600 000 U. It is also used as an orally active benzylpenicillin because it is only slightly affected by a patient's meal [241].

1017

Phenoxymethylpenicillin [*87-08-1*], (**102**), penicillin V, $C_{16}H_{18}N_2O_5S$, M_r 350.40.

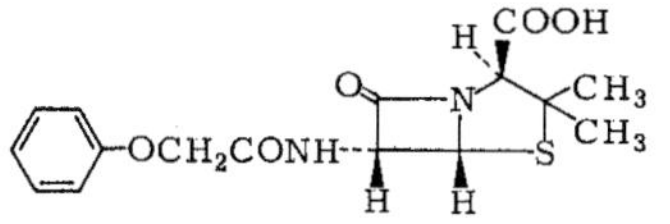

structure 102

Phenoxymethylpenicillin was produced in the culture broth of *Penicillium chrysogenum* when phenoxyacetic acid was added to the medium at Biochemie in 1953 [242]. It is more stable against acid than benzylpenicillin and is used as an orally active penicillin. Its therapeutic applications are the same as those of benzylpenicillin.

Phenoxymethylpenicillin potassium [*132-98-9*], (**103**), penicillin V potassium, $C_{16}H_{17}KN_2O_5S$, M_r 388.49.

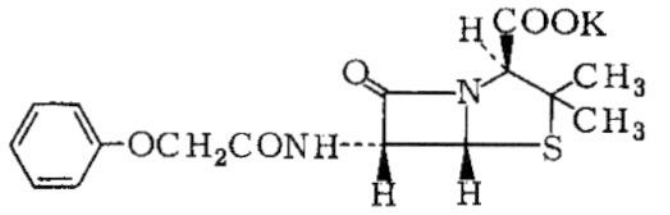

structure 103

Phenoxymethylpenicillin potassium was first obtained at Lilly Research Laboratories in 1948, following the addition of phenoxyacetic acid to a culture of *Penicillium chrysogenum*. Its industrial-scale production was established in 1953 [243]. Its usefulness as an orally active penicillin was suggested by SPITZY et al. in 1955. It is less hygroscopic and much more stable against gastric acid than benzylpenicillin, and it has been used orally for therapy of infections caused by *Streptococcus*, *Staphylococcus*, and other gram-positive bacteria as well as *Neisseria* and *Leptospira*.

Phenethicillin potassium [*132-93-4*], (**104**), $C_{17}H_{19}KN_2O_5S$, M_r 402.52.

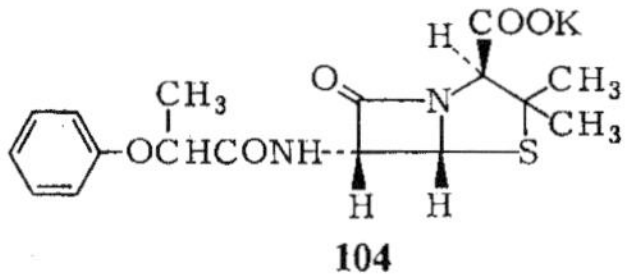

structure 104

Phenethicillin was the first member of the semisynthetic penicillin class of antibiotics to be introduced in clinics. This drug was synthesized in 1960 by Bristol-Myers in collaboration with Beecham Research Laboratories starting with 6-aminopenicillanic acid [244]. It is as stable against gastric acid as phenoxymethylpenicillin and is used orally for therapy of gram-positive and *Neisseria* infections.

Propicillin [551-27-9], (**105**), $C_{18}H_{22}N_2O_5S$, M_r 378.45.

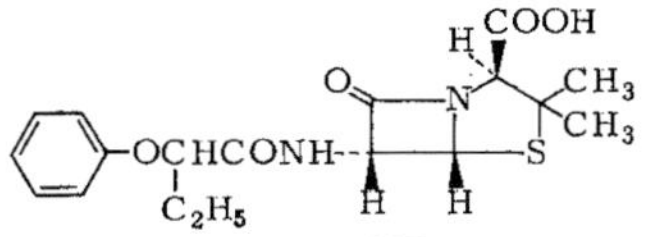

105

Propicillin was synthesized by a collaboration of Beecham Research Laboratories with Bristol-Myers Laboratories in 1961 starting with 6-aminopenicillanic acid [245]. It is stable against gastric acid and shows three- to fourfold higher serum concentrations than phenoxymethyl-penicillin when administered orally. Its antibacterial spectrum and activity are almost the same as those of phenoxymethylpenicillin. Propicillin has been used for therapy of infections caused by *Streptococcus*, *Staphylococcus*, and *Neisseria*.

Carbenicillin disodium [4800-94-6], (**106**), $C_{17}H_{16}N_2Na_2O_6S$, M_r 422.37.

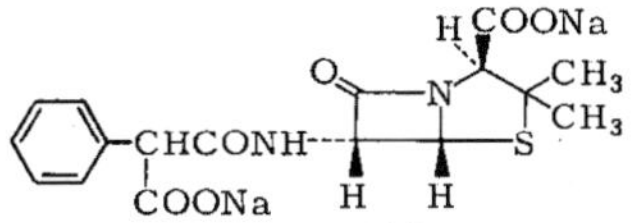

106

Carbenicillin was synthesized by BRAIN et al. of Beecham Research Laboratories in 1965. It was the first synthetic penicillin to show activity against *Pseudomonas aeruginosa* [246]. Although its activity against the microorganism is not strong (MIC = 25 – 100 µg/mL), it is widely used against *P. aeruginosa* infections because of its low toxicity and the lack of other antibiotics suitable for use against this microorganism. Carbenicillin is mainly used clinically to treat urinary tract and respiratory tract infections and sepsis caused by *Proteus*, *Escherichia coli*, *Klebsiella*, and *Pseudomonas aeruginosa*.

Carbenicillin phenyl sodium [21649-57-0], (**107**), carfecillin, $C_{23}H_{21}N_2NaO_6S$, M_r 476.49.

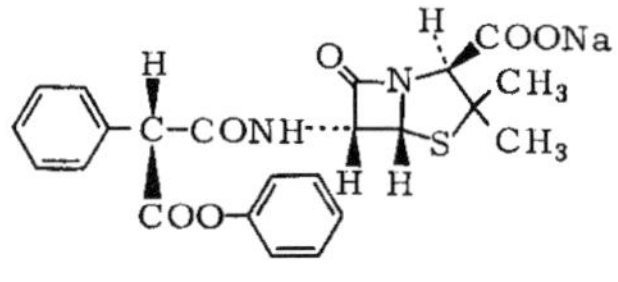

107

Carbenicillin phenyl was synthesized by Beecham Research Laboratories in 1966 as an orally active carbenicillin [247]. It is hydrolyzed to carbenicillin by intestinal esterase and thus acts the same when administered orally. The phenol produced by the hydrolysis is conjugated and excreted in urine.

Carbenicillin indanyl sodium [*26605-69-6*], (**108**), carindacillin, $C_{26}H_{25}N_2NaO_6S$, M_r 516.55.

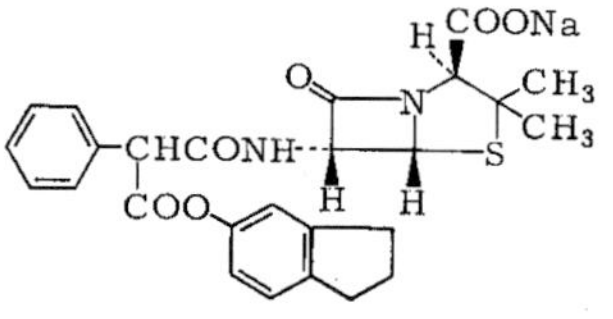

108

Carbenicillin indanyl was synthesized by Pfizer in 1972 as an orally active carbenicillin [248]. It shows strong activity against a variety of bacteria in vitro, and, when administered orally, it behaves as carbenicillin after being hydrolyzed by intestinal esterase.

Sulbenicillin [*34779-28-7*], (**109**), $C_{16}H_{18}N_2O_7S_2$, M_r 414.46.

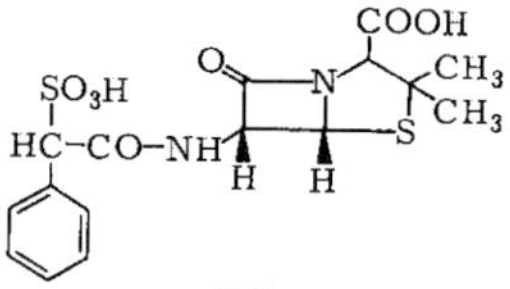

109

Sulbenicillin was synthesized by Takeda Chemical Industries in 1968 [249]. It shows slightly lower activity against gram-positive bacteria, almost the same activity against gram-negative bacteria, and slightly higher activity against anaerobic bacteria in comparison with carbenicillin. Sulbenicillin has been used by intravenous and intramuscular administration for therapy of sepsis, bacterial endocarditis, pyoderma, urinary tract and respiratory tract infections, and other infections caused by *Staphylococcus, Streptococcus, Klebsiella, Proteus, Enterobacter, Citrobacter, Escherichia coli* and *Haemophilus influenzae*.

Ticarcillin disodium [*4697-14-7*], (**110**), $C_{15}H_{14}N_2Na_2O_6S_2$, M_r 428.40.

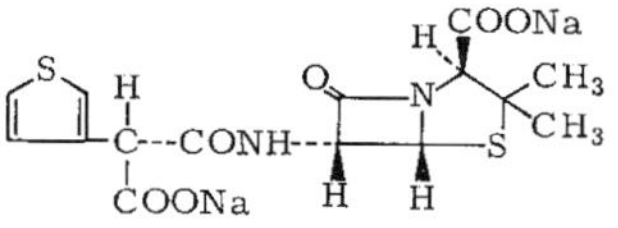

110

Ticarcillin was synthesized by Beecham Research Laboratories in 1971. The phenyl residue of carbenicillin was replaced by the 3-thienyl moiety [250]. Ticarcillin shows almost the same activity as carbenicillin against gram-positive bacteria and a twofold higher activity against gram-negative bacteria. Its in vivo activity against *Pseudomonas aeruginosa* infections in mice is fourfold higher than that of sulbenicillin. Ticarcillin is used to treat sepsis, urinary tract infections, and serious *Pseudomonas aeruginosa*,

Escherichia coli, Proteus, and *Enterobacter* infections in leukemic and other cancer patients.

Methicillin sodium [7246-14-2], (**111**), meticillin, $C_{17}H_{19}N_2NaO_6S \cdot H_2O$, M_r 420.42.

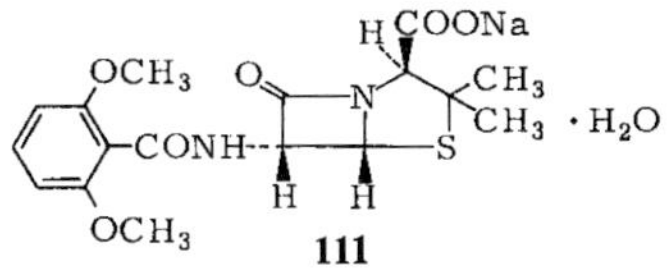

111

Methicillin was synthesized independently by Bristol-Myers Laboratories and Beecham Research Laboratories in 1960 [251]. It was the first member of the penicillinase-stable semisynthetic penicillin class of antibiotics to be introduced clinically. This antibiotic is a parenteral penicillin having an antibacterial spectrum similar to that of benzylpenicillin. Although its activity against benzylpenicillin-sensitive bacteria is about 1/3 to 1/50 that of benzylpenicillin, it shows strong activity against benzylpenicillin-resistant strains because of its stability toward penicillinase. Methicillin has been used for therapy of respiratory tract and urinary tract infections, sepsis, and gynecological and other infections caused by benzylpenicillin-resistant bacteria.

Oxacillin sodium [7240-38-2], (**112**), $C_{19}H_{18}N_3NaO_5S \cdot H_2O$, M_r 441.43.

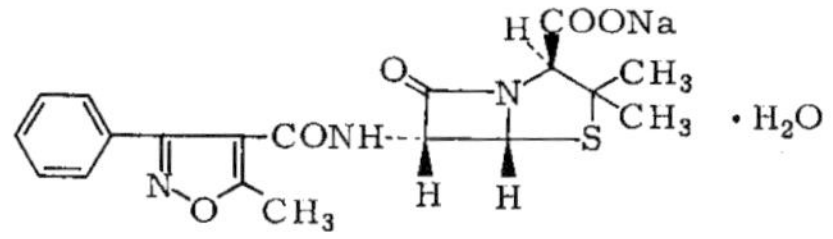

112

Oxacillin was synthesized by Bristol-Myers Laboratories in 1961 starting with 6-aminopenicillanic acid [252]. It was the first orally active and penicillinase-stable semisynthetic penicillin to be introduced clinically. Oxacillin is slightly less stable against gastric acid and shows a lower serum concentration than phenoxymethylpenicillin, but it is highly active against phenoxymethylpenicillin-resistant *Staphylococcus aureus*. Oxacillin has been used by oral and intramuscular administration for therapy of respiratory tract, urinary tract, gynecological, and other infections caused by benzylpenicillin-resistant bacteria.

Cloxacillin sodium [7081-44-9], (**113**), $C_{19}H_{17}ClN_3NaO_5S \cdot H_2O$, M_r 475.89.

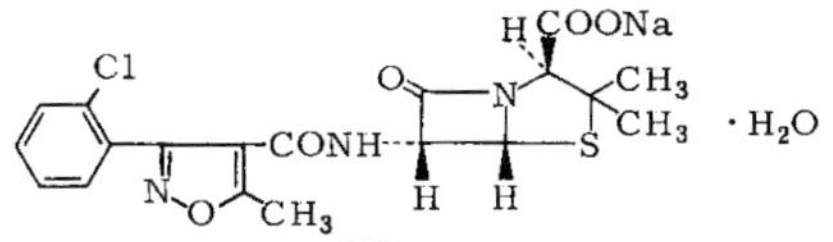

113

Cloxacillin was synthesized by Beecham Research Laboratories in 1962 starting with 6-aminopenicillanic acid [253]. It was the first semisynthetic penicillin with a halogen atom in the side chain to be used clinically. It shows 5- to 20-fold stronger activity than methicillin and is twice as active as oxacillin against *Staphylococcus aureus*, including benzylpenicillin-resistant strains. Cloxacillin is highly stable against penicillinase and well absorbed by oral administration. It has been widely used, by oral and intramuscular administration, to treat internal, surgical, gynecological, and other infections caused by both benzylpenicillin-sensitive and benzylpenicillin-resistant bacteria.

Dicloxacillin [*3116-76-5*], (**114**), $C_{19}H_{17}Cl_2N_3O_5S$, M_r 470.33.

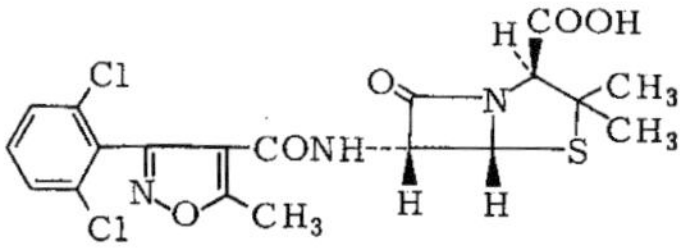

Dicloxacillin was synthesized by Bayer in 1965 starting with 6-aminopenicillanic acid [254]. It is a penicillinase-stable and orally active semisynthetic penicillin and shows higher and longer serum concentrations than cloxacillin when administered orally. Dicloxacillin is used orally, either alone or in combination with ampicillin, to treat various infections, including those caused by benzylpenicillin-resistant bacteria.

Flucloxacillin [*5250-39-5*], (**115**), floxacillin, $C_{19}H_{17}ClFN_3O_5S$, M_r 453.88.

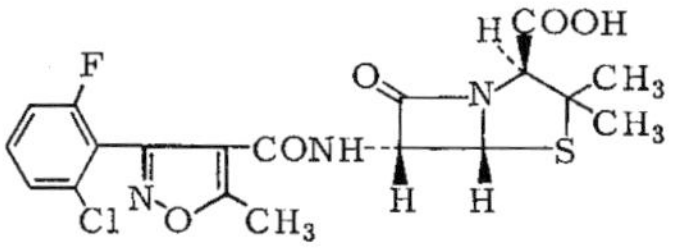

Flucloxacillin was synthesized by Beecham Research Laboratories in 1962 as a penicillinase-stable and orally active semisynthetic penicillin [255]. It shows almost the same activity as dicloxacillin, and it has slightly higher serum and tissue concentrations than dicloxacillin. This drug has been used to treat pyoderma, sepsis, and postoperative infections as well as ear and nose, respiratory tract, and other infections caused by *Staphylococcus* and *Streptococcus*, including benzylpenicillin-resistant strains.

Ampicillin [*69-53-4*], (**116**), aminobenzyl penicillin, $C_{16}H_{19}N_3O_4S$, M_r 349.41.

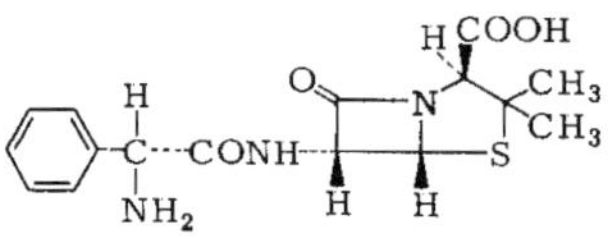

Ampicillin was synthesized by Beecham Research Laboratories in 1961 and evaluated for its anti-gram-negative activity by ROLINSON et al. [256]. It was the first semisynthetic penicillin showing strong activity against gram-negative bacilli. Although it is hydrolyzed by bacterial penicillinase, it is active against *Escherichia coli, Shigella, Proteus mirabilis*, and *Haemophilus influenzae* and is used very widely as an oral antibiotic. Ampicillin also is used by injection for serious infections.

Mezlocillin [*51481-65-3*], (**117**), $C_{21}H_{25}N_5O_8S_2$, M_r 539.59.

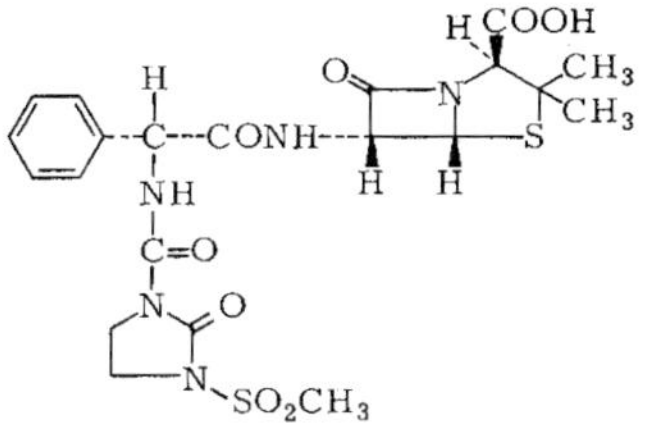

117

Mezlocillin was synthesized by Bayer in 1974 [257]. The NH_2 residue of ampicillin was acylated. Mezlocillin is a member of the so-called ureidopenicillins and shows two- to eightfold greater activity against *Citrobacter, Enterobacter, Klebsiella, Escherichia coli*, and *Haemophilus influenzae* than ampicillin, carbenicillin, or sulbenicillin. Its activity against gram-positive bacteria is almost the same as that of carbenicillin. Mezlocillin is used by intravenous administration for therapy of sepsis, meningitis, and respiratory tract, urinary tract, and abdominal infections.

Piperacillin sodium [*59703-84-3*], (**118**), $C_{23}H_{26}N_5NaO_7S$, M_r 539.55.

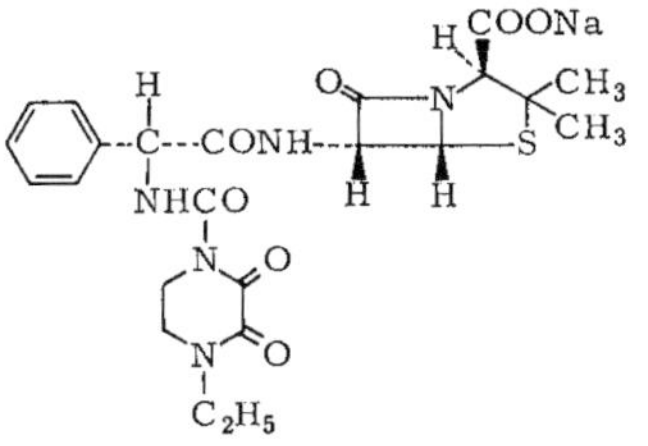

118

Piperacillin was synthesized by Toyama Chemicals Company in 1976 by acylation of the amino residue of ampicillin [258]. It shows excellent activity against a wide range of gram-positive and gram-negative bacteria and has been used as the most potent semisynthetic penicillin. Piperacillin is administered intravenously or intramuscularly for therapy of sepsis, meningitis, respiratory and urinary tract infections, and for abdominal infections caused by *Staphylococcus, Streptococcus, Escherichia coli, Klebsiella, Haemophilus influenzae, Serratia marcescens, Pseudomonas aeruginosa, Proteus, Enterobacter, Citrobacter*, and *Bacteroides*.

Amoxicillin [61336-70-7], (**119**), amoxycillin, $C_{16}H_{19}N_3O_5S \cdot 3\,H_2O$, M_r 419.46.

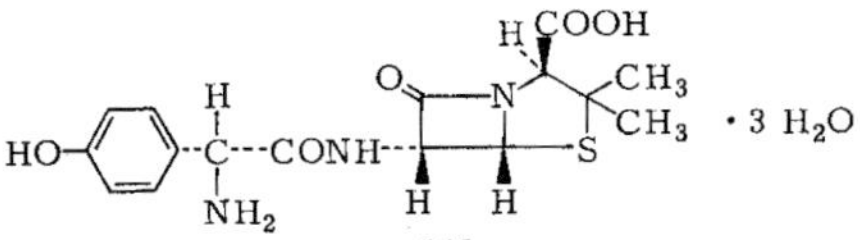

119

Amoxicillin was synthesized by Beecham Research Laboratories in 1968 [259]. A hydroxyl group was introduced on the benzene ring of ampicillin. Amoxicillin shows about a twofold higher serum concentration than ampicillin when administered orally. It was shown by double-blind comparative studies with ampicillin that amoxicillin was as effective as ampicillin when administered at half the dose of ampicillin.

Ciclacillin [3485-14-1], (**120**), cyclacillin, $C_{15}H_{23}N_3O_4S$, M_r 341.43.

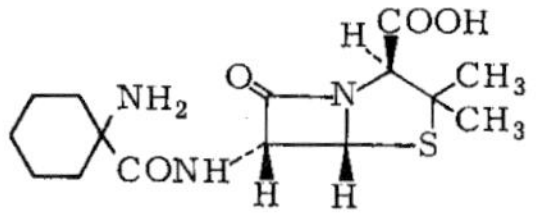

120

Ciclacillin was synthesized by Wyeth Laboratories in 1967 in the course of studies on the improvement of oral absorption of ampicillin [260]. Although its antibacterial activity is one-sixteenth to one-half that of ampicillin, it shows a four to tenfold higher oral absorption and higher urinary excretion. Ciclacillin shows less tendency than ampicillin to cause diarrhea and is used for therapy of pyoderma, wound infection, respiratory and urinary tract infections, as well as ear and nose, and other infections caused by *Staphylococcus*, *Streptococcus*, *Escherichia coli*, *Citrobacter*, *Klebsiella*, *Proteus*, and *Haemophilus influenzae*.

Hetacillin [3511-16-8], (**121**), $C_{19}H_{23}N_3O_4S$, M_r 389.48.

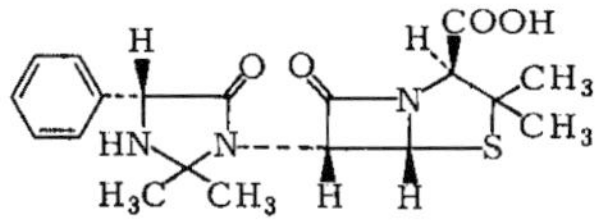

121

Hetacillin was synthesized by Bristol-Myers Laboratories in 1964. It shows activity as strong as that of ampicillin against a variety of gram-positive and gram-negative bacteria [261]. Hetacillin shows almost the same pharmacokinetic properties as ampicillin when administered orally and a later serum peak when given by intramuscular administration. When administered either orally or intramuscularly, it is hydrolyzed in vivo into ampicillin. Therefore, the clinical applications of hetacillin are the same as those of ampicillin.

Sulbactam sodium [*69388-84-7*], (**122**), $C_8H_{10}NNaO_5S$, M_r 255.22.

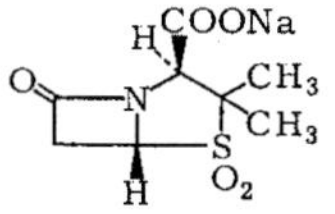

122

Sulbactam was synthesized by Pfizer Research Laboratories in 1977 in the course of screening for β-lactamase inhibitors [262]. It shows strong activity against penicillinase and moderate activity against cephalosporinase. Sulbactam itself shows activity against some gram-negative bacteria but no activity against most pathogenic bacteria. The use of sulbactam in combination with cefoperazone, which is partially hydrolyzed by penicillinase, is under study along with its use as an esterified complex with ampicillin (sultamicillin) for therapy of cefoperazone-ampicillin-resistant infections.

Talampicillin hydrochloride [*39878-70-1*], (**123**), ampicillin phthalidyl, $C_{24}H_{23}N_3O_6S \cdot HCl$, M_r 517.99.

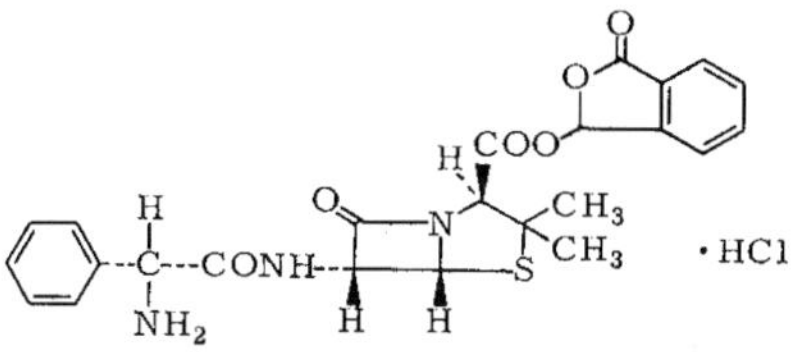

123

Talampicillin was synthesized by Yamanouchi Pharmaceutical Co. in 1971 by esterifying the carboxylic acid group of ampicillin in order to improve the oral absorption [263]. When administered orally, it is hydrolyzed to ampicillin by intestinal esterase. Its bioavailability is two or more times as high as that of ampicillin, and it is used to treat the same infections as those for which ampicillin is used orally but in doses only half or one-third as large.

Bacampicillin hydrochloride [*37661-08-8*], (**124**), ampicillin ethoxycarbonyloxyethyl hydrochloride, $C_{21}H_{27}N_3O_7S \cdot HCl$, M_r 501.99.

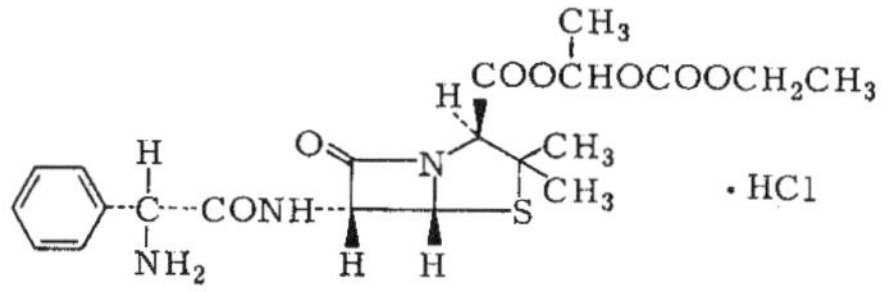

124

Bacampicillin was synthesized by EKSTRÖM et al. of Astra Läkemedel in 1975 [264]. It is more stable against acid than ampicillin and more rapidly absorbed orally; its absorption is less affected by a patient's most recent meal than is the case for ampicillin. Bacampicillin is hydrolyzed by intestinal esterase after oral administration and then behaves the same as ampicillin. In double-blind comparison studies, bacampicillin was shown to be as effective as ampicillin when administered at half the dose.

Pivmecillinam [*32886-97-8*], (**125**), amdinocillin pivoxil, $C_{21}H_{33}N_3O_5S$, M_r 439.58.

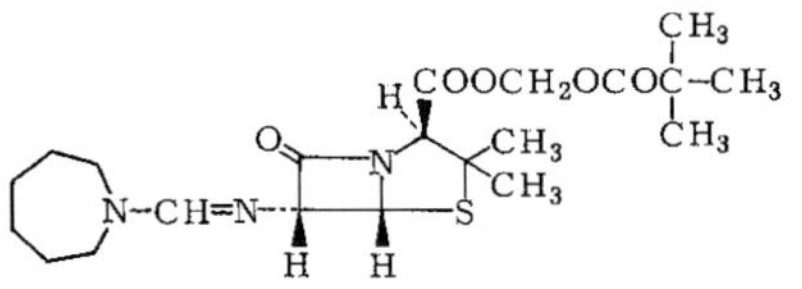

125

Pivmecillinam was found by a series of studies at Leo in 1969 [265]. It is a derivative of 6β-formamidinopenicillanic acid and shows strong activity against gram-negative bacilli. Pivmecillinam is hydrolyzed by intestinal esterase and acts similarly to mecillinam after oral administration. Mecillinam shows weaker activity than ampicillin against gram-positive cocci but much stronger activity against a wide range of gram-negative bacilli [266].

Cephalexin [*23325-78-2*], (**126**), cefalexin, $C_{16}H_{17}N_3O_4S \cdot H_2O$, M_r 365.41.

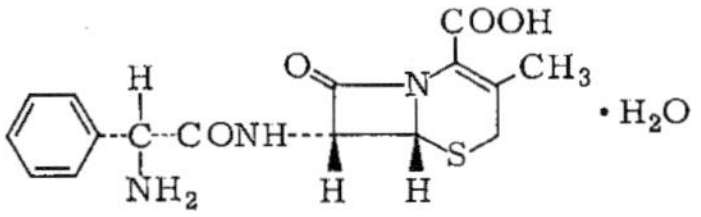

126

Cephalexin was first synthesized in 1967 by Glaxo Research Laboratories [267] and first produced on an industrial scale by Eli Lilly & Co. in the same year. It is a deacetoxylated derivative of cephaloglycin that is not metabolized in vivo. When administered orally, it shows a much higher serum concentration and much lower tendency to induce diarrhea than cephaloglycin. Cephalexin has been used widely and is the most popular orally active antibiotic in the world for treatment of respiratory tract, urinary tract, surgical, ear and nose, and other infections caused by *Staphylococcus*, *Streptococcus*, *Escherichia coli*, *Klebsiella*, *Enterobacter*, and *Proteus*.

Cefaclor [*70356-03-5*], (**127**), $C_{15}H_{14}ClN_3O_4S \cdot H_2O$, M_r 385.83.

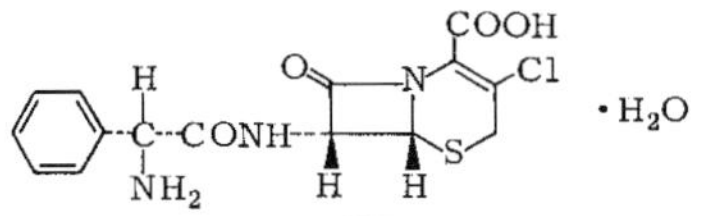

127

Cefaclor was found by Eli Lilly & Co. in 1976 in the course of studies on the improvement of synthetic procedures for cephalexin [268]. The intermediates with chlorine at the 3 position of the cephem nucleus have excellent antibacterial activity.

Cefaclor was selected from among the various derivatives that differed at the 7 position because of its strong activity and high bioavailability following oral administration.

Cephaloglycin [22202-75-1], (**128**), cefaloglycin, $C_{18}H_{19}N_3O_6S \cdot 2\ H_2O$, M_r 441.46.

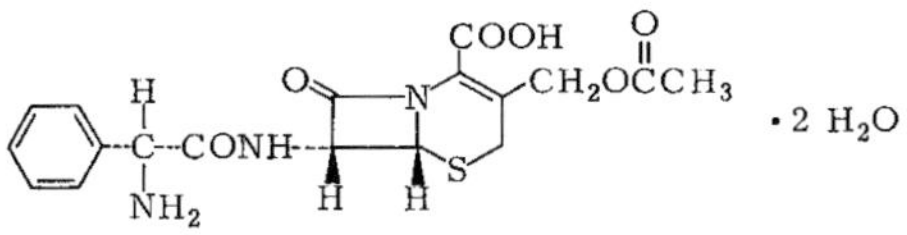

128

Cephaloglycin was synthesized by Lilly Research Laboratories in 1966 as the first member of the orally active cephalosporin C class of antibiotics [269]. Its activity against *Staphylococcus* and *Streptococcus* is one-fourth to one-half that of cephaloridin, a parenteral antibiotic, but against gram-negative bacteria it is almost the equal of cephaloridin. Cephaloglycin is readily absorbed by oral administration but is partially metabolized in vivo into deacetylcephaloglycin, which shows one-tenth to one-half as much activity. This antibiotic has been used in therapy of urinary tract infections and pyoderma, but it is being replaced by the newer orally active cephalosporins. Cephaloglycin often causes diarrhea, especially in children, and pediatric use is not allowed.

Cefadroxil [66592-87-8], (**129**), $C_{16}H_{17}N_3O_5S \cdot H_2O$, M_r 381.41.

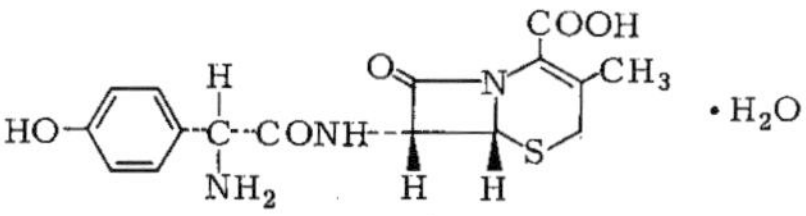

129

Cefadroxil was found by Bristol-Myers Co. in 1976 [270]. A hydroxyl group was attached to the benzene ring of cephalexin. Like amoxicillin and ampicillin, cefadroxil shows almost the same antibacterial activity spectrum as cephalexin and superior oral absorption. Its in vivo activity is four to six times greater than that of cephalexin, and its half-life in serum is twice as long.

Cephradine [38821-53-3], (**130**), cefradine, $C_{16}H_{19}N_3O_4S$, M_r 349.41.

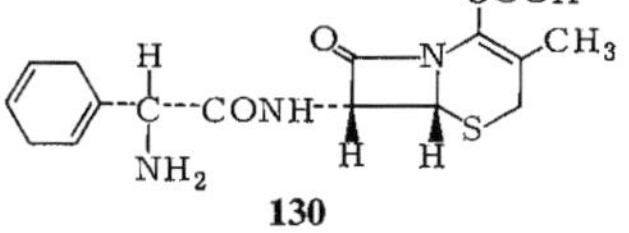

130

Cephradine was synthesized by the Squibb Institute of Medical Research in 1971 [271]. It shows almost the same antibacterial activity and pharmacokinetic properties as cephalexin. Cephradine has been used for therapy of urinary and respiratory tract infections caused by *Staphylococcus*, *Streptococcus*, *Escherichia coli*, *Klebsiella*, and *Proteus mirabilis*.

Cefroxadine [51762-05-1], (**131**), $C_{16}H_{19}N_3O_5S$, M_r 365.41.

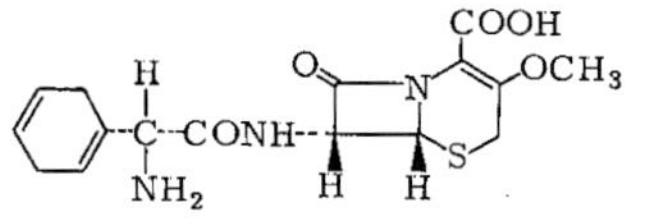

131

Cefroxadine was synthesized by Ciba-Geigy in 1972. A methoxyl group replaced the methyl group of cephradine at the 3 position of the cephem nucleus [272]. Cefroxadine shows stronger activities than cephalexin, especially bactericidal and bacteriolytic activities, and it has better oral absorption that is less affected by a recent meal. Cefroxadine shows less renal toxicity than cephalexin in toxicological studies using animals.

Cephapirin sodium [24356-60-3], (**132**), cefapirin, $C_{17}H_{16}N_3NaO_6S_2$, M_r 445.45.

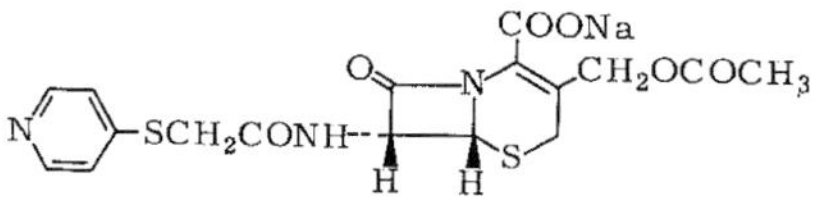

132

Cephapirin was synthesized by Bristol-Myers Laboratories in 1970 [273]. It shows almost the same in vitro antibacterial activity as cephalothin, but its in vivo effects are slightly greater than those of cephalothin. Like cephalothin, it is metabolized in vivo, and its deacetylated metabolite shows almost the same activity against gram-positive bacteria as cephalothin, but weaker activity against gram-negative bacteria. Cephapirin has been used for therapy of urinary tract infections and osteomyelitis caused by *Staphylococcus*, *Streptococcus*, and *Escherichia coli*.

Cephalothin sodium [58-71-9], (**133**), $C_{16}H_{15}N_2NaO_6S_2$, M_r 418.43.

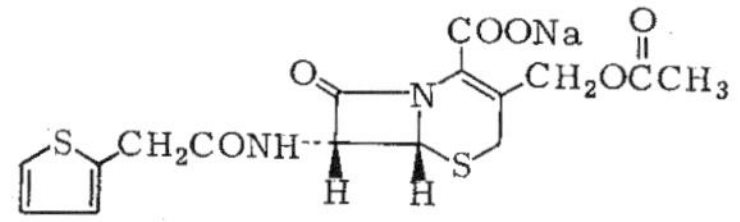

133

Cephalothin, along with cephaloridine, was the first of the synthetic cephalosporin C class antibiotics to be introduced clinically. It was synthesized from 7-amino-cephalosporanic acid by Lilly Research Laboratories in 1962 [274]. Cephalothin shows strong activity against gram-positive and gram-negative bacteria and *Leptospira*, including benzylpenicillin-resistant strains. It has been used intravenously and intramuscularly to treat a variety of infections caused by *Staphylococcus*, *Streptococcus*, *Escherichia coli*, and *Neisseria*. The drug is metabolized in vivo, and the metabolite, deacetylcephalothin, is almost inactive.

Cephacetrile sodium [23239-41-0], (**134**), cefacetrile, $C_{13}H_{12}N_3NaO_6S$, M_r 361.31.

```
              COONa
      O       CH2OCOCH3
       \  N /
CH2CONH--|  |
 |       |  S
 CN      H  H
         134
```

Cephacetrile was synthesized by Ciba-Geigy in 1970 [275]. It shows almost the same activity against gram-negative bacteria as cephalothin, but it has a higher activity against β-lactamase-producing *Escherichia coli*. Like cephalothin, it is metabolized partially by deacetylation in vivo. Cephacetrile has been used to treat sepsis and abdominal and respiratory tract infections caused by *Staphylococcus*, *Streptococcus*, *Escherichia coli*, and *Klebsiella pneumoniae*, but it is gradually being replaced clinically by the newer and more active cephalosporins.

Cefsulodin sodium [52152-93-9], (**135**), $C_{22}H_{19}N_4NaO_8S_2$, M_r 554.54.

```
                 COO-
        H    O       CH2-N+⟨  ⟩-CONH2
   ⟨  ⟩-C--CONH--|  |
        |        |  S
        SO3Na    H  H
                 135
```

Cefsulodin was synthesized by Takeda Chemicals Industries in 1974 by introducing the sulfobenzyl group, the same moiety as in sulbenicillin, at the 7 position of the cephem nucleus [276]. Its side chain at the 3 position is similar to that of cephaloridine except for the carbamoyl group. The introduction of these hydrophilic groups increases the activity against *Pseudomonas aeruginosa*, but it markedly decreases it against gram-positive and other gram-negative bacteria. Therefore cefsulodin is used as a specific antibiotic against infections caused by the opportunistic pathogen *P. aeruginosa*.

Cephaloridine [50-59-9], (**136**), cefaloridine, $C_{19}H_{17}N_3O_4S_2$, M_r 415.49.

```
             COO-
      O          CH2N+⟨  ⟩
       \  N /
 ⟨S⟩-CH2CONH--|  |
              |  S
              H  H
              136
```

Cephaloridine, along with cephalothin, was the first member of the synthetic cephalosporin C class of antibiotic, to be introduced clinically. It was synthesized, starting with 7-aminocephalosporanic acid, by Glaxo Research Laboratories in 1962 [277]. This drug shows strong activity against gram-positive and gram-negative bacteria and *Leptospira*, including benzylpenicillin-resistant strains. It has been given widely by intravenous, intramuscular, and intraspinal injection to treat a variety of infections caused by *Staphylococcus*, *Streptococcus*, *Neisseria*, *Klebsiella*, *Escherichia coli*, and other pathogens. Because of its renal toxicity and the development of newer and more active synthetic cephalosporins, its use is declining.

Cefatrizine [51627-14-6], (**137**), $C_{18}H_{18}N_6O_5S_2$, M_r 462.51.

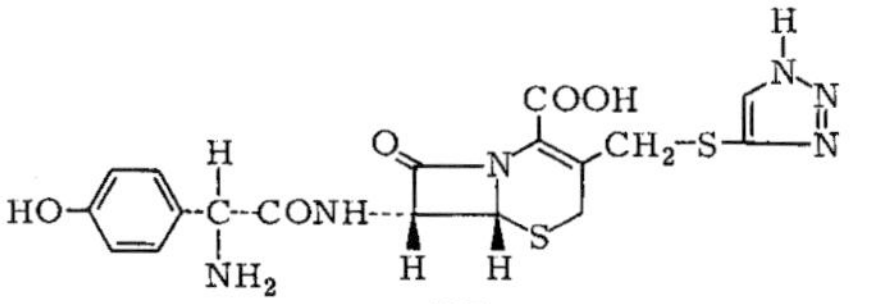

(Structure 137)

Cefatrizine was synthesized by Bristol-Myers Co. and Smith Klein & French Laboratories in 1974 [278]. It shows two to four times higher activity against gram-positive and four to eight times higher activity against gram-negative bacteria than cephalexin. Cefatrizine also shows excellent oral absorption, and its in vivo activity is 30 to 500 times higher than that of cephalexin.

Cefoperazone sodium [62893-20-3], (**138**), $C_{25}H_{26}N_9NaO_8S_2$, M_r 667.66.

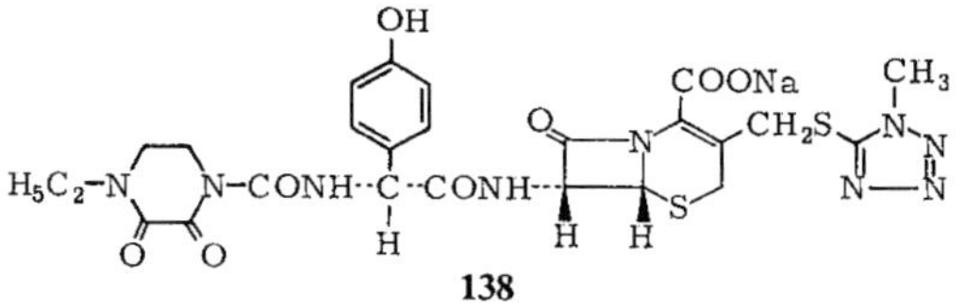

(Structure 138)

Cefoperazone was synthesized by Toyama Chemicals Co. in 1978. Except for the hydroxyl group, the side chain attached to the cephem nucleus is the same as that of piperacillin [279]. Cefoperazone shows excellent activity against gram-positive (except *Staphylococcus*) and gram- negative bacteria, including *Pseudomonas aeruginosa*. Its pharmacological characteristics are unique. Cefoperazone is excreted mainly in bile, and a concentration five to tenfold higher in bile than in serum is obtained. The transfer into cerebrospinal fluid is 10–30% of the serum concentration; the half-life in serum is 2.0–2.6 h, and the degree of binding with serum protein is as high as 86.6%.

Cefamandole [34444-01-4], (**139**), $C_{18}H_{18}N_6O_5S_2$, M_r 462.51.

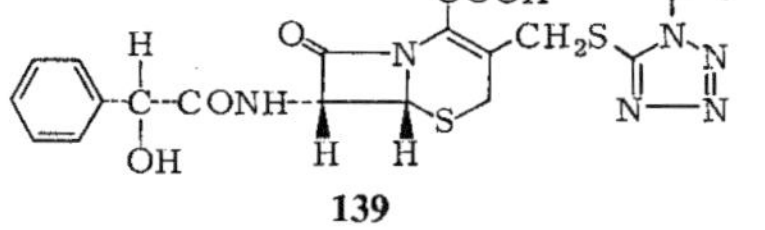

(Structure 139)

Cefamandole was synthesized by Eli Lilly & Co. in 1972 [280]. It shows strong activity against *Proteus* (indole-positive) species, *Enterobacter*, and *Citrobacter*, against which the earlier cephalosporins, such as cephalothin and cefazolin, are inactive. The nafate (sodium salt of the O-formyl ester) has been used in the United States and Europe, and the sodium salt of cefamandole has been used in Japan by injection.

Cefotiam hydrochloride [66309-69-1], (**140**), $C_{18}H_{23}N_9O_4S_3 \cdot 2$ HCl, M_r 598.56.

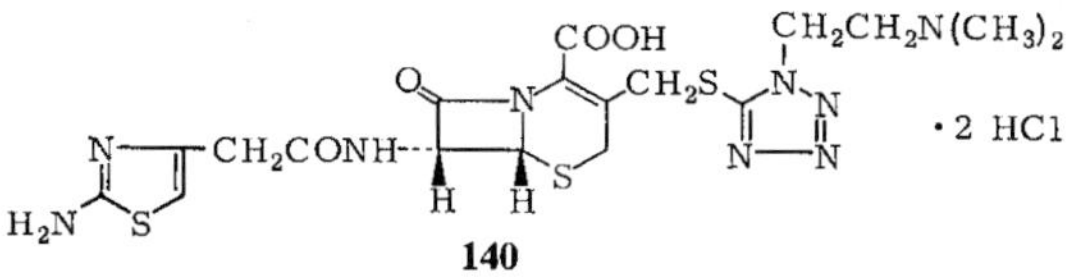

140

Cefotiam was synthesized by Takeda Chemicals Industries in 1977 as the first cephem derivative to introduce the aminothiazol group in the 7 side chain and the alkylated tetrazole group at the 3 side chain [281]. This drug shows greater antibacterial activity against gram-negative bacteria than the earlier cephalosporins, including *Enterobacter*, *Citrobacter*, and indole-positive *Proteus*, and it retains the same activity as they have against gram-positive organisms. Its excellent activity against *Klebsiella* and *Escherichia coli* is based on its ability to penetrate the cell. This is 3- to 70-fold greater than that of cephalothin or cefazolin.

Cefazolin sodium [27164-46-1], (**141**), $C_{14}H_{13}N_8NaO_4S_3$, M_r 476.50.

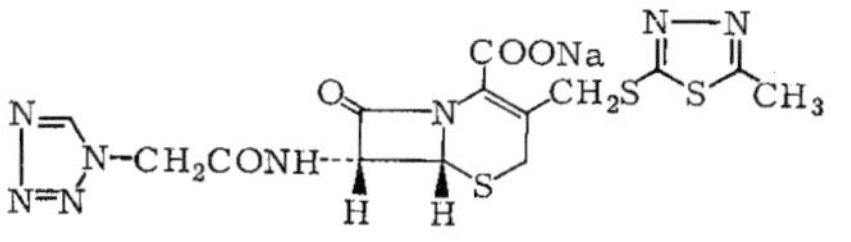

141

Cefazolin was synthesized by Fujisawa Pharmaceutical Co. in 1969 [282]. It was the first of the cephem antibiotics to introduce a thiadiazolylthiomethyl group at the 3 position and a tetrazole group at the 7 position in the side chain. Cefazolin is a parenteral cephem antibiotic showing better activity against gram-negative bacteria than cephalothin or cephaloridine. Its bactericidal activity, tissue distribution, and urinary excretion are excellent and it has wide clinical use.

Ceftizoxime sodium [68401-82-1], (**142**), $C_{13}H_{12}N_5NaO_5S_2$, M_r 405.39.

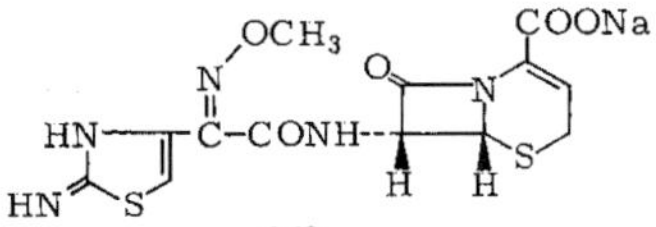

142

Ceftizoxime was synthesized by Fujisawa Pharmaceutical Industries in 1979 [283]. It possesses the (iminothiazolyl)methoxy-iminomethyl group at the 7 position of the cephem nucleus, but there is no side chain at the 3 position. The compound shows excellent activity against gram-positive and gram-negative bacteria, behavior similar to that of cefotaxime. Unlike cefotaxime, however, it is not metabolized in vivo.

Cefotaxime sodium [64485-93-4], (**143**), $C_{16}H_{16}N_5NaO_7S_2$, M_r 477.45.

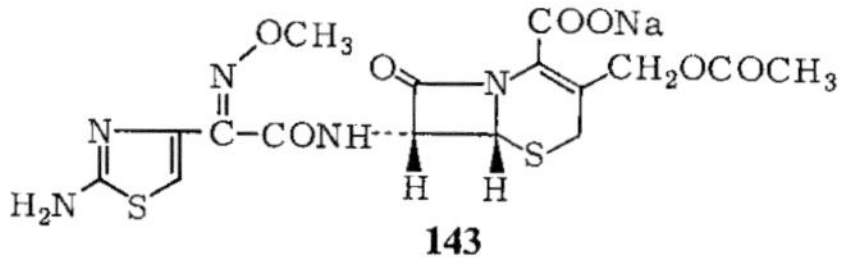

[Structure 143: Cefotaxime sodium — aminothiazole ring with H_2N and S, linked via $C(=N-OCH_3)-CONH-$ to cephem nucleus bearing COONa and CH_2OCOCH_3]

Cefotaxime was synthesized by Hoechst and Roussel-Uclaf in 1977 [284]. It was the first derivative of cephalosporin to introduce the methoxyimino and aminothiazole groups at the 7 position of the cephem nucleus. Although it shows unexpectedly low oral absorption, its excellent activity against a wide range of gram-positive and gram-negative organisms, including *Serratia, Enterobacter, Citrobacter,* and anaerobes, guided research and development of the newer synthetic cephems, the so-called third-generation cephalosporins.

Cefmenoxime hydrochloride [75738-58-8], (**144**), $(C_{16}H_{17}N_9O_5S_3)_2 \cdot HCl$, M_r 1059.59.

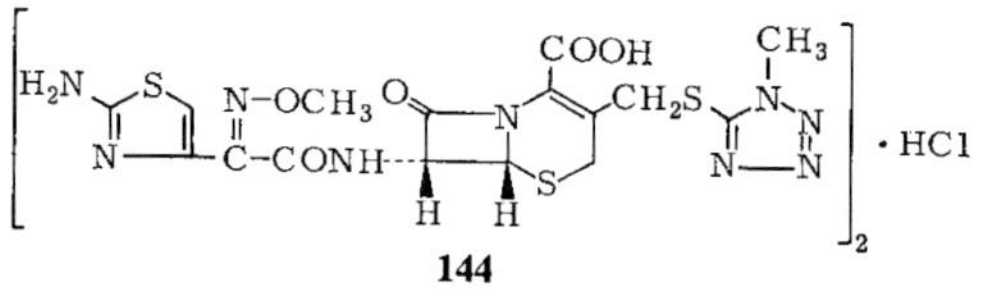

[Structure 144: Cefmenoxime hydrochloride — aminothiazole-methoxyimino acyl group attached to cephem nucleus with COOH and CH_2S-methyltetrazole group; shown as dimer · HCl]

Cefmenoxime was synthesized by Takeda Pharmaceutical Co. in 1978 [285]. It contains aminothiazole and methoxyimino groups in the 7-acyl position and a methyl-tetrazole group at the 3 position of the cephem nucleus. Cefmenoxime has a wide activity spectrum against gram-positive and gram-negative bacteria, including anaerobes, except *Staphylococcus aureus*. Cefmenoxime is a member of the so-called third-generation cephalosporins, a recently introduced class of clinically important antibiotics.

Cefuroxime [55268-75-2], (**145**), $C_{16}H_{16}N_4O_8S$, M_r 424.39.

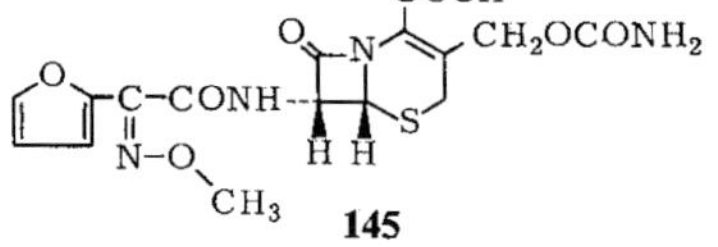

[Structure 145: Cefuroxime — furyl-C(=N-OCH$_3$)-CONH- attached to cephem nucleus with COOH and CH_2OCONH_2]

Cefuroxime was synthesized by Glaxo Laboratories in 1975 as the first cephem antibiotic with the methoxyimino group at the 7 position [286]. It is highly resistant to hydrolysis by cephalosporinase and is active against a variety of gram-negative bacteria, including indole-positive *Proteus, Enterobacter,* and *Citrobacter*. Cefuroxime is considered to be one of the so-called second-generation cephalosporins.

Ceftriaxone sodium [74578-69-1] (**146**), $C_{18}H_{16}N_8Na_2O_7S_3$, M_r 598.55.

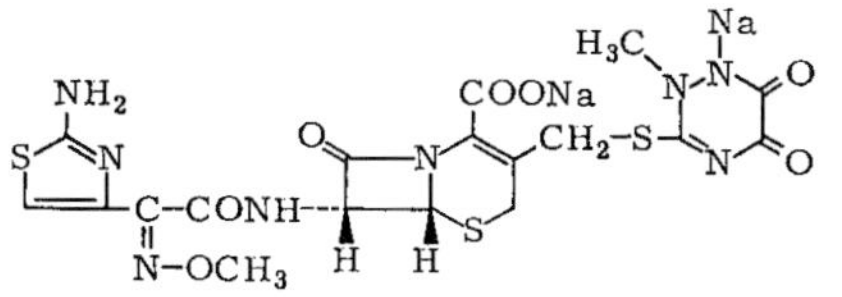

Ceftriaxone was synthesized by Hoffmann-La Roche in 1981 [287]. The triazinyl moiety was introduced at the 3 position of the cephem nucleus. The same side chain as possessed by cefotaxime and the other so-called third-generation cephalosporins was retained at the 7 position. The antibacterial activity of ceftiaxone is almost the same as that of cefotaxime in vitro, but its in vivo activity is 10 to 100 times higher. Its most characteristic property is its seven to eight hour half-life in serum, the longest among the known cephem antibiotics.

Ceftazidime [72558-82-8], (**147**), $C_{22}H_{22}N_6O_7S_2$, M_r 546.58.

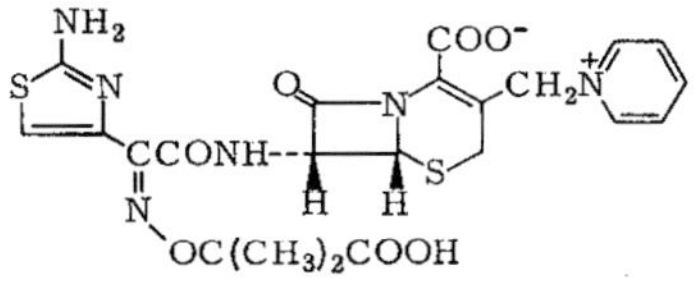

Ceftazidime was synthesized by Glaxo Laboratories in 1979 [288]. A new side chain, a carboxypropyloxyimino group with an aminothiazolyl group, was attached at the 7 position of the cephem nucleus. The compound shows excellent activity against *Pseudomonas aeruginosa*, indole-positive *Proteus* spp., *Enterobacter cloacae*, and *Serratia*, which are called opportunistic pathogens. Its activity against clinically isolated *P. aeruginosa* strains was found to be twofold higher than that of cefsulodin, a specific antibiotic widely used against that microorganism. Unlike cefsulodin, it shows balanced activity against other gram-negative bacteria and *Streptococcus*.

Cefoxitin [35607-66-0], (**148**), $C_{16}H_{17}N_3O_7S_2$, M_r 427.46.

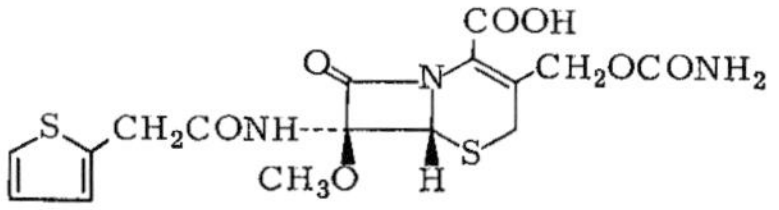

Cefoxitin was synthesized by Merck Sharp & Dohme Research Laboratories in 1972 as the first clinically applicable cephamycin; it was derived from a cephamycin component obtained from *Streptomyces lactamdurans* [289]. The drug is slightly hydrolyzed by β-lactamases and shows great activity against gram-negative bacteria and anaerobes.

Cefmetazole [56796-20-4], (**149**), $C_{15}H_{17}N_7O_5S_3$, M_r 471.54.

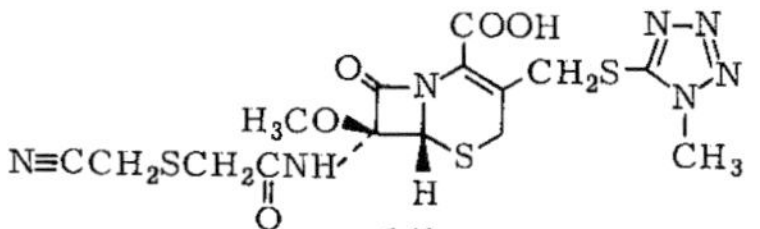

Cefmetazole was synthesized by Sankyo Co. in 1976 starting with a biosynthetic cephamycin [290]. It shows excellent activity against *Serratia* and *Proteus*, against which cefazolin is not active, and it has stronger activity than cefoxitin, another derivative of cephamycin. Cefmetazole is active against anaerobes and resistant to β-lactamase, but it is not active against *Pseudomonas aeruginosa*.

Cefotetan [69712-56-7] (**150**), $C_{17}H_{17}N_7O_8S_4$, M_r 575.62.

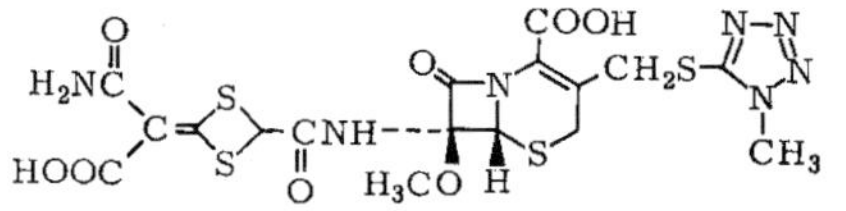

Cefotetan was synthesized by Yamanouchi Pharmaceutical Co. in 1979 starting with oganomycin, a cephamycin, produced by *Streptomyces oganoensis* YG19Z [291]. Its 7β side chain, 1,3-dithietan, is unique and contributes greatly to its strong activity against gram-negative bacteria, including *Serratia, Citrobacter, Enterobacter,* indole-positive *Proteus,* and anaerobes. Its half-life in the serum is as long as three hours, and about 90 % of it is excreted in the urine.

Latamoxef [64953-12-4], (**151**), disodium moxalactam, $C_{20}H_{18}N_6Na_2O_9S$, M_r 564.44.

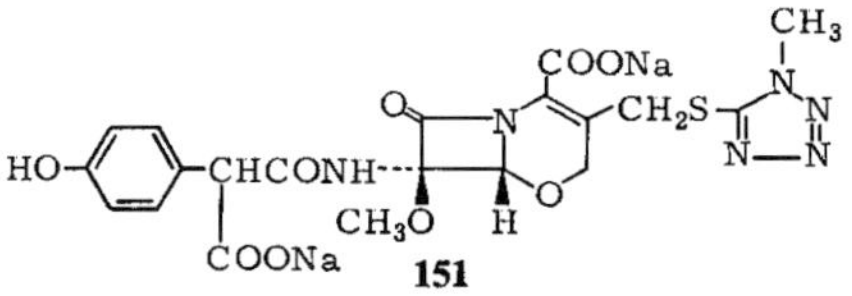

Latamoxef was synthesized by Shionogi Pharmaceuticals in 1975 starting with benzylpenicillin and using a novel drug design [292]. The oxacephem nucleus, in which the sulfur atom had been replaced by oxygen, was substituted with a methoxyl group at the 7α position, as in the cephamycins. A carboxyl moiety and a hydroxybenzyl group were added at the 7β position, as in carbenicillin, and a methyltetrazolylthiomethyl group was attached at the 3 position. These substitutions resulted in a strong activity against gram-negative bacteria and a high resistance to the action of β-lactamase, along with excellent activity against *Pseudomonas aeruginosa*, even though the compound has no activity against *Staphylococcus aureus*.

Clavulanic acid [58001-44-8], (**152**), $C_8H_9NO_5$, M_r 199.16.

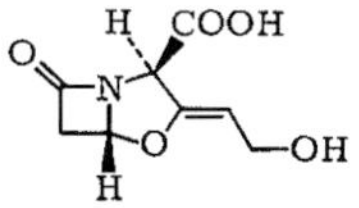

152

Clavulanic acid was found in the culture broth of *Streptomyces clavuligerus* by Beecham Research Laboratories in 1976 [293]. It was the first β-lactamase inhibitor. This antibiotic shows weak antibacterial activity against gram-positive and gram-negative organisms but strong inhibitory activity against the β-lactamase produced by ampicillin-resistant bacteria. Clavulanic acid is used orally in combination with amoxicillin and with ticarcillin by injection to enhance the activities of these antibiotics against resistant infections.

Imipenem [64221-86-9], (**153**), *N*-formimidoylthienamycin.

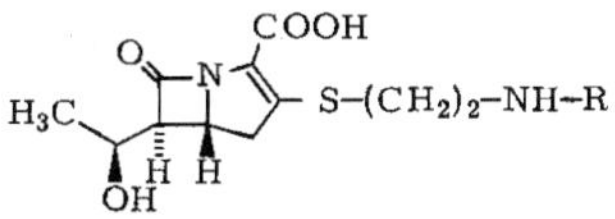

Thienamycin: R = H
$C_{11}H_{16}N_2O_4S$, M_r 272.33
N-Formimidoylthienamycin: R = -CH=NH
$C_{12}H_{17}N_3O_4S$, M_r 299.35

153

Thienamycin was found in the culture broth of *Streptomyces cattleya* by Merck Sharp & Dohme in 1976, as a very unstable substance. It has a unique carbapenem structure, like that of the olivanic acids found in *S. olivaceus* by Beecham Research Laboratories in 1979 [294]. Thienamycin shows excellent activity against a variety of pathogenic bacteria, including *Pseudomonas aeruginosa*. Its chemical stability has been improved by derivatization with the formimidoyl group, and its biological stability has been improved by combining it with cilastatin, an inhibitor of kidney dihydropeptidase. The combination drug imipenem–cilastatin is now under study to evaluate its clinical efficacy and safety.

Aztreonam [78110-38-0], (**154**), azthreonam, $C_{13}H_{17}N_5O_8S_2$, M_r 435.44.

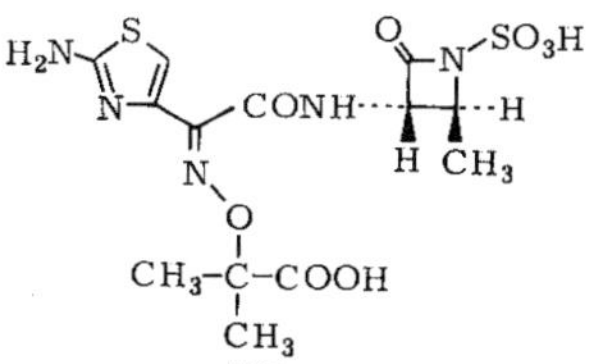

154

Aztreonam was synthesized by the Squibb Institute for Medical Research in 1981 starting with L-threonine. The synthesis was based on findings about bacterial β-lactam compounds of a monocyclic nature [295]. The β-lactam compounds, called monobactams, were isolated from *Chromobacterium violaceum, Agrobacterium radiobacter*, etc. Such monocyclic β-lactams of bacterial origin had previously been found independently in 1981 by Takeda Chemicals Industries in the culture broths of *Pseudomonas acidophila* and *P. mesoacidophila* and named sulfazecin and isosulfazecin, respectively [82] (see p. 976). Aztreonam was selected from among hundreds of derivatives as a candidate for clinical trials because of its unique antibacterial spectrum and strong activity. This antibiotic shows excellent activity against a variety of gram-negative aerobic bacteria but no activity against gram-positive bacteria or anaerobes. Its efficacy and safety are now being clinically evaluated.

Tetracycline [*60-54-8*], (**155**), $C_{22}H_{24}N_2O_8$, M_r 444.45.

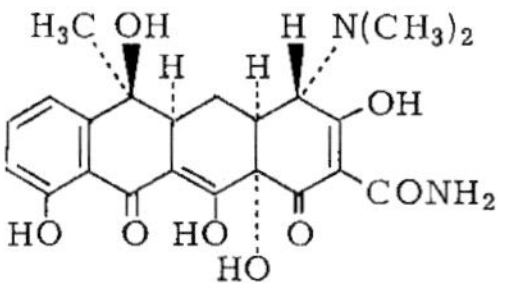

Tetracycline was first obtained from chlortetracycline by reductive dehalogenation at Lederle Laboratories and Pfizer in 1953 [296]. It was also obtained either by fermentation of the chlortetracycline-producing organism, *Streptomyces aureofaciens*, under conditions of chlorine limitation, or by fermentation of a mutant of the organism lacking the chlorinating enzyme. Tetracycline is more stable than chlortetracycline in aqueous solution. Its antimicrobial activity is the same as that of chlortetracycline and oxytetracycline, but its serum concentration after oral administration is considerably higher and more enduring.

Chlortetracycline hydrochloride [*64-72-2*], (**156**), $C_{22}H_{23}ClN_2O_8 \cdot HCl$, M_r 515.35.

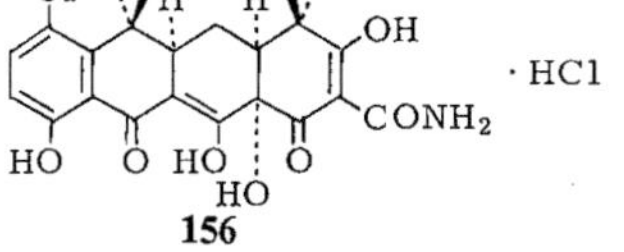

Chlortetracycline was found in the culture broth of *Streptomyces aureofaciens* by DUGGAR et al. of Lederle Laboratories in 1948 and named aureomycin [85]. It shows a wider range of antibiotic activity than the earlier antibiotics, penicillins, and streptomycins and as great as that of chloramphenicol. Its activity covers gram-positive and gram-negative bacteria as well as *Rickettsia* and *Chlamydiae*. Chlortetracycline has been replaced by other tetracyclines in clinical use and is used now used as a feed additive to promote the growth of livestock.

Demethylchlortetracycline [*127-33-3*], (**157**), demeclocycline, $C_{21}H_{21}ClN_2O_8$, M_r 464.86.

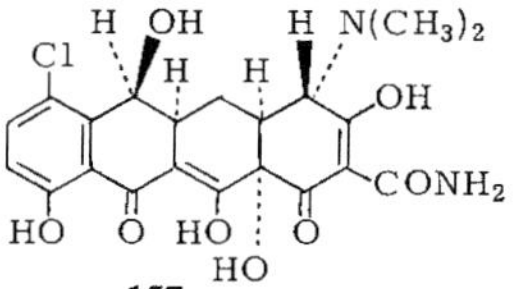

157

Demethylchlortetracycline was isolated from the culture broth of a mutant of *Streptomyces aureofaciens*, the chlortetracycline-producing strain, by Lederle Research Laboratories in 1957 [297]. It shows one and one-half to two times as much in vitro antimicrobial activity and in vivo protective effect as tetracycline. Its base and hydrochloride have been used orally and by topical application to treat infections caused by *Staphylococcus, Streptococcus, Rickettsia, Chlamydiae, Neisseria, Klebsiella, Proteus, Escherichia coli,* and *Haemophilus influenzae.*

Oxytetracycline [*6153-64-6*], (**158**), $C_{22}H_{24}N_2O_9 \cdot 2\,H_2O$, M_r 496.48.

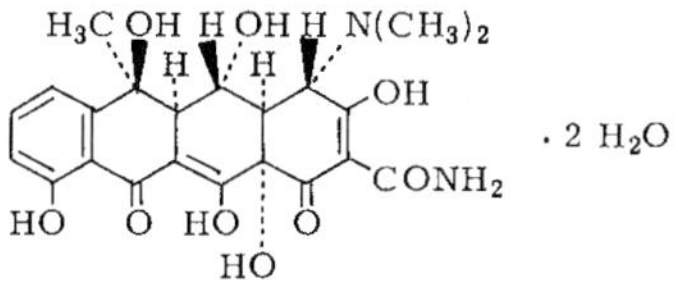

158

Oxytetracycline was found in the culture broth of *Streptomyces rimosus* by Pfizer in 1950 and named terramycin [298]. It is closely related to chlortetracycline (aureomycin), and its structure was elucidated in 1952. Oxytetracycline has had wide clinical use as a substitute for chlortetracycline, even after the introduction of tetracycline. Oxytetracycline is administered orally, topically, and parenterally.

Methacycline [*914-00-1*], (**159**), metacycline, $C_{22}H_{22}N_2O_8$, M_r 442.43.

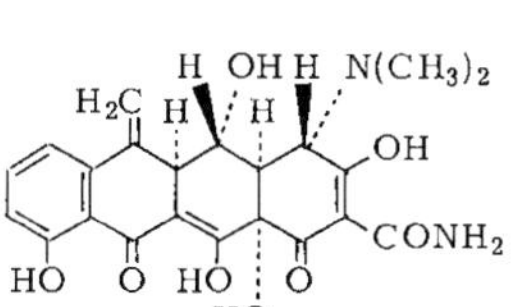

159

Methacycline was synthesized by Pfizer Research Laboratories in 1961 starting with oxytetracycline [299]. It shows two to fivefold greater in vitro antibacterial activity and in vivo protective activity than tetracycline. Methacycline has been given orally to treat infections by *Rickettsia, Chlamydiae, Staphyloccus, Streptococcus, Neisseria, Klebsiella, Proteus, Escherichia coli,* and *Haemophilus influenzae.*

Doxycycline [*17086-28-1*], (**160**), $C_{22}H_{24}N_2O_8 \cdot H_2O$, M_r 462.46.

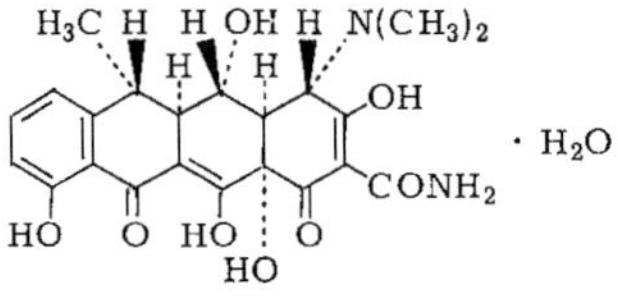

160

Doxycycline was synthesized by Pfizer starting with either tetracycline or oxytetracycline in 1958 [300]. It shows about fourfold higher antibacterial activity than tetracycline against a variety of pathogens. Doxycycline shows higher oral absorption than tetracycline, and its concentration in tissue is higher and maintained longer. Hyclate is the preparation of doxycycline currently in use; it contains one mole of hydrochloric acid and one-half mole each of ethanol and water per mole of doxycycline.

Rolitetracycline [*751-97-3*], (**161**), $C_{27}H_{33}N_3O_8$, M_r 527.58.

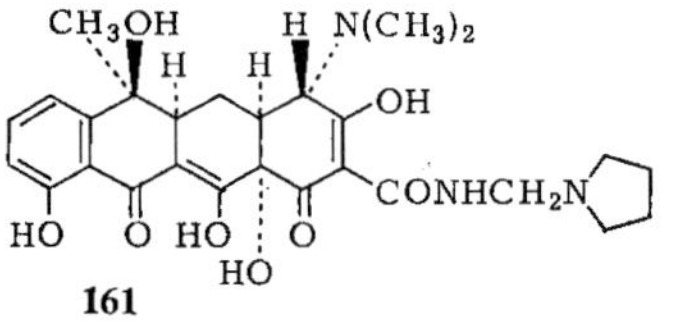

161

Rolitetracycline was synthesized in 1958. Bristol-Myers Laboratories prepared it from tetracycline by introducing a pyrrolidinylmethyl moiety [301]. This antibiotic is very soluble in water and more stable than tetracycline under acidic conditions. Rolitetracycline is used by intravenous injection, and its nitrate is used by both intravenous and intramuscular injections for therapy of the same infections as those treated by tetracycline.

Minocycline [*10118-90-8*], (**162**), $C_{23}H_{27}N_3O_7$, M_r 457.49.

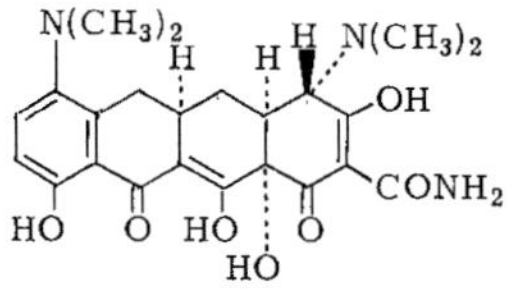

162

Minocycline was synthesized by American Cyanamid in 1966 in the course of studies on derivatives of 6-deoxytetracyclines [302]. It shows activity against tetracycline-resistant bacteria and higher activity than tetracycline against a variety of pathogens. Minocycline has a higher oral absorption than tetracycline, and its concentrations in several tissues are four to tenfold higher than that of tetracycline. It is widely used by oral administration, and it is also administered by drip infusion for serious infections.

Daunorubicin hydrochloride [23541-50-6], (**163**), $C_{27}H_{29}NO_{10} \cdot HCl$, M_r 563.99.

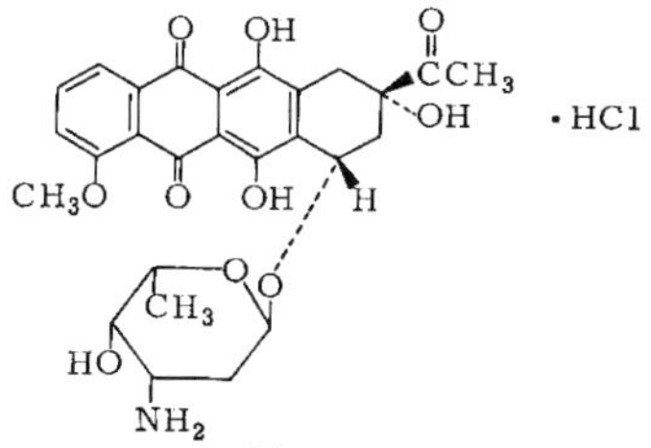

163

Daunorubicin was found independently by Farmitalia and Rhône Poulenc in 1963 in the mycelium of *Streptomyces peucetius* and the culture broth of *Streptomyces coeruleorubidus*, respectively [303]. It was the first anthracycline antibiotic clinically used for therapy of cancers, especially leukemia. Daunorubicin is used in combination with other anticancer drugs.

Doxorubicin [23214-92-8], (**164**), $C_{27}H_{29}NO_{11}$, M_r 543.53.

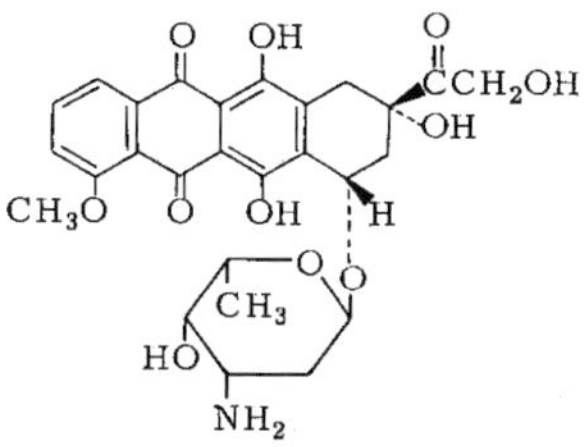

164

Doxorubicin was found in the culture broth of *Streptomyces peucetius* var. *cesius* by Farmitalia in 1967 in the course of studies of anthracycline antibiotics [304]. It shows stronger activity against a variety of tumors and leukemia than daunorubicin, and its clinical application in the therapy of cancer is wider than that of daunorubicin. Doxorubicin is sometimes called by its old generic name, adriamycin, especially in the medical field.

Aclarubicin [57576-44-0], (**165**), $C_{42}H_{53}NO_{15}$, M_r 811.89.

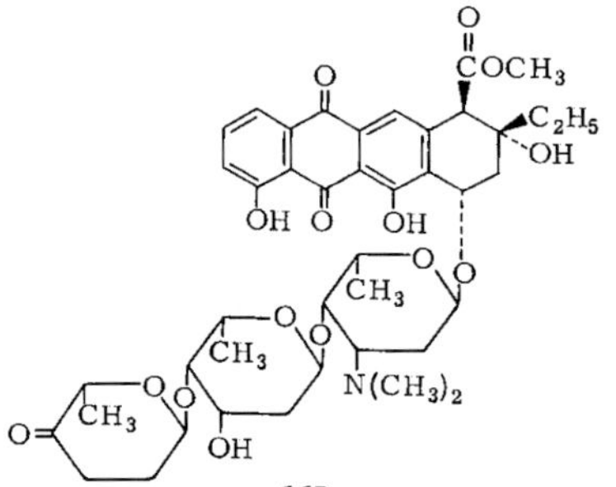

165

Aclarubicin was found in the culture broth of *Streptomyces galilaeus* MA144-M1 by UMEZAWA et al. of the Institute of Microbial Chemistry in 1975 [305]. It was produced along with structurally related compounds showing antileukemic activity and named aclacinomycin A. Sanraku-Ocean cooperated in isolating aclacinomycin A as a yellow crystalline powder and evaluated its strong antileukemic activity and low cardiac toxicity. Its generic name was changed to aclarubicin on the recommendation of the World Health Organization.

Kanamycin sulfate [25389-94-0], (**166**), $C_{18}H_{36}N_4O_{11} \cdot H_2SO_4$, M_r 582.59.

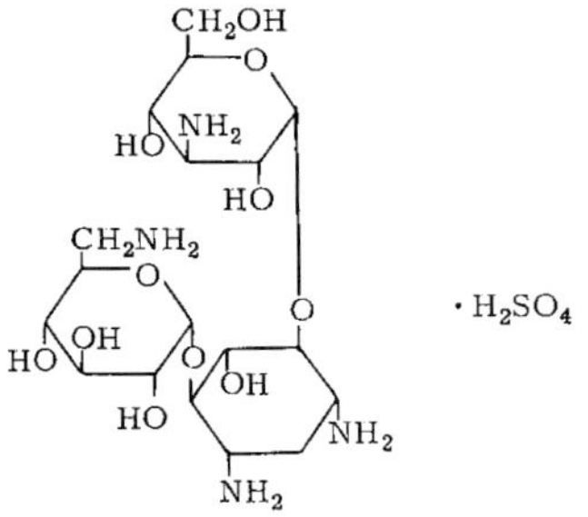

166

Kanamycin was found by UMEZAWA et al. in the culture broth of *Streptomyces kanamyceticus* in 1957 [306]. It is produced with other components, kanamycins B (bekanamycin) and C, which are also separated during the purification. The compound shows much lower toxicity than the earlier aminoglycosides, streptomycin and neomycin, and strong activity against a wide range of gram-positive and gram-negative bacteria, including *Mycobacterium*. Kanamycin has been used clinically for treatment of such serious infections as dysentery, salmonellosis, and tuberculosis.

Bekanamycin [*4696–76–8*], (**167**), kanamycin B, $C_{18}H_{37}N_5O_{10}$, M_r 483.52.

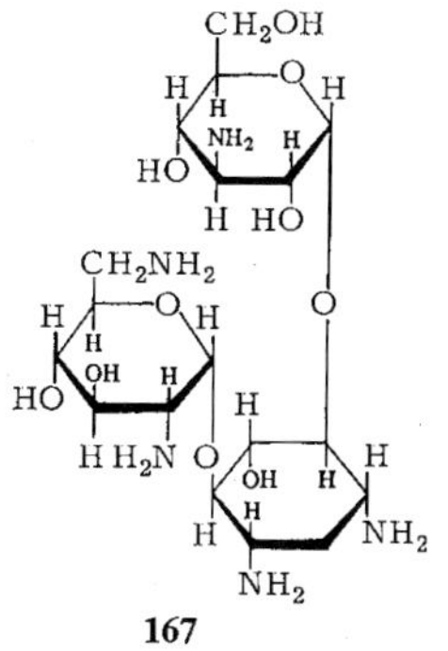

167

Bekanamycin, kanamycin B, was found in the culture broth of *Streptomyces kanamyceticus* by UMEZAWA et al. in 1957 [307]. It shows the same antibacterial spectrum as kanamycin but with stronger activity. The total synthesis of bekanamycin was completed by UMEZAWA et al. in 1968 and the knowledge gained from its synthesis was successfully applied to the synthesis of dibekacin.

Tobramycin [*32986-56-4*], (**168**), $C_{18}H_{37}N_5O_9$, M_r 467.52.

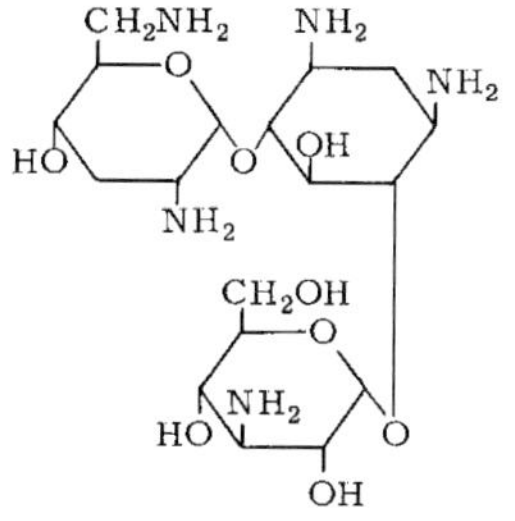

168

Tobramycin was found in 1967 by Eli Lilly Co. in the nebramycins complex that was produced in the culture broth of *Streptomyces tenebrarius* [308]. Structurally it is closely related to kanamycin, a naturally produced 3′-deoxy derivative of bekanamycin. The 3′-hydroxyl group was found to be the target of enzymatic phosphorylation by resistant bacteria. As expected, tobramycin shows strong activity against resistant bacteria, including *Pseudomonas aeruginosa*, having this phosphorylating enzyme.

Gentamicin sulfate [1405-41-0], (**169**), gentamicin.
Gentamicin C_1, $C_{21}H_{43}N_5O_7$, M_r 477.61.
Gentamicin C_2, $C_{20}H_{41}N_5O_7$, M_r 463.58.
Gentamicin C_{1a}, $C_{19}H_{39}N_5O_7$, M_r 449.55.

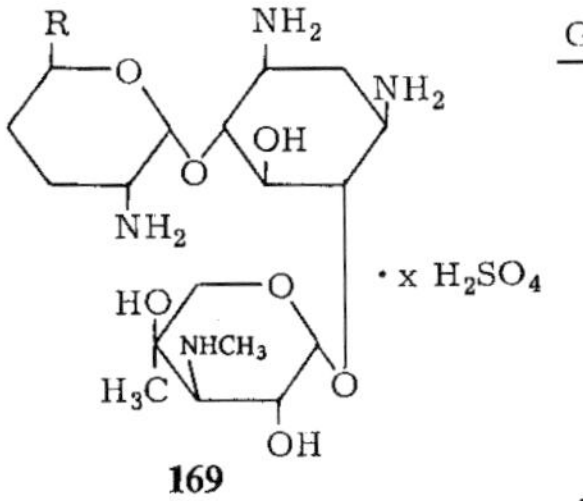

Gentamicin	R
C_1	$H_3C-HN-\overset{CH_3}{\underset{|}{C}}-H$
C_2	$H_2N-\overset{CH_3}{\underset{|}{C}}-H$
C_{1a}	$-CH_2NH_2$

Gentamicin was found by Schering Plough Co. in the culture broths of *Micromonospora purpurea* and *M. echinospora* in 1963 [309]. It is a mixture of at least 16 structurally related components. The major components used in clinical preparations are gentamicins C_{1a}, C_1, and C_2. Gentamicin shows high activity against a variety of gram-positive and gram-negative bacteria, including *Pseudomonas aeruginosa*, *Proteus*, and *Serratia*. It has been widely used for the clinical treatment of serious infections. Gentamicin is used alone or in combination with β-lactam antibiotics and is being replaced gradually by the newer, less toxic aminoglycosides.

Dibekacin [34493-98-6], (**170**), $C_{18}H_{37}N_5O_8$, M_r 451.52.

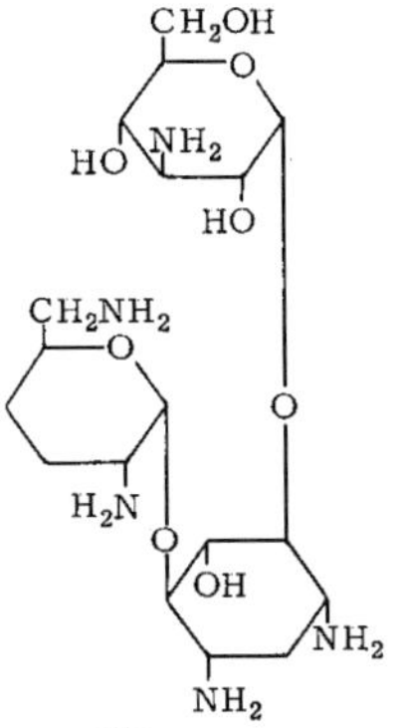

Dibekacin was synthesized in 1967 by UMEZAWA et al. by the removal of the 3′- and 4′-hydroxyl groups of kanamycin B [310]. Studies by the same workers on the mechanisms of bacterial resistance to kanamycin-group antibiotics preceded the discovery. Dibekacin shows excellent activity, as expected, against a variety of bacteria, including kanamycin-resistant strains. It shows higher activity than kanamycin against *Pseudomonas aeruginosa*, *Proteus*, and other pathogens.

Amikacin [*37517-28-5*], (**171**), $C_{22}H_{43}N_5O_{13}$, M_r 585.61.

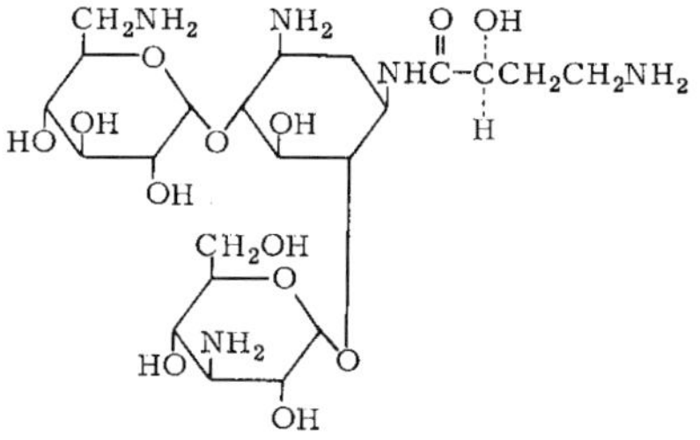

171

Amikacin was synthesized by KAWAGUCHI et al. of the Bristol-Banyu Research Institute in 1970 starting with kanamycin and the acyl moiety of butirosin [311]. Its design is based on knowledge of the mechanisms of bacterial resistance to kanamycin and related compounds in which the 3′-hydroxyl group of the antibiotic is phosphorylated enzymatically. The acyl moiety in butirosin prevents this enzymatic inactivation.

Micronomicin [*52093-21-7*], (**172**), $C_{20}H_{41}N_5O_7$, M_r 463.58.

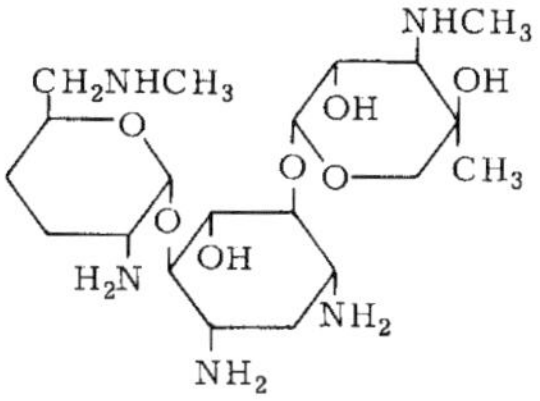

172

Micronomicin, formerly called sagamicin, was found in the culture broth of *Micromonospora sagamiensis* var. *nonreducans* by Kyowa Hakko Kogyo Company and Abbott Laboratories in 1974 [312]. The compound is identical with one of the minor components contained in gentamicin, gentamicin C_{2b}, but it is produced as a single component. Its amino group at the 6′-position is methylated and is not subject to the enzymatic acetylation caused by resistant bacteria, including *Pseudomonas aeruginosa*. Micronomicin is less toxic than gentamicin to the renal and aural systems.

Ribostamycin [25546-65-0], (**173**), $C_{17}H_{34}N_4O_{10}$, M_r 454.48.

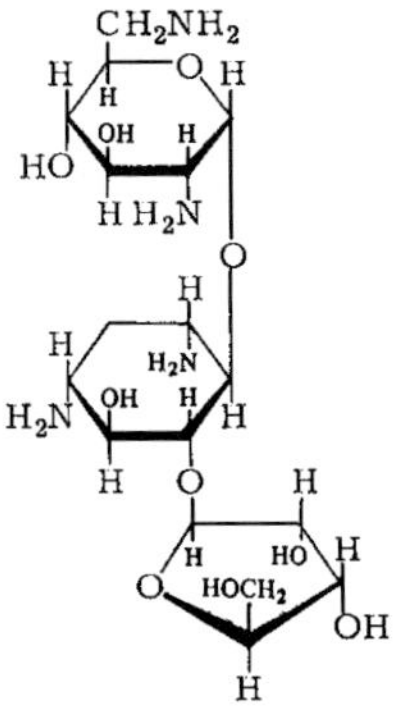

173

Ribostamycin was found in the culture broth of *Streptomyces ribosidificus* by Meiji Seika Kaisha in 1970 in the course of screening aminoglycoside antibiotics [313]. It is structurally related to neomycin but lacks the diaminoidose (glucose) moiety substituted on the ribose moiety. Ribostamycin is much less toxic than neomycin and shows strong activity against a variety of gram-positive and gram-negative bacteria, except *Pseudomonas aeruginosa*. It is used parenterally for therapy of urinary tract, respiratory tract, surgical, and other infections.

Neomycin sulfate [1405-10-3], (**174**), fradiomycin, $C_{23}H_{46}O_{13}N_6$, M_r 614.66.

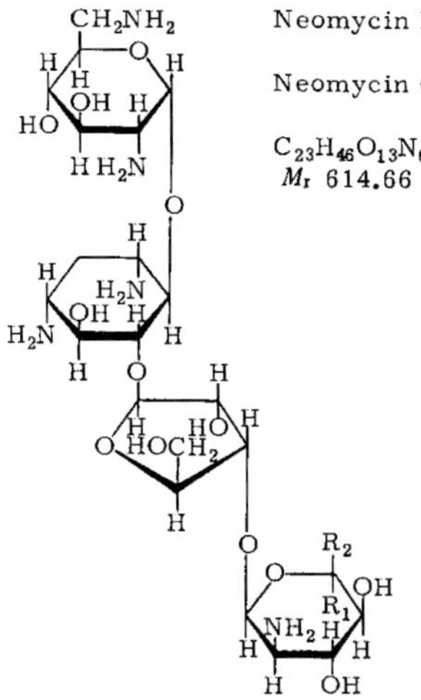

Neomycin B: R_1 = H
R_2 = CH_2NH_2
Neomycin C: R_1 = CH_2NH_2
R_2 = H

$C_{23}H_{46}O_{13}N_6$
M_r 614.66

174

Neomycin was found independently in 1949, by UMEZAWA et al. and WAKSMAN et al. in the culture broth of *Streptomyces fradiae* [314]. It consists of two closely related components, B and C, and shows strong activity against a wide range of gram-positive and gram-negative bacteria, including *Serratia* and *Pseudomonas aeruginosa*. Because of its renal and ototoxicity, it is given orally or by topical application. Because it is not absorbed orally, as are other aminoglycoside antibiotics, it is used orally only for the purpose of suppressing intestinal flora, i.e., in treating dysentery, salmonellosis, and diarrhea in the pediatric field. Neomycin also has been used topically in the treatment of bacterial infections of the eye and skin.

Paromomycin sulfate [1263-89-4], (**175**), $C_{23}H_{45}N_5O_{14} \cdot xH_2SO_4$, M_r 615.64 (base).

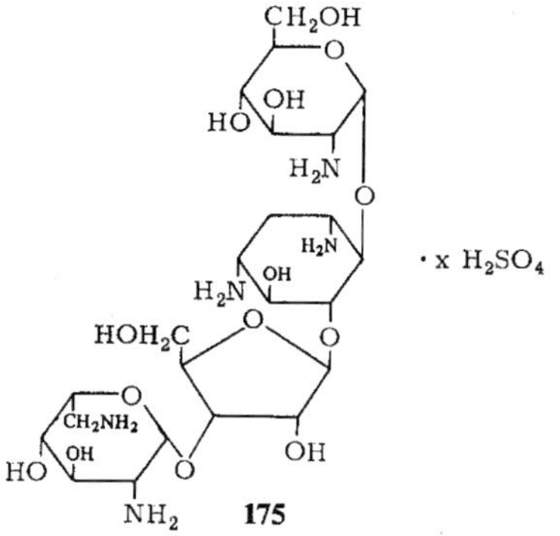

Paromomycin was found in the culture broth of *Streptomyces rimosus* forma *paromomycinus* by Parke Davis & Co. in 1959 [315]. In the same year it was found in the culture broth of *S. Crestomyceticus* by Farmitalia and named amminosidin. Paromomycin is related structurally to neomycin, but it has a hydroxyl group at the 6′ position, whereas neomycin has an amino group. Its antibacterial activity is weaker than that of neomycin, but its toxicity is much less. Paromomycin is used by intramuscular injection for therapy of respiratory, urinary, and surgical infections and by oral administration to treat dysentery and salmonellosis.

Streptomycin sulfate [3810-74-0], (**176**), $C_{42}H_{84}N_{14}O_{36}S_3$, M_r 1457.40.

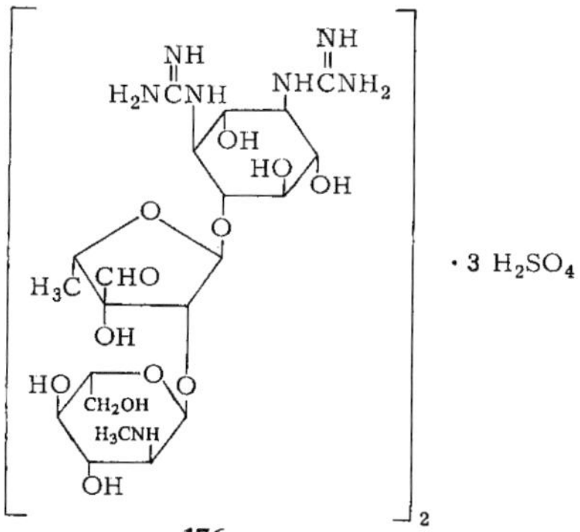

Streptomycin was found in the culture broth of *Streptomyces griseus* by S. A. WAKSMAN of Rutgers University in 1944; it was the second antibiotic introduced clinically (after penicillin) [316]. This drug is a water soluble, basic substance having an aminoglycoside structure and showing strong activity against a wide range of gram-positive and gram-negative bacteria including *Mycobacterium*. Streptomycin is the first choice among antituberculotic antibiotics and has been used for the therapy of *Spirochaeta* and *Treponema* infections.

Dihydrostreptomycin [*128-46-1*], (**177**), $C_{21}H_{41}N_7O_{12}$, M_r 583.60.

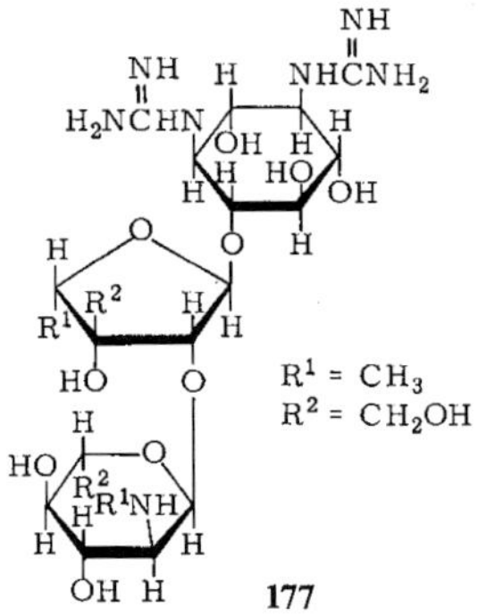

Dihydrostreptomycin was first synthesized by Parke Davis Co. in 1946 by the reduction of streptomycin [317]. Naturally occurring dihydrostreptomycin was found in the culture broth of *Streptomyces humidus* by Takeda Chemicals Industries in 1957. Its hydrochloride or sulfate is more easily cristallized and more stable under alkaline conditions than streptomycin. Dihydrostreptomycin has been used for therapy of tuberculosis, but because it has a higher ototoxicity than streptomycin its use is now restricted to animal therapy.

Destomycin A [*14918-35-5*], (**178**), $C_{20}H_{37}N_3O_{13}$, M_r 527.53.

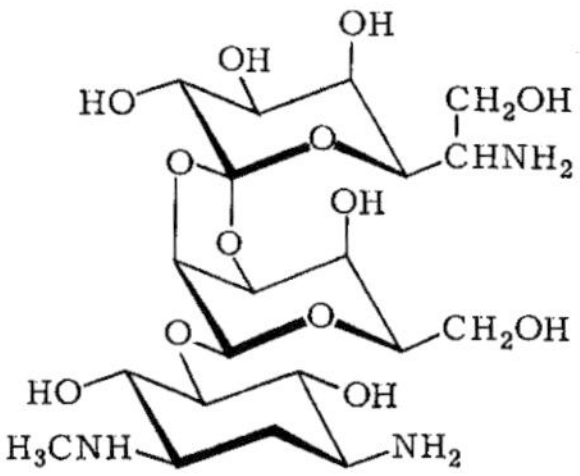

Destomycin A was found in the culture broth of *Streptomyces rimofaciens* by Meiji Seika Kaisha in 1965 in the course of screening for water-soluble and basic antibiotics, such as aminoglycosides [318]. It shows activity against a variety of gram-positive and gram-negative bacteria as well as fungi and, more interestingly, against helminths. Destomycin A has been used to treat helminth infections in swine and poultry.

Hygromycin B [31282-04-9], (**179**), $C_{20}H_{37}N_3O_{13}$, M_r 527.53.

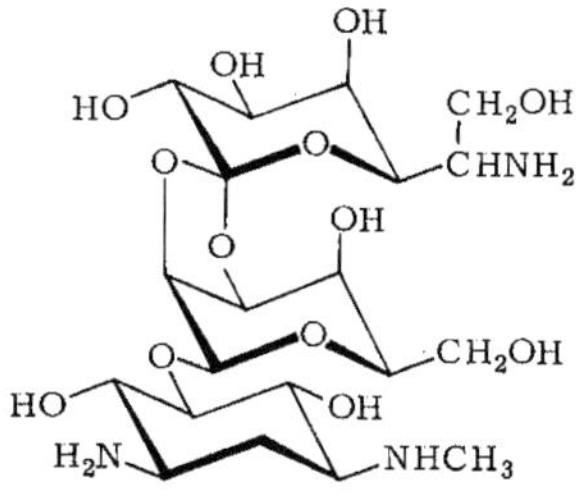

179

Hygromycin B was found in the culture broth of *Streptomyces hygroscopicus* by Lilly Research Laboratories in 1953, as a mixture with hygromycin A [319]. It showed activity against a variety of gram-positive and gram-negative bacteria as well as fungi. Hygromycin B has been used for therapy of helminthic infections in swine and poultry.

Apramycin [41194-16-5], (**180**), $C_{21}H_{41}N_5O_{11}$, M_r 539.59.

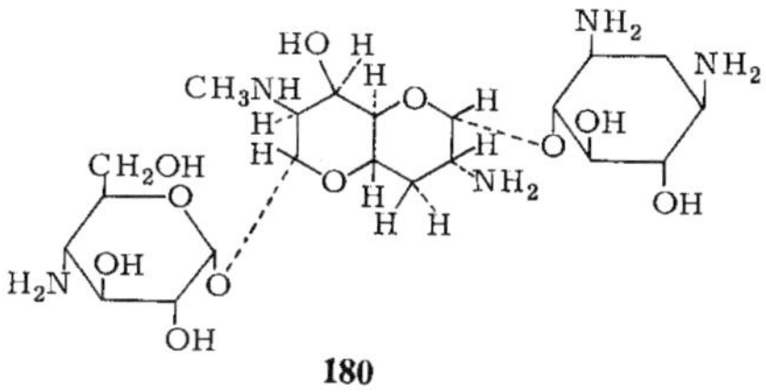

180

Apramycin was found in the culture broth of *Streptomyces tenebrarius* by Eli Lilly in 1967 as one of the eight components of the nebramycin complex [320]. Apramycin is used in feed additives.

Sisomicin [32385-11-8], (**181**), $C_{19}H_{37}N_5O_7$, M_r 447.54.

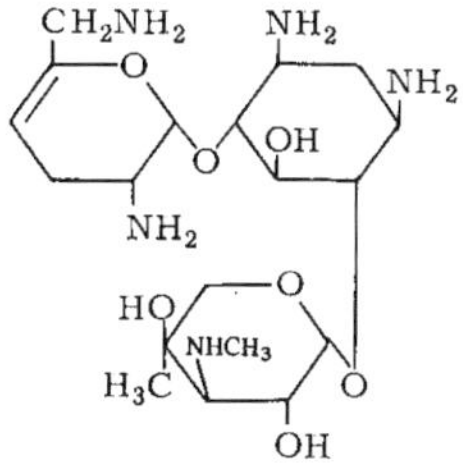

181

Sisomicin was found in the culture broth of *Micromonospora inyoensis* by Schering-Plough Co. in 1970, following the discovery of gentamicin by the same research group [321]. The structure and activity of sisomicin are very similar to those of gentamicin C_{1a}, the major component of the gentamicin complex. Sisomicin shows stronger bacterial activity and lower renal and ototoxicity than gentamicin C_{1a}.

Netilmicin sulfate [56391-57-2], (**182**), $(C_{21}H_{41}N_5O_7)_2 \cdot 5\ H_2SO_4$, M_r 1441.57.

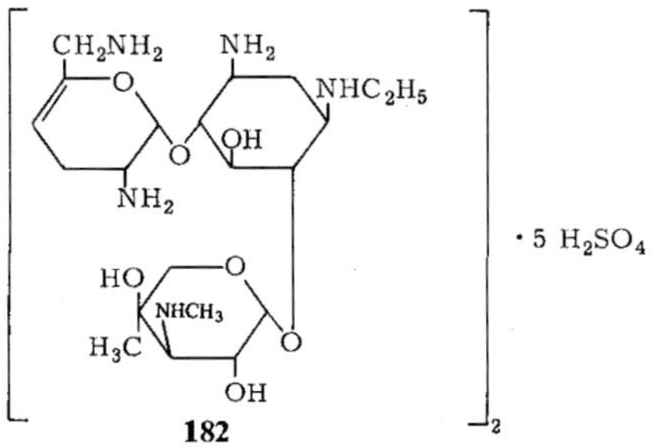

Netilmicin was synthesized by Schering-Plough Co. in 1976 [322]. The ethyl moiety was introduced at the 1-amino group of sisomicin. Its design was based on an understanding of the biochemical mechanisms of bacterial resistance to the gentamicin–sisomicin antibiotic group. The modification at the 1-amino group is known to prevent adenylation at the 2″-hydroxyl group and acetylation at the 3-amino group, and to deter acetylation at the 6′-amino residue. Netilmicin shows almost the same activity against a variety of gram-positive and gram-negative bacteria as sisomicin and strong activity against gentamicin–sisomicin-resistant bacteria. It is now under study for evaluation of its clinical efficacy and safety.

Spectinomycin hydrochloride [22189-32-8], (**183**), $C_{14}H_{24}N_2O_7 \cdot 2\ HCl \cdot 5\ H_2O$, M_r 495.36.

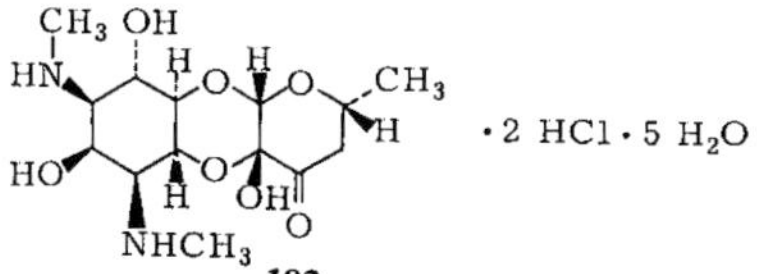

Spectinomycin was found in the culture broth of *Streptomyces spectabilis* by Upjohn Co. in 1961 [323]. This antibiotic was intended for use in treating various infections, as are the other aminoglycoside antibiotics, but its activity is insufficient for clinical efficacy except against gonorrhea. Spectinomycin does not develop crossresistance with any other antibiotic and shows low toxicity; it is used by deep intramuscular injection for "single-session," bolus-injection therapy of gonorrhea.

Astromicin sulfate [72275-67-3], (**184**), fortimicin A, $C_{17}H_{35}N_5O_6 \cdot 2\,H_2SO_4$, M_r 601.65.

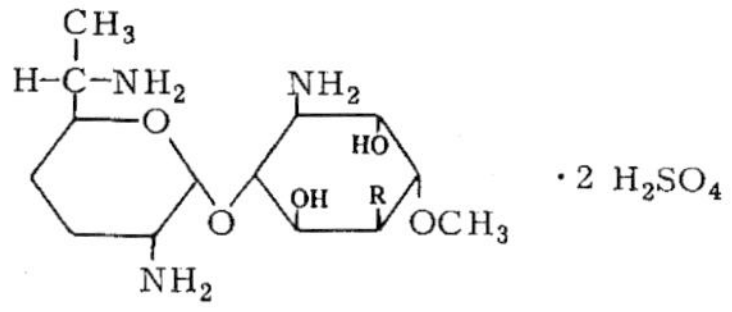

[Structure of compound 184]
R = CH$_3$–N–COCH$_2$NH$_2$
184

Astromicin was found in the culture broth of *Micromonospora olivoasterospora* by NARA et al. of Kyowa Hakko Kogyo Co. in 1976 [324]. It has a unique conformation with an acylated diamino inositol moiety different from other aminoglycoside antibiotics. Astromicin is produced with one major byproduct, fortimicin B, and several minor components.

Validamycin [37248-47-8], (**185**), validamycin A, $C_{20}H_{35}NO_{13}$, M_r 497.50.

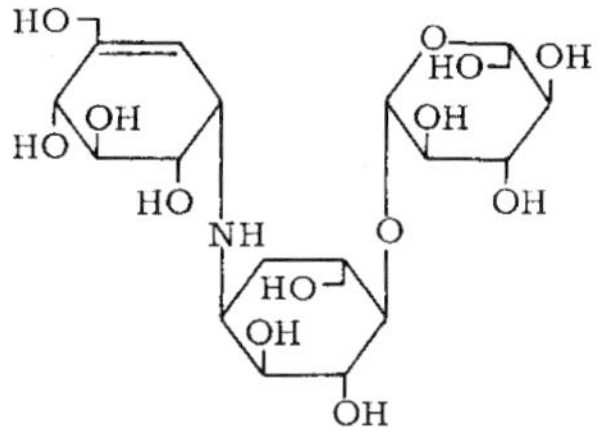

[Structure of Validamycin A]
Validamycin A
185

Validamycin was found in the culture broth of *Streptomyces hygroscopicus* var. *limoneus* by Takeda Chemicals Industries in 1971 in the course of screening for substances active against sheath blight in the rice plant [325]. It consists of five components. The major component, validamycin A, was found to be the major contributor to the activity. Validamycin is active against a variety of phytopathogenic fungi, especially *Pellicularia sasakii*, and it has been used to protect rice plants against sheath blight.

Kasugamycin [6980-18-3], (**186**), $C_{14}H_{25}O_9N_3$, M_r 379.37.

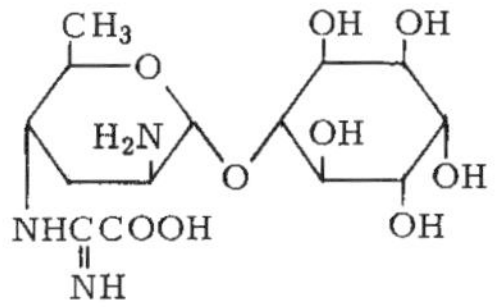

[Structure of compound 186]
186

Kasugamycin was found in the culture broth of *Streptomyces kasugaensis* by UMEZAWA et al. in 1965 [326]. It has an aminocyclitol structure and shows strong activity against phytopathogenic fungi, especially *Pericularia oryzae*, the pathogen causing rice blast.

This drug also shows activity against *Pseudomonas,* and its toxicity is very low; no mice died following intravenous injection of doses as high as two grams per kilogram. Kasugamycin has been used to protect rice plants against rice blast and for animal infections.

Polyoxin [*11113-80-7*], (**187**), polyoxin A [*19396-03-3*], $C_{23}H_{32}N_6O_{14}$, M_r 616.54, poly-oxin B, [*19396-06-6*], $C_{17}H_{25}N_5O_{13}$, M_r 507.41.

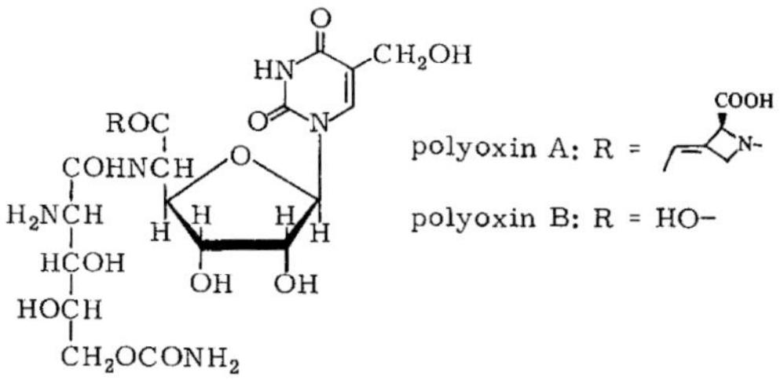

187

Polyoxin was found in the culture broth of *Streptomyces cacaoi* var. *asoensis* by SUZUKI et al. in 1965 [145]–[147]. It consists of several closely related components, A through O, and shows activity against phytopathogenic fungi by inhibition of cell-wall chitin synthesis. Polyoxin has been used in agriculture against fungal infections, especially *Alternaria* leaf spot in vegetables and fruits.

Blasticidin S [*2079-00-7*], (**34**), $C_{17}H_{26}N_8O_5$, M_r 422.45 (for structure, see p. 995).

Blasticidin S was found in the culture broth of *Streptomyces griseochromogenes* by YONEHARA et al. of the University of Tokyo in 1958 [157]. It has a nucleoside-analogue structure and shows strong activity against phytopathogenic fungi, especially *Pericularia oryzae,* the pathogen causing rice blast. Blasticidin S has been used to protect rice plants.

Erythromycin [*114-07-8*], (**188**), $C_{37}H_{67}NO_{13}$, M_r 733.95.

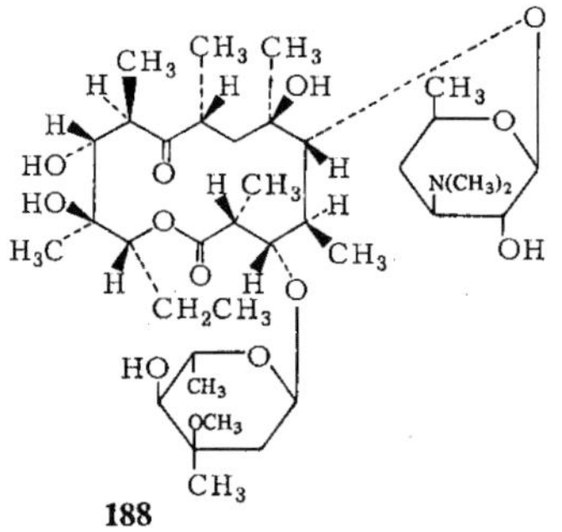

188

Erythromycin, the first of the macrolide antibiotics, was found in the culture broth of *Streptomyces erythreus* by Eli Lilly Co. in 1952 [327]. Its chemical structure and synthesis have been studied extensively by the WOODWARD group at Harvard University. Erythromycin shows activity against gram-positive bacteria and gram-negative cocci as well as *Mycoplasma* and *Leptospira*. Oral administration of its base and esters is used widely in clinics to treat respiratory tract and other infections. An ester, lactobionate, is used by drip infusion for serious infections (see next two paragraphs).

Erythromycin estolate [*3521-62-8*], (**189**), erythromycin 2'-propionate dodecyl sulfate (salt), $C_{40}H_{71}NO_{14} \cdot C_{12}H_{26}O_4S$, M_r 1056.41.

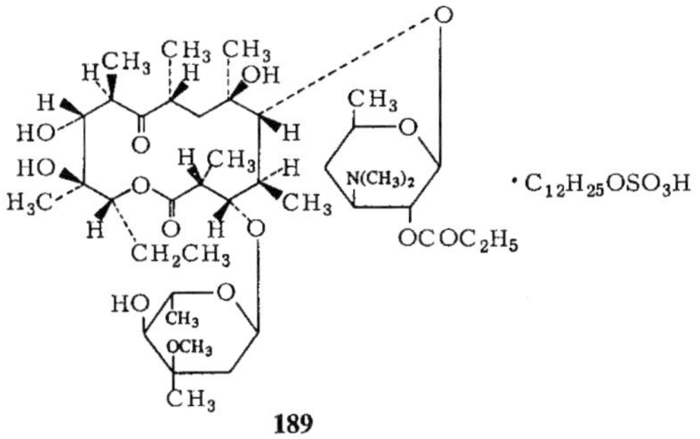

189

Erythromycin estolate was synthesized by the Lilly Research Laboratories in 1959 [328]. It shows a higher and longer lasting serum level than erythromycin and the other esters, ethylsuccinate and stearate, when administered orally. This antibiotic is stable against gastric acids, and after oral absorption it is hydrolyzed gradually into erythromycin. Long-term usage of erythromycin estolate is toxic to the liver. Erythromycin ethylsuccinate [*41342-53-4*], ($C_{43}H_{75}NO_{16}$, M_r 862.07) and erythromycin stearate [*643-22-1*] ($C_{37}H_{67}NO_{13} \cdot C_{18}H_{36}O_2$, M_r 1018.43) also have been used orally, and erythromycin lactobionate [*3847-29-8*] ($C_{37}H_{67}NO_{13} \cdot C_{12}H_{22}NO_{12}$, M_r 1092.25) has been used by intravenous administration and opthalmic topical application.

Oleandomycin phosphate [*7060-74-4*], (**190**), $C_{35}H_{61}NO_{12} \cdot H_3PO_4$, M_r 785.87.

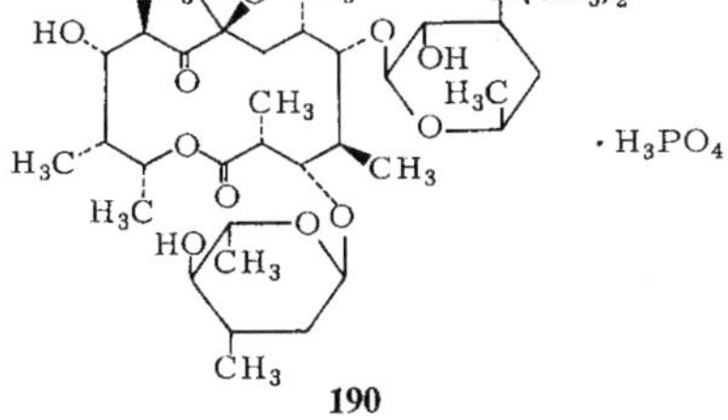

190

Oleandomycin is a macrolide antibiotic with a 14-membered lactone constituent. It was found in the culture broth of *Streptomyces antibioticus* by Pfizer Research Laboratories in 1954 [329]. Oleandomycin shows almost the same antimicrobial spectrum and activity as the other macrolide antibiotics. Its serum concentration is low, but its tissue concentration is high enough to provide a therapeutic effect. Oleandomycin phosphate is used orally or intravenously to treat pyoderma, sepsis, meningitis, surgical and abdominal infections, respiratory tract and urinary tract infections, and other infections caused by *Staphylococci, Streptococci, Corynebacterium, Neisseria,* and *Mycoplasma.* Oleandomycin is also used as a feed additive and therapeutic agent for animals.

Triacetyloleandomycin [*2751-09-9*], (**191**), troleandomycin, $C_{41}H_{67}NO_{15}$, M_r 813.99.

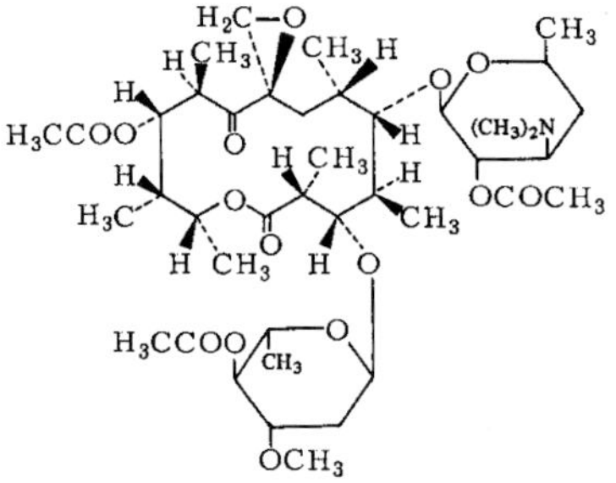

191

Triacetyloleandomycin was synthesized by Pfizer Research Laboratories in 1958 [330]. It shows a higher and longer-lasting serum level than oleandomycin when administered orally. Triacetyloleandomycin behaves as oleandomycin in vivo, following its hydrolysis by intestinal esterase. However, considerable amounts of the intermediates, the monoacetates and diacetates, are detected in the serum and urine. Prolonged use of triacetyloleandomycin causes damage to the liver as does erythromycin estolate.

Kitasamycin [*1392-21-8*], (**192**),

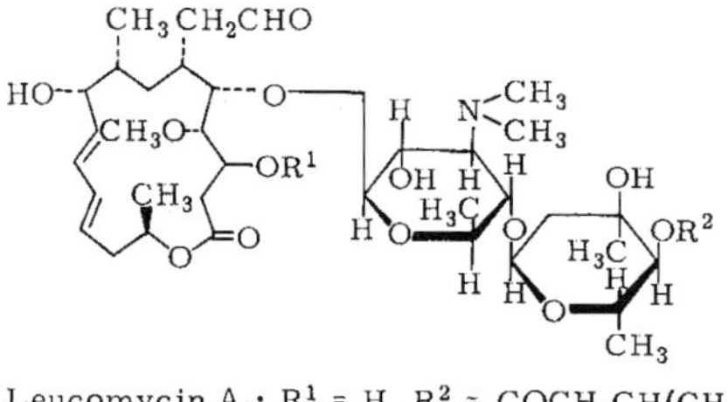

Leucomycin A_1: R^1 = H, R^2 = $COCH_2CH(CH_3)_2$
Leucomycin A_3: R^1 = $COCH_3$, R^2 = $COCH_2CH(CH_3)_2$
Leucomycin A_4: R^1 = $COCH_3$, R^2 = $COCH_2CH_2CH_3$
Leucomycin A_5: R^1 = H, R^2 = $COCH_2CH_2CH_3$
Leucomycin A_6: R^1 = $COCH_3$, R^2 = $COCH_2CH_3$
Leucomycin A_7: R^1 = H, R^2 = $COCH_2CH_3$

192

Kitasamycin, formerly called leucomycin, was found in the culture broth of *Streptoverticillium kitasatoensis* by HATA et al. of the Kitasato Institute, Japan, in 1953 [166]. This was the first macrolide antibiotic with a 16-membered lactone constituent. Kitasamycin is a complex of eight leucomycin A components and is used in its base and ester forms to treat gram-positive bacterial and gram-negative coccal infections, as well as infections of *Mycoplasma*, *Spirochaeta*, and *Treponema* by injection or by oral topical administration.

Josamycin [56689-45-3], (**193**), $C_{42}H_{69}NO_{15}$, M_r 828.02.

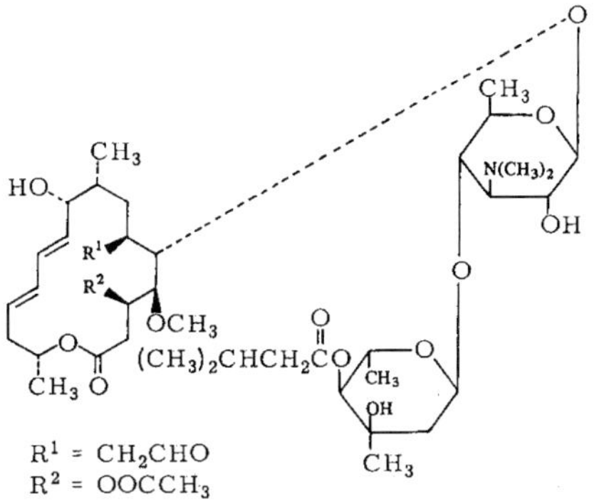

R¹ = CH₂CHO
R² = OOCCH₃
193

Josamycin was found in the culture broth of *Streptomyces narboensis* var. *josamyceticus* by Yamanouchi Pharmaceutical Co. and the Institute of Microbial Chemistry in 1964 [331]. It was the first macrolide antibiotic with a 16-membered lactone constituent to be prepared as a single compound. Its lactone structure was first thought to be 17-membered, but later this idea was revised and its identity with leucomycin A_3, a component of kitasamycin, was confirmed. Clinically, josamycin is given to children in the form of a base or propionate ester to avoid its bitter taste.

Spiramycin [8025-81-8], (**194**).

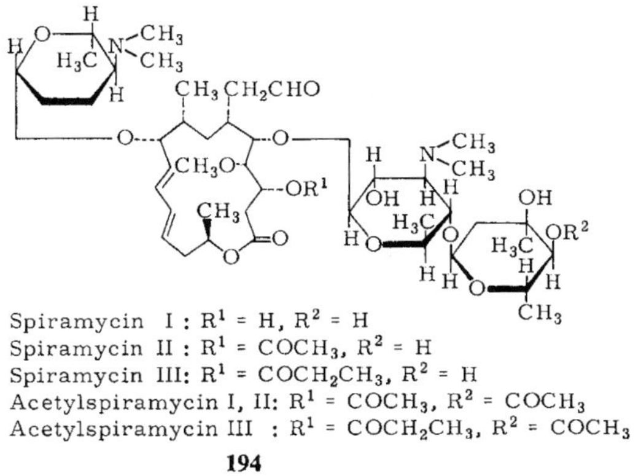

Spiramycin I : R¹ = H, R² = H
Spiramycin II : R¹ = COCH₃, R² = H
Spiramycin III: R¹ = COCH₂CH₃, R² = H
Acetylspiramycin I, II: R¹ = COCH₃, R² = COCH₃
Acetylspiramycin III : R¹ = COCH₂CH₃, R² = COCH₃
194

Spiramycin was found in the culture broth of *Streptomyces ambofaciens* by Rhône Poulenc in 1954 [168] and its acetate was synthesized by Kyowa Hakko Kogyo Co. in 1965. This antibiotic consists of three closely related macrolide components, I, II, and III, whose ratio is 2:1:1. Spiramycin shows almost the same antimicrobial spectrum and activity as the other macrolides, but its acetate, acetylspiramycin, has much better pharmacokinetic properties and activity in vivo.

Tylosin [*1401-69-0*], (**195**), $C_{46}H_{77}NO_{17}$, M_r 916.12.

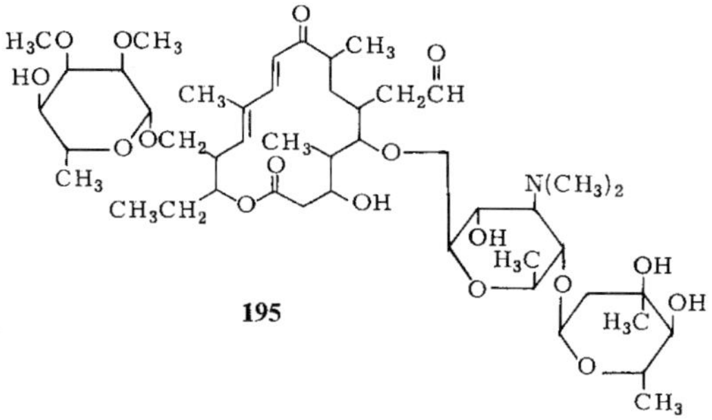

[Structure 195]

Tylosin was found in the culture broth of *Streptomyces fradiae* by Lilly Research Laboratories in 1959 [179]–[181]. It shows the same antimicrobial spectrum as the other macrolide antibiotics, and its activity against *Mycoplasma* is the widest and highest of any member of its group. Against gram-positive bacteria it is slightly weaker than erythromycin. Tylosin shows a very high and long-term tissue concentration when administered subcutaneously. It is used by injection or oral administration to treat *Mycoplasma* and gram-positive bacterial infections in poultry, swine, and other livestock.

Ivermectin [*70288-86-7*], (**196**), component B_{1a} [*70161-11-4*], component B_{1b} [*70209-81-3*].

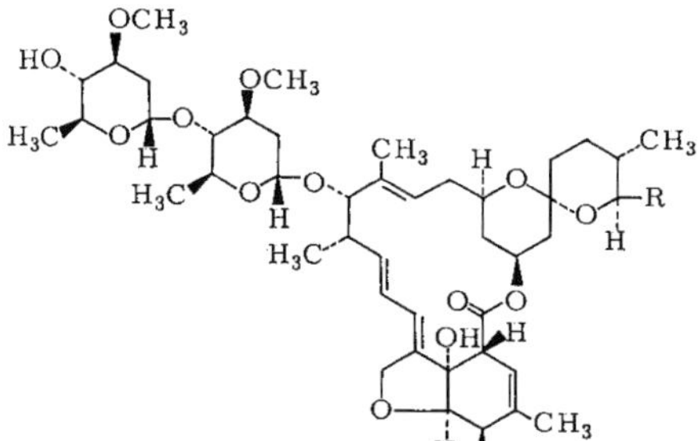

[Structure 196]

Ivermectin B_{1a}: R = $-\overset{|}{C}-CH_2CH_3$
Ivermectin B_{1b}: R = $-CH(CH_3)_2$

Ivermectin was synthesized starting with avermectin, which was found by Merck Sharp & Dohme in the culture broth of *Streptomyces avermitilis* in 1977 [332]. Unlike other antibiotics, its activity is strictly against insects, mites, and animal parasites. Ivermectin has been used against pests, mites, and other parasites in domestic animals and livestock.

Midecamycin [*35457-80-8*], (**197**), $C_{41}H_{67}NO_{15}$, M_r 813.99.

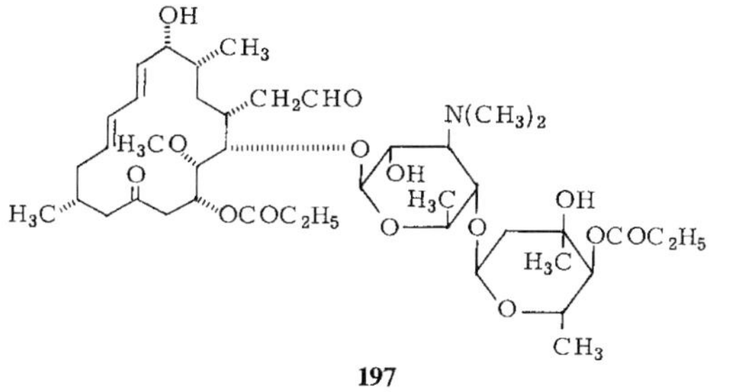

197

Midecamycin, a macrolide antibiotic having a 16-membered lactone ring, was found in the culture broth of *Streptomyces mycarofaciens* by Meiji Seika Kaisha in 1971 [333]. Under specific culture conditions, it is produced by the organism as a single component. Midecamycin shows almost the same antimicrobial spectrum and activity as kitasamycin. Although its serum and urine concentrations are low, it distributes in tissues at high concentration following oral administration.

Rifampicin [*13292-46-1*], (**198**), rifampin, $C_{43}H_{58}N_4O_{12}$, M_r 822.96.

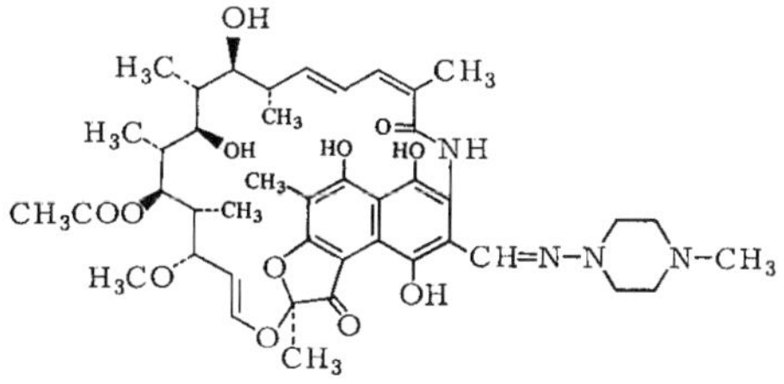

198

Rifampicin was synthesized from rifamycin O obtained from a culture of *Nocardia mediterranei* by Lepetit and Ciba Geigy in 1967 [334]. It shows strong activity against gram-positive bacteria and gram-negative cocci, including *Mycobacterium*, and against viruses. Its clinical application has been restricted to the therapy of tuberculosis caused by streptomycin-resistant tuberculotic bacteria, but recently its usefulness against infections by *Legionella* has been considered.

Bleomycin sulfate [*9041-93-4*], (**199**), variable composition.

Bleomycin was found by UMEZAWA et al. in the culture broth of *Streptomyces verticillus* B80-Z2 in 1965 as a mixture of 13 closely related components [193]. It is produced as a complex with copper (II) ion showing a blue color. Removal of the copper ion with 8-hydroxyquinoline gives a slightly yellowish powder that has the same anticancer activity as the complex but is less toxic. The major component of bleomycin sulfate is bleomycin A_2, which shows strong activity and low toxicity. It is effective against skin, head, neck, lung, and other cancers and malignant lymphoma.

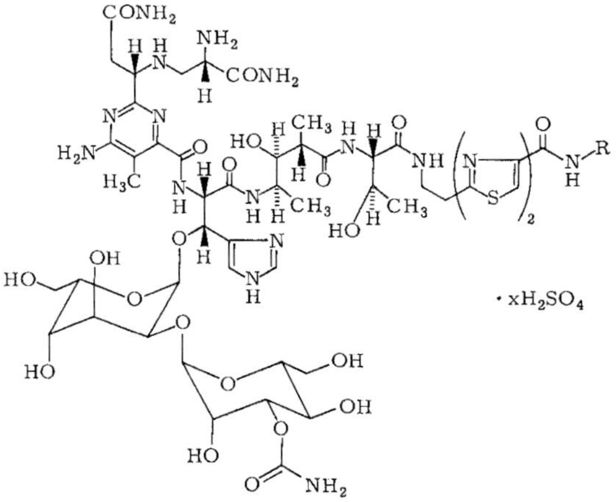

(Main component: Bleomycin A₂, in which R is (CH₃)₂S⁺CH₂CH₂CH₂–)

199

Peplomycin sulfate [*70384-29-1*], (**200**), pepleomycin, $C_{61}H_{88}N_{18}O_{21}S_2 \cdot H_2SO_4$, M_r 1571.70.

Peplomycin was derived biosynthetically from *Streptomyces verticillus*, the bleomycin-producing organism, in the course of efforts to obtain bleomycin derivatives with high activity and low toxicity. This antibiotic was found by Nippon Kayaku Co. and the Institute of Microbial Chemistry, Japan, in 1974 [335]. Its anticancer effects appear earlier than those of bleomycin; consequently, the period of therapy is shortened. Peplomycin is also active against metastatic cancer of the lymph nodes and shows less pneumotoxicity than bleomycin.

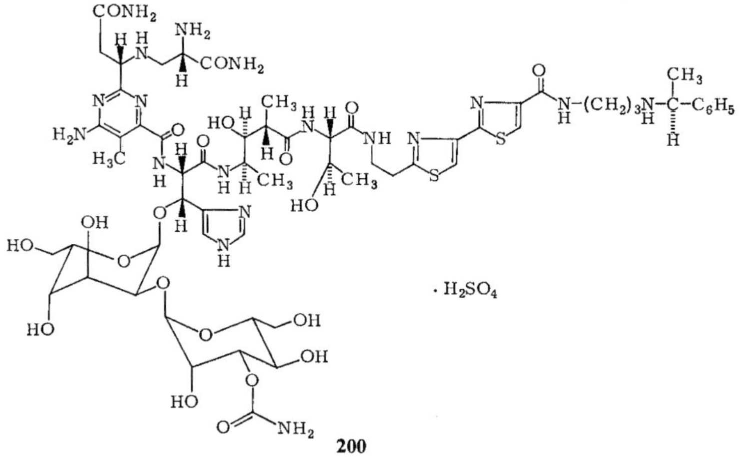

200

Gramicidin S [*113-73-5*], (**201**), $C_{60}H_{92}N_{12}O_{10}$, M_r 1141.48.

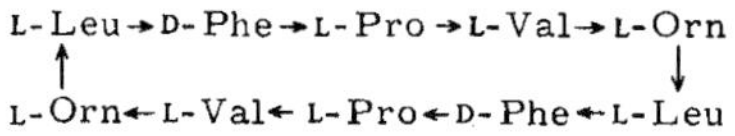

```
L-Leu→D-Phe→L-Pro→L-Val→L-Orn
 ↑                          ↓
L-Orn←L-Val←L-Pro←D-Phe←L-Leu
              201
```

Gramicidin S was found in the culture broth of *Bacillus brevis* by GAUSE of the USSR Academy of Medical Sciences in 1944 [199]. It was later found by OTANI of Osaka University in the cells of the *Nagano* strain of the same species. The antibiotic is a basic peptide showing strong activity against gram-positive bacteria and considerable activity against gram-negative cocci and *Mycobacterium*. Gramicidin S is rather toxic but is not absorbed orally; it is used topically as an ointment or as eye or ear drops in combination with other antibacterial drugs.

Polymyxin B [*1404-26-8*], (**202**).

Polymyxin B was isolated by Wellcome Research Laboratories, in 1949, from the mixture of polymyxins A, B, C, and D produced by *Bacillus polymyxa* [336]. It was later separated into the major component, B_1, and the minor component, B_2. Polymyxin B is a basic polypeptide and shows strong activity against gram-negative bacteria, but its activity against gram-positive bacteria, *Mycobacterium*, and fungi is weak. Because of its toxicity, it is used carefully by intramuscular injection for resistant *Pseudomonas aeruginosa* infections, e.g., sepsis. Polymyxin B is used orally to sterilize the gut in leukemic patients, intraspinally for meningitis, or topically.

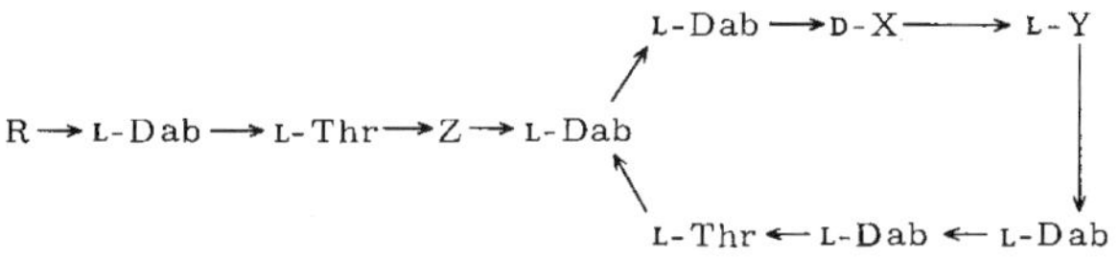

```
                        L-Dab→D-X────→L-Y
                       ↗                ↓
R→L-Dab→L-Thr→Z→L-Dab
                       ↖                ↓
                        L-Thr←L-Dab←L-Dab
```

Dab = α, γ-diaminobutyric acid

Polymyxin B_1, [*4135-11-9*], $C_{56}H_{98}N_{16}O_{13}$: M_r 1203.50, R = (+)-6-methyloctanoyl; X = phenylalanine; Y = leucine; Z = L-Dab

Polymyxin B_2, [*34503-87-2*], $C_{55}H_{96}N_{16}O_{13}$: M_r 1189.48, R = 6-methylheptanoyl; X = phenylalanine; Y = leucine; Z = L-Dab

202

Bacitracin [*1405-87-4*], bacitracin A (the major component) [*22601-59-8*] (**203**), $C_{66}H_{103}N_{17}O_{16}S$, M_r 1422.73.

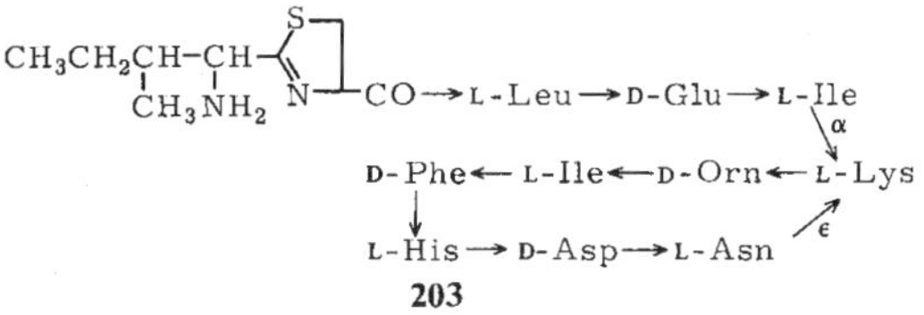

```
CH₃CH₂CH-CH⟨S⟩
     |    |    ⟩
    CH₃  NH₂  N─CO→L-Leu→D-Glu→L-Ile
                                    \α
              D-Phe←L-Ile←D-Orn← L-Lys
               ↓                   /ε
              L-His→D-Asp→L-Asn
                     203
```

Bacitracin was found as a polypeptide complex in the culture broth of *Bacillus subtilis* and *B. licheniformis* by JOHNSON of Columbia University in 1945 [337]. It was first used as a mixture of at least nine bacitracin components. The structure of bacitracin A was determined in 1966 by RESSLER et al [203].

Colistin sulfate [*1264-72-8*], (**204**), $C_{45}H_{85}O_{10}N_{13} \cdot 2.5\ H_2SO_4$, M_r 1213.45.

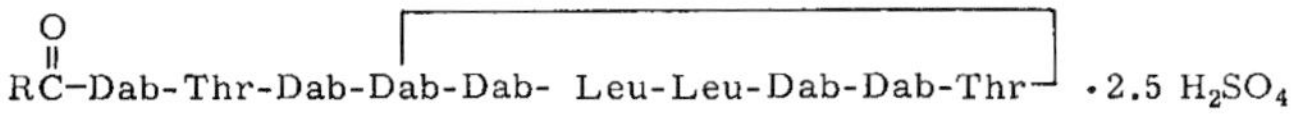

$$\overset{O}{\overset{\|}{R C}}-Dab-Thr-Dab-Dab-Dab-\ Leu-Leu-Dab-Dab-Thr \cdot 2.5\ H_2SO_4$$

Dab is L-α,γ-diaminobutyric acid; R is 5-methylheptyl in colistin A and 5-methylhexyl in colistin B

204

Colistin was found in the culture broth of *Bacillus polymyxa* var. *colistinus* by Kayaku Antibiotics Research in 1950 [338]. It is closely related to polymyxin and shows strong activity against gram-negative bacteria, including *Pseudomonas aeruginosa*. Colistin and its methanesulfonic acid derivative have been used to treat urinary tract infections caused by *Escherichia coli* and *P. aeruginosa*. They have also been used parenterally to treat dysentery and abdominal infections and topically for ophthalmic and otorhinolaryngological infections.

Colistinmethanesulfonate sodium (**205**) [*8068-28-8*], colistimethate sodium, $C_{58}H_{107}N_{16}Na_5O_{28}S_5$ (colistin A component), M_r 1749.84; [*21362-08-3*] $C_{57}H_{103}N_{16}Na_5O_{28}S_5$ (colistin B component), M_r 1735.82.

Colistinmethansulfonate was synthesized by Kayaku Antibiotics Research Co. in 1955 [339]. It is much less toxic than colistin and has been used intramuscularly and orally to treat infections caused by *Pseudomonas aeruginosa, Escherichia coli, Klebsiella pneumoniae, Enterobacter,* and *Shigella*.

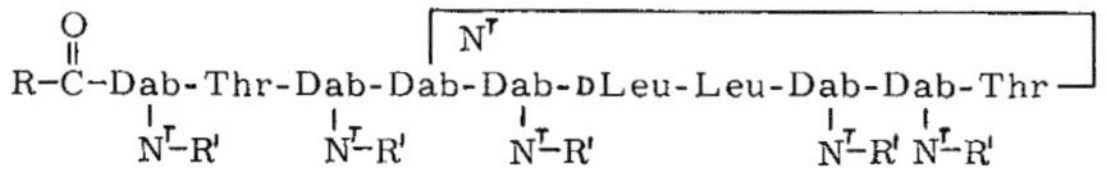

$$R-\overset{O}{\overset{\|}{C}}-Dab-Thr-Dab-Dab-Dab-DLeu-Leu-Dab-Dab-Thr$$

Dab is L-α, γ-diaminobutyric acid; R is $CH_3CH_2\overset{CH_3}{\underset{|}{C}H}(CH_2)_4$ in the colistin A component and $CH_3\overset{CH_3}{\underset{|}{C}H}(CH_2)_4$ in the colistin B component; R' is CH_2SO_3Na

205

Enramycin [*11115-82-5*], (**206**), enduracidin, A: $C_{107}H_{138}N_{25}O_{31}Cl_2$, M_r 2355.35; B: $C_{108}H_{140}N_{26}O_{31}Cl_2$, M_r 2369.38.

Enramycin, formerly called enduracidin, was found in the culture broth of *Streptomyces fungicidicus* by Takeda Chemicals Industries in 1967 [340]. This antibiotic shows strong activity against gram-positive bacteria but not against gram-negative bacteria. It is not subject to any of the mechanisms of resistance against other antibiotics and shows activity against otherwise resistant strains. Enramycin was evaluated for human therapy, but it is now used as a feed additive for growth promotion in poultry.

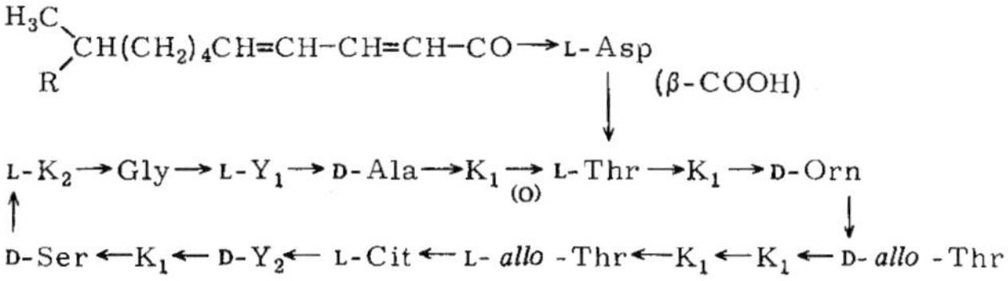

```
H₃C
   CH(CH₂)₄CH=CH-CH=CH-CO ⟶ L-Asp
R                              │ (β-COOH)
                               ↓
L-K₂ ⟶ Gly ⟶ L-Y₁ ⟶ D-Ala ⟶ K₁ ⟶ L-Thr ⟶ K₁ ⟶ D-Orn
↑                              (O)              │
D-Ser ⟵ K₁ ⟵ D-Y₂ ⟵ L-Cit ⟵ L-allo-Thr ⟵ K₁ ⟵ K₁ ⟵ D-allo-Thr
```

Enramycin A: R = CH₃
Enramycin B: R = C₂H₅
K₁: α-Amino-4-hydroxyphenylacetic acid
K₂: α-Amino-3,5-dichloro-4-hydroxyphenylacetic acid
Y₁: α(S)-Amino-β-4(R)-(2-iminoimidazolidinyl)propionic acid
Y₂: α(R)-Amino-β-4(R)-(2-iminoimidazolidinyl)propionic acid

206

Mikamycin [*11006-76-1*], mikamycin B [*3131-03-1*] (**207**), $C_{45}H_{54}N_8O_{10}$, M_r 866.98.

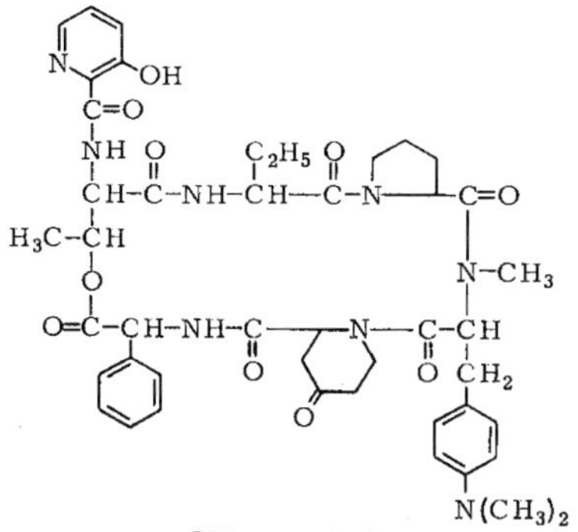

Mikamycin B
207

Mikamycin was found in the culture broth of *Streptomyces mitakaensis* by UMEZAWA et al. of the University of Tokyo in 1956 [341]. It is a mixture of two components, A and B, showing synergistic effects on each other. Staphylomycin, which was found by DE SOMER of Leuven University in 1955, is a closely related compound. Mikamycin shows strong activity against gram-positive bacteria, including strains resistant to other antibiotics. It has been used topically to treat *Staphylococcus* or *Streptococcus* infections of the skin, including burns.

Virginiamycin [11006-76-1], (208),

Factor M_1, $C_{28}H_{35}N_3O_7$, M_r 525.61

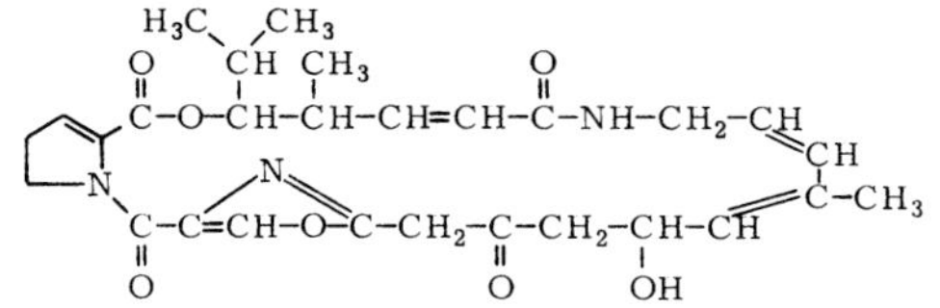

Factor S, $C_{42}H_{47}N_7O_{10}$, M_r 809.88

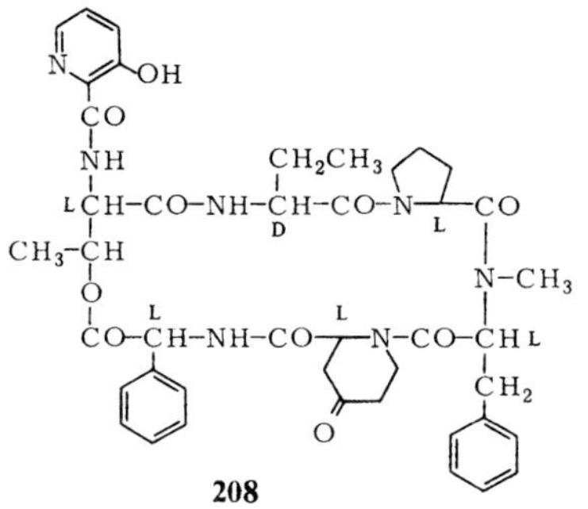

208

Virginiamycin, formerly called staphylomycin, was found in the culture broth of *Streptomyces virginiae* by DE SOMER et al. of Leuven University in 1955 [342]. The commercial product contains about 75% of the fraction M_1 and 5% of the fraction S. The structure of the minor component, M_2, has not yet been elucidated. Virginiamycin has very low toxicity and is not absorbed from the intestine when administered orally. It shows activity against gram-positive bacteria, and it has been used as a feed additive for growth promotion in swine, cattle, and other livestock.

Capreomycin sulfate [1405-37-4], (209),

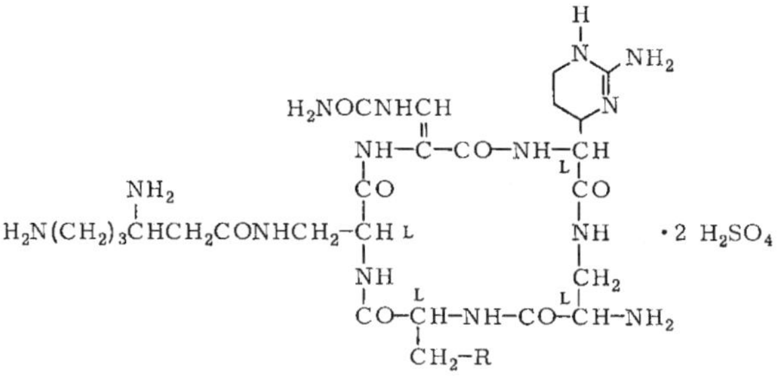

Capreomycin I A : R = OH; $C_{25}H_{44}N_{14}O_8 \cdot 2\ H_2SO_4$, M_r 864.87
Capreomycin I B : R = H; $C_{25}H_{44}N_{14}O_7 \cdot 2\ H_2SO_4$, M_r 848.87

209

Capreomycin was found as an antituberculotic antibiotic by HERR et al. of Eli Lilly & Co. in the culture broth of *Streptomyces capreolus* NRRL2773 in 1962 [343]. It shows no cross-resistance with streptomycin and is used against tuberculosis caused by streptomycinresistant *Mycobacterium* or for patients with adverse side effects caused by streptomycin.

Viomycin [*32988-50-4*], (**210**), $C_{25}H_{43}N_{13}O_{10}$, M_r 685.70.

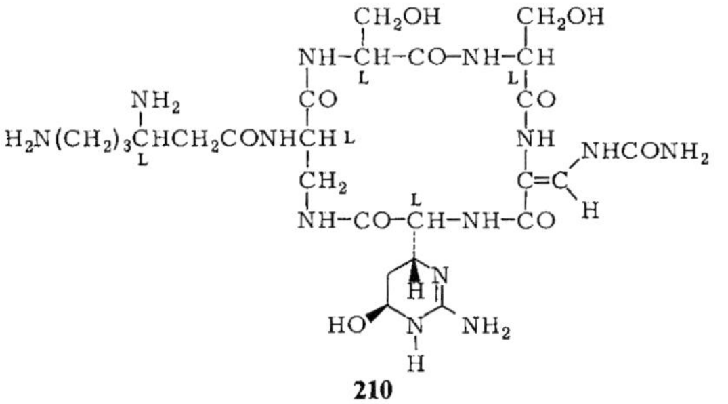

$$\begin{array}{c}
\text{structure 210}
\end{array}$$

Viomycin was found independently in 1951, in the culture broth of *Streptomyces puniceus* by Pfizer Research Laboratories and in that of *S. floridae* by Parke Davis Co. [344]. It shows strong activity against *Mycobacterium* species and moderate activity against gram-positive and gram-negative bacteria. Viomycin has been used by intramuscular administration to treat tuberculosis, but because of its ototoxicity and renal toxicity it is being replaced by more active and less toxic drugs, such as enviomycin and other antituberculotic antibiotics.

Enviomycin [*33103-22-9*], (**211**), tuberactinomycin N, $C_{25}H_{43}N_{13}O_{10}$, M_r 685.70.

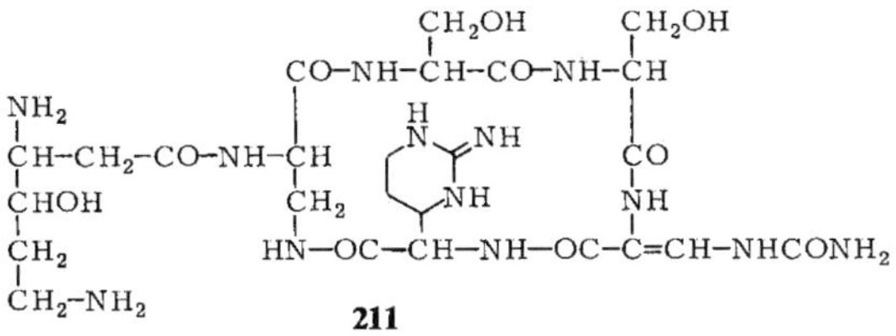

$$\begin{array}{c}
\text{structure 211}
\end{array}$$

Enviomycin was found in the culture broth of *Streptomyces griseoverticillatus* var. *tuberacticus* by Toyo Jozo Co. in 1966 as a mixture of tuberactinomycins N and O [345]. It is a water-soluble, basic peptide closely related to viomycin and showing selective activity against *Mycobacterium* species about twice as strong as that of viomycin. Enviomycin, in combination with other antituberculotic drugs, is used to treat tuberculosis caused by streptomycin-resistant *Mycobacterium*.

Vancomycin [*1404-90-6*], (**212**), $C_{66}H_{75}Cl_2N_9O_{24}$, M_r 1449.29.

Vancomycin was found by Lilly Research Laboratories in 1956 in the culture broth of *Streptomyces orientalis* [346]. It is a large, molecular glycopeptide showing bactericidal activity against gram-positive bacteria and anaerobes. It is used to treat resistant infections of *Staphylococcus*, to sterilize the gut in the perioperation of bone-marrow transplantation, and in leukemic patients. Recently its efficacy has been demonstrated against pseudomembrane colitis, which is caused by *Clostridium difficile*.

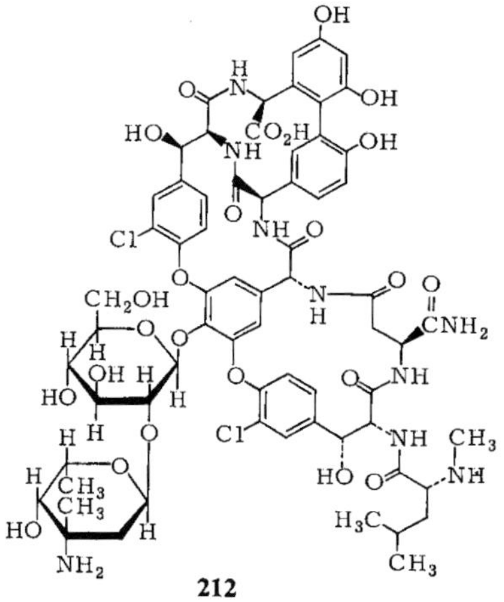

212

Actinomycin D [50-76-0], (**213**), dactinomycin, $C_{62}H_{86}N_{12}O_{16}$, M_r 1255.45.

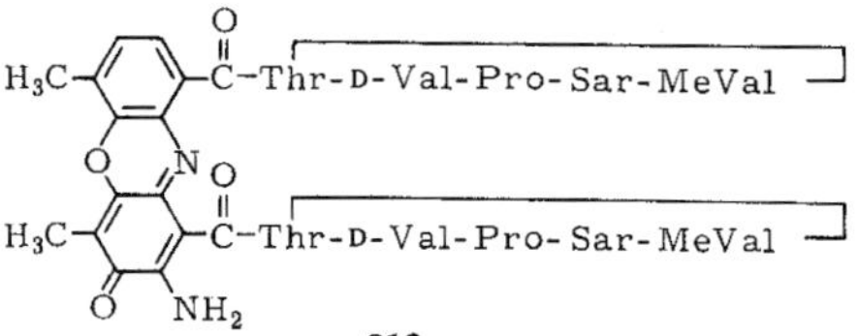

213

Actinomycins (mixtures of A, B, C, and other components) were first found by WAKSMAN et al. of Rutgers University in 1940 in the culture broth of *Streptomyces antibioticus* [347]. The mixture of the C components has been used as cactinomycin for anticancer therapy. Actinomycin D was found by the same group in 1954, obtained as a single component from *S. parvullus*. Its strong activity against Wilm's tumor and other cancers has been evaluated. Dactinomycin is the name recommended by INN and used in the United States.

Neocarzinostatin [9014-02-2], zinostatin.

Ala – Ala – Pro – Thr – Ala – Thr – Val – Thr – Pro – Ser – Ser –
Gly – Leu – Ser – Asp – Gly – Thr – Val – Val – Lys – Val – Ala –
Gly – Ala – Gly – Leu – Gln – Ala – Gly – Thr – Ala – Tyr – Asp –
Val – Gly – Gln – Cys – Ala – Ser – Val – Asn – Thr – Gly – Val –
Leu – Trp – Asn – Ser – Val – Thr – Ala – Ala – Gly – Ser – Ala –
Cys – Asx – Pro – Ala – Asn – Phe – Ser – Leu – Thr – Val – Arg –
Arg – Ser – Phe – Glu – Gly – Phe – Leu – Phe – Asp – Gly – Thr –
Arg – Trp – Gly – Thr – Val – Asx – Cys – Thr – Thr – Ala – Ala –
Cys – Gln – Val – Gly – Leu – Ser – Asp – Ala – Ala – Gly – Asp –
Gly – Glu – Pro – Gly – Val – Ala – Ile – Ser – Phe – Asn

Neocarzinostatin was found in the culture broth of *Streptomyces carzinostaticus* by ISHIDA et al. of Tohoku University in 1957 [348]. It showed strong cytotoxicity against sarcoma 180 ascites, tumor cells, and leukemia SN-36. Neocarzinostatin was found to be an acidic peptide of M_r 10700, consisting of 109 amino acid residues. Recently the principal agent of anticancer and antibacterial activity in neocarzinostatin was found to be a small molecular chromophore with the peptide component playing a role by stabilizing the chromophore in vivo. Neocarzinostatin is used to treat leukemia and gastric and pancreatic cancer.

Bestatin [*58970-76-6*], (**214**), $C_{16}H_{24}N_2O_4$, M_r 308.38.

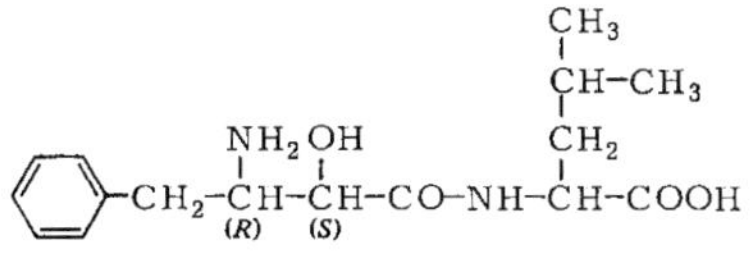

$$\underset{\underset{214}{}}{\underset{(R)\quad(S)}{\text{Ph-CH}_2\text{-}\overset{\overset{NH_2}{|}}{C}H\text{-}\overset{\overset{OH}{|}}{C}H\text{-CO-NH-}\overset{\overset{CH_2\text{-CH(CH}_3)_2}{|}}{C}H\text{-COOH}}}$$

Bestatin was found in the culture broth of *Streptomyces olivoreticuli* in 1976 by UMEZAWA et al. in the course of screening specific enzyme inhibitors of microbial origin [349]. It shows inhibitory activity against specific exopeptidases, e.g., aminopeptidase B and leucine aminopeptidase. However, it does not act against aminopeptidase A, carboxypeptidases A and B, or endopeptidases. Bestatin acts as a bioresponse modifier and shows antitumor activity via stimulation of immuno response in the host. In combination with cytotoxic anticancer drugs, it is now under clinical evaluation for use in cancer therapy.

Pepstatin [*26305-03-3*], (**215**), $C_{34}H_{63}N_5O_9$, M_r 685.91.

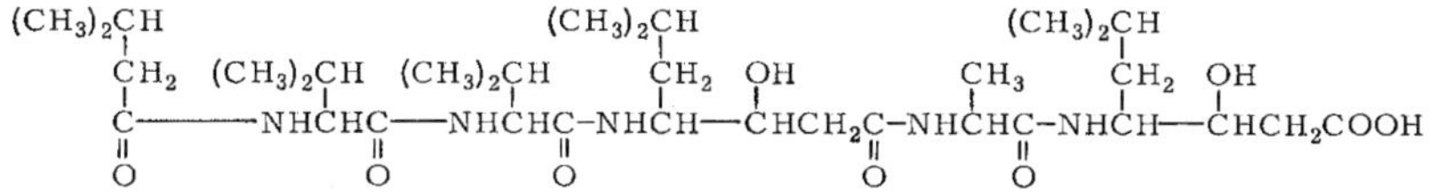

215

Pepstatin was found in the culture broth of *Streptomyces testaceus* by UMEZAWA of the Institute of Microbial Chemistry in 1970 in the course of screening for specific enzyme inhibitors [350]. It shows strong inhibitory activity against several proteases, such as pepsin, cathepsin D, and renin, but not against trypsin, chymotrypsin, papain, etc. Its potential use against ulcers in humans and animals has been studied.

Monensin [*17090-79-8*], (**216**), $C_{36}H_{62}O_{11}$, M_r 670.89.

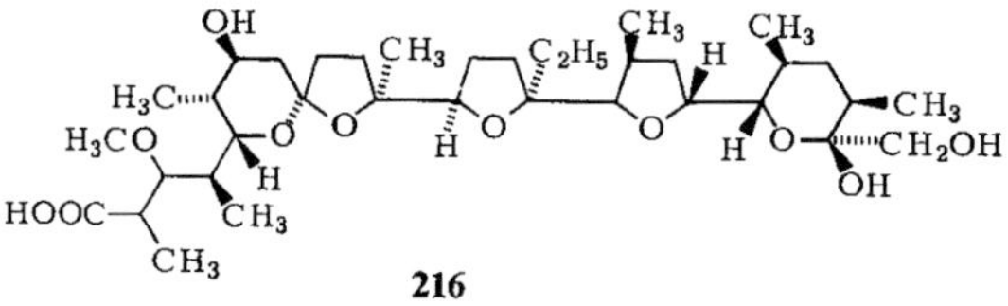

216

Monensin was found in the mycelium of *Streptomyces cinnamonensis* by Eli Lilly & Co. in 1967 in the course of screening for anticoccidial and growth-promoting substances for cattle, swine, and chickens [351]. It shows antiprotozoal, antibacterial, and antifungal activities, as well as anticoccidial activity. Monensin is widely used to treat such infections and as a feed additive to promote the growth of livestock.

Lasalocid [*25999-31-9*], (**217**), $C_{34}H_{54}O_8$, M_r 590.80.

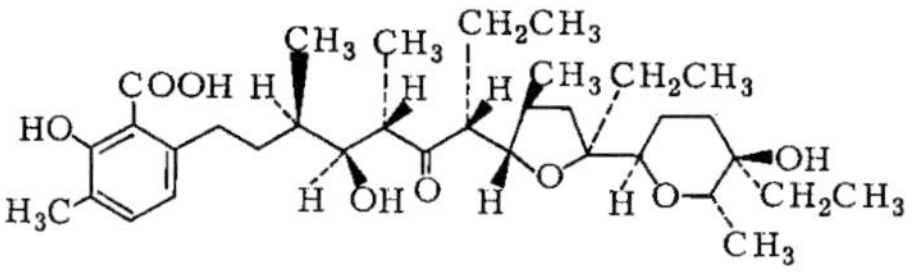

217

Lasalocid is one of the polyether ionophore antibiotics and was found in the culture broth of *Streptomyces lasaliensis* by Hoffmann-La Roche in 1951 [352]. It shows activity against gram-positive bacteria, including *Mycobacterium* and *Streptomyces*, and also against anaerobic bacteria, but it has no activity against gram-negative bacteria or fungi. Lasalocid also shows strong anticoccidial activity, and it stimulates propionic acid formation in the rumen. It is used for prophylaxis and for therapy of coccidial infections in poultry and as a feed additive for growth promotion in cattle.

Salinomycin [*53003-10-4*], (**218**), $C_{42}H_{70}O_{11}$, M_r 751.02.

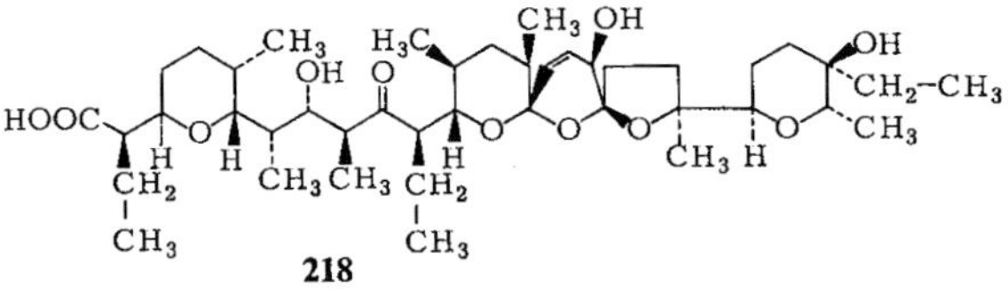

218

Salinomycin was found in the culture mycelium of *Streptomyces albus* by OTAKE et al. of the University of Tokyo in 1973 [353]. Like other polyether ionophore antibiotics, it shows activity against gram-positive bacteria, fungi, and *Coccidium*. Salinomycin has been used as a feed additive to protect poultry against coccidial infections.

Amphotericin B [*1397-89-3*], (**219**), $C_{47}H_{73}NO_{17}$, M_r 924.10.

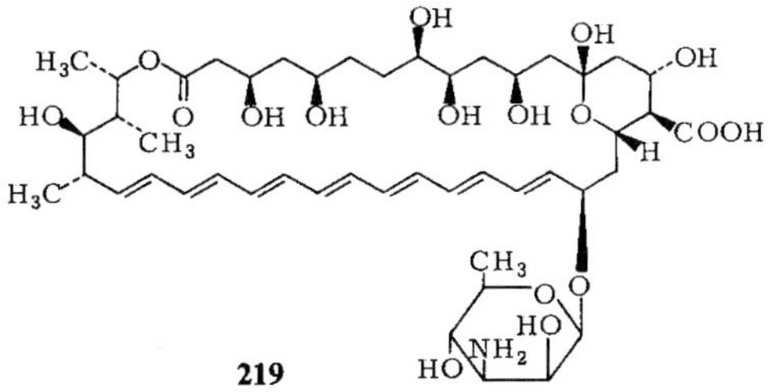

Amphotericin B was found in the mycelium of *Streptomyces nodosus* M-4575 by GOLD et al. of the Squibb Institute of Medical Research in 1956 [354]. It is produced with another polyene macrolide antibiotic, amphotericin A, and separated by solvent extraction. Amphotericin B shows strong antimycotic activity against a variety of fungi and pathogenic yeasts (*Candida*) and is used by injection and as a vaginal suppository.

Nystatin [*1400-61-9*], (**220**).
Nystatin A_1 [*34786-70-4*], $C_{47}H_{75}NO_{17}$, M_r 926.12.

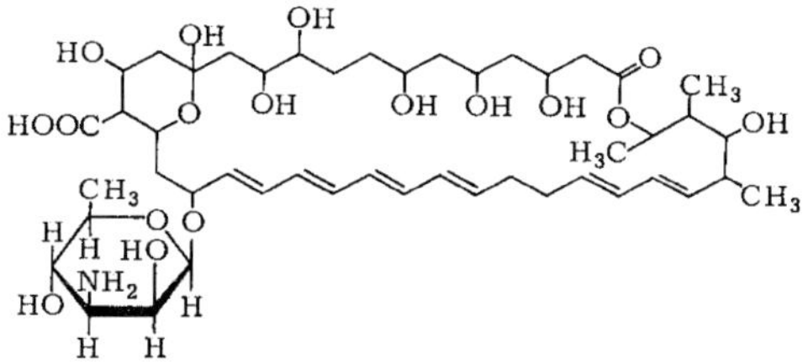

Nystatin was found in the mycelium of *Streptomyces noursei* in 1950 [355] and produced on an industrial scale by Squibb & Sons Co. in 1954. This antibiotic was used orally and topically as the first clinically applied polyene macrolide with antifungal properties. Nystatin shows activity against *Candida* and filamentous fungi and is used to treat *Candida* infections of the mouth, digestive organs, and vagina. The application of nystatin in combination with gentamicin and vancomycin to sterilize the gut in perioperation of bone-marrow transplantation has been developed recently.

Natamycin [*7681-93-8*], (**221**), pimaricin, $C_{33}H_{47}NO_{13}$, M_r 665.74.

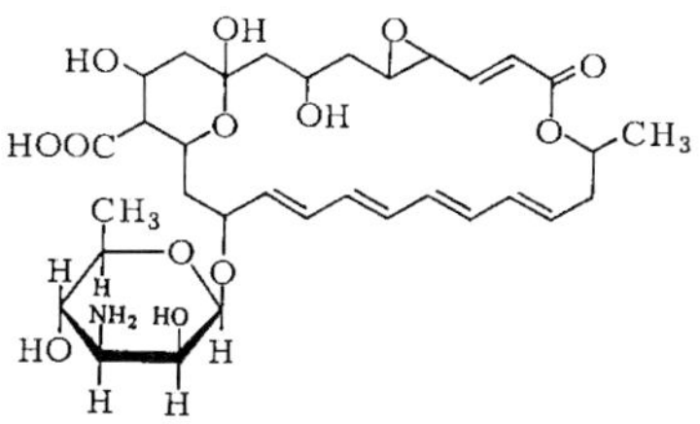

Natamycin was found in the culture mycelium of *Streptomyces natalensis* by Royal Dutch Yeast & Fermentation Industries and in that of *S. gilvosporeus* by American Cyanamid, independently, in 1957 [356]. It has a tetraene structure, and like other polyene macrolide antibiotics, it shows activity against pathogenic fungi. Natamycin has been used as a vaginal suppository for therapy of *Candida* infections.

Trichomycin [1394-02-1], (**222**), hachimycin.

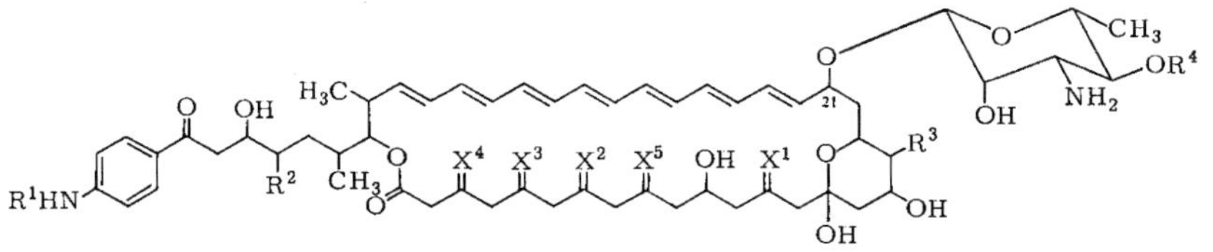

Antibiotic:	R^1	R^2	R^3	R^4	X^1	X^2	X^3	X^4	X^5
Trichomycin B:	H	H	COOH	H	H, OH	O	H, H	O	H, OH
Levorin A₂ (candi= cidin D):	H	CH₃	COOH	H	H, OH	O	H, H	O	H, OH
Hamycin (67-121 B):	H	H	COOH	H	H, OH	H, OH	H, OH	H, OH	H, OH

222

Trichomycin (hachimycin in the INN) was found in the mycelium of *Streptomyces hachijoensis* by HOSOYA et al. of the University of Tokyo in 1952 [357]. It has a heptaene structure and once was considered identical with two other polyene macrolide antibiotics, candicidin (found by WAKSMAN et al. of Rutgers University in 1953 in *S. griseus*) and hamycin (found by THIRUMALACHAR et al. of Hindustan Antibiotics Research Institute in 1961 in *S. pimprina*). Its difference from them was shown by an HPLC comparison in 1980. Trichomycin shows activity against pathogenic *Candida, Trichomonas*, and *Trypanosoma* and has been used in vaginal applications to treat *Trichomonas vaginalis* infections.

Mithramycin [18378-89-7], (**223**), plicamycin, $C_{52}H_{76}O_{24}$, M_r 1085.17.

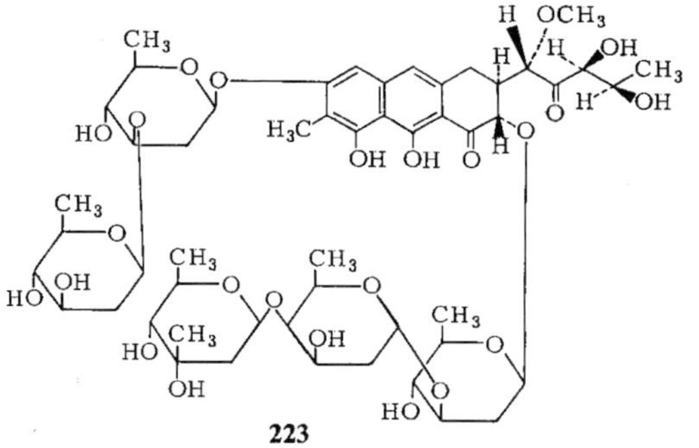

223

Mithramycin, recently renamed plicamycin, was found in the culture broth of *Streptomyces argillaceus* and *S. tanashiensis* by Abbott Laboratories in 1952 [358]. It is structurally related to chromomycin A_3. Mithramycin shows strong inhibitory activity against malignant cells of human origin. It acts by inhibition of the DNA-directed RNA synthesis through binding with DNA. Mithramycin is used intravenously to treat cancers of the embryonal cells, seminoma, choriocarcinoma, etc.

Lincomycin [154-21-2], (**224**), $C_{18}H_{34}N_2O_6S$, M_r 406.55.

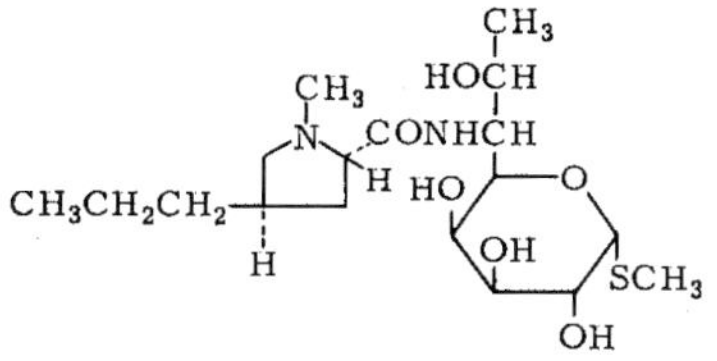

224

Lincomycin was found in the culture broth of *Streptomyces lincolnensis* var. *lincolnensis* by the Upjohn Co. in 1962 [359]. It shows antibacterial activity similar to that of the macrolide antibiotics and also shows excellent activity against anaerobic bacteria. Lincomycin is used clinically in combination with other classes of antibiotics for postoperative, gynecological, urinary tract, ear and nose, and other infections.

Clindamycin [18323-44-9], (**225**), 7(*S*)-chloro7-deoxylincomycin, $C_{18}H_{33}ClN_2O_5S$, M_r 424.99.

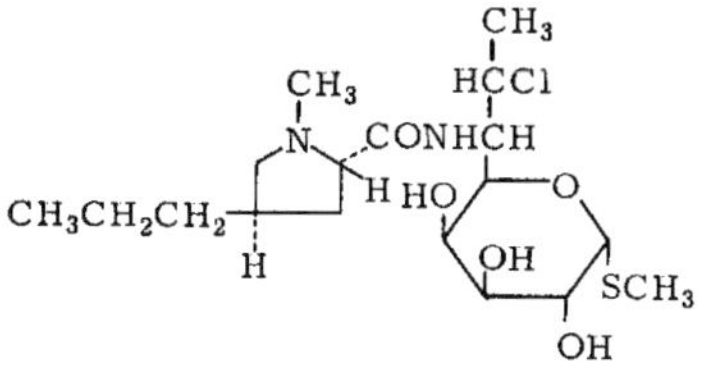

225

Clindamycin was synthesized at the Upjohn Co. in 1966 by introducing a chlorine atom at the 7 position of lincomycin [360]. It is used as an orally active lincomycin-group antibiotic with the same range of activity as lincomycin. Clindamycin shows excellent activity against gram-positive bacteria and anaerobes, and it is used with other antibiotics for complicated infections.

Clindamycin palmitate hydrochloride [*25507-04-4*], (**226**), $C_{34}H_{63}ClN_2O_6S \cdot HCl$, M_r 699.87.

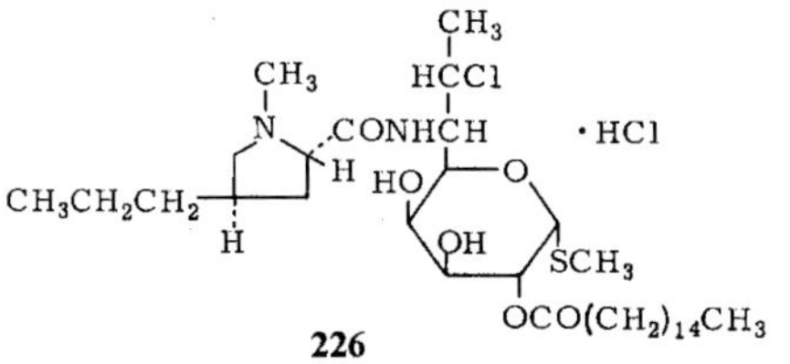

226

Clindamycin palmitate was synthesized by the Upjohn Co. in 1968 [361]. It has a much less bitter taste than clindamycin and is suitable for oral preparations for children. Clindamycin palmitate is very rapidly hydrolyzed into clindamycin after oral administration. Clindamycin phosphate [*24729-96-2*], $C_{18}H_{34}N_2O_8ClPS$, M_r 504.97, has been introduced clinically as an injectable clindamycin.

Flavophospholipol [*11015-37-5*], (**227**), moenomycin, bambermycin, $C_{69}H_{107}N_4O_{35}P$, M_r 1583.60.

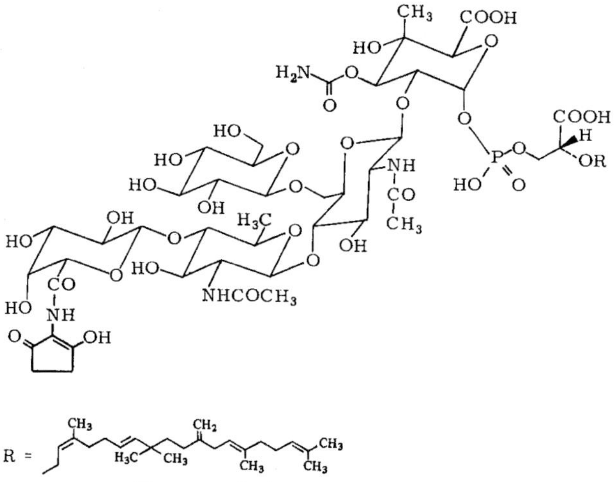

Moenomycin A

227

Flavophospholipol, formerly called moenomycin, was found in the culture broth of *Streptomyces bambergiensis* by Hoechst in 1955 [362]. This drug is called bambermycin by INN and in the United States. It shows strong activity against gram-positive bacteria and weak activity against gram-negative bacteria. Flavophospholipol has been used as a feed additive for growth promotion in poultry.

Cycloserine [68-41-7], (**228**), $C_3H_6N_2O_2$, M_r 102.09.

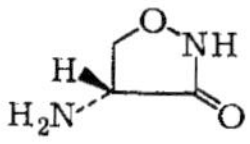

228

Cycloserine was found in the culture broth of *Streptomyces orchidaceus* by Commercial Solvent Co. [363]. It was developed in collaboration with Eli Lilly & Co. in 1955 and is manufactured synthetically. Cycloserine shows weak activity against gram-positive and gram-negative bacteria, including *Mycobacterium*. Its activity against clinically isolated tuberculotic bacteria, including streptomycin- and viomycin-resistant strains, is five to ten micrograms per millimeter. Cycloserine is used to treat tuberculosis caused by organisms resistant to other antituberculotic antibiotics.

Pecilocin [19504-77-9], (**229**), variotin, $C_{17}H_{25}NO_3$, M_r 291.39.

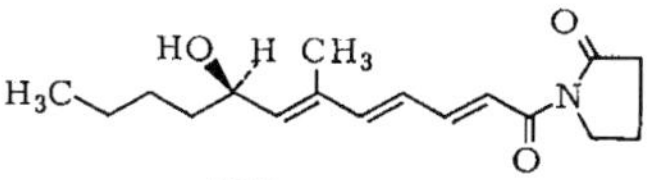

229

Pecilocin was discovered in the culture broth and mycelium of *Paecilomyces varioti* var. *antibioticus* by YONEHARA et al. of the University of Tokyo in 1959 [364]. It has an oily nature and shows activity against specific filamentous fungi. Pecilocin is used as an ointment or in ethanolic solution for the treatment of dermatomycoses caused by *Trichophyton*, *Microsporum*, and *Epidermophyton*.

Griseofulvin [126-07-8], (**230**), $C_{17}H_{17}ClO_6$, M_r 352.77.

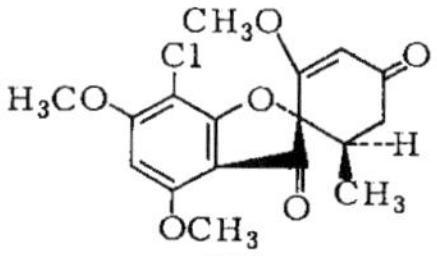

230

Griseofulvin was discovered in the mycelium of *Penicillium griseofulvum* and other *Penicillium* species by OXFORD et al. in 1939 [210]. Its application in the treatment of fungal infections was initiated by ICI in 1946. Griseofulvin is used clinically by topical and oral administration for therapy of dermatomycoses caused by *Trichophyton* and *Microsporum* species.

Chloramphenicol [56-75-7], (**231**), $C_{11}H_{12}Cl_2N_2O_5$, M_r 323.13.

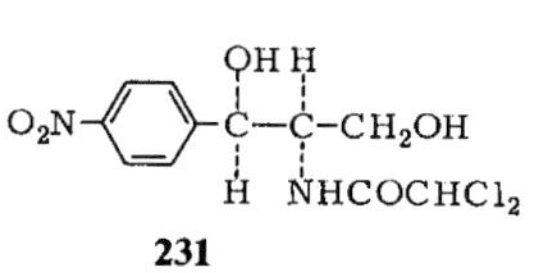

231

Chloramphenicol was found in the culture broth of *Streptomyces venezuelae* by EHRLICH et al. of Parke Davis & Co. in 1947 [212]. The effective industrial synthesis was developed in 1949, and the compound is manufactured synthetically. This was the third antibiotic to be introduced clinically and showed a wide range of activity against grampositive and gram-negative bacteria, as well as *Treponema*, *Rickettsia*, and *Chlamydiae*. Chloramphenicol shows excellent activity in the treatment of dysentery and salmonellosis, and it is widely used against these diseases.

Chloramphenicol palmitate [530-43-8], (**232**), $C_{27}H_{42}Cl_2N_2O_6$, M_r 561.54.

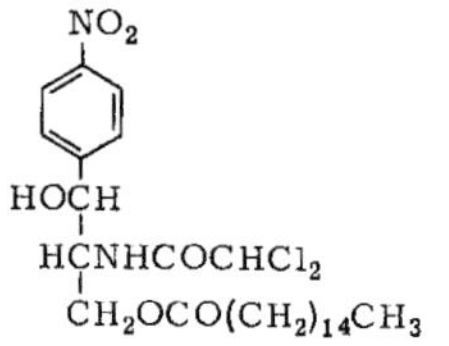

NO$_2$
⬡
HOCH
|
HCNHCOCHCl$_2$
|
CH$_2$OCO(CH$_2$)$_{14}$CH$_3$

232

Chloramphenicol palmitate was synthesized by Parke Davis Co. in 1952 [365]. It has a much less bitter taste than chloramphenicol and is suitable for oral administration, especially for children. The palmitate shows a higher serum level after oral administration than chloramphenicol does and acts the same as chloramphenicol in vivo.

Chloramphenicol sodium succinate [982-57-0], $C_{15}H_{15}N_2O_8Cl_2Na$, M_r 445.19, is highly soluble in H$_2$O and can be given by intravenous injection.

Mitomycin C [50-07-7], (**233**), $C_{15}H_{18}N_4O_5$, M_r 334.33.

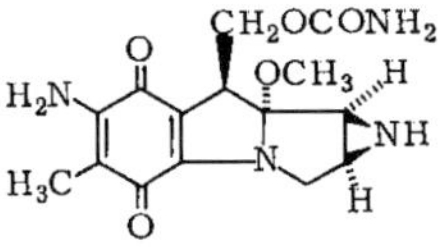

233

Mitomycin C was found in the culture broth of *Streptomyces caespitosus* by WAKAKI in 1958 [216]. This followed the discovery of mitomycins A and B, derived from the same organism by HATA of the Kitasato Institute in 1956. Mitomycin C shows strong activity against a variety of tumors and bacteria. Because it decreases the concentration of leukocytes, its use had been limited, but after 1975 its low-dosage use in anticancer therapy in combination with other anticancer drugs increased.

Pyrrolnitrin [1018-71-9], (**234**), $C_{10}H_6Cl_2N_2O_2$, M_r 257.08.

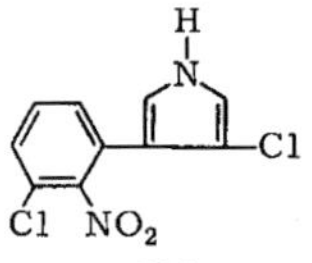

234

Pyrrolnitrin was found in the cells of *Pseudomonas pyrrocinia* grown in a medium containing a high concentration of inorganic phosphate. It was discovered by ARIMA of the University of Tokyo in collaboration with Fujisawa Pharmaceutical Industries Co. in 1965 [366]. The same compound was found soon after that by Eli Lilly & Co. in the cells of *P. aureofaciens*. Pyrrolnitrin shows strong activity against a variety of fungi and weak activity against gram-positive bacteria. It is used topically to treat dermatomycoses caused by *Trichophyton, Epidermophyton,* and *Microsporum*. The antibiotic is unstable in sunlight and cannot be used to protect plants against phytopathogenic fungi.

Fosfomycin [*23155-02-4*], (**235**), $C_3H_7O_4P$, M_r 138.06.

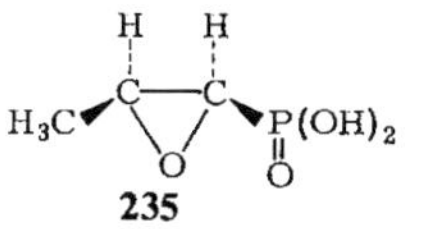

Fosfomycin was found in the culture broth of *Streptomyces fradiae* by Compania Espanola de Penicilina y Antibióticos and Merck Sharp & Dohme in 1967 [219]. Its chemical structure is simple and unique among antibiotics in having a C–P bond. Fosfomycin shows antibacterial activity against gram-positive and gram-negative organisms, including *Pseudomonas aeruginosa* and *Serratia marcescens,* and β-lactam-resistant *Staphylococcus aureus*. Its mechanism of action is probably the inhibition of cell-wall synthesis. It shows no cross-resistance with other classes of antibiotics.

Fusidic acid [*6990-06-3*], (**236**), $C_{31}H_{48}O_6$, M_r 516.72.

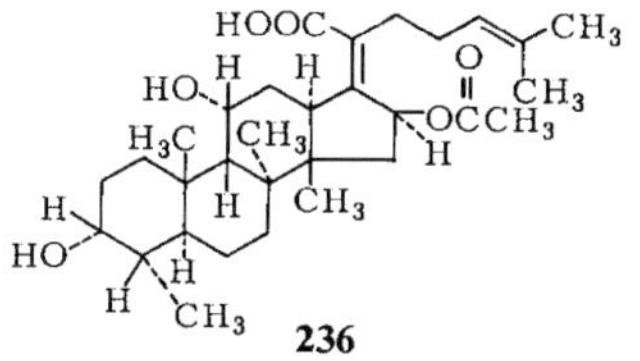

Fusidic acid was found in the culture broth of a fungus imperfectus, *Fusidium coccineum,* by Leo in 1962 [220]. It has a steroid structure but shows no hormonal activity. Fusidic acid shows very strong activity against *Staphylococcus aureus* and weak activity against other gram-positive bacteria and gram-negative cocci and *Mycobacterium*. Its clinical use is restricted to staphylococcal infections resistant to other classes of antibiotics.

Bicozamycin [*38129-37-2*], (**237**), bicyclomycin, $C_{12}H_{18}N_2O_7$, M_r 302.29.

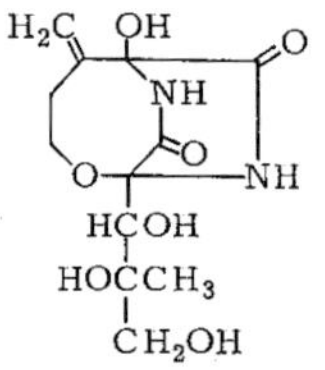

237

Bicozamycin, formerly called bicyclomycin, was found independently in 1972, in the culture broth of *Streptomyces sapporonensis* by Fujisawa Pharmaceuticals Industries and in that of *S. aizuensis* by MIYAMURA et al. of Niigata University [367]. It has a unique bicyclic structure and shows activity against *Klebsiella*, *Salmonella*, and *Shigella* but none against other gram-negative bacteria or gram-positive microorganisms. Bicozamycin is not absorbed orally and shows very low toxicity; no mice died following its intravenous injection at a dose as high as two grams per kilogram. This drug is used orally to treat salmonellosis and dysentery.

Tiamulin [*55297-95-5*], (**238**), $C_{28}H_{47}NO_4S$, M_r 493.76.

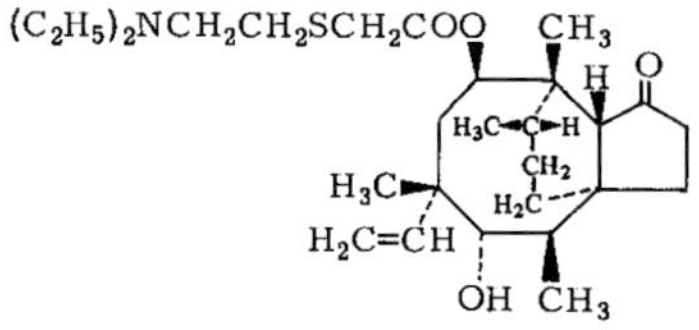

238

Tiamulin was synthesized by Sandoz Co. in 1973, starting from pleuromutilin produced by *Pleurotus mutilus* [368]. It has been developed by that company in collaboration with E. R. Squibb & Sons for use against animal infections. Tiamulin is used against dysentery and *Mycoplasma* infections in sheep, swine, cattle, and chickens.

Siccanin [*22733-60-4*], (**239**), $C_{22}H_{30}O_3$, M_r 342.48.

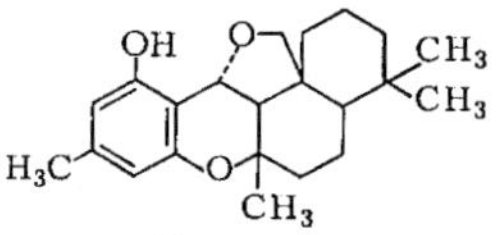

239

Siccanin was found in the culture broth of *Helminthosporium siccans* by Sankyo Co. in 1962 [369]. It shows specific and strong activity against the dermatophytes *Trichophyton*, *Epidermophyton*, and *Microsporum*, but it has almost no activity against other fungi. Siccanin has been used as a topical ointment or solution to treat dermatomycoses.

5. Fermentation

Fermentation is considered here from the following points of view:

1) Biological development, which includes screening and selection, mutation, and maintenance of the strain.
2) Process development leading to large-scale manufacture.
3) Improvements in fermentation technology.

5.1. Screening

Technical developments in the production of penicillin have given the field new momentum and have stimulated the search, not only for more efficient strains, but also for microorganisms that produce completely different antibiotics. This process is called screening because valuable antibiotic producers are separated from the large number of organisms found in nature. The screening process and the expected results are influenced by several factors.

Source of Sample. Worldwide screening endeavors to isolate the individual microorganisms not only from soil samples from different sources, but also from other microbe-containing materials. Samples from unusual sources often show the occurrence of selection and adaptation. For example, thermophilic microorganisms are examined in samples taken from deep caves, the sea bottom, hot springs, or geysers.

Examination Technique. There are several factors that determine the conditions under which a certain microorganism not only lives and grows but also efficiently produces its antibiotic. These factors include the composition and pH of the culture medium, the additives, the air supply, and the temperature. These factors are also of prime importance for any later industrial fermentation. The isolation and testing of the new antibiotic, first in vitro and then in animal experiments, and the indisputable proof that the new compound is not identical to one of the numerous known antibiotics are part of the examination technique.

Purpose of the Examination. In the early days of antibiotic screening, any organism that showed antibiotic activity was screened, but later definite objectives were set and appropriate examination techniques were developed. The factors mentioned in the previous paragraph narrow the choice to certain bacteria and fungi. Further restrictions are brought about by the selection of organisms used to test the efficacy of the antibiotic. Such specific screening methods are used to find, e.g., antifungal agents, antibiotics active against cancer or viruses, or antibiotics effective against bacteria resistant to other antibiotics.

A general overview of the successes and failures experienced during the search for antibiotics has been presented [370]. GOULDEN [371] reported that in the United States from 1955 to 1966, about 90000 synthetic compounds, 20000 plant extracts, and 120000 culture solutions of microorganisms were tested against different types of neoplasms. About 1600 substances showed sufficient activity to justify their purification; 31 fermentation products reached the first clinical test, but only five of them got as far as the second step. Of these only two products, mithramycin and streptonigrin, are clinically used today. It can be concluded that, starting with a limited number of samples, the probability of obtaining a therapeutic agent is extremely low. This also applies to other screening objectives, e.g., the search for antibiotics more effective against tuberculosis, resistant microbes, or fungi and yeasts. In order to realize a definite goal, new test methods had to be introduced. Asteromycin was discovered in the process of introducing new tests against mycoplasmas. A search for antibiotics active against bacteriophages led to the discovery of a strain producing dextrochrysin. Dienomycin was found when testing nucleotides with Wood's reagent. The leucopeptines, which are active against phytopathogenic microorganisms, gram-positive bacteria, and mycobacteria, were discovered as a result of their antiplasmin activity. Although all these antibiotics have not been approved for use, they show the importance of new screening methods.

5.2. Selection, Mutation, and Maintenance of Strains

The biological production of antibiotics is carried out predominantly by microorganisms. The discovery and isolation of the microorganism are the first steps of a long process leading to the production of the antibiotic. A yield-improvment program, a very time-consuming process, is needed to raise the yield of the strain to an economic level [372]. This is done primarily by developing optimal cultivation conditions, keeping in mind that the deep-tank and submerged methods are the only ones technically applicable. Even so, the concentration of antibiotic in the culture medium is generally not enough to start production.

For this reason *selection* is necessary. A large number of single individuals belonging to a strain are isolated. These are bred, and the antibiotic production in the cultures is quantitatively measured. New individuals with good, average, poor, or even no productivity usually develop. Hence, selection is carried out from generation to generation in an effort to develop a strain with as high an antibiotic productivity as possible, one that produces few interfering byproducts (dyes, toxins, other antibiotics, etc.), and one that remains stable over a long period of time, i.e., one whose antibiotic production does not decline.

Another technique used to obtain improved strains is *mutation*. Cultures are exposed systematically to mutagens, such as ultraviolet radiation or specific chemical compounds. The dosage is chosen so that of a very large number of treated individuals only a few survive, and these are genetically altered. The mutants obtained in this way are generally valueless. In a few cases, however, it is possible to separate a single organism that possesses properties, such as increased antibiotic production and strain stability, superior to those of the untreated strain.

Penicillin production has been perfected to such an extent that today's industrial strains produce at least 35000 U/mL of culture medium. On a smaller scale, still higher production rates have been obtained. The *maintenance* of the strain, i.e., the production, choice, testing, and storage of efficient antibiotic producers, plays a very important part in the manufacture of antibiotics. The yield of antibiotic tends to decrease through many successive rounds of selection. This tendency must be monitored using tests in culture plates, which include methods using the agar diffusion test, photocytometry, and tests in shaken erlenmeyer flasks or in small fermenters with volumes ranging from 10 to 3000 L.

5.3. Process Development Leading to Large-Scale Production

After an efficient microorganism has been found in the laboratory, the strain must be brought into large-scale production. This process, known as scaling up, is undertaken in steps and poses tremendous technical problems. Several factors are important in scaling up.

Transition to Larger Volumes. Microbial growth, begun in the laboratory on culture plates, transferred subsequently to shaking flasks having a maximum volume of one liter, and later to small industrial fermenters having a working volume of 3000 L, must ultimately be carried out in production fermenters having a total capacity of about 150 m^3 or more. The main problem involved in this process of scaling up is to modify the fermentation conditions in such a way that the same yields are obtained in the larger fermenters as in the smaller ones.

Changes in the Fermenter Geometry. The structure of the fermenter, its construction, and its dimensions, greatly affect the yield obtained. The height versus the diameter of the fermenter, the stirring and aeration systems, the cooling system (jacket, spiral, or inserted cooling), and the protection of the inlets and outlets of the fermenter against infection (pressure sealing) are all important factors in the fermentation process.

Experiments in large-scale fermenters are time consuming and expensive. Critical comparisons of similar experiments conducted by different institutes or companies must take into account the fact that the research laboratories have only a few types of fermenters at their disposal. In addition, almost every researcher handles similar fermentation problems using different strains of a microorganism. Hence, a real comparison cannot be made and only reserved conclusions can be drawn.

Variations in the Culture Medium or Nutrient Solution. The culture medium influences the growth of the microorganism and, independently of this, the amount of antibiotic it produces. The growth media or nutrient solutions required for the prefermentation treatment, which primarily must support the rapid multiplication of the microorganisms as a monoculture, have a different composition than the nutrient solution used during fermentation. For example, in fermentation the carbon source should not be too plentiful. As a result of rapid consumption, nutrients must be resupplied to prevent a nutrient shortage.

In the choice and supply of nutrients, factors other than the achievement of an optimal antibiotic production must be considered. A nutrient suitable for improving the yield of an antibiotic may simultaneously hinder its recovery. Only an accurate comparison of the yield with the effort required, from the prefermentation treatment to the final product, can decide whether an apparently good fermentation raw material is also suitable for production. The addition or removal of certain substances has a direct effect on the antibiotic production. In the manufacture of penicillin, the addition of building blocks or precursors to the fermentation broth causes, depending on the addition, a preferred production of benzylpenicillin (addition of phenylacetic acid) or of phenoxymethylpenicillin (addition of phenoxyacetic acid). On the other hand, if chloride ions are largely removed from the culture medium, e.g., by pretreatment with silver salts [373], [374], or by ion exchange [375], the production of chlortetracycline is suppressed in favor of tetracycline. Certain inhibitors, e.g., inorganic additives, such as bromides and thiocyanates [376], and a great number of organic compounds, also suppress the production of chlortetracycline.

Variation of Other Fermentation Conditions. Strict control must be maintained during fermentation. The temperature, the pH (including the effects of nutrients and additives on the pH), the stirrer speed, the air supply, and the duration of fermentation must be monitored constantly.

Control of Fermentation by Means of Additives. A resting surface culture or a simple shaken culture contains a definite nutrient medium, which is required to support the growth of the microorganism. The growing culture eventually slows down and ceases growth, usually because the medium is spent. A submerged fermenter allows the sterile addition of additives during the course of fermentation. In this way important changes can be made. A *sterile air* supply and its generally continuous distribution are vital to all aerobic microorganisms. Any interruption of the air supply must be avoided and the air must be evenly distributed throughout the fermenter. This is achieved by a mixing and air-distributing system. The air supply is often limited in large fermenters as a result of high viscosity and foaming, which makes the addition of *antifoam substances* necessary (oils, silicones). Mechanical foam destroyers can also be used but they consume large amounts of energy and are not applicable in production plants. The addition of *nutrients*, in portions or continuously, permits the supply of nutrients to be adjusted at each stage in the fermentation process. The added nutrients may be organic (e.g., sugar as the carbon source) or inorganic (e.g., ammonia as the nitrogen source).

The pH can be adjusted during fermentation by the addition of *acid or base*. However, experience has shown that the addition of certain nutrients causes a simultaneous change in the pH. Slower adjustment is physiologically preferable in this case. The addition of sugar often causes a fall in pH because of carbon dioxide formation; peptides and amino acids cause an increase in pH (via the formation of ammonia or other nitrogen bases). The addition of *building blocks, precursors, and inhibitors* during the course of fermentation has proved useful, especially for the production of penicillins and tetracyclines. In such cases, a single addition at the beginning of the fermentation procedure leads to a concentration toxic to the fungus, but because of its rapid consumption, the precursor concentration should be maintained at a certain level.

Measurement and Control Techniques; Analytical Measurements during Fermentation. The process of fermentation is relatively long, and the antibiotic production is very sensitive to disturbances. Precise analytical measurements and rapid and accurate control mechanisms are therefore required.

The monitoring of conditions during the course of fermentation can be divided into direct measurements in or at the fermenter and indirect testing in the laboratory of samples withdrawn at regular intervals and under sterile conditions.

The direct measurements are immediate and can sometimes be automated. For example, an electrode could be installed to monitor the foam level and automatically release an antifoam additive as required. The formation of foam is influenced, within certain limits, by changes in the air supply. Equipment to monitor and control the temperature (by adjusting the amount of cooling water) and the pH of the medium is common. Some other direct measurements are the determination of the weight of the full fermenter, e.g., with the help of a pressure gage (especially at the start of fermentation and at harvesting), the measurement and control of the stirring speed and of the air supply, and the determination of the partial pressures of oxygen and carbon dioxide in the fermenter and their concentrations in the exhaust gas.

Computer monitoring is of great help in industrial production because of the large number and size of the fermenters. The results of measurements on indirect samples, which are available after hours or a few days, also are fed into the computer.

The tendency today is to analyze numerous fermentation samples, taken at as short intervals as possible, using automatic analyzing instruments for sugar, nitrogen, phosphates, biomass, product, etc. Methods that permit the fully automatic withdrawal of samples during fermentation and the automatic transfer of the samples to different analyzers have been perfected.

Scale Down. When a fermentation procedure is carried out for the first time on an industrial scale, scale-up problems occur. After their start-up problems have been overcome, large fermenters often produce disproportionately larger amounts than the previously used smaller fermenters. It is difficult to explain this phenomenon when the same strain and approximately the same fermentation conditions are used. This leads to the scale-down problem, i.e., the problem of increasing the yield obtained from the smaller fermenter to that of the larger.

Solution of the scale-down problem is very important for the further development of a strain. A new, promising mutant or variant developed during a large-scale fermentation must first be evaluated in a small fermenter. The yield thus obtained must be comparable to the yield obtained using the original production strain, also in a small fermenter.

Continuous Fermentation. Continuous fermentation has proved to be feasible in breeding yeast, in the activated-sludge purification of waste water, and in the brewing of beer [377], [381]. Continuous industrial production of antibiotics suffers from several difficulties, and it has made little progress in displacing batch methods.

1) It is very difficult to keep the yield constant. The highly efficient strains used today tend to degenerate; i.e., the antibiotic production declines. This process makes the maintenance of the strain very important.
2) Maintaining the sterility of the fermentation environment and the additives is much more difficult than in a batch process.
3) A purely technical problem is the continuous accumulation of culture solutions; this generally necessitates continuous further processing. It therefore becomes necessary to convert a factory normally working only days into one working round-the-clock shifts.
4) The main saving in introducing a continuous process lies in the reduction of the volume of the equipment. However, the extent of space saving is directly proportional to the rate of the reaction. When the time required for fermentation is two weeks or less, then the volume of equipment required for a continuous process is scarcely less than that needed for a batch process. In addition, the construction of the equipment required for a continuous process is more complicated and more expensive.

Even in the case of antibiotics produced by fast-growing bacteria, e.g., tyrothricin [382] and gramicidin S [383], reasons 1, 2, and 3, along with higher equipment and factory costs, speak against continuous production.

5.4. Fermentation Technology

The essential prerequisites for the production of antibiotics using either submerged fermentation or other fermentation methods are the same.

1) A strain of microorganism should produce the desired antibiotic in satisfactory amounts and, as far as possible, without unwanted byproducts that are difficult to remove. It strain should be as stable as possible; i.e., the production of antibiotic should not decrease with time. The strain should also be resistant to other microorganisms, phages, etc.
2) Complete industrial facilities, which include laboratories for the preparation of inoculum and for the maintenance of the strain and vessels for the prefermentation treatment and for fermentation, must be available. The vessels must be equipped with devices such as temperature regulators, automatic foam destroyers, and appliances for the addition of nutrients and for the supply of sterile air. In addition, a sufficiently large recovery plant and enough storage space for raw materials, fermentation aids, intermediates, and finished products must be available.
3) The fermentation process and its optimal operation, the properties of the antibiotic formed, its isolation, and its efficient purification must be known in detail.
4) Analytical equipment and methods to monitor the operation of the fermentation and recovery processes and to control the raw materials, intermediates, and end products must be available.

Figure 1 shows a schematic outline of the large-scale production of penicillin, an example of a fermentation process.

5.4.1. Maintenance of the Strain and Production of Inoculum

The strain of microorganism is maintained as a pure culture in a microbiological laboratory. The underlying principle is preservation; i.e., a form of the microorganism that is as stable as possible must always be available. The microorganism is stored in a large number of small, separate ampules or vials that are used successively. Cultures of good colonies are constantly restarted so that the strain is never depleted.

If the microorganism forms spores, its storage is relatively easy. The spores, a resting form, are dried, usually mixed with sterile soil, and stored in ampules. Spores can be stored for months or years.

The application of frozen inoculum, stored in the vegetative state, has advantages. This form is easy to prepare in large amounts and germination is no longer necessary. However, the storage times are limited.

The inoculum for penicillin fermentation is produced by placing the spore-containing soil in a sterile agar nutrient medium in Roux bottles and incubating at 24 °C. The

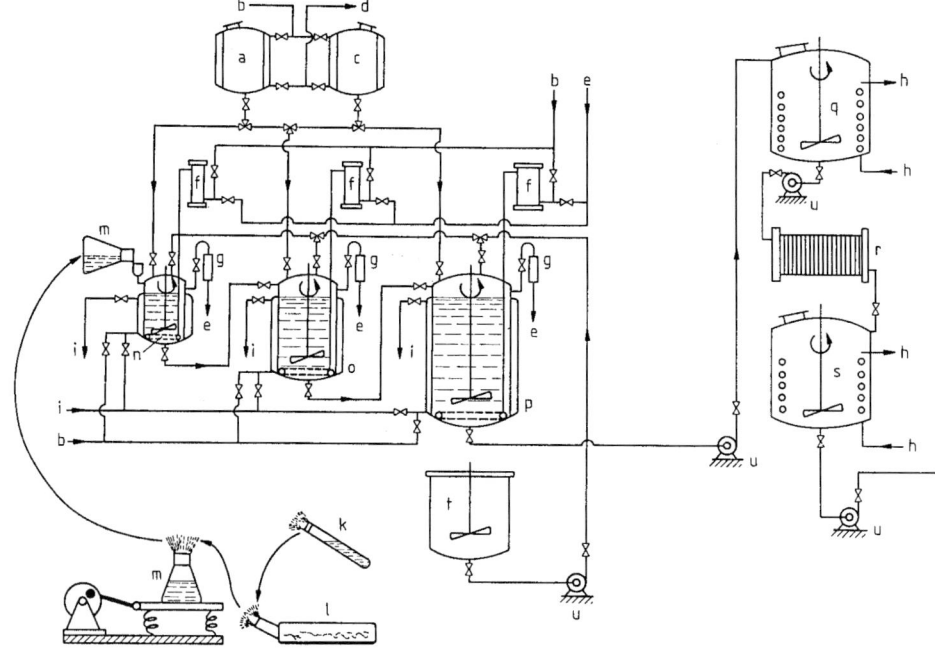

Figure 1. Schematic outline of the manufacture of penicillin
a) Antifoam substance; b) Steam; c) Precursors for penicillin formation; d) Condensate; e) Air; f) Air filter; g) Airflow recorder; h) Cooling brine; i) Cooling water; k) Spore culture (filled into l); l) Fungal culture with spores; m) Spore suspension; n) Prefermenter (inoculation culture for o); o) Intermediate fermenter (inoculation culture for p); p) Production fermenter; q) Cooling tank; r) Filtration unit; s) Filtrate container; t) Starting vessel for nutrient solution; u) Pump.

spores germinate in one or two days (vegetative form). A rich mycelium network is formed, from which new spores develop in a few days. These young, freshly formed spores are removed from the fungal network under sterile conditions and with water or normal saline. They are then transferred to erlenmeyer flasks containing a suitable sterile nutrient solution. The suspension of spores is shaken at 24 °C, enabling them to undergo multiplication.

The inoculum is transferred to another shaken flask and is allowed to grow in nutrient solution until a submerged culture can be started. The next steps (Fig. 1) lead to rapid growth and an increase in volume until finally enough mass of mycelium is obtained to inoculate the production fermenter.

Besides breeding the inoculation material, the microbiology laboratory has the equally important task of guaranteeing the maintenance and care of the strain, which insures a steady supply of a microorganism with a constant efficiency. If the laboratory limited itself to breeding, storage, and regular reinoculation, the antibiotic activity of the fermentation cultures would very soon decrease because these highly productive strains tend to mutate and degenerate. To avoid a decrease in the antibiotic production

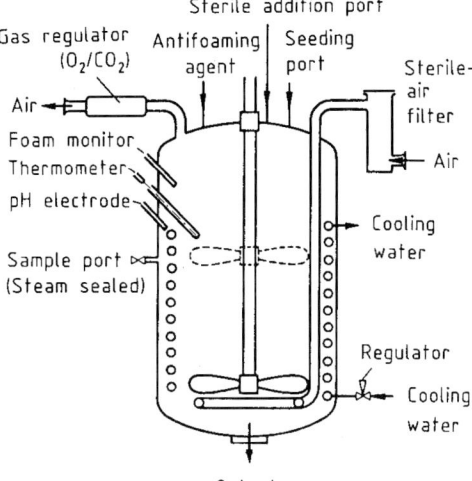

Figure 2. Schematic outline of a fermenter

in industrial fermentation, efficient strains must be subjected annually to several thousand single selections, and the resultant colonies must be tested. The single strains thereby isolated show considerable differences in their stability, i.e., in their tendency to develop into good or bad producers or even into nonproducers. Only after the minimum number of selections necessary for the maintenance of a strain has been exceeded, do the chances of surpassing the efficiency of the original strain increase. Then an improvement in the factory productivity becomes possible.

5.4.2. Treatment Before and During Fermentation

The manufacture of antibiotics by means of fermentation is always carried out in closed, sterile vessels constructed of stainless steel or of steel lined with stainless steel. Figure 2 shows the type of construction generally used today.

The supply of sterile air for fermentation is very important. Foreign organisms are filtered from the air by means of glass wool, a filtering candle, or other methods.

The composition of the nutrient solutions must meet the nutritional needs of the microorganism; these needs vary depending on the stage of fermentation. The solutions are produced in separate vessels and are sterilized therein, in the fermenter itself, or in a continuous-flow heat exchanger. This heater is also used to sterilize the additives [380].

The amounts of raw material required may only be transported and stored in silos. Hydraulic transport and weighing with a pressure gage have replaced conventional methods, and only minor additives, e.g., trace element salts, are weighed in the normal way.

Table 1. Balance of energy and materials

	Mass, kg	Mass fraction, %	Energy distribution*, MJ
Raw materials	2350	–	–
Products			
benzylpenicillin sodium	100	4	2453
Fungal mycelium	825	35	11744
Remaining substance			
in culture medium	660	28	8051
Carbon dioxide **	765	33	13176
Total			35424

* as heat of formation or combustion.
** Heat of combustion released during transition to carbon dioxide (carried off with the spent air or cooling water).

Balance Studies. The balance of energy and materials in the particular case of benzylpenicillin has been described [384] The manufacture of 100 kg of the sodium salt of penicillin requires 1.2 t carbohydrates, 60 kg animal and vegetable fats, 770 kg cornsteep liquor, 220 kg inorganic compounds (buffer, sources of sulfur and phosphorus), and 100 kg phenylacetic acid as precursors. The amount of product and the distribution of energy are shown in Table 1.

The energy requirements for the production of 100 kg of benzylpenicillin sodium are:

Electrical energy:	10.8 GJ (mainly for stirring)
Steam:	4 t (sterilization, sealing)
Fermentation air:	50000 m^3 (at STP)
Cooling water:	900 m^3

Waste Materials. The accumulation of substantial amounts of fermentation waste materials, such as the fungal mycelium and the culture solution freed of antibiotic, is a real problem. After the removal of organic solutions, e.g., by distillation (stripping), the spent medium must be fed into a biological water treatment plant. Seepage is no longer allowed.

The fungal mycelium can be processed or disposed of in several ways.

1) It can be fed to animals, directly or after drying. The proceeds cover only a part of the costs, especially if the mycelium has been dried. Also, the presence of antibiotics in the filter cake must be avoided, and filtering aids, e.g., kieselguhr or activated charcoal, may not be used.
2) It can be incinerated after the addition of liquid fuel; this is a clean but very expensive procedure.
3) It can be disposed of by dumping and humus production along with sludge depositing. This alternative often must be considered, although it entails high transportation costs.
4) It can be recycled. The mycelium can be used, directly or after intermediate processing, as a raw material for further fermentation. Mycelium is usually used as a nutrient for another microorganism. This approach is economically attractive (see Table 2 [385] – [392]).

Table 2. Fermentation residues used as raw material for further fermentation

Fermentation residue	Condition	Raw material for subsequent fermentation	Reference
Penicillium (from benzylpenicillin)	moist mycelium	oxytetracycline	[385]
	moist mycelium	chlortetracycline	[386]
	moist mycelium	streptomycin, vitamin B_{12}, or riboflavin	[387]
	mycelium hydrolyzate	phenoxymethylpenicillin	
Penicillium or other mold mycelia from fermentation	mycelium hydrolyzate	nutrient medium (e.g., for Lactobacillus bifidus)	[388]
Penicillium (from benzylpenicillin)	moist mycelium	calcium gluconate	[389]
Streptomyces (from tetracyclines)	moist mycelium	tetracyclines, etc.	[390]
Penicillium, Aspergillus, Rhizopus, or yeast	moist mycelium	nisin, in connection with lactic acid fermentation (silage)	[391]
Penicillium, Aspergillus, Actinomyces, Rhizopus, yeasts, and activated sludge (from water treatment plants)	moist or dry mycelium	moenomycin (flavophospholipol)	[392]

Control of the Fermentation Process. At regular intervals of several hours, samples of the culture solution are withdrawn through the sample port for analysis. Important data are obtained by means of chemical, physical, and biological tests. The values are plotted and curves that present a good picture of the fermentation process are obtained.

The most important analyses are:

1) Determination of the amount of antibiotic. (Biological assay is described on 7.)
2) Determination of the weight of the mycelium as an indication of the growth of the microorganism. After inoculation of the fermenter and an initial slow phase, rapid multiplication occurs. This slows down later and finally almost comes to a standstill. The point in time at which the antibiotic production decreases and falls short of economic viability can be empirically determined. At about this point the fermentation is interrupted and the culture harvested.
3) Microscopic control of the growth of the microorganism.
4) Sterility tests, i.e., tests for the absence of foreign microorganisms.
5) Measurement and correction of the pH of the culture.
6) Determination of sugar. Figures 3 A and 3 B graphically show two possibilities. Figure 3 A shows the sugar consumption. The amount of sugar consumed, plotted against time, is the difference between the total amount of sugar added and the analytically determined sugar concentration at that particular time. Figure 3 B shows the sugar content of the nutrient solution. Here, the shape of the curve is a measure of the sugar consumption (provided no more sugar is added).
7) Determination of nitrogen in the mycelium and in the culture solution, possibly combined with the addition of a nitrogen-containing nutrient solution.

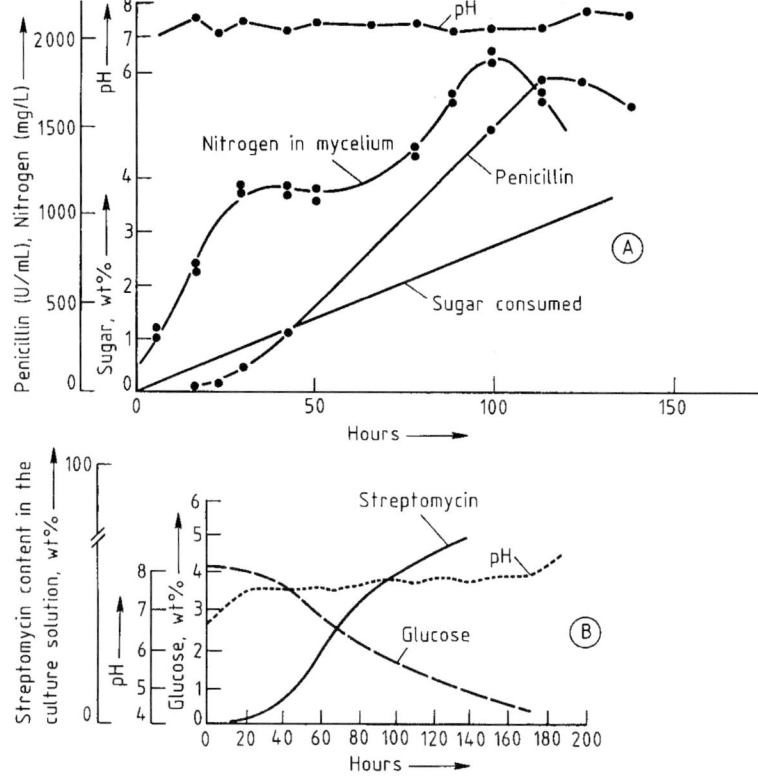

Figure 3. A. Penicillin formation with continuous glucose addition after development of the mycelium [393]
B. Streptomycin formation by *Streptomyces griseus* [394]

6. Isolation and Purification of Antibiotics; Quality Specifications

6.1. Isolation

When fermentation is completed, i.e., when a sufficiently high amount of antibiotic has accumulated in the culture solution, the antibiotic must be separated from the spent medium. The contents of the fermenter are transferred to a harvesting tank so that the fermenter can be turned immediately to the production of the next batch. The aeration is stopped, the solution cooled if necessary, and recovery of the product is begun as soon as possible. If permitted, a disinfectant, e.g., formaldehyde, is added or heat is applied to prevent further proliferation of the microorganism.

The recovery of the antibiotic can be carried out in several different ways, depending not only on its properties, but also on its subsequent processing.

Drying Process. Technically speaking, the easiest and cheapest process is to dry the entire culture, the culture filtrate, or the filter cake. Drying is employed on a large scale only in the manufacture of antibiotics used to supplement animal feed, e.g., tetracyclines and moenomycin (flavophospholipol, flavomycin) or salinomycin. Spray drying is the method most often used. Other methods, such as roller drying (possibly under vacuum) are also used. It is advisable to concentrate the solution, e.g., using a downdraft evaporator, before it is actually dried. In any case, the antibiotic must be resistant to higher temperatures. Because of its very short heating time, the spray-drying method is one of the most gentle procedures.

Filtration Followed by Extraction and Precipitation. The mycelium is separated from the liquid medium by passing the entire culture solution through a filter press, using filtering aids if necessary. A rotating filter can also be used e.g., an Oliver filter, which has three zones, intended for suction, washing, and peeling. If the culture solutions contain small amounts of mycelium, separation can also be carried out in a centrifuge.

Extraction is the method used to separate most antibiotics contained in the filtrate. A classic example is the extraction of benzylpenicillin (and phenoxymethylpenicillin) with butyl acetate (Fig. 4). It leads to a 120- to 150-fold enrichment. The penicillins are then precipitated from the extract as salts. Only those organic bases that preferentially form sparingly soluble salts with penicillin G or V but highly soluble salts with other penicillins can be used for precipitation from either water or organic solvents.

Tertiary morpholines, *N*-ethylhexahydropicoline and *N*-ethylpyrrolidine, besides *N*-ethylpiperidine, can be used for precipitation from butyl acetate, amyl acetate, and similar esters. *N*-Ethylhexamethyleneimine is used for precipitation from chloroform.

The salts thus obtained generally are easily crystallized easily and are quite stable in the dry state. They can be stored until further production steps are carried out to give the product that is used in clinical practice.

Filtration and Direct Precipitation. After filtration the aqueous culture filtrate can be subjected to direct precipitation. This method was once important in the isolation of streptomycin as a highly insoluble, colored salt, but this use has long been abandoned. Direct precipitation from the culture filtrate is of interest now because the amount of antibiotic produced by today's highly developed, efficient strains is so large that the traces of antibiotic remaining in the aqueous solution after precipitation are negligible. This method has acquired importance in the isolation of, e.g., tetracycline, which can be precipitated at its isoelectric point (pH 4.8) [395]. Another example is the direct precipitation of 5-hydroxytetracycline using a long-chain quaternary ammonium salt [396].

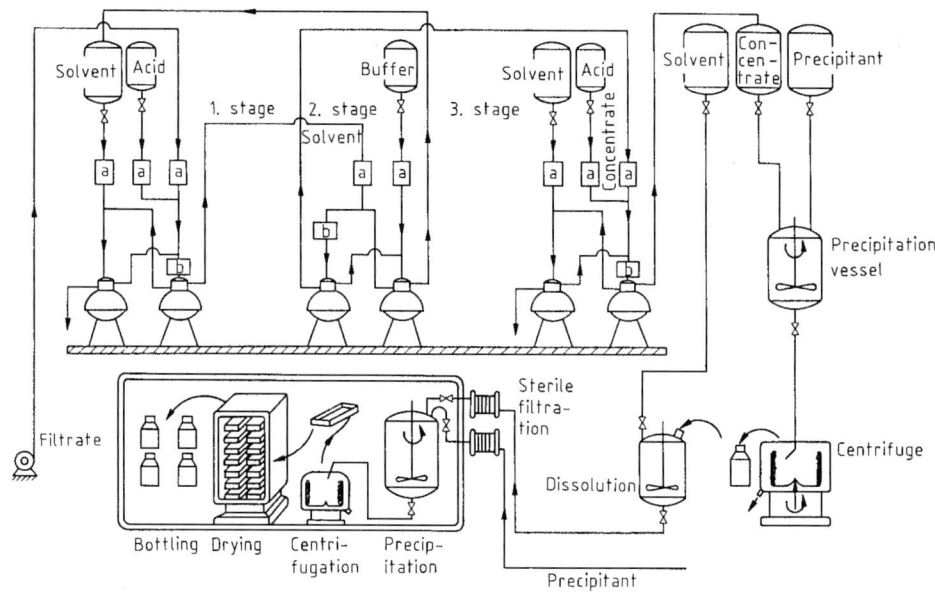

Figure 4. Penicillin extraction and the sterile final stage
a) Rotameter; b) Mixer

Filtration Followed by Extraction from the Filter Cake. If the antibiotic is present entirely or almost entirely in the mycelium, its isolation is greatly facilitated by filtration, which causes a considerable decrease in the volume of material. The moist or dry filter cake bearing the product is extracted with a solvent. The resulting solution is filtered again and processed further [397]. Examples are griseofulvin and moenomycin (flavophospholipol, flavomycin).

Adsorption Methods. Adsorption on activated charcoal following filtration is no longer used industrially. However, this method is used with some success in the developmental stages of new antibiotics.

Direct adsorption, e.g., on resins, without prior filtration, is still industrially important for the separation of such basic antibiotics as streptomycin, kanamycin, neomycin, and paromomycin. Filtration of the culture solution, especially because of the slimy substances produced by actinomycetes, is laborious and requireds large amounts of filtering aids. The real breakthrough came when the adsorption material (cation exchanger) was brought into contact with an ascending stream of the culture solution, without prior filtration. In this case the antibiotic molecules leave the solution and attach themselves to the surface of the adsorption material.

6.2. Purification Techniques, Sterile End Products, Official Regulations

Antibiotics are fermentation products and are isolated either as unfinished products or as intermediates, generally solid substances of limited stability. They are purified by methods normally employed in organic chemistry, which include chromatography, crystallization, and precipitation.

A major requirement is that the antibiotics intended for parenteral administration be free of pyrogens (fever-producing substances) and histamine-like compounds. These unwanted substances can be carried over from the fermentation, but such impurities must not appear in the final product. Special purification steps, e.g., treatment with elemental chlorine (destruction of pyrogens associated with streptomycin), filtration through activated charcoal, or deep-bed adsorption filtration, must be carried out. Precise tests must confirm the absence of all unwanted substances.

In the manufacture, purification, and preparation of antibiotics, special measures also must be taken to prevent penicillin contamination. In the processing of active substances to give pharmaceuticals, a strict spatial separation is necessary to avoid mutual contamination through the air.

Even minute amounts of antibiotics, especially penicillin, can lead to sensitization in humans and to the formation of resistant microorganisms. These antibiotics are then ineffective in the treatment of diseases because either the patient is allergic or the pathogens have become resistant. Therefore, during the processing and filtering operations the air supply must be monitored very carefully to insure the protection of the operating personnel and to guarantee that no patient unintentionally receives even traces of penicillin with another drug.

Production of Sterile Bulk Drug Substances. If sterile bulk drug substances are required, then aseptic conditions must be maintained from the start of the process. Many antibiotics are chemically unstable and cannot tolerate sterilization by heat or other agents. Generally, the solution is sterilized before the final crystallization, precipitation, spray drying, or freeze drying. This is done by filtering through porcelain filters, sintered metal filters, layers of filter paper, or graded-porosity films.

Work is carried out in specially equipped sterile rooms, which are fully air conditioned with practically germ-free air. Air-filtering devices similar to those used for the production of fermentation air are used. The rooms are disinfected using, e.g., gaseous formaldehyde, before work is commenced. The floor is kept damp with a solution of a disinfectant (phenol, quaternary ammonium bases) mixed with glycerol, to control dust. The air pressure in these sterile rooms is higher than atmospheric pressure. This prevents the entrance of unclean air. One enters the sterile rooms only after carefully washing and donning sterile clothes and through an airlock equipped with UV lamps and foot mats soaked in disinfectant. Small objects (tools, etc.) are brought into the sterile room through smaller air locks, in which they are disinfected using intense UV radiation. Containers, e.g., stainless steel cans, are first washed thoroughly and then sterilized in an autoclave. The autoclave is provided with one door that leads to the unsterile washing room and a second door leading to the sterile room.

The different steps in the course of further processing the antibiotics, such as centrifugation, drying, pulverization, sieving, and packaging, are performed in sterile glove boxes, which are provided with sterile air at a slightly elevated pressure and are equipped with UV irradiators.

Laminar-Flow Technique. In this technique, only one part of a clean or sterile room is maintained as a clean bench. This area is surrounded by a sterile box provided with a working access. To avoid the penetration of unclean air into this confined clean space, a displacing, turbulence-free air stream is created in the box. A continuous stream of sterile, filtered air enters the box from the top or from one side at a fixed speed. This air is distributed uniformly and is then sucked out the opposite side at the same speed. Special devices insure that there is minimum turbulence in the area of contact between the flowing, sterile air and the stationary, unclean air. Laminar-flow (LF) units can also be used for work with substances that should not escape to the outside world. In this case the air is recirculated through a filter.

Clean Packing. Antibiotics packed in bulk, in large containers, and intended for therapeutic use, usually are present in a sterile, highly pure state. Products such as tetracycline hydrochloride, intended for oral administration, must be very pure but not necessarily sterile.

Official Quality Requirements, Pharmacopoeias. The production process and the quality of the end product are subjected to rigorous official controls. The requirements have long been stipulated in the pharmacopoeia of each country and generally possess legal authority. The pharmacopoeias *European Pharmacopoeia, International Pharmacopoeia,* and *Compendium Medicamentorum* (standard pharmacopoeia for all Comecon states) are each valid in several countries.

In the United States, the influence exerted by the federal government goes beyond the determination of minimum quality standards for drugs. The Food and Drug Administration (FDA), attached to the U.S. Department of Health and Human Services, has published a Code of Federal Regulations (CFR) that is continuously supplemented with new regulations and reissued annually. The demands on the quality of a drug are substantially stricter and more comprehensive than those stipulated in the *U.S. Pharmacopeia*. In addition, detailed requirements have been established for the production and encapsulation techniques and rooms, the documentation, and the storage of raw materials, additives, intermediates, and end products. A detailed analysis of the starting materials, process controls, and tests of the end products and preparations must also be carried out. When a product not yet approved by the FDA is to be registered, preliminary tests must be conducted as well.

Food and Drug Administration officials have the right to inspect production plants regularly. Complaints can lead to the temporary closing of the plant and to the recall of certain preparations or particular batches of a preparation. In order to achieve a uniform standard in the production of drugs, the FDA [398] and the WHO [399]–[402] have elaborated and published fundamental rules that are now internationally called Regulations for Good Manufacturing Practices (GMPs). Drugs manufactured in countries outside the United States but imported into the United States are also subject to FDA regulations, including the GMPs, and FDA inspections. The FDA has published detailed rules in the *Federal Regulations* [403], especially for the registration of imported products.

Officials in other countries also demand a detailed description of the manufacture, quality, and safety of any drug they import. In many cases, sales depend on a prerequisite inspection of the factory, similar to the one conducted by the FDA. Canadian officials, American military forces purchasers, and the British Department of Health and Social Security all have this requirement.

7. Analytical Measurements and Quality Control

The analyses of antibiotics can be divided into two basic groups:

1) Tests during production, usually process surveillance and control.
2) Quality control, practiced as required by the WHO and the FDA. These tests have been routinely conducted by independent laboratories for a long time.

The end product, raw materials, and intermediates all are tested. The analytical measurements can be divided into chemical and physical tests, on the one hand, and biological tests, on the other. For the former, general methods used in synthetic organic chemistry are applied, along with some special methods that have been worked out for antibiotics. Fully or partially automatic techniques have been introduced to handle the large number of samples.

7.1. Microbiological Analysis

Biological Assay. Numerous microbiological methods are available to determine the amount of antibiotic present in a sample.

Agar Diffusion Test (Cylinder-Plate Method). See [404] for the original method; improvements are described in [405], [406].

A standard method has been published by the FDA [407]. For supplements, see [408] and [409]. Twenty milliliters of nutrient agar is placed in a flat-bottomed petri dish. After this solidifies, four milliliters of a second nutrient solution, seeded with the test bacteria, is poured evenly onto the first layer (at 48 °C). As soon as the second layer has solidified, six sterile stainless steel cylinders are placed on the agar, preferably using a cylinder-placing machine. To the open cylinders are added equal amounts of a standard penicillin solution containing 2.0, 1.5, 1.0, 0.5, and 0.25 U/mL. Samples of the test antibiotic solution are deposited analogously on other petri dishes.

The dishes are incubated at 37 °C for 16–18 h. During this time the penicillin diffuses out of the cylinders into the surrounding agar and suppresses the growth of the test organism. Thus, the cylinders are surrounded by clear zones, free of bacteria. The diameter of each zone provides an index of the activity of the penicillin preparation. The mean values obtained from 10–20 standard plates are used to draw a calibration curve, and the biological activity of the test solution in international units is determined using conversion tables.

Antibiotic Disk Method. This is a modification of the diffusion test. The method is widely used to determine whether a definite strain or a mixture of different microorganisms is sensitive or resistant to a given antibiotic [410]–[415].

The pathogen is freshly isolated from patients and used to inoculate a suitable nutrient agar plate. Filter paper disks 6 or 9 mm in diameter are placed on the petri dish before incubation. These disks are impregnated with a solution of the antibiotic. The amount is chosen so that the concentration of active substance present after diffusion into the agar medium corresponds to the level attainable in the patient (blood or tissue level). Test doses of 0.5 to 20 U are normal for a disk test of penicillins G or V.

In order to maintain a certain uniformity in the production of the nutrient medium for the disk method, the following directions for the preparation of peptone–casein agar have proved useful:

Peptone	6.0 g
Pancreatic–digested casein	4.0 g
Yeast extract	3.0 g
Meat extract	1.5 g
Dextrose	1.0 g
Agar	15.0 g

The components are dissolved in 1000 mL of distilled water and the pH of the liquid agar is adjusted to 6.55 after sterilization. If an agar plate thus prepared is incubated, the growth of the microorganisms seeded on the plate can be observed from the turbidity of the agar surface. If the antibiotic on the disk is effective, a clear zone of inhibition forms surrounding the disk (Fig. 5). Table 3 shows the experimental values for the diameters of the zones of inhibition for *Staphylococcus aureus* ATTC 6538P, using the pH 6.55 nutrient medium described above.

Tube Dilution Method. Three milliliters of a nutrient solution is put into each of a row of tubes. Three milliliters of a penicillin solution with a dilution of 1:100 is pipetted into the first tube. After thorough mixing, three millimeters is removed and added to the next tube. After mixing, three millimeters is removed and added to the third tube, and so on. The tubes contain successively lower concentrations of the drug. If the initial penicillin solution had a 1:100 dilution, the first tube now contains a 1:200 dilution, the second a 1:400, and so on.

Each tube is inoculated with one drop of a day-old *staphylococcus* culture (test bacteria). After incubation for one day at 37 °C, the end point is determined, i.e., the lowest concentration of penicillin that prevents the development of turbidity. If tube 3, for example, is clear but tube 4 is turbid, the end

Table 3. Diameter of zones of inhibition for *Staphylococcus aureus* ATTC 6538 P

Antibiotic	Concentration per disk	Inhibition zone
Ampicillin	5 µg	26
Chloramphenicol	10 µg	20
Lincomycin	2 µg	19
Methicillin	10 µg	26
Novobiocin	10 µg	26
Oxacillin	10 µg	30
Penicillin (P)	0.5 I.U.	26
Penicillin (*P)[a]	20 I.U.	>40
Streptomycin[b]	10 µg	16
Tetracycline	10 µg	27

[a] Massive dose of penicillin (see Fig. 5);
[b] A pH of 8.0 is required for the optimal evaluation of the substance. Under these conditions the diameter of this zone is also 26 mm.

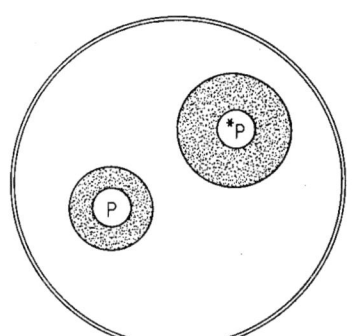

Figure 5. Antibiotic disk test
The diameter of the clear zone depends on the test dose of antibiotic: disk P (left) contained 0.5 U, disk *P (right) 20 U benzylpenicillin sodium.

point is calculated by multiplying the starting dilution of 1 : 100 by 2^3. Thus the bacteriostatic units are obtained, here 800 Bact. U. These can be converted into international units with the help of the penicillin sensitivity factor. For instance, if this factor is 0.04, then 25 Bact. U = 1 I. U. and the solution contains 800/25 = 32 I. U. [416] – [419].

Slight turbidity also can be caused by protein precipitation resulting from a change in pH. For this reason, it is advisable to add a pH indicator to the nutrient broth. Changes in the pH are then clearly visible. Phenol red and bromothymol blue generally are used. A critical assessment of the test is given in [408].

The tube dilution method is also used to determine the sensitivity of freshly isolated single strains or mixtures of pathogens to antibiotics. A comparison of the lowest inhibitory concentration of an antibiotic with the serum levels attainable in vivo indicates which antibiotic is most suitable for clinical administration.

Automatic Analyses. Automatic sample removal and preparation have speeded up analyses. However, in cases that involve special preparatory steps, such as dissolution, filtration, or extraction, full automation is not yet possible. Many references discussing automatic biotesting, are available [408], [409], [420] – [425].

The measurement of the clear zones in the agar diffusion test, which formerly was conducted visually, can now be performed objectively and automatically using commercially available instruments (scanning analysis systems). By connecting a laboratory calculator (e.g., HP 85) to such an instrument, the amount of antibiotic in a test solution can be calculated using reference standards and printed out directly.

7.2. Isotopically Labeled Antibiotics

Antibiotics containing a radioisotope at a definite position in the molecule are very important for scientific studies. Labeled substances can be used to trace:

1) Accumulation in specific tissues, e.g., tumors, for diagnostic purposes.
2) Metabolism of an antibiotic, i.e., tracing the metabolites and cleavage products in animal and human organs or excretions.
3) Determining the location and partial degradation of the antibiotic during further fermentation, processing, and purification steps.

Isotopically labeled antibiotics can be manufactured in two ways.

Fermentation Production. If appropriate isotopically labeled compounds are added to the culture solution during fermentation, a corresponding amount of labeled antibiotic is produced and can be isolated with the unlabeled antibiotic. This technique also is suitable for following labeled precursors, nutrients, salts, etc., during the fermentation procedure. In this way, insights into the mechanism of formation of the antibiotic in a microorganism or the mechanism of formation in the presence of an isolated enzyme system can be obtained. Benzylpenicillin, phenoxymethylpenicillin, and 6-aminopenicillanic acid have been labeled with ^{14}C and ^{35}S in this way [395, vol. 4, pp. 266, 296], [426]. Streptomycin has been labeled with ^{14}C and ^{3}H [395, vol. 4, p. 349].

Labeling. Acylation of 6-aminopenicillanic acid with a ^{14}C-labeled acid yields a product labeled in the side chain [427]. A subsequent isotopic exchange within the antibiotic molecule is also possible. However, only the easily removed ^{1}H atoms can be replaced by ^{2}H or ^{3}H.

8. Economic Aspects

Antibiotics find widespread use in human and veterinary medicine. As yet, agricultural usage is low and generally confined to Asia where they are used for antifungal treatment of rice plants, etc. Over 30 kt/a of antibiotics was produced worldwide in 1984. There are six main categories of antibiotics:

1) β-Lactam
2) Tetracycline
3) Macrolide
4) Peptide and glycopeptide
5) Aminoglycoside
6) Polyether

At least twenty other commercial antibiotics are not included within these six categories. They belong to a variety of chemical types, i.e., polyene, ansamycin, anthracycline, nucleoside, etc.

Over the past ten years, output has grown by approximately 4% per year, with the most rapid growth in the β-lactams, macrolides, and polyethers. On the other hand, tetracyclines have presented a static or declining market, particularly for human therapy. Dollar volume sales have grown correspondingly, with a successful new human antibiotic product being defined as one commanding minimal sales of $ 100 000 000 worldwide. β-Lactam sales account for at least half of the total human antibiotic market, which exceeds $ 5 000 000 000.

All the categories except the polyethers find use in human medicine. The most important veterinary antibiotics belong to the tetracycline, macrolide, peptide, and polyether families although some β-lactams, aminoglycosides, and other antibiotics also have veterinary markets.

Worldwide, there are over seventy primary producers of antibiotics by fermentation. If companies involved in producing semisynthetic penicillins and cephalosporins from purchased parent antibiotics are included, the number is well over one hundred. Some companies specialize in the production of a single antibiotic, but more generally a number of different antibiotics are produced, e.g., benzylpenicillin, phenoxymethylpenicillin, cephalosporin C, oxytetracycline, and streptomycin. Large multinational pharmaceutical companies frequently operate a number of separate antibiotic fermentation plants in one or more countries. For technical reasons, it may not be possible to produce two different antibiotics in the same plant.

United States and European companies are active in all categories of antibiotics; Japanese, Chinese, and Korean producers have tended to specialize in the aminoglycosides, macrolides, anticancer drugs, semisynthetic second- and third-generation β-lactams, and agricultural antibiotics.

Some old antibiotics, which are no longer protected by patents, are traded in bulk at the prices quoted in the following paragraphs. The bulk products are purchased for use in specialities or conversion into semisynthetic drugs by companies that do not have their own fermentation facilities or whose fermentation capacities are not adequate to supply their growing needs. Bulk antibiotics are also purchased on tender by government agencies, charities, etc., for use in developing countries. There is usually a significant difference between a bulk price and that of a finished (branded or generic) speciality.

In the following paragraphs the estimated worldwide antibiotic output for the year 1985 is listed. The antibiotics are grouped into the six main categories plus "other antibiotics." The production figures include antibiotics for human and veterinary applications. The specific compounds are arranged alphabetically, not according to their commercial importance. The bulk prices are quoted only for the antibiotics that are traded; this price is much lower than the price of the finished specialty product.

β-Lactams. Total output is 10–20 kt/a. There are over 50 producers. The bulk price for benzylpenicillin is 25–30 $/kg. The following compounds are included: ampicillin, amoxycillin, carbenicillin, cefaclor, cefamandole, cefazolin, cefoperazone, cefotaxime, cefoxitin, ceftazidime, cefuroxime, cephadroxil, cephalexin, cephalosporin C, cephalothin, cephamycin C, cephradine, clavulanic acid, cloxacillin, dicloxacillin, flucloxacillin, oxacillin, benzylpenicillin, phenoxymethylpenicillin, piperacillin, and ticarcillin.

Tetracyclines. Total output is 5–10 kt/a. There are 30–40 producers. The bulk price for oxytetracycline is 25–30 $/kg. The following compounds are included: chlortetracycline, democlocycline, doxycycline, methacycline, minocycline, oxytetracycline, and tetracycline.

Macrolides. Total output is 3–5 kt/a. There are 20–30 producers. The bulk price for erythromycin base is 100–120 $/kg. The following compounds are included: erythromycin, ivermectin, josamycin, kitasamycin, midecamycin, milbemycin, miocamycin, oleandomycin, spiramycin, and tylosin.

Peptides and Glycopeptides. Total output is 2–3 kt/a. There are 10–20 producers. The bulk price for bacitracin is 15 $/kg. The following compounds are included: avoparcin, bacitracin, colistin, enramycin, gramicidin, nisin, polymixin, and thiopeptin.

Aminoglycosides. Total output was 1–2 kt/a. There are 20–30 producers. The bulk price for streptomycin is 30 $/kg. The following compounds are included: amikacin, apramycin, dibekacin, dihydrostreptomycin, gentamicin, hygromycin, kanamycin, lincomycin, neomycin, netilmicin, paromomycin, ribostamycin, sagamicin, sisomicin, streptomycin, and tobramycin.

Polyethers. Total output is 3–5 kt/a. There are 5–10 producers. The following compounds are included: laidlomycin, lasalocid, maduromycin, moenomycin, monensin, narasin, and salinomycin.

Other Antibiotics. Total output was 1–2 kt/a. There are 30–40 producers. The following compounds are included: amphotericin, anticancer (including bleomycin, daunorubicin, doxorubicin, epirubicin, and mitomycin), blasticidin, clindamycin, cycloserine, flavomycin, fusidic acid, griseofulvin, novobiocin, nystatin, pimaricin, pleuromutilin, pyrrolnitrin, rifampicin, spectinomycin, vancomycin, viomycin and virginiamycin.

9. References

General References

[1] D. Gottlieb, P. D. Shaw (eds.): *Antibiotics,* vol. **I**. Mechanism of Action, Springer-Verlag, Berlin–Heidelberg–New York 1967.

[2] D. Gottlieb, P. D. Shaw (eds.): *Antibiotics,* vol. **II**, Biosynthesis, Springer-Verlag, Berlin–Heidelberg–New York 1967.

[3] J. W. Corcoran, F. E. Hahn (eds.): *Antibiotics,* vol. **III**, Mechanism of Action of Antimicrobial and Antitumor Agents, Springer-Verlag, Berlin–Heidelberg–New York 1975.

[4] J. W. Corcoran (ed.): *Antibiotics,* vol. **IV**, Biosynthesis, Springer-Verlag, Berlin–Heidelberg–New York 1981.

[5] F. E. Hahn (ed.): *Antibiotics,* vol. **V**, part 1, Mechanisms of Action of Antibiotic Agents, Springer-Verlag, Berlin–Heidelberg–New York 1979.

[6] F. E. Hahn (ed.): *Antibiotics,* vol. **V**, part 2, Mechanisms of Action of Antieukaryotic and Antiviral Compounds, Springer-Verlag, Berlin–Heidelberg–New York, 1979.

[7] J. S. Glasby: *Encyclopedia of Antibiotics,* 2nd ed., J. Wiley & Sons, Chichester–New York–Brisbane–Toronto 1979.

[8] G. Lancini, F. Parenti: *Antibiotics. An Integrated View,* Springer-Verlag, New York–Heidelberg–Berlin 1982.

[9] L. P. Garrod, H. P. Lambert, F. O'Grady: *Antibiotic and Chemotherapy,* Churchill Livingstone, Edinburgh–London–Melbourne–New York 1981.

[10] E. F. Gale, E. Cundliffe, P. E. Reynolds, M. H. Richmond, M. J. Waring: *The Molecular Basis of Antibiotic Action,* 2nd ed., J. Wiley & Sons, London–New York–Sydney–Toronto 1981.

[11] *Kirk-Othmer,* 3rd ed., vol. **2, 3**.

[12] J. Bérdy, A. Aszalos, M. Bostian, K. L. McNitt, *CRC Handbook of Antibiotic Compounds,* vol. **I–X**, CRC Press Inc., Boca Raton, Florida, 1980.

[13] H. Umezawa (ed.): *Index of Antibiotics from Actinomycetes,* vol. **I, II**, Japan Scientific Societies Press, Tokyo, University Park Press, Baltimore 1978.

[14] T. Korzylski, Z. Kowszyk-Gindifer, W. Kurylowicz: *Antibiotics Origin, Nature, and Properties,* American Society for Microbiology, Washington D.C., 1978.

[15] M. J. Weinstein, G. H. Wagman: "Antibiotics Isolation, Separation and Purification," *J. Chromatogr. Libr.* **15** (1978).

[16] R. Reiner: *Antibiotics,* Georg Thieme Verlag, Stuttgart 1982.

[17] W. Kurylowicz (ed.): *Antibiotics (a Critical Review),* Polish Medical Publishers, Warsaw 1976.

[18] P. Sammes (ed.): *Topics in Antibiotic Chemistry,* vol. **I–VI**, Ellis Horwood Ltd, 1980.

[19] H. P. Kuemmerle (ed.): *Clinical Chemotherapy,* vol. **I, II, III**, Thieme-Stratton Inc., New York 1984.

Specific References

[20] S. A. Waksman, *Antibiot. Chemother. (Basel, 1954–70)* **6** (1956) 90.
[21] S. A. Waksman, *The Antibiotic Era*, The Waksman Foundation of Japan Inc., Tokyo 1975.
[22] A. Fleming, *Br. J. Exp. Pathol.* **10** (1929) 226.
[23] G. Domagk, *Dtsch. Med. Wochenschr.*, **61** (1935) 250.
[24] E. B. Chain et al., *Lancet II* (1940) 226.
[25] E. P. Abraham, P. B. Loder, in E. H. Flynn (ed.): *Cephalosporins and Penicillins*, Academic Press, New York 1972, p. 1.
[26] W. Dürckheimer et al., *Angew. Chem. Int. Ed. Engl.* **24** (1985) 180.
[27] R. D. G. Cooper in P. G. Sammes (ed.): *Topics in Antibiotic Chemistry*, vol. **3,** Ellis Horwood Limited, Chichester 1980, p. 39.
[28] F. A. Jung, W. R. Pilgrim, J. P. Poyser, P. J. Siret in P. G. Sammes (ed.): *Topics in Antibiotic Chemistry*, vol. **4,** Ellis Horwood Limited, Chichester 1980, p. 11.
[29] G. Albers-Schönberg et al. in R. B. Morin, M. Gorman (eds.): *Chemistry and Biology of β-Lactam Antibiotics*, vol. 1, vol. 2, vol. 3, Academic Press, New York 1982.
[30] L. P. Garrod, H. P. Lambert, F. O'Grady: *Antibiotic and Chemotherapy*, 4th ed., Churchill Livingstone, Edinburgh 1973.
[31] S. A. Waksman, *Jpn. J. Microbiol.* **10** (1966) 129.
[32] H. Umezawa: *Enzyme Inhibitors of Microbial Origin*, University of Tokyo Press, Tokyo 1972.
[33] E. P. Abraham in S. Mitsuhashi (ed.): *Beta-Lactam Antibiotics*, Japan Scientific Societies Press, Tokyo, and Springer-Verlag, Berlin–Heidelberg–New York 1981, p. 3.
[34] D. C. Hodgkin, *Advancemt Sci.* **6** (1949) 85.
[35] J. C. Sheehan, K. R. Henery-Logan, *J. Am. Chem. Soc.* **79** (1957) 1262; **81** (1959) 3089.
[36] S. Nakatsuka, H. Tanino, Y. Kishi, *J. Am. Chem. Soc.* **97** (1975) 5008, 5010.
[37] J. E. Baldwin, M. A. Christie, S. B. Haber, L. I. Kruse, *J. Am. Chem. Soc.* **98** (1976) 3045.
[38] R. B. Woodward et al., *J. Am. Chem. Soc.* **88** (1966) 852.
[39] E. O. Stapley, J. Birnbaum in M. Salton, G. D. Shockman (eds): *β-Lactam Antibiotics*, Academic Press, New York 1981, p. 327.
[40] S. Karady et al., *Tetrahedron Lett.* 1976, 2401.
[41] S. Karady et al., *J. Am. Chem. Soc.* **94** (1972) 1410.
[42] L. M. Weinstock et al. *Tetrahedron Lett.* **1975**, 3979.
[43] M. Shinozaki, N. Ishida, K. Iino, T. Hiraoka, *J. Chem. Soc. Chem. Commun.* 1978, 517.
[44] Y. Hamashima, H. Matsumura, S. Matsuura, W. Nagara, M. Narisada, T. Yoshida in G. I. Gregory (ed.): *Recent Advances in the Chemistry of β-Lactam Antibiotics*, Special Publ. no. 38, The Royal Society of Chemistry, London 1981, p. 57.
[45] W. Nagata in H. Nozaki (ed.): *Current Trends in Organic Synthesis*, Pergamon Press, Oxford 1983, p. 83.
[46] A. G. Brown et al., *J. Antibiot.* **29** (1976) 668.
[47] A. G. Brown et al., in J. Elks (ed.): *Recent Advances in the Chemistry of β-Lactam Antibiotics*, Special Publ. no. 28, The Chemical Society, London 1977, p. 295.
[48] P. H. Bentley, P. D. Berry, G. Brooks, M. L. Gilpin, E. Hunt, I. I. Zomaya, *J. Chem. Soc. Chem. Comm.* 1977, 748.
[49] M. Lang, K. Prasad, W. Holick, J. Gosteli, I. Ernest, R. B. Woodward, *J. Am. Chem. Soc.* **101** (1979) 6296.
[50] J. S. Kahan et al., *Intersci. Conf. Antimicrob. Agents Chemother.*, 16th, Chicago, Pap. no. 227 (1976).

[51] A. G. Brown, D. F. Corbett, A. J. Eglington, T. T. Howarth, *J. Antibiot.* **32** (1979) 961.
[52] S. G. Box, J. D. Hood, S. R. Spear, *J. Antibiot.* **32** (1979) 1239.
[53] D. F. Corbett, A. J. Eglington, T. T. Howarth, *J. Chem. Soc. Chem. Commun.* 1977, 953.
[54] A. G. Brown, D. F. Corbett, A. J. Eglington, T. T. Howarth, *J. Chem. Soc. Chem. Comm.* 1977, 523.
[55] E. O. Stapley et al., *Intersci. Conf. Antimicrob. Agents Chemother.*, 17th, New York, Pap. no. 80 (1977).
[56] E. O. Stapley et al., *J. Antibiot.* **34** (1981) 628.
[57] P. J. Cassidy et al., *Intersci. Conf. Antimicrob. Agents Chemother.*, 17th, New York, Pap. no. 81 (1977).
[58] P. J. Cassidy et al., *J. Antibiot.* **34** (1981) 637.
[59] K. Maeda et al., *J. Antibiot.* **30** (1977) 770.
[60] K. Okamura et al., *J. Antibiot.* **31** (1978) 480.
[61] K. Okamura et al., *J. Antibiot.* **32** (1979) 262.
[62] K. Yamamoto, T. Yoshioka, Y. Kato, N. Shibamoto, Y. Shimauchi, T. Ishikura, *J. Antibiot.* **33** (1980) 796.
[63] N. Shibamoto et al., *J. Antibiot.* **33** (1980) 1128.
[64] T. Mori et al., *Intersci. Conf. Antimicrob. Agents Chemother.*, 20th, New Orleans, Pap. no. 165 (1980).
[65] M. Nakayama et al., *J. Antibiot.* **33** (1980) 1388.
[66] K. Tanaka et al. *J. Antibiot.* **34** (1981) 909.
[67] D. B. R. Johnston, S. M. Schmitt, F. A. Bouffard, B. G. Christensen, *J. Am. Chem. Soc.* **100** (1978) 313.
[68] F. A. Bouffard, D. B. R. Johnston, B. G. Christensen, *J. Org. Chem.* **45** (1980) 1130.
[69] S. M. Schmitt, D. B. R. Johnston, B. G. Christensen, *J. Org. Chem.* **45** (1980) 1135; **45** (1980) 1142.
[70] T. Kametani, S. P. Huang, S. Yokohama, Y. Suzuki, M. Ihara, *J. Am. Chem. Soc.* **102** (1980) 2060.
[71] T. Kametani, T. Nagahara, Y. Suzuki, S. Yokohama, S. P. Huang, M. Ihara, *Heterocycles* **14** (1980) 403.
[72] T. N. Salzmann, R. W. Ratcliffe, B. G. Christensen, F. A. Bouffard, *J. Am. Chem. Soc.* **102** (1980) 6161.
[73] D. G. Melillo, I. Shinkai, T. Liu, K. Ryan, M. Sletzinger, *Tetrahedron Lett.* 1980, no. 21, 2783.
[74] D. G. Melillo, T. Liu, K. Ryan, M. Sletzinger, I. Shinkai, *Tetrahedron Lett.* 1981, no. 22, 913.
[75] R. J. Ponsford, R. Southgate, *J. Chem. Soc. Chem. Commun.* 1979, 846; 1980, 1085.
[76] M. Ohno, S. Kobayashi, T. Iimori, Y.-F. Wang, T. Izawa, *J. Am. Chem. Soc.* **103** (1981) 2405.
[77] H. Kotani et al., *Agric. Biol. Chem.* **47** (1983) 1363.
[78] S. Kobayashi, T. Iimori, T. Izawa, M. Ohno, *J. Am. Chem. Soc.* **103** (1981) 2406.
[79] K. Okano, T. Izawa, M. Ohno, *Tetrahedron Lett.* 1983, no. 24, 217.
[80] T. Iimori, Y. Takahashi, T. Izawa, S. Kobayashi, M. Ohno, *J. Am. Chem. Soc.* **105** (1983) 1659.
[81] H. Aoki et al., *J. Antibiot.* **29** (1976) 492.
[82] A. Imada, K. Kitano, K. Kintaka, M. Muroi, M. Asai, *Nature (London)* **289** (1981) 590.
[83] R. B. Sykes et al. *Nature (London)* **291** (1981) 489.
[84] C. M. Cimarusti et al., *J. Org. Chem.* **47** (1982) 179.
[85] B. M. Duggar, *Ann. N.Y. Acad. Sci.* **51** (1948) 177.
[86] T. Money, A. I. Scott, *Prog. Org. Chem.* **7** (1968) 1.
[87] D. L. J. Clive, *Quart. Rev.* **22** (1968) 435.

[88] C. W. Waller et al., *J. Am. Chem. Soc.* **74** (1952) 4980.
[89] H. Muxfeldt, *Chem. Ber.* **92** (1959) 3122.
[90] J. J. Goodman et al. *J. Bacteriol.* **69** (1955) 70.
[91] A. Mitscher et al., *J. Am. Chem. Soc.* **92** (1970) 6070.
[92] F. Arcamone, A. D. Di Marco, M. Gaetani, T. Scotti, *G. Microbiol.* **9** (1961) 83.
[93] F. Arcamone in P. G. Sammes (ed.): *Topics in Antibiotic Chemistry,* vol. **2,** Ellis Horwood Ltd., Chichester 1978, p. 99.
[94] S. Neidle in P. G. Sammes (ed.): *Topics in Antibiotic Chemistry,* vol. **2,** Ellis Horwood Ltd., Chichester 1978, p. 240.
[95] S. A. Waksman: *Streptomycin,* The Williams and Wilkins Co., Baltimore 1949; *Neomycin,* Rutgers Univ. Press, New Brunswick, New Jersey, 1953.
[96] L. Weinstein, N. J. Ehrenkranz in: H. Welch, F. Marti-Ibánez (eds.): *Streptomycin and DHS, Antibiotic Monographs,* no. 10, Medical Encyclopedia, New York 1958.
[97] H. Schmidt et al., *Pharmazie* **23** (1968) 161.
[98] J. S. Brimacombe, *Angew. Chem.* **83** (1971) 261.
[99] H. Grisebach, R. Schmid, *Angew. Chem.* **84** (1972) 192.
[100] H. Umezawa et al., *J. Antibiot. Ser. A* **20** (1967) 136.
[101] H. Umezawa et al., *Science (Washington, D.C.)* **157** (1967) 1559.
[102] H. Umezawa et al., *J. Antibiot.* **24** (1971) 485.
[103] I. R. Hooper in H. Umezawa, I. R. Hooper (eds.): *Handbook of Experimental Pharmacology,* vol. **62,** Springer Verlag, Heidelberg 1982.
[104] S. Umezawa in S. Mitsuhashi (ed.): *Drug Action and Drug Resistance in Bacteria,* vol. **2,** University of Tokyo Press, Tokyo 1975, p. 3.
[105] J. Davies, D. I. Smith in M. Starr (ed.): *Annual Review of Microbiology,* vol. **32,** Annual Reviews Inc., California 1978, p. 469.
[106] H. Umezawa in R. S. Tipson, D. Horton (eds.): *Advances in Carbohydrate Chemistry and Biochemistry,* vol. **30,** Academic Press, New York 1974, p. 183.
[107] H. Umezawa, in S. Mitsuhashi (ed.): *Drug Action and Drug Resistance in Bacteria,* vol. **2,** University of Tokyo Press, Tokyo 1975, p. 211.
[108] H. Umezawa, S. Kondo in H. Umezawa, I. R. Hooper (eds.): *Handbook of Experimental Pharmacology,* vol. 62, Springer Verlag, Heidelberg 1982.
[109] R. U. Lemieux, M. L. Wolfrom in W. W. Pigman, M. L. Wolfrom (eds.): *Advances in Carbohydrate Chemistry,* vol. **3,** Academic Press, New York 1948, p. 337.
[110] S. Neidle, D. Rogers, M. B. Hursthouse, *Tetrahedron Lett.* 1968, 4725.
[111] S. Umezawa et al., *J. Antibiot.* **27** (1974) 997.
[112] T. G. Cochran, D. J. Abraham, L. L. Martin, *J. Chem. Soc. Chem. Commun.* 1972, 494.
[113] S. Hanessian, R. Roy, *J. Am. Chem. Soc.* **101** (1979) 5839.
[114] D. R. White et al., *Tetrahedron Lett.* 1979, 2737.
[115] K. L. Rinehart Jr.: *The Neomycins and Related Antibiotics,* J. Wiley and Sons, New York 1964.
[116] S. Umezawa, Y. Nishimura, *J. Antibiot.* **30** (1977) 189.
[117] M. Hichens, K. L. Rinehart Jr., *J. Am. Chem. Soc.* **85** (1963) 1547.
[118] S. Tatsuoka, S. Horii, *Proc. Jpn. Acad.* **39** (1963) 314.
[119] T. Ito et al., *Agric. Biol. Chem.* **34** (1970) 980.
[120] G. Koyama et al., *Tetrahedron Lett.* 1968, 1875.
[121] S. Umezawa et al., *J. Antibiot.* **21** (1968) 367.
[122] D. J. Cooper et al., *J. Chem. Soc. C* 1971 3126.
[123] H. Reimann et al., *J. Org. Chem.,* **39** (1974) 1451.

[124] S. Fukatsu, *J. Antibiot.* **32** (suppl), (1979) S-178.
[125] H. Kawaguchi et al., *J. Antibiot.* **25** (1972) 695.
[126] T. L. Nagabushan et al., *J. Am. Chem. Soc.* **100** (1978) 5253.
[127] T. Tsuchiya, Y. Takagi, S. Umezawa, *Tetrahedron Lett.* 1979, 4951.
[128] T. Deushi et al., *J. Antibiot.* **32** (1979) 173; **32** (1979) 1061.
[129] S. Inouye et al., *J. Antibiot.* **32** (1979) 1354.
[130] T. Nara et al., *J. Antibiot.* **30** (1977) 533.
[131] Y. Okami et al., *J. Antibiot.* **32** (1979) 964.
[132] A. Bloch in E. J. Ariëns (ed.): *Drug Design*, vol. **4,** Academic Press, New York 1973, p. 286.
[133] M. Ohno in J. M. Cassady, J. D. Douros (eds.): *Anticancer Agents Based on Natural Product Models*, Academic Press, New York 1980, p. 73.
[134] A. Bloch in M. Grayson (ed.): *Antibiotics, Chemo-therapeutics, and Antibacterial Agents for Disease Control*, J. Wiley & Sons, New York 1982.
[135] H. E. Bergy, R. R. Herr, *Antimicrob. Agents Chemother. (1961–70)* 1966, 625.
[136] K. Mizuno, M. Tsujino, M. Takada, M. Hayashi, K. Atsumi, K. Asano, T. Matsuda, *J. Antibiot.* **27** (1974) 775.
[137] T. Tsuruoka et al., *Meiji Seika Kenkyu Nempo* **9** (1967) 17.
[138] P. W. K. Woo, H. W. Dion, et al., *J. Heterocycl. Chem.* **11** (1974) no. 4, 641–3.
[139] H. Nakamura, G. Koyama, Y. Iitaka, M. Ohno, N. Yagisawa, S. Kondo, K. Maeda, H. Umezawa, *J. Am. Chem. Soc.* **96** (1974) 4327.
[140] E. Cherbuliez, K. Bernhard, *Helv. Chim. Acta* **15** (1932) 464.
[141] L. Ehrenberg et al., *Sven. Kem. Tidskr.* **58** (1946) 269.
[142] K. Isono, S. Suzuki, *J. Antibiot. Ser. A* **13** (1960) 270.
[143] H. Nishimura, K. Katagiri, K. Sato, M. Mayama, N. Shimaoka, *J. Antibiot. Ser. A* **9** (1956) 60.
[144] K. G. Cunningham, S. A. Hutchinson, W. Manson, F. S. Spring, *J. Chem. Soc.* 1951, 2299.
[145] K. Isono, K. Asahi, S. Suzuki, *J. Am. Chem. Soc.* **91** (1969) 7490.
[146] K. Isono, S. Suzuki, *Heterocycles* **13** (1979) 333.
[147] M. Uramoto, M. Matsuoka, J. G. Liehr, J. A. McCloskey, K. Isono, *Agric. Biol. Chem.* **45** (1981) 1901.
[148] M. Hori, E. Ito, T. Takita, G. Koyama, T. Takeuchi, H. Umezawa, *J. Antibiot. Ser. A* **17** (1964) 96.
[149] S. Aizawa, T. Hidaka, N. Otake, H. Yonehara, K. Isono, N. Igarashi, S. Suzuki, *Agric. Biol. Chem.* **29** (1965) 375.
[150] G. Koyama, H. Umezawa, *J. Antibiot. Ser. A.* **18** (1965) 175.
[151] T. Sawa, Y. Fukagawa, I. Homma, T. Wakashiro, T. Takeuchi, M. Hori, T. Komai, *J. Antibiot.* **21** (1968) 334.
[152] H. Nishimura, M. Mayama, Y. Komatsu, H. Kato, N. Shimaoka, Y. Tanaka, *J. Antibiot. Ser. A* **17** (1964) 148.
[153] T. Haneishi et al., *J. Antibiot.* **24** (1971) 797.
[154] T. Kusaka, H. Yamamoto, M. Shibata, M. Muroi, T. Kishi, K. Mizuno, *J. Antibiot.* **21** (1968) 255.
[155] S. Yaginuma, N. Muto, M. Tsujino, Y. Sudate, M. Hayashi, M. Otani, *J. Antibiot.* **34** (1981) 359.
[156] M. Arita, K. Adachi, Y. Ito, H. Sawai, M. Ohno, *J. Am. Chem. Soc.* **105** (1983) 4049.
[157] S. Takeuchi et al. *J. Antibiot. Ser. A* **11** (1958) 1.
[158] S. Masamune, G. S. Bates, J. W. Corcoran, *Angew. Chem. Int. Ed. Engl.* **16** (1977) 585.
[159] S. Masamune, J. W. Ellingboe, W. Choy, *J. Am. Chem. Soc.* **104** (1982) 5526.

[160] M. N. Donin et al., *Antibiot. Annu.* 1953–1954, 179.
[161] C. Djerassi, J. A. Zderic, *J. Am. Chem. Soc.* **78** (1956) 2907.
[162] J. M. McGuire et al., *Antibiot. Chemother. (Washington D.C.)* **2** (1952) 281.
[163] R. A. LeMahieu, J. F. Blount, R. W. Kierstead, *J. Antibiot.* **28** (1975) 705, and references cited therein.
[164] S. Omura, A. Nakagawa, *J. Antibiot.* **28** (1975) 401.
[165] S. Omura, A. Nakagawa, M. Machida, H. Imai, *Tetrahedron Lett.* 1977, 1045.
[166] T. Hata et al., *J. Antibiot. Ser. A* **6** (1953) 87.
[167] T. Watanabe, T. Fujii, K. Satake, *J. Biochem. (Tokyo)* **50** (1961) 197.
[168] S. Pinnert-Sindico et al., *Antibiot. Annu.* **2** (1954–55) 274.
[169] G. H. Wagman, J. A. Waitz, J. Marquez, A. Murawski, E. M. Oden, R. T. Testa, M. J. Weinstein, *J. Antibiot.* **25** (1972) 641.
[170] H. Reimann, R. S. Jaret, *J. Chem. Soc. Chem. Commun.* 1972, 1270.
[171] H. Koshiyama, H. Tsukiura, K. Fujisawa, M. Konishi, M. Hatori, K. Tomita, H. Kawaguchi, *J. Antibiot.* **22** (1969) 61.
[172] H. Tsukiura, M. Konishi, M. Saka, T. Naito, H. Kawaguchi, *J. Antibiot.* **22** (1969) 89.
[173] K. Hatano, E. Higashide, M. Shibata, *J. Antibiot.* **29** (1976) 1163.
[174] T. Kishi, S. Harada, H. Yamana, A. Miyake, *J. Antibiot.* **29** (1976) 1171.
[175] Y. Shimauchi, K. Kubo, K. Osumi, K. Okamura, Y. Fukagawa, T. Ishikura, *J. Antibiot.* **32** (1979) 878.
[176] R. B. Woodward, *Angew. Chem.* **69** (1957) 50.
[177] R. B. Woodward, L. S. Weiler, P. C. Dutta, *J. Am. Chem. Soc.* **87** (1965) 4662.
[178] A. Kinumaki, M. Suzuki, *J. Antibiot.* **25** (1972) 480.
[179] R. B. Morin, M. Gorman, P. L. Hamill, P. V. Demarco, *Tetrahedron Lett.* 1970, 4737.
[180] S. Bhuwapathanapun, P. Gray, *J. Antibiot.* **30** (1977) 673.
[181] S. Omura, H. Matsubara, A. Nakagawa, A. Furusaki, T. Matsumoto, *J. Antibiot.* **33** (1980) 915.
[182] M. Hayashi, M. Ohno, S. Satoi, *J. Chem. Soc. Chem. Commun.* 1980, 119.
[183] P. W. K. Woo, H. W. Dion, Q. R. Bartz, *Chem. Comun.* **86** (1964) 2724, 2726.
[184] F. J. Antosz in M. Grayson (ed.): *Antibiotics, Chemotherapeutics and Antibacterial Agents for Disease Control*, J. Wiley & Sons, New York 1982, p. 76.
[185] A. Lüttringhaus, H. Gralheer, *Justus Liebigs Ann. Chem.* **550** (1942) 67.
[186] K. L. Rinehart Jr., *Acc. Chem. Res.* **5** (1972) 57.
[187] P. Sensi, *Chim. Ind. (Milan)* **51** (1969) 811.
[188] P. Sensi, *Res. Prog. Org. Biol. Med. Chem.* **1** (1964) 337.
[189] M. T. Timbal, *Antibiot. Annu.* 1959–1960, 271.
[190] N. Bergamini, G. Fowst, *Arzneim. Forsch.* **15** (1965) 951.
[191] S. Riva, L. G. Silvestri, *Annu. Rev. Microbiol.* **26** (1972) 199.
[192] D. Perlman in M. Grayson (ed.): *Antibiotics, Chemotherapeutics, and Antibacterial Agents for Disease Control*, J. Wiley&Sons, New York 1982, p. 210.
[193] H. Umezawa, K. Maeda, T. Takeuchi, Y. Okami, *J. Antibiot. Ser. A.* **19** (1966) 200.
[194] T. Takita, Y. Muraoka, T. Nakatani, A. Fujii, Y. Umezawa, H. Naganawa, H. Umezawa, *J. Antibiot.* **31** (1978) 801.
[195] T. Takita et al., *Tetrahedron Lett.* 1982, no. 23, 521, and references cited therein.
[196] H. Umezawa in J. M. Cassady, J. D. Douros (ed.): *Anticancer Agents Based on Natural Product Models*, Academic Press, New York 1980, p. 147.
[197] H. Kawaguchi et al., *J. Antibiot.* **30** (1977) 779.
[198] M. Konishi et al., *J. Antibiot.* **30** (1977) 789.

[199] G. F. Gauze, M. G. Brazhnikova, *Lancet* **247** (1944) 715.
[200] T. S. G. Jones, *Ann. N.Y. Acad. Sci.* **51** (1949) 909.
[201] I. M. Lockhart, E. P. Abraham, *Biochem. J.* **58** (1954) 633.
[202] A. Arriagada et al., *Br. J. Exp. Pathol.* **30** (1949) 425.
[203] C. Ressler, D. V. Kashelikar, *J. Am. Chem. Soc.* **88** (1966) 2025.
[204] H. Brockmann, *Angew. Chem.* **72** (1960) 939.
[205] A. J. Whiffen, J. N. Bohono, R. L. Emerson, *J. Bacteriol.* **52** (1946) 610.
[206] G. M. Schull, J. L. Sardinas, *Antibiot. Chemother.* **5** (1955) 398.
[207] S. Takeuchi, H. Yonehara, H. Umezawa, *J. Antibiot. Ser. A* **12** (1959) 195.
[208] H. Umezawa et al., *J. Antibiot. Ser. A* **6** (1953) 101.
[209] H. Hoeksema, J. L. Johnson, J. W. Hinman, *J. Am. Chem. Soc.* **77** (1955) 6710.
[210] A. E. Oxford, H. Raistrick, P. Simonart, *Biochem. J.* **33** (1939) 240.
[211] J. F. Grove, J. MacMillan, T. P. C. Mulholl, M. A. T. Rogers, *J. Chem. Soc.* 1952, 3977.
[212] J. Ehrlich, Q. R. Bartz, R. M. Smith, D. A. Joslyn, P. R. Burkholder, *Science (Washington D.C.)* **106** (1947) 417.
[213] G. Keiser, *Dtsch. Med. Wochenschr.* **96** (1971) 1544.
[214] K. Shirahata, N. Hirayama, *J. Am. Chem. Soc.* **105** (1983) 7199.
[215] T. Hata, Y. Sano, R. Sugawara, A. Matsume, K. Kanamori, T. Shima, T. Hoshi, *J. Antibiot. Ser. A* **9** (1956) 141.
[216] S. Wakaki, H. Marumo, K. Tomioka, G. Shimizu, E. Kato, H. Kamada, S. Kudo, Y. Fujimoto, *Antibiot. Chemother. (Washington D.C.)* **8** (1958) 228.
[217] J. Vodrazka, *Pharm. Ind.* **32** (1970) 951.
[218] A. Agtarap et al., *J. Am. Chem. Soc.* **89** (1967) 5737.
[219] D. Hendlin et al., *Science (Washington D.C.)* **166** (1969) 122.
[220] W. O. Godtfredsen, S. Jahnsen, H. Lorck, K. Roholt, L. Tybring, *Nature (London)* **193** (1962) 987.
[221] S. Takahashi, T. Ohashi, Y. Kii, H. Kumagai, H. Yamada, *J. Ferment. Technol.* **57** (1979) 328.
[222] T. Ohashi, S. Takahashi, T. Nagamichi, K. Yoneda, H. Yamada, *Agric. Biol. Chem.* **45** (1981) 831.
[223] K. C. Nicolaou, W.-M. Dai, *Angew. Chem. Int. Ed. Engl.* **30** (1990) 1387.
[224] N. Ishida, K. Miyazaki, K. Kumagai, M. Rikimaru, *J. Antibiot.* **18** (1965) 68.
[225] K. Edo et al., *Tetrahedron Lett.* **26** (1985) 331.
[226] M. D. Lee et al., *J. Am. Chem. Soc.* **109** (1987) 3464.
[227] M. D. Lee et al., *J. Am. Chem. Soc.* **109** (1987) 3466.
[228] J. Golik et al., *J. Am. Chem. Soc.* **109** (1987) 3461.
[229] J. Golik et al., *J. Am. Chem. Soc.* **109** (1987) 3462.
[230] M. Konishi et al., *J. Antibiot.* **42** (1989) 1449.
[231] J. Hu et al., *J. Antibiot.* **41** (1988) 1575.
[232] T. Otani, Y. Minami, T. Marunaka, R. Zhang, M.-Y. Xie, *J. Antibiot.* **41** (1988) 1580.
[233] K. S. Lam et al., *J. Antibiot.* **44** (1991) 472.
[234] S. J. Hofsteader, J. A. Matson, A. R. Malacko, H. Marquardt, *J. Antibiot.* **45** (1992) 1250.
[235] N. Zein et al., *Proc. Natl. Acad. Sci. USA* **90** (1993) 8009.
[236] J. E. Leet et al., *J. Am. Chem. Soc.* **115** (1993) 8432.
[237] M. Hanada et al., *J. Antibiot.* **44** (1991) 403.
[238] A. G. Meyer, *Tetrahedron Lett.* **28** (1987) 4493.
[239] H. Maeda, K. Edo, N. Ishida (eds.): *Neocarzinostatin. The Past, Presend, and Future of an Anticancer Drug*, Springer-Verlag, Heidelberg 1997.

[240] N. P. Sullivan et al., *Science (Washington, D.C.)* **107** (1948) 169.
[241] J. L. Szabo et al., *Antibiot. Chemother. (Washington, D.C.)* **1** (1951) 499.
[242] E. Brandl et al., *Wien. Med. Wochenschr.* 1953, 602.
[243] K. H. Spitzy, *Wien. Klin. Wochenschr.* **65** (1953) 583.
[244] Y. G. Perron et al., *Antibiot. Ann.* 1959–1960, 107; *J. Am. Chem. Soc.* **82** (1960) 3934.
[245] G. M. Williamson et al., *Lancet 1961,* vol. **I**, 847;
M. Nagley, *Lancet 1961,* vol. **I**, 851.
[246] E. T. Kundsen et al., *Br. Med. J.* **3** (1967) 75.
[247] J. P. Clayton et al., *J. Med. Chem.* **18** (1975) 172.
[248] K. Butler, *Del. Med. J.* **43** (1971) 366. A. R. English et al., *Antimicrob. Agents Chemother.* **1** (1972) 185.
[249] K. Tsuchiya et al., *J. Antibiot.* **24** (1971) 607.
[250] H. Neu, E. B. Windshell, *Appl. Microbiol.* **21** (1971) 66.
[251] E. T. Kundsen, G. N. Rolinson, *Br. Med. J.* **2** (1960) 700.
[252] F. P. Doyle et al., *Nature (London)* **192** (1961) 1183.
[253] E. T. Kundsen et al., *Lancet 1962,* vol. **II**, 632.
[254] H. Yoshioka et al., *J. Antibiot. Ser. B* **20** (1967) 34. I. P. Fomina et al., *Antibiotiki (Moscow)* **16** (1971) 153.
[255] R. Sutherland et al., *Br. Med. J.* **4** (1970) 455.
[256] G. N. Rolinson, S. Stevens, *Br. Med. J.* **2** (1961) 191.
[257] H. B. Koenig et al., *Infection (Munich)* **5** (1977) 170.
[258] K. Ueo et al., *Antimicrob. Agents Chemother.* **12** (1977) 455.
[259] R. Sutherland et al., *Antimicrob. Agents Chemother. (1961–70)* 1971, 411. R. Sutherland et al., *Br. Med. J.* **3** (1972) 13.
[260] A. J. Gonzaga et al., *J. Infect. Dis.* **129** (1974) 545. H. C. Neu, *Drugs* **9** (1975) 81; *Int. J. Clin. Pharmacol. Biopharm.* **11** (1975) 132.
[261] R. Sutherland et al., *Br. Med. J.* **2** (1967) 804. R. Kahrimanis, P. Pierpaoli, *N. Engl. J. Med.* **285** (1971) 236.
[262] C. N. Baker et al., *Intersci. Conf. Antimicrob. Agents Chemother. 18th,* Abstract 293, Atlanta 1978.
[263] Y. Shiobara et al., *J. Antibiot.* **27** (1974) 665.
[264] M. Ehrnebo et al., *J. Pharmacokinet. Biopharm.* **7** (1979) 429.
[265] K. Roholt, *J. Antimicrob. Chemother. (Suppl. B)* **3** (1977) 71.
[266] F. Lund, L. Tybring, *Nature (London)* **236** (1972) 135.
[267] C. W. Ryan et al., *J. Med. Chem.* **12** (1969) 310.
[268] R. R. Chauvette, P. A. Pennington, *J. Med. Chem.* **18** (1975) 403.
[269] J. L. Spencer et al., *J. Med. Chem.* **9** (1966) 746.
[270] R. F. Buck, K. E. Price, *Antimicrob. Agents Chemother.* **11** (1977) 324.
[271] J. E. Dolfini et al., *J. Med. Chem.* **14** (1971) 117.
[272] O. Zak et al., *J. Antibiot.* **29** (1976) 653.
[273] L. B. Crast et al., *J. Med. Chem.* **16** (1973) 1413.
[274] R. R. Chauvette et al., *J. Am. Chem. Soc.* **84** (1962) 3401.
[275] F. Knüsel et al., *Antimicrob. Agents Chemother. (1961–70)* 1970, 140. H. C. Neu et al., *J. Antibiot.* **25** (1972) 400.
[276] H. Nomura et al., *J. Med. Chem.* **17** (1974) 1312. K. Tsuchiya et al., *Antimicrob. Agents Chemother.* **13** (1978) 137.
[277] G. T. Stewart et al., *Lancet (1964),* vol. **II**, 1305.

[278] R. Del Busto et al., *Antimicrob. Agents Chemother.* **9** (1976) 397.
[279] I. Saikawa et al., *Yakugaku Zasshi* **99** (1979) 929.
[280] W. E. Wick, D. A. Preston, *Antimicrob. Agents Chemother.* **1** (1972) 221.
[281] M. Numata et al., *J. Antibiot.* **31** (1978) 1262.
[282] K. Kariyone et al., *J. Antibiot.* **23** (1970) 131.
[283] T. Kamimura et al., *Antimicrob. Agents Chemother.* **16** (1979) 540.
[284] R. Heymes et al., *Infection (Munich)* **5** (1977) 529.
[285] K. Tsuchiya et al., *Antimicrob. Agents Chemother.* **19** (1981) 56.
[286] R. N. Brogden et al., *Drugs* **17** (1979) 223. C. H. O'Callaghan et al., *Antimicrob. Agents Chemother.* **9** (1976) 511.
[287] R. Reiner et al., *J. Antibiot.* **33** (1980) 783.
[288] C. H. O'Callaghan et al., *Antimicrob. Agents Chemother.* **17** (1980) 876.
[289] E. O. Stapley et al., *Antimicrob. Agents Chemother.* **2** (1972) 122.
[290] H. Nakao et al., *J. Antibiot.* **29** (1976) 554.
[291] M. Iwanami et al., *Chem. Pharm. Bull.* **28** (1980) 2629.
[292] M. Narisada et al., *J. Med. Chem.* **22** (1979) 757.
[293] T. T. Howarth et al., *J. Chem. Soc. Chem. Commun.* 1976, 266.
[294] W. J. Leanza et al., *J. Med. Chem.* **22** (1979) 1435.
[295] R. B. Sykes, I. Phillips, *J. Antimicrob. Chemother. (Suppl. E)* **8** (1981) 1.
[296] J. H. Boothe et al., *J. Am. Chem. Soc.* **75** (1953) 4621. L. H. Conover et al., *J. Am. Chem. Soc.* **75** (1953) 4622.
[297] J. R. D. McCormick et al., *J. Am. Chem. Soc.* **79** (1957) 4561.
[298] A. C. Finlay et al., *Science (Washington, D.C.)* **111** (1950) 85. P. P. Regna, I. A. Solomons, *Ann. N.Y. Acad. Sci.* **53** (1950) 221.
[299] R. K. Blackwood et al., *J. Am. Chem. Soc.* **83** (1961) 2773; **85** (1963) 3943.
[300] C. R. Stephens et al., *J. Am. Chem. Soc.* **80** (1958) 5324. J. R. D. McCormick et al., *J. Am. Chem. Soc.* **82** (1960) 3381.
[301] W. J. Gottstein et al., *J. Am. Chem. Soc.* **81** (1959) 1198.
[302] G. S. Redin, *Antimicrob. Agents Chemother. (1961–70)* 1967, 371.
[303] A. D. Di Marco et al., *Nature (London)* **201** (1964) 706. F. Arcamone et al., *J. Am. Chem. Soc.* **86** (1964) 5334.
[304] F. Arcamone et al., *Tetrahedron Lett.* 1969, 1007.
[305] T. Oki et al., *J. Antibiot.* **28** (1975) 830.
[306] H. Umezawa et al., *J. Antibiot. Ser. A* **10** (1957) 181.
[307] T. Wakazawa et al., *J. Antibiot. Ser. A* **14** (1961) 180, 187.
[308] D. A. Preston, W. E. Wick, *Antimicrob. Agents Chemother.* 1971, 322.
[309] M. J. Weinstein et al., *Antimicrob. Agents Chemother. (1961–70)* 1964, 1. J. Black et al., *Antimicrob. Agents Chemother. (1961–70)* 1964, 138.
[310] H. Umezawa et al., *J. Antibiot.* **24** (1971) 485.
[311] H. Kawaguchi, *J. Infect. Dis. (Suppl.)* **134** (1976) 242.
[312] R. Okachi et al., *J. Antibiot.* **27** (1974) 793.
[313] T. Shomura et al., *J. Antibiot.* **23** (1970) 155.
[314] S. A. Waksman, H. A. Lechevalier, *Science (Washington, D.C.)* **109** (1949) 305.
[315] T. H. Haskell et al., *J. Am. Chem. Soc.* **81** (1959) 3480, 3482.
[316] A. Schatz, E. Bugie, S. A. Waksman, *Proc. Soc. Exp. Biol. Med.* **55** (1944) 66.
[317] R. L. Peck et al., *J. Am. Chem. Soc.* **68** (1946) 1390.
[318] S. Kondo et al., *J. Antibiot. Ser. A* **18** (1965) 38.

[319] R. L. Mann, W. W. Broomer, *J. Am. Chem. Soc.* **80** (1958) 2714.
[320] R. Q. Thompson, E. A. Presti, *Antimicrob. Agents Chemother. (1961–70)* 1967, 332.
[321] J. A. Waitz et al., *Antimicrob. Agents Chemother.* **2** (1972) 431. C. C. Crowe, E. Sanders, *Antimicrob. Agents Chemother.* **3** (1973) 24.
[322] S. A. Kabins et al., *Antimicrob. Agents Chemother.* **10** (1976) 139. J. J. Rahal et al., *Antimicrob. Agents Chemother.* **9** (1976) 595. V. Dhawan et al., *Antimicrob. Agents Chemother.* **11** (1977) 64.
[323] D. J. Mason et al., *Antibiot. Chemother. (Washington, D.C.)* **11** (1961) 118.
[324] R. Okachi et al., *J. Antibiot.* **30** (1977) 541.
[325] S. Horii et al., *J. Antibiot.* **24** (1971) 57, 59.
[326] H. Umezawa et al., *J. Antibiot. Ser. A* **18** (1965) 101.
[327] J. M. McGuire et al., *Antibiot. Chemother. (Washington, D.C.)* **2** (1952) 281.
[328] V. C. Stephens et al., *J. Am. Pharm. Assoc. Sci. Ed.* **48** (1959) 620.
[329] B. A. Sobin et al., *Antibiot. Annu.* 1954–1955, 87.
[330] W. D. Celmer et al., *Antibiot. Annu.* 1957–1958, 476.
[331] T. Osono et al., *J. Antibiot.* **20** (1967) 174. K. Nitta et al., *J. Antibiot.* **20** (1967) 181.
[332] J. C. Chabala et al., *J. Med. Chem.* **23** (1980) 1134.
[333] T. Tsuruoka et al., *J. Antibiot.* **24** (1971) 319, 452.
[334] P. Sensi et. al., *Antimicrob. Agents Chemother. (1961–70)* 1967, 699.
[335] W. Tanaka et al., *Heterocycles* **13** (1979) 469.
[336] W. Hausmann et al., *J. Am. Chem. Soc.* **76** (1954) 4892.
[337] B. A. Johnson et al., *Science (Washington, D.C.)* **102** (1945) 376.
[338] Y. Koyama et al., *J. Antibiot.* **3** (1950) 457.
[339] S. Benjamin et al., *Antibiot. Annu.* 1959–1960, 41.
[340] S. Tanayama et al., *J. Antibiot.* **21** (1968) 313.
[341] M. Arai et al., *J. Antibiot. Ser. A* **9** (1956) 193.
[342] P. J. Van Dijck et al., *Chemotherapy (Basel)* **14** (1969) 109. P. Crooy et al., *J. Antibiot.* **25** (1972) 371.
[343] E. B. Herr Jr. et al., *Ann. N.Y. Acad. Sci.* **135** (1966) 940.
[344] A. C. Finlay et al., *Am. Rev. Tuberc.* **63** (1951) 1. Q. R. Bartz et al., *Am. Rev. Tuberc.* **63** (1951) 4.
[345] H. Yoshioka et al., *Tetrahedron Lett.* 1971, 2043.
[346] M. H. McCormick et al., *Antibiot. Annu.* 1955–1956, 606.
[347] S. A. Waksman et al., *Proc. Soc. Exp. Biol. Med.* **45** (1940) 609.
[348] N. Ishida et al., *J. Antibiot. Ser. A* **18** (1965) 68.
[349] H. Umezawa et al., *J. Antibiot.* **29** (1976) 97.
[350] H. Umezawa et al., *J. Antibiot.* **23** (1970) 259.
[351] M. E. Haney Jr. et al., *Antimicrob. Agents Chemother. (1961–70)* 1967, 349.
[352] J. Berger et al., *J. Am. Chem. Soc.* **73** (1951) 5295.
[353] Y. Miyazaki, *J. Antibiot.* **27** (1974) 814.
[354] W. Gold et al., *Antibiot. Annu.* 1955–1956, 579.
[355] E. L. Hazen, *Science (Washington, D.C.)* **112** (1950) 423.
[356] A. P. Struyk et al., *Antibiot. Annu.* 1957–1958, 878.
[357] S. Hosoya et al., *Jpn. J. Exptl. Med.* **22** (1952) 505. S. Hosoya, *Chemotherapy (Tokyo)* **2** (1954) 1.
[358] W. E. Grundy et al., *Antibiot. Chemother. (Washington, D.C.)* **3** (1953) 1215.
[359] C. Lewis et al., *Antimicrob. Agents Chemother. (1961–70)* 1963, 570.
[360] B. J. Magerlein, *Antimicrob. Agents Chemother. (1961–70)* 1967, 727.
[361] Upjohn Co., ZA 6802283, 1968. (W. Morozowich, A. A. Sinkula)
[362] K. H. Wallhäuser et al., *Antimicrob. Agents Chemother. (1961–70)* 1965, 734.

[363] R. L. Harned et al., *Antibiot. Chemother. (Washington, D.C.)* **5** (1955) 204.
[364] S. Takeuchi et al. *J. Antibiot. Ser. A.* **12** (1959) 109.
[365] A. J. Glazko et al., *Antibiot. Chemother. (Washington, D.C.)* **2** (1952) 234.
[366] K. Arima et al., *Agric. Biol. Chem.* **28** (1964) 575. K. Arima et al., *J. Antibiot. Ser. A* **18** (1965) 201.
[367] T. Kamiya et al., *J. Antibiot.* **25** (1972) 576.
[368] H. Egger, H. Reinshagen, *J. Antibiot.* **29** (1976) 915.
[369] K. Ishibashi, *J. Antibiot. Ser. A* **15** (1962) 161.
[370] S. A. Waksman, *Dtsch. Apoth. Ztg.* **109** (1969) 1019.
[371] S. A. Goulden, *Manuf. Chem. Aerosol News* **36** (1965) no. 4, 45.
[372] C. T. Calan, *Process Biochem.* **7** (1972) no. 7, 29.
[373] American Cyanamid, GB 773453, 1954.
[374] Bristol-Myers, US 2970946, 1960.
[375] American Cyanamid, US 2734018, 1953 (P. P. Minieri, H. Sokol, M. C. Firman).
[376] Bristol Lab., US 2739924, 1953.
[377] C. G. T. Evans, *Manuf. Chem.* **31** (1960) 5–9.
[378] G. D. Wilkin, *Manuf. Chem.* **31** (1960) 329.
[379] J. Málek, J. Hospodka, *Folia Microbiol. (Prague)* **5** (1960) 120.
[380] J. Málek, Z. Fencl: *Theoretical and Methodological Basis of Continuous Culture of Microorganisms*, Publ. House of the Czechosl. Acad. of Sciences, Prague, and Academic Press, New York – London, Engl. ed. 1966.
[381] A. L. Demain, C. L. Cooney, *Process Biochem.* **7** (1972) no. 7, 21.
[382] W. Oberzill, N. Matsché, *Chem. Ing. Tech.* **43** (1971) 83.
[383] H. W. Blanch, P. L. Rogers, *Biotechnol. Bioeng.* **13** (1971) 843.
[384] R. Kreutzfeldt, *Angew. Chem. Int. Ed. Engl.* **6** (1967) 470.
[385] M. J. Thirumalachar, R. S. Sukapure, P. W. Rahalkar, K. S. Gopalkrishnan, *Hind. Antibiot. Bull.* **5** (1962) 1.
[386] B. N. Ganguli, V. M. Doctor, *Hind. Antibiot. Bull.* **7** (1964) no. 2, 85.
[387] Distillers, GB 649818, 1948.
[388] Benckiser, DE 1000572, 1952.
[389] G. R. Ambekar, S. B. Thadani, *Appl. Microbiol.* **13** (1965) 713.
[390] Ankerfarm, DE 1467764, 1965.
[391] Benckiser, DE 1103735, 1958.
[392] Hoechst, DE 1247549, 1965; NL-A 6602132, 1966; BE 677053, 1966.
[393] P. Hosler, M. J. Johnson, *Ind. Eng. Chem.* **45** (1953) 871.
[394] C. Rainbow, A. H. Rose: *Biochemistry in Industrial Microbiology,* Academic Press, New York 1963, p. 254.
[395] A. Söder: "Tetracycline," in G. Ehrhart, H. Ruschig (eds.): *Arzneimittel, Entwicklung, Wirkung, Darstellung,* 2nd ed., vol. **4**, Verlag Chemie, Weinheim 1972, p. 368.
[396] R. V. Reeves, *Chem. Eng. (N.Y)* **59** (1952) Jan., 145.
[397] W. Dürckheimer in [395] vol. **5**, p. 302.
[398] Bundesverband der Pharmazeutischen Industrie: "GMP-Regulations of Food and Drug Administration/USA," 15. Jan. 1971, including amendment of 2. Mar. 1971, *Pharm. Ind.* **33** (1971) 364.
[399] WHO: "Draft Requirements for Good Manufacturing Practice in the Manufacture and Quality Control of Drugs and Pharmaceutical Specialities," World Health Organisation, Tech. Rep., Ser. 1969, no. 418, Annex 2.

[400] WHO: "Quality and Control of Drugs," Official Records of the World Health Organisation, no. 176, Dec. 1972, Annex 12.
[401] R. Marris, *Pharm. Ind.* **33** (1971) 749.
[402] *Pharm. Ind.* **32** (1970) 813–819.
[403] FDA, *Fed. Regist.* **37** (1972) no. 101, 10510.
[404] E. P. Abraham, E. Chain et al., *Lancet 1941,* vol. **II,** 177, 189.
[405] W. H. Schmidt et al., *J. Bacteriol.* **47** (1944) 199.
[406] M. D. Reeves et al., *J. Bacteriol.* **49** (1945) 395.
[407] FDA, *Fed. Regist.* **12** (1947), 4th Apr., 2215, 2217–2226.
[408] H. Seyfarth, O. P. Ewald, *Pharm. Ind.* **34** (1972) 40.
[409] K. H. Wallhäusser, *Pharm. Ind.* **34** (1972) 23.
[410] T. Dimmling, *Ärztl. Wochenschr.* **8** (1953) no. 27, 633.
[411] P. Naumann, *Antibiot. Chemother. (Basel)* **10** (1962) 1–93.
[412] A. L. Barry, *Am. J. Med. Technol.* **30** (1964) 153, 333.
[413] L. J. Griffith, C. G. Mullins, *Appl. Microbiol.* **16** (1968) 656.
[414] L. G. Wayland, P. J. Weiss, *J. Pharm. Sci.* **57** (1968) 806.
[415] K. H. Wallhäusser, *Ärztl. Lab.* **16** (1970) 150.
[416] H. Knöll, *Pharmazie* **2** (1947) 392.
[417] W. Irmer, *Dtsch. Med. Rundsch.* 1949, 123.
[418] K. Irrgang, *Z. Naturforsch. B: Anorg. Chem. Org. Chem. Biochem. Biophys. Biol.* **5 B** (1950) 155.
[419] G. Dorner, T. Lammers, *Med. Klin. (Munich)* **46** (1951) 522.
[420] W. H. C. Shaw, R. E. Duncombe, *Analyst (London)* **88** (1963) 694.
[421] T. A. Haney et al., *Ann. N.Y. Acad. Sci.* **87** (1960) 782; **93** (1962) 627.
[422] K. Heil, V. Beitz, *Pharm. Ind.* **34** (1972) 37.
[423] W. H. Shaw et al., *Ann. N.Y. Acad. Sci.* **130** (1965) 647.
[424] D. A. Burns et al., *Biotechnol. Bioeng.* **11** (1969) 1011.
[425] R. E. Hone, C. T. Rhodes, *Process Biochem.* **7** (1972) Feb., 27.
[426] E. P. Abraham, G. G. F. Newton, *Adv. Chemother.* **2** (1966) 23.
[427] D. E. Nettleton et al., *Int. J. Appl. Radiat. Isot.* **13** (1962) 259.

Synthetic Chemotherapeutic Agents

Separate keywords: → Anthelmintics, → Antibiotics.

PAUL ACTOR, Smith Kline & French Laboratories, Philadelphia, Pennsylvania 19101, United States (Chaps. 2, 3)

ALFRED W. CHOW, Smith Kline & French Laboratories, Philadelphia, Pennsylvania 19101, United States (Chaps. 2, 3)

FRANK J. DUTKO, Sterling-Winthrop Research Institute, Rensselaer, New York 12144, United States (Chap. 4)

MARK A. MCKINLAY, Sterling-Winthrop Research Institute, Rensselaer, New York 12144, United States (Chap. 4)

1.	**Introduction**		1108
2.	**Chemotherapy of Bacterial Infections**		1109
2.1.	Classification of Bacteria Causing Disease		1109
2.2.	Emergence of New Bacterial Pathogens		1109
2.3.	Antimicrobial Resistance		1111
2.4.	Pathogenesis and Virulence Factors		1112
2.5.	Selection of an Appropriate Antimicrobial Agent		1113
2.6.	Chemoprophylaxis		1114
2.7.	Chemotherapy		1115
2.7.1.	Biochemical Targets for Antimicrobial Agents		1115
2.7.2.	Quinolone Antibacterial Agents		1115
2.7.2.1.	Structure Function		1116
2.7.2.2.	Mechanism of Action		1117
2.7.2.3.	Nalidixic Acid and First-Generation Quinolones		1117
2.7.2.4.	Second-Generation Fluoroquinolones		1121
2.7.3.	Sulfa Drugs		1128
2.7.3.1.	Biological Activity and Medical Uses		1129
2.7.3.2.	Mechanism of Action and Antimicrobial Resistance		1129
2.7.3.3.	Structure – Function Relationships		1130
2.7.3.4.	Pharmacokinetics		1131
2.7.3.5.	Toxicity and Drug Interactions		1131
2.7.3.6.	Combination Therapy		1132
2.7.3.7.	Rapidly Absorbed Short- and Medium-Acting Sulfa Drugs		1132
2.7.3.8.	Long-Acting Sulfonamides		1136
2.7.3.9.	Sulfonamides for Use in the Gastrointestinal Tract		1136
2.7.4.	Agents for Treating Mycobacterial Infections		1138
2.7.4.1.	Antituberculosis Agents		1139
2.7.4.2.	Antileprosy Agents		1142
2.7.5.	Miscellaneous Nitroheterocycles Used to Treat Bacterial Infection		1144
3.	**Chemotherapy of Protozoan Infections**		1146
3.1.	Classification of Pathogenic Protozoa		1146
3.2.	Flagellates		1146
3.2.1.	Hemoflagellates		1146
3.2.1.1.	African Trypanosomiasis		1148
3.2.1.2.	American Trypanosomiasis		1151
3.2.1.3.	Leishmaniasis		1153

3.2.2.	Intestinal and Urogenital Flagellates	1155	4.2.	**Classification of Viruses**	1173
3.2.2.1.	Trichomonas Vaginalis	1155	4.3.	**Assessment of Antiviral Activity in Cell Culture**	1175
3.2.2.2.	*Giardia Lamblia*	1157	4.4.	**Animal Models of Virus Infection**	1175
3.3.	**Sporozoans**	1159			
3.3.1.	Plasmodia	1159			
3.3.1.1.	Biology and Epidemiology	1159	4.5.	**Rationale for Chemotherapy of Viral Infections**	1176
3.3.1.2.	Chemotherapy	1160			
3.3.2.	Babesia	1166	4.6.	**Chemotherapeutic Agents**	1177
3.3.3.	Isosporiasis	1166	4.6.1.	Nucleoside Analogues	1177
3.3.4.	Toxoplasmosis	1167	4.6.2.	Phosphonoacetate and Phosphonoformate	1183
3.3.5.	Cryptosporidium	1167			
3.3.6.	*Pneumocystis Carinii*	1168	4.6.3.	Amantadine and Rimantadine	1184
3.4.	**Ciliates**	1168	4.6.4.	Enviroxime	1185
3.5.	**Amebas**	1168	4.6.5.	4',6-Dichloroflavan	1186
3.5.1.	Biology and Epidemiology	1168	4.6.6.	Chalcone Ro 09–0410	1186
3.5.2.	Chemotherapy	1169	4.6.7.	Arildone and Disoxaril	1187
4.	**Chemotherapy of Viral Infections**	1172	4.6.8.	3'-Azidothymidine	1188
			4.6.9.	Suramin	1188
4.1.	**Physical and Biological Characteristics**	1172	4.6.10.	HPA 23	1189
			5.	**References**	1189

1. Introduction

This article discusses the synthetic agents that are effective against pathogenic bacteria, protozoa, and viruses. Although a large percentage of the agents employed to treat these infections, especially bacterial infections, are natural products, i.e., antibiotics, many synthetic compounds continue to provide useful alternatives, and some are the agents of choice for the treatment of specific clinical entities.

2. Chemotherapy of Bacterial Infections

2.1. Classification of Bacteria Causing Disease

A broad variety of organisms are primary human pathogens. No universal agreement exists for the classification of these bacteria, and it is beyond the scope of this chapter to discuss the arguments inherent in attempting to create a universally acceptable and useful system. The definitive work in bacterial classification is Bergy's *Manual of Determinative Bacteriology* [1]. Bacterial species still are arbitrarily defined by descriptive features. In the approximate sequence of importance, some of the major features employed include the following:

1) *morphological appearance of the cell,* including shape, size, flagellar pattern if motile, capsule occurrence, colonial morphology, and pigmentation
2) *staining properties,* including the gram capsule, spore, and acid fast
3) *metabolic patterns*
4) *macromolecular composition* and *structure*
5) *ecological habitat*
6) ability to be *pathogenic*

The major families of bacteria causing human infection are shown in Table 1. However, many bacteria that are not primary pathogens are capable of causing clinical infections in immunocompromised hosts. Furthermore, transfer of genetic material between bacterial species, e.g., via plasmids, allows saprophytic bacteria to acquire virulence factors and become pathogenic.

2.2. Emergence of New Bacterial Pathogens

The emergence and recognition of organisms as important human pathogens are strongly influenced by three major factors: (1) the transfer of virulence factors and antibiotic resistance between bacterial species by extrachromosomal elements, (2) opportunistic infection of immunocompromised hosts by saprophytic bacteria, and (3) improvement and widespread use in the microbiological clinical laboratory of new diagnostic procedures for isolating and identifying pathogens.

Opportunistic infections, particularly in hospitalized individuals, have become major infectious disease problems. This problem is likely to continue because of the increasing use of instrumentation, antibiotics, and drugs that either bypass or reduce the level of

Table 1. The major groups of pathogenic bacteria

Morphological type	Family	Genus	Gram stain	Oxygen utilization*
Cocci	Micrococcaceae	*Micrococcus*	+	A
		Staphylococcus	+	A/F – AN
	Streptococcaceae	*Streptococcus*	+	F – AN
	Peptococcaceae	*Peptococcus*	+	AN
		Peptostreptococcus	+	AN
	Neisseriaceae	*Neisseria*	–	A
	Veillonellaceae	*Veillonella*	–	AN
Rods	Enterobacteriaceae	*Escherichia*	–	F – AN
		Shigella	–	F – AN
		Salmonella	–	F – AN
		Citrobacter	–	F – AN
		Klebsiella	–	F – AN
		Enterobacter	–	F – AN
		Erwinia	–	F – AN
		Serratia	–	F – AN
		Hafnia	–	F – AN
		Edwardsiella	–	F – AN
		Proteus	–	F – AN
		Providencia	–	F – AN
		Morganella	–	F – AN
		Yersinia	–	F – AN
	Vibrionaceae	*Vibrio*	–	F – AN
		Aeromonas	–	F – AN
	Pasteurellaceae	*Pasteurella*	–	F – AN
		Haemophilus	–	F – AN
	Pseudomonadaceae	*Pseudomonas*	–	A
	Legionellaceae	*Legionella*	–	A
	Neisseriaceae	*Moraxella*	–	A
		Acinetobacter	–	A
		Brucella	–	A
		Bordetella	–	A
	Bacteroidaceae	*Bacteroides*	–	AN
		Fusobacterium	–	AN
		Leptotricha	–	AN
	Bacillaceae	*Bacillus*	+	AN
		Clostridium	+	AN
		Listeria	+	A
		Erysipelothrix	+	A
Actinomycetes and related organisms		*Corynebacterium*	+	A/F – AN
	Propionibacteriaceae	*Propionibacterium*	+	AN
		Eubacterium	+	AN
	Actinomycetaceae	*Actinomyces*	+	FA
	Mycobacteriaceae	*Mycobacterium*	+	A
	Nocardiaceae	*Nocardia*	+	A
Rickettsias and Chlamydias	Rickettsiaceae	*Rickettsia*	–	P
		Coxiella	–	P
	Bartonellaceae	*Bartonella*	–	P
	Anaplasmataceae	*Grahamella*	–	P
		Anaplasma	NR**	P

Table 1. (continued)

Morphological type	Family	Genus	Gram stain	Oxygen utilization*
		Haemobartonella	–	P
		Eperythrozoan	–	P
	Chlamydiaceae	*Chlamydia*	–	P
Mycoplasmas	Mycoplasmataceae	*Mycoplasma*	–	FA
		Ureaplasma	–	FA
	Acholeplasmataceae	*Acholoplasma*	NR**	FA
Spirochetes	Spirochaetaceae	*Treponema*	–	AN
		Borrelia	–	AN
		Leptospira		A

* A = Aerobic; AN = Anaerobic; F – AN = Facultative anaerobic; A/F – AN = Aerobic or facultative anaerobic growth; P = Parasitic usually require host cells for growth.
** NR = Gram stain not revealed.

natural resistance or the specific immune mechanisms of the host. Thus, microorganisms previously considered as innocuous commensals or contaminants are now able to invade and cause disease. Some of the more important bacterial pathogens that have created significant problems include *Staphylococcus epidermidis, Bacteroides fragilis* and other *Bacteroides* species, *Clostridium difficile,* the gram-negative rods including *Acinetobacter* species, *Serratia, Citrobacter* species, *Yersinia, Moraxella,* and the atypical mycobacteria.

Since the mid-1960s, the role of *S. epidermidis* has changed from a bothersome contaminant to a major pathogen causing infections associated with foreign bodies (prosthetic valvular endocarditis and infections of CNS shunts and of joint and vascular prostheses). In addition, this organism can cause infections not associated with foreign bodies, such as endocarditis and urinary tract infections. *Bacteroides fragilis,* the preeminent anaerobic human pathogen, has evolved as a problem. Information as to the importance and widespread incidence of this organism only became available when techniques were established for its facile isolation and rapid diagnosis. Another important anaerobe is *Clostridium difficile,* a component of the normal gut flora, which overgrows to large populations in the presence of antibiotics. Toxigenic stains can cause pseudomembranous colitis, a severe necrotizing disease of the large intestine. With the waning of *M. tuberculosis* infections in the United States, other mycobacterioses, the so-called atypical mycobacteria, have become increasingly important.

2.3. Antimicrobial Resistance

Mechanisms of Resistance. The widespread clinical use of antimicrobics has resulted in the emergence of many strains of bacteria resistant to one or more of these agents. In most cases in which adequate studies have been done, the role of antimicrobial agents apparently is to exert selective pressure, resulting in the emergence of resistant organisms. In some instances, the organisms are *naturally resistant* to the antibiotic used. In other cases, the resistant bacteria may have acquired *R-factors* or *plasmids.* These extrachromosomal agents may provide the organism with the ability to synthesize enzymes that modify or inactivate the antimicrobial agent. The plasmids may also cause changes in the organism's ability to accumulate the antimicrobial agent, or they may stimulate the cell to produce or overproduce metabolic enzymes resistant to inhibition by the antimicrobial agent. Additionally, alterations in the permeability of the bacterial cell envelope could result in drug resistance.

Chromosomal resistance develops as a result of spontaneous mutation in a locus that controls susceptibility to a given antimicrobial agent. The presence of the drug serves as a selection mechanism for drug-resistant mutants. Chromosomal mutants are most commonly resistant by virtue of an alteration in a structural receptor for the drug. Cross resistance is frequently observed between chemically related drugs showing a similar mechanism of action, but may also exist for unrelated chemicals.

Clinical Implications of Resistance. Bacterial resistance problems have resulted in the continuing need for new antimicrobics or modification in the ways in which we treat patients with the available antimicrobials. Important examples with the synthetic antimicrobials include the sulfonamides, which were only active against a small percentage of gonococcal strains 6 years after they were first successfully employed for the treatment of gonorrhea. The sulfonamides also have lost their usefulness in the prevention and treatment of meningococcal infection because of drug resistance. Drug-resistant mutants have arisen in tuberculosis, and naturally resistant species of mycobacteria have become clinically important especially in immunocompromised hosts. The emergence of drug-resistant bacteria in the hospital setting has led to restricted use of certain valuable drugs in hospitals. Other strategies that have been employed to minimize the drug resistance problem include treatment regimes to maintain high drug levels in tissues and the simultaneous administration of two drugs, each of which delays emergence of resistance to the other drug (e.g., sulfonamides and trimethoprim).

2.4. Pathogenesis and Virulence Factors

All organ systems are subject to the pathogenic effects of bacteria, and the resulting infections range from the trivial to the fatal. The major organ systems involved are skin and soft tissues, urinary and reproductive tissues, the respiratory tract, and the central nervous, digestive, and cardiovascular systems. Infections accompanying other medical problems that result in a breach of the anatomical and immunological barriers to infection are particularly difficult to treat, e.g., *Pseudomonas* infection in burn patients or those with cystic fibrosis. Mixed infections involving more than one organism are common.

Although humans are continually exposed to many different environmental microorganisms, only a small percentage of these have the capacity to produce disease. The production of *virulence factors* by microorganisms is the important determinant in disease. Virulence factors are those components of the microbe that are essential for the establishment of infection and the development of disease in the host. Usually more than one factor is involved in the disease process. The adherence of microorganisms to host tissues is determined by highly specific host receptors for bacterial surface components and is a necessary prerequisite for infection. Some bacteria, e.g., *Vibrio cholerae* and *Escherichia coli*, colonize mucosal surfaces and cause damage by elaboration of a toxin. Other organisms, e.g., *Salmonella* and *Shigella*, can invade following attachment and either enter host cells or disseminate throughout the body. The invasion is facilitated by the production of enzymatic substances that circumvent anatomical barriers. To survive, these organisms may have specialized virulence factors that enable them to avoid or disarm host defenses. Survival and continued proliferation of the invading organism are often accompanied by the production of *toxins*. These toxins are proteinaceous substances capable of producing adverse biological effects.

2.5. Selection of an Appropriate Antimicrobial Agent

The selection of a specific agent for treating infection involves consideration of the infecting organism, the status of the host, and the specific attributes of the antimicrobic. The key factors to be considered are the drug's antimicrobial spectrum and potency, its physical characteristics (solubility and stability in body fluids), safety profile, pharmacokinetics, compatibility with other drugs, and cost. The choice of an appropriate agent involves identifying the infecting organism, obtaining information as to its susceptibility, and assessing various host factors.

Identification of the Microorganism. Rapid identification can often be made by *microscopic examination* of gram-stained specimens. A number of immunologic procedures can be employed, such as *latex agglutination tests* or the *enzyme-linked immunoabsorbent assay* (ELISA). Much progress has been made with assays employing *DNA probes* but these are not generally available to the clinical laboratory. *Cultural techniques* currently represent the major means for identification of microbial pathogens. A number of automated and semiautomated tests and kits facilitate this identification and may combine the capability of antimicrobial susceptibility testing.

Determination of Antimicrobial Susceptibility. Several methods for determination of antimicrobial susceptibility are available. The *disk-diffusion method* is widely employed and gives semiquantitative information that is clinically useful. Methods employing *serial dilution* of the drug in culture broth or agar give quantitative data. This is usually expressed as the minimal inhibitory concentration (MIC). A subculture onto an antibiotic-free medium of broth cultures from the MIC tube showing no visible growth can allow for the determination of the minimal bactericidal concentration (MBC), which is usually defined as the concentration that causes decline in colony count of 99.9% or more.

Host Factors. The clinician must consider a number of host factors in selecting the drug. A history of adverse reactions to antimicrobials, the patient's age, genetic or metabolic abnormalities, pregnancy, renal or hepatic function, and the site of infection all influence this choice.

Combination Therapy. At times combination therapy offers advantages over treatment with a single antimicrobial agent; however, most infections can be treated with one drug. Combination therapy may be valuable in prevention of the emergence of resistance or for polymicrobial infections. When the initial diagnosis is unclear, a combination of antimicrobial agents may be needed.

2.6. Chemoprophylaxis

Chemoprophylaxis is the administration of drugs prior to or shortly after exposure to an infectious agent to prevent infection. For example, isoniazid may be given to individuals who show recent change from negative to positive skin tests for tuberculosis. The risk of infection must be balanced against toxicity, cost, efficacy, and inconvenience of taking the proposed chemoprophylaxis.

Antibacterial chemoprophylaxis is an accepted clinical procedure for preventing group A *Streptococcus* and *Meningococcus,* and plague infections. Prophylactic drugs have also been used to prevent infective endocarditis, postcoital cystitis, and exacerbations in chronic bronchitis in high-risk individuals. Immunosuppressed patients are often given antimicrobials to prevent the complications resulting from the spread of endogenous bacteria. In some surgical procedures, prophylactic administration of antimicrobials is valuable in preventing infection.

2.7. Chemotherapy

2.7.1. Biochemical Targets for Antimicrobial Agents

A chemical affects cell growth if it hinders the cell's endogenous biosynthetic processes by restricting the availability of building materials or catalytic enzymes or interferes with the supply of usable energy. Considerable advances have been made in identifying several general biochemical targets (i.e., peptidoglycan synthesis, biosynthetic enzymes, and ribosomal function), although in many cases the final molecular site of action is uncertain. At the cellular and subcellular level, most known antimicrobial agents function in one of four major ways:

1) inhibition of bacterial cell wall synthesis
2) alteration of cell membrane permeability or inhibition of active transport across cell membranes
3) inhibition of nucleic acid synthesis
4) inhibition of protein synthesis

The specific mechanisms of action for the individual classes of antimicrobials are discussed in the following sections [2].

2.7.2. Quinolone Antibacterial Agents

Historical Background. Following the discovery of nalidixic acid, numerous antibacterial quinolones were synthesized [3]. These early quinolone carboxylic acids were absorbed fairly well, but suffered from poor pharmacokinetics and tissue penetration, which relegated them to use only in urinary tract infections caused by gram-negative bacteria. In addition to poor pharmacokinetics, these compounds also showed CNS side effects and rapid development of resistant microorganisms. Since the initial introduction of the early compounds, many second- and third-generation compounds have been discovered with increased potency and antibacterial spectrum against gram-negative bacteria, including *P. aeruginosa*. In addition, some of the newer compounds are active against gram-positive bacteria and anaerobes and appear to have fewer adverse effects, as well as improved pharmacokinetics and tissue penetration. Several have shown promise when administered parenterally and thus offer therapeutic potential outside the urinary tract. The early compounds have been reviewed comprehensively, especially with respect to synthetic methods, microbiology, and structure activity relationships [4]. The newer compounds have been reviewed in several recent publications [5]–[7].

2.7.2.1. Structure Function

Nalidixic acid, the prototype of the series, is a naphthyridine derivative; however, the newer, more potent congeners are not derived from the naphthyridine skeleton. The prototype compound of the quinoline class is oxolinic acid. Although all congeners of nalidixic acid can be named as derivatives of quinoline, such a system is cumbersome because the common skeleton is termed 4-oxo-1,4-dihydroquinoline. A simpler term, 4-quinolone (**1**), has been suggested as a generic name for the agents.

1
4-Quinolones

X = N, Naphthyridine
X = CH, Quinolone

The second- and third-generation 4-quinolones have been modified at C-6, C-7, and C-8 (**1**). The position 1 is vital to antimicrobial activity. The presence of a two-carbon fragment or a spatial equivalent is essential and almost all marketed compounds in this class have an ethyl group at position 1. Ciprofloxacin has a cyclopropyl group at this site, which fills the spatial requirements of the two-carbon chain. Very few modifications at C-2 have been studied, although cinoxacin, which substitutes a nitrogen for the carbon at this position, has improved absorption properties. Relative to its parent compound, oxolinic acid, cinoxacin is somewhat less active in vitro.

Positions C-3 and C-4 are the most structurally critical, and substitution of other groups at these positions results in loss of activity. Modification of C-5 also offers few advantages. Addition of a nitrogen within the ring may result in an improved use profile. Cinoxacin, mentioned previously, is an example of improved pharmacokinetics resulting from such a change at C-2; however, substitution of a ring nitrogen for C-5, C-6, C-7, and C-8 generally decreases or abolishes activity.

Addition of a fluorine atom to C-6 results in a dramatic increase in the antibacterial spectrum and potency of these compounds [8]. Consequently, all of the newer quinolones have C-6-substituted fluorine as a component of the molecule. These structures have been given the name fluoroquinolones [5].

An aromatic group at C-7 may tend to increase CNS side effects (e.g., rosoxacin), but the piperazine ring may not produce this enhanced effect. The piperazine ring at C-7 also seems to achieve the necessary balance of enhancing penetration capabilities and in vivo activity. The activity of the 4-quinolones against gram-positive and anaerobic bacteria is enhanced through substitutions at C-8 with short one-to-three-atom chains (e.g., ofloxacin). Increased activity against gram-positive bacteria is achieved by the addition of a fluorine atom at C-8 (e.g., CI–934).

2.7.2.2. Mechanism of Action

The quinolone antibacterial agents act by inhibiting bacterial DNA-gyrase (topoisomerase II). This essential bacterial enzyme, discovered in 1976 in *Escherichia coli* [9], contains two A-subunits and two B-subunits. The *B-subunit*, which is the site of action of coumermycin–novobiocin type antibiotics, is responsible for supplying the energy necessary for the supercoiling of DNA. The *A-subunit*, the site of action for the quinolone type antibiotics, performs the physical act of supercoiling the bacterial DNA molecule in conjunction with the B-subunit [10].

2.7.2.3. Nalidixic Acid and First-Generation Quinolones

Nalidixic acid and the related first-generation 4-quinolones are orally administered agents that concentrate primarily in the urinary tract. Because effective plasma concentrations are not obtained with safe doses of these agents, they cannot be used for the treatment of systemic infections; thus, their use is for the most part limited to infections of the urinary tract.

Nalidixic acid (International Nonproprietary Name – INN, *United States Pharmacopeial – U.S.P.*) [*389-08-2*], 1-ethyl-1,4-dihydro-7-methyl-4-oxo-1,8-naphthyridine-3-carboxylic acid, $C_{12}H_{12}N_2O_3$, M_r 232.24, *mp* 229–230 °C, is a pale buff crystalline powder.

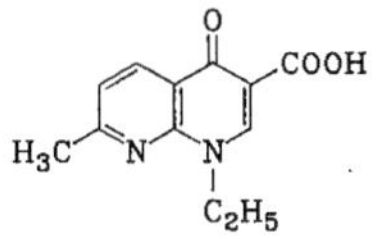

Nalidixic acid is more active against gram-negative bacteria than against gram-positive bacteria. It is active against most of the members of the Enterobacteriaceae, including ca. 99% of the strains of *Escherichia coli*, 98% of *Proteus mirabilis*, 75–97% of other *Proteus* species, 92% of *Klebsiella* and *Enterobacter* species, and 80% of other coliform bacteria.

These organisms are susceptible to a urinary concentration of 16 µg/mL or less [3]. Some strains of *Salmonella* and *Shigella* are also sensitive. *Pseudomonas* species are resistant, as are most of the important species of gram-positive clinical pathogens, including *Staphylococcus* species, *Streptococcus pneumoniae*, and *Streptococcus faecalis*. Resistance can be induced by in vitro passage [11]; however, surveys of resistance in clinically isolated uropathogens have shown that the incidence of isolates resistant to nalidixic acid remained surprisingly low despite extensive clinical use [12].

Nalidixic acid administered orally is 96% absorbed from the gastrointestinal tract [13]. It is rapidly metabolized in the liver. Plasma levels of 20–50 µg/mL may be obtained 2 h after a 1-g dose [14]; however, the drug does not accumulate in the tissues even after prolonged administration, and the kidney is the only organ in which tissue concentrations may exceed plasma levels [13]. Approximately 85 % of the drug excreted in the urine is the inactive conjugated form; the remainder is the hydroxynalidixic acid metabolite, which is 16 times more active than the parent compound [15]. Urinary concentrations range from 25 – 250 µg/mL following a single oral dose of 0.5 – 1 g and remain between 100 and 500 µg/mL with a 1-g dose administered every 6 h [14], [16].

Oral nalidixic acid is generally well tolerated; however, various *adverse reactions* have been reported. Gastrointestinal side effects include nausea, vomiting, diarrhea, and abdominal pain. Dermatological reactions and photosensitivity have also been reported. Additionally, a range of reversible central nervous system reactions are observed. The drug is contraindicated in infants and during early pregnancy; it should not be used in children.

Synthesis: Condensation of 2-amino-6-methylpyridine with diethyl ethoxymethylenemalonate and thermal cyclization gives 4-hydroxy-7-methyl-1,8-naphthyridine-3-carboxylate. Alkaline saponification of the ester group and alkylation of the nitrogen atom in one step yields nalidixic acid [17].

Trade Names: Cybis (Sterling-Winthrop), NegGram, and Wintomylon (Sterling-Winthrop).

Oxolinic acid (INN) [*14698-29-4*], 5-ethyl-5,8-dihydro-8-oxo-1,3-dioxolo[4.5-*g*]quinoline-7-carboxylic acid, $C_{13}H_{11}NO_5$, M_r 261.23, *mp* 314 – 316 °C (decomp.), occurs as crystals.

Oxolinic acid shares a similar antibacterial spectrum of activity with nalidixic acid, i.e., it is active against gram-negative rods with the exception of *Pseudomonas* species. Its in vitro potency, however, is significantly greater and it is more active in vivo [18]. Except for *S. aureus* strains that are inhibited at concentrations of 6.25 µg/mL, oxolinic acid does not inhibit gram-positive bacteria. Cross resistance with nalidixic acid is observed [19]. As with nalidixic acid, emergence of resistance during treatment of patients with bacteriuria has been reported.

Oxolinic acid achieves good urinary concentrations within 4 h following oral administration. An oral dosage of 2 g/d produces an average urinary concentration range of 16 – 64 µg/mL in 24-h urinary collections. Elimination of the drug is via urine and feces with low or borderline plasma levels against susceptible bacteria.

Adverse Effects: Central nervous system toxicity has been observed with this drug, and the potential is increased in elderly patients, in which CNS side effects are more common than with nalidixic acid.

Synthesis: Condensation of 3,4-methylenedioxyaniline with diethyl ethoxymethylenemalonate in boiling diphenyl ether and alkylation of the resultant 1,4-dihydroxy-6,7-

methylenedioxy-4-oxoquinoline-3-carboxylate with ethyl iodide in the presence of caustic soda – DMF (*N,N*-dimethylformamide) gives the precursor ester. Saponification with dilute sodium hydroxide and neutralization yields oxolinic acid [20].
Trade Name: Utibid (Parke-Davis).

Cinoxacin (INN, U.S.P.) [*28657-80-9*], 1-ethyl-1,4-dihydro-4-oxo[1,3]dioxolo[4,5-g]-cinnoline-3-carboxylic acid, $C_{12}H_{10}N_2O_5$, M_r 262.22, *mp* 261 – 262 °C (decomp.), occurs as tan crystals.

The antibacterial activity spectrum of cinoxacin resembles that of nalidixic acid, although it is more potent against selected species [21]. Generally its potency lies between that of nalidixic acid and oxolinic acid. Cinoxacin is most active against *E. coli*, of which > 90 % of the strains are inhibited at 16 µg/mL. Its activity is mainly against gram-negative rods; a broad spectrum of these organisms are susceptible to cinoxacin [21], [22]. Strains that are resistant to cinoxacin are cross-resistant with nalidixic acid and oxolinic acid [22], [23]. Similar to observations with the other quinolones, no evidence exists that plasmids play a role in resistance.

Excellent absorption is observed on oral administration. Peak plasma levels occur at 2 – 3 h after administration of a 250- or 500-mg dose. Serum binding is ca. 70 % and the serum half-life is ca. 1 h. Food may delay and reduce absorption. Peak urine concentrations range from 88 to 925 µg/mL within 4 – 6 h after dosing. Cinoxacin concentrates in renal tissue, where levels may exceed serum levels [24]. Approximately 60 % of the parent compound is excreted in the urine along with at least four microbiologically inactive metabolites [25].

Low frequencies of *adverse effects* (4.4 %) are reported on oral administration of cinoxacin with gastrointestinal reactions being the most common. Central nervous system side effects are < 1 %.

Synthesis: 2-Nitro-4,5-methylenedioxyacetophenone is catalytically reduced to the corresponding amine, which is cyclized by treatment with $NaNO_2$ and HCl in water to give 6,7-methylenedioxycinnolin-4-ol. Bromination followed by treatment with cuprous cyanide in refluxing DMF gives 4-hydroxy-6,7-methylenedioxycinnolin-3-carbonitrile; hydrolysis with HCl in refluxing acetic acid gives cinoxacin [26].
Trade Name: Cinubac (Eli Lilly).

Pipemidic Acid and Piromidic Acid. Both pipemidic and piromidic acid have been marketed outside of the United States as urinary tract antiseptics. These compounds previously have been reviewed in detail [27], [28]. Generally they offer no significant advantages over the other available 4-quinolones and are cross-resistant with these quinolones. They are of historical interest because they are the *forerunners of the second-*

generation fluoroquinolones. The addition of a piperazine ring at C-7 of the 4-quinolone skeleton (**1**) in pipemidic acid seems to enhance in vivo performance through improved pharmacokinetics. Pipemidic acid has been reported to be orally effective against experimental *P. aeruginosa* infections in mice; however, its clinical utility against this organism is not established.

Pipemidic acid (INN, dénomination commune Francaise, French approved name – DCF, *Merck Index* – MI) [*51940-44-4*], 8-ethyl-5,8-dihydro-5-oxo-2-(1-piperazinyl)pyrido[2,3-*d*]pyrimidine-6-carboxylic acid, $C_{14}H_{17}N_5O_3$, M_r 303.32, *mp* 253–255 °C, is a yellowish-white, odorless, bitter-tasting crystal.

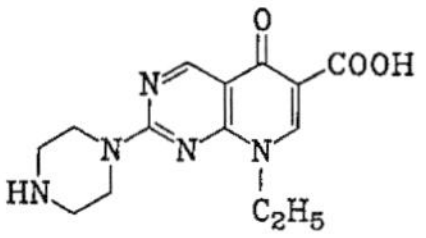

Synthesis: Ethyl 8-ethyl-5,8-dihydro-2-methylthio-5-oxopyrido[2,3-*d*]pyrimidine-6-carboxylate is heated with excess piperazine at 90–110 °C in DMSO (dimethyl sulfoxide) for 3 h [29].

Piromidic acid (INN, National Formulary Name – NFN, MI 9) [*19562-30-2*], 8-ethyl-5,8-dihydro-5-oxo-2-(1-pyrrolidinyl)pyrido[2,3-*d*]pyrimidine-6-carboxylic acid, $C_{14}H_{16}N_4O_3$, M_r 288.31, *mp* 314–316 °C, is crystalline in form.

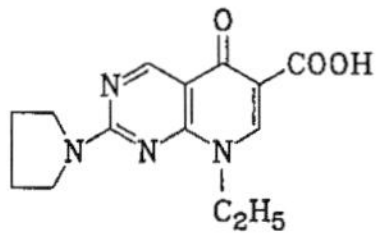

Synthesis: Ethyl 8-ethyl-5,8-dihydro-2-methylthio-5-oxopyrido[2,3-*d*]pyrimidine-6-carboxylate is heated in a pressure tube with pyrrolidine at 95 °C in ethanol for 6 h to yield piromidic acid [30].

Rosoxacin (INN) [*40034-42-2*], 1-ethyl-1,4-dihydro-4-oxo-7-(4-pyridyl)-3-quinolinecarboxylic acid, $C_{17}H_{14}N_2O_3$, M_r 294.31, *mp* 290 °C, forms yellow crystals, is stable in dry heat at 70 °C, and is sensitive to light.

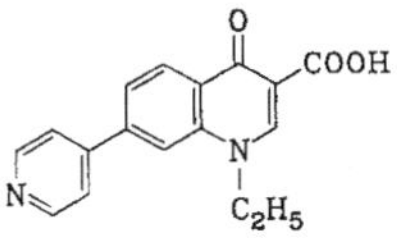

Rosoxacin is a potent first-generation quinolone with minimum inhibitory concentrations (MIC) ranging from 0.1 µg/mL for the most susceptible *E. coli* to 1.6 µg/mL for *Providencia* and *Klebsiella* species [31]. Some strains of *Pseudomonas* are susceptible. *Neisseria gonorrhoeae, Neisseria meningitidis,* and *Haemophilus influenzae* are particularly susceptible to the action of rosoxacin (MIC, 0.02 µg/mL); low plasma levels are achieved in humans after oral administration. A dose of 250 mg produces a peak plasma level of 6.4 µg/mL 2 h after administration.

This quinolone has been marketed as an antigonorrheal agent. Single oral doses are reported to produce high cure rates in patients with acute uncomplicated gonorrhea. *Adverse effects* include dizziness and itching.

Synthesis: A five-step synthesis starting with 3-nitrobenzaldehyde yields 4-(3-aminophenyl)-pyridine, which is condensed with diethyl ethoxymethylenemalonate and then cyclized thermally to yield ethyl 1,4-dihydro-4-oxo-7-(4-pyridyl)quinoline-3-carboxylate. Ethylation followed by saponification and neutralization yields rosoxacin [32].
Trade Name: Roxadyl (Sterling-Winthrop).

Miloxacin (INN) [*37065-29-5*], 5,8-dihydro-5-methoxy-8-oxo-1,3-dioxolo[4,5-*g*]quinoline-7-carboxylic acid, $C_{12}H_9NO_6$, M_r 263.21, *mp* 264 °C (decomp.), occurs as colorless prisms from DMF.

Miloxacin, which is currently marketed in Japan, has an antimicrobial spectrum and potency similar to that of oxolinic acid. Clinically, it does not seem to have any advantages over oxolinic acid; however, it appears to offer somewhat superior absorption and higher urinary excretion over oxolinic acid [33], [34]. A peak serum concentration of 7.7 µg/mL is achieved in humans 1–2 h following a 500-mg oral dose [35]. The drug is highly metabolized (87%) to a biologically inactive glucuronide form, and only 5.4% is found in the urine 8 h after dosing.
Synthesis: [36].

2.7.2.4. Second-Generation Fluoroquinolones

The second-generation fluoroquinolones are distinguished by their excellent and broad antibacterial activity, their relatively fewer adverse effects when compared with nalidixic acid, and their low propensity for inducing bacterial resistance. Some of these compounds can be administered parenterally and are effective in systemic infections outside of the urinary tract.

Their *activity* seems scarcely affected by inoculum size, type of medium, or presence of serum [37]–[40]. Some of the compounds lose activity in the presence of urine and when the pH of the medium is < 5.0 [38]–[41]. Generally these drugs are highly active against the enteric gram-negative bacilli and cocci. They are also active against other gram-negative bacteria, including *P. aeruginosa* (but less so against other *Pseudomonas* species), *Aeromonas hydrophila*, *H. influenzae*, and *Legionella pneumophilia*. They have excellent activity against pathogens of the gastrointestinal tract, including *E. coli*, *Salmonella* species, *Shigella* species, *Yersinia enterocolitica*, *Camplobacter jejuni*, and

Vibro species. Their activity against gram-positive species is poorer than against gram-negative ones, but it is still within a potential therapeutic range especially for ofloxacin and ciprofloxacin. Activity against anaerobic bacteria is marginal. Cross resistance among the various quinolones is observed [41], [42]. The fluoroquinolones are active in vitro against multiple antibiotic-resistant bacteria. They are rapidly bactericidal. Other organisms reported to be susceptible to one or more of these compounds include *Chlamydia trachomatis, Ureaplasma urealyticum, Mycoplasma hominis, Mycoplasma pneumoniae,* and *Mycobacterium tuberculosis.* Fungi are not susceptible to the fluoroquinolones.

Although differences in potency between the fluoroquinolones exist for the various pathogens, their clinical usefulness will be predicted best in conjunction with key parameters, such as pharmacokinetic properties and toxicities of individual drugs [5].

Norfloxacin (INN, BAN – British Approved Name) [*70458-96-7*], 1-ethyl-6-fluoro-1,4-dihydro-4-oxo-7-(1-piperazinyl)-3-quinolinecarboxylic acid, $C_{16}H_{18}FN_3O_3$, M_r 319.34, *mp* 227–228 °C, is crystalline in form.

Norfloxacin has an in vitro spectrum of antibacterial activity that includes most gram-negative organisms with MIC_{90} values in the range of 1 µg/mL or less [43], [44]. Norfloxacin is about 100 times more active than nalidixic acid, and its spectrum of activity includes enterococci and staphylococci as well as *Pseudomonas* [43]. Norfloxacin is also active against *H. influenzae* at a concentration of 0.12 µg/mL and *N. gonorrhoeae* at 0.016 µg/mL. The MIC_{90} for *S. aureus* strains is 1.6 µg/mL, with the streptococci being even less sensitive [44]. Members of the *Bacteroides fragilis* group of anaerobes are relatively resistant to norfloxacin (MIC 8–128 µg/mL), as are other anaerobic bacteria. Norfloxacin is more potent against sexually transmitted diseases than available quinolone chemotherapeutics [45]. It is less active than other fluoroquinolones against *Ureaplasma* and shows only moderate activity against *Chlamydia trachomatis* [46].

Absorption of norfloxacin is poorer than that of enoxacin or ciprofloxacin [47]–[50]. Prostatic tissue levels in humans exceed serum levels, whereas blister fluid levels are 70% of serum levels [48], [51]. Approximately 30% of the administered dose is excreted in urine; 80% of the excreted dose is parent compound [52]. Oral administration of 400 mg BID (twice daily) for 10 d was as effective as trimethoprim–sulfamethoxazole in treatment of patients with urinary tract infections or acute pyelonephritis [53]. *Adverse effects,* which included dizziness and nausea, were low. Norfloxacin administered in a single oral dose of 600 mg gave 100% cure of gonococcal infections in men [45].

Synthesis: by condensation of 1-ethyl-6-fluoro-7-chloro-1,4-dihydro-4-oxo-3-quinolinecarboxylic acid with piperazine at 170 °C in water in a pressure tube [54].

Trade (or code) names: Noroxin, Baccidal, MK−0366, and AM−715 (Kyorin, Merck).

Ciprofloxacin (INN, BAN) [*85721-33-1*], 1-cyclopropyl-6-fluoro-1,4-dihydro-4-oxo-7-(1-piperazinyl)-3-quinolinecarboxylic acid, $C_{17}H_{18}FN_3O_5$, M_r 331.35, *mp* 255−257 °C, is crystalline in form.

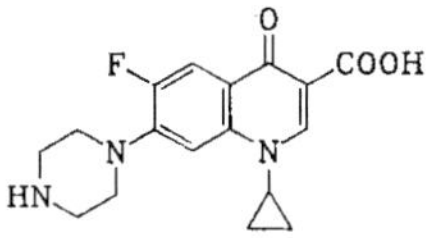

Ciprofloxacin has the most potent in vitro activity of the fluoroquinolones reported to date. It is active against most bacterial strains that cause urinary tract infections at concentrations well in excess of those observed in the urine [55]−[57]. The minimum inhibitory concentrations of ciprofloxacin that inhibit 90% of clinical isolates (MIC_{90}), in μg/mL, are as follows.

Pathogen	MIC_{90}, μg/mL
Escherichia coli	0.06
Klebsiella species	0.25
Salmonella species	0.015
Shigella species	0.008
Citrobacter diversus	0.03
Citrobacter freundii	0.125
Enterobacter cloacae	0.03
Enterobacter aerogenes	0.06
Proteus mirabilis	0.06
Proteus vulgaris	0.06
Morganella morganii	0.016
Providencia stuartii	0.50
Pseudomonas aeruginosa	0.25
Pseudomonas maltophilia	4.0
Pseudomonas cepacia	8.0
Serratia marcescens	0.13
Acinetobacter calcoaceticus	0.5
Yersinia enterocoliticia	0.06
Seromonas hydrophilia	0.008
Pasteurella multocida	0.016
Haemophilus influenzae	0.015
Neisseria gonorrhoeae	0.004
Bacteroides fragilis	4.00
Staphylococcus aureus	0.50
Staphylococcus epidermidis	0.25
Streptococcus pyogenes	0.20
Streptococcus agalactiae	1.00
Streptococcus faecalis	2.00
Streptococcus pneumoniae	2.00
Streptococcus viridans	4.00
Listeria monocyogenes	1.00
Chlamydia trachomatis	1.00

As can be seen from these values, most organisms are highly susceptible to ciprofloxacin, although the gram-positive bacteria are much less susceptible than the gram-negative organisms.

Oral doses of ciprofloxacin are rapidly absorbed and peak plasma levels of 2–3 µg/mL are observed 1–1.5 h after a 500-mg dose. The serum half-life is 3.9–4.9 h [50], [58]. Approximately 20% of the administered dose can be recovered in the urine as active drug during the first 4 h, with a total of 30–40% recovered in 24 h. Ciprofloxacin can also be recovered from blister fluid, where it achieves 57% of the serum concentration [50]. A parenteral formulation is available. An intravenous dose results in a serum half-life of 4 h, with ca. 76% of the dose recovered in the urine [49]. Clinical trials have shown the drug to be efficacious for the treatment of urinary tract infections as well as systemic bacterial infections.

Synthesis: Condensation of 2,4-dichloro-5-fluorobenzoylchloride with diethyl malonate by means of magnesium ethoxide in ether gives diethyl 2,4-dichloro-5-fluorobenzoylmalonate, which is partially hydrolyzed and decarboxylated with *p*-toluenesulfonic acid–water, yielding ethyl 2,4-dichloro-5-fluorobenzoylacetate. Condensation of this with triethyl orthoformate in refluxing acetic anhydride affords ethyl 2-(2,4-dichloro-5-fluorobenzoyl)-3-ethoxyacrylate, which is treated with cyclopropylamine in ethanol to give ethyl 2-(2,4-dichloro-5-fluorobenzoyl)-3-cyclopropylaminoacrylate. Cyclization with NaH in refluxing dioxane then yields 7-chloro-1-cyclopropyl-6-fluoro-1,4-dihydro-4-oxoquinoline-3-carboxylic acid, which is finally condensed with piperazine in hot DMSO to yield ciprofloxacin [59].

Trade Name: Bay 09867 (Bayer).

Enoxacin (INN) [*74011-58-8*], 1-ethyl-6-fluoro-1,4-dihydro-4-oxo-7-(1-piperazinyl)-1,8-naphthyridine-3-carboxylic acid, $C_{15}H_{17}FN_4O_3$, M_r 320.32, *mp* 220–224 °C, occurs as crystals; HCl salt *mp* 300 °C.

Enoxacin is currently under late clinical development worldwide. Its spectrum of activity is similar to that of norfloxacin, but it appears to be somewhat less potent [41]. The antibiotic is well-absorbed in humans and animals [47], [49], [60]. In experimental animals, enoxacin produced greater tissue concentrations than did norfloxacin when administered orally. In humans, a 600-mg oral dose of enoxacin shows superior oral absorption to norfloxacin, a higher peak serum concentration c_{max} (3.7 µg/mL), a longer serum half-life (6.2 h), greater urinary recovery (67% in 0–24 h), and greater blister fluid penetration [49]. This quinolone is effective in treating urinary tract and respiratory infections in humans, with a low incidence of side effects [49]. Clinical success has also been reported in the treatment of uncomplicated gonorrhea and systemic *P. aeruginosa* infections [61], [62]. A parenteral formulation is available and is currently being tested clinically.

Synthesis: Condensation of 2,3-difluoro-6-nitrophenol with chloroacetone yields 2-acetonyloxy-3,4-difluorobenzene, which is cyclized (H_2/Raney Ni) to give 7,8-difluoro-2,3-dihydro-3-methyl-4*H*-benzoxazine. Condensation with diethyl ethoxymethylenemalonate followed by cyclization (ethyl polyphosphate, 145 °C) yields 9,10-difluoro-3-methyl-7-oxo-2,3-dihydro-7*H*-pyrido[1,2,3-*d*]1,4-benzoxazine-6-carboxylate. Hydrolysis with HCl gives enoxacin [63].

Trade (or code) names: CI–919, AT–2266, Flumark, and PD–107779 (Dainippon).

Pefloxacin (INN) [*70458-92-3*], 1-ethyl-6-fluoro-7-(4-methyl-1-piperazinyl)-4-oxo-1,4-di-hydroquinoline-3-carboxylic acid, $C_{17}H_{20}FN_3O_3$, M_r 333.36, *mp* 270–272 °C. The mesylate salt, *mp* 284–286 °C (decomp.), is a yellowish-white crystalline powder.

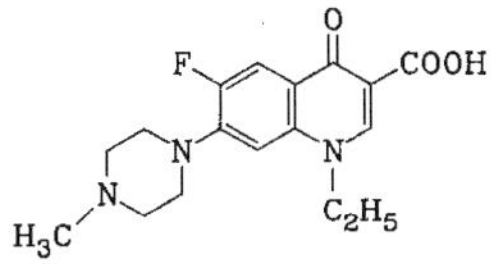

Pefloxacin was recently marketed in France and is under early development in the United States. The in vitro spectrum and potency of this fluoroquinolone are similar to norfloxacin but pefloxacin appears to have improved activity against gram-positive bacteria [64]. In experimental immunocompromised animals with *Pseudomonas* infection, pefloxacin showed better activity than did ciprofloxacin. In humans, renal plasma levels were 3.8 µg/mL following a 400-mg oral dose, and the plasma half-life (9–10 h) exceeded that for most quinolones currently under development [65]. Approximately 10 % is recovered in the urine after an 800-mg oral dose. One of the active metabolites of this compound is norfloxacin. Pefloxacin has been reported to be efficacious in treating meningitis in humans [66].

An intravenous preparation is available, and significant bone levels are achieved after i.v. dosing. Pefloxacin has been reported active clinically in the treatment of staphylococcal osteomyelitis when administered along with rifampicin [67].

Synthesis: 3-Chloro-4-fluoroaniline is condensed with diethyl ethoxymethylenemalonate and then cyclized thermally to give ethyl 6-fluoro-7-chloro-4-hydroxyquinoline-3-carboxylate. Subsequent alkylation with ethyl iodide leads to the ester, which is then saponified and neutralized to give pefloxacin [68].

Trade (or code) names: 1589–RB, Eu–5306 (Rhône–Poulenc).

Ofloxacin (INN) [*83380-47-6*], (±)9-fluoro-3-methyl-10-(4-methyl-1-piperazinyl)-7-oxo-2,3-dihydro-7*H*-pyrido[1,2,3-*de*]-1,4-benzoxazine-6-carboxylic acid, $C_{18}H_{20}FN_3O_4$, M_r 361.16, *mp* 250–257 °C (decomp.), occurs as colorless needles.

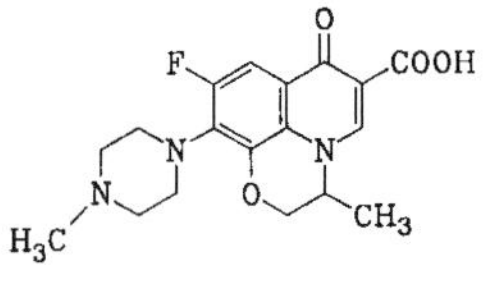

Ofloxacin has no spectrum or potency advantages over the other newer quinolones. Its in vitro microbiological activity is comparable to that of norfloxacin, but it has somewhat better activity against gram-positive cocci [37], [64]. The major attribute of this fluoroquinolone lies in its superior pharmacokinetics. A peak serum level of 10.7 μg/mL at 1.2 h after a 600-mg oral dose was achieved in human volunteers [69]. A long serum half-life (7.0 h) and excellent urinary recovery (73%) make this compound one of the more interesting members of the group of fluoroquinolones under worldwide development. Ofloxacin has been reported to be efficacious in a variety of clinical indications after oral dosing of 100–600 mg/d [60].

Synthesis: prepared from 2,3,4-trifluoronitrobenzene by a seven-step synthesis [70].
Trade (or code) names: DL–8280, Hoe–280, Ru–43–280, and Tarivid (Daiichi Seiyaku).

Amifloxacin (INN) [*86393-37-5*], 6-fluoro-1,4-dihydro-1-(methylamino)-7-(4-methyl-1-piperazinyl)-4-oxo-3-quinolinecarboxylic acid, $C_{16}H_{19}FN_4O_3$, M_r 334.35, mp 299–301 °C (decomp.), occurs as crystals.

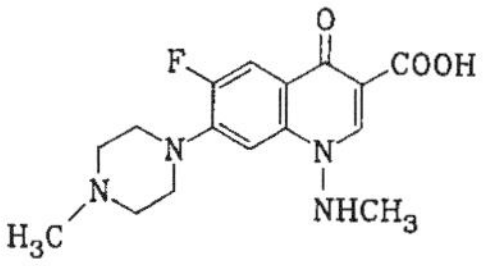

Amifloxacin has broad spectrum in vitro activity similar to that of the other quinolones, but with poorer potency than ciprofloxacin and little activity against anaerobic bacteria [71], [72]. The MIC_{90} for various Enterobacteriaceae, including *E. coli, Klebsiella, Proteus, Enterobacter, Citrobacter, Salmonella,* and *Shigella*, was comparable to that of norfloxacin and enoxacin, but was 2- to 8-fold less than that of ciprofloxacin [73]. Amifloxacin was more active than norfloxacin and enoxacin against *P. aeruginosa*, but less active than ciprofloxacin.

Amifloxacin is well-absorbed on oral administration and is excreted at high urinary levels in experimental animals [74]. A parenteral dosage form is available for humans, and early phase 1 studies have been initiated.

Synthesis: prepared from ethyl 6-fluoro-7-chloro-1,4-dihydro-4-oxo-3-quinolonecarboxylate through a five-step synthesis [75].
Trade (or code) name: WIN 49375 (Sterling-Winthrop).

Other Fluoroquinolones. A number of interesting fluoroquinolones are currently under study in various countries. They include flumequine (R–802), A–56620,

AM–833, and CI–934. The most advanced of these is flumequine, which is available for human and animal use in Europe.

Compound A–56620 (difloxacin), like the other fluoroquinolones, has potent antibacterial activity, but also has good activity against staphylococci, streptococci, and *B. fragilis* [76], [77]. Its potency is not appreciably affected by inoculum size. In dogs, the half-life is extended (8.2 h), and the compound undergoes enterohepatic circulation. The compound is in phase I development.

Fluoroquinolone AM–833 has in vitro microbiological activity similar to that of norfloxacin, but improved activity against staphylococci. It is more active in animal infections when administered orally than is norfloxacin, probably because of superior pharmacokinetics [78]. Peak serum levels and serum half-life in dogs are clearly superior to the values obtained with norfloxacin [79].

Compound CI–934 has broad gram-negative activity, but is less potent against these organisms than is norfloxacin or ciprofloxacin. It does have superior activity against gram-positive cocci when compared with the other quinolones [80]–[82].

Flumequine (INN, BAN) [42835-25-6], 9-fluoro-6,7-dihydroxy-5-methyl-1-oxo-1H,5H-benzo[i,j]quinolizine-2-carboxylic acid, $C_{14}H_{12}FNO_3$, M_r 261.15, mp 253–255 °C, is a white crystalline powder.

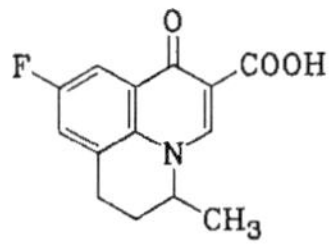

Synthesis: Condensation of 5-fluoro-2-methyltetrahydroquinoline with diethyl ethoxymethylenemalonate followed by thermal cyclization gives ethyl 6,7-dihydro-9-fluoro-5-methyl-1-oxo-1H,5H-benzo[i,j]quinolizine-2-carboxylate, which is saponified with sodium hydroxide to give flumequine [83].

A–56620 [98105-99-8], 1-(p-fluorophenyl)-6-fluoro-1,4-dihydro-4-oxo-7-(1-piperazinyl)-quinoline-3-carboxylic acid hydrochloride, $C_{20}H_{17}F_2N_3O_3 \cdot HCl$, M_r 421.83, mp 275 °C.

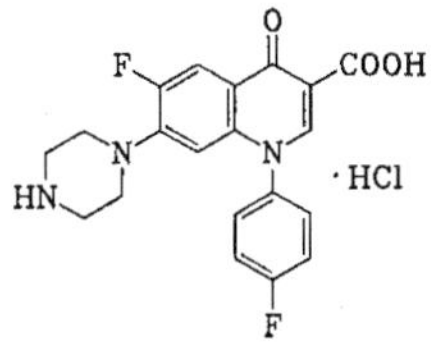

Synthesis: Condensation of 7-chloro-1-(p-fluorophenyl)-6-fluoro-1,4-dihydro-4-oxoquinoline-3-carboxylic acid with N-carboethoxypiperazine in hot 1-methyl-2-pyrrolidinone yields the N-carboethoxy derivative of A–56620. Hydrolysis of the derivative with sodium hydroxide in aqueous ethanol followed by treatment with dilute hydrochloric acid gives A–56620 [84].

AM – 833 [*79660-72-3*], 6,8-difluoro-1-(2-fluoroethyl)-1,4-dihydro-7-(4-methyl-1-piperazinyl)-4-oxo-3-quinolinecarboxylic acid, $C_{17}H_{18}F_3N_3O_3$, M_r 369.34; HCl salt [*79660-53-0*], mp 269–271 °C.

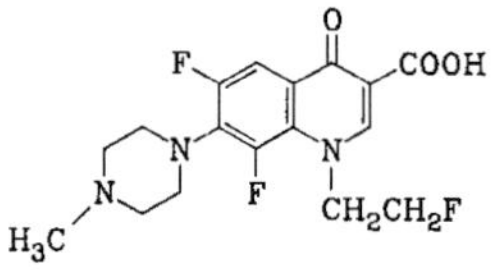

Synthesis: The reaction of 6,7,8-trifluoro-1,4-dihydro-4-oxoquinoline-3-carboxylic acid with 1-bromo-2-fluoroethane by means of NaI in DMF gives 6,7,8-trifluoro-1-(2-fluoroethyl)-1,4-dihydro-4-oxoquinoline-3-carboxylic acid, which is then condensed with *N*-methylpiperazine in refluxing pyridine to give AM – 833.
Manufacturer: Kyorin.

CI – 934 [*91188-00-0*], 1-ethyl-7-{3-(ethylamino)methyl-1-pyrrolidinyl}-6,8-difluoro-1,4-dihydro-4-oxo-3-quinolinecarboxylic acid, $C_{19}H_{23}F_2N_3O_3$, M_r 379.41, mp 208–210 °C.

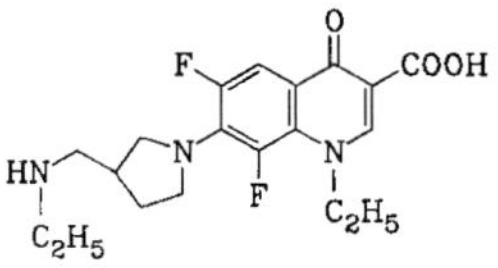

Synthesis: Refluxing 1-ethyl-1,4-dihydro-4-oxo-6,7,8-trifluoro-3-quinolinecarboxylic acid and *N*-ethyl-3-pyrrolidinemethanamine and 1,8-diazabicyclo[5.4.0]undec-7-ene in acetonitrile gives CI – 934 [85].

2.7.3. Sulfa Drugs

The sulfonamides continue to maintain a niche in human and animal medicine even 50 years after their discovery. More than 5000 sulfonamides have been synthesized with a broad diversity of antimicrobial activity and pharmacokinetic profiles. Approximately 13 compounds continue to be used in human medicine; however, their use is limited by the emergence of resistant organisms. The sulfonamide prontosil, one of a group of dyes synthesized in 1935, was the first clinically effective compound. Prontosil protected mice and rabbits against bacterial infections, but had no in vitro activity. Subsequently, prontosil was found to be, in effect, a prodrug, which on metabolic cleavage produced 4-aminobenzenesulfonamide (sulfonalamide) [86]. Because of their ease of synthesis, low cost, relative safety, and broad efficacy in humans, many research organizations embarked on broad synthesis programs designed to produce compounds with improved potency, spectrum of activity, and pharmacokinetic properties. Interest shifted from the sulfonamides in 1945 after the introduction of penicillin, but was revived again in the late 1950s.

2.7.3.1. Biological Activity and Medical Uses

Sulfonamides have broad antimicrobial activity that includes many species of bacteria and protozoa, such as the following:

Gram-negative bacteria	Gram-positive bacteria	Others
Escherichia coli	*Bacillus anthracis*	*Chlamydia trachomatis*
Proteus species	(some strains)	Trachoma virus
Klebsiella species	*Staphylococcus aureus*	Lymphogranuloma
Salmonella species	*Streptococcus pyogenes*	venereum virus
Shigella	*Clostridium welchii*	*Plasmodium falciparum*
Vibrio cholerae	*Clostridium tetani*	*Plasmodium malariae*
Neisseria species		*Toxoplasma*
Haemophilus species		*Nocardia* species
Pseudomonas (some strains)		
Calymmatobacterium granulomatis		
Legionella pneumophila		

The common gram-positive and gram-negative bacteria are susceptible, as are the protozoan organisms that cause malaria and toxoplasmosis. The lymphogranuloma venereum and trachoma viruses are reported to be susceptible as well. A high percentage of bacterial organisms have developed resistance to sulfonamides; and thus, these drugs are indicated only in a few diseases, including urinary tract infections, chancroid, inclusion conjunctivitis, trachoma, and nocardia infections.

The sulfonamides have been classified on the basis of their serum half-lives in humans and their topical or gastrointestinal uses. Those compounds with a half-life of < 10 h are termed short-acting, and those with a half-life between 10 and 24 h are medium-acting; long-acting sulfonamides have a half-life > 24 h.

2.7.3.2. Mechanism of Action and Antimicrobial Resistance

4-Aminobenzoic acid (**3**) is an essential metabolite for many microorganisms.

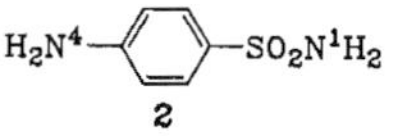

$H_2N^4\text{—}\langle\text{—}\rangle\text{—}SO_2N^1H_2$ $H_2N^4\text{—}\langle\text{—}\rangle\text{—}COOH$
                2   **3**

4-Aminobenzenesulfonamide   4-Aminobenzoic acid
[63-74-1]   [150-13-0]

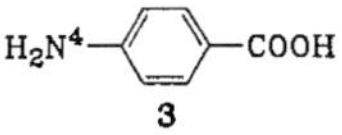

This compound is used by bacteria as a precursor in the synthesis of folic acid [59-30-3], which, in turn, serves as an important intermediate in nucleic acid synthesis. *Sulfonamides*, as structural analogues of **3**, interfere with microbial folic acid synthesis by inhibiting the adenosine triphosphate dependent condensation of a pteridine with **3** to yield dihydropteroic acid, which is subsequently converted to folic acid [87]–[89]. As a result, nonfunctional analogues of folic acid are formed that do not allow bacterial cells to grow. Because mammalian cells cannot synthesize folic acid, the activity of sulfonamides is selective for bacteria that are capable of producing it. Tubercule bacilli are poorly inhibited by sulfonamides, but their growth is inhibited by 4-aminosalicylic acid [65-49-6], whereas most sulfonamide-sensitive bacteria are resistant to this acid. Thus, the receptor site for 4-aminobenzoic acid apparently differs in different types of organisms.

Trimethoprim [738-70-5] inhibits dihydrofolic acid reductase 10 000 times more efficiently in bacteria than in mammalian cells. This enzyme is important in the folic acid pathway leading to the synthesis of purines and ultimately of DNA. The combination of sulfonamides with trimethoprim acts synergistically to inhibit bacteria by sequential blockage (see Section 2.7.3.6).

Pyrimethamine [58-14-0] (Daraprim) also inhibits dihydrofolate reductase, but it is more active against the enzyme in mammalian cells, and, thus more toxic than trimethoprim. Toxoplasma and other protozoal infections can be successfully treated with pyrimethamine – sulfonamide combinations.

The usefulness of sulfa drugs has been limited by the development of *bacterial resistance*. At least three mechanisms are well documented: (1) alteration of cell wall permeability, (2) increased production of an essential metabolite, and (3) a mechanism involving dihydropteroic acid synthetase. Two enzymes have been reported for *E. coli* with less affinity for sulfonamides than enzymes from susceptible strains, but with no change in 4-aminobenzoic acid affinity. Resistant bacteria, e.g., staphylococci, synthesize excess 4-aminobenzoic acid, which can antagonize sulfa drugs. Plasmids may code for drug-resistant enzymes [90] or decreased bacterial cell permeability to sulfonamides [91]. Multiple resistance mechanisms in the same organism are possible [92]. Sulfonamide resistance is found in 20–40% of community and nosocomial strains of bacteria, including *Staphylococcus*, Enterobacteriaceae, *Neisseria meningitidis,* and *Pseudomonas* species [93].

2.7.3.3. Structure – Function Relationships

All clinically useful sulfa drugs are derived from sulfonalamide (**2**), an analogue of 4-aminobenzoic acid (**3**). The synthesis of thousands of derivatives of **2** has allowed for excellent structure – function determinations, which have been reviewed in detail [94]. A free amino group at N-4 is essential for microbiological activity. Acyl substitution, such as phthalyl or succinyl, at N-4 is acceptable, provided the substituent is hydrolyzed in the body to the free amine (e.g., phthalylsulfathiazole). Replacement of the benzene ring by a heterocycle leads to loss in activity; thus, only the acid portion of the molecule is amenable to substitution. Substitution at the sulfonyl radical can result in marked changes in pharmacologic properties, such as absorption, solubility, pharmacokinetics, and gastrointestinal tolerance.

Monosubstitution at N-1, especially when the substituent is a heterocycle, has led to most of the clinically useful sulfa drugs. Derivatives containing five- or six-membered heterocyclic rings, including oxazole, thiazole, pyridine, pyrimidine, and many others, have been most successful.

2.7.3.4. Pharmacokinetics

Sulfa drugs are commonly administered orally; absorption is generally rapid. Systemic derivatives, such as sulfisoxazole and sulfadiazine, are available, but are rarely used in clinical medicine. An intravenous formulation of trimethoprim and sulfamethoxazole is used to treat *Pneumocystis carinii* pneumonitis, shigellosis, and urinary tract infections caused by susceptible organisms. In addition various topical preparations are available for treating ophthalmic and vaginal infections and burn patients.

Generally the available sulfa drugs are rapidly absorbed from the gastrointestinal tract. Their excretion rates vary widely, depending on their physical and chemical properties. Several derivatives of **2** substituted at N-4 are poorly absorbed and are employed for diseases of the gastrointestinal tract. Peak blood levels following oral absorption are usually observed at 2–4 h. A wide variation in half-lives from 2.5 to 150 h for the commercially available agents has been reported, but most agents are excreted over a 24-h period.

Sulfonamides are usually well-distributed in the body and achieve tissue levels of ca. 80% of those observed in the serum [95]. The degrees of protein binding and lipid solubility play a major role in determining serum half-life and distribution in the body. The primary route of metabolism is via the liver, where acetylation or glucuronidation occur. Excretion is mainly via the kidney, where most drugs are removed by glomerular filtration, although partial reabsorption and active tubular secretion also are involved. Sulfa drugs bind reversibly to serum proteins, primarily albumin, which tends to slow metabolism by the liver [96].

2.7.3.5. Toxicity and Drug Interactions

Significant hypersensitivity reactions have been associated with the sulfonamides; the long-acting derivatives, especially, have caused fatal hypersensitivity reactions. Other hypersensitivity reactions include erythema nodosum and erythema multiforme, including Stevens–Johnson syndrome. Photosensitization, itching, and rash usually require cessation of therapy. With earlier drugs, crystalluria and related renal damage occurred, primarily with the less water-soluble compounds, such as sulfadiazine, sulfamerizine, and sulfapyridine. Other adverse effects observed less commonly are nausea, vomiting, diarrhea, hepatic toxicity, and a syndrome that resembles serum sickness. Serious adverse reactions include hemolytic anemia, aplastic anemia, agranulocytosis, thrombocytopenia, and leukopenia.

Sulfonamides interact with other drugs and compete with drugs, such as warfarin and methotrexate, for albumin-binding sites resulting in increased toxicity of the displaced drugs. Concomitant administration of probenecid, which results in decreased renal tubular secretion, prolongs sulfa drug levels and may increase toxicity.

2.7.3.6. Combination Therapy

Both oral and intravenous formulations of sulfamethoxazole in combination with trimethoprim are available. This synergistic mixture is used orally to treat urinary tract and certain other infections. In the United States, trimethoprim is also available as a single agent for the treatment of urinary tract infections. Combinations of sulfa drugs and various antibiotics for both oral and topical use are also available.

Systemic formulations in combination with 2,6-diamino-3-phenylazopyridine · hydrochloride, an azo dye that is a mild local analgesic, have been employed to reduce pain of inflammation that results from urinary tract infections.

2.7.3.7. Rapidly Absorbed Short- and Medium-Acting Sulfa Drugs

Sulfamethoxazole (INN, U.S.P.) [723-46-6], N^1-(5-methyl-3-isoxazolyl)sulfanilamide, $C_{10}H_{11}N_3O_3S$, M_r 253.28, mp 170 °C, is an odorless and colorless crystalline powder.

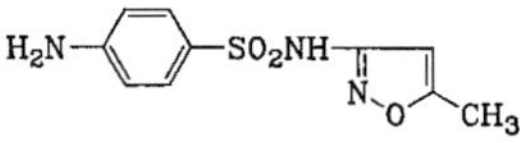

Sulfamethoxazole is rapidly absorbed from the gastrointestinal tract and has a moderate elimination rate. Peak blood levels of 80–100 µg/mL are observed 1–4 h following a 2-g oral dose. The plasma half-life is 9–11 h and the drug is 70% protein-bound; 80–90% of the unbound compound remains unacylated and biologically active in the blood. About 60% of the drug is excreted in the urine, and half of this is N-4 acylated or conjugated. Urine levels are 3 times blood levels.

The major indication for sulfamethoxazole is treatment of urinary tract infections caused by susceptible strains of *E. coli*, *Klebsiella*, *Proteus* species, and staphylococci. Although the compound has broader microbiological activity, it generally is not used for infections outside of the urinary tract. Sulfamethoxazole is also indicated in the treatment of nocardiosis, inclusion conjunctivitis, trachoma, chancroid, and malaria caused by chloroquine-resistant strains, and is used in combination with pyrimethamine for toxoplasmosis.

The usual dose for mild infections is 2 g followed by 1 g every 12 h for 4 days. For severe infections, the drug is administered 3 times a day at twice this dose. Pediatric dose is 50–60 mg/kg followed by 12-h doses of 25–30 mg/kg.

Synthesis: prepared from ethyl 5-methylisoxazole-3-carbamate [97].

Trimethoprim (INN, U.S.P.) [*738-70-5*], 2,4-diamino-5-(3,4,5-trimethoxybenzyl)pyrimidine, $C_{14}H_{18}N_4O_3$, M_r 290.32, *mp* 199–203 °C, is a white-to-cream-colored bitter, crystalline powder.

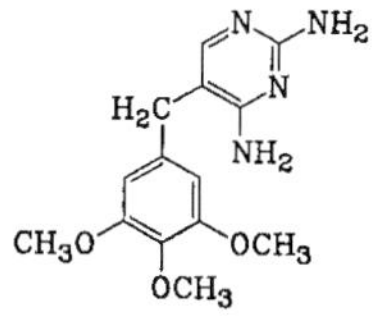

Synthesis: prepared from guanidine and β-ethoxy-3,4,5-trimethoxybenzylbenzalnitrile [98].

Trade Names: Cofrim (Lemmon), Bactrim (Hoffman-La Roche), Septa (Burroughs Wellcome), Proloprim (Burroughs Wellcome), and Trimpex (Hoffman-La Roche, trimethoprim only).

Sulfamethoxazole – Trimethoprim. Sulfamethoxazole is usually administered along with trimethoprim in a fixed ratio of 5:1. This synergistic combination results in a sequential blockage of folic acid synthesis. Synergistic antibacterial activity has been demonstrated in both in vitro and in vivo systems [99], [100].

The pharmacokinetic profiles of sulfamethoxazole and trimethoprim fit well, both compounds showing peak blood levels 1–4 h after oral administration. With administration at 12-h intervals, a 20:1 ratio of unbound sulfa drug: trimethoprim is achieved in the plasma, an optimal ratio for antibacterial activity. Trimethoprim is 44% protein bound and has a serum half-life of 8–10 h. Both drugs are metabolized by the liver and excreted by the kidneys.

Trimethoprim has a broad spectrum of antimicrobial activity and extends the spectrum of sulfamethoxazole to include many urinary and systemic pathogens untreatable by this sulfa drug alone. The major use of the drug combination is for urinary tract infections, in which case the drugs are administered orally for 10–14 days every 12 h. Additionally, the combination is indicated for chronic bronchitis in adults and otitis media in children for susceptible strains of *Streptococcus pneumoniae* and *Haemophilus influenzae*. Bacillary dysentery caused by *Shigella* species also responds to oral treatment.

An intravenous preparation is available for treatment of severe infections, as well as pneumonitis caused by the protozoan, *Pneumocystis carinii*. For treatment of *Pneumocystis* pneumonia, the recommended i.v. dose is 15–20 mg/kg, based on the trimethoprim component, given in 3 or 4 equally divided doses every 6–8 h for up to 14 days [101]. For severe urinary tract infections and shigellosis, the total daily dose is 8–10 mg/kg, based on the trimethoprim component, given in 2–4 equally divided doses every 6, 8, or 12 h for up to 14 days for urinary tract infections and 5 days for shigellosis.

Sulfisoxazole [127-69-5], sulfafurazole (U.S.P., INN), N^1-(3,4-dimethyl-5-isoxazolyl)sulfanilamide, $C_{11}H_{13}N_3O_3S$, M_r 267.30, *mp* 192 °C, occurs as colorless and odorless prisms.

Sulfisoxazole has antimicrobial activity somewhat less than that of sulfadiazine; however, it is more easily tolerated because of its high solubility and rapid excretion. It is clinically useful in urinary tract infections and is especially potent against *E. coli* and *Prot. vulgaris*. Later derivatives of this compound, i.e., sulfamethoxazole, have superior solubility properties. At 2 h after oral administration of a 2-g dose, peak serum levels of ca. 150 µg/mL are achieved. The drug is 85% protein-bound, and about 30% of the unbound portion is in the acetylated form.

Excretion is primarily by the kidneys with ca. 95% of a single oral dose eliminated within 24 h; 70% of the excreted material is in the active form. The acetyl sulfisoxazole is metabolized in the gastrointestinal tract to release sulfisoxazole, which results in delayed absorption.

Although this sulfa drug is still used clinically, its effectiveness is limited mainly because of high protein binding. The usual adult dose is 2–4 g in a loading dose followed by 4–8 g/d divided in 4–6 doses. The compound is also formulated in vaginal preparations for the treatment of vaginitis caused by *Haemophilus vaginalis*. This sulfa drug is also available as the water soluble diolamine salt in 40% aqueous solution for dilution with sterile water as a parenteral preparation.

Synthesis: p-Acetaminobenzenesulfonyl chloride is reacted with 3,4-dimethyl-5-aminoisoxazole followed by deacetylation [102].

Trade Names: Gantrisin (Hoffmann-La Roche), SK-Soxazole (Smith Kline & French).

Sulfadiazine (INN, U.S.P.) [68-35-9], N^1-2-pyrimidinylsulfanilamide, $C_{10}H_{10}N_4O_2S$, M_r 250.27, *mp* 252–256 °C, is a white or pale yellow crystalline powder.

Sulfadiazine is classified as a rapidly absorbed agent with a moderate rate of excretion. It is the sulfa drug of choice for treating CNS infections caused by antibiotic-resistant organisms and for treating nocardia infections. Sulfadiazine is also used to treat urinary tract infections; however, crystalluria is a problem because of its poor water solubility. The usual precautions of urine alkalinization and forced fluid intake are recommended for patients taking sulfadiazine.

Sulfadiazine is absorbed rapidly from the gastrointestinal tract after oral administration. The peak blood level of free drug following a 2-g dose is 30–60 µg/mL with about 45% protein-binding. The drug achieves excellent cerebrospinal fluid levels, which are 5–13% of the blood levels. The plasma half-life is 17 h. About 20% of an oral dose is excreted in the urine as the free form. The major metabolite is the N-4-acetylated compound.

This sulfa drug can be administered parenterally; however, because its alkaline solution is irritating, the i.v. route is preferred. The usual oral dose for adults is 2 – 4 g initially, followed by 2 – 4 g/d in 3 – 6 divided doses. When given intravenously for adults and children over 2 months of age, the drug is administered at a loading dose of 50 mg/kg or 1.25 g/m^2 followed by a total daily dose of 100 mg/kg or 2.25 g/m^2 administered 4 times a day.

A silver salt of sulfadiazine is available for topical treatment of burn patients. In this case the sulfonamide acts primarily as a vehicle for release of silver ions, which exert an antibacterial effect. Resistant organisms have been reported in burn units.

Synthesis: prepared by condensing 2-aminopyrimidine with acetylsulfaninyl chloride followed by hydrolysis with NaOH [103].

Trade Name: Suladyne (Reid-Provident).

Sulfacytine (INN) [*17784-12-2*], N^1-(1-ethyl-1,2-dihydro-2-oxo-4-pyrimidinyl)sulfanilamide, C$_{12}$H$_{14}$N$_4$O$_3$S, M_r 294.33, *mp* 166.5 – 168 °C, is crystalline in form.

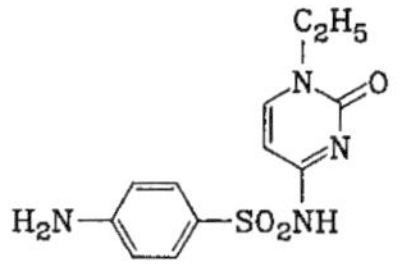

Sulfacytine is absorbed rapidly from the GI (gastrointestinal) tract with a high rate of urinary excretion. Its major use is for treating urinary tract infections. It is active against the common urinary tract pathogens, such as *E. coli, Klebsiella* species, *Proteus* species, as well as *S. aureus*. Urinary levels far in excess of the MIC are found following a daily dose of 1 g. The usual dose is 500 mg initially followed by 250 mg 4 times a day for 10 days. Peak blood levels occur within 2 – 3 h after administration, with ca. 90 % of the oral dose being recovered in the urine as the free form. High urinary concentrations are observed, but the drug is bound 86 % to serum proteins. The serum half-life is 4.5 h.

Synthesis: [104].

Trade Name: Renoquid (Parke-Davis).

Sulfamethizole (INN, U.S.P.) [*144-82-1*], N^1-(5-methyl-1,3,4-thiadiazol-2-yl)sulfanilamide, C$_9$H$_{10}$N$_4$O$_2$S$_2$, M_r 270.32, *mp* 208 °C, is a colorless crystal.

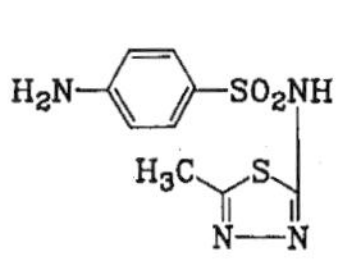

Sulfamethizole is well-absorbed and rapidly excreted. Its major indications are confined to therapy of urinary tract infections caused by susceptible organisms. It has the same therapeutic indications as sulfisoxazole; however, it is less bound to serum proteins than is sulfisoxazole. The usual adult dose is 2 – 4 g followed by 2 – 4 g/d administered in 6 divided doses.

Synthesis: prepared by the reaction of acetaldehyde thiosemicarbazone with *p*-acetylaminobenzenesulfonyl chloride in pyridine [105].

Trade Names: Thiosnifil-A (Ayerst), Proklar (O'Neal, Jones & Feldman), component of Azotrex (Bristol), and component of Suladyne (Reid-Provident).

2.7.3.8. Long-Acting Sulfonamides

Sulfadoxine (INN, U.S.P.) [*2447-57-6*], N'-(5,6-dimethoxy-4-pyrimidinyl)sulfanilamide, $C_{12}H_{14}N_4O_4S$, M_r 310.33, *mp* 201 – 202 °C, forms colorless crystals.

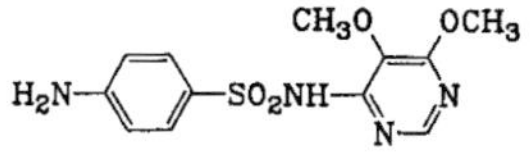

Sulfadoxine is the only long-acting agent available for clinical use in the United States. All of the long-acting sulfonamides are associated with serious hypersensitivity reactions, such as Stevens – Johnson syndrome. Sulfadoxine, originally known as sulfamethoxine, is combined with pyrimethamine (Fansidar). It has a half-life of 200 – 230 h and a peak serum level of 51 – 76 µg/mL at 2.5 – 6 h following an oral dose of 0.5 g. The major indication is for the treatment and prophylaxis of chloroquine-resistant *Plasmodium falciparum* malaria.

Synthesis: [106].

Trade Name: Fanasil (Hoffmann-La Roche International).

2.7.3.9. Sulfonamides for Use in the Gastrointestinal Tract

Sulfasalazine (INN, U.S.P.) [*599-79-1*], 5-{[*p*-(2-pyridylsulfamoyl)phenyl]azo}salicylicacid, $C_{18}H_{14}N_4O_5S$, M_r 398.39, *mp* 240 – 245 °C, is a brownish-yellow powder.

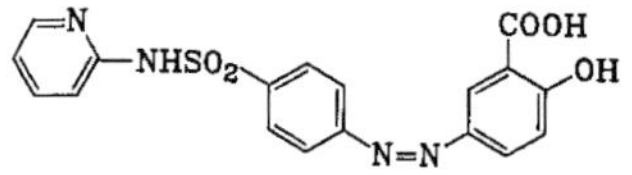

Sulfasalazine, used for the treatment of ulcerative colitis, is a prodrug of sulfapyridine. The compound is a combination of sulfapyridine and 5-aminosalicylic acid joined by an azo link, which is cleaved in the lower intestinal tract by bacterial enzymes. The release of 5-aminosalicylic acid is believed to produce a local antiinflammatory action in the colon; this is the mechanism by which the drug is effective in ulcerative colitis.

Sulfasalazine is absorbed only 10 – 15 % from the upper gastrointestinal tract and gives peak blood levels of 14 µg/mL 2 – 4 h after a 2-g oral dose. The drug is excreted in the bile and reabsorbed from the intestine as its metabolite, sulfapyridine. Peak blood levels of sulfapyridine have been reported at 13 µg/mL 6 – 24 h after dosing. The major urinary metabolite of sulfasalazine is sulfapyridine and its metabolites (60%); however, small amounts of the unchanged drug (15%) and 5-aminosalicylic acid and its metabolites (20 – 33%) are also observed.

Synthesis: prepared by coupling diazotized 2-sulfanilamidopyridine with salicylic acid [107].

Sulfaguanidine, Sulfasuxidine, Sulfathalidine. Sulfaguanidine, sulfasuxidine, and sulfathalidine are all absorbed poorly from the GI tract following an oral dose. They have been employed as prophylactic agents prior to bowel surgery.

Sulfaguanidine (INN, *National Formulary* – NF **11**, MI) [*57-67-0*], N^1-(diaminomethylene)-sulfanilamide, $C_7H_{10}N_4O_2S$, M_r 214.24, *mp* 190 – 193 °C, occurs as colorless crystals.

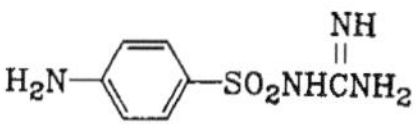

Synthesis: prepared by fusing N^4-acetylsulfanilamide and dicyanodiamide [108].

Sulfasuxidine, succinylsulfathiazole (U.S.P. **18,** MI) [*116-43-8*], 4′-(2-thiazolylsulfamoyl)-succinanilic acid monohydrate, $C_{13}H_{13}N_3O_5S_2 \cdot H_2O$, M_r 373.41, *mp* reported as 184 – 186 °C and as 192 – 195 °C, occurs as crystals.

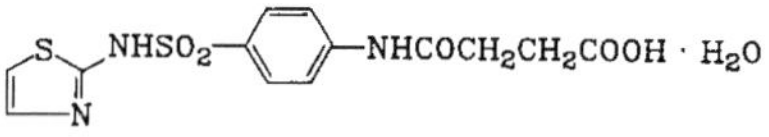

Synthesis: prepared by refluxing sulfathiazole with a slight excess of succinic anhydride in alcohol [109].

Sulfathalidine [*85-73-4*], phthalylsulfathiazole (INN, U.S.P.), 4′-(2-thiazolylsulfamoyl)-phthalanilic acid, $C_{17}H_{13}N_3O_5S_2$, M_r 403.43, *mp* 270 °C, is a white crystalline powder.

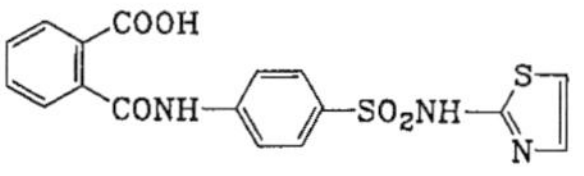

Synthesis: prepared by refluxing sulfathiazole with a slight excess of succinic anhydride in alcohol [110].

2.7.4. Agents for Treating Mycobacterial Infections

The mycobacteria are a group of rod-shaped bacteria that include many pathogenic species [111]–[113]. They do not stain readily, but once stained, they resist decoloration by acid or alcohol; therefore, they are called acid-fast bacilli. They cause chronic diseases producing granulomatous lesions. The major organisms causing human disease are *Mycobacterium tuberculosis, M. leprae,* and the so-called atypical mycobacteria.

Mycobacterium tuberculosis and *M. bovis,* the causative agents of tuberculosis, produce no known toxins. The organisms are usually taken in via the respiratory tract, where they establish pulmonary infections. Resistance and hypersensitivity of the host generally influence the development of the disease. Once established in the tissues, they reside in phagocytic cells, where their intracellular location makes chemotherapy difficult. The organisms spread in the host by direct extension, through the lymphatics and bloodstream, and via bronchi and the gastrointestinal tract. Every organ system can be involved. Chronic pulmonary infections are usually established; however, meningitis or urinary tract involvement can occur in the absence of other symptoms of tuberculosis. Dissemination via the bloodstream leads to miliary tuberculosis involving many organs and a high fatality rate.

Treatment of *M. tuberculosis* infections usually involves combinations of chemotherapeutic agents. The most widely used antituberculosis drugs are isoniazid, pyrazinamide, ethambutol, and the antibiotics rifampin and streptomycin (→ Antibiotics). Other drugs, such as ethionamide, 4-aminosalicylic acid, viomycin, and cycloserine, have severe adverse effects and are employed less frequently. Treatment is for long periods of time and clinical cure can be achieved in 6–18 months. Drug-resistant strains to all of these agents emerge rapidly; thus, greater success is achieved when the drugs are administered concomitantly.

There has been a increase in the incidence of atypical mycobacterial infection usually observed in patients whose immunity has been compromised by other factors. Many of these species of *Mycobacterium* cause severe disease closely resembling tuberculosis. These atypical mycobacteria, e.g., *M. avium – intracellularis* group, are generally much less sensitive to the available antituberculosis chemotherapeutics, especially streptomycin and isoniazid. Other organisms respond to the antitubercular chemotherapeutic agents, whereas many are sensitive to antibiotics employed for treating other bacterial infections (aminoglycosides and tetracyclines). A summary of the chemotherapeutic agents used to treat atypical mycobacterial infections is shown in Table 2 [114].

Mycobacterium leprae, the etiological agent of *leprosy,* is a disease involving the cooler tissues of the body, i.e., skin, superficial nerves, nose, pharynx, larynx, eyes, and testicles. In untreated cases, severe disfiguration is observed because of skin infiltration and nerve involvement. There are two distinct types of leprosy, lepromatous and tuberculoid. The more severe, lepromatous type is progressive and leads to lesions involving skin and nerves and to bacteremia. The tuberculoid type is benign and

Table 2. Antimicrobials used to treat atypical mycobacteria *

Mycobacterial species	Primary drug	Secondary drug
M. kansasii	isoniazid with rifampin, with or without ethambutol	ethambutol, streptomycin, pyrazinamide, PAS, cycloserine, ethionamide, kanamycin, and capreomycin
M. fortuitum complex	amikacin and doxycycline	cefoxitin, rifampin, erythromycin, and a sulfonamide
M. avium – intracellularis – scrofulaceum	isoniazid, rifampin, ethambutol, and streptomycin	clofazimine, capreomycin, ethionamide, cycloserine, ansamycin, imipenem, and amikacin
M. marinum (balnei)	minocycline	trimethoprim – sulfamethoxazole, rifampin, and cycloserine

* Adapted from [114].

nonprogressive. Treatment usually is lengthy and requires administration of one of the sulfones and the semisynthetic antibiotic, rifampin. Drug resistance to the sulfones has emerged.

2.7.4.1. Antituberculosis Agents

Isoniazid (INN, U.S.P.) [*54-85-3*], isonicotinic acid hydrazide, $C_6H_7N_3O$, M_r 137.14, mp 171.4 °C, occurs as crystals (alc).

In 1952 isoniazid was demonstrated to be valuable in treating tuberculosis [115]. It kills *M. tuberculosis* organisms by inhibiting mycolic acid synthesis [116]. At higher concentrations, it may have another mechanism of action against atypical bacteria. The MIC of isoniazid for susceptible *M. tuberculosis* strains ranges from 0.025 to 0.05 µg/mL. Emergence of resistance is observed when the drug is administered alone [115].

Isoniazid is well-absorbed after oral or intramuscular administration and is subsequently distributed throughout the body. The drug is acetylated in the liver, with the rate of acetylation being separated into genetic populations of slow and normal acetylators [116].

Isoniazid is generally well-tolerated. An infrequent major toxicity is hepatitis [117]. Elevations in serum glutamic-oxaloacetic transaminase (SGOT), which are observed in 15% of patients, disappear with continued therapy. Peripheral neuropathy occurs in 17% of people receiving 6 mg kg^{-1} d^{-1} of isoniazid. Manifestations of central nervous system toxicity have been reported, as have hypersensitivity reactions. Isoniazid potentiates dilantin toxicity, especially in slow acetylators.

Isoniazid is indicated for all forms of tuberculosis. It is administered in combination with one or more other antitubercular agents, the usual dose being 5–10 mg kg^{-1} d^{-1}. Higher doses of the drug (15 mg/kg, orally) can be administered twice weekly along with other agents. This twice-weekly regime can be started after an initial period of daily drug administration and serves to encourage compliance and reduce cost [118].

Synthesis: prepared by condensing pyridine-4-carboxyethylate and hydrazine [119].
Trade Names: Cotinazin (Pfizer), Dinacrin (Sterling-Winthrop), INH (Ciba-Geigy), and Nydrazid (Squibb).

Ethambutol hydrochloride (INN, U.S.P.) [*1070-11-7*], [*74-55-5*], (+)-2,2′-(ethylenediimino)-di-1-butanol dihydrochloride, $C_{10}H_{24}N_2O_2 \cdot 2$ HCl, M_r 277.23, *mp* 194.5 °C.

$$CH_3CH_2-\underset{H}{\overset{CH_2OH}{C}}-NHCH_2CH_2NH-\underset{CH_2OH}{\overset{H}{C}}-CH_2CH_3 \cdot 2\ HCl$$

Ethambutol is a tuberculostatic agent active against most human strains of tubercle bacilli at 8 μg/mL, with ca. 75% of the strains susceptible at 1 μg/mL. It is not cross-resistant with isoniazid-resistant strains. Resistance develops stepwise when the drug is administered alone. The major use of the drug is in combination with other antitubercular agents to limit the emergence of resistance. Ethambutol has replaced 4-aminosalicylic acid as the companion drug for isoniazid because of its greater potency and lower incidence of adverse effects.

The usual dose is 15–25 mg kg^{-1} d^{-1} initially, followed after 60 days by 15 mg kg^{-1} d^{-1} as a single dose. Ethambutol is 75–80% absorbed after an oral dose with peak plasma levels of 5 μg/mL after a 25-mg/kg dose. It is well-distributed in the body and reached CNS levels of 10–50% of serum levels in patients with meningeal inflammation. The drug is excreted mainly unchanged in the urine with ca. 15% being converted to inactive metabolites following absorption. Neuropathic toxicities are the major *adverse effects* of this drug, especially retrobulbar neuritis, which is common at a dose of 50 mg kg^{-1} d^{-1}.

Synthesis: prepared by warming (+)-2-amino-1-butanol with ethylene bromide at 110–115 °C, liberation of the base with potassium hydroxide, and conversion to the dihydrochloride [120].
Trade Name: Myambutol (Lederle).

Pyrazinamide (INN, U.S.P.) [*98-96-4*], pyrazinecarboxamide, $C_5H_5N_3O$, M_r 123.11, *mp* 189–191 °C, occurs as crystals.

Pyrazinamide is a bactericidal antitubercular agent that is used in combination with other antitubercular agents, especially in developing nations [121]. It is a nicotinamide analogue with an unknown mechanism of action. When used alone, resistance develops rapidly. Because of its toxicity, it is used for short-course therapy regimes. Treatment with pyrazinamide can result in hepatotoxicity, especially in the extended dose regimes employed in the early clinical trials, i.e., 40 mg kg^{-1} d^{-1}, but a current recommended dose of 20–35 mg kg^{-1} d^{-1} appears to be safe [122].

In the United States, pyrazinamide is generally reserved for combination use in patients infected with drug-resistant strains. Mainly because of low cost and convenient once-weekly administration, its use is favored in developing nations, especially in areas of high incidence of drug resistance [123].

Pyrazinamide is well-absorbed when administered orally and reaches levels in body fluids in excess of the MIC for susceptible organisms. It is metabolized by the liver and excreted in the urine. The usual total daily dose is 1.5–2.0 g divided into 2–4 dosing intervals. Higher levels administered once or twice weekly appear to be clinically effective with no observable liver toxicity.

Synthesis: prepared by ammonolysis of methyl pyrazinoate (from quinoxaline) [124].
Manufacturer: Lederle.

Ethionamide (INN, U.S.P.) [*536-33-4*], 2-ethylthioisonicotinamide, $C_8H_{10}N_2S$, M_r 166.24, *mp* 164–166 °C (decomp.), occurs as yellow crystals.

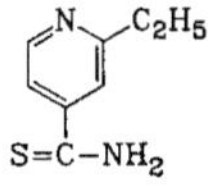

This nicotinic acid analogue is considered a secondary agent for treating drug-resistant tuberculosis. It is tuberculostatic for most susceptible strains at 0.6–2.5 µg/mL. The usual dose is 1 g/d, starting at 250 mg in divided doses and increasing by 125 mg kg^{-1} d^{-1} until the desired dose is reached. A frequent *adverse effect* is gastrointestinal irritation accompanied by nausea and vomiting. Central nervous system side effects, including psychiatric disturbances and peripheral neuropathy, have been reported. In about 5% of the patients, a reversible hepatotoxicity is observed. Ethionamide is absorbed well from the gastrointestinal tract, giving plasma concentrations of 20 µg/mL. It penetrates both normal and inflamed meninges, giving high CNS levels. Ethionamide metabolism is in the liver, and the metabolites are excreted in the urine.

Synthesis: prepared by addition of hydrogen sulfide to 2-ethylnicotinonitrile in the presence of triethanolamine [125].
Trade Name: Trecator–SC (Ives).

p-Aminosalicylic acid [*65-49-6*], PAS (U.S.P.), 4-aminosalicylic acid, $C_7H_7NO_3$, M_r 153.14, *mp* 150–151 °C (effervescence), is in the form of minute crystals (alcohol).

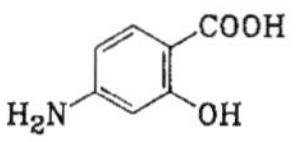

4-Aminosalicylic acid is an inhibitor of mycobacterial growth by impairing of folate synthesis. It has largely been replaced by ethambutol in the standard combination therapy for *M. tuberculosis*. The compound is poorly absorbed when administered orally; thus, a 4-g dose results in peak plasma concentrations of only 7–8 µg/mL. The metabolites of PAS are excreted in the urine. The major *adverse effect* is gastrointestinal intolerance, which frequently results in poor patient compliance. 4-Aminosalicylic acid has been reported to produce a lupus-like syndrome. Hypersensitivity reactions are frequent, occurring at a 5–10% rate. The usual adult dose is 10–12 g/d in three or four divided doses and 200–300 mg kg^{-1} d^{-1} in divided doses for children.

Synthesis: prepared by carboxylation of 3-aminophenol with ammonium carbonate solution at 110 °C under pressure [126].

Trade Name: PAS – Heyl (Heyl).

Amithiozone [*104-06-3*], thiacetazone (INN, MI), 4′-formylacetanilide thiosemicarbazone, $C_{10}H_{12}N_4OS$, M_r 236.29, mp 225–230 °C (decomp.), occurs as minute, pale yellow crystals that darken on exposure to light.

CH₃CONH—⟨ ⟩—CH=NNHCNH₂
                      ‖
                      S

Amithiozone is a second-line drug for the treatment of tuberculosis. This thiosemicarbazone is active against *M. tuberculosis*, inhibiting most strains at 1 µg/mL [127]. The drug is administered orally at 150 mg/d or 450 mg twice weekly. Peak serum concentrations are 1–2 µg/mL. Amithiozone should be used in combination with other drugs because resistance develops readily when the drug is used alone. The drug is not available in the United States and has only limited use in Europe because of gastrointestinal irritation and bone-marrow suppression. It is, however, used as a first-line drug in East Africa, where apparent lower incidence of adverse effects in Africans and low cost favor its use.

Synthesis: prepared by treating 4-acetamidobenzaldehyde with thiosemicarbazide in alcohol [128].

2.7.4.2. Antileprosy Agents

Dapsone (U.S.P.) [*80-08-0*], 4,4′-sulfonyldianiline, $C_{12}H_{12}N_2O_2S$, M_r 248.30, mp 175–176 °C (also recorded 180.5 °C), occurs as crystals (alc).

H₂N—⟨ ⟩—SO₂—⟨ ⟩—NH₂

Dapsone is the basic therapeutic agent for the treatment of leprosy. It is used either alone or more frequently as a component of multidrug programs. Dapsone has been employed, primarily as a single chemotherapeutic agent, since the mid-1940s with satisfactory clinical efficacy. Both secondary and primary drug resistance have emerged and are of worldwide concern; thus, combination therapy with the antibiotic rifampin or with clofazimine is recommended. Many dapsone derivatives, all of which share many pharmacological properties, have been synthesized; however, dapsone remains the most clinically useful agent for the treatment of leprosy.

Sulfones share a mechanism of action similar to that of the sulfonamides and are bactericidal for *M. leprae* [129]. Dapsone is slowly and completely absorbed from the gastrointestinal tract with peak levels achieved in 1–3 h. The half-life varies over a broad range with a mean of 28 h. The drug is 50% bound to plasma proteins [130]. Tissue distribution is broad and reabsorption of bile-secreted drug from the gastrointestinal tract tends to extend the residence time of the drug in the circulation. The drug is excreted in the urine.

Daily therapy with 50 mg of dapsone has been successful in adults. The daily dose can be increased to 100 mg if necessary, and twice-weekly doses of 100–400 mg have been clinically successful. Therapy is gradually increased to the effective level and should continue for at least 2 years, but may be necessary for the lifetime of the patient. The most common *adverse effect* is hemolysis, generally observed in patients dosed with levels greater than 100 mg/d. Anorexia, nausea, and vomiting may result from oral administration.

Synthesis: [131].
Trade Name: Avlosulfon (Ayerst).

Sulfoxone, sodium (INN, U.S.P.) [*144-75-2*], [*144-76-3*] free acid, disodium sulfonylbis(*p*-phenyleneimino)dimethanesulfinate, $C_{14}H_{14}N_2Na_2O_6S_3$, M_r 448.43.

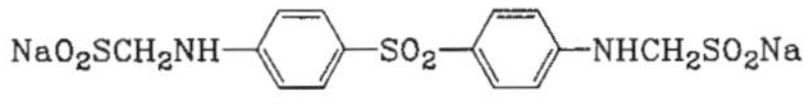

NaO$_2$SCH$_2$NH—⟨ ⟩—SO$_2$—⟨ ⟩—NHCH$_2$SO$_2$Na

Sulfoxone has been substituted for dapsone in cases in which poor gastrointestinal tolerance restricts dapsone therapy. The maximum daily oral dose of sulfoxone is 660 mg. Sulfoxone is incompletely absorbed from the gastrointestinal tract and large amounts are excreted in the feces. Generally the distribution and excretion of sulfoxone in humans after oral absorption is similar to that observed for dapsone.

Synthesis: prepared by combining 4,4′-sulfonyldianiline with sodium formaldehyde sulfoxylate in acetic acid or alcohol [132].
Trade Name: Diasone Sodium Enterab (Abbott).

Acedapsone (INN) [*77-46-3*], 4,4′-sulfonylbis(acetanilide), $C_{16}H_{16}N_2O_4S$, M_r 332.37, mp 289–292 °C, is a crystalline solid.

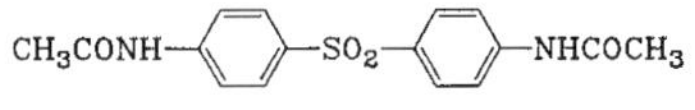

CH$_3$CONH—⟨ ⟩—SO$_2$—⟨ ⟩—NHCOCH$_3$

Acedapsone is a long-acting injectable repository derivative of dapsone. It is slowly absorbed when administered intramuscularly. Peak serum concentrations occur 3–5 weeks after administration. Acedapsone is metabolized to dapsone in the body, where it

has a serum half-life exceeding 40 d [133]. Injections administered 5 times yearly have been employed with promising results [134].
Synthesis: [135].

Clofazimine (INN) [*2030-63-9*], 3-(*p*-chloroanilino)-10-(*p*-chlorophenyl)-2,10-dihydro-2-(isopropylimino)phenazine, $C_{27}H_{22}Cl_2N_4$, M_r 473.40, mp 210–212 °C, occurs as dark red crystals.

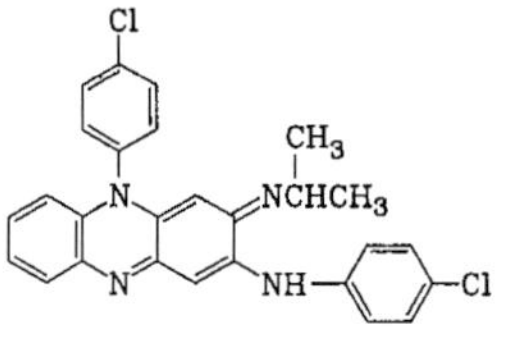

Clofazimine is a phenazine dye with weak bactericidal activity against *M. leprae* [136]. Its main use has been for treating sulfone-resistant infections and people who cannot tolerate sulfones. It also exerts an antiinflammatory effect and prevents the development of erythema nodosum leprosum. The drug is absorbed from the gastrointestinal tract and appears to accumulate in the tissues. Appreciable clinical effects are not observed until 50 days after initiation of therapy. The dose of clofazimine used is 100–300 mg spaced at 2-week intervals [137]–[139]. The major *adverse effects* are skin pigmentation and mild gastrointestinal intolerance; other effects are negligible.

Synthesis: by heating the hydrochloride of the corresponding 2-imino compound with isopropylamine at 80 °C [140].
Trade Name: Lampren (Ciba-Geigy).

2.7.5. Miscellaneous Nitroheterocycles Used to Treat Bacterial Infection

Metronidazole (INN, U.S.P.) [*443-48-1*], 2-methyl-5-nitroimidazole-1-ethanol, $C_6H_9N_3O_3$, M_r 171.16, mp 158–160 °C, occurs as cream-colored crystals.

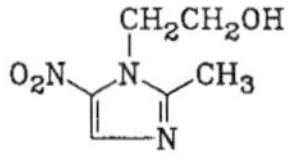

Metronidazole was originally introduced as an antitrichomonal agent and was later found to be useful in treating infections caused by several other protozoal organisms as well as by anaerobic bacteria.

Metronidazole inhibits most anaerobic bacteria at in vitro concentrations of 16 µg/mL or less; however, some organisms are less susceptible [141], [142].

This drug is useful clinically for treating the majority of anaerobic infections with the exception of actinomycosis. A major use of metronidazole is to treat infections caused by the *Bacteroides fragilis* group, the most common etiological agent of serious anaerobic infections. The emergence of resistance to clindamycin, an antibiotic useful in the treatment of anaerobic bacteria, makes metronidazole an even more important agent.

Metronidazole is absorbed rapidly after oral dosing and the serum levels obtained during the elimination phase are similar to those observed when an equivalent dose is administered intravenously. The drug is distributed widely in the body and diffuses well into all tissues including the central nervous system. Five major metabolites are found in the urine, with the hydroxy metabolite being the most important. The usual intravenous treatment regime for the susceptible anaerobic bacteria is a loading dose of 15 mg/kg followed by 7.5 mg/kg 4 times a day. The oral dose is 1–2 g/d in 2–4 doses at intervals of 6 or 12 h [143]. The maximum daily dosage is 4 g. Generally the drug is well-tolerated; however, major *adverse reactions* involving the central nervous system have been reported. In patients ingesting alcohol, metronidazole may cause reactions similar to those observed with disulfiram.

Synthesis: prepared by nitrating 2-methyl-5-nitroimidazole followed by alkylation with 2-chloroethyl alcohol [144].

Trade Names: Flagyl (Searle), Metro I.V. (American McGaw), Satric (Savage), Metronid (Asher), Metryl (Lemmon), and SK-Metronidazol (Smith Kline & French).

Nitrofurantoin (INN, U.S.P.) [67-20-9], [17140-81-7], monohydrate, 1-[(5-nitrofurfurylidene)amino]hydantoin, $C_8H_6N_4O_5$, M_r 238.16, *mp* 270–272 °C (decomp.), occurs as orange-yellow needles (dilute acetic acid).

Nitrofurantoin is used to prevent and treat urinary tract infections caused by susceptible bacteria. The drug is bacteriostatic for most susceptible organisms at 32 µg/mL or less. Nitrofurantoin is active against many strains of *E. coli*; however, most *Proteus* and *Pseudomonas* species and many of the *Enterobacter* and *Klebsiella* species are resistant [145]. Other bacteria commonly found outside the urinary tract are susceptible in vitro to nitrofurantoin, but their susceptibility is of little practical significance [146]. Microorganisms initially sensitive to nitrofurantoin generally do not become resistant during therapy.

Nitrofurantoin is absorbed rapidly and completely after oral administration with therapeutically active concentrations found only in the urinary tract. Approximately one-third of the drug is rapidly excreted in the urine by both glomerular filtration and tubular secretion, with significant reabsorption when the urine is acid [147]. The macrocrystalline form of the drug is absorbed and excreted more slowly. The average dose gives urinary concentrations of ca. 200 µg/mL. The usual adult dose for treating acute and recurrent uncomplicated urinary tract infections is 50 or 100 mg 4 times per day for 2 weeks. The most common *adverse effects* are nausea, vomiting, and diarrhea. Various hypersensitivity reactions have been reported. Other rare, but serious side effects involving various organ systems occur.

Synthesis: prepared by reaction of 5-nitrofurfural diacetate with 1-aminohydantoin in acid solution [148].
Trade Names: Furadantin (Norwich Eaton), Macrodantin (Norwich Eaton), and the sodium salt is Dantrirem (Norwich Eaton).

3. Chemotherapy of Protozoan Infections

3.1. Classification of Pathogenic Protozoa

The human parasites that are members of the phylum Protozoa are a diverse group of eukaryotic single-cell organisms. They can be divided into four principal groups:

1) flagellates (Mastigophora)
2) sporozoans (Sporozoa),
3) ciliates (Ciliata)
4) amebas (Sarcodina)

These organisms continue to be among the leading causes of infectious diseases in underdeveloped countries and are now encountered more frequently throughout the world. The ease of travel and the increased use of immunosuppressive drugs have increased the incidence of protozoal infections in the temperate climates. Table 3 lists the major parasitic protozoal human diseases and their causative agents. Excellent reviews and textbooks, which detail the epidemiology, pathogenesis, and chemotherapy of protozoal infections, are available [149]–[163].

3.2. Flagellates

3.2.1. Hemoflagellates

The hemoflagellates of humans include the genera *Trypanosoma* and *Leishmania* (Table 3). There are two distinct types of human trypanosomiasis: (1) African (or sleeping sickness), caused by *T. rhodesiense* and *T. gambiense*, and transmitted by tsetse flies; and (2) American, caused by *T. cruzi*, which is the agent of Chagas' disease and is transmitted by cone-nosed bugs. The *Leishmania* include four species and several subspecies, all of which are transmitted by sand flies.

Table 3. Major p causing human disease

Disease	Causative organism
	Flagellates
African trypanosomiasis (sleeping sickness)	*Trypanosoma gambiense* *Trypanosoma rhodesiense*
American trypanosomiasis (Chagas' disease)	*Trypanosoma cruzi*
Visceral leishmaniasis	*Leishmania donovani*
Mucocutaneous leishmaniasis	*Leishmania braziliensis*
Cutaneous leishmaniasis	*Leishmania tropica* *Leishmania mexicana*
Trichomoniasis	*Trichomonas vaginalis*
Lambliasis	*Giardia lamblia*
	Sporozoans
Malaria	*Plasmodium falciparum* *Plasmodium vivax* *Plasmodium malariae* *Plasmodium ovale*
Babesiosis	*Babesia* species
Isosporiasis (human coccidiosis)	*Isospora belli*
Toxoplasmosis	*Toxoplasma gondii*
Cryptosporidium	*Cryptosporidium* species
Pneumocystosis	*Pneumocystis carinii*
	Ciliates
Balantidiasis	*Balantidium coli*
	Amebas
Amebiasis	*Entamoeba histolytica* *Entamoeba polechi* *Dientamoeba fragilis*

Trypanosomes appear in the blood as trypanomastigotes, an elongated form with a free flagellum that is an extension of a lateral undulating membrane. Other developmental forms among the hemoflagellates include the amastigote, a leishmanial rounded intracellular stage; the promastigote (formerly called a leptomonad), an elongated flagellated extracellular form without an undulating membrane; and an epimastigote (formerly called the crithidial stage), a flagellated form with a short undulating membrane.

In *Leishmania* only the amastigote and promastigote stages are found, the latter being restricted to the insect. In African trypanosomes only the trypanomastigote is found in humans, whereas the other flagellated stages appear in the tsetse fly. In *Trypanosoma cruzi* all three stages appear in humans, whereas only the trypanomastigote and epimastigote stages are found in the insect.

3.2.1.1. African Trypanosomiasis

3.2.1.1.1. Biology and Epidemiology

The two human parasites causing African trypanosomiasis, i.e., *Trypanosoma rhodesiense* and *T. gambiense*, are believed to have evolved from *T. brucei*, a pathogen of livestock and game animals, and are considered by some to be subspecies of this animal trypanosome. The three forms are indistinguishable morphologically but differ ecologically and epidemiologically.

These parasites are introduced through the bite of the tsetse fly (*Glossina*) and the disease is generally restricted to tsetse fly areas. *Trypanosoma gambiense*, transmitted primarily by *Glossina palpalis*, extends from west to central Africa and produces a relatively chronic infection with progressive central nervous system disease. *Glossina morsitans* transmits *T. rhodesiense* in more restricted geographical areas, primarily to the south and west of Lake Tanganyika (Africa). This disease is generally more acute, leading to death in a matter of weeks to months.

After introduction of the parasite by the tsetse fly bite, a primary lesion is formed; the parasite may spread to lymph nodes and the bloodstream and in terminal stages to the central nervous system, where it produces the syndrome of sleeping sickness. Trypanosomes appear in the blood and tissues as trypanomastigotes. When trypanosomes are ingested with a blood-meal, they undergo a developmental cycle in the tsetse fly, producing metacyclic forms in the salivary glands that are reintroduced with the next blood-meal.

3.2.1.1.2. Chemotherapy

The chemotherapy of African trypanosomiasis currently centers on three key drugs: (1) pentamidine for chemoprophylaxis, (2) suramin for treatment of the early stages of the disease, and (3) melarsoprol for treatment of the late stages when trypanosomes are present in the central nervous system [164]. A polyamine biosynthesis inhibitor, α-difluoromethylornithine, recently has been reported to be effective clinically in the treatment of African trypanosomiasis [165]. Detailed reviews have been published on the chemotherapy of the African trypanosomes [166]–[172].

Drug resistance is not a serious clinical problem despite the fact that producing resistance to the available agents in the laboratory is relatively easy [166]–[169]. Suramin and pentamidine do not penetrate the central nervous system, a serious defect that limits the clinician to the more toxic arsenic compounds for treating late-stage disease. In addition lack of oral activity, adverse effects, and lack of efficacy against all stages of disease and all strains and species of pathogens limit the available agents.

Suramin (see Section 4.6.9). Suramin is a sulfated naphthylamine that is the drug of choice for the treatment of early stage hemolymphatic African trypanosomiasis. The usual adult dose is a test dose of 100–200 mg followed by 5 intravenous doses of 1 g

administered on days 1, 3, 7, 14, and 21. Initial plasma concentrations fall rapidly and the drug, tightly bound to serum proteins, remains in low concentration for up to 3 months. Suramin does not penetrate the central nervous system and, thus, does not cure infections of the central nervous system. A transient albuminuria is often observed during treatment, as are shock, febrile reactions, skin lesions, and other toxic effects.
Trade Names: Antrypol, Bayer 205, Belganyl, Fourneau 309, Germanin, Moranyl, Naganol, and Naphuride.

Pentamidine (INN, BAN, DCF, NFN) [*100-33-4*], 4,4'-(pentamethylenedioxy)dibenzamidine, $C_{19}H_{24}N_4O_2$, M_r 340.42, *mp* 186 °C (decomp.); pentamidine isethionate (M & B 800) [*140-64-7*], $C_{23}H_{36}N_4O_{10}S_2$, M_r 592.28, *mp* ca. 180 °C, occurs as very bitter crystals.

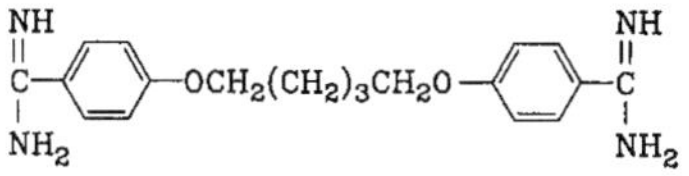

Two salts of this aromatic diamidine are in use: pentamidine isethionate and pentamidine methanesulfonate (lomidine). The doses and adverse effects of the two salts are similar and are discussed together. Treatment usually involves an intramuscular dose of 4 mg kg^{-1} d^{-1} for 10 days. For chemoprophylaxis, the usual dose is 4 mg/kg given intramuscularly every 3–6 months. Generally, this drug is more effective against the Gambian disease than the Rhodesian disease. Abdominal cramping may occur during treatment. Pentamidine can cause hypoglycemia and renal damage.

Pentamidine has been shown to be effective in the treatment of pneumonia due to *Pneumocystis carinii*, an opportunistic protozoal pathogen, and as a secondary drug for treating visceral leishmaniasis.

Synthesis: Saturating an anhydrous alcoholic solution of 4,4'-dicyanodiphenoxypentane with dry hydrogen chloride and allowing it to stand gives pentamidine [173].
Trade Names: Lomidine, dimethanesulfonate salt (Specia, France); pentamidine isethionate B.p., diisethionate salt (May & Baker, United Kingdom).

Melarsoprol (INN, BAN, DCF, NFN, MI 9) [*494-79-1*], 2-[4-(4,6-diamino-1,3,5-triazin-2-yl-amino)phenyl]-1,3,2-dithiarsolane-4-methanol, $C_{12}H_{15}AsN_6OS_2$, M_r 398.33.

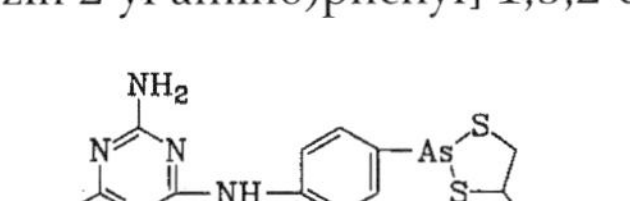

Melarsoprol and the other arsenic compounds discussed in this section are rarely used because of their extreme toxicity. The use of these arsenic compounds had been restricted to patients with overt central nervous system symptoms; however, it is now realized that invasion of the central nervous system occurs early in the course of infection and that late relapses after treatment with drugs other than the arsenic compounds are due to failure to kill organisms present in the central nervous system.

Thus, a regime of intravenous suramin followed by one or more courses of intramuscular injections of melarsoprol has been proposed to prevent relapse [155]. The recommended treatment for central nervous system infections is a daily intravenous dose of 2 – 3.6 mg/kg for 3 days followed after a week with 3.6 mg/kg for 3 days and repeated after 10 – 21 days.

Adverse effects with melarsoprol and other arsenic compounds are common and can be quite severe. These include various neuropathies, including the optic nerve, skin rashes, and a syndrome that resembles acute encephalitis and is believed to be due to a reaction that destroys parasites in the brain, which occurs in up to 10% of the patients. Treatment with arsenic compounds is also complicated by the emergence of drug-resistant strains.

Synthesis: [174].

Tryparsamide (INN; U.S.P. **17,** MI 9) [*554-72-3*], monosodium *N*-(carbamoylmethyl)arsanilate, $C_8H_{10}AsN_2NaO_4 \cdot 1/2\,H_2O$, M_r 305.10, occurs as hemihydrate, platelets, is slowly affected by light and stable to air.

$$\text{ONa} \atop \text{O=As}-\!\!\!\diagdown\!\!\!\diagup\!\!\!-\text{NHCH}_2\text{CNH}_2 \atop \text{OH}$$

Tryparsamide is a pentavalent arsenic compound that has no in vitro activity, but is most probably metabolized to the trivalent arsenic form after injection. This drug is active only against *T. gambiense,* which limits its use. It crosses the blood – brain barrier, and, thus can be used to treat late-stage trypanosomiasis; however, adverse effects are such that melarsoprol is preferred. Tryparsamide has been employed as an alternate drug of choice in tandem with suramin for treating late-stage disease with central nervous system involvement. The regime employed for tryparsamide is intravenous administration of 30 mg/kg given every 5 days for a total of 12 injections. The regime may be repeated after 1 month. Suramin should be administered intravenously at 10 mg/kg at the same dose regime employed for tryparsamide.

Synthesis: reaction of arsanilic acid with chloracetamide in the presence of sodium hydroxide and sodium carbonate [175].

Trade Names: Tryparsam, Tryparsamidium, Glyphenarsine, and Tryparsone.

Berenil [*908-54-3*], diminazene aceturate (INN, BAN, NFN), 4,4′-(1-triazene-1,3-diyl)bis(benzenecarboximidamide)bis(*N*-acetylglycinate), $C_{22}H_{29}N_9O_6$, M_r 515.54, *mp* 217 °C (decomp.), is a yellow solid; free base [*536-71-0*].

$$\text{HN}\!\!=\!\!\text{C}-\!\!\!\diagdown\!\!\!\diagup\!\!\!-\text{N=NNH}-\!\!\!\diagdown\!\!\!\diagup\!\!\!-\text{C}\!\!=\!\!\text{NH} \atop \text{H}_2\text{N}\qquad\qquad\qquad\qquad\qquad\qquad\text{NH}_2$$
$$\cdot\,2\;(\text{HOOCCH}_2\text{NHCCH}_3)$$

This aromatic diamidine, closely related to pentamidine, was developed originally as a cattle trypanocide. It has been employed effectively in an intravenous dose to treat both Gambian and Rhodesian forms of trypanosomiasis. Recently, the drug was also found to be active on oral administration, with the effective dose being 5 mg/kg taken at 2-day intervals for a total of 3 doses [164]. The drug appears to be relatively well tolerated; however, a persistent albuminuria has been observed in humans following therapy.

Synthesis: [176].

Mel W [*13355-00-5*], melarsonyl potassium (INN, BAN, DCF), melarsenoxide potassium dimercaptosuccinate, dipotassium 2{4-[(4,6-diamino-1,3,5-triazin-2-yl)amino]phenyl}-1,3,2-dithiarsolane-4,5-dicarboxylate, $C_{13}H_{11}AsK_2N_6O_4S_2$, M_r 532.51; free acid [*37526-80-0*].

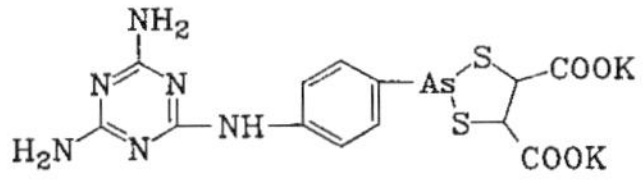

Mel W is an arsenic compound that was developed as a possible replacement for melarsoprol. Similar to other arsenic compound, it crosses the blood–brain barrier and thus can be employed to treat late-stage Gambian and Rhodesian trypanosomiasis. Although it is more water soluble than melarsoprol, Mel W has similar adverse effects to those of melarsoprol. It is almost never employed clinically because of lack of obvious efficacy and lack of safety advantages.

3.2.1.2. American Trypanosomiasis

3.2.1.2.1. Biology and Epidemiology

Trypanosoma cruzi, which causes Chagas' disease, is a parasite of humans and many small animals. This disease is found mainly in Central and South America, but several cases have been reported as far north as the southern United States. This organism is transmitted by triatomid bugs of the family Reduviidae, of which at least 36 species have been found to be naturally infected.

The trypanosomes are ingested with a blood-meal and undergo development in the intestine of the bug. They eventually give rise to infective metacyclic trypanosomes, which resemble those in vertebrate blood. Transmission usually results from contamination of mucous membranes or skin with infected insect excreta.

After the organisms gain access to the host, they usually multiply in lymphoid tissue. From there, all cells of the body may be infected, but the reticuloendothelial system, the central nervous system, and cardiac muscle are most likely involved. The digestive tract also may be infected, especially the esophagus and colon, leading to development of megaesophagus or megacolon.

The amastigote (leishmania) form is found intracellularly where it multiplies, leading to the formation of pseudocysts, which rupture and release trypanosomes capable of invading other cells or circulating in the blood. Inflammatory reactions resulting from cell damage and infected cells produce many of the disease symptoms.

Acute Chagas' disease occurs mainly in children and only in a small percentage of those people who become infected in endemic areas. Chronic disease develops slowly, with cardiac involvement observed in 20–40% of patients.

3.2.1.2.2. Chemotherapy

A great need exists for agents to treat Chagas' disease. Only two drugs have been reported to have any significant efficacy for treatment (nifurtimox and benznidazole) and none are available for chemoprophylaxis [169], [177]–[181].

Nifurtimox [*23256-30-6*], 4-[(5-nitrofurfurylidene)amino]-3-methyl-4-thiomorpholine-1,1-dioxide, $C_{10}H_{13}N_3O_5S$, M_r 287.29, *mp* 180–182 °C, occurs as orange red crystals.

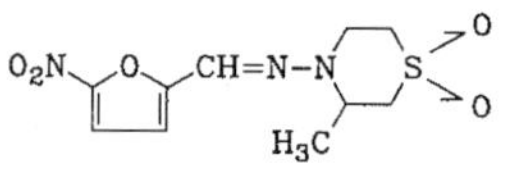

This nitrofuran has been available for 10 years for treating the acute stage of the disease. Its efficacy in chronic Chagas' disease is open to question. Apparently there are variations in response by various strains of the parasite. The usual dose is 8–10 mg kg^{-1} d^{-1} in 4 divided oral doses given for 120 days. It is considered by some to be the only available agent for the treatment of Chagas' disease. Nifurtimox is not well tolerated by patients over the 120-day recommended treatment period. *Adverse effects*, such as nausea, weight loss, and memory and sleep disorders, are common, and few patients complete the full treatment period.

Synthesis: prepared from 5-nitrofurfural and 5-amino-3-methyltetrahydro-1,4-thiazine-1,1-dioxide [182].
Trade Names: Bayer 2502, Lampit.

Benznidazole [*22994-85-0*], *N*-benzyl-2-nitroimidazole-1-acetamide, $C_{12}H_{12}N_4O_3$, M_r 260.26, *mp* 188.5–190 °C, is crystalline in form.

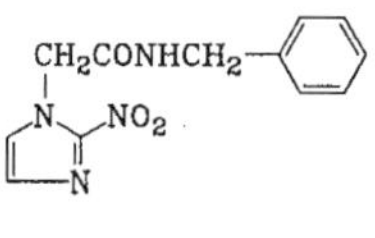

This orally administered nitroimidazole is claimed to be effective for both acute and chronic Chagas' disease. The recommended dose is 5 mg kg^{-1} d^{-1} for 60 days. No evidence has shown that this drug has any significant advantages over nifurtimox. In doses employed in the initial clinical trials in excess of 5 mg kg^{-1} d^{-1}, serious *adverse effects*, such as polyneuropathy, were observed.
Synthesis: [183].
Trade Names: R07-1051, Radanil and Rochagan.

3.2.1.3. Leishmaniasis

3.2.1.3.1. Biology and Epidemiology

Leishmania infections are caused by four species of *Leishmania* (*L. donovani, L. tropica, L. braziliensis,* and *L. mexicana*). These are obligate intracellular parasites during their amastigote (nonflagellar) stage in the human host. One species, *L. donovani*, mainly invades the internal organs whereas the others invade the skin. The Central and South American forms of leishmaniasis, caused by *L. braziliensis* and *L. mexicana* involve the mucous membranes of the nose, mouth, and pharynx, in addition to the skin. *Leishmania* are transmitted by sand flies of the genus *Phlebotomus*. The sand fly acquires the parasite with a blood-meal or from infected skin. In the sand fly, the organism is transformed into the flagellated promastigote stage, which multiplies and is introduced into the skin of a host when the fly feeds.

A broad spectrum of pathology is observed, partly dependent on the site of invasion of the specific parasite and the host's inflammatory response to the organism. The visceral form of the disease, kala-azar, caused by *L. donovani*, produces a progressive disease mainly of the reticuloendothelial system (spleen, liver, and bone marrow). The disease usually results in death if untreated. The cutaneous forms vary from self-healing sores to diffuse progressive disfiguring lesions that can lead to a disseminated disease in hosts with defective cellular immunity.

3.2.1.3.2. Chemotherapy

Two groups of compounds are used to treat visceral and cutaneous forms of leishmaniasis: pentavalent organic antimony compounds and aromatic diamidines. The antifungal antibiotic, amphotericin B, has been shown to be effective in selected situations.

Pentavalent antimony compounds have been used to treat kala-azar since the 1920s. Their introduction was preceded, for a few years, by the use of trivalent antimony compounds, especially tartar emetic. With correct use, cure rates exceeding 90% have been obtained with pentavalent antimony compounds [184]. Response to therapy for visceral leishmaniasis varies in different geographical locations, with more resistance to therapy being observed in the Sudan than in India.

Pentamidine, a drug used for treating African trypanosomiasis, has been employed for the therapy of visceral leishmaniasis after initial treatment with pentavalent antimony compounds has failed. Although this compound, which now is the only readily available diamidine for use in humans, is effective, the patients may have a high relapse rate. The recommended dose of pentamidine is 2–4 mg/kg intramuscular or slow intravenous injections administered weekly or, at most, 3 times weekly, until a clinical and parasitological cure is achieved, usually involving many months of therapy.

Sodium stibogluconate (INN, BAN, DCF) [*16037-91-5*], antimony(V) derivative of sodium gluconate, 2′,4′-O-(oxydistibylidyne)bis(D-gluconic acid)-2,4-Sb,Sb′-dioxide trisodium salt · nonahydrate, $C_{12}H_{17}Na_3O_{17}Sb_2 \cdot 9\,H_2O$, M_r 907.6, water-soluble amorphous powder.

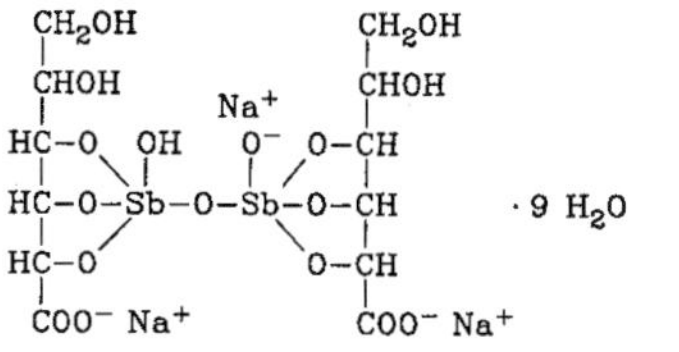

[Structural formula]

Sodium stibogluconate is the drug of first choice for the treatment of all forms of leishmaniasis. Some evidence suggests that the antiprotozoal activity of this compound may depend on its reduction to the trivalent antimony compound after treatment. The drug can be administered by either the intravenous or the intramuscular routes.

A large percentage of patients are cured with a single course of therapy consisting of 20 mg kg^{-1} d^{-1} up to a maximum of 800 mg/d. Sodium stibogluconate is tolerated relatively well; however, a disturbing aspect of therapy is the occurrence of sudden death, although it is difficult to determine whether the deaths are related to the clinical disease, the drug therapy, or the interaction of disease and drug.

Synthesis: heating gluconic acid with freshly made antimonic acid paste until it is completely dissolved and then neutralized with sodium hydroxide.

Trade Name: Pentostam (Burroughs Wellcome, United Kingdom).

Glucantime [*133-51-7*], meglumine antimonate, 1-deoxy-1-(methylamino)-D-glucitol trioxoantimonate, $C_7H_{18}NO_8Sb$, M_r 365.91.

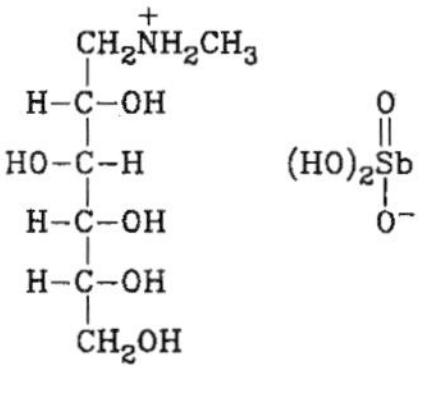

[Structural formula]

Glucantime is a pentavalent antimony compound related to pentostam. The former drug has no obvious advantage over the latter one and is used interchangeably with pentostam as the drug of choice in some countries. It can be administered intramuscularly or intravenously. The usual dose is 10 or 20 mL of a 30% solution on alternate days for a total of 200–250 mL. As has been observed with pentostam, viable organisms can be recovered from treated patients both during and even after therapy.

Trade Names: Protosib, 2168-RP (Rhône-Poulenc).

3.2.2. Intestinal and Urogenital Flagellates

Although the human alimentary and urogenital tracts are colonized by seven species of flagellate protozoa, only two, *Trichomonas vaginalis* and *Giardia lamblia*, are generally considered to be pathogens. Reviews covering broad biological data for these two protozoans are available [185] – [191].

3.2.2.1. Trichomonas Vaginalis

3.2.2.1.1. Biology and Epidemiology

Of the three species of *Trichomonas* that infect humans, only *T. vaginalis* is pathogenic, causing trichomoniasis. *Trichomonas vaginalis* is a common pathogen in the female genitourinary tract, where it exists only as the trophozoite form. In the female, the infection is normally limited to the vulva, vagina, and cervix. In the male, where infection occurs less often, the prostate, seminal vesicles, and urethra may be involved. Signs and symptoms in the female include profuse vaginal discharge in addition to local tenderness, vulval pruritus, and burning. About 10% of infected males have a thick, white urethral discharge.

Trichomonas infection is typically transmitted during sexual intercouse, although nonvenereal routes of transmission cannot be excluded. Infection rates vary greatly, but may be quite high in some populations, e.g., 38 – 56% of symptomatic women attending a venereal disease clinic. Assessing the prevalence of trichomoniasis in men is difficult because most infections in men are asymptomatic. An estimated 3×10^6 American women contract trichomoniasis every year.

3.2.2.1.2. Chemotherapy

Successful treatment of vaginal infections requires the destruction of the trichomonads. Numerous topical preparations are available for treatment; however, they suffer from failure to completely eliminate the parasite in female infections and cannot be applied to infected males. Systemically active drugs are, therefore, the normal treatment of choice. Treatment is necessary both in symptomatic patients and simultaneously in asymptomatic sexual consorts to prevent reinfection or spread of infection.

The effective agents for the systemic treatment of trichomoniasis are all related to the 5-nitroimidazole, metronidazole (see Section 2.7.5). The drugs include tinidazole, nimorazole, ornidazole, secnidazole, and carnidazole. None of these compounds are approved for use in the treatment of trichomoniasis in the United States, and not all are available in every country. No clear clinical advantage has been shown for any of these 5-nitroimidazoles over the others. Cure rates for trichomoniasis are ca. 85 – 95%, with symptoms usually relieved within a few days. The original recommended regimes for the various compounds for therapy called for oral doses of up to 1 g/d in divided doses for 4 – 10 days; e.g., the recommended dose for metronidazole is 250 mg given 3 times each day for 7 days. Patients now generally receive a shortened regime or a single large oral dose of 2 g. A single dose is as efficacious as multiple-dose treatment and may be superior with some imidazoles. Treatment failures with the 5-imidazoles may be associated with poor serum and local tissue levels or resistant isolates. Cross resistance with metronidazole and the other nitroimidazoles, has been reported. Generally, the 5-nitroimidazoles are free of acute side effects. The most common *adverse effects* are gastrointestinal, and these generally are transient and mild. This group of compounds is mutagenic in bacteria, and tumors have been observed in rodents fed high doses of metronidazole for long periods.

Metronidazole: see Section 2.7.5.

Tinidazole (INN) [*19387-91-8*], 1-[2-(ethylsulfonyl)ethyl]-2-methyl-5-nitroimidazole, $C_8H_{13}N_3O_4S$, M_r 247.27, *mp* 127–128 °C, occurs as colorless crystals (benzene).

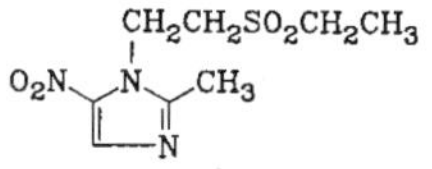

Synthesis: heating 2-methyl-5-nitroimidazole with 2-ethylsulfonylethyl-4-toluenesulfonate for 4 h [192].
Trade Names: Fasigyn (Pfizer, USA), Simplotan (Pfizer, FRG).

Nimorazole (INN, BAN, NFN, MI) [*6506-37-2*], 4-[2-(5-nitroimidazolyl-1-yl)ethyl]-morpholine, $C_9H_{14}N_4O_3$, M_r 226.23, *mp* 110–111 °C, occurs as crystals (water).

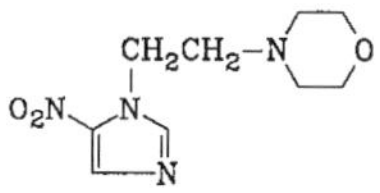

Synthesis: condensing the sodium salt of 5-nitroimidazole with 4-(2-chloroethyl)morpholine in boiling acetone [193].
Trade Name: Naxogin (Carlo Erba, Italy).

Ornidazole (INN) [*16773-42-5*], α-(chloromethyl)-2-methyl-5-nitro-1*H*-imidazole-1-ethanol, $C_7H_{10}ClN_3O_3$, M_r 219.63, *mp* 77–78 °C, is crystalline in form.

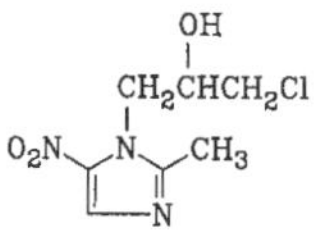

Synthesis: [194].

Secnidazole (INN, DCF, NFN, BAN) [*3366-95-8*], α-2-dimethyl-5-nitro-1*H*-imidazole-1-ethanol, $C_7H_{11}N_3O_3$, M_r 185.18, *mp* 76 °C, occurs as crystals.

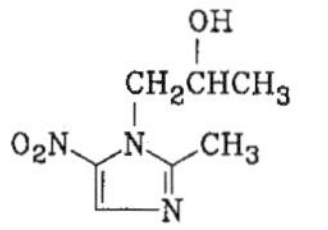

Synthesis: [195].
Code names: PM 185184, RP 14539.

Carnidazole (INN) [*42116-76-7*], *O*-methyl [2-(2-methyl-5-nitro-1*H*-imidazol-1-yl)ethyl]thiocarbamate, $C_8H_{12}N_4O_3S$, M_r 244.27.

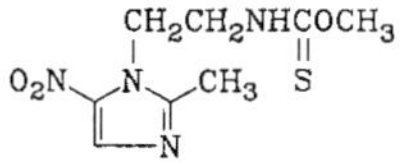

Synthesis: [196].
Code names: R28096 (as hydrochloride), R25831 (as the free base).

3.2.2.2. Giardia Lamblia

3.2.2.2.1. Biology and Epidemiology

Giardia lamblia is a flagellated protozoan residing in humans in the upper portion of the small intestine. The trophozoite form of the parasite is converted to the infective cyst stage on its passage to the colon. Ingestion of cysts and the subsequent formation of motile trophozoite forms in the upper small intestine complete the life cycle. This parasite, which was long considered a harmless commensal, is now regarded as the most common intestinal parasite in the United States and United Kingdom.

Endemic infections with *Giardia* are found in every country of the world, but the incidence varies from one country to another. Giardiasis occurs more commonly in children. Transmission is either by direct fecal–oral contamination or by indirect transfer of cysts in food or water. Diarrhea is the most frequent symptom. The disease may be acute or chronic. Anorexia, abdominal cramps, bloating, and weight loss are common symptoms. Most infected individuals do not present evidence of the disease.

3.2.2.2.2. Chemotherapy

Several reviews describe in detail the treatment regimes for *Giardia* infection [191], [197], [198]. Although there is not general agreement as to the optimal chemotherapeutic agent, all three of the following chemical compounds have been reported to give cure rates $\geq 90\%$ with oral administration: metronidazole (and other 5-nitroimidazoles), quinacrine, and furazolidone. In the United States, quinacrine is the most commonly used drug, whereas in many other countries the 5-nitroimidazoles are preferred. Furazolidone is not widely used except as a suspension for children. This agent is generally considered less efficacious than the nitroimidazoles.

Metronidazole has been used to treat giardiasis since 1961, but the original dose regime has been modified. The recommended oral dose for metronidazole is 250 mg 3 times a day for 5 days, although shortened dosing regimes similar to those employed in the treatment of trichomoniasis have been reported to be successful. Other 5-nitroimidazoles that are employed include tinidazole, nimorazole, and ornidazole. Similar to metronidazole, these compounds are generally used with good clinical success in the same dose regimes as have been employed for treating trichomoniasis.

Metronidazole: see Section 3.2.2.1.1.

Tinidazole: see Section 3.2.2.1.2.

Nimorazole: see Section 3.2.2.1.2.

Ornidazole: see Section 3.2.2.1.2.

Quinacrine hydrochloride (U.S.P.) [*6151-30-0*], 6-chloro-9-{[4-(diethylamino)-1-methylbutyl]-amino}-2-methoxyacridine dihydrochloride dihydrate, $C_{23}H_{30}ClN_3O \cdot 2\,HCl \cdot 2\,H_2O$, M_r 508.91, mp 248–250 °C (*mp* poorly discernable), is a bitter, bright yellow crystal; anhydrous quinacrine [*83-89-6*], atabrine, mepacrine.

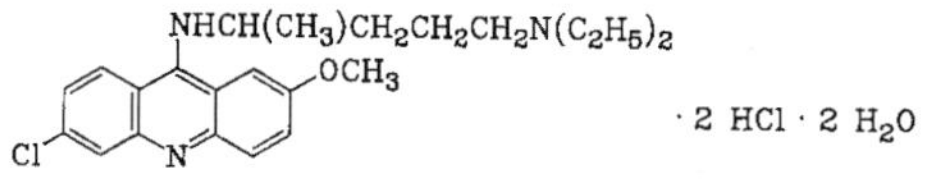

Quinacrine is an acridine derivative previously employed to treat malaria and tapeworm infections. The usual adult dose is 100 mg 3 time a day given for 5 days. The pediatric dose is 2 mg/kg 3 times a day for 5 days with a maximum of 300 mg/d. Gastrointestinal *adverse effects* as well as malaise and headache, occur with administration of this compound. Quinacrine is excreted in the urine, imparting a deep yellow color. Approximately 4–5% of the patients develop yellow skin staining at the dose used for giardiasis. Quinacrine can induce red blood cell hemolysis in G-6-PD deficient patients. Prolonged high-dose therapy with quinacrine has caused rare instances of retinopathy similar to that observed with chloroquine.

Synthesis: condensing 1-(diethylamino)-4-aminopentane with 3,9-dichloro-7-methoxyacridine [199].

Furazolidone (INN, BAN, U.S.P.) [*67-45-8*], furoxone, 3-[(5-nitrofurfurylidene)-amino]-2-oxazolidinone, $C_8H_7N_3O_5$, M_r 225.16, mp 275 °C (decomp.), occurs as yellow crystals.

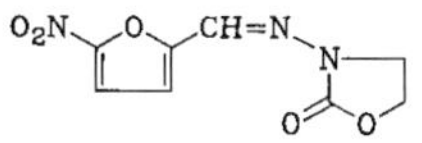

Furazolidone is a nitrofuran that is active against *G. lamblia* and various gram-negative bacteria, including *Salmonella, Shigella,* and *Vibrio cholerae.* The recommended adult oral dose is 100 mg 4 times a day for 7–10 days. The pediatric dose is 1.25 mg/kg 4 times a day for 7–10 days. Compared with other agents used for treating *Giardia* infections, furazolidone is somewhat less active. Furazolidone treatment can result in gastrointestinal and central nervous system *adverse effects.* Furazolidone turns the urine brown and can cause red blood cell hemolysis and mild, reversible anemia in individuals with G-6-PD deficiency. Furazolidone, similar to other nitrofurans, has carcinogenic potential. Adenocarcinoma of the lung in mice and of the mammary gland in female rats has been reported with long-term administration of furazolidone.

Synthesis: [200].
Trade Names: Furox (Smith Kline Beckman), Furoxone (Norwich Eaton).

3.3. Sporozoans

Sporozoans have a complex life cycle that often involves two hosts (e.g., arthropod and human). The Coccidia, a subclass of essentially intestinal Sporozoa, and the hemosporidians, which include the malaria parasite, are animal and human parasites. *Toxoplasma,* a parasite of cats, is a common human parasite. *Babesia,* a tick-borne protozoan and a common animal parasite, is a rare human pathogen (these organisms are no longer considered sporozoans, but will be covered in this section). *Pneumocystis carinii* and *Cryptosporidium* have emerged as important human pathogens, especially in immunocompromised patients.

3.3.1. Plasmodia

3.3.1.1. Biology and Epidemiology

Malaria infections are distributed widely in countries of Africa, Asia, and Latin America, with an estimated worldwide prevalence of 100×10^6 cases associated with ca. 1×10^6 deaths each year. At least five species of *Plasmodium* can infect humans: *P. vivax, P. ovale, P. malaria, P. falciparum,* and *P. knowlesi* (only in Malaysia). In addition, at least two species of nonhuman primate plasmodia are transmissible to humans experimentally. Diagnosis of infection is made on the basis of morphological characteristics of specific species. The identification of the infecting organism is important in determination of the chemotherapeutic approach. Equally important in the therapy and prophylaxis of infection is an understanding of the life cycle and course of infection of the malaria parasites.

Transmission to humans is by the bloodsucking bite of various species of mosquitoes of the genus *Anopheles,* in which the sexual or sporogonic cycle of development occurs. The asexual cycle (schizogany) takes place in humans. The mosquito introduces the infective sporozoites, which quickly invade the liver parenchymal cells (the preerythrocytic cycle). After further development, numerous asexual progeny (the merozoites) enter the bloodstream and invade the erythrocytes. Multiplication in the red blood cell is characteristic for each species, resulting in synchronous destruction of host cells (the erythrocytic cycle). Successive production of merozoites occurs every 48 h (*P. vivax, P. ovale,* and *P. falciparum*) or at 72-h intervals (*P. malariae*). The *P. falciparum* infections are generally confined to the red blood cells after the first liver cycle; thus, untreated infections will terminate spontaneously, usually in 6–8 months, or end fatally. The other three species continue to multiply in liver cells, and persistence of the parasite in these cells long after the parasites have disappeared from the bloodstream is observed. During the erythrocytic cycle, certain merozoites become differentiated as male or female gametocytes. Thus, the sexual cycle begins in humans, but for continuation of the cycle, the gametocytes must be ingested by mosquitoes. Development time of the various stages in the mosquito's stomach wall ranges from 8 to 14 days, depending on the malaria species. The infective stage is the sporozoite, which migrates to the salivary gland of the mosquito and is injected with a blood-meal.

The *pathogenic mechanisms* resulting in clinical illness in humans can be divided into four processes: (1) fever, (2) anemia, (3) tissue hypoxia, and (4) immunopathologic events. Cyclic fever and its physiological consequences involve rupture of erythrocytes and release of schizonts. Anemia is a common complication of malaria and mainly results from the rupture of red blood cells during schizogony, but other factors, such as autoimmune mechanisms, may contribute to the anemia. Tissue hypoxia, resulting from alterations in microcirculation and anemia, may cause serious complications including renal failure, pulmonary edema, and cerebral dysfunction in *P. falciparum* infections. Immunologic response to malaria infection can result in clinical disease. Immune-complex glomerulonephritis and greatly enlarged spleens are two examples of this phenomenon.

3.3.1.2. Chemotherapy

Antimalarial drugs can be divided into two groups on the basis of their mechanism of action: (1) The aminoquinolines, such as chloroquine, apparently exert their effect by intercalation into parasite DNA. This is not the only mechanism of action of these compounds, because mefloquine does not intercalate DNA. (2) The other group of compounds inhibits the synthesis of folic acid from 4-aminobenzoic acid. This group includes chloroguanide, pyrimethamine, and their derivatives, as well as the sulfonamides and sulfones. Quinine, a natural product and a mainstay of therapy for more than 60 years, still remains an important chemotherapeutic agent, especially for treating malaria parasites that are resistant to the newer synthetic agents.

Antimalarial drugs have been grouped in at least six different categories according to use, including (1) causal prophylaxis, (2) suppressive treatment, (3) clinical cure or treatment of the acute attack, (4) radical cure, (5) suppressive cure, and (6) gametocidal therapy [201]. True *causal prophylactic agents* should be capable of killing sporozoites prior to their entry into red blood cells. Although no such agent is available, several agents, e.g., primaquine and chloroguanide, have activity against the preerythrocytic stages of *P. falciparum. Suppressive therapy* inhibits development of the erythrocytic stages, thus preventing clinical symptoms. Chloroquine, chloroguanide, and pyrimethamine are suppressive; however, insensitivity or drug resistance in malarial strains in certain localities has created problems. A *clinical cure* can be achieved by agents that interrupt the development of the intracellular schizont in the red blood cell. The 4-aminoquinoline derivatives, chloroquine and amodiaquin, are the major drugs in this category, although the slower-acting chloroguanide and pyrimethamine are also highly active schizonticides. *Radical cure* involves elimination of both erythrocytic and exoerythrocytic parasites. Vivax malaria can be treated with the 8-aminoquinoline derivatives, of which only primaquine is used currently. Radical cure of falciparum malaria is relatively easy to achieve by continuation of treatment. A *suppressive cure* involves complete elimination of the malaria parasites by treatment that exceeds the life span of the parasite. *Drugs that kill gametocytes* directly are not available, but chloroguanide and pyrimethamine prevent the development of gametocytes in mosquitoes.

Drug resistance to all of the synthetic agents used in the treatment and prophylaxis of malaria has been observed in many areas of the world. *Plasmodium falciparum* is resistant to amodiaquine and chloroquine and to the combination of sulfadoxine and pyrimethamine. This therapeutic problem extends to South America, 13 African nations, south of the Equator, and to the South Asian continent, from Pakistan to eastern India. When used alone, quinine has lost some of its activity against falciparum malaria in those areas where it has been used indiscriminately or with poor compliance. Treatment of the other malarial species with 4-aminoquinoline derivatives, such as amodiaquine and chloroquine, still remains effective.

Chloroquine (INN, U.S.P.) [*54-05-7*], 7-chloro-4-(4-diethylamino-1-methylbutyl-amino)quinoline; $C_{18}H_{26}ClN_3$, M_r 319.88, *mp* 87 °C; diphosphate, *mp* 193 – 195 °C or 215 – 218 °C (two modifications); sulfate, *mp* ca. 207 °C.

Cl—[quinoline ring]—N
   |
   HNCH(CH$_2$)$_3$N(C$_2$H$_5$)$_2$
   |
   CH$_3$

Chloroquine, a 4-aminoquinoline, is highly effective against the asexual erythrocytic forms of *P. vivax* and *P. falciparum*, and gametocytes of *P. vivax*. In human vivax malarias, chloroquine has no prophylactic or radically curative value; however, it is effective in terminating acute attacks of vivax malaria. When administered continuously for long periods, it acts as a suppressive agent. It is highly effective in controlling acute attacks of falciparum malaria and generally cures the disease. In certain parts of the world, drug-resistant strains of *P. falciparum* limit the use of chloroquine.

Chloroquine is well absorbed when administered orally. It is highly bound to tissues, especially liver, spleen, kidney, and lung; about 55% of the drug is bound to plasma proteins. The drug is excreted slowly in the urine, with ca. 70% being recovered as parent compound. Because of high tissue binding, a loading dose is necessary to achieve adequate plasma concentration. After a single dose, the half-life of the drug in plasma is 3 days, with a longer half-life achieved on multiple doses. Chloroquine at the dose employed for prophylaxis of acute malarial attacks causes relatively few *adverse effects*, mainly pruritus and gastrointestinal discomfort.

The oral dose for suppression of sensitive malarial strains in adults is 300 mg of the free base weekly, starting 2 weeks prior to and while present in a malarious area and for 6 weeks after leaving the area. For oral treatment of acute malaria in adults, 600 mg of base is given initially, followed by 300 mg 6 h later and 300 mg at 24 and 48 h. Primaquine should be given after chloroquine doses when treating malaria due to *P. vivax* and *P. ovale* or for radical cure following exposure.

Chloroquine is of use in treating extraintestinal amebiasis, and is also reported to be of some value in the treatment of *Giardia* and *Babesia* infections.

Synthesis: condensation of 4,7-dichloroquinoline with 1-(diethylamino)-4-aminopentane [202].

Trade Names: Aralen hydrochloride, Aralen phosphate (Sterling Winthrop), Resochin (diphosphate, Bayer), Nivaquine (sulfate, Specia, France).

Amodiaquine (INN, U.S.P.) [*86-42-0*], 4-[(7-chloro-4-quinolyl)amino]-α-(diethylamino)-*o*-cresol, $C_{20}H_{22}ClN_3O$, M_r 358.87, *mp* 208 °C, is crystalline in form; dihydrochloride dihydrate, *mp* 150–160 °C (decomp.), is a yellow, bitter crystal.

Amodiaquine, a congener of chloroquine, is claimed to have superior activity against some strains of *P. falciparum* that are partially resistant to chloroquine. It is used for treating overt malaria attacks and for suppression. The observed *adverse effects* and the frequency of these effects are similar to chloroquine, i.e., diarrhea, vomiting, and vertigo. Dosing also is similar to that employed for chloroquine. For suppressive therapy, the unit dose is 400 mg of the free base (520 mg of dihydrochloride). For treatment of acute attacks, 600 mg of the base is given initially with subsequent daily doses of 400 mg for 2 days.

Synthesis: Condensation of 4,7-dichloroquinoline and 4-acetamido-α-diethylamino-*o*-cresol gives amodiaquin dihydrochloride dihydrate [203].

Trade Name: Camoquin hydrochloride (Parke-Davis).

Hydroxychloroquine sulfate (U.S.P.), hydroxyquinine (INN) [*747-36-4*], 2-({4-[(7-chloro-4-quinolyl)amino]pentyl}ethylamino)ethanolsulfate (1:1 salt), $C_{18}H_{26}ClN_3O \cdot H_2SO_4$, M_r 439.95, *mp* (usual form) ca. 240 °C, *mp* (other form) ca. 198 °C, is a white crystalline odorless powder with a bitter taste; hydroxychloroquine [*118-42-3*].

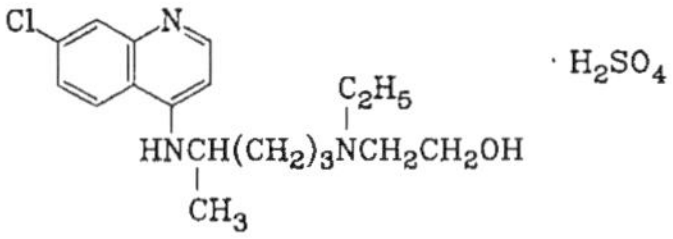

Hydroxychloroquine is an *N*-ethyl-*β*-hydroxylated chloroquine. This compound has an activity and safety profile that is similar to chloroquine. A dose of 400 mg of hydroxychloroquine sulfate is equivalent to 500 mg of chloroquine phosphate.

Synthesis: prepared by reacting a mixture of 4,7-dichloroquinoline, phenol, and *N'*-ethyl-*N'*-*β*-hydroxyethyl-1,4-pentadiamine at 125 – 130 °C [204].

Trade Name: Plaquenil sulfate (Sterling-Winthrop, USA).

Primaquine (INN) [*90-34-6*]; primaquine phosphate (U.S.P.) [*63-45-6*], 8-[(4-amino-1-methylbutyl)amino]-6-methoxyquinoline phosphate (1:2 salt), $C_{15}H_{21}N_3O \cdot 2\ H_3PO_4$, M_r 455.34, *mp* 197 – 198 °C, occurs as yellow crystals (90% ethanol).

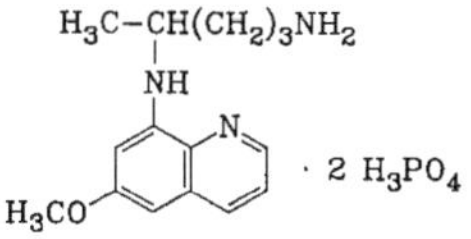

Primaquine is an α-aminoquinoline used for prevention of malarial relapses and for radical cure of *P. vivax* and *P. ovale* malaria by acting on the exoerythrocytic stage of the parasite. It also destroys the gametocytes of these species. It has no significant activity against the asexual blood forms of *P. falciparum*.

Primaquine, similar to all of the other α-aminoquinolines, is absorbed rapidly when administered orally and is rapidly metabolized. Plasma levels peak at 6 h. Minor *adverse effects* include nausea, abdominal discomfort, and headache. Severe hemolytic reactions in people with a glucose-6-phosphate dehydrogenase deficiency of the type found in those of Mediterranean ancestry have been observed. Most blacks have this enzyme deficiency, but it is usually confined to older erythrocytes and thus the hemolytic reaction is less severe.

Primaquine is supplied as the phosphate salt with 26.3 mg being equivalent to 15 mg of free base. For prevention of attack after departure from areas where *P. vivax* and *P. ovale* are endemic, a daily oral dose of 15 mg of base along with the last 2 weeks of chloroquine prophylaxis is recommended. For prevention of relapses to vivax and ovale malaria, a daily dose of 15 mg/d for 14 days or a weekly dose of 45 mg of base for 8 weeks is employed.

Synthesis: 6-Methoxy-8-aminoquinoline is condensed with 1-phthalimido-4-bromopentane, and the phthalyl group is cleaved by heating with hydrazine in ethanol and hydrochloric acid [205].

Pyrimethamine (INN, U.S.P.) [*58-14-0*], 2,4-diamino-5-(4-chlorophenyl)-6-ethylpyrimidine, $C_{12}H_{13}ClN_4$, M_r 248.71, *mp* 233–234 °C, is crystalline in form.

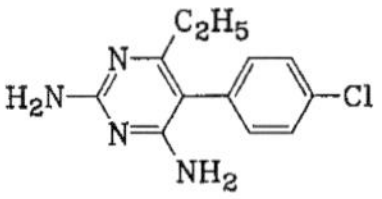

Pyrimethamine is a 2,4-diaminopyrimidine that has high affinity for the enzyme dihydrofolate reductase from the malaria parasite and thus interferes with folate synthesis. Trimethoprim, a related compound with good antibacterial activity, also has antimalarial activity. Pyrimethamine by itself has little value in treating a primary attack of malaria. However, combined with sulfadoxine (Fansidar), malaria has been prevented in subjects in areas where there is drug-resistant falciparum malaria. The combination oral product of 25 mg of pyrimethamine and 500 mg of sulfadoxine is given once weekly during exposure. It is also employed in combination with quinine sulfate for treatment of chloroquine-resistant falciparum malaria. The dose employed for the combination is 650 mg of quinine 3 times daily and 25 mg of pyrimethamine twice daily for 3 days. Pyrimethamine is well absorbed when administered orally. It is eliminated slowly with a plasma half-life of 4 days. The dose of 25 mg weekly produces few adverse effects. Excessive doses may result in a reversible megaloblastic anemia resembling folic acid deficiency.

Synthesis: prepared by condensing 3-isobutoxy-2-(4-chlorophenyl)pent-2-enonitrile with guanidine nitrate in the presence of sodium methylate [206].
Trade Name: Daraprim (Burroughs Wellcome).

Chloroguanide [*500-92-5*]; chloroguanide hydrochloride (U.S.P. **4**) [*637-32-1*], proguanil (INN), 1-(4-chlorophenyl)-5-isopropylbiguanide hydrochloride, $C_{11}H_{16}ClN_5 \cdot HCl$, M_r 290.20, *mp* 243–244 °C, is a white powder with a bitter taste.

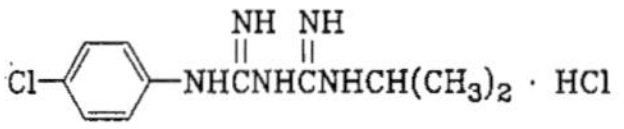

Chloroguanide is used to treat overt clinical vivax and falciparum malaria; however, response to treatment is slower than that observed with most other antimalarial agents. Chloroguanide is not recommended for treatment of an acute attack of falciparum malaria. Although it is of use in acute vivax malaria, it offers no advantages over other available agents. It is a causal prophylactic, suppressive, and radical cure agent in falciparum malaria. Chloroguanide is active against developing preerythrocytic stages of some malarias and is reported to sterilize gametocytes. Resistance to chloroguanide greatly compromises its usefulness.

Chloroguanide is converted metabolically to a triazine derivative that inhibits the enzyme dihydrofolate reductase. It is slowly absorbed from the gastrointestinal tract with peak serum concentrations achieved at 2–4 h after oral administration. About 50% of the drug is excreted in the urine and 10% directly into the feces. The prophylactic dose is 100–200 mg daily for non-immune subjects. Less drug (300 mg weekly) is employed for partly immune individuals. For treatment of acute attack of vivax malaria, an initial dose of 300–600 mg is followed by a daily dose of 300 mg, usually for 5–10 days. Chloroguanide is a relatively safe drug at the doses employed for malaria treatment. Large daily doses of 1 g may cause vomiting, abdominal pain, and diarrhea.

Synthesis: prepared by heating 4-chlorophenyldicyandiamide with isopropylamine [207].

Trade Name: Paludrine (ICI, United Kingdom).

Mefloquine (INN) [53230-10-7], (DL-*erythro*-α-2-piperidyl-2,8-bis(trifluoromethyl)-4-quinolinemethanol, $C_{17}H_{16}F_6N_2O$, M_r 378.32, *mp* 178–178.5 °C; mefloquine · HCl [51773-92-3], *mp* 259–260 °C.

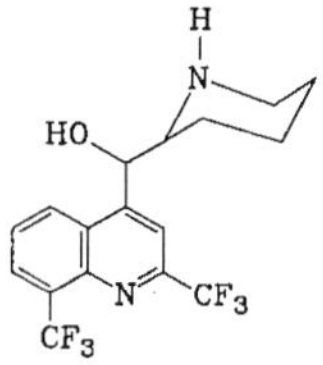

Mefloquine is a 4-quinolinemethanol that is currently well along into clinical trials, especially in Asia and South America, where multiple drug-resistant *P. falciparum* is a major problem [208]–[209]. This compound has an unusually long half-life in humans (17 days) and may be efficacious for use as a chemoprophylactic when given weekly. It has proved highly effective in curing infections caused by *P. falciparum* when administered orally. Liver concentrations are high and prolonged, and the drug is excreted mainly in the feces. A quinate salt is available with a markedly increased water solubility and an improved pharmacokinetic profile.

Synthesis: [210].

Trade Name: Mefloquine Quinate, $C_{17}H_{16}F_6N_{20} \cdot C_7H_{12}O_6$ (Smith Kline Beckman).

Halofantrine [69756-53-2]; [±]-halofantrine [66051-63-6]; halofantrine hydrochloride (INN) [36167-63-2], 1,3-dichloro-α-[2-(dibutylamino)-ethyl]-6-(trifluoromethyl)-9-phenanthrenemethanol hydrochloride, $C_{26}H_{30}Cl_2F_3NO \cdot HCl$, M_r 536.89, *mp* 93–96 °C, *mp* 203–204 °C, occurs as crystals.

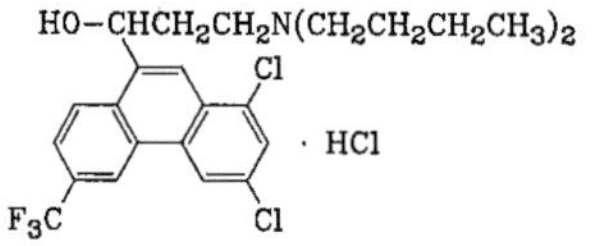

HO−CHCH$_2$CH$_2$N(CH$_2$CH$_2$CH$_2$CH$_3$)$_2$

[Structure of Halofantrine with Cl, Cl, F$_3$C substituents on phenanthrene] · HCl

Halofantrine is a 9-phenanthrene methanol currently under development for the treatment of drug-resistant falciparum malaria. It is curative with minimal adverse effects in nonimmune subjects infected with Vietnam or Cambodian strains of *P. falciparum* or Chesson strain of *P. vivax*. The oral dose employed in these studies was 250 mg every 6 h for 12 doses [211]. The compound is rapidly but incompletely absorbed when administered orally.

Synthesis: [212].
Trade Name: Halofantrine-β-glycerophosphate, $C_{26}H_{30}Cl_2F_3NO \cdot C_3H_9O_6P$ (Smith Kline Beckman).

3.3.2. Babesia

Babesia are intraerythrocytic protozoan parasites that are transmitted by ticks. They are mainly parasites of domestic and wild animals, although on occasion they can be transmitted to humans. Infections tend to be self-limiting and are characterized by fever, hemolytic anemia, and hemoglobinuria. No effective therapy exists for these infections.

3.3.3. Isosporiasis

The sporozoan order Coccidia contains a number of parasites that invade the intestinal mucosa at one stage in their life cycle. *Isospora belli,* usually acquired by fecal contamination of food or water, is a member of this group. Although a rare human pathogen, *T. belli* can cause serious gastrointestinal disease. Treatment is usually with antifolates, such as pyrimethamine and sulfadiazine. The drug of choice is trimethoprim–sulfamethoxazole. The usual oral dose of this combination is 160 mg of trimethoprim and 800 mg of sulfamethoxazole four times daily for 10 days then twice a day for 3 weeks. *Isospora hominis,* now called *Sarcocystis hominis,* is a related parasite that can cause intestinal infection in humans. This organism, similar to toxoplasma, has an asexual stage in the muscles of many mammals. Human infection is acquired from eating improperly cooked beef or pork containing *Sarcocystis*. Treatment is similar to that employed for *I. belli*.

3.3.4. Toxoplasmosis

Toxoplasma gondii is a sporozoan of the order Coccidia. The parasite is an obligate intracellular organism that exists in three forms including the tachyzoites (formerly trophozoites), tissue cysts, and oocysts. *Toxoplasma gondii* is a ubiquitous parasite with a worldwide distribution and a capability of infecting a wide range of animals and birds. The final hosts are members of the cat family. Infection is by ingestion of either tissue cysts or oocysts, releasing viable organisms that invade the epithelial cells of the intestine, where they undergo an asexual cycle and a sexual cycle in the cat. Many oocysts are shed in the feces and sporulate outside the cat.

Ingestion of primarily lamb or pork containing tissue cysts or other food products contaminated with oocysts is the major means of transmission. Active infection during gestation can result in congenital infection. Toxoplasma infection can be acute or chronic, symptomatic or asymptomatic. Most infected adults show no symptoms. Chronic infections can be latent and later exacerbate in immunocompromised individuals, resulting in severe disease, such as encephalitis, myocarditis, and pneumonitis. Congenital infection may lead to subsequent disease, such as impaired vision, hearing loss, or neurologic disorders.

The most effective treatment for toxoplasmosis is a combination of pyrimethamine and sulfadiazine. Although the tachyzoites are susceptible to these agents, the tissue cyst is resistant to available agents. A loading regime of pyrimethamine at 100 mg $kg^{-1} d^{-1}$ twice a day for 2 days followed by 25 mg every other day is the recommended treatment for adults with significant infection. A synergistic sulfa drug, such as sulfadiazine, sulfamethazine, or sulfamerazine, is usually given in combination with pyrimethamine. The antibiotics spiramycin and clindamycin or trimethoprim–sulfamethoxazole are less active than the combination of pyrimethamine and a sulfa drug.

3.3.5. Cryptosporidium

This coccidian protozoa is a significant cause of death in immunocompromised patients. The parasite, long known in domestic animals, is the same organism that causes disease in humans. Infection is acquired by ingestion of oocysts, which excyst and release sporozoites that invade and replicate in the intestinal microvilli. Male and female gametes are produced, initiating the sexual cycle. Oocysts are then formed and can be passed in the feces or can release sporozoites to cause autoinfection. In individuals with normal host defenses, the organism may not cause symptoms or may cause a transient diarrhea and gastrointestinal disease that is generally self-limiting. The disease is more serious in immunocompromised individuals and is a major contribution to death in patients with acquired immune deficiency syndrome (AIDS). No effective chemotherapy exists for cryptosporidiosis.

3.3.6. *Pneumocystis Carinii*

This protozoan causes infection in the immunocompromised host and is the most common pathogen of AIDS patients, accounting for 43% of reported opportunistic diseases. The disease is usually limited to the lungs, where it causes a diffuse pneumonitis. The parasite has been found in the lungs of a wide variety of animals and is distributed globally. Subclinical infections in normal individuals are probably common. The untreated infection in the immunocompromised patient is 90–100% fatal. The drug of choice for treating pneumocystis infection is the combination of trimethoprim and sulfamethoxazole. The recommended dose is 20 mg kg^{-1} d^{-1} of trimethoprim with 100 mg kg^{-1} d^{-1} of sulfamethoxazole given orally or intravenously in 4 doses for 14 days. This combination therapy results in a high incidence of rash, neutropenia, fever, and diarrhea in AIDS patients. Pentamidine isethionate is an alternative choice, but is associated with significant adverse effects. The recommended dose is 4 mg kg^{-1} d^{-1} administered intramuscularly for 12–13 days.

3.4. Ciliates

Infections in humans caused by ciliated protozoans are rare. Only one parasite, *Balantidium coli*, causes disease in humans. The ciliated form of this parasite penetrates the mucosa of the colon and multiplies in the submucosal tissues. This results in colitis, with mucus and blood in the feces. The parasite can form cysts when conditions are not ideal for penetration and it is the cyst that allows for transmission of this disease, usually from animals (swine) to humans. The disease is treated with the antibiotics tetracycline or paromomycin or with metronidazole (750 mg twice a day for 5 days).

3.5. Amebas

3.5.1. Biology and Epidemiology

The disease amebiasis is worldwide in distribution and is almost always associated with poor sanitary conditions. The etiological agent, *Entamoeba histolytica*, occurs in three stages: (1) the inactive cyst, (2) the intermediate precyst, and (3) the ameboid trophozoite, which is the only stage found in the tissues. The ameboid form is also found in liquid feces during amebic dysentery. Multiplication among trophozoites occurs by binary fission. After ingestion of infective cysts in food or water contaminated with feces, the cysts are activated in the stomach, and development takes place during passage to the large intestine.

A population of lumen-dwelling trophozoites, capable of invading the intestinal epithelium, emerge from the cyst. Most infections are without symptoms, although when tissue invasion occurs (in ca. 10% of the cases), there is disease associated with the infection. Asymptomatic infected persons harbor lumen-dwelling amebas that produce cysts, which are passed in the feces. Diseased individuals usually pass trophozoites and cysts in the feces. Extraintestinal amebiasis is observed in ca. 4% or more of clinical infections, usually taking the form of amebic hepatitis or liver abscess. Abscesses may also occur rarely in other areas, including the lungs, brain, and spleen.

Entamoeba histolytica must be distinguished from four other amebas that are also intestinal parasites of humans. These include the common ameba *Ent. coli; Dientamoeba fragilis,* the only intestinal ameba other than *Ent. histolytica* suspected of causing diarrhea, but not of invading the tissues; *Iodamoeba butschlii;* and *Endolimax nana.*

Primary amebic meningoencephalitis has been reported in less than 100 cases with amebas living free in soil and in water. Two genera of ameba, *Naegleria* and *Acanthamoeba,* have been associated with pathogenicity. Most cases of infection have developed in children who were swimming in contaminated outdoor pools. *Naegleria* meningoencephalitis infection is rapidly fatal and does not respond to the available amebicides. A few patients have been reported to respond to intravenous and intrathecally administered amphotericin B.

3.5.2. Chemotherapy

The approach to treating amebiasis is in part based on the location of the amebic organism and includes treatment of the asymptomatic individual who is passing cysts. The criterion for cure in intestinal amebiasis is the elimination of the organism. This is usually achieved with tetracycline and an 8-hydroxyquinoline derivative or diloxanide furoate. The antibiotic paromomycin also may be employed. For those infections involving only the bowel wall and resulting in acute amebic dysentery, a variety of agents are employed. Symptomatic amebiasis involving the intestine can be treated with a nitroimidazole, such as metronidazole or tinidazole, a course of therapy usually resulting in 90% cure. Dehydroemetine, a semisynthetic derivative of the alkaloid emetine, can be used for the rapid relief of symptoms in severely ill patients. As an alternative to metronidazole, combination therapy of dehydroemetine with tetracycline or paromomycin has been employed. For extraintestinal infections, use of metronidazole is recommended, usually in combination with diiodohydroxyquin or diloxanide furoate, to prevent continued intraluminal infection. In seriously ill patients with complicated amebic infections, parenteral emetine in combination with iodoquinol may be employed.

8-Hydroxyquinolines. A number of 8-hydroxyquinolines are available for treating amebiasis. Iodoquinol, the most widely used drug, is described in this chapter; however, other derivatives have been applied in specific areas of the world. These include clioquinol (iodochlorhydroxyquin), broxyquinoline, chlorquinaldol, and chiniofon.

Iodoquinol [*83-73-8*], diiodohydroxyquinoline (INN, U.S.P.), 5,7-diiodo-8-quinolinol, $C_9H_5I_2NO$, M_r 396.95, *mp* 200 – 215 °C (extensive decomp.), occurs as crystals or yellowish brown powder.

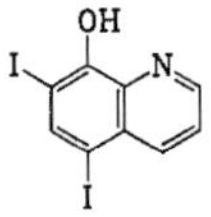

Iodoquinol is an 8-hydroxyquinoline that is directly amebicidal. It is active only on amebas in the intestinal tract and, like the other members of this class of compounds, is ineffective against amebic abscess and hepatitis. The drug is partly absorbed when it is administered orally, with ca. 25 % of the drug recovered in the urine as the glucuronide. Although the 8-hydroxyquinolines were originally thought to be of low toxicity, a number of *toxic reactions* are now known to result from their use, the most significant being a subacute myelooptic neuropathy, which is particularly observed with clioquinol. The usual dose of iodoquinol for treating asymptomatic carriers of *Ent. histolytica* is 650 mg 3 times a day for 20 days. This same dose may be employed in combination therapy with metronidazole in treating frank intestinal disease or in hepatic abscess, where it serves to limit reoccurrence of the intestinal forms.

Synthesis: prepared by the action of iodine monochloride on 8-hydroxyquinoline [213].
Trade Name: Yodoxin (Glenwood).

Clioquinol [*130-26-7*], iodochlorhydroxyquin (INN, U.S.P., BAN), 5-chloro-8-hydroxy-7-iodoquinoline, C_9H_5ClINO, M_r 305.50, *mp* ca. 178 – 179 °C (decomp.), is a brownish-yellow bulky powder.

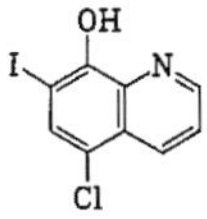

Synthesis: [214].
Trade Names: Vioform (Ciba-Geigy), Rheaform, veterinary (Squibb).

Broxyquinoline (INN, DCF, MI) [*521-74-4*], 5,7-dibromo-8-hydroxyquinoline, $C_9H_5Br_2NO$, M_r 302.95, *mp* 196 °C, occurs as monoclinic needles.

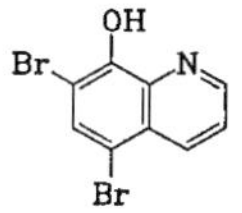

[Structure: 5,7-dibromo-8-hydroxyquinoline]

Synthesis: prepared by bromination of 8-quinolinol [215].

Chiniofon (INN, DCF, NFN, NF **5**) [*8002-90-2*], mixture of four parts (by mass) 8-hydroxy-7-iodo-5-quinolinesulfonic acid and one part sodium hydrogencarbonate.

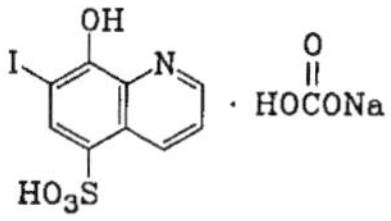

[Structure: 8-hydroxy-7-iodo-5-quinolinesulfonic acid · HOCONa]

Synthesis: of 8-hydroxy-7-iodo-5-quinolinesulfonic acid [216].

Diloxanide furoate [*3736-81-0*], 4-[(dichloroacetyl)methylamino]phenyl-2-furancarboxylate, $C_{14}H_{11}Cl_2NO_4$, M_r 328.15, *mp* 112.5 – 114 °C.

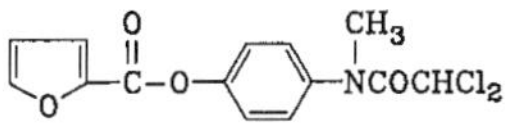

[Structure: diloxanide furoate]

The furoate ester of diloxanide is one of the agents of choice in the treatment of persons who are asymptomatic passers of cysts or for invasive and extraintestinal amebiasis (administered with other appropriate drugs). The recommended oral dose is 500 mg 3 times daily given for 10 days; a second course may be necessary. When administered orally, the ester is hydrolyzed in the lumen or mucosa of the intestine, resulting in diloxanide and furoic acid. Most of the oral dose is excreted in the urine within 48 h, and peak drug concentrations appear in the plasma in 1 h. There are few *adverse effects* reported with this compound, the most common being flatulence or mild abdominal discomfort.

Diloxanide (INN, BAN, DCF, NFN, MI) [*579-38-4*], 2,2-dichloro-4′-hydroxy-*N*-methylacetanilide, $C_9H_9Cl_2NO_2$, M_r 234.08, *mp* 175 °C, is crystalline in form.

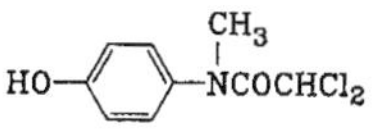

[Structure: diloxanide]

Synthesis: Diloxanide is prepared from 3,3-dianisyl-4-hexanone [217]. Diloxanide furoate is obtained by reacting diloxanide with 2-furoyl chloride in pyridine.

4. Chemotherapy of Viral Infections

4.1. Physical and Biological Characteristics

Obligate Intracellular Parasites. Viruses are among the smallest of all life forms. Unlike bacteria, viruses are obligate intracellular parasites that are metabolically inert in the extracellular state. Even though viruses depend to varying degrees on the host cell's metabolic machinery, the differences existing between the metabolic processes of the host cell and those specified by viruses have been exploited in the search for antiviral drugs that selectively inhibit virus replication.

Nucleic Acid. Unlike other microorganisms, viruses contain only one form (double-stranded or single-stranded) of only one type (RNA or DNA) of nucleic acid as their genetic material. The polarity of the genetic material determines whether transcriptase enzymes (enzymes not found in eukaryotic cells) must be present within virions. For example, a negative-stranded RNA genome is complementary to messenger RNA (mRNA) and must be transcribed by a virion transcriptase enzyme to produce viral mRNA necessary for initiation of the infection process. Other virion transcriptases, such as the reverse transcriptase present in retroviruses, catalyze the formation of DNA from viral RNA. The unique reverse transcriptase enzyme of the human T-cell lymphotrophic virus type III, lymphadenopathy associated virus, AIDS-related virus, or human immunodeficiency virus (HIV), which are various names for the virus implicated in AIDS (acquired immune deficiency syndrome), is the target of several antiviral drugs.

Viral Proteins. The functions of the structural virion proteins are to protect the genetic material from degradation and also to determine what types of host cells can be infected by a particular virus (host range). The latter function is determined by envelope proteins, which may be glycosylated, because these proteins are the first to contact the host cell. In viruses that lack an envelope, capsid proteins protect the genetic material and determine host range. The capsid is the protein coat that is complexed with the nucleic acid. Nonstructural proteins, such as the virion transcriptases, are also present within the virion.

Lipid Envelope. The lipids present in enveloped viruses resemble, in composition, those lipids of the host-cell membranes. Certain viral proteins may be embedded in the lipid bilayer of virions.

Laboratory Diagnosis of Viral Infections. Rapid detection of viral infections is critical for rational chemotherapeutic intervention [218]. Currently, detection methods include serological tests, as well as direct examination of clinical specimens for virus. Serological tests measure the rise in serum antibody late in the disease course (convalescent) vs. serum antibody levels found early in the course of the disease (acute). The presence of antibodies specific for a particular virus may be difficult to assess if antibodies to related viruses are present or if the virus infection represents reactivation or reinfection of the host with the same virus. The immunoglobulin antibody of class IgM, which is produced early in infections, can indicate the presence of active virus replication. This IgM antibody can be detected by using IgM capture techniques in which serum interacts with anti-IgM on an immobilized solid surface. Other serological techniques use enzyme- (peroxidase), radioactive-, or fluorescent-labeled antibodies.

Direct observation using an electron microscope can rapidly detect some virus infections. This technique works well for examining fecal contents in acute gastroenteritis caused by rotaviruses, caliciviruses, astroviruses, and the Norwalk agent, in which sufficiently large numbers of morphologically distinct virions are present. In these cases, the viruses do not grow well in cell culture.

The classic method, and still the standard, is to isolate and identify viruses by growing them from clinical specimens in a cell culture. Several different cell cultures must be used because no one culture can support the replication of all viruses. Preliminary virus identification can be obtained by examination of infected cells for the cytopathic effect (CPE) that is characteristic for a virus group. In the absence of a characteristic CPE, serological tests are used for final identification of viruses.

Recently, many viruses have been cloned by using recombinant DNA techniques. Probes produced from these cloned virus DNAs can be used to detect certain viruses in clinical samples. Eventually, nonradioactive probes that use biotin or fluorescent tags may be widely used in clinical laboratories.

4.2. Classification of Viruses

The International Committee on Taxonomy of Viruses (ICTV) published a report in 1982 [219] that is summarized in Table 4. A total of 16 of the 55 virus families recognized by the ICTV contain viruses that are human pathogens. The distinguishing characteristics are the type of nucleic acid, the presence or absence of an envelope, the replication strategy of the genome, whether it is a positive-, negative-, or ambisense genome, and genome segmentation.

Table 4. Virus families

Virus family	Distinguishing characteristics *	Human pathogens *
Poxviridae	DS DNA, enveloped	smallpox, vaccinia
Herpesviridae	DS DNA, enveloped	HSV, VZV, CMV, EBV
Adenoviridae	DS DNA, nonenveloped	human adenoviruses
Papovaviridae	DS DNA, nonenveloped	papilloma, BK and JC
Hepadnaviridae **	DS DNA, nonenveloped	hepatitis B virus
Parvoviridae	SS DNA, nonenveloped	parvoviruses, Norwalk (?)
Reoviridae	DS RNA, nonenveloped	reoviruses, rotaviruses
Togaviridae	SS RNA, enveloped, no DNA step, (+) sense	EEE, WEE, VEE, rubella, yellow fever, dengue
Coronaviridae	SS RNA, enveloped, no DNA step, (+) sense	coronaviruses
Paramyxoviridae	SS RNA, enveloped, no DNA step, (−) sense, nonsegmented	parainfluenza, measles, mumps, RSV
Rhabdoviridae	SS RNA, enveloped, no DNA step, (−) sense, nonsegmented	VSV, rabies
Filoviridae **	SS RNA, enveloped, no DNA step, (−) sense, nonsegmented	Marburg, Ebola
Orthomyxoviridae	SS RNA, enveloped, no DNA step, (−) sense, segmented	influenza A and B
Bunyaviridae	SS RNA, enveloped, no DNA step, segmented, ambisense	Bunyamwera, Rift Valley fever
Arenaviridae	SS RNA, enveloped, no DNA step, segmented, ambisense	Lassa, Machupo, Junin, LCMV
Retroviridae	SS RNA, enveloped, DNA step	human T-lymphotrophic viruses
Picornaviridae	SS RNA, nonenveloped	poliovirus, Coxsackievirus, echovirus, hepatitis A, rhinovirus
Caliciviridae	SS RNA, nonenveloped	Norwalk (?), caliciviruses

* DS = double-stranded; SS = single-stranded; HSV = herpes simplex virus; VZV = varicella–zoster virus; CMV = cytomegalovirus; EBV = Epstein–Barr virus; EEE = eastern equine encephalitis virus; WEE = western equine encephalitis virus; VEE = Venezuelan equine encephalitis virus; RSV = respiratory syncytial virus; VSV = vesicular stomatitis virus; LCMV = lymphocytic choriomeningitis virus; (+) sense = genomes containing nucleotide sequences translated by ribosomes; (−) sense = genomes composed of nucleotide sequences complementary to (+) sense; ambisense = genome composed of (−) sense covalently attached to (+) sense.
** Not yet approved by ICTV.

4.3. Assessment of Antiviral Activity in Cell Culture

In vitro cell culture systems for determining the antiviral effect of compounds have been useful, even though in vitro activity does not always predict in vivo or clinical efficacy. Plaque reduction [220], [221], yield reduction, cell growth, and macromolecular synthesis assays are some of the methods currently used to quantify the effects of antiviral candidates on virus replication and host cell metabolism. In the plaque reduction assay, the concentration of compound that reduces the number of plaques to a defined end point (usually 50 % or 90 %) is determined. In the yield reduction assay, the concentration of compound that reduces the yield of infectious virus by 50 % or 90 % is determined. In the cell growth or macromolecular synthesis assay, the effect of a compound on the increase in cell number or the incorporation of radioactive precursors into DNA, RNA, and protein, respectively, is determined. A compound is a viral-specific inhibitor if the antiviral concentration in the plaque reduction or virus yield assay does not affect cell growth or macromolecular synthesis.

4.4. Animal Models of Virus Infection

In addition to the in vitro cell culture systems described in Section 4.3, animal models of human virus infections are used to assess the systemic efficacy of candidate antiviral compounds. A compound that is a potent inhibitor of virus replication in a cell culture is commonly completely ineffective in preventing disease when administered to a virus-infected animal. These results are usually explained by poor absorption or rapid metabolism of the compound, resulting in a concentration of the compound that is too low to inhibit virus replication in the critical tissues.

Results in animal models of human viral infections often differ significantly from the natural course of the human infection, even when the identical virus is used. For example, mice infected intracerebrally with a mouse-adapted human poliovirus develop flaccid limb paralysis and succumb to the infection within 24 h of the onset of paralysis [222]. The severity of the poliovirus infection in mice is greater than that usually observed in humans, presumably because the route of inoculation in mice does not allow the animal to develop a protective immune response prior to the onset of paralysis. In humans, poliovirus replicates first in the Peyer's patches in the gut, which stimulates an immune response prior to the spread of virus to the central nervous system.

Mouse models of herpes simplex virus (HSV) infections are used to assess the efficacy of antiherpetic compounds; however, the disease in mice is more severe than the acute infection in humans. Mice generally succumb to encephalitis following intraperitoneal or intravaginal inoculation of virus. The severity of the disease in the animal model used in evaluating an antiviral compound should be considered when assessing potency and projecting dose levels in humans.

4.5. Rationale for Chemotherapy of Viral Infections

The rationale for the discovery of specific chemotherapeutic agents effective against viruses is based on the knowledge that certain virus functions are distinct from cellular functions (for reviews of antiviral chemotherapeutics, see [223]). A number of unique virus-coded enzymes are known to be critical for virus replication. These enzymes (e.g., deoxypyrimidine kinase, also known as thymidine kinase, and ribonucleotide reductase from HSV) have different substrate specificities compared to cellular enzymes and, therefore, can be used as targets for antiviral chemotherapy. The antiherpetic drug, acyclovir (see Section 4.6.1), is an excellent example of the exploitation of the differences between the viral and cellular enzyme. Acyclovir is activated by deoxypyrimidine kinase to eventually form acyclo-GTP, which inhibits the viral DNA polymerase. This process does not occur to a significant extent in uninfected cells. A second example is the reverse transcriptase enzyme (from the retroviruses implicated in AIDS), which presumably does not have a normal cell counterpart. This enzyme has been the target of several agents that specifically inhibit the replication of the AIDS viruses.

Virus replication can also be inhibited by agents, such as rhodanine or arildone, that specifically bind with high affinity to the picornaviral capsid proteins and stabilize virion capsid conformation. This stabilization prevents cell-induced uncoating of the virion nucleic acid, thereby halting the virus infection at the uncoating step.

The host's immune response to a viral infection works in concert with antiviral drugs to clear the infection. The virus infection of a host can be viewed as a race between replication of the virus and the mounting immune response to the virus. If the extent of virus replication and resultant tissue destruction exceeds a threshold value, disease occurs. If the extent of virus replication is limited by the immune response or by chemotherapeutic intervention, then the disease course will be milder or asymptomatic. Thus, the role of antiviral chemotherapy in the treatment of viral disease is to inhibit virus replication sufficiently to enable the host's immune system to overcome the virus infection.

4.6. Chemotherapeutic Agents

4.6.1. Nucleoside Analogues

Acyclovir (INN) [59277-89-3], 9-[(2-hydroxyethoxy)methyl]guanine, $C_8H_{11}N_5O_3 \cdot 1/2\,H_2O$, M_r 234.2, mp > 200 °C.

The antiherpetic activity of acyclovir, the acyclic analogue of guanosine, was first reported along with its synthesis in 1978 [224]. The rationale for this synthesis was the previous observation that the acyclic analogue of adenosine was a substrate for adenosine deaminase [225]. Acyclovir has potent activity against several herpesviruses (herpes simplex virus types 1 and 2, Epstein–Barr virus, and varicella–zoster virus), but is not as potent against human cytomegalovirus.

Acyclovir is clinically effective in the treatment of herpesvirus infections in humans. Acyclovir was first reported to be clinically efficacious in the topical treatment of herpetic keratitis [226]. Acyclovir was subsequently shown to be as effective as vidarabine [227] or as trifluorothymidine [228] against herpetic keratitis. Topical acyclovir cream was found to be effective against herpes labialis [229] and primary genital herpes [230], but was not effective against recurrent genital herpes [231]. Unlike the topical cream, orally administered acyclovir did reduce the duration of viral shedding, as well as the time to healing in recurrent genital herpes. Oral acyclovir, however, had no significant effect on symptoms of recurrent genital herpes [232]–[234]. Recently, treatment protocols using daily administration of prophylactic oral acyclovir for 4-month periods have been described for patients with frequent recurrences [235], [236]. During treatment with acyclovir, the patients had significantly fewer recurrences than those receiving placebo. However, after treatment with acyclovir was stopped, the patients' recurrence rates returned to pretreatment frequencies. While on treatment with acyclovir, many patients reported the symptoms of a prodrome that did not progress to a complete lesion [237].

Adverse Effects. Several problems exist with the chronic use of acyclovir in patients. The major question is one of long-term safety. First, because acyclovir is a nucleoside analogue, there is concern it may be mutagenic. Thus far, however, acyclovir has not been shown to cause cancer in laboratory animals. Furthermore, it is not known whether long-term prophylactic use of acyclovir is more likely to result in development of resistant viruses. Other minor concerns with acyclovir include reversible toxicity to the bone marrow, reversible kidney damage due to acyclovir crystals, and diarrhea. Because only 15–20 % of acyclovir is absorbed after oral administration, a prodrug of acyclovir activated by xanthine oxidase has recently been developed [238].

Synthesis: Reaction of 2,6-dichloro-9-(2-benzoyloxyethoxymethyl) purine with methanolic ammonia, followed by treatment with nitrous acid and then with methanolic ammonia yields acyclovir [224].

Trade Name: Zovirax (Burroughs Wellcome).

Idoxuidine [*54-42-2*], IDU (INN, U.S.P.), 2′-deoxy-5-iodouridine, $C_9H_{11}IN_2O_5$, M_r 354.12, *mp* 160 °C (decomp.).

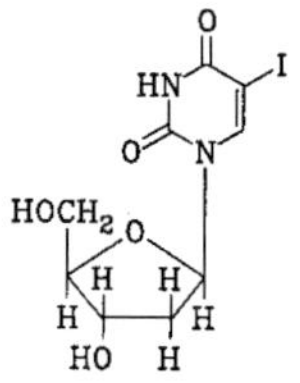

Idoxuidine was the first clinically effective nucleoside analogue. It was synthesized in 1959 [239] as part of a program aimed at developing synthetic nucleosides as DNA synthesis inhibitors to treat cancer. In effect, antiviral chemotherapy became a reality in 1962 when IDU was reported to cure herpes simplex keratitis in humans [240]. Although IDU was effective topically against herpes simplex keratitis, it was not active when administered systemically.

Adverse Effects. Because of the insolubility and toxicity of IDU, other drugs, such as trifluorothymidine, are currently used for the topical treatment of ocular herpetic infections [241].

Synthesis: obtained by refluxing uracildeoxyriboside, iodine, chloroform, and HNO_3 [242].

Trade Names: Stoxil (Smith Kline).

Trifluridine (INN) [*70-00-8*], 5′-trifluoromethyl-2′-deoxyuridine, trifluorothymidine, TFT, $C_{10}H_{11}F_3N_2O_5$, M_r 296.2, *mp* 186–189 °C.

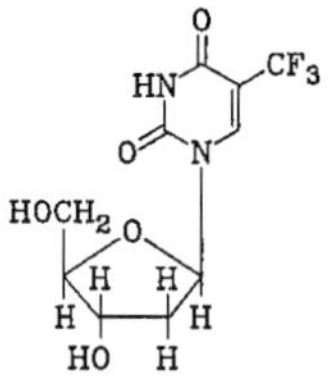

Trifluridine was first synthesized in 1964 [243]. Its use is limited to the topical treatment of herpetic keratitis. It was more effective than IDU in patients with dendritic or ameboid herpetic keratitis [244].

The *antiviral activity* and the *mechanisms of action* of TFT have been reviewed [245], [246]. It is converted to the active triphosphate form by cellular and viral thymidine kinases. The triphosphate form of TFT is then preferentially incorporated into viral DNA, which results in inhibition of the transcription of late viral mRNAs.

Adverse Effects. Because TFT can also be incorporated into cellular DNA, TFT is toxic to uninfected cells; thus, TFT cannot be used systemically.

Synthesis: prepared by treatment of 5-trifluoromethyluracil with a bacterial enzyme [243].

Trade Name: Viroptic (Burroughs Wellcome).

Vidarabine [*24356-66-9*], ara-A (INN, U.S.P.), 9-β-D-arabinofuranosyladenine, $C_{10}H_{13}N_5O_4 \cdot 4\,H_2O$, M_r 339.36, *mp* 257 °C.

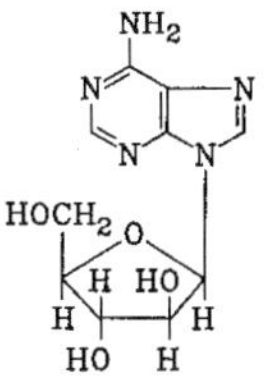

Ara-A was initially synthesized in the 1960s as a potential anticancer agent [247], [248]. Ara-A was reported to inhibit herpesvirus, vaccinia virus, cytomegalovirus, and varicella–zoster viruses [249], [250]. Topical ara-A was as effective against herpetic keratitis as IDU [251]. Intravenous ara-A was also effective against herpes zoster in immunosuppressed patients [252]. Ara-A was sufficiently nontoxic to uninfected cells to permit systemic administration to patients with herpes encephalitis. Ara-A reduced mortality from 70 to 28 % in a study of 28 patients with herpes encephalitis [253]. Recently, acyclovir was shown to be superior to ara-A for herpes encephalitis [254]. Topical ara-A and ara-A monophosphate have no effect in patients with genital or labial herpes [255], [256].

Mechanism of Action. Clearly, ara-A has multiple sites of inhibition of HSV. Ara-A is phosphorylated to give ara-A triphosphate (ara-ATP). Ara-ATP is a selective inhibitor of ribonucleoside diphosphate reductase and HSV DNA polymerase, as well as of the addition of poly-A to viral mRNA. Ara-A also slows viral DNA elongation, inhibits terminal deoxynucleotidyl transferase, and inhibits *S*-adenosylmethionine-dependent methylation (capping) of viral mRNA [257].

Synthesis: Ara-A is prepared by treating 9-(3′,5′-*O*-isopropylidene-β-D-xylofuranosyl)-adenine with methanesulfonyl chloride. The crystalline 9-(3′,5′-*O*-isopropylidene-2′-*O*-methanesulfonyl-β-D-xylofuranosyl)adenine is exposed to 90 % aqueous acetic acid at 100 °C for 5 h. The epoxide is formed with the use of methanolic sodium hydroxide. The epoxide is converted to the arabinoside by reaction with sodium benzoate or sodium acetate in 95 % aqueous *N*,*N*-dimethylformamide [248].

Trade Name: Vira-A (Parke-Davis).

Bromovinyldeoxuridine [*73110-56-2*], BVDU, (*E*)-5-(2-bromoethenyl)-2′-deoxyuridine, $C_{11}H_{13}BrN_2O_5$, M_r 331.6.

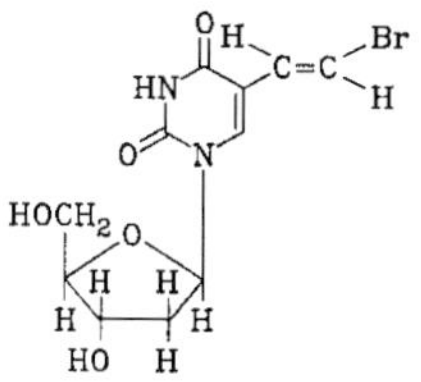

The antiherpetic activity of BVDU was first reported in 1979 [258]. This pyrimidine derivative is about 100 times more active against HSV−1 in cell culture than against HSV−2 and is also active against varicella−zoster virus.

The clinical efficacy of BVDU has been demonstrated in several uncontrolled (open) trials. First, because BVDU was well-absorbed when given orally, the progression of varicella−zoster virus in immunocompromised patients was retarded by BVDU [259]. Second, topical administration of BVDU was effective against herpetic keratitis [260]. Placebo-controlled, double-blinded clinical trials have been initiated.

The *mechanism of action* of BVDU is similar to that of other selective nucleoside analogues having antiherpetic activity. It is converted to the triphosphate by the viral and cellular thymidine kinases. The 5′-triphosphate is then incorporated into viral DNA in virally infected cells.

Adverse Effects. There are several problems with the use of BVDU. First, BVDU is rapidly degraded by thymidine phosphorylase. Second, herpesviruses can readily develop resistance to BVDU by lowering the virus content of thymidine kinase activity. Finally, the development of BVDU by G. D. Searle has been halted because of increased incidence of tumors in animals dosed for long periods [261].

Synthesis: Condensation of the trimethylsilyl derivative of (*E*)-5-(2-bromovinyl)uracil with 2′-deoxy-3,5-di-*O*-*p*-toluoyl-α-D-erythropentofuranosyl chloride gives a mixture of the α- and β-anomers of the protected nucleoside. The *p*-toluoyl protecting groups are removed by treatment with sodium methoxide in methanol. The β-anomer, but not the α-anomer, is biologically active [258].

FIAC [*69124-05-6*], 1-(2′-fluoro-2′-deoxy-β′-D-arabinofuranosyl)-5-iodocytosine hydrochloride, $C_9H_{11}FIN_3O_4 \cdot HCl$, M_r 407.58, *mp* 177−181 °C.

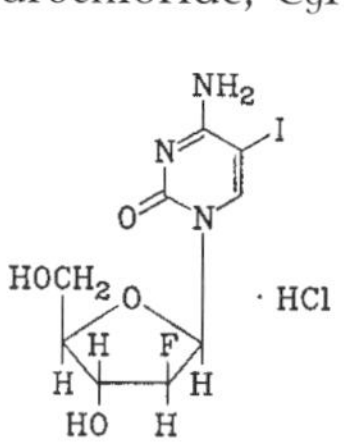

The synthesis and antiherpetic activity of a series of 2′-fluoro-5-substituted arabinofuranosylcytosines and uracils have been recently reported [262], [263]. Several compounds in this series are potent inhibitors of HSV−1, HSV−2, varicella−zoster virus, cytomegalovirus, and Epstein−Barr virus. For example, 1-(2′-fluoro-2′-deoxy-β-D-arabinofuranosyl)-5-iodocytosine (FIAC) was active systemically after i.v. administration to immunocompromised patients infected with varicella–zoster virus [264].

The mode of action of these 2'-fluoro-5-substituted nucleoside analogues is similar to that of acyclovir. The analogues are substrates for the herpesvirus-coded thymidine kinase (deoxypyrimidine kinase) [265]. Viral DNA polymerase is more sensitive to inhibition by the 5'-triphosphate of FIAC than is the host cell DNA polymerase [266].

A related compound, 1-(2'-fluoro-2'-deoxy-β'-D-arabinofuranosyl)-5-iodouracil (FIAU), has recently been reported to have some toxic effects in humans. FIAU caused cardiac fibrosis at a dose of 150 mg/kg in a long-term study [267].

Synthesis: FIAC is prepared by condensation of 3-O-acetyl-5-O-benzoyl-2'-deoxy-2'-fluoro-D-arabinofuranosyl bromide with trimethylsilylated cytosines to yield the blocked nucleosides. After deprotection by saponification to the 2'-fluoroarabinofuranosylcytosine nucleoside, the 5-iodo analogue (FIAC) is obtained by iodination [262].

Ribavirin [*36791-04-5*], Virazole (INN, British Approved Name – BAN), 1-β-D-ribofuranosyl-1 H-1,2,4-triazole-3-carboxamide, $C_8H_{12}N_4O_5$, M_r 244.2, mp 166–168 °C.

Ribavirin was initially described in 1972 as the result of a program to synthesize broad spectrum antiviral agents [268]. Ribavirin is active in cell culture against most DNA and RNA viruses [269].

Ribavirin probably has several *mechanisms of action*. Ribavirin is phosphorylated by cellular enzymes, and both the monophosphate (RMP) and the triphosphate (RTP) have antiviral activity. The monophosphate inhibits synthesis of guanosine 5'-monophosphate (GMP) [270]. Thus, the supply of guanosine triphosphate is depleted and nucleic acid synthesis is inhibited. In addition RTP inhibits the virus-specific mRNA capping enzymes of vaccinia virus [271]. The ability of RTP to inhibit the capping of viral-specific mRNA could explain how ribavirin could be active against both DNA and RNA viruses. An interesting finding is that ribavirin was active in cell culture against the virus which causes AIDS [272].

Clinical trials with ribavirin have been performed with respiratory syncytial virus (RSV), Lassa virus, and influenza viruses. Aerosolized ribavirin was effective in reducing fever and symptoms in adults experimentally infected with RSV [273]. Similar findings were observed in infants treated with aerosolized ribavirin after natural RSV infection [274]. Intravenous ribavirin reduced mortality in comatose patients infected with Lassa virus [275]. Conflicting results have been found with influenza virus and oral ribavirin. Several studies demonstrated decreased severity of illness and decreased viral shedding with oral ribavirin [276], [277]. Other studies demonstrated no therapeutic effect of oral ribavirin against influenza virus [278], [279]. However, aerosolized ribavirin has been shown to reduce fever and systemic illness in patients infected with influenza virus [280], [281].

Synthesis: Acid-catalyzed fusion of methyl 1,2,4-triazole-3-carboxylate and 1,2,3,5-tetra-*O*-acetyl-*β*-D-ribofuranose or 1-*O*-acetyl-2,3,5-tri-*O*-benzoyl-*β*-D-ribofuranose yields blocked methyl ester nucleosides. Ribavirin is obtained after treatment of the blocked nucleoside with methanolic NH_3 [268].

Trade Name: Virazole (ICN).

Cyclaradine [*69979-46-0*], carbocyclic arabinofuranosyladenine, [±]-9-[2α,3β-dihydroxy-4α-(hydroxymethyl)cyclopent-1α-yl]adenine, $C_{11}H_{15}N_5O_3 \cdot H_2O$, M_r 283.33, *mp* 253–255 °C.

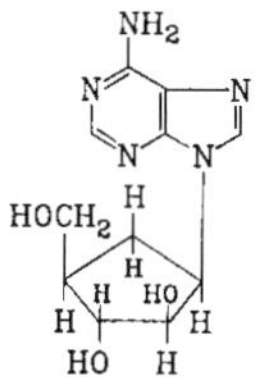

Cyclaradine is a carbocyclic derivative of ara-A. Cyclaradine was synthesized in 1977 in an effort to obtain a derivative of ara-A that is resistant to adenosine deaminase [282]. The deamination product of ara-A is 9-*β*-D-arabinofuranosylhypoxanthine, which is considerably less active than ara-A against herpesviruses. Cyclaradine is resistant to adenosine deaminase.

Cyclaradine is active in cell culture against herpes simplex virus type 1 [282]. Clinical studies in humans have not yet been conducted with cyclaradine. However, cyclaradine is effective against HSV–1 induced encephalitis in mice [283] and HSV–2-induced genital infections in guinea pigs [284].

Synthesis: Reaction of 5-amino-4-*N*-[2α,3β-dihydroxy-4α-(hydroxymethyl)cyclopent-1α-yl]-amino-6-chloropyrimidine with diethoxymethyl acetate gives the 6-chloropurine. Treatment of the 6-chloropurine with liquid ammonia gives cyclaradine [282].

DHPG [*82410-32-0*], 2-amino-1,9-dihydro{[2-hydroxy-1-(hydroxymethyl)ethoxy]-methyl}-6*H*-purin-6-one, BIOLF–62 [285], BW B759U, 2′-nor-2′-deoxyguanosine (2′-NDG) [286], $C_9H_{13}N_5O_4$, M_r 255.3, *mp* > 300 °C.

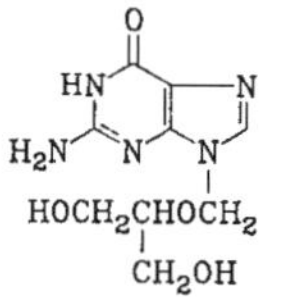

DHPG is an analogue of acyclovir; its potency is equal to or greater than that of acyclovir against some herpesviruses. In comparison with acyclovir, DHPG is at least

10-fold more active against CMV and Epstein–Barr virus in cell culture and about 50-fold more active than acyclovir in animal models of herpesvirus infection [287].

Several preliminary clinical studies with DHPG have been reported. It was effective against CMV pneumonia in transplant recipients [288] and against CMV retinitis [289]. DHPG has also been used for CMV infections in AIDS patients [290].

Synthesis: 2-O-(Acetoxymethyl)-1,3-di-O-benzylglycerol is condensed with N^2,9-diacetylguanine in the presence of a catalytic amount of *p*-toluenesulfonic acid in sulfolane. The desired isomer is crystallized from toluene, debenzylated over 20% palladium hydroxide, and deacetylated with concentrated NH_4OH/methanol [291].

Buciclovir [*86304-28-1*], (*R*)-9-(3,4-dihydroxybutyl)guanine, (*R*)-DHBG, (*R*)-2-amino-9-(3,4-dihydroxybutyl)-1,9-dihydro-6*H*-purin-6-one, $C_9H_{13}N_5O_3$, M_r 239.23, mp 260–261 °C (decomp.).

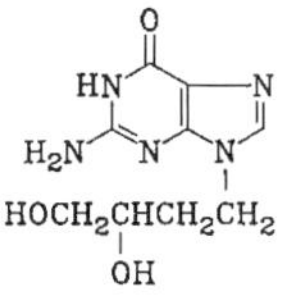

(*RS*)-9-(3,4-Dihydroxybutyl)guanine, which is a close analogue of acyclovir and DHPG, has recently been shown to inhibit HSV–1 and HSV–2 both in cell culture and in animal models of virus infection [292]. The (R)-enantiomer of DHBG is about 5-fold more active against HSV–2 in plaque reduction tests than the (S)-enantiomer. The mode of action of DHBG is similar to that of acyclovir [293], [294]. DHPG is phosphorylated by the HSV-induced thymidine kinase; presumably it is phosphorylated by cellular enzymes to give the 5′-triphosphate derivative of DHPG, which selectively inhibits viral DNA synthesis.

No clinical data on DHBG are available at the present time.

Synthesis: Condensation of 2-amino-6-chloropurine with 4-bromo-2-hydroxybutyrate and subsequent hydrolysis yields 4-(9-guanyl)-2-hydroxybutyric acid, which is esterified and reduced to buciclovir [295].

4.6.2. Phosphonoacetate and Phosphonoformate

Phosphonoacetic acid (PAA) [*4408-78-0*], $C_2H_5O_5P$, M_r 136.01, was discovered by routine screening methods in 1973 to inhibit HSV [296]. The sodium salt is called phosphonoacetate. PAA has been shown to inhibit the replication in cell culture of HSV, cytomegalovirus, Epstein–Barr virus, varicella–zoster virus, vaccinia virus, avian herpesvirus, and equine abortion virus. PAA inhibits the replication of herpesvirus by interacting directly with the virus-induced DNA polymerase at the pyrophosphate binding site [297]. Host cell polymerases are less sensitive to PAA [298]. Preclinical studies in rats showed that intravenous PAA was deposited in bone [299]. This result prevented clinical studies from being performed with PAA. Structure–activity studies have been reported recently for 100 analogues of PAA, but all analogues were less active than PAA [300].

Phosphonoformate [63585-09-1], foscarnet sodium, phosphonoformate sodium, PFA (INN), phosphonoformic acid (trisodium salt), foscarnet, CNa_3O_5P, M_r 191.95, mp >250 °C.

$$\text{NaO}-\underset{\text{ONa}}{\overset{\text{O}}{\underset{\|}{P}}}-\overset{\text{O}}{\underset{\|}{C}}-\text{ONa}$$

Phosphonoformate is closely related to PAA and also directly inhibits the virus-induced DNA polymerase [301]. Phosphonoformate is deposited in bone, but to a lesser extent than PAA [302]. Clinical trials have indicated that topical treatment of patients with recurrent genital herpesvirus infection was effective in shortening the time to healing, but this effect was only observed in male patients [303]. Intravenous phosphonoformate was effective against CMV in allograft recipients [304].

Phosphonoformate has also shown activity in cell culture against HTLV–III (the virus implicated in AIDS) [305].

Synthesis: Phosphonoformate is prepared by saponification of the triethyl ester of phosphonoformate with NaOH. Trisodium phosphonoformate hexahydrate is recrystallized from H_2O [301].

Trade Name: Foscarnet (Astra).

4.6.3. Amantadine and Rimantadine

Amantadine is presently the only chemotherapeutic agent accepted for clinical use against influenza virus. Amantadine was discovered in 1964 and was found to be active against influenza A viruses [306]. Administered orally, subcutaneously, or intraperitoneally, amantadine reduced the mortality in mice infected intranasally with influenza A virus. The precise *mechanism of action* of amantadine is not defined although most studies indicate that amantadine inhibits uncoating (the release of infectious nucleic acid into the cell's cytoplasm) [307]. The first clinical trials in volunteers lacking antibody showed that prophylactic amantadine reduced the incidence of influenza [308]. These results were confirmed in a natural outbreak of influenza virus [309]. The use of amantadine did not become widespread because until September, 1976; the FDA would not approve its use against any strain of influenza except for the original A2/Asian strain. Thus, amantadine could not be used against the major Hong Kong influenza virus epidemic in the late 1960s.

In comparison studies with amantadine, rimantadine is equally effective prophylactically with fewer *adverse effects* (nervousness, difficulty in concentrating, and insomnia)

[310]. Rimantadine has similar therapeutic efficacy as amantadine with respect to faster resolution of symptoms in patients with influenza [311]. Recently, a randomized, placebo-controlled, double-blind trial demonstrated the therapeutic efficacy of oral rimantadine given once daily to patients with influenza A virus infection [312].

Amantadine (INN) [*665-66-7*], 1-adamantanamine hydrochloride, $C_{10}H_{17}N \cdot HCl$, M_r 187.74, *mp* 206–208 °C.

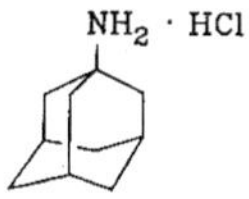

Synthesis: 1-Bromoadamantane is reacted with acetonitrile in the presence of concentrated sulfuric acid. The resulting 1-acetaminoadamantane is hydrolyzed with NaOH in boiling diethylene glycol.
Trade Name: Symmetrel (Du Pont).

Rimantadine (INN) [*13392-28-4*], [*1501-84-4*], α-methyl-tricyclo[3.3.1.1^{3,7}]decane-1-methan-amine, $C_{12}H_{21}N$, M_r 179.31, *mp* 373–375 °C, an analogue of amantadine, was also described in 1964 [306]; HCl salt [*1501-84-4*].

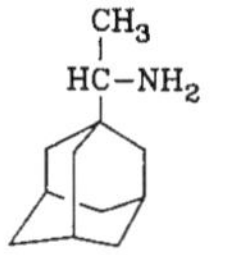

4.6.4. Enviroxime

Enviroxime (INN) [*72301-79-2*], (*E*)-2-amino-6-benzoyl-1-(isopropylsulfonyl)benzimidazoleoxime, 6-{[(hydroxyimino)phenyl]methyl}-1-[(1-methylethyl)sulfonyl]-1*H*-benzimidazol-2-amine, $C_{17}H_{18}N_4O_3S$, anti-isomer, M_r 358.45, *mp* 198–199 °C.

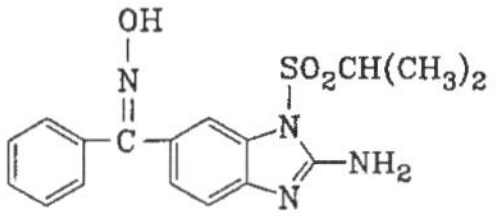

Enviroxime is an extremely potent inhibitor of human rhinoviruses in cell culture. A total of 15 different rhinovirus serotypes were inhibited in plaque reduction assays by enviroxime concentrations ranging from 0.04 μg/mL to < 0.01 μg/mL [313], [314]. However, the results of five clinical trials showed that enviroxime had only a modest, clinically significant, beneficial effect on rhinovirus infections. Prophylactic intranasal enviroxime administered prior to virus challenge in volunteers resulted in a statistically significant reduction in rhinorrhea, with no significant difference detected in infection rate or in the quantity of secreted virus. In addition, oral administration of enviroxime caused nausea and vomiting [315]–[317]. Therapeutically, intranasal administration of enviroxime lessened symptoms, but only on the fifth (last) day of treatment [318]. Recently, an open-field trial of intranasally applied enviroxime against natural rhinovirus infections showed no statistically significant differences compared to placebo [319]. Eli Lilly has terminated all studies on enviroxime because of the marginal efficacy observed in the clinic.

Synthesis: Reaction of cyanogen bromide and 3,4-diaminobenzophenone gives 2-amino-1-(isopropylsulfonyl)-6-benzimidazolyl phenylketone. The benzimidazole is sulfonated by using isopropylsulfonyl chloride and sodium hydride in dimethoxyethane. The anti- and syn-isomers of enviroximes are separated by high-performance liquid chromatography or by fractional crystallization [313].

4.6.5. 4',6-Dichloroflavan

4',6-Dichloroflavan [*73110-56-2*], BW-683C, 6-chloro-2-(4-chlorophenyl)-3,4-dihydro-4',6-di-chloroflavan, $C_{15}H_{12}Cl_2O$, M_r 279.18.

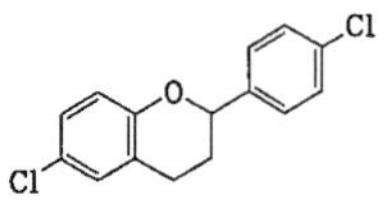

Dichloroflavan, like enviroxime, is a potent inhibitor of the replication of human rhinoviruses in cell culture [320]. Dichloroflavan appears to bind to virion capsid proteins, thereby inhibiting uncoating of the viral RNA [321], [322]. Only one clinical trial with oral dichloroflavan has been reported. In this double-blind, placebo-controlled trial, orally administered dichloroflavan was given to volunteers before and after challenge with rhinovirus 9. Dichloroflavan had no effect on the incidence or course of rhinovirus infection in this study [323].

Chemical synthetic details are not available.

4.6.6. Chalcone Ro 09-0410

Chalcone Ro 09-0410 [*76554-66-0*], 1-(4-ethoxy-2-hydroxy-6-methoxyphenyl)-3-(4-methoxyphenyl)-2-propen-1-one, 4'-ethoxy-2'-hydroxy-4,6'-dimethoxychalcone, Ro 09-0410, $C_{19}H_{20}O_5$, M_r 328.37.

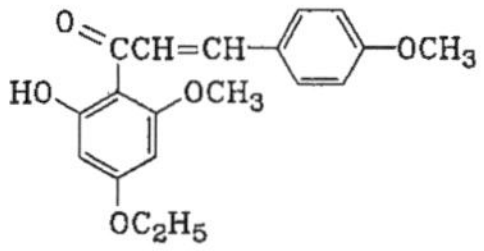

Chalcone Ro 09-0410 is a potent inhibitor of human rhinovirus in cell culture. Only 0.002 µg of Ro 09-0410 per mL is sufficient to inhibit some serotypes of human rhinovirus. Other serotypes of human rhinovirus require as much as 2 µg of the chalcone per mL for inhibition to occur [324]. The mode of action of Ro 09–0410 is similar to dichloroflavan in that it prevents uncoating by irreversible binding to viral capsid proteins. The chalcone could, however, be removed from virus particles by extraction with a nonpolar solvent [325].

A phosphate ester prodrug of the chalcone has been tested in volunteers because the chalcone itself was not well-absorbed orally and was irritating intranasally. The prodrug did not have any effect on the course of the rhinovirus infection, probably because even though the prodrug was well-absorbed, the chalcone did not reach the nasal mucosa [326].

Chemical synthetic details are not available.

4.6.7. Arildone and Disoxaril

Arildone [327] and disoxaril [328] have been shown to be active in cell culture, as well as in experimental animals infected with human picornaviruses [329]–[331]. *Mechanism of action* studies indicated that arildone blocked poliovirus uncoating, but did not inhibit adsorption or penetration [332]. Further work has shown that both arildone and disoxaril interact directly with picornavirion capsid proteins to prevent uncoating of virions.

Arildone (INN) [*56219-57-9*], 4-[6-(2-chloro-4-methoxy)phenoxy]hexyl-3,5-heptanedione, $C_{20}H_{29}ClO_4$, M_r 368.94, *bp* 180 °C.

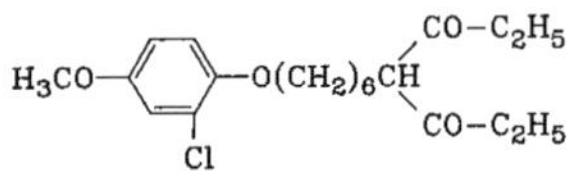

Synthesis: Treatment of 1-(2-chloro-4-methoxyphenoxy)-6-iodohexane and lithio-3,5-heptanedione in dimethylformamide for 48 h at 60 °C gives arildone [327].

Disoxaril (INN), Win 51711 [*87495-31-6*], 5-{7-[4-(4,5-dihydro-2-oxazolyl)phenoxy]-heptyl}-3-methylisoxazole, $C_{20}H_{26}N_2O_3$, M_r 342.42, *mp* 86–89 °C.

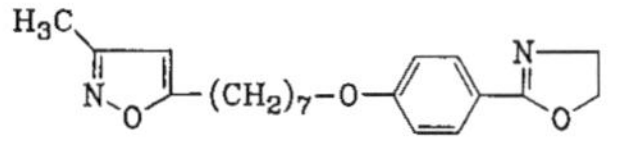

4.6.8. 3'-Azidothymidine

3'-Azidothymidine [*30516-87-1*], AZT, 3'-azido-3'-deoxythymidine, $C_{10}H_{13}N_5O_4$, M_r 267.3, *mp* 119 – 121 °C.

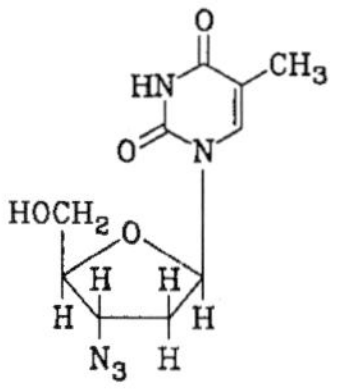

3'-Azidothymidine was initially synthesized in 1979 [333]. This compound is active against the viruses responsible for AIDS (HTLV – III); it blocks the cytopathic effects of HTLV – III in cell culture [334]. 3'-Azidothymidine entered phase II clinical trials against AIDS in January, 1986 [335]; it has shown some positive results in preventing mortality because of the AIDS virus in patients treated with AZT for a few months.

Synthesis: prepared by detritylation of the 3'-azidoderivative of 1-(2'-deoxy-5'-O-trityl-β-D-lyxosyl)thymine [333], [336].

4.6.9. Suramin

Suramin (INN, U.S.P. **12**, DCF) [*145-63-1*], 8,8'{carbonylbis[imino-3,1-phenylenecarbonyl-imino(4-methyl-3,1-phenylene)carbonylimino]}-bis(1,3,5-naphthalene)trisulfonic acid; suramin sodium (hexasodium salt) [*129-46-4*], $C_{51}H_{34}N_6Na_6O_{23}S_6$, M_r 1429.15.

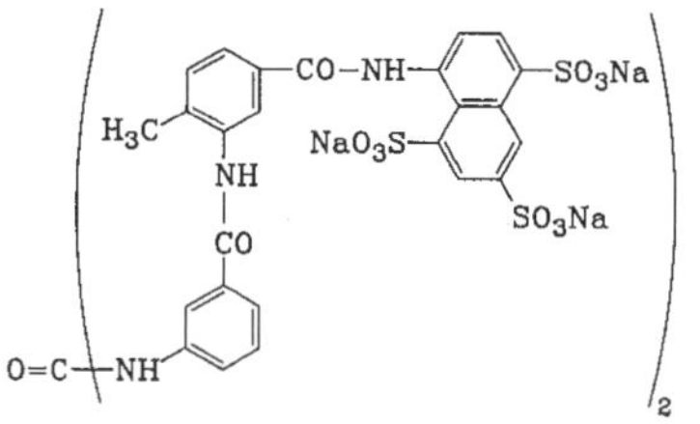

Suramin sodium is the hexasodium salt of 8,8'-{carbonylbis[imino-3,1-phenylenecarbonyl-imino(4-methyl-3,1-phenylene)carbonylimino]}-bis(1,3,5-naphthalene)trisulfonic acid. It is the drug of choice for the therapy of early African trypanosomiasis [337], and recently has been the subject of renewed interest as an inhibitor of the reverse transcriptase of the virus that causes AIDS [338], [339]. Preliminary findings in a few AIDS patients showed that suramin sodium was beneficial after short-term intravenous treatment [340].

Synthesis: Suramin sodium is prepared by condensing 1-naphthylamine-4,6,8-trisulfonic acid with 3-nitro-4-methylbenzoyl chloride, reducing the product, condensing with 3-nitrobenzyl chloride, reducing again, and then treating with carbonyl chloride and neutralizing with sodium hydroxide [341].

4.6.10. HPA 23

HPA 23 [*89899-81-0*], ammonium 21-tungsto-9-antimonate, 5′-tungsto-2-antimonate, $(NH_4)_{17}Na/NaW_{21}Sb_9O_{86} \cdot 14 H_2 O$, M_r 6937.39.

HPA 23 is a tungsto-antimonate compound that is a competitive inhibitor of the reverse transcriptase of murine oncornavirus [342] and of the AIDS viruses. In a preliminary uncontrolled clinical trial, HPA 23 was able to inhibit AIDS virus replication in patients. The virus, however, reappeared when therapy was discontinued [343].

Synthesis: Addition of a hot hydrochloric acid solution of antimony trioxide to an aqueous solution of sodium tungstate and NH_3 gives HPA 23 [344].

5. References

[1] R. E. Buchanan, N. E. Gibbons, (eds.): *Bergey's Manual of Determinative Bacteriology,* 8th ed., The Williams & Wilkins Co., Baltimore 1974.
[2] E. F. Gale, E. Cundliffe, P. E. Reynolds, M. H. Richmond, M. J. Waring in *The Molecular Basis of Antibiotic Actions,* 2nd ed., J. Wiley & Sons, London 1972.
[3] G. Y. Lesher, E. J. Froelich, M. D. Gruett, J. H. Bailey et al., *J. Med. Pharm. Chem.* **5** (1962) 1063.
[4] R. Albrecht, *Prog. Drug Res.* **21** (1977) 9.
[5] J. S. Wolfson, D. C. Hooper, *Antimicrob. Agents Chemother.* **28** (1985) 581–585.
[6] M. P. Wentland, J. B. Cornett in D. M. Bailey (ed.): *Annual Reports in Medicinal Chemistry,* vol. **20,** Academic Press, New York 1985, pp. 145–153.
[7] D. C. Hooper, J. S. Wolfson, *Antimicrob. Agents Chemother.* **29** (1986).
[8] H. Koga, A. Itoh, S. Murayama, S. Suzae, T. Irikura, *J. Med. Chem.* **23** (1980) 1358–1364.
[9] M. Gellert, K. Mizuuchi, M. H. O'Dea, H. A. Nash, *Proc. Natl. Acad. Sci.* **73** (1976) 3872–3878.
[10] A. Sugino, C. L. Peebles, K. N. Kruezer, N. R. Cozzar, *Proc. Natl. Acad. Sci.* **47** (1977) 4767–4771.
[11] A. M. Barlow, *Br. Med. J.* **2** (1963) 1308.
[12] T. A. McAllister, A. Percival, J. G. Alexander, J. G. Boyce et al., *Postgrad. Med. J.* **47** (1971, Sept.) Suppl., 7.
[13] E. W. McChesney, E. J. Froelich, G. Y. Lesher et al., *Toxicol. Appl. Pharmacol.* **6** (1964) 292.
[14] T. A. Stamey, N. J. Nemoy, M. Higgins, *Invest. Urol.* **6** (1969) 582.
[15] G. A. Portmann, E. W. McChesney, H. Stander et al., *J. Pharm. Sci.* **55** (1966) 72.
[16] M. Buchbinder, J. C. Webb, L. Anderson, W. R. McCabe, *Antimicrob. Agents Chemother.* 1962, 308.
[17] Sterling Drug, US 3 149 104, 1964 (G. Y. Lesher, M. D. Gruett).
[18] F. J. Turner, S. M. Ringel, J. F. Martin et al., *Antimicrob. Agents Chemother.* 1967, 475.
[19] S. M. Ringel, F. J. Turner, F. L. Lindo et al., *Antimicrob. Agents Chemother.* 1967, 480.
[20] D. Kaminsky, R. I. Meltzer, *J. Med. Chem.* **11** (1968) 160.
[21] W. E. Wick, D. S. Preston, W. A. White et al., *Antimicrob. Agents Chemother.* **4** (1973) 415.

[22] R. G. Gordon, L. I. Stevens, C. E. Edmiston et al., *Antimicrob. Agents Chemother.* **10** (1976) 918.
[23] W. A. Goss, W. H. Dietz, T. M. Book, *J. Bacteriol.* **10** (1965) 918.
[24] R. A. P. Burt, T. Morgan, J. P. Payne et al., *Br. J. Urol.* **49** (1977) 147.
[25] H. R. Black, K. S. Israel, R. L. Wolen et al., *Antimicrob. Agents Chemother.* **15** (1979) 165.
[26] Eli Lilly, DE 2 065 719, 1975 (W. A. White).
[27] J. R. Prous (ed.), *Ann. Drug Data Rep.* **2** (1979/1980) 177–180.
[28] J. R. Prous (ed.), *Ann. Drug Data Rep.* **2** (1979/1980) 181–184.
[29] J. Matsumoto, S. Minami, *J. Med. Chem.* **18** (1975) 74.
[30] Dainippon, JP 25 912, 1967 (Minami et al.).
[31] J. R. O'Connor, R. A. Dobson, P. E. Came, R. B. Wagner in J. D. Nelson, C. Grassi (eds.): "Current Chem. and Infect. Dis.," *Proc. 11th ICC and 19th ICAAC*, vol. **1**, The American Society for Microbiology, Washington, D.C., 1980, pp. 440–442.
[32] Sterling Drug, US 3 753 993, 1973 (G. Y. Lesher, P. M. Carabateas).
[33] A. Izawa, Y. Kisaki, K. Irie, Y. Eda et al., *Antimicrob. Agents Chemother.* **18** (1980) 37–40.
[34] A. Izawa, A. Yoshitake, T. Komatsu, *Antimicrob. Agents Chemother.* **18** (1980) 41–44.
[35] A. Yoshitake, K. Kawahara, F. Shono, I. Umeda et al., *Antimicrob. Agents Chemother.* **18** (1980) 45–49.
[36] Sumitomo, US 3 799 930, 1974 (T. Nakagome et al.).
[37] K. Sato, Y. Matsuura, M. Inoue, T. Une et al., *Antimicrob. Agents Chemother.* **22** (1982) 548–553.
[38] N. X. Chin, H. C. Neu, *Antimicrob. Agents Chemother.* **25** (1984) 319–326.
[39] D. S. Reeves, M. J. Bywater, H. A. Holt, L. O. White, *J. Antimicrob. Chemother.* **13** (1984) 333–346.
[40] J. B. Cornett, R. B. Wagner, R. A. Dobson, M. P. Wentland et al., *Antimicrob. Agents Chemother.* **27** (1985) 4–10.
[41] N. X. Chin, H. C. Neu, *Antimicrob. Agents Chemother.* **24** (1983) 754–763.
[42] A. L. Barry, R. N. Jones, *Antimicrob. Agents Chemother.* **25** (1984) 775–777.
[43] A. Ito, K. Hirai, M. Inoue et al., *Antimicrob. Agents Chemother.* **17** (1980) 103.
[44] H. C. Neu, P. Labthavikul, *Antimicrob. Agents Chemother.* **22** (1982) 23.
[45] S. R. Crider, S. D. Colby, L. K. Miller, W. O. Harrison et al., *N. Eng. J. Med.* **311** (1984) 137.
[46] P. Cantet, H. Ranaudin, C. Quentin, C. Babear, *Pathol. Biol. (Paris)* **31** (1983) 501.
[47] S. Nakamura, N. Kurobe, S. Kashimoto, T. Ohue et al., *Antimicrob. Agents Chemother.* **24** (1983) 54.
[48] Z. N. Adhami, R. Wise, D. Wetson, B. Crump, *Antimicrob. Chemother.* **13** (1984) 87.
[49] R. Wise, R. Lockley, J. Dent, M. Webberly, *Antimicrob. Agents Chemother.* **26** (1984) 17.
[50] B. Crump, R. Wise, J. Dent, *Antimicrob. Agents Chemother.* **24** (1983) 784.
[51] M. Bologna, L. Vaggii, C. M. Forchetti, E. Martine, *Lancet* **2** (1983) 280.
[52] B. N. Swanson, V. K. Boppana, P. H. Vlasses, H. M. Rotmensch et al., *Antimicrob. Agents Chemother.* **23** (1983) 284.
[53] G. Panichi, A. Pantosti, G. P. Testore, *J. Antimicrob. Chemother.* **11** (1983) 589.
[54] Kyorin, BE 870 917, 1979.
[55] R. Wise, J. M. Andrew, L. J. Edwards, *Antimicrob. Agents Chemother.* **23** (1983) 559.
[56] H. L. Muytjens, J. van der Ros-van de Repe, G. van Veldhuizen, *Antimicrob. Agents Chemother.* **24** (1983) 302.
[57] R. J. Fass, *Antimicrob. Agents Chemother.* **24** (1983) 568.
[58] W. Brumfitt, I. Franklin, D. Grady, J. M. T. Hamilton-Miller et al., *Antimicrob. Agents Chemother.* **26** (1984) 757.

[59] Bayer, EP-A 78 362, 1983 (K. Grohe, H. J. Zeiler, K. G. Metzger).
[60] S. Nakamura, K. Nakata, H. Katae, A. Minami et al., *Antimicrob. Agents Chemother.* **23** (1983) 742.
[61] A. Notowicz, E. Stolz, B. van Klingeren, *J. Antimicrob. Chemother.* **14** (1984) Suppl. C, 91.
[62] J. M. Hubrechts, R. Vanhoof, J. Servais, R. Toen et al., *Lancet* **1** (1984) 860.
[63] Dainippon, EP-A 9 425, 1980 (J. I. Matsumoto, Y. Takase, Y. Nishimura).
[64] D. L. van Caekenberghe, S. R. Pattyn, *Antimicrob. Agents Chemother.* **25** (1984) 518.
[65] G. Montay, Y. Gouffon, R. Roquet, *Antimicrob. Agents Chemother.* **25** (1984) 463.
[66] M. Wolff, B. Regnier, C. Daldoss, M. Nkam et al., *Antimicrob. Agents Chemother.* **26** (1984) 289.
[67] N. Desplaces, L. Gutmann, J. F. Acar, *24th ICAAC*, Washington, D.C., 1984, article no. 279.
[68] F. B. McGillion, *Drugs of the Future* **7** (1982) 946.
[69] M. R. Lockley, R. Wise, J. Dent, *Antimicrob. Agents Chemother.* **14** (1984) 647.
[70] Daiichi Seiyaku, EP-A 47 005, 1982 (I. Hayakawa, T. D. S. Hiramitsu, Y. Tanaka).
[71] N. V. Jacobus, F. P. Tally, M. Barza, *Antimicrob. Agents Chemother.* **26** (1984) 104–107.
[72] I. Garcia, G. P. Bodey, V. Fainstein, D. Hsitto et al., *Antimicrob. Agents Chemother.* **26** (1984) 421–423.
[73] H. C. Neu, P. Labthavikul, *24th ICAAC*, Washington, D.C., 1984, article no. 402.
[74] R. B. Wagner, R. A. Dobson, M. P. Wentland, D.M. Bailey et al., *23rd ICAAC*, Las Vegas, Nev., 1983, article no. 378.
[75] Sterling Drug, EP-A 90 424, 1983 (M. P. Wentland, D. M. Bailey).
[76] P. B. Fernandes, D. Chu, R. Bowen, N. Shipkowitz et al., *24th ICAAC*, Washington, D.C., 1984, article no. 78.
[77] P. B. Fernandes, N. Shipkowitz, D. Chu, L. Doen et al., *24th ICAAC*, Washington, D.C., 1984, article no. 79.
[78] N. X. Chin, D. C. Brittain, H. C. Neu, *24th ICAAC*, Washington, D.C., 1984, article no. 399.
[79] K. Takagi, M. Hosaka, H. Kusajima, Y. Oomori et al., *23rd ICAAC*, Las Vegas, Nev., 1983, article no. 659.
[80] M. A. Cohen, P. A. Bien, T. J. Griffin, C. L. Heifetz, *24th ICAAC*, Washington, D.C., 1984, article no. 81.
[81] J. C. Sesnie, M. A. Shapiro, C. L. Heifetz, T. F. Mich, *24th ICAAC*, Washington, D.C., 1984, article no. 82.
[82] J. M. Domagala, J. B. Nichols, C. L. Heifetz, T. F. Mich, *24th ICAAC*, Washington, D.C., 1984, article no. 80.
[83] Riker Lab, US 3 896 131, 1975 (J. F. Gerster).
[84] D. T. W. Chu, G. R. Granneman, P. B. Fernandes, *Drugs of the Future* **10** (1985) 546.
[85] Warner Lambert, AU 8 318 698, 1984 (T. P. Culbertson, J. M. Domatgala, T. F. Michh, J. B. Nickols).
[86] J. Trefouël, J. Trefouël, F. Nitti, D. Bovet, *C. R. Seances Soc. Biol. Ses Fil.* **120** (1935) 756.
[87] D. D. Woods, *Br. J. Exp. Pathol.* **21** (1940) 74.
[88] P. Fildes, *Lancet* **1** (1940) 955.
[89] A. K. Miller, P. Bruno, R. M. Berglund, *J. Bacteriol.* **54** (1947) 9.
[90] O. Sköld, *Antimicrob. Agents Chemother.* **9** (1976) 49.
[91] S. A. Kabins, M. V. Panse, S. Cohen, *J. Infect. Dis.* **123** (1971) 158.
[92] R. L. Then, *Rev. Infect. Dis.* **4** (1982) 261.
[93] J. M. J. Hamilton-Miller, *J. Antimicrob. Chemother.* **5** Suppl. B (1979) 61.
[94] J. K. Seydel, *J. Pharm. Sci.* **57** (1968) 1455.

[95] N. Anand: "Sulfonamides and Sulfones," in M. E. Wolff (ed.): *Burger's Medicinal Chemistry*, 4th ed., J. Wiley & Sons, New York 1979, Part II, Chap. 13.
[96] T. Fujita, C. Hanch, *J. Med. Chem.* **10** (1967) 991.
[97] H. Kano, US 2 888 455, 1959.
[98] B. Roth et al., *J. Med. Chem.* **23** (1980) 535.
[99] S. R. M. Bushby, *J. Infect. Dis.* **128** Suppl. (1973) S443.
[100] E. Grunberg, *J. Infect. Dis.* **128** Suppl. (1973) S478.
[101] D. J. Winston, W. K. Lau, R. P. Gale, L. S. Young, *Ann. Intern. Med.* **92** (1980) 762.
[102] J. Wüst, *Antimicrob. Agents Chemother.* **11** (1977) 631.
[103] R. O. Roblin, *J. Am. Chem. Soc.* **62** (1940) 2002.
[104] L. Doub, *J. Med. Chem.* **13** (1970) 242.
[105] H. Iundbeck & Co., US 2 447 702, 1948 (O. Hübner).
[106] H. Bretschneider, US 3 132 139, 1962.
[107] Aktiebolaget Pharmacia, US 2 396 145, 1946 (A. Askëlof).
[108] ICI, GB 551 513, 1943 (E. Haworth).
[109] M. L. Moore, *J. Am. Chem. Soc.* **64** (1942) 1572.
[110] Sharpe & Dohme, US 2 324 013–15, 1943 (M. L. Moore).
[111] R. M. Des Prez, R. A. Goodwin, Jr. in G. L. Mandell, R. G. Douglas, Jr., J. E. Bennett (eds.): *Principles and Practice of Infectious Diseases,* 2nd ed., J. Wiley & Sons, New York 1985, p. 1383.
[112] W. E. Bullock, [111] p. 1406.
[113] W. E. Sanders, Jr., [111] p. 1413.
[114] *Med. Lett.* **28** (1986, Mar. 28) 710.
[115] Tuberculosis Chemotherapy Trials Committee: Interim Report to the Medical Research Council, "The Treatment of Pulmonary Tuberculosis with Isoniazid," *Br. Med. J.* **2** (1952) 735.
[116] W. B. Pratt in *Chemotherapy of Infection,* Oxford University Press, New York 1977, p. 231.
[117] D. E. Kopanoff, D. E. Spider, G. J. Caras, *Am. Rev. Respir. Dis.* **117** (1978) 991.
[118] American Thoracic Society, *Am. Rev. Respir. Dis.* **110** (1974) 374.
[119] Distillers, US 2 280 994, 1958.
[120] R. G. Wilkinson et al., *J. Am. Chem. Soc.* **83** (1961) 2212.
[121] W. W. Stead, A. K. Dutt, *Am. Rev. Respir. Dis.* **125** Suppl. (1982) no. 3, 94.
[122] D. J. Girling, *Tubercule* **59** (1978) 13.
[123] A. K. Dutt, W. W. Stead, *J. Infect. Dis.* **146** (1982) 698.
[124] Merck, US 2 149 279, 1939 (O. Dalmer).
[125] D. Lieberman, *Bull. Soc. Chim. Fr.* (1958) 687.
[126] J. T. Sheehan, *J. Am. Chem. Soc.* **70** (1948) 1665.
[127] G. L. Mandell, M. A. Sande in A. G. Gilman, L. S. Goodman, A. Gilman (eds.): *Drugs Used in the Chemotherapy of Tuberculosis and Leprosy, The Pharmacological Basis of Therapeutics,* 6th ed., Macmillan Publ. Co., New York 1980, p. 1200.
[128] G. Domagk et al., *Naturwissenschaften* **33** (1946) 315.
[129] C. C. Shepard, L. Levy, P. Fasal, *Am. J. Trop. Med. Hyg.* **18** (1969) 258.
[130] R. W. Riley, L. Levy, *Proc. Soc. Exp. Biol. Med.* **142** (1973) 1168.
[131] I.G. Farbenind, FR 829 926, 1938.
[132] H. Bauer, *J. Am. Chem. Soc.* **61** (1939) 617.
[133] J. H. Peters, J. F. Murray, G. R. Gordon et al., *Am. J. Trop. Med. Hyg.* **26** (1977) 127.
[134] D. A. Russell, C. C. Shepard, D. H. McRae et al., *Am. J. Trop. Med. Hyg.* **24** (1975) 485.
[135] E. Eslarger et al., *J. Med. Chem.* **12** (1969) 357.
[136] L. Levy, *Am. J. Trop. Med. Hyg.* **23** (1974) 1097.

[137] J. Convit, S. G. Browne, J. Languillon et al., *WHO Bull.* **42** (1970) 667.
[138] C. C. Shepard, L. L. Walker, R. M. Van Lindingham et al., *Proc. Soc. Exp. Biol. Med.* **137** (1971) 725.
[139] C. C. Shepard, L. L. Walker, R. M. Van Lindingham et al., *Proc. Soc. Exp. Biol. Med.* **137** (1971) 728.
[140] V. C. Barry, *J. Chem. Soc.* (1958) 859.
[141] V. L. Sutter in S. M. Finegold, W. L. George, R. D. Rolfe (eds.): *First United States Metronidazole Conference, Proc. Symp.*, Tarpon Springs, Fla., Biomedical Information Corp., New York 1982, p. 61.
[142] Hoffmann La Roche, US 2 430 094, 1947 (H. M. Wuest).
[143] S. M. Finegold, [111] p. 220.
[144] R. M. Jacob, Rhône-Poulenc, US 2 944 061, 1960.
[145] M. Turck, A. R. Ronald, R. G. Petersdorf, *Antimicrob. Agents Chemother.* 1966, 446.
[146] A. Kucers, N. M. Bennett in A. Kucers, N. M. Bennett (eds.): *The Use of Antibiotics*, J. B. Lippincott Co., Philadelphia 1979, p. 749.
[147] H. K. Reckendorf, R. G. Castringius, H. K. Spingler, *Antimicrob. Agents Chemother.* 1962, 531.
[148] Eaton Labs., US 2 610 181, 1952 (K. J. Hayes).
[149] R. A. Marcial-Rojas (ed.): *Pathology of Protozoal and Helminth Diseases, With Clinical Correlation*, Williams & Wilkins, New York 1971.
[150] H. Van de Bossche, *Nature London* **273** (1978) 626.
[151] T. C. Jones, [111] pp. 1505, 1513, 1560.
[152] D. J. Wyler, [111] p. 1514.
[153] D. R. Hill, [111] p. 1552.
[154] J. J. Marr, [111] p. 286.
[155] [111] p. 1537.
[156] J. I. Ravdin, T. C. Jones, [111] p. 1506.
[157] W. T. Hughes, [111] p. 1549.
[158] L. V. Kirchoff, F. Neva, [111] p. 1531.
[159] M. R. Rein, [111] p. 1556.
[160] T. K. Ruebush, II, [111] p. 1559.
[161] H. W. Brown, *Basic Clinical Parasitology*, 4th ed., Appleton-Century-Crofts, New York 1976.
[162] R. D. Pearson, A. de Queiroz Sousa, [111] p. 1522.
[163] R. E. McCabe, J. E. Remington, [111] p. 1540.
[164] W. E. Gutteridge, *Br. Med. Bull.* **41** (1985) 162.
[165] A. Sjoerdsma, J. A. Golden, P. J. Schechter et al., *Clin. Res.* **32** (1984) 559 A.
[166] J. Williamson, *Exp. Parasitol.* **12** (1962) 274.
[167] J. Williamson in H. W. Mulligan (ed.): *The African Trypanosomiases*, Allen & Unwin, London 1970, p. 125.
[168] J. Williamson, *Trop. Dis. Bull.* **73** (1976) 531.
[169] F. Hawking in R. J. Schnitzer, F. Hawking (eds.): *Experimental Chemotherapy*, vol. **1**, Academic Press, New York 1963, p. 129.
[170] B. B. Waddy, [167] p. 711.
[171] F. I. C. Apted, *Pharmacol. Ther.* **11** (1980) 391.
[172] D. A. Evans, *Antibiot. Chemother.* **30** (1981) 272.
[173] May & Baker, US 2 410 796, 1946 (G. Neuberry).
[174] E. Friedheim, *Chem. Abstr.* **47** (1953) 144.
[175] W. A. Jacobs, *J. Am. Chem. Soc.* (1919) 1590.

[176] Hoechst, US 2 838 485, 1958 (R. Brodersten).
[177] Z. Brenner, *Adv. Pharmacol. Chemother.* **13** (1975) 1.
[178] Z. Brenner, *Pharmacol. Ther.* **7** (1979) 71.
[179] W. E. Gutteridge, *Trop. Dis. Bull.* **73** (1976) 699.
[180] W. E. Gutteridge in J. R. Baker (ed.): "Perspectives in Trypanosomiasis Research," *Proceedings of the Twenty-first Trypanosomiasis Seminar*, Research Studies Press, Cinchester 1982, p. 47.
[181] R. Ribeiro-dos-Santos, A. Rassi, F. Koberle, *Antibiot. Chemother.* **30** (1980) 115.
[182] Bayer, US 3 262 930, 1964 (H. Herlinger).
[183] Hoffmann-La Roche, *Chem. Abstr.* **71** (1969) 3383.
[184] P. H. Rees, P. A. Kager, T. Ogada, J. K. M. Eeftinck Schattenkerk, *Trop. Geogr. Med.* **37** (1985) 37.
[185] L. L. Branborg, R. Owen, R. Fogel, H. Goldberg et al., *Gastroenterology* **78** (1980) 1602.
[186] B. M. Honigberg in L. Kreiser, *Parasitic Protozoa*, vol. **2**, Academic Press, New York 1978, p. 275.
[187] O. Jirovec, M. Petru, *Adv. Parasitol.* **6** (1968) 117.
[188] R. Knight, *Clin. Gastroenterol.* **7** (1978) 47.
[189] E. A. Meyer, S. Radulescu, *Adv. Parasitol.* **17** (1979) 1.
[190] J. W. Smith, M. S. Wolfe, *Ann. Rev. Med.* **31** (1980) 373.
[191] J. G. Meingassner, P. G. Heyworth, *Antibiot. Chemother.* **30** (1981) 163.
[192] K. Butler, *Chem. Abstr.* **71** (1964) 3384 e.
[193] Carlo Erba, US 3 399 193, 1965 (P. N. Giraldi).
[194] Hoffmann La Roche, US 3 435 049, 1966 (M. Hoffer).
[195] Rhône-Poulenc, *Chem. Abstr.* **63** (1965) 11 571 d.
[196] J. Heeres, *Eur. J. Med. Chem.-Chim. Ther. (Paris)* **11** (1976) 237.
[197] S. J. Lerman, R. A. Walker, *Clin. Pediatr. (Phila.)* **21** (1982) 409.
[198] M. S. Wolfe, *J. Am. Med. Assoc.* **233** (1975) 1362.
[199] B. R. Brown, *J. Chem. Soc.* 1948, 99.
[200] G. D. Drake, *Chem. Abstr.* **51** (1957) 2051 e.
[201] I. M. Rollo in A. F. Goodman, L. S. Goodman, A. Gilman (eds.): *The Pharmacological Basis of Therapeutics*, 6th ed., Macmillan Publ. Co., New York 1980, p. 1061.
[202] A. R. Surrey, *J. Am. Chem. Soc.* **68** (1946) 113.
[203] J. H. Burckhalter, *J. Am. Chem. Soc.* **70** (1948) 1363.
[204] A. R. Surrey, *J. Am. Chem. Soc.* **72** (1950) 1814.
[205] R. C. Elderfield, *J. Am. Chem. Soc.* **68** (1946) 1525.
[206] P. B. Russel, *J. Am. Chem. Soc.* **73** (1951) 3763.
[207] J. Souza, *WHO Bull.* **61** (1983) 815.
[208] F. H. S. Curd, *J. Chem. Soc.* (1948) 1630.
[209] T. Harinasuta, D. Bunnag, W. H. Wernsdorfer, *WHO Bull.* **61** (1983) 299.
[210] F. I. Carroll, *J. Med. Chem.* **17** (1974) 210.
[211] J. Rinehart, *Am. J. Trop. Med. Hyg.* **25** (1976) 769.
[212] K. Killer, *Drugs of the Future* **5** (1980) 547.
[213] V. Papesch, *J. Am. Chem. Soc.* **58** (1936) 1314.
[214] A. Das, *J. Org. Chem.* **22** (1957) 1111.
[215] A. Luis, *Chem. Abstr.* **47** (1953) 10 533 d.
[216] I.G. Farbenind, DE 545 915, 1930 (K. Schranz).
[217] Boots Pure Drug, US 2 912 438, 1959 (P. Oxley).
[218] T. H. Flewett, *Br. Med. Bull.* **41** (1985) 315–321.

[219] R. E. F. Matthews, *Intervirology* **17** (1982) 1–199.
[220] B. Rada, D. Blaskovic, F. Sorm, J. Skoda, *Experientia* **16** (1960) 487.
[221] E. C. Herrmann, *Proc. Soc. Exp. Biol. Med.* **107** (1961) 142–145.
[222] B. Jubelt, O. Narajan, R. T. Johnson, *J. Neuropathol. Exp. Neurol.* **39** (1980) 149–159.
[223] B. Clement, *Pharm. Unserer Zeit* **15** (1986) 72–84.
[224] H. J. Schaeffer, L. Beauchamp, P. deMiranda, G. B. Elion et al., *Nature (London)* **272** (1978) 583–585.
[225] H. J. Schaeffer, S. Gurwara, R. Vince, S. Bittner, *J. Med. Chem.* **14** (1971) 367–369.
[226] B. R. Jones, D. J. Coster, P. N. Fison, G. M. Thompson et al., *Lancet* **1** (1979) 243–244.
[227] P. R. Laibson, D. Pavan-Langston, W. R. Yeakley, J. Lass, *Am. J. Med.* **73 A** (1982) 281–285.
[228] C. La Lau, J. A. Oosterhuis, J. Versteeg, G. Van Rij et al., *Am. J. Med.* **73 A** (1982) 305–306.
[229] J. M. Yeo, A. P. Fiddian, *J. Antimicrob. Chemother.* **12** (1983) Suppl. B, 95–103.
[230] L. Corey, A. J. Nahmias, M. E. Guinan, J. K. Benedetti et al., *New. Engl. J. Med.* **306** (1982) 1313–1319.
[231] L. Corey, J. K. Benedetti, C. W. Critchlow, M. R. Remington et al., *Am. J. Med.* **73** (1982) 326–334.
[232] A. E. Nilsen, T. Aasen, A. M. Halsos, B. R. Kinge et al., *Lancet* **2** (1982) 571–573.
[233] O. P. Salo, A. Lassus, T. Hovi, A. P. Fiddian, *Eur. J. Sex. Trans. Dis.* **1** (1983) 95–98.
[234] R. C. Reichman, G. J. Badger, G. J. Mertz, L. Corey et al., *J. Am. Med. Assoc.* **251** (1984) 2103–2107.
[235] J. M. Douglas, C. Critchlow, J. Benedetti, G. J. Mertz et al., *New. Engl. J. Med.* **310** (1984) 1551–1556.
[236] S. E. Straus, H. E. Takiff, M. Seidlin, S. Bachrach et al., *New Engl. J. Med.* **310** (1984) 1545–1550.
[237] D. Brigden, *Br. Med. Bull.* **41** (1985) 357–360.
[238] T. A. Krenitsky, W. W. Hall, P. deMiranda, L. M. Beauchamp et al., *Proc. Natl. Acad. Sci. U.S.A.* **81** (1984) 3209–3213.
[239] G. R. Revankar, J. H. Huffman, L. B. Allen, R. W. Sidwell et al., *J. Med. Chem.* **18** (1975) 721–726.
[240] H. E. Kaufman, *Proc. Soc. Exp. Biol. Med.* **109** (1962) 251–252.
[241] H. E. Kaufman, E. D. Varnell, Y. M. Centifanto Fitzgerald, J. G. Sanitato, *Antiviral Res.* **4** (1984) 333–338.
[242] W. H. Prusoff, *Biochim. Biophys. Acta* **32** (1959) 295–296.
[243] P. C. Heidelberger, D. G. Parsons, D. C. Remy, *J. Med. Chem.* **7** (1964) 1–5.
[244] C. Wellings, P. N. Awdry, F. H. Bors, B. R. Jones et al., *Am. J. Ophthalmol.* **73** (1972) 932–942.
[245] P. C. Heidelberger, D. King, *Pharmacol. Ther.* **6** (1979) 427–442.
[246] P. C. Heidelberger, *Ann. N. Y. Acad. Sci.* **255** (1975) 317–325.
[247] W. W. Lee, A. Benitez, L. Goodman, B. R. Baker, *J. Am. Chem. Soc.* **82** (1960) 2648–2649.
[248] E. J. Reist, A. Benitez, L. Goodman, B. R. Baker et al., *J. Org. Chem.* **27** (1962) 3274–3279.
[249] J. DeRudder, M. Privat de Garlihe, *Antimicrob. Agents Chemother.* **5** (1965) 578–584.
[250] F. M. Schabel, *Chemotherapy* **13** (1968) 321–338.
[251] D. Pavan-Langston, C. H. Dohiman, *Am. J. Ophthalmol.* **74** (1972) 81–88.
[252] R. J. Whitley, L. T. Ch'ien, R. Dolin, G. J. Galasso et al., *New. Engl. J. Med.* **294** (1976) 1193–1199.
[253] R. J. Whitley, S.-J. Soong, R. Dolin, G. J. Galasso et al., *New. Engl. J. Med.* **297** (1977) 289–294.
[254] B. Sköldenberg, M. Forsgren, K. Alestig, T. Bergström et al., *Lancet* **2** (1984) 707–711.

[255] H. G. Adams, E. A. Benson, E. R. Alexander, L. A. Vontver et al., *J. Infect. Dis.* **133** (1976) Suppl. A, 151–159.

[256] S. L. Spruance, C. S. Crumpacker, H. Haines, C. Bader et al., *New. Engl. J. Med.* **300** (1979) 1180–1184.

[257] W. H. Prusoff, M. Zucker, W. R. Mancini, M. J. Otto et al., *Antiviral Res.* (1985) Suppl. 1, 1–10.

[258] E. DeClercq, J. Descamps, P. DeSomer, P. J. Barr et al., *Proc. Natl. Acad. Sci.* **76** (1979) 2947–2951.

[259] E. DeClercq, H. Degreef, J. Wildiers, G. DeJonge et al., *Br. Med. J.* **281** (1980) 1178.

[260] P. C. Maudgal, E. De Clercq, L. Missotten, *Antiviral Res.* **4** (1984) 281–291.

[261] R. K. Robins, *Chem. Eng. News* **64** (1986) 28–40.

[262] K. A. Watanabe, U. Reichman, K. Hirota, C. Lopez et al., *J. Med. Chem.* **22** (1979) 21–24.

[263] K. A. Watanabe, T.-L. Su, R. S. Klein, C. K. Chu, *J. Med. Chem.* **26** (1983) 152–156.

[264] C. W. Young, R. Schneider, B. Leyland-Jones, D. Armstrong et al., *Cancer Res.* **43** (1983) 5006–5009.

[265] Y.-C. Cheng, G. Dutschman, J. J. Fox, K. A. Watanabe et al., *Antimicrob. Agents Chemother.* **20** (1981) 420–423.

[266] H. S. Allaudeen, J. Descamps, R. K. Sehgal, J. J. Fox, *J. Biol. Chem.* **257** (1982) 11 879–11 882.

[267] C. McLaren, M. S. Chen, R. H. Barbhaiya, R. A. Buroker et al. in R. Kono (ed.): *Herpesviruses and Virus Chemotherapy,* Elsevier Science Publ., New York 1985, pp. 57–61.

[268] J. T. Witkowski, R. K. Robins, R. W. Sidwell, L. N. Simon, *J. Med. Chem.* **15** (1972) 1150–1154.

[269] R. W. Sidwell, R. K. Robins, I. W. Hillyard, *Pharmacol. Ther.* **6** (1979) 123–146.

[270] D. G. Streeter, J. T. Witkowski, G. P. Khare, R. W. Sidwell et al., *Proc. Natl. Acad. Sci.* **70** (1973) 1174–1178.

[271] B. B. Goswami, E. Borek, O. K. Sharma, J. Fujitaki et al., *Biochem. Biophys. Res. Commun.* **89** (1979) 830–836.

[272] J. B. McCormick, J. P. Getchell, S. W. Mitchell, D. R. Hicks, *Lancet* **2** (1984) 1367–1369.

[273] C. B. Hall, J. T. McBride, E. E. Walsh, D. M. Bell et al., *New Engl. J. Med.* **308** (1983) 1443–1447.

[274] C. B. Hall, J. T. McBride, C. L. Gala, S. W. Hildreth et al., *J. Am. Med. Assoc.* **254** (1985) 3047–3051.

[275] J. B. McCormick, I. J. King, P. A. Webb, C. L. Scribner et al., *New Engl. J. Med.* **314** (1986) 20–26.

[276] C. R. Magnussen, J. R. Douglas, R. F. Betts, F. K. Roth et al., *Antimicrob. Agents Chemother.* **12** (1977) 498–502.

[277] F. Salido-Rengell, H. Nasser-Quinones, B. Briseno Garcia, *Ann. N.Y. Acad. Sci.* **284** (1977) 272–277.

[278] C. B. Smith, R. P. Charette, J. P. Fox, M. K. Cooney et al. in R. A. Smith, W. Kirkpatrick (eds.): *Ribavirin: A Broad Spectrum Antiviral Agent,* Academic Press, New York 1980, pp. 147–164.

[279] Y. Togo, E. A. McCracken, *J. Infect. Dis.* **133** (1976) Suppl. A, 109–113.

[280] V. Knight, S. Wilson, J. M. Quarles, S. E. Greggs et al., *Lancet* **2** (1981) 945–949.

[281] H. W. McClung, V. Knight, B. E. Gilbert, S. Z. Wilson et al., *J. Am. Med. Assoc.* **249** (1983) 2671–2674.

[282] R. Vince, S. Daluge, *J. Med. Chem.* **20** (1977) 612–613.

[283] W. M. Shannon, L. Westbrook, G. Arnett, S. Daluge et al., *Antimicrob. Agents Chemother.* **24** (1983) 538–543.

[284] R. Vince, S. Daluge, H. Lee, W. M. Shannon et al., *Science* **221** (1983) 1405–1406.
[285] K. K. Ogilvie, U. O. Cheriyan, B. K. Radatus, K. O. Smith et al., *Can. J. Chem.* **60** (1982) 3005–3010.
[286] W. T. Ashton, J. D. Karkas, A. K. Field, R. L. Tolman, *Biochem. Biophys. Res. Commun.* **108** (1982) 1716–1721.
[287] A. K. Field, M. E. Davies, C. DeWitt, H. C. Perry et al., *Proc. Natl. Acad. Sci.* **80** (1983) 4139–4143.
[288] D. H. Shepp, P. S. Dandliker, P. deMiranda, T. C. Burnette et al., *Ann. Intern. Med.* **103** (1985) 368–373.
[289] D. Felsenstein, D. J. D'Amico, M. S. Hirsch, D. A. Neumeyer et al., *Ann. Intern. Med.* **103** (1985) 377–380.
[290] M. C. Bach, S. P. Bagwell, N. P. Knapp, K. M. Davis et al., *Ann. Intern. Med.* **103** (1985) 381–382.
[291] J. C. Martin, C. A. Dvorak, D. F. Smee, T. R. Matthews et al., *J. Med. Chem.* **26** (1983) 759–761.
[292] A. Larsson, B. Öberg, S. Alenius, C.-E. Hagberg et al., *Antimicrob. Agents Chemother.* **23** (1983) 664–670.
[293] A. Larsson, P.-Z. Tao, *Antimicrob. Agents Chemother.* **25** (1984) 524–526.
[294] A. Larsson, S. Alenius, N.-G. Johansson, B. Öberg, *Antiviral Res.* **3** (1983) 77–86.
[295] Astra, EP-A 55 239, 1982 (C. E. Hagberg, K. N. G. Johansson, Z. M. I. Kovacs, G. B. Stening).
[296] N. L. Shipkowitz, R. R. Bower, R. N. Appell, C. W. Nordeen et al., *Appl. Microbiol.* **26** (1973) 264–267.
[297] J. C.-H. Mao, E. E. Robishaw, *Biochemistry* **14** (1975) 5475–5479.
[298] C. L. K. Sabourin, J. M. Reno, J. A. Boezi, *Arch. Biochem. Biophys.* **187** (1978) 96–101.
[299] B. A. Bopp, C. B. Estep, D. J. Anderson, *Fed. Proc.* **36** (1977) 939.
[300] J. C.-H. Mao, E. R. Otis, A. M. von Esch, T. R. Herrin et al., *Antimicrob. Agents Chemother.* **27** (1985) 197–202.
[301] J. M. Reno, L. F. Lee, J. A. Boezi, *Antimicrob. Agents Chemother.* **13** (1978) 188–192.
[302] E. Helgstrand, H. Flodh, J.-O. Lernestedt, J. Lundstrom et al. in L. H. Collier, J. Oxford (eds.): *Developments in Antiviral Therapy*, Academic Press, London 1980, pp. 63–83.
[303] J. Wallin, J.-O. Lernestedt, S. Ogenstad, E. Lycke, *Scand. J. Infect. Dis.* **17** (1985) 165–172.
[304] G. Klintmalm, B. Lönngvist, B. Öberg, G. Gahrton et al., *Scand. J. Infect. Dis.* **17** (1985) 157–163.
[305] E. G. Sandstrom, J. C. Kaplan, R. E. Byington, M. S. Hirsch, *Lancet* **1** (1985) 1480–1482.
[306] W. L. Davies, R. R. Grunert, R. F. Haff, J. W. McGahen et al., *Science* **144** (1964) 862–863.
[307] N. Kato, H. J. Eggers, *Virology* **37** (1969) 632–641.
[308] G. G. Jackson, R. L. Muldoon, L. W. Akers, *Antimicrob. Agents Chemother.* **3** (1963) 703–707.
[309] A. W. Galbraith, J. S. Oxford, G. C. Schild, G. I. Watson, *Lancet* **2** (1969) 1026–1028.
[310] R. Dolin, R. C. Reichman, H. P. Madore, R. Maynard et al., *New Engl. J. Med.* **307** (1982) 580–584.
[311] L. P. VanVoris, R. F. Betts, F. G. Hayden, W. A. Christmas et al., *J. Am. Med. Assoc.* **245** (1981) 1128–1131.
[312] F. G. Hayden, A. S. Monto, *Antimicrob. Agents Chemother.* **29** (1986) 339–341.
[313] J. H. Wikel, C. J. Paget, D. C. DeLong, J. D. Nelson et al., *J. Med. Chem.* **23** (1980) 368–372.
[314] D. C. DeLong, S. E. Reed, *J. Infect. Dis.* **141** (1980) 87–91.
[315] R. J. Phillpotts, D. C. DeLong, J. Wallace, R. W. Jones et al., *Lancet* **1** (1981) 1342–1344.
[316] F. G. Hayden, J. M. Gwaltney, *Antimicrob. Agents Chemother.* **21** (1982) 892–897.

[317] R. A. Levandowski, C. T. Pachucki, M. Rubenis, G. G. Jackson, *Antimicrob. Agents Chemother.* **22** (1982) 1004–1007.
[318] R. J. Phillpotts, J. Wallace, D. A. J. Tyrrell, V. B. Tagart, *Antimicrob. Agents Chemother.* **23** (1983) 671–675.
[319] F. D. Miller, A. S. Monto, D. C. DeLong, A. Exelby et al., *Antimicrob. Agents Chemother.* **27** (1985) 102–106.
[320] D. J. Bauer, J. W. T. Selway, J. F. Batchelor, M. Tisdale et al., *Nature (London)* **292** (1981) 369–370.
[321] M. Tisdale, J. W. T. Selway, *J. Gen. Virol.* **64** (1983) 795–803.
[322] M. Tisdale, J. W. T. Selway, *J. Antimicrob. Chem.* **14** (1984) Suppl. A, 97–105.
[323] R. J. Phillpotts, J. Wallace, D. A. J. Tyrrell, D. S. Freestone et al., *Arch. Virol.* **75** (1983) 115–121.
[324] H. Ishitsuka, Y. T. Ninomiya, C. Ohsawa, M. Fujiu et al., *Antimicrob. Agents Chemother.* **22** (1982) 617–621.
[325] Y. Ninomiya, C. Ohsawa, M. Aoyama, I. Umeda et al., *Virology* **134** (1984) 269–276.
[326] R. J. Phillpotts, P. G. Higgins, J. S. Willman, D. A. J. Tyrrell et al., *J. Antimicrob. Chemother.* **14** (1984) 403–409.
[327] G. D. Diana, U. J. Salvador, E. G. Zalay, R. E. Johnson et al., *J. Med. Chem.* **20** (1977) 750–756.
[328] G. D. Diana, M. A. McKinlay, M. J. Otto, V. Akullian et al., *J. Med. Chem.* **28** (1985) 1906–1910.
[329] M. A. McKinlay, J. V. Miralles, C. J. Brisson, F. Pancic, *Antimicrob. Agents Chemother.* **22** (1982) 1022–1025.
[330] M. J. Otto, M. P. Fox, M. J. Fancher, M. F. Kuhrt et al., *Antimicrob. Agents Chemother.* **27** (1985) 883–886.
[331] M. A. McKinlay, B. A. Steinberg, *Antimicrob. Agents Chemother.* **29** (1986) 30–32.
[332] J. J. McSharry, L. A. Caliguiri, H. J. Eggers, *Virology* **97** (1979) 307–315.
[333] T.-S. Lin, W. H. Prusoff, *J. Med. Chem.* **21** (1978) 109–112.
[334] C. Norman, *Science* **230** (1985) 1355–1358.
[335] *FDC Reports*, Mar. 3, 1986.
[336] J. P. Horwitz, J. Chua, M. Noel, *J. Org. Chem.* **29** (1964) 2076–2078.
[337] R. S. Goldsmith in B. G. Katzung (ed.): *Basic Clinical Pharmacology*, 2nd ed., Lange Medical Publications, Los Altos, Calif., 1984, pp. 648–675.
[338] E. DeClercq, *Cancer Lett.* **8** (1979) 9–22.
[339] H. Mitsuya, M. Popovic, R. Yarchoan, S. Matsushita et al., *Science* **226** (1984) 172–174.
[340] D. Rouvroy, J. Bogaerts, J.-B. Habyarimana, D. Nzaramba et al., *Lancet* **1** (1985) 878–879.
[341] M. Windholz, S. Budavari, R. F. Blumetti, E. S. Otterbein (eds.): *The Merck Index*, 10th ed., Merck & Co., Rahway, N.J., 1983, p. 1294.
[342] J. C. Chermann, F. C. Sinoussi, C. Jasmin, *Biochem. Biophys. Res. Commun.* **65** (1975) 1229–1236.
[343] W. Rozenbaum, D. Dormont, B. Spire, E. Vilmer et al., *Lancet* **1** (1985) 450–451.
[344] M. Michelon, G. Herve, *C. R. Acad. Sci. Paris* **274** (1972) 209–212.

Antimycotics

Separate keyword: → *Drugs Used in Dermatology.*

Axel Schmidt, Bayer AG, Wuppertal, Federal Republic of Germany
Frank-Ulrich Geschke, Bayer AG, Wuppertal, Federal Republic of Germany

1.	Introduction	1199	3.	Polyene Antimycotics	1232
2.	Azole Antimycotics	1201	3.1.	Amphotericin B	1232
2.1.	Azole Antimycotics for Systemic Treatment	1201	3.2.	Natamycin (Pimaricin)	1235
			3.3.	Nystatin A1	1237
2.1.1.	Fluconazole	1201	4.	Flucytosine	1238
2.1.2.	Itraconazole	1203	5.	Griseofulvin	1241
2.1.3.	Ketoconazole	1206	6.	Ciclopirox	1244
2.1.4.	Miconazole	1209	7.	Thiocarbamates	1246
2.2.	Azole Antimycotics for Topical Treatment	1212	7.1.	Tolciclate	1246
			7.2.	Tolnaftate	1247
2.2.1.	Bifonazole	1212	8.	Allylamines	1248
2.2.2.	Butoconazole	1214	8.1.	Naftifine	1249
2.2.3.	Clotrimazole	1216	8.2.	Terbinafine	1250
2.2.4.	Croconazole	1218	9.	Amorolfine	1252
2.2.5.	Econazole	1219	10.	Unspecific Topical Antimycotics	1254
2.2.6.	Fenticonazole	1220			
2.2.7.	Isoconazole	1222	11.	Recent Developments and Outlook	1256
2.2.8.	Omoconazole	1223			
2.2.9.	Oxiconazole	1224	12.	References	1256
2.2.10.	Sertaconazole	1225			
2.2.11.	Sulconazole	1226			
2.2.12.	Terconazole	1228			
2.2.13.	Tioconazole	1230			

1. Introduction

The significant advances in antimycotic chemotherapy in the last decades are mainly due to the introduction of new, potent, and broadly active antimycotics to the market. The substances can be classified according to their mode of application into topical and systemic antimycotics. Topical antimycotics are applied to the skin, mucous membranes, and intravaginally in order to treat superficial mycoses locally. Peroral therapy and prophylaxis of intraluminal colonizations or infections of the gastrointestinal tract, e.g., with drops, syrups, or lozenges containing antimycotics that are not absorbed or are absorbed to an extremely small extent from the gastrointestinal tract (no or extremely low bioavailability) must also be regarded as topical therapy. Systemic antimycotics are applied parenterally and/or orally and show moderate to good bioavailability.

At present, numerous drugs for the topical treatment of superficial mycoses such as cutaneous and mucocutaneous infections are available, but there are only a limited number of drugs for the treatment of systemic mycoses. Additionally, resistance problems are arising, especially in the antimycotic chemotherapy of systemic mycoses, so that the medical need for new substances for this indication is great. Several "historical" compounds that show a relatively unselective antifungal activity, such as dyes, organic acids, mercury salts, and quaternary ammonium compounds, are sometimes still used in the treatment of superficial mycoses. Their therapeutic efficacy, tolerance, and patient acceptance are rated as low nowadays, and these substances are of decreasing relevance as much more selective antimycotics are available.

Modern antimycotics are expected to be effective in vitro, in animal models, and in infected humans and animals against one or more of the following classes of pathogenic fungi: dermatophytes (e.g., *Trichophyton* and *Microsporum* species, *Epidermophyton floccosum*), yeasts (e.g., *Candida* and *Torulopsis* species, *Cryptococcus neoformans*), moulds (e.g., *Aspergillus fumigatus* and other *Aspergillus* species, *Fusarium* species, zygomycetes such as *Mucor, Rhizopas,* and *Absidia* species), and dimorphic (biphasic) fungi, which can express a yeast phase, which is mostly the pathogenic form in the infected host, and a mould phase, which is mostly the infective, saprophytic form in nature, e.g., in soils [1].

In addition to the very common superficial mycoses of the skin and mucous membranes, yeasts, moulds, and dimorphic fungi can also induce serious, life-threatening generalized opportunistic infections, especially in immunocompromised hosts. As the number of patients at high risk of systemic mycoses (e.g., patients after transplantations, under aggressive antineoplastic chemotherapy or in intensive care, AIDS patients) is increasing steadily, these infections are much more common nowadays [1]. Dermatophytes solely cause infections of the skin, hair, and nails, and are unable to induce systemic, generalized mycoses, even in immunocompromised hosts.

The in vitro antimycotic activity of the substances is assessed by the determination of MIC values (minimal inhibitory concentrations). This is the lowest concentration of a substance that inhibits the growth of a fungal isolate in vitro. The MICs of antimycotics can vary considerably — up to as much as several decades — depending on culture conditions such as medium composition, incubation temperature, inoculum size and preparation method. Therefore, a high degree of standardization must be maintained to obtain comparable results. Furthermore, there is often only a poor or no in vitro/in vivo correlation of MIC values with the clinical outcome.

Some antimycotics, including azole agents, also show a moderate to mostly low in vitro antibacterial activity which does not seem to be relevant for the clinical situation during therapy. Therefore, these minor antibacterial properties of the substances are not discussed in this chapter. Many formulations, especially of topically applicable antimycotics, contain other compounds such as urea, antibiotics, and steroids. These combinations are not extensively discussed here.

2. Azole Antimycotics

Azoles with antimycotic properties can be devided into imidazole and triazole derivatives. Both groups have the same mechanism of antifungal action. Imidazole derivatives with antimycotic activity were first synthesized at Bayer and Janssen and described in 1968/1969. Serval important commercial products have been developed for human and veterinary medicine and plant protection [2].

Low doses of azole antimycotics produce morphological changes in sensitive fungi. These are reflected in changes in the cell volume and plasma membrane [3] and are probably due to the interaction of the azoles with ergosterol biosynthesis [4]–[7]. Biochemically, azoles specifically inhibit fungal ergosterol biosynthesis by interference with fungal specific isoenzymes of the cytochrome P450-dependent 14α-demethylation of lanosterol or 24-methylenedihydrolanosterol, which is a key step in the biosynthesis of ergosterol. The accumulation of membrane-disturbing 14α-methylsterols seems to be the origin of the antifungal properties of azole compounds, as known from in vitro studies [8]. Ergosterol is an essential constituent of the fungal plasma membrane, helping to maintain the integrity of the cell membrane which has a barrier function. Ergosterol deficiency leads to loss of essential cytoplasmatic constituents through the plasma membrane, and results in cessation of growth and even death of the fungus. In all studies, an accumulation of sterols with a methyl group on C14, which are unsuitable as membrane components, was observed [9], [10]. At higher doses, azoles also interfere with synthesis of fatty acids and triglyceride [4]. The weak antibacterial activity of these substances seems to be based on another mechanism. However, MIC values of azole antimycotics strongly depend on the test method and are mostly of limited use in predicting clinical efficacy.

2.1. Azole Antimycotics for Systemic Treatment

2.1.1. Fluconazole

Fluconazole [*86386-73-4*], α-(2,4-difluorophenyl)-α-(1*H*-1,2,4-triazol-1-ylmethyl)-1*H*-1,2,4-triazole-1-ethanol, $C_{13}H_{12}F_2N_6O$, M_r 306.27, *mp* 138–140 °C, crystals from ethyl acetate/hexane.

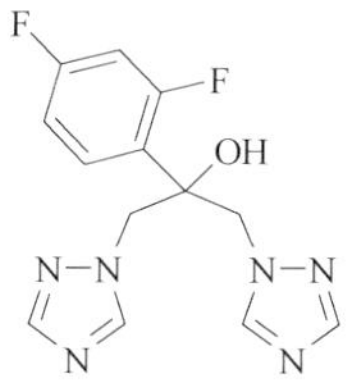

Table 1. MIC values for fluoconazole

Fungus	No. isolates tested	MIC range, µg/mL
Candida albicans	443	0.125 –> 80
Candida kefyr	22	32 – 64
Candida lusitaniae	28	0.25 – 32
Candida parapsilosis	68	0.5 – 4
Candida tropicalis	88	0.5 – >64
Trichosporon beigelii	15	1 – 64

Synonym(s). UK 49858.

Solubility. Soluble in water.

Stability. The pure solid form is stable under normal storage conditions.

Description. The triazole antimycotic fluconazole was first synthesized at Pfizer Pharmaceuticals [11], [12].

Formulations. Capsules à 0.05, 0.1, 0.15, and 0.2 g. Suspensions (0.5 and 1%) for oral application. Infusion bottles à 0.1, 0.2, and 0.4 g.

Trade Names: Biozolene (Bioindustria), Diflucan (Mason, Pfizer, Roerig), Dimycon (Alkaloid), Elazor (Sigma-Tau), Flavizol (Gador), Fungata (Mack), Mutum (Raffo), Triflucan (Pfizer).

Antimycotic Properties. Fluconazole is a broad spectrum triazole antimycotic with good in vitro and in vivo activity against pathogenic *Candida* species (apart from *Candida krusei*, which is a primary resistant *Candida* species), *Cryptococcus neoformans*, and dermatophytes [13]. *Aspergillus* species mostly show primary resistance towards fluconazole. Secondary resistance development can be observed after prolonged or repeated therapy of *Candida*-associated mycoses, especially in AIDS patients. In vitro susceptibilities of yeasts against fluconazole are reported in Table 1 [14].

Pharmacokinetics. The bioavailability after oral application ranges around 90% [15]. The plasma concentrations and AUC (Area Under the Curve) are dose-proportional in the dose range of 50 – 400 mg/d [15]. After a single oral dose of 100 mg, a plasma concentration of 1.7 µg/L is achieved (6.7 µg/L after 400 mg) [16]. A steady state is mostly achieved after a therapy duration of 6 – 10 d. The serum protein binding of fluconazole is 12% [17], and the plasma half-life 25 h. In *Cryptococcus neoformans* meningitis in AIDS patients, the liquor concentration was 80% of the plasma concentration [18]. In vaginal secretions, the concentration of fluconazole was similar to the plasma concentration [19]; 60 – 75% of the substance is eliminated by the kidneys in unmetabolized form. The elimination half-life is between 22 and 37 h [17].

Indications. Vaginal candidiasis [20] and systemic and mucocutaneous candidiasis [21]–[24]. In *Cryptococcus neoformans* meningitis the substance is mostly used in combination with amphotericin B and flucytosine [25]. Prophylaxis and preemptive therapy in AIDS patients, patients with leukemia, and patients with bone marrow transplants [26], [27].

Side Effects. Fluconazole is well tolerated.
CNS. Vertigo: 3.7 %, headache: 1.9 %.
GIT. Stomachache: 1.7 %, vomiting: 1.7 %, diarrhoea: 1.5 %.
Liver. Rise in transaminases (> 8 % of normal level): 1 %.
Skin. Exanthema: 1.8 %.
A Stevens – Johnson syndrome was observed after fluconazole therapy in an AIDS patient.

2.1.2. Itraconazole

Itraconazole [*84625-61-6*], 4-[4-[4-[4-[[2,4-Dichlorophenyl)-2-(1*H*-1,2,4-triazol-1-yl-methyl)-1,3-dioxolan-4-yl]-methoxy]phenyl]-1-piperazinyl]phenyl]-2,4-dihydro-2-(1-methylpropyl)-3*H*-1,2,4-triazol-3-one, $C_{35}H_{38}Cl_2N_8O_4$, M_r 705.64, *mp* 166.2 °C, crystals from toluene.

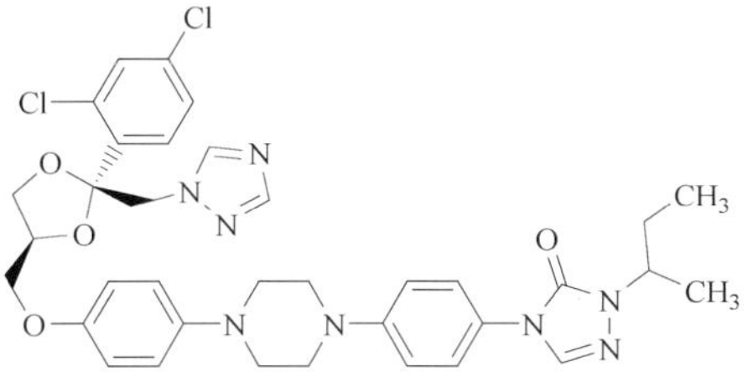

Synonym(s). R 51211.

Solubility. Itraconazole is a lipophilic substance which is practically insoluble in water and dilute acidic solutions. Partition coefficient in *n*-octanol/aqueous buffer of pH 8.1: 5.66.

Stability. The pure solid compund is stable under normal storage conditions.

Synthesis. Itraconazole is synthesized as follows [28], [29]:

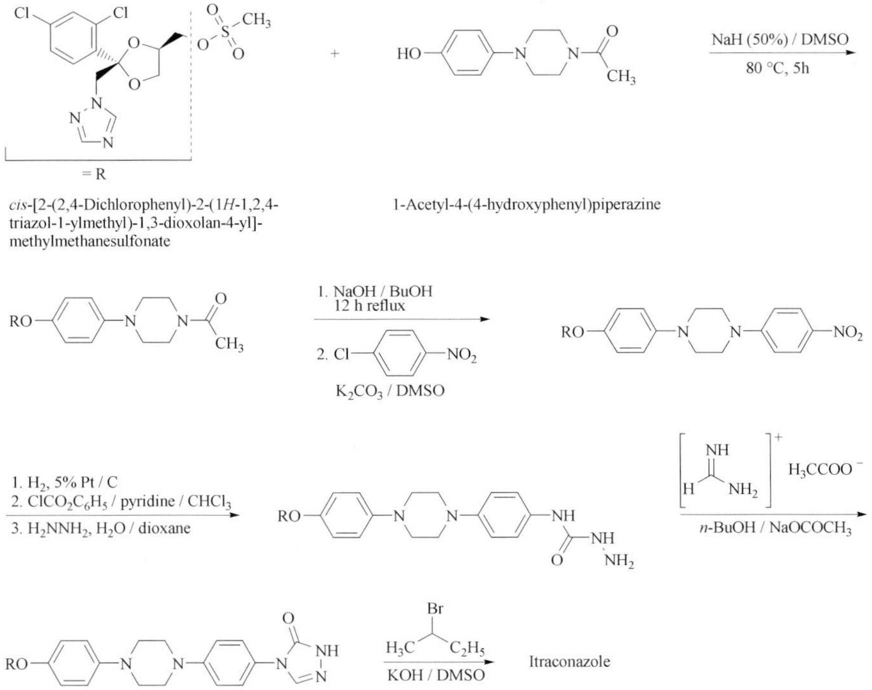

Description. The triazole antimycotic itraconazole was first synthesized at Janssen Pharmaceuticals [30], [31].

Formulations. Capsules à 0.1 g.

Trade Names: Itrizole (Cilag), Sempera (Janssen), Siros (Cilag), Sporanox (Edward Keller, Janssen), Triasporin (Lifepharma).

Antimycotic Properties. Itraconazole is a broad-spectrum triazole antimycotic with good in vitro and in vivo activity against pathogenic dermatophytes, moulds (including *Aspergillus* species), *Candida* species, and dimorphic fungi such as *Blastomyces dermatitidis, Coccidioides immitis, Histoplasma capsulatum, Cryptococcus neoformans,* and *Sporothrix schenckii* [32]–[40]. MIC values of itraconazole are given in Table 2 [14].

Pharmacokinetics. The substance has a bioavailability after oral application of 40 % if taken on an empty stomach to 55 % if taken simultaneously with food [41]. The absorption of itraconazole after oral application is slow; peak plasma concentrations are reached 3–6 h after intake [41], [42]. Plasma concentrations are low, and a peak plasma concentration of 0.1 µg/L was observed after a single oral dose of 100 mg [42]. The serum protein binding of itraconazole is high; more than 99 % of the substance is bound to plasma proteins, mostly to albumin. The substance shows good tissue penetration, especially into the skin and nails [41]. Itraconazole is extensively meta-

Table 2. MIC values for itraconazole

Fungus	No. isolates tested	MIC range, µg/mL
Aspergillus fumigatus	96	0.001 – 10
Candida albicans	1605	0.063 – 128
Candida (Torulopsis) glabrata	4	2 – 128
Candida krusei	15	0.125 – 0.5
Candida lusitaniae	15	$\leq$ 0.03 – 0.25
Candida parapsilosis	46	0.063 – > 128
Candida tropicalis	50	0.13 – > 128
Cryptococcus neoformans	118	0.001 – 0.5
Epidermophyton floccosum	5	0.063
Fusarium species	44	8 – > 32
Histoplasma capsulatum	10	0.063
Malassezia furfur	35	0.001 – 1
Microsporum species	10	0.063 – 0.25
Sporothrix schenckii	33	0.001 – 4
Trichophyton species	15	0.063 – 64
Trichosporon beigelii	15	$\leq$ 0.03 – 0.25

bolized in the liver, and more than 30 metabolites have been identified so far [43]. The elimination of the substance mostly occurs by the biliary-fecal route and renally in a metabolized form [44]. The elimination half-life ranges between 15 and 42 h, depending on the applied dose [41].

Indications. Therapy of dermatophytoses and onychomycoses [45] – [47] with a dosing regimen of 100 to 200 mg per day. Vaginal candidosis with 200 mg/d for 3 days [48]. Therapy of systemic infections such as candidosis and aspergillosis including disseminated infections, blastomycosis, coccidioidomycosis, histoplasmosis, and cryptococcosis with a dosing regimen of 100 to 400 (or up to 600) mg/d [49] – [51]. Especially in aspergillosis, an early onset of therapy is important for the clinical outcome.

Side Effects. Itraconazole is generally well tolerated.
CNS. Vertigo and headache: 10 %.
GIT. Stomachache, vomiting, and diarrhoea were observed.
Liver. Rise in transaminases and other reversible liver affections: 7 %.
Metabolism. Hypokaliemia and rise in serum triglycerides [52], [53].

2.1.3. Ketoconazole

Ketoconazole [65277-42-1], cis-1-acetyl-4-[4-[[2-(2,4-dichlorophenyl)-2-(1H-imidazol-1-ylmethyl)-1,3-dioxolan-4-yl]methoxy]phenyl]piperazine, $C_{26}H_{28}Cl_2N_4O_4$, M_r 531.4, mp 146 °C, colorless crystals.

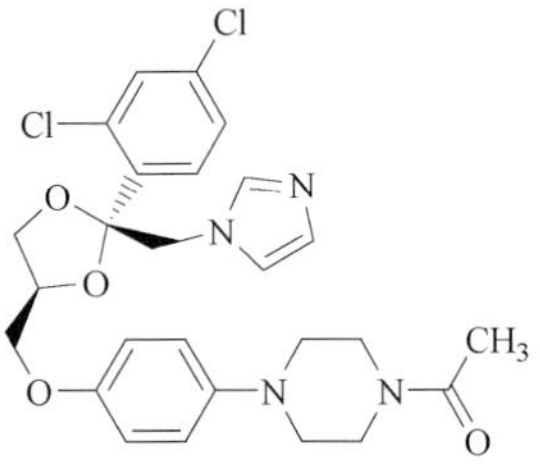

Synonym(s). R 41400.

Solubility. Ketoconazole is soluble in methanol, ethanol, and chloroform; slightly soluble in dilute acids; and barely soluble in water.

Stability. The pure solid compund is stable under normal storage conditions.

Synthesis. The synthesis of ketoconazole begins with the ketalization of 2,4-dichloroacetophenone (**1**) with glycerol. The intermediate **2** is not isolated but treated with bromine to form the bromide **3**. The benzoxylation of **3** gives **4** in the form of a cis–trans mixture. The cis form can be separated by crystallization. The reaction of **4** with imidazole in dimethylacetamide gives **5**, which is then hydrolyzed to **6**, which in turn is converted with methanesulfonyl chloride into the sulfonate **7**. The reaction of **7** with the sodium salt of N-acetyl-N-p-hydroxyphenylpiperazine in DMSO gives ketoconazole.

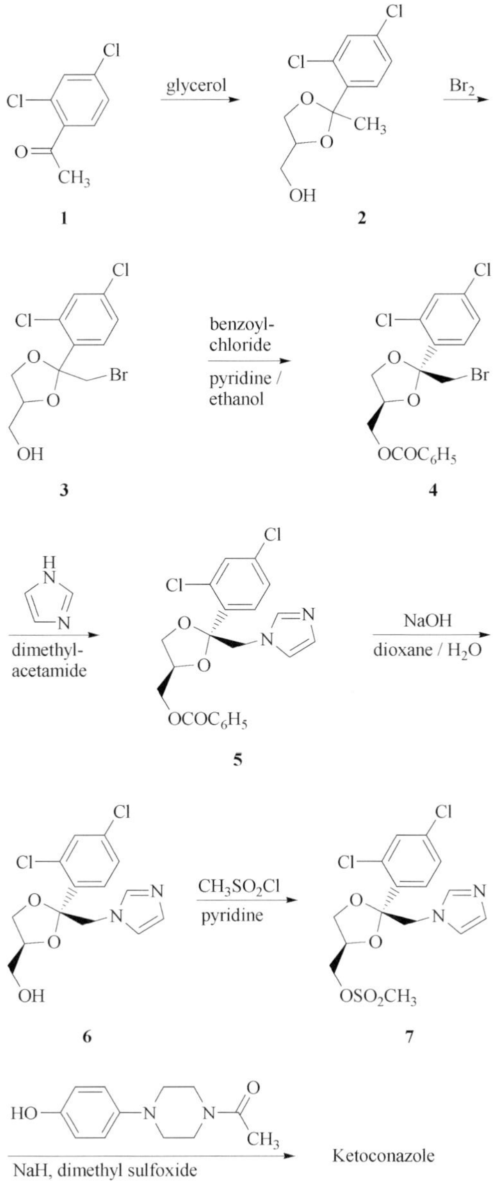

Description. The imidazole antimycotic ketoconazole was first synthesized at Janssen Pharmaceuticals [54]–[58].

Formulations. Tablets à 0.2 g; 2 % solutions, ointments, and cream preparations.

Table 3. MIC values for ketoconazole

Fungus	MIC range, µg/mL
Dermatophytes	0.04 – 10
Candida species	0.04 – 40
Cryptococcus neoformans	0.08 – 5
Aspergillus species	1.25 – 40
Dimorphic fungi	0.04 – 5
Chromomycetes	0.04 – 20

Trade Names: Candoral (Aché), Cetonax (Cilag), Fitonal (Disprovent), Fungarest (Janssen), Fungicil (Labinca), Fungo-Hubber (Hubber), Fungoral (Ilsan, Janssen, Leo), Ketazol (Exa), Ketoderm (Janssen), Ketoisdin (Isdin), Ketonan (IMA), Ketoral (Bilim), Micoral (Cassara), Micotek (Kressfor), Micoticum (Vita), Nizoral (Abic, Edward Keller, Janssen), Nizshampoo (Janssen), Oromycosal (Alkaloid), Oronazol (Krka), Panfungol (Esteve), Rofenid (Rhône-Poulenc), Terzolin (Janssen).

Antimycotic Properties. Ketoconazole shows a good efficacy against dermatophytes, dimorphic fungi (*Blastomyces dermatitidis, Coccidioides immitis, Histoplasma capsulatum*), and pathogenic *Candida* species [59], [60]. The activity against hyphomycetes and *Cryptococcus neoformans* is limited. MIC values of ketoconazole are summarized in Table 3. The in vitro activity of ketoconazole, like that of the other azoles, is strongly dependent on the test method used.

Pharmacokinetics. After oral administration, ketaconazole shows a variable degree of absorption of up to 70 %. The degree of absorption depends on the pH of the gastric fluid [61], [62]. Antacids and H_2-antagonists reduce the absorption of the drug. Peak serum concentrations are reached 2 – 4 h after administration and are between 2.0 and 3,5 µg/mL for a dose of 200 mg (3 mg/kg), and between 3.5 and 7.0 µg/mL for a 400 mg dose. Serum concentrations of 10 – 20 µg/mL can be reached after administration of a single dose of 1200 mg [63]. The substance shows a binding potency to plasma proteins and erythrocytes of 99 % [61]. After vaginal application, 1 % of the substance is absorbed; after cutaneous application, no absorption could be measured [64]. Ketoconazole penetrates well into the stratum corneum of the skin and vaginal secretions [61]. The concentration of ketoconazole in urine is low, with less than 1 µg/mL after a daily dose of 400 mg. The drug diffuses only slowly into the cerebrospinal fluid, where it reaches about 5 % of the serum concentration. The elimination half-life ranges from 4 to 9 h. The substance is mainly eliminated by the biliary – fecal route in a metabolized, inactive form.

Indications. Systemic therapy of dermatomycoses, blastomycosis, coccidioidomycosis, and histoplasmosis. Because of the side effects of ketoconazole, the newer azoles such as fluconazole and itraconazole should be considered instead of ketoconazole for systemic treatment. Ketoconazole is especially indicated in topical therapy of dermatomycoses, pityriasis versicolor, seborrheoic dermatitis, and dandruff.

Side Effects. After oral administraition of ketaconazole, side effects are seen at up to 10 % (400 mg/d) to > 50 % (800 mg/d) [65].

CNS. Vertigo, dizziness, and headache.

GIT. Somachache, loss of appetite, vomiting, and diarrhoea.

Liver. Rise in transaminases and other liver affections which are mostly reversible, although fulminant hepatitis was observed in 1 in 10 000 cases.

Vessels. Angioedema. Further, skin reactions, insomnia, and lethargy can be observed. Gynaecomastia, oligospermia, loss of libido, and alopecia, which may occur, are due to an inhibition of testosterone and cortisol biosynthesis [66]–[69]. Daily doses of higher than 400 mg reduce the testosterone serum concentration.

2.1.4. Miconazole

Miconazole [*22916-47-8*], 1-[2-(2,4-dichlorophenyl)-2-[(2,4-dichlorophenyl)methoxy]ethyl]-1*H*-imidazole, $C_{18}H_{14}Cl_4N_2O$, M_r 415.92.

Miconazole nitrate, [*22832-87-7*], $C_{18}H_{15}Cl_4N_3O_4$, M_r 479.15, *mp* 184 – 185 °C, colorless crystals.

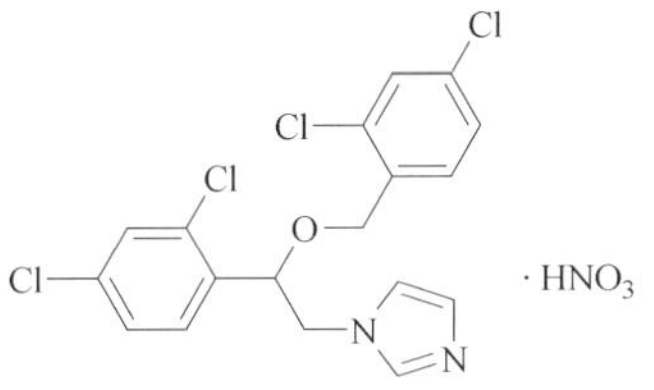

Synonym(s). R 18134; R 14889 (mononitrate).

Solubility. Miconazole is barely soluble in water (0.3 mg/mL) and slightly soluble to soluble in the many organic solvents and dilute inorganic acids.

Stability. The pure solid compound is stable under normal storage conditions.

Synthesis. The synthesis [70], [71] begins with 2,4-dichloroacetophenone, which is first brominated to form the phenacyl bromide and then treated with imidazole to form the phenacylimidazole. Reduction with sodium borohydride and reaction of the resulting alcohol with 2,4-dichlorobenzyl chloride yields miconazole.

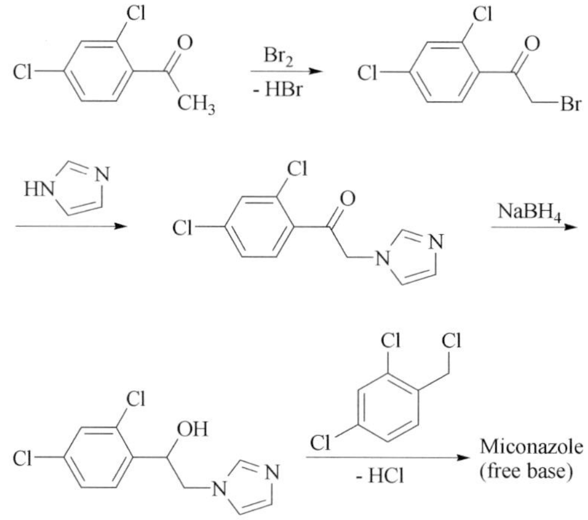

Description. The imidazole antimycotic miconazole was first synthesized at Janssen Pharmaceuticals [72], [73].

Formulations. Ampulles à 0.2 g; lozenges à 0.25 g; 2% solutions, lotions, powders, gels, ointments, and cream preparations; vaginal ovula.

Trade Names (Miconazole): Andergin (Pierell), Brentan (Janssen), Daktar (Janssen), Daktarin (Abic, Edward Keller, Janssen), Dumicoat (Dumex), Femeron (Janssen), Micofim (Elofar), Micotar (Dermapharm), Micotef (LPB), Monistat (Cilag, Ethnor, Janssen), Monistat I.V. (Janssen), Vodol (Searle).

Trade Names (Miconazole Nitrate): Aflorix (Gerardo Ramon), Albistat (Cilag), Aloid (Janssen), Andergin (Pierrel), Anotit (Janssen), Brenazol (DuraScan), Brentan (Janssen), Britane (Janssen), Canofite (Janssen), Conoderm (C-Vet), Conofite (Janssen, Pitman-Moore), Daktar (Janssen), Daktarin (Abic, Janssen), Decomyc (Merck), Deralbine (Andromaco), Dermacure (Janssen), Dermonistat (Ortho), Epi-Monistat (Cilag, Janssen), Florid (Janssen, Mochida), Fungiderm (Janssen), Funginazol (Morgens), Fungisdin (Esteve, Janssen), Fungucit (Iltas), Fungur (Salutas), Gyno-Daktar (Janssen), Gyno-Daktarin (Edward Keller, Janssen, Johnson & Johnson), Gyno-Monistat (Cilag, Janssen), Ipec (Byk-Gulden), Micatin (McNeil, Ortho), Micoderm (Italsuisse), Micogel (Cipla), Micogyn (Elofar), Miconal (Ecobi), Micotef (LPB), Monistat (Cilag, Janssen, Ortho), Monostat 7 (Ortho), Mykotral (Chephasaar), Prilagin (Gambar), Surolan (Janssen), Vodol (Searle).

Antimycotic Properties. Miconazole is a broad-spectrum imidazole antimycotic that generally shows activity against pathogenic dermatophytes (*Epidermophyton floccosum; Trichophyton* and *Microsporum* species), *Candida* species, moulds including *Asper-*

Table 4. MIC values for miconazole

Fungus	No. isolates tested	MIC range µg/mL
Aspergillus fumigatus	74	0.5–64
Blastomyces dermatitidis	57	0.001–2
Candida albicans	1815	0.016–100
Candida (Torulopsis) glabrata	224	0.016–64
Candida kefyr	33	0.016–4
Candida krusei	40	< 0.063–6.25
Candida lusitaniae	7	0.18–6.3
Candida parapsilosis	54	0.016–32
Candida tropicalis	172	0.016–33
Coccidioides immitis	99	0.063–6
Cryptococcus neoformans	69	0.063–25
Epidermophyton floccosum	19	0.06
Histoplasma capsulatum	48	0.05–2
Malassezia furfur	44	0.2–50
Microsporum canis	18	0.1–4
Pseudoallescheria boydii	42	< 0.016–4
Sporothrix schenckii	65	0.5–16
Trichophyton mentagrophytes	28	0.1–4
Trichophyton rubrum	63	0.001–32

gillus species, and pathogenic dimorphic fungi [74]–[78]. MIC values are summarized in Table 4.

Phamacokinetics. Miconazole is incompletely absorbed after oral application with an average bioavailability of 27% [79]. The active drug is dissolved in Cremophor EL for intravenous administration. After intravenous administration of a single dose of 200 mg, a peak plasma concentration of 1.6 mg/L was observed. Infusion of 500 mg of miconazole briefly produces serum concentrations of 2–9 µg/mL, which, however, rapidly fall to under 0.2 µg/mL. Miconazole is barely absorbed after cutaneous and/or mucocutaneous application. The substance shows a plasma protein binding of 90% and poor penetration into the liquor cerebrospinalis and other body fluids. Miconazole is extensively metabolized in the liver. The substance is eliminated predominantly by the biliary–fecal route as inactive metabolites. Elimination occurs in phases with a terminal elimination half-life of 24 h [79].

Indications. Especially topical therapy of skin infections caused by dermatophytes and yeasts, and mucocutaneous infections including *Candida* vulvovaginitis. Systemic, intravenous therapy especially of infections caused by *Pseudoallescheria boydii* (dosage up to 1.2 g 3× per day) [80]. For treatment of other generalized mycoses, mostly the newer, more effective, and less toxic azole antimycotics for systemic use, such as fluconazole and itraconazole, are indicated.

Side Effects. Multiple side effects have been observed after intravenous infusion of miconazole, which seem to be partly due to the formulation of the solvent, which contains the detergent Cremophor EL; mostly anaphylactic reactions, pruritus, dizziness, tachyarrhythmia, hyperaggregation of erythrocytes, anemia, thromocytosis, phlebitis, nausea, vomiting, hyperlipidemia, hyponatriemia, arthralgia, and flush [81].

2.2. Azole Antimycotics for Topical Treatment

2.2.1. Bifonazole

Bifonazole [*60628-96-8*], 1-([1,1'-biphenyl]-4-ylphenylmethyl)-1*H*-imidazole, $C_{22}H_{18}N_2$, M_r 310.4, *mp* 147 – 148 °C, colorless, odorles, tasteless crystals.

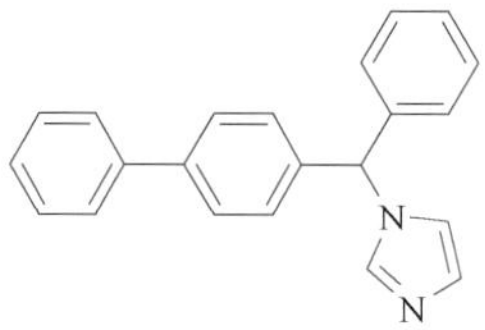

Synonym(s). Trifonazole, Bay h 4502.

Solubility. Bifonazole is soluble in glacial acetic acid (833 mg/mL), chloroform (344 mg/mL), benzene (38.2 mg/mL), methanol (36.5 mg/mL), ethanol (24.6 mg/mL), ethyl acetate (17.1 mg/mL), and acetonitrile (10.4 mg/mL). It is practically insoluble in water, but in Walpole's buffer, solubility is 3.75 mg/mL at pH 2.

Stability. Pure bifonazole is stable under normal storage conditions. In solution, the substance is stable under neutral or slightly alkaline conditions. It slowly decomposes in strongly alkaline or strongly acidic media. Bifonazole solutions are sensitive to light.

Table 5. MIC values for clotrimazole and bifonazole

Fungus	MIC range, µg/mL	
	clotrimazole	bifonazole
Dermatophytes	<0.01–2	0.01–1
Candida species	0.04–10	0.08–10
Cryptococcus neoformans	0.04–10	0.04–2.5
Pityrosporum species	2–4	0.5–1
Aspergillus species	0.1–1.25	0.5–1
Chromomycetes	0.04–2.5	<0.04–0.31
Dimorphic fungi	0.04–0.31	<0.04–0.08

Synthesis. Bifonazole is synthesized as follows:

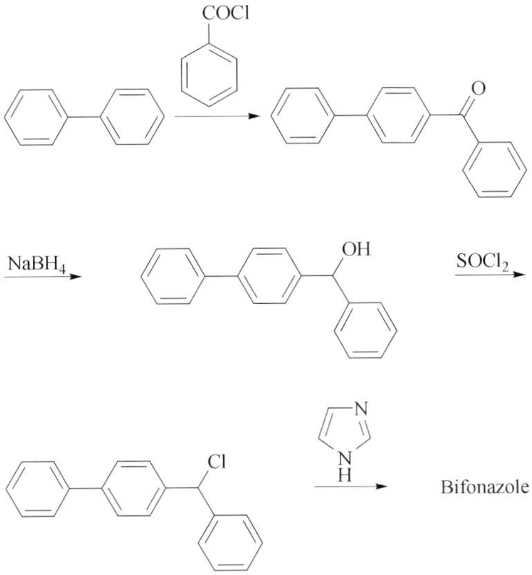

Description. The imidazole antimycotic bifonazole was first synthesized at Bayer Pharmaceuticals in 1974 [82]–[86].

Formulations. 1% solutions, powders, gels, ointments, and cream preparations.

Trade Names: Amycor (Lipha), Azolmen (Menarini), Bifonazol (Bayropharm), Micofun (Incobra), Mycospor (Bayer, Kai Cheong), Mycosporin (Bayer).

Antimycotic Properties. Bifonazole is a broad-spectrum imidazole antimycotic with excellent in vitro activity against pathogenic dermatophytes, yeasts, moulds, and dimorphic fungi. MIC values of bifonazole are summarized in Table 5.

Pharmacokinetics. Bifonazole is absorbed adequately after oral administration to humans and animals. The substance quickly induces ribosomal liver enzymes and is primarily metabolized to metabolites without antimycotic activity. Therefore, bifonazole can only be used as a topical antimycotic [87], [88]. The substance is minimally absorbed after topical application of 1% formulations. Concentrations in blood serum remain below 5 ng/mL, even after long term treatment of a 200 cm^2 skin area. Concentrations in the stratum papillare of the skin reach 2–3 µg/g [88].

Indications. Topical treatment of mycoses of the skin induced or sustained by fungi such as dermatophytes, yeasts, and also chromomycetes.

Side Effects. Topical applications, including intravaginal application, are well tolerated. Skin irritations and allergic reactions are rarely observed and seem to be mainly due to the galenic formulation (e.g., fatty preparations on acute, inflammatory lesions). Allergic reactions are rarely observed [89].

2.2.2. Butoconazole

Butoconazole [64872-76-0], (±)-1-[4-(4-chlorophenyl)-2-[(2,6-dichlorphenyl)thio]butyl]-1H-imidazole, $C_{19}H_{17}Cl_3N_2S$, M_r 411.78, mp 68–70.5 °C, crystals from cyclohexane.

Butoconazole nitrate [64872-77-1], $C_{19}H_{18}Cl_3N_2O_3S$, M_r 475.01, mp 162–163 °C, colorless blades from acetone/ethyl acetate.

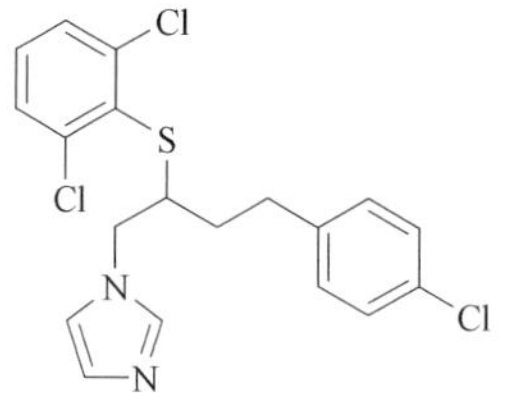

Solubility. For further information, see [90].

Stability. The pure solid compound is stable under normal storage conditions.

Synthesis [91]:

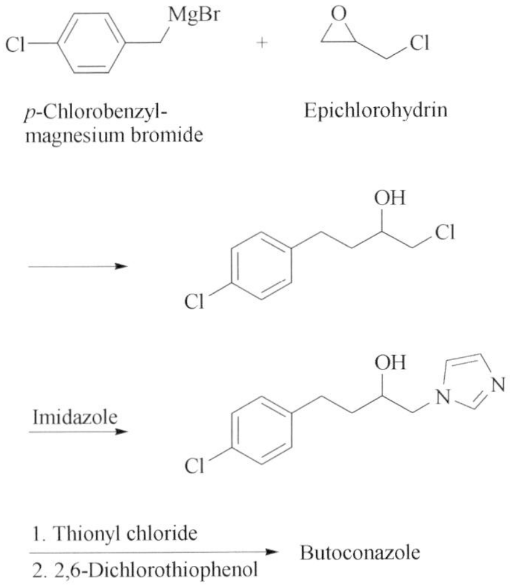

p-Chlorobenzyl-
magnesium bromide + Epichlorohydrin

→ [intermediate: 4-(4-chlorophenyl)-1-chloro-2-butanol]

Imidazole → [intermediate: 1-[4-(4-chlorophenyl)-2-hydroxybutyl]imidazole]

1. Thionyl chloride
2. 2,6-Dichlorothiophenol
→ Butoconazole

Description. The imidazole antimycotic butoconazole was first synthesized at Syntex Pharmaceuticals [92].

Formulations. 2% cream preparations.
Trade Names (Butoconazole Nitrate): Femcosyn (Protochemie), Femstat (Syntex), Gynomyk (Cassenne).

Antimycotic Properties. Broad-spectrum imidazole antimycotic with in vitro and in vivo activitiy against pathogenic yeasts, dermatophytes, moulds, and dimorphic fungi [93], [94].

Pharmacokinetics. After intravaginal application of 100 mg of butoconazole nitrate, 5% of the dose was found in the urine and feces [95].

Indications. Butoconazole is generally indicated for therapy of vulvovaginal mycoses caused by *Candida* species and *Candida (Torulopsis) glabrata* [96], [97].

Side Effects. Local irritations such as itching and burning sensations and allergic reactions may occur in rare cases and are mainly due to the galenic formulation [98].

2.2.3. Clotrimazole

Clotrimazole [23593-75-1], 1-[(2-chlorophenyl)diphenylmethyl]-1H-imidazole, $C_{22}H_{17}ClN_2$, M_r 344.8, mp 143–144 °C, colorless, tasteless crystals.

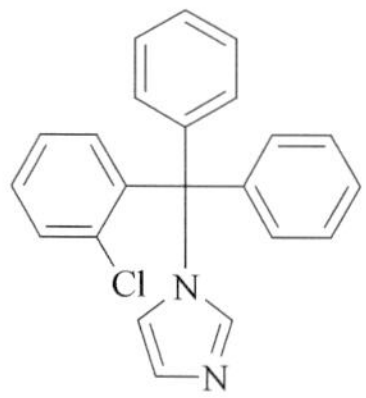

Synonym(s). Chlortritylimidazole, Bay 5097, FB b 5097, PCPIM.

Solubility. Clotrimazole is readily soluble in methanol, chloroform, and DMF (all > 100 mg/mL); slightly soluble in acetone (50 mg/mL), ethyl acetate (45 mg/mL), ethanol (95 mg/mL), and PEG 400 (60 mg/mL); and practically insoluble in water (< 0.01 mg/mL).

Stability. Pure clotrimazole is stable under normal storage conditions. The stability of the solution is pH dependent. Above pH 7 the solution is stable. In acid solution, clotrimazole undergoes gradual hydrolysis to form 2-chlorophenylbisphenylmethanol and imidazole. The hydrolysis half-life at pH 1 is about 170 h in ethanol/water (1/1) at room temperature.

Synthesis. The synthesis of clotrimazole begins with 2-chlorobenzyl chloride, which is converted to benzotrichloride. Friedel–Crafts reaction with benzene followed by reaction with imidazole forms clotrimazole [99].

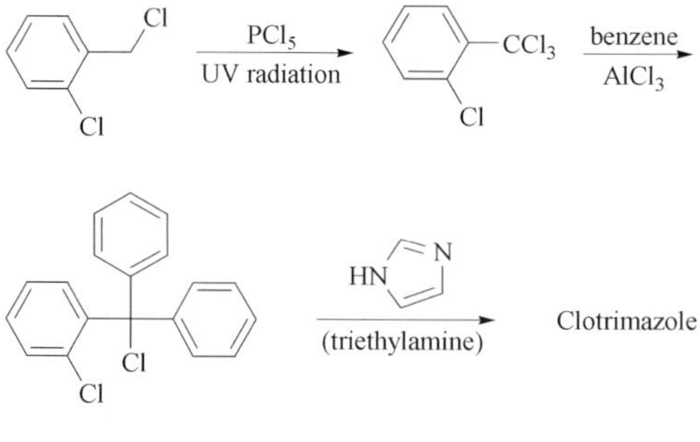

Description. The imidazole antimycotic clotrimazole was first synthesized at Bayer Pharmaceuticals in 1967 [100]. Ist antimycotic activity was first reported in 1969 [101], [102] – [104].

Formulations. Solutions, powders, gels, ointments, and cream preparations; lozenges and vaginal tablets.

Trade Names: Agisten (Agis), Akneclor (Spirig), Antifungol (Hexal), Antimicotico (Savoma), Antimyk Neu (Pfleger), Apocanda (Apogepha), Azutrimazol (Azuchemie), Canesten (Bayer, Bayropharm, Kai Cheong), Canifug (Wolf), Clazol (Ilsan), Clocim (Cimex), Clot-basan (Sagitta), Clotrifug (Wolff), Clotrimaderm (Taro), Clotrimix (Hosbon), clotri OPT (Optimed), Clotrizol (Jossa), Clozole (Jean-Marie), Contrafungin (PharmaGalen), Cutistad (Stada), Dignotrimazol (Dignos), Dolexalan (Wölfer), Durafungol (Durachemie), Empecid (Bayer, Yoshitomi), Eurosan (Mepha), Femcare (Schering-Plough), Fungiframan (Oftalmiso), Fungi-med (Permamed), Fungizid-ratiopharm (ratiopharm), Fungosten (Mulda), Fungotox (Mepha), Gilt (Lyssia), Gino-Canesten (Bayer), Gino-Clotrimix (Hosbon), Gino-Lotramina (Schering Corp.), Gromazol (Grossman), Gyne-Lotremin (Mason, Schering Corp., Schering-Plough), Jenamazol (Jenapharm), KadeFungin (Kade), Localicid (Dermapharm), Lotramina (Schering Corp.), Lotremin (Mason, Schering Corp.), Lotrimin (Schering Corp., Schering-Plough), Micomisan (Hosbon), Micoter (Cusi), Mono-Baycuten (Bayropharm), Mycelex (Bayer, Dome), Mycelex-7 (Bayer), Mycelex-G (Bayer, Dome), Myclo (Boehringer Ingelheim), Mycofug (Hermal), Myco-Hermal (Hermal, Jebsen), Mycoril (Remedica), Mycosporin (Bayer), Mycotrim (Lagap), Myko Crdes (Ichthyol), Mykofungin (Wyeth), Mykohaug C (Salutas), Ovis Neu (Aldenylchemie), Peckle (Hisamitsu), Pedisafe (Sagitta), Plimycol (Pliva), SD-Hermal (Hermal), Stiemazol (Stiefel), Tibatin (DAK), Tricosten (Farmion), Trimysten (Bellon), Uromykol (Hoyer).

Antimycotic Properties. Broad-spectrum imidazole antimycotic with good activity against yeasts, dermatophytes, moulds, and dimorphic fungi. MIC values of clotrimazole are summarized in Table 5.

Pharmacokinetics. Clotrimazole is absorbed adequately after oral administration to humans and animals. The substance quickly induces ribosomal liver enzymes and is primarily metabolized to metabolites without antimycotic activity. Therefore, clotrimazole can only be used as a topical antimycotic [105], [106]. The substance is minimally absorbed after the topical application of 1% formulations. Concentrations in blood serum remain below 5 ng/mL, even after long-term treatment of a 200 cm^2 skin area. Concentrations in the stratum papillare of the skin reach 2 – 3 µg/g [106].

Indications. Topical treatment of mycoses of the skin induced or sustained by fungi such as dermatophytes, yeasts, and chromomycetes. In addition, clotrimazole is indicated for therapy of vulvovaginal mycoses caused by *Candida* species and *Candida (Torulopsis) glabrata*.

Side Effects. Topical applications, including intravaginal application, are well tolerated. Skin irritations and allergic reactions are rarely observed and seem to be mainly due to the galenic formulation (e.g. fatty preparations on acute, inflammative lesions). Allergic reactions are rarely observed [107].

2.2.4. Croconazole

Croconazole [77175-51-0], 1-[1-[2-(3-chlorobenzyloxy)phenyl]vinyl]imidazole, $C_{18}H_{15}ClN_2O$, M_r 310.78, mp 72–73 °C, white crystals.

Croconazole hydrochloride [77174-66-4], $C_{18}H_{16}Cl_2N_2O$, M_r 347.24, mp 148.5–150 °C, white crystals.

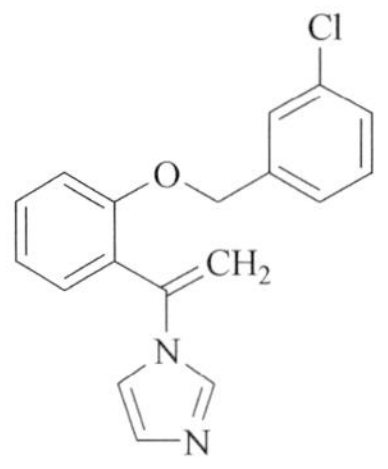

Synonym(s). Croconazole (free base); S 710674 (hydrochloride).

Solubility. Soluble in ethyl acetate.

Stability. The pure solid compound is stable under normal storage conditions.

Description. The imidazole antimycotic croconazole was first synthesized at Shionogi Pharmaceuticals [108], [109].

Formulations. 1% creams and gel preparations.

Trade Names (Croconazole Hydrochloride). Pilzcin (Merz, Merz + Schoeller, Shionogi).

Antimycotic Properties. Broad-spectrum imidazole antimycotic with activity against almost all pathogenic fungi.

Pharmacokinetics. Croconazole is not absorbed after cutaneous application.

Indications. Topical treatment of mycoses of the skin induced or sustained by fungi such as dermatophytes and yeasts.

Side Effects. Local irritations and allergic reactions after topical treatment may occur in rare cases and are mainly due to the galenic formulation.

2.2.5. Econazole

Econazole [*27220-47-9*], 1-[2-[(4-chlorphenyl)methoxy]-2-(2,4-dichlorophenyl)ethyl]-1*H*-imidazole, $C_{18}H_{15}Cl_3N_2O$, M_r 381.69, *mp* 86.8 °C, colorless crystals.

Econazole nitrate [*68797-31-9*], $C_{18}H_{16}Cl_3N_3O_4$, M_r 444.7, *mp* 164–165 °C, colorless crystals.

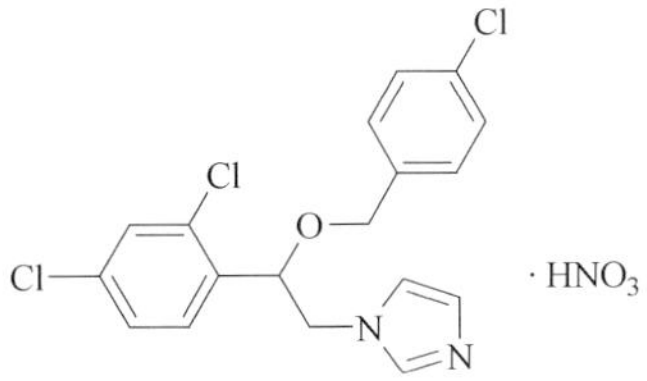

Synonym(s). R 14827 (mononitrate).

Solubility. Econazole is soluble in methanol, acetic acid, and PEG 400; less soluble in ethanol, acetone, chloroform, and butanol; and barely soluble in water, ether, cyclohexane, and hexane [110].

Stability. The solid substance is stable under usual storage conditions but should be stored in the dark.

Synthesis. The synthesis is almost the same as that of miconazole (see Section 2.1.4) [70], [71], [111], differing only in the last stage, in which the 1-[2,4-dichloro[β-hydroxy)phenethyl]imidazole intermediate is treated with 4-chlorobenzyl chloride.

Description. The imidazole antimycotic econazole was first synthesized at Janssen Pharmaceuticals [111] in 1969 [112], [113].

Formulations. 1 % solutions, powders, ointments, and cream preparations; vaginal ovula (150 mg).

Trade Names (Econazole): Pevalip (Janssen), Pevaryl (Cilag), Pevaryl Lipogel (Cilag, Janssen), Pevaryl lotion (Cilag), Pevaryl P.V. (Cilag).

Trade Names (Econazole Nitrate): Amicel (Salus), Chemionazolo (Brocchieri), Dermazol (CT), Dermazole (Lision Hong), Ecalin (JAKA-80), Ecodergin (Leben's), Eco Mi (Geymonat), Ecorex (Tosi), Ecostatin (Squibb, Westwood-Squibb), Ecotam (Alacan), Epi-Pevaryl (Cilag), Etramon (Johnson & Johnson), Fitonax (Cilag), Gyno-Pevaryl (Cilag, Edward Keller), Ifenec Zilliken), Micocert (Dexter), Micoespec (Centrum), Micofungal (Labopharma), Micogin (Crosara), Micogyn (Crosara), Micos (AGIPS), Micosten (Bergamon), Micostyl (Stiefel), Mitekol (Lek), Mykopevaryl (Cilag), Pargin (Gibipharma),

Table 6. MIC values for econazole

Fungus	No. isolates tested	MIC range µg/mL
Aspergillus fumigatus	24	0.15 – 25
Candida albicans	283	0.016 – 25
Candida (Torulopsis) glabrata	38	0.016 – 25
Candida kefyr	33	0.016 – 6.25
Candida krusei	34	0.125 – 12.5
Candida parapsilosis	32	0.016 – 25
Candida tropicalis	52	0.016 – 25
Malassezia furfur	19	0.0125 – 25

Pevaryl (Cilag, Edward Keller, Smith Kline Beecham), Pevaryl P.V. (Edward Keller), Polycain (Taiho), Skilar (Bonomelli), Spectazole (Ortho).

Antimycotic Properties. Broad-spectrum imidazole antimycotic with activity against almost all species of pathogenic fungi [114], [115]. MIC values of econazole are summarized in Table 6 [14].

Pharmacokinetics. After vaginal application, 1 – 27 % of the applied dose was found in the urine [116], [117].

Indications. Topical treatment of mycoses of the skin induced or sustained by fungi such as dermatophytes and yeasts [118]. In addition, econazole is indicated for therapy of vulvovaginal mycoses caused by *Candida* species and *Candida (Torulopsis) glabrata* [119], [120].

Side Effects. Local irritations such as itching and burning sensations and allergic reactions may occur in rare cases after topical therapy and are mainly due to the galenic formulation [121].

2.2.6. Fenticonazole

Fenticonazole [*72479-26-6*], 1-[2-(2,4-dichlorophenyl)-2-[[4-(phenylthio)phenyl]-methoxy]ethyl]-1*H*-imidazole, $C_{24}H_{20}Cl_2N_2OS$, M_r 455.4.

Fenticonazole nitrate [*73151-29-8*], $C_{24}H_{21}Cl_2N_3O_4S$, M_r 518.63, *mp* 136 °C, odorless white crystalline powder.

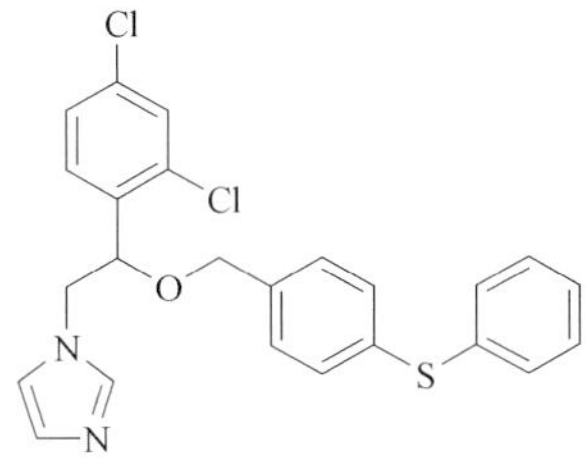

Synonym(s). Rec 15/1476 (mononitrate).

Solubility. Solubility at 20 °C: water < 0.1 mg/mL, diethyl ether < 0.1 mg/mL, ethanol 30 mg/mL, methanol 100 mg/mL, chloroform 300 mg/mL, DMF 600 mg/mL [122].

Stability. The pure solid compund is stable under normal storage conditions.

Description. The imidazole antimycotic fenticonazole was first synthesized at Recordati Pharmaceuticals [123], [124].

Formulations. 2% solutions, ointments, and cream formulations; vaginal ovula (600 mg).

Trade Names (Fenticonazole Nitrate). Falvin (Farmades), Fenizolan (Organon), Fentiderm (Zyma), Fentigyn (Novartis), Lomexin (Grünenthal, Recordati), Mycodermil (Vifor Fribourg).

Antimycotic Properties. Broad-spectrum imidazole antimycotic with activity against almost all species of pathogenic fungi.

Pharmacokinetics. After topical application, the substance is not absorbed.

Indications. Topical treatment of mycoses of the skin induced or sustained by fungi such as dermatophytes and yeasts [125], [126]. In addition, fenticonazole is indicated for therapy of vulvovaginal mycoses caused by *Candida* species and *Candida (Torulopsis) glabrata* [127].

Side Effects. Local irritations such as itching and burning sensations and allergic reactions may occur in rare cases and are mainly due to the galenic formulation.

2.2.7. Isoconazole

Isoconazole [*27523-40-6*], 1-[2-(2,4-dichlorophenyl)-2-[(2,6-dichlorophenyl)methoxy]ethyl]-1*H*-imidazole, $C_{18}H_{14}Cl_4N_2O$, M_r 416.1, white crystals.

Isoconazole nitrate [*24168-96-5*], $C_{18}H_{15}Cl_4N_3O_4$, M_r 479.1, *mp* 182 – 183 °C, colorless crystals.

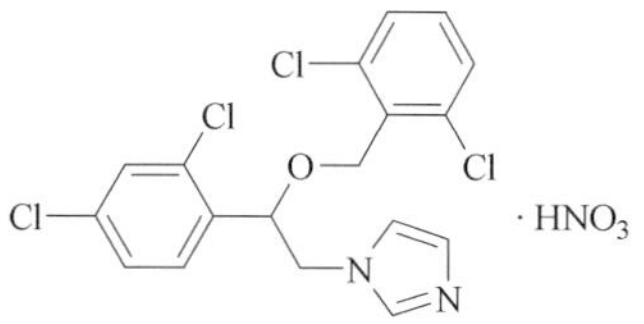

Synonym(s). R 15454 (mononitrate).

Solubility. Isoconazole is soluble in methanol, acetic acid, and PEG 400; less soluble in ethanol, acetone, chloroform, and butanol; and barely soluble in water, diethyl ether, cyclohexane, and hexane.

Stability. The solid substance is stable under normal storage conditions.

Synthesis. The synthesis of isoconazole [128] differs from that of miconazole (see Section 2.1.4) only in the last stage, in which 1-[2,4-dichloro(β-hydroxy)phenethyl]-imidazole is treated with 2,6-dichlorobenzyl chloride in hot benzene – DMF in the presence of sodium hydride.

Description. The imidazole antimycotic isoconazole was first synthesized at Janssen Pharmaceuticals [128] – [131].

Formulations. 1 % solutions, ointments, and cream formulations.

Trade Names (Isoconazole Nitrate). Fazol (Bellon), Gyno-Travogen (Jebsen, Schering), Icaden (Schering), Isogyn Ginecologico (Crosara), Mupaten (Schering-Plough), Travogen (Jebsen, Schering), Travogyn (Schering).

Antimicrobial Properties. Broad-spectrum azole antimycotic with activity against almost all species of pathogenic fungi [132], [133].

Pharmacokinetics. After topical application, isoconazole remains practically unabsorbed. Application in an ethanol/propyleneglycol vehicle increases penetration into the stratum corneum of the skin significantly [134].

Indications. Topical treatment of mycoses of the skin induced or sustained by fungi such as dermatophytes and yeasts. In addition, isoconazole is indicated for therapy of vulvovaginal mycoses causes by *Candida* species and *Candida (Torulopsis) glabrata* [135].

Side Effects. Local irritations such as itching and burning sensations and allergic reactions may occur in rare cases and are mainly due to the galenic formulation.

2.2.8. Omoconazole

Omoconazole [*74512-12-2*], (*Z*)-1-[2-[2-(4-chlorophenoxy)ethoxy]-2-(2,4-dichlorophenyl)-1-methylethenyl]-1*H*-imidazole, $C_{20}H_{17}C_{13}N_2O_2$, M_r 423.73, *mp* 89 – 90 °C, crystals from ethyl acetate/hexane (1/4).

Omoconazole nitrate [*83621-06-1*], $C_{20}H_{18}Cl_3N_3O_5$, M_r 485.96, *mp* 118 – 120 °C (Büchi) / 122.5 (Mettler), crystals from ethyl acetate/ethanol.

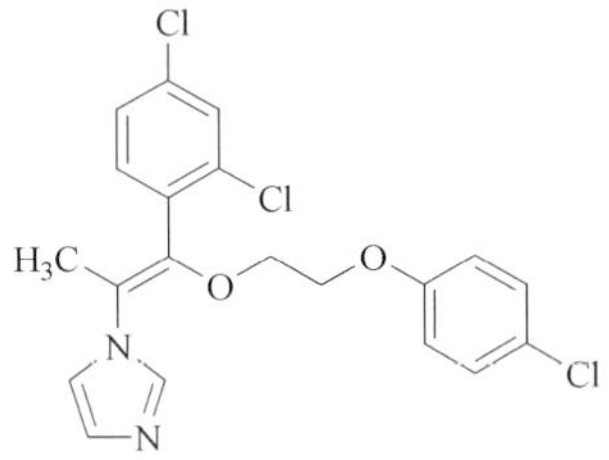

Synonym(s). CM 8282 (mononitrate).

Stability. The pure solid substance is stable under normal storage conditions.

Description. The imidazole antimycotic omoconazole was first synthesized at Siegfried Pharmaceuticals [136], [137]. Stereospecific synthesis [138].

Formulations. 1% cream formulations.

Trade Names (Omoconazole Nitrate). Azameno (Kwizda, Wyeth), Fongamil (Biorga), Fongarex (Sanofi), Fongorex (Siegfried), Melur (Siegfried).

Antimycotic properties. Broad-spectrum imidazole antimycotic with activity against almost all pathogenic fungi [139].

Pharmacokinetics. After topical application, omoconazole is not absorbed.

Indications. Topical treatment of mycoses of the skin induced or sustained by fungi such as dermatophytes and yeasts.

Side Effects. Local irritations such as itching and burning sensations and allergic reactions may occur in rare cases and are mainly due to the galenic formulation.

2.2.9. Oxiconazole

Oxiconazole [64211-45-6], (Z)-1-(2,4-dichlorophenyl)-2-(1H-imidazol-1-yl)ethanone O-[(2,4-dichlorophenyl)methyl]oxime, $C_{18}H_{13}Cl_4N_3O$, M_r 429.92.

Oxiconazole nitrate [64211-46-7], $C_{18}H_{14}Cl_4N_4O_4$, M_r 492.15, mp 137–138 °C, crystals from ethanol.

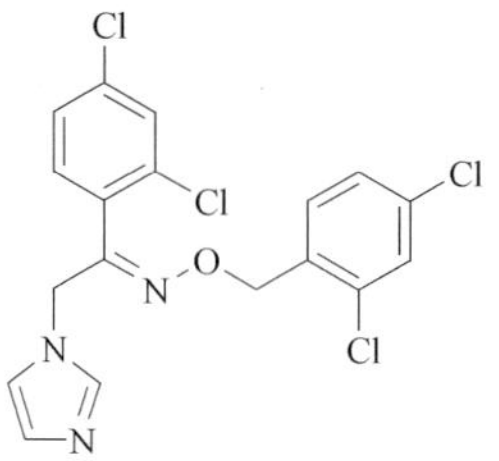

Synonym(s). Ro 13-8996/000 (free base); Ro 13-8996/001, Sgd 301-76 (mononitrate).

Stability. The pure solid substance is stable under normal storage conditions.

Synthesis. Oxiconazole is synthesized as follows [140], [141]:

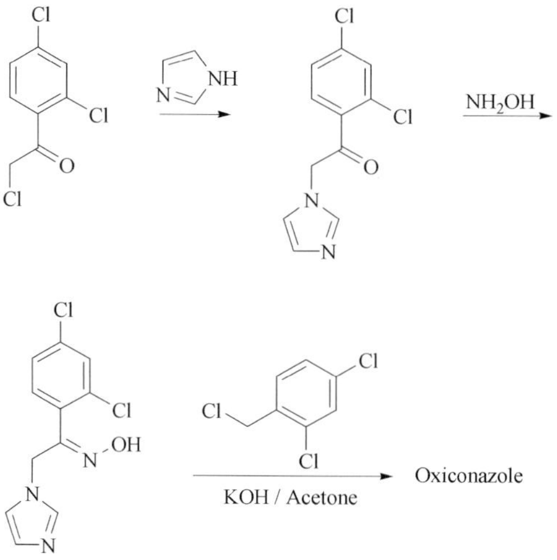

Description. The imidazole antimycotic oxiconazole was synthesized at Roche and Siegfried Pharmaceuticals [142], [143].

Formulations. 1 % solutions, powders, ointments, and cream formulations. Vaginal tablets (600 mg).

Trade Names (Oxiconazole Nitrate): Gyno-Myfungar (Klinge), Myfungar (Klinge, Siegfried, Wyeth), Oceral (Roche), Oxistat (Glaxo).

Antimycotic Properties. Broad-specktrum imidazole antimycotic with activity against almost all species of pathogenic fungi [144].

Pharmacokinetics. After local application, oxiconazole penetrates well into the stratum corneum of the skin and the hair follicle [145].

Indications. Topical treatment of mycoses of the skin induced or sustained by fungi such as dermatophytes and yeasts. In addition, oxiconazole is indicated for therapy of vulvovaginal mycoses caused by *Candida* species and *Candida (Torulopsis) glabrata* [146].

Side Effects. Local irritations such as itching and burning sensations, and allergic reactions may occur in rare cases and are mainly due to the galenic formulation.

2.2.10. Sertaconazole

Sertaconazole [*99592-32-2*], (±)-1-[2,4-dichloro-β-[(7-chlorobenzo[*b*]thien-3-yl)-methoxy]phenethyl]imidazole, $C_{20}H_{15}Cl_3N_2OS$, M_r 437.78.

Sertaconazole nitrate [*99592-39-9*], $C_{20}H_{15}Cl_3N_3O_4S$, M_r 501.01, *mp* 158 – 160 °C, odorless, white crystalline powder.

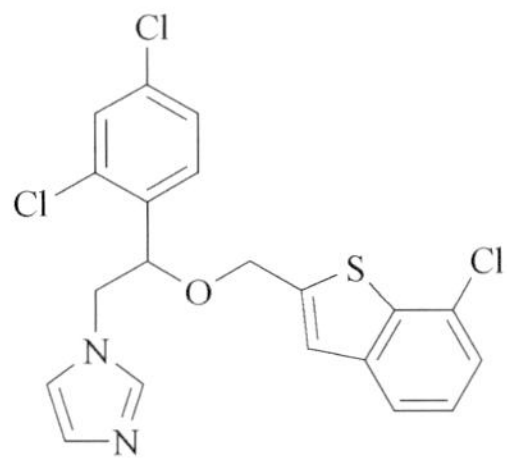

Synonym(s). FI-7045 (free base); FI-7056 (mononitrate).

Solubility. Fairly soluble in ethanol (1.7 %), chloroform (1.5 %); slightly soluble in acetone (0.95 %); very slightly soluble in *n*-octanol (0.069 %). Practically insoluble in water (< 0.01 %).

Stability. The pure solid substance is stable under normal storage conditions.

Synthesis. The Synthesis is described in [147].

Description. See [148], [149].

Formulations. 1 % cream formulations, gels, and ointments. New formulations for the treatment of vaginal candidiasis and oral formulations for the prophylaxis and treatment of mucocutaneous buccopharyngeal candidiasis are being developed.

Trade Names (Sertaconazole Nitrate). Dermofix (Ferrer), Dermoseptic (Smith Kline Beecham), Zalain (Robert).

Antimycotic Properties. Sertaconazole is a rather new broad-spectrum imidazole antimycotic with activity against almost all species of pathogenic fungi. It also has excellent activity against pathogenic yeasts [150] – [153].

Parmacokinetics. The substance is not absorbed after topical administration.

Indications. Topical treatment of mycoses of the skin induced or sustained by fungi such as yeasts and dermatophytes [154]. New formulations for the treatment of vaginal mycoses are in development.

Side Effects. Local irritations and allergic reactions may occur in rare cases and are mainly due to the galenic formulation.

2.2.11. Sulconazole

Sulconazole [*61318-90-9*], 1-[2-[[(4-chlorophenyl)methyl]thio]-2-(2,4-dichlorophenyl)-ethyl]-1*H*-imidazole $C_{18}H_{15}Cl_3N_2S$, M_r 460.8.

Sulconazole nitrate [*61318-91-0*], $C_{18}H_{15}Cl_3N_3O_3S$, M_r 524.03, *mp* 130.5 – 132.0 °C, white or off-white crystalline powder.

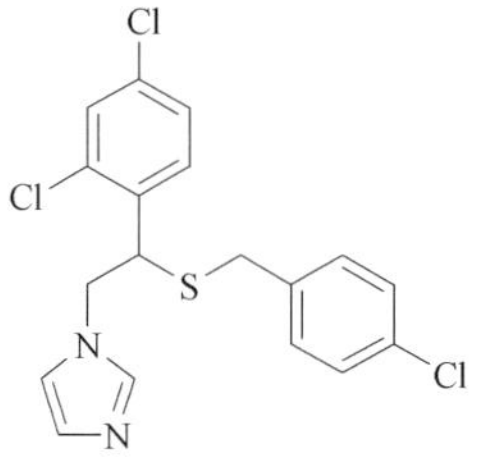

Synonym(s). RS-44872, RS-44872-00-10-3 (both mononitrate).

Solubility. 1 part in 3,333 (water), 1 part in 100 (ethanol), 1 part in 130 (acetone), 1 part in 333 (chloroform), 1 part in 2000 (dioxan), 1 part in 71 (methanol), 1 part in 286 (chloromethane), 1 part in 10 (pyridine), 1 part in 2000 (toluene).

Stability. The pure solid substance is stable under normal storage conditions, but shoud be protected from light.

Synthesis. [155]:

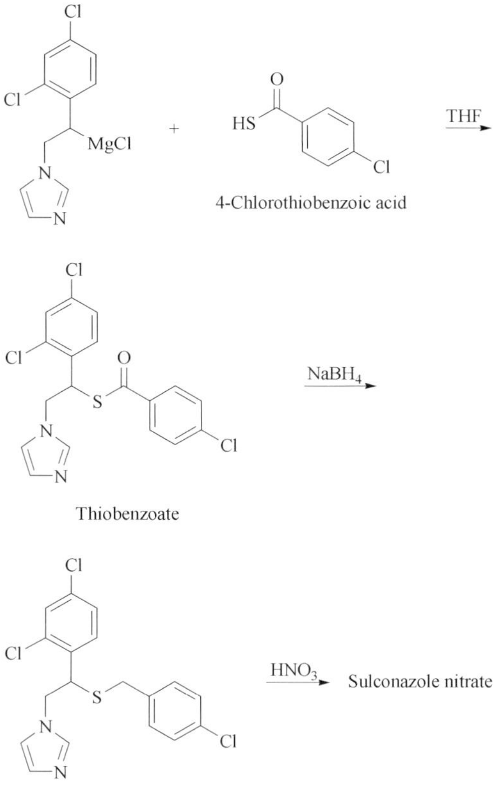

Description. See [156]–[158].

Formulations. 1% solutions, creams, and ointments.

Trade Names (Sulconazole Mononitrate): Exelderm (Schwarz), MYK (Cassenne), MYK-1 (Will), Sulcosyn (Syntex), Suldicyn (Syntex).

Antimycotic Properties. Sulconazole is a broad-spectrum imidazole antimycotic with activity against almost all pathogenic fungi [159]–[161].

Pharmacokinetics. The substance is not absorbed after topical administration.

Indications. Topical treatment of mycoses of the skin induced or sustained by fungi such as dermatophytes and yeasts.

Side Effects. Local irritations and allergic reactions may occur in rare cases and are mainly due to the galenic formulation.

2.2.12. Terconazole

Terconazole [*67915-31-5*], *cis*-1-[4-[[2-(2,4-dichlorophenyl)-2-(1*H*-1,2,4-triazol-1-ylmethyl)-1,3-dioxolan-4-yl]methoxy]phenyl]-4-(1-methylethyl)piperazine, $C_{26}H_{31}Cl_2N_5O_3$, M_r 532.48, *mp* 126.3 °C, crystals from isopropyl ether.

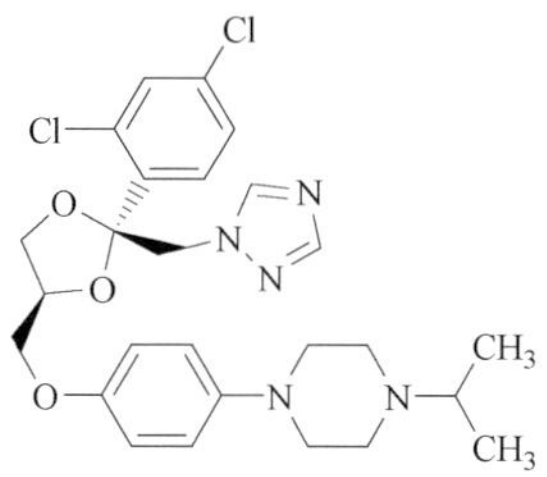

Synonym(s). Triaconazole, R 42470.

Stability. The pure solid substance is stable under normal storage conditions.

Synthesis. [162], [163]:

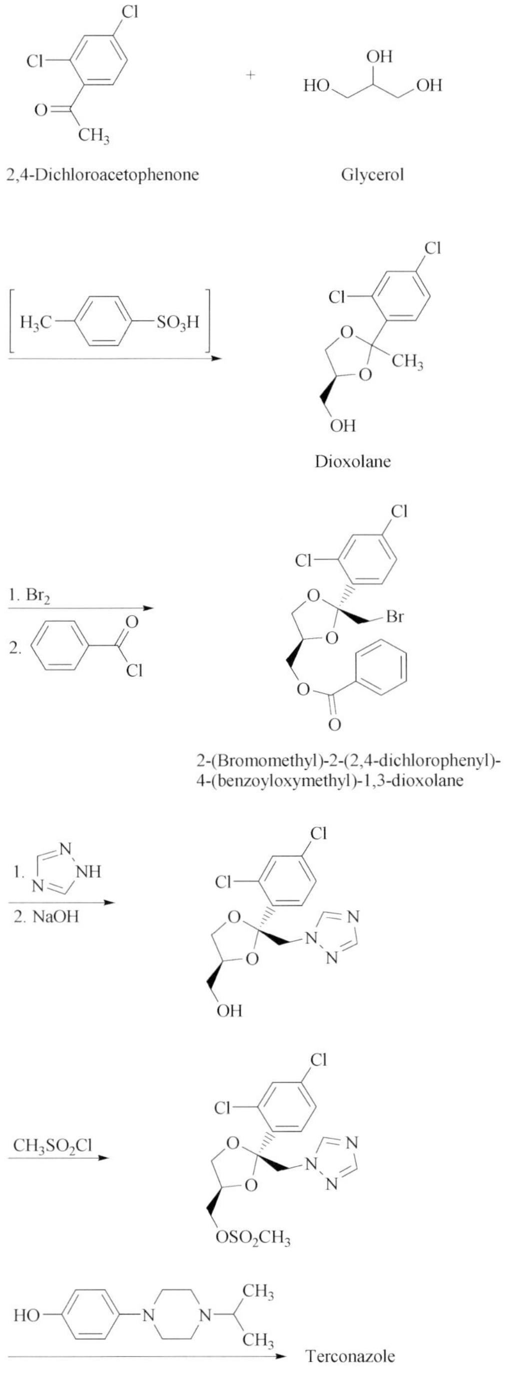

2,4-Dichloroacetophenone + Glycerol

[H₃C—C₆H₄—SO₃H] →

Dioxolane

1. Br₂
2. PhC(O)Cl
→

2-(Bromomethyl)-2-(2,4-dichlorophenyl)-4-(benzoyloxymethyl)-1,3-dioxolane

1. 1,2,4-triazole (NH)
2. NaOH
→

CH₃SO₂Cl →

4-(4-hydroxyphenyl)-1-isopropylpiperazine → Terconazole

Description. The triazole antimycotic terconazole was first synthesized at Janssen Pharmaceuticals [164]–[166].

Formulations. 1% solutions, powders, ointments, and cream formulations; vaginal pessar (80 mg).

Trade Names: Fungistat (Janssen), Gyno-Terazol (Cilag), Panlomyc (Janssen), Terazol (Cilag, Organon, Ortho), Terconal (Fisons), Tercospor (Cilag).

Antimycotic Properties. Terconazole is a broad-spectrum triazole antimycotic with activity against most species of pathogenic fungi [163].

Pharmacokinetics. After intravaginal application, 5–16% of the dose is aborbed. The substance is extensively metabolized and eliminated via the urine and feces [163].

Indications. Topical treatment of mycoses of the skin induced or sustained by fungi such as dermatophytes and yeasts. In addition, terconazole is indicated for theraphy of vulvovaginal mycoses caused by *Candida* species and *Candida (Torulopsis) glabrata* [167], [168].

Side Effects. *CNS.* After intravaginal application of terconazole, a flu syndrome (fever, headache, hypotension) was observed [169].
Local irritations and allergic reactions may occur in rare cases and are mainly due to the galenic formulation.

2.2.13. Tioconazole

Tioconazole [*65899-73-2*], 1-[2-[(2-chloro-3-thienyl)methyloxy]-2-(2,4-dichlorophenyl)-ethyl]-1*H*-imidazole, $C_{16}H_{13}Cl_3N_2OS$, M_r 387.70, crystals.

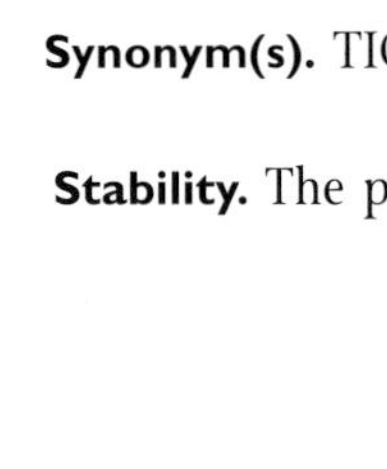

Synonym(s). TIO, IK 20 349.

Stability. The pure solid substance is stable under normal storage conditions.

Synthesis. [170]:

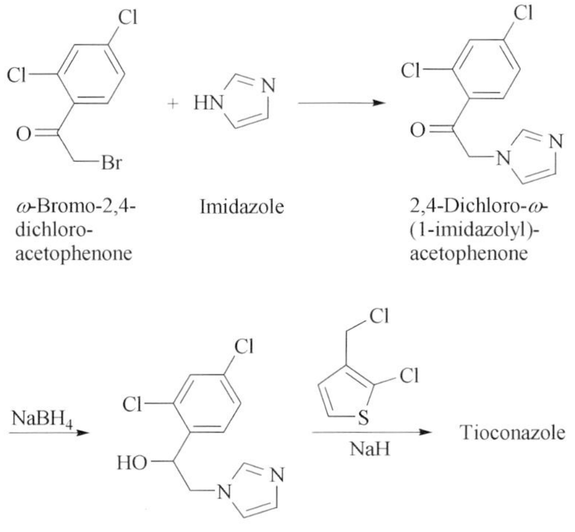

ω-Bromo-2,4-dichloro-acetophenone + Imidazole → 2,4-Dichloro-*ω*-(1-imidazolyl)-acetophenone

NaBH₄ → (alcohol intermediate) → (with 3-chloromethyl-2-chlorothiophene, NaH) → Tioconazole

Description. The imidazole antimycotic tioconazole was first synthesized at Pfizer Pharmaceuticals [171], [172].

Formulations. 1% solutions, lotions, ointments, and cream preparations for therapy of dermatomycoses; 28% solutions for onychomycoses; 2–6% cream preparations for vaginal candidiasis.

Trade Names: Dermo-Trosyd (Pfizer), Fungibacid (Asche), Gino-Tralen (Pfizer), Gyno-Trosyd (Mason, Pfizer), Tralen (Pfizer), Trosid (Pfizer), Trosyd (Mason, Pfizer, Roerig), Trosyl (Roerig, Novex, Pfizer), Vagistat (Bristol-Myers-Sqibb), Zoniden (Irbi).

Antimycotic Properties. Tioconazole is a broad-spectrum imidazole antimycotic with activity against almost all species of pathogenic fungi [173]–[175].

Pharmacokinetics. Tioconazole is normally not absorbed after topical application to the skin [173]. On vaginal application of an ovulum containing 300 mg tioconazole, a serum concentration of 21 µg/mL was observed after 8 h [174].

Indications. Topical treatment of mycoses of the skin induced or sustained by fungi such as dermatophytes and yeasts [176]. In addition, tioconazole is indicated for therapy of vulvovaginal mycoses caused by *Candida* species and *Candida (Torulopsis) glabrata*.

Side Effects. Local irritations such as itching, burning sensations, erythema, and dermatitis, and allergic reactions may occur in rare cases and are mainly due to the galenic formulation [177], [178].

3. Polyene Antimycotics

The mechanism of action of polyene antimycotics is based on the formation of a complex with ergosterol in the fungal plasma membrane. This complex causes changes in the permeability of the plasma membrane, loss of ions and low molcular mass cytoplasmatic components, and inhibition of glycolysis [179]. The effect of polyenes on fungi is antagonized by Ca^{2+}, Mg^{2+}, and sterols in vitro. Further, an antagonism between polyene and azole antimycotics has been observed in several cases, both in vitro and in vivo. This antagonism can be explained in terms of both substance classes interfering with ergosterol biosynthesis.

The polyenes are primarily fungistatic at concentrations around the MIC level. In vitro fungicidal effects are achieved with 2, 4 or up to 10 times the MIC. The three relevant polyene antimycotics amphotericin B, natamycin (pimaricin), and nystatin show broad-spectrum in vitro antimycotic activity. Amphotericin B is the most potent of these three substances. However, in vivo activity is confined to yeasts, dimorphic (biphasic) fungi, and moulds, including *Aspergillus* species. Generally, the polyenes are less active to mostly ineffective and therefore generally not indicated in the treatment of dermatophytoses.

Polyenes are not absorbed sufficiently, if at all, on oral administration. Parenteral administration causes serious local inflammations. Nystatin and pimaricin are therefore unsuitable for intravenous administration.

3.1. Amphotericin B

Amphotericin B [*1397-89-3*], natural product, 38-membered cyclic lactone, isolated, e.g., from culture filtrates of *Streptomyces nodosus*. There are seven double bonds in the molecule. The keto group in position 1 is cyclized with the hydoxl group in position 35 to form a hemiacetal. The carbon atom at position 33 is atached to the dideoxy sugar mycosamine by a β-glucoside bond. $C_{47}H_{73}NO_{17}$, M_r 924.1, *mp*: does not have an exact melting point and decomposes above 170 °C, amorphous yellow powder.

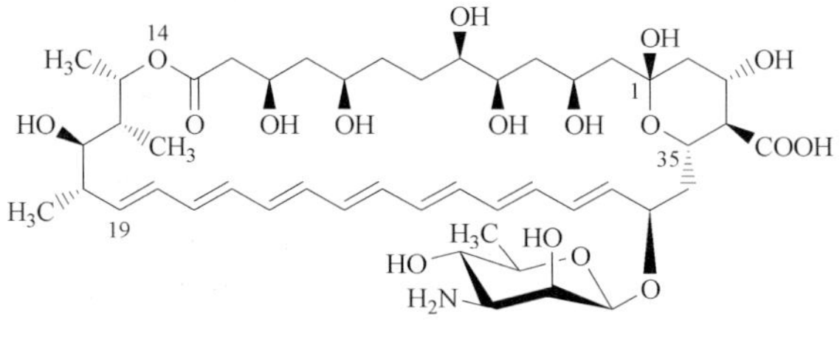

Synonym(s). RP 17774.

Solubility. Insoluble in water at pH 6–7. Solubility in water at pH 2 or pH 11: 0.1 mg/mL. Soluble in aqueous ethanol; DMF: 2–4 mg/mL; DMF+HCl: 60–80 mg/mL; DMSO: 30–40 mg/mL; insoluble in ether, ethyl acetate, or benzene. Soluble in aqueous sodium desoxycholate.

Stability. Amphotericin B is unstable to heat and prolonged exposure to light. It should be stored in the dark and kept refrigerated. In solution it decomposes at pH < 4 and > 10 [180]. Solutions show an optimum stability in a citrate–phosphate buffer at pH 5–7 [181]–[183].

Synthesis. Mostly fermentative from *Streptomyces* species [184]. First isolated from a culture of *Streptomyces nodosus* M4575, an isolate obtained from soil of the Orinoco river region of Venezuela.

Description. The first gereal characterization appeared in 1961 [185]. After some partial structures were determined, the complete structure was published in 1970 [186], and the total synthesis in 1987 [187], [188].

Formulations. For systemic treatment: Ampules à 0.05 g. Liposomal formulations with reduced toxicity [189]–[191] are available for systemic treatment. For local therapy, amphotericin B formulations such as cream preparations and ointments, lozenge formulations (à 0.1 and 0.01 g), and suspensions (100 mg/mL) are used. Also topical cream formulations additionally containing antibiotics (e.g., neomycine) and/or steroids (e.g., triamcinolone) are on the market.

Trade Names: AmBisome (Vestar), Ampho-Moronal (Bristol-Myers Squibb, Heyden, Squibb), Funganiline (Med. y Prod. Quim.), Fungilin (Bristol-Myers Squibb, Novo Nordisk, Squibb), Fungizone (Bristol-Myers Squibb, Squibb).

Antimycotic Properties. Amphotericin B is a broad-spectrum polyene antimycotic with activity against pathogenic yeasts, moulds, dimorphic fungi, and in vitro also against dermatophytes [192], [193]. MIC values of amphotericin B are summarized in Table 7.

Pharmacokinetics. Amphotericin B is scarcely absorbed after oral application. Therefore, amphtericin B has to be applied by the intravenous route for the treatment of systemic fungal infections [194]. Intravenous administration of 5–75 mg/d leads to serum concentrations of 0.14–2.39 mg/L 4 h after infusion [194]. The extravascular tissue distribution of amphotericin B is minimal because of cell membrane and lipoprotein binding [195]. The highest amphotericin B tissue concentrations are achieved in the liver and spleen, and also in the lungs and kidneys [196], [197]. Because of its poor penetration of the blood–brain barrier, meningitis and infections of the central nervous system can only be treated intralumbally and/or intrathecally [194].

Table 7. MIC values for amphotericin B, nystatin, and pimaricin

Fungus	MIC range, µg/mL		
	amphotericin B	nystatin	pimaricin
Candida species	0.1 – 0.5	1 – 5	1.5 – 10
Cryptococcus neoformans	0.1 – 0.5	0.5 – 4	6 – 12.5
Aspergillus species	0.1 – 0.5	1 – 10	3 – 12
Coccidioides immitis	0.1 – 0.5	0.5 – 1.5	2.5 – 25
Histoplasma capsulatum	0.1 – 1	0.5 – 1	3
Blastomyces species	0.1 – 0.5	0.5 – 1	1.5
Dermatophytes	2 – 4	2 – 10	1.2 – 100

The metabolization of amphotericin B is still unclear, and no metabolites have been charcterized so far [196]. Only 3 % of the administered dose is found in the urine [198], and 1 – 15 % of the daily dose is eliminated by the biliary – fecal route [197]. Much of the substance is trapped in the organs. Elimination occurs in two phases. The initial half-life is between 1 and 2 d, with a terminal half-life of 15 d [194].

Indications. Amphotericin B is effective in therapy of *Candida* infections, also including generalized and disseminated forms, caused by *Candida albicans* and other pathogenic *Candida* species [199]. Further, amphotericin B is indicated in the therapy of mycotic diseases such as histoplasmosis, sporotrichosis, cryptococcosis, blastomycosis, aspergillosis, zygomycosis, and coccidioidomycosis [200]. Although the substance shows an in vitro activity against dermatophytes, amphotericin B is not indicated in the therapy of dermatophytoses, mostly due to toxicity and stability problems and high cost. In the treatment of generalized mycoses, often a synergistic effect between amphotericin B and fluorocytosine [201], [202], and an antagonistic effect between amphotericin B and azole antimycotics is observed (the latter with the exception of CNS cryptococcosis due to *Cryptococcus neoformans*). Indications and efficacy of amphotericin B are summarized in Table 8. Amphotericin B can be given as a continuous drip infusion, even though serious, and in some cases, irreparable side effects, predominantly to the kidney tubulus, have been observed. Therefore, systemic treatment should be initiated only after exact diagnosis, and the patient should be kept under strict supervision. The optimum dosage for systemic therapy is 1 mg/kg every 24 to 48 h. Despite serious side effects, amphotericin B still remains a drug of choice for treating systemic mycoses.

Side Effects. The following side effects have been observed under systemic therapy:
CNS. Fever, vomiting, and chills are observed in about 50 % and seem to be mediated by prostaglandines.
Heart. Rarely, arrythmia is observed under infusion [203].
Circulation. Collapse.
Vessels. Thrombophlebitis at the injection site is requently observed, liposomal formulations show a lower phlebotoxicity [204].

Table 8. Indications and efficacy of amphotericin B at optimal dosage

Indication	Pathogen	Clinical efficacy
Candidosis	*Candida* species, mostly *C. albicans*	effective in > 50% of cases
Aspergillosis	*Aspergillus* species except *A. nidulans*	uncertain
Zygomycosis	*Mucor* and *Absidia* species	uncertain
Blastomycoses	*Blastomyces dermatidis*, *Bl. brasiliensis*	effective in ≈ 50% of cases
Coccidioidomycosis	*Coccidioides immitis*	effective
Histoplasmosis	*Histoplasma capsulatum*	usually effective

Blood/Spleen. Anemia and thrombocytopenia [204] are rarely observed.

Liver. In some cases, hepatotoxicity was observed.

Kidney. Amphotericin B shows a high degree of nephrotoxicity. Cumulative doses of 4–5 g generally cause reversible damage to the kidneys [205]. Higher doses mostly cause an irreversible kidney damage with persisting hypopotassemia, hematuria, proteinuria, and azotemia [203]–[206]. Therefore, serum potassium, magnesium, and creatinine levels have to be frequently monitored.

In the case of intralumbal and/or intrathecal administration, severe side effects of the central nervous system are common. Liposomal formulation of amphotericin B can reduce toxicity problems [207].

Topical application of amphotericin B is highly effective and well tolerated in the treatment of cutaneous and mucocutaneous candidoses. This is also true for inhalation and instillation. Local irritations and allergic reactions may occur in rare cases and are partly due to the galenic formulation.

3.2. Natamycin (Pimaricin)

Natamycin [*7681-93-8*], natural product isolated from culture filtrates of *Streptomyces natalensis* (original isolate from soil of Pietermaritzburg, South Africa) and *Streptomyces chattanoogensis* [208], [209], contains mycosamine as sugar component, like amphotericin B and nystatin. $C_{33}H_{47}NO_{13}$, M_r 665.7, *mp*: does not have an exact melting point and decomposes above 200 °C, colorless substance.

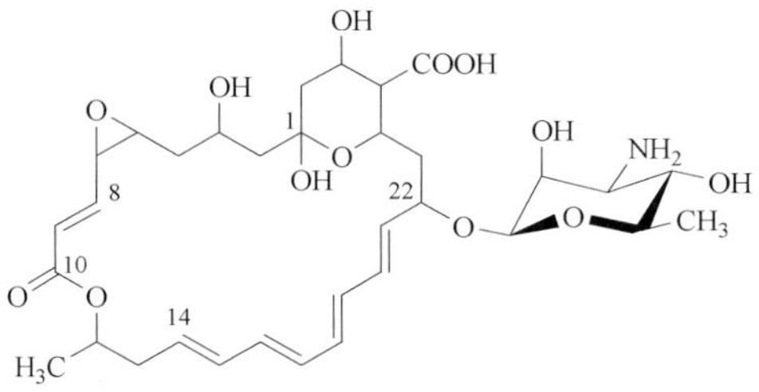

Synonym(s). Pimaricin, antibiotic A 5283, CL 12625, tennecetin.

Solubility. Natamycin is soluble in propylene glycol (20 mg/mL), DMF (50 mg/mL), and *N*-methylpyrrolidone (120 mg/mL); slightly soluble in methanol (2 mg/mL); and practically insoluble in water (0.05 mg/mL). Natamycin is amphoteric, dissolving in dilute acids and bases. However, these solutions are unstable.

Stability. Solutions or suspensions of natamycin are stable for several weeks at pH 5–7. The solutions can even be sterilized by heating at 110 °C for 20 min. However, they must be protected from light, which causes rapid decomposition [208], [210].

Synthesis. Fermentative from *Streptomyces* species.

Description. First isolated in 1957 [208], [211], [212].

Formulations. For therapy of human infections: Ointments, powders, creams, lotions, lozenges, dragees, suspensions, and vaginal tablets of different concentration. Also in combination with antibiotics and/or steroids (e.g., neomycine, hydrocortisone).

Trade Names: Deronga (Basotherm), Mycophyt (Gist-Brocades), Myprozine (Lederle), Natacyn (Alcon), Natafucin (Brocades), Pima-Biciron (Basotherm), Pimafucin (Basotherm, Beytout, Byk Procienx, Gist-Brocades), Pimagyn (Doetsch-Grether), Synogil (Basotherm).

Antimycotic Properties. Natamycin is a broad-spectrum polyene antimycotic, MIC values are summarized in Table 7.

Phamacokinetics. After topical application, no systemic absorption was detectable.

Indications. Cutaneous and mucocutaneous candidosis; in some cases, the substance was also successful for therapy of human dermatophytoses [213]–[215]. In veterinary medicin, natamycin is used for the therapy of dermatophytoses in cows (generally caused by *Trichophyton verrucosum*) and horses (generally caused by *Trichophyton equi-*

num) [216], [217]. As natamycin also has antibacterial properties, it is also used as a food preservative.

Side Effects. When applied topically, natamycin is highly effective and well tolerated in the treatment of cutaneous and mucocutaneous candidoses. Local irritations and allergic reactions may occur in rare cases and are mostly due to the galenic formulation. Especially in veterinary medicine, phototoxic effects were observed.

3.3. Nystatin A1

Nystatin (A1) [*1400-61-9*], three biologically active components — Nystatin A1, A2, and A3 — have been described. Nystatin A1 is a natural product isolated from culture filtrates of *Streptomyces noursei* [218], *Streptomyces aureus*, and other *Streptomyces* species. Like amphtericin B it is a 38-membered macrocyclic lactone, from which it differs solely in the position of the hydroxyl groups and in the absence of a double bond at C23 [219]. $C_{47}H_{75}NO_{17}$, M_r 926.1, *mp*: does not have a melting point, begins to decompose above 160 °C and decomposes above 250 °C without melting, yellow powder.

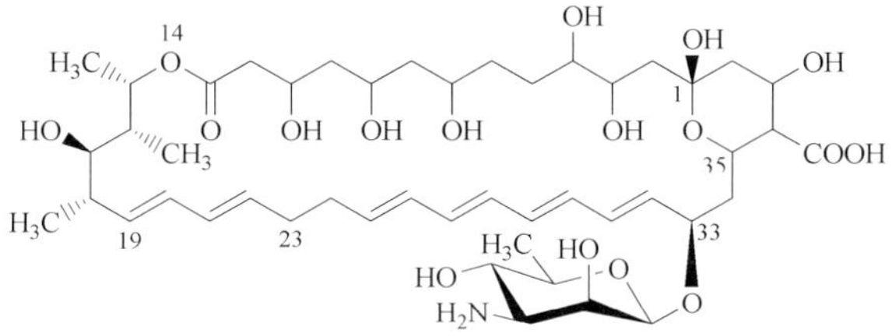

Synonym(s). Polyfungin A1.

Solubility. Nystatin A1 is soluble in pyridine, DMSO, DMF, ethylene glycol (8.7 µg/mL), methanol (11 µg/mL), ethanol (1.2 µg/mL), butanol, dioxane, and water (4.0 µg/mL). The substance is hygroscopic [220].

Stability. The substance is amphoteric, but aqueous and alkaline solutions are unstable. Nystatin shows optimum stability in phosphate – citrate buffers at pH 5 – 7. If kept refrigerated, the pure substance can be stored for several months without loss of activity. For stability, see [221].

Synthesis. Fermentative from *Streptomyces* species.

Description. See [222], [223].

Formulations. Tablets, dragées, lozenges, suspensions, drops, ointments, powders, gels, creams, ovulas, and vaginal tablets of different concentrations. There are also formulations available in combination with antibiotics and/or steroids such as tetracycline, neomycin, gramicidin, and cortisone. Dosages are often expressed in IE (100 000 IE = 22.73 mg nystatin A1).

Trade Names: Adiclair (Ardeypharm), Biofanal (Pfleger), Candex (Dome), Candio-Hermal (Hermal, Merck), Canstat (Lederle), Diastatin (Pfizer), Fungicidin (Spofa), Fungireduct (Azupharma), Herniocid (Mayrhofer), Korostatin (Holland-Rantos), Lederlind (Lederle), Lystin (Mekim), Mikostatin (Squibb), Moronal (Heyden, Squibb), Multilind (Bristol-Myers Squibb, Fair), Mycostatin (Bristol-Myers Squibb, Heyden, Sanofi Winthrop, Westwood-Squibb), Mykinac (NMC), MycoPosterine N (Kade), Mykundex (Jossa), Nadostine (Pan-Well), Nilstat (Lederle), Nyaderm (Taro), Nysert (Norwich Eaton), Nystacid (Farmos Group), Nystaderm (Dermapharm), Nysta-Dome (Dome), Nystan (Squibb), Nystat-Rx (Pharma-Tek), Nystavescent (Squibb), Nystex (Savage), Oranyst (Taro), O-V Statin (Squibb), Restatin (Remedica), Rivostatin (Rivopharm) Stereomycin (Medica).

Antimycotic Properties. Nystatin A1 is a broad-spectrum polyene antimycotic with in vitro activity against pathogenic yeasts, moulds, dermatophytes, and dimophic fungi. MIC values of nystatin are summarized in Table 12.

Pharmacokinetics. After topical application, no systemic absorption was detectable.

Indications. Nystatin A1 is especially indicated in the therapy of cutaneous and mucocutaneous infections caused by pathogenic yeasts.

Side Effects. When applied topically, nystatin A1 is highly effective and well tolerated in the treatment of cutaneous and mucocutaneous candidoses. Local irritations and allergic reactions may occur in rare cases and are mostly due to the galenic formulation. If topically applied in the gastrointestinal tract, nausea, vomiting, and diarrhoea are rarely observed.

4. Flucytosine

Flucytosine [*2022-85-7*], 4-amino-5-fluoro-2(1*H*)-pyrimidinone, $C_4H_4FN_3O$, M_r 129.03, *mp* 297 °C (decomp.), colorless crystals.

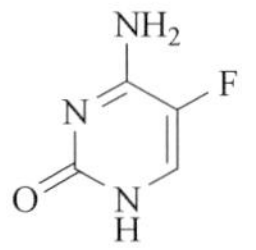

Synonym(s). Fluorocytosine, 5-Flurocytosine, 5-FC, Ro 2-9915.

Solubility. Flucytosine is readily soluble in water. It is basic and forms salts with acids.

Stability. Pure flucytosine as well as aqueous solutions, both acidic and basic, are stable. The substance should be stored in the dark.

Synthesis. The synthesis begins with *S*-ethylisothioureahydrobromide and ethyl-2-fluoro-3-methoxy-acrylate [224], [225]:

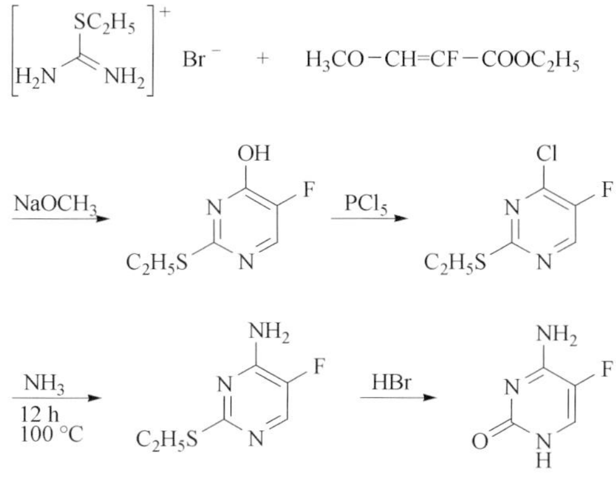

Description. The substance was initially synthesized at Hoffmann-La Roche Pharmaceuticals as an antimetabolite [224], [226]–[228].

Formulations. Infusion bottles à 0.1, 0.2, and 0.4 g. Capsules à 0.05, 0.1, 0.15, and 0.2 g. 0.5 and 1% suspensions for oral application, 10% suspension creams.

Trade Names: Alcoban (Roche), Alcobon (Roche), Ancobon (Roche), Ancotil (Edward Keller, Hoffmann-La Roche, Roche), Cocol (Horita, Tobishi).

Mechanism of Action. The mechanism of action of flucytosine is based on the incorporation of 5-fluorouracil, which is relatively selectively formed in fungal cells by deamination of flucytosine by the fungal cytosine-desaminase, into the messenger RNA and ribosomal RNA of the fungal cell. Fungal RNA containing 5-fluorouracil-riboside increasingly inhibits fungal protein biosynthesis [231]–[234].

Antimycotic Properties. Flucytosine has a narrow in vitro antimycotic activity spectrum that includes yeasts, *Cryptococcus neoformans*, as well as some *Aspergillus* strains and some chromomycetes. The activity spectrum and MIC values of flucytosine are summarized in Table 9 [231], [232]. Because the antimycotic action of flucytosine is antagonized by purines, pyrimidines, peptides, and some amino acids in vitro, only media free from these antagonists, e.g., yeast–nitrogen base medium containing dextrose, can be used for in vitro testing. Of the fungus strains affected by flucytosine,

Table 9. MIC values for flucytosine

Fungus	MIC range, µg/mL
Candida albicans	0.1–10 (>100)
Candida tropicalis	0.1–100
Candida parapsilosis	0.1–0.5
Candida pseudotropicalis	0.1
Candida krusei	0.5–1
Candida (Torulopsis) glabrata	0.1–1
Cryptococcus neoformans	0.5–4 (16)
Aspergillus fumigatus	0.5–10
Aspergillus niger	0.5–1
Aspergillus flavus	1
Aspergillus nidulans	100
Phialophora species	1–10
Cladosporium species	1–10

more than 10 % can be expected to show primary resistance. The development of secondary resistance by sensitive fungi during treatment also occurs, especially in cases of underdosage or prolonged treatment [232], [234]. In vitro fungicidal effects can be obtained with >10 times the MIC and a contact time of > 150 h.

Pharmacokinetics. Flucytosine can be administered orally, parenterally, and topically. After oral administration, the drug is absorbed rapidly and almost completely in the intestine (80–90 %) [235]. Average serum concentrations of 30–60 µg/mL are normally reached with a single oral dose of 35 mg/kg. In patients with normal kidney function, optimal continuous serum levels between 25 and 100 mg/L are achieved with a dose of 50 mg/kg every 6 h [236]. Flucytosine penetrates readily into the tissues, body fluids, and cerebrospinal fluid (CSF) [237]. The CSF level reaches approximately 70–80 % of blood levels [236]. The plasma protein binding of flucytosine is low (ca. 4 %) [238]. More than 90 % of the dose is eliminated via the kidneys in unchanged form. The elimination half-life is 3–6 h. In patients with severely impaired kidney function, the elimination half-life can rise to 4 d. In patients with restricted renal function, the daily dose of flucytosine therefore must be adjusted according to creatinine elimination. The compound can be readily eliminated by hemodialysis [231]–[233]. A small amount of flucytosine is metabolized to 5-fluorouracil, e.g., by the bacterial flora of the gut, which seems to be a reason for the hematotoxicity of flucytos e [239], [240].

Indications. Flucytosine has proved very effective in the treatment of *Candida* fungemia and systemic candidoses (but not mucocutaneous candidiasis) in combination with amphotericin B. Also, *Cryptococcus* infections, including *Cryptococcus neoformans* meningoencephalitis in immunocompromised patients (e.g., AIDS patients), can be treated with flucytosine in combination with amphotericin B (and in the case of *Cryptococcus neoformans* associated infections, additionally fluconazole). Further, flucytosine is indicated in the therapy of chromomycoses [241]. There can also be some activity of flucytosine in disseminated aspergilloses. Because of the rapid development of resistance, the substance should only be given in combination with other antimycotic agents, mostly amphotericin B [242].

Side Effects. Flucytosine is normally well tolerated provided dosage instructions are observed and precautionary measures are taken [232]. The following side effects have been observed during treatment with flucytosine: *Blood/spleen.* Neutropenia, leukopenia, thrombopenia [243].

GIT. Nausea, vomiting, and diarrhoea were observed in rare cases [243].

Liver. Liver malfunctions with reversible hepatomegalia and increase in serum transaminases; an extensive liver cell necrosis was reported after flucytosine therapy in two patients [243].

Skin. Allergic skin reactions were observed under therapy. Changes in blood counts can have serious consequences. Blood counts and liver function should therefore be checked regularly during therapy. Hematotoxic side effects were especially observed if serum levels exceeded 100 mg/L. Flucytosine can prolong engraftment in bone marrow transplantation patients.

5. Griseofulvin

Griseofulvin [*126-07-8*], 7-chloro-4,6-trimethoxy-6'-methylspiro[benzofuran-2(3*H*),1'-[2]cyclohexene]-3,4'-dione, $C_{17}H_{17}ClO_6$, M_r 352.8, *mp* 222 °C, colorless crystals.

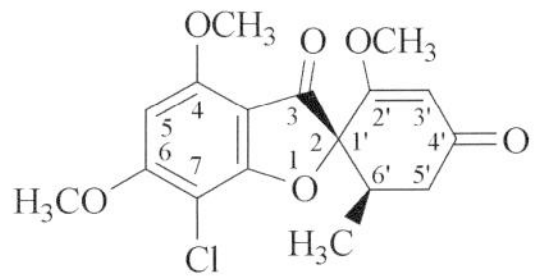

Synonym(s). Amudane, Curling Factor.

Solubility. Griseofulvin is soluble in methanol (2.5 g/mL), ethanol (1.5 g/mL), ether, and glacial acetic acid; quite soluble in DMF (150 mg/mL); slightly soluble in chloroform (25 mg/mL), ethyl acetate, acetone (33 mg/mL), and toluene; and insoluble in petroleum ether. The solubility in water is 40 µg/mL.

Stability. Griseofulvin is stable under normal storage conditions.

Synthesis. Griseofulvin is mostly produced by fermentation with strains of *Penicillium patulum*. After the first total synthesis in 1960 [244], other processes were published, but none of them has acquired industrial significance up to now.

Description. Griseofulvin was isolated from culture broths of *Penicillium griseofulvum* in 1938 and later also found in culture filtrates of *Penicillium nigricans* (= *P. janczewskii*),

P. patulum, and other *Penicillium* species. The structure was determined in 1952. Griseofulvin was first used in plant protection and was introduced to medicine as an antimycotic in 1958 [245].

Formulations. Tablets à 125 and 500 mg. 5% cream formulations.

Trade Names: B-GF (Biokema), Delmofulvina (Coli), Fulcin (ICI, Mason, Zeneca), Fulcin S (ICI), Fulvicin (Schering), Fulvicina (Interpharma), Fulvicin P/G (Schering-Plough), Fungivin (Nycomed), Gefulvine (Dif-Dogu), Greosin (Glaxo), Gricin (Arzneimittelwerke Dresden), Grifulin (Teva), Grifulvin V (McNeil, Ortho), Grisactin (Ayerst, Wyeth), Griséfuline (Sanofi Winthrop), Griseo (Ayerst), Griseoderm (Trinity), Griseomed (Waldheim), Griseostatin (Mason, Schering Corp.), Griseo von ct (ct-Arzneimittel), Grisfulvin (Protea), Grisol (Gebro), Grisovin (Glaxo), Grisovina Fp (Glaxo, Teofarma), Grisovin-FP (Glaxo), Grivate (Fujisawa), Grysio (Ayerst), Lamoryl (Leo, Lövens), Likuden (Hoechst), Norofulvin (Norbrook), Polygris (Schering Corp.), Sulvina (Dibios), Ultragris (Sidmak).

Mechanism of Action. In addition to the inhibition of protein synthesis resulting from the interference of griseofulvin with guanine during RNA synthesis, there may also be a specific inhibition of chitin synthesis. Chitin is an essential cell-wall constituent in dermatophytes. Inhibition of its synthesis would explain the characteristic morphological changes in the hyphae, called curling, that occur in dermatophytes in the presence of griseofulvine [231], [232], [246].

Antimycotic Properties. Griseofulvin has a narrow spectrum of antimycotic activity, including only dermatophytes. *Trichophyton* and *Microsporum* species and *Epidermophyton floccosum* are inhibited by concentrations of 0.1–0.5 µg/mL. The in vitro effect of griseofulvin depends primarily on the test system used and is confined to proliferating hyphae. Griseofulvin is fungistatic. Purines and pyrimidines are antagonistic in vitro [232], [247]. Acquired in vitro resistance towards griseofulvin develops gradually following a number of fungal passages [248], [249]. However, increases in resistance during therapy are rarely observed, even during long-term treatment. Although dermatophyte variants showing primary resistance have been described, they are very rare [250]. MIC values for griseofulvin are summarized in Table 10 [14].

Pharmacokinetics. The degree of absorption of orally administered griseofulvin varies from individual to individual from 25 to 75% of the administered dose. The absorption is strongly dependent on the degree of micronization of the drug, a high fat content in the food also increases absorption. After the administration of a single dose of 0.25–0.5 g, serum concentrations reach 0.6 to 1.5 µg/mL. The serum half-life is about 20 h [231], [232], [251]. Griseofulvin accumulates during keratinization in human epidermis cells, hair, and nails. In the course of long-term griseofulvin treatment, concentrations up to 25 µg/g can be detected in the stratum corneum of the skin

Table 10. MIC values for griseofulvine

Fungus	No. isolates tested	MIC range µg/mL
Epidermophyton floccosum	6	3 – > 5
Microsporum canis	19	0.1 – > 18
Microsporum gypseum	6	0.5 – 3.13
Trichophyton mentagrophytes	57	0.1 – > 30
Trichophyton rubrum	558	0.1 – > 30
Trichophyton schoenleinii	170	0.1 – 25
Trichophyton tonsurans	25	0.1 – > 18
Trichophyton violaceum	345	0.1 – 21

[232]. Griseofulvin concentrations of 0.2 to 0.3 µg/mL are found in the sweat [252]. Further, the substance shows high concentrations in liver, body fat, and muscles. Griseofulvin is metabolized in the liver and is mostly eliminated by the kidneys in a metabolized, inactive form. Up to 75 % of the quantity absorbed is excreted in the urine as inactive 6-demethylgriseofulvin. The concentration of unchanged griseofulvin in the urine is 1 – 2 µg/mL, corresponding to approximately 1 % of the dose [232], [253]. Elimination occurs biphasically with a terminal elimination half-life of 10 – 40 h [254].

Indications. The indications for griseofulvin are dermatophytoses and onychomycoses which do not respond to topical therapy. The optimal daily dose is 5 – 10 mg/kg body weight. The duration of treatment ranges from 3 to 5 weeks for dermatophytoses of the trunk and extremities, and from 5 to 10 weeks for infections of palms, soles, and scalp, but is more than 6 months and even up to 18 months for onychomycoses [231], [232]. Additional topical therapy increases the cure rate and reduces duration of treatment. Due to better tolerability and efficacy, newer antimycotic compounds such as azole antimycotics and allylamines have mostly replaced griseofulvin in the therapy of dermatomycoses and onychomycoses. Griseofulvin is inactive in mycoses caused by fungi other than dermatophytes.

Side Effects. *CNS.* Headache, depression, dizziness, sleeplessness.

Nerves. Paresthesia and neuropathia [255].

GIT. Dryness of the mouth, changes in taste, and other unspecific gastrointestinal disorders.

Skin. Minor allergic skin reactions, Lyell syndrome [256], erythema multiforme, exfoliative dermatitis, angioedema, photosensitization.

Occasionally, reversible hematopathies, proteinuria, and, especially in case of previous hepatopathy, increases in the liver enzyme values in the blood may be observed. A carcinogenic and embryotoxic potential of griseofulvin was observed in laboratory animals such as rats.

6. Ciclopirox

Ciclopirox [29342-05-0], 6-cyclohexyl-1-hydroxy-4-methyl-2(1H)-pyridinone, $C_{12}H_{17}NO_2$, M_r 207.27, mp 144 °C, colorless crystals.

Ciclopirox olamine [41621-49-2], 6-cyclohexyl-1-hydroxy-4-methyl-2(1H)-pyridinone olamine, $C_{14}H_{24}N_2O_3$, M_r 268.4, mp 143 °C, colorless crystals.

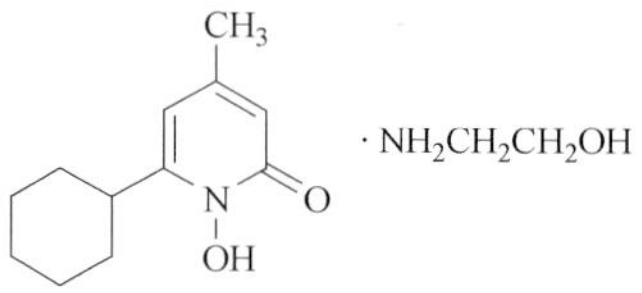

Synonym(s). HOE 296b (free base); Ciclopirox ethanolamine, Hoe 296 (olamine salt).

Stability. The substance is stable under normal storage conditions.

Synthesis. Base **8** is produced by heating 4-methyl-6-cyclohexyl-2-pyrone (**9**) [257] for 8 h at 80 °C with hdroxylamine hydrochloride and 2-aminopyridine [258] or from 5-cyclohexyl-3-methyl-5-oxopentenoate (**10**) [258] by reaction with hydroxylamine hydrochloride via the oxime **11**, followed by cyclization. The final step is to form the salt with ethanolamine.

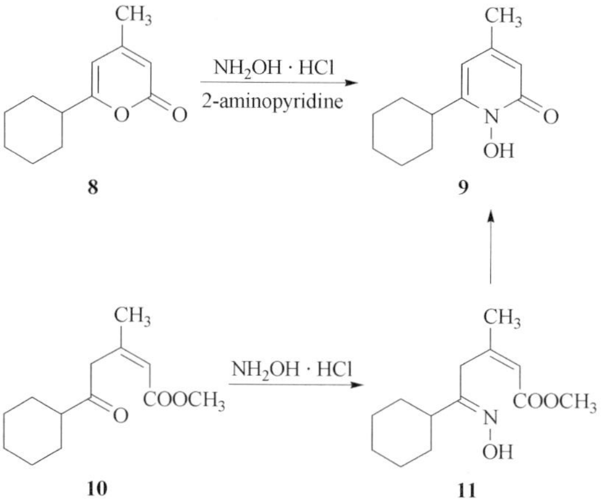

Description. The substance was first synthesized by Hoechst Pharmaceuticals in 1973 [258]–[262].

Formulations. 1% solutions, ointments, and cream preparations.

Trade Names (Ciclopirox): Nagel Batrafen (Casella-Riedel).

Table 11. MIC values for ciclopirox olamine

Species	MIC range, µg/mL
Dermatophytes	0.5 – 5
Candida species	0.5 – 5
Moulds	0.5 – 30

Trade Names (Ciclopiroxolamine): Batrafen (Casella-Riedel, Hoechst, Knoll), Brumixol (Bruschettini), Ciclochem (Novag), Dafnegin (Poli), Fungiderm (Heilmittelwerke Wien), Fungowas (Wassermann), Loporox (Hoechst-Roussel), Miclast (Ellem), Micomicen (Vita), Micoxolamin (Delalande), Mycoster (Sinbio), Obytin (Hoechst, Jugoremedija).

Mechanism of Action. Ciclopirox is a substituted pyridone derivative. Experiments show that the mechanism of action involves a disruption of transport mechanisms into the cell plasma [263], [264], especially the uptake of leucine [265]. Further, the uptake of other amino acids such as phenylalanine and lysine, as well as the uptake of potassium and phosphate ions is impaired. This all probably involves a change in membrane permeability, as reported for some other antimycotics, although ciclopirox does not show lytic activity [263], [264]. Consecutively, the substance significantly reduces fungal growth rates [266].

Antimycotic Properties. Ciclopirox has a broad in vitro antifungal activity spectrum, including dermatophytes, yeasts, and moulds. The most important fungi within the activity spectrum of ciclopirox and the corresponding MIC values are listed in Table 11 [263], [264], [267]. Ciclopirox is primarily fungistatic, but if the contact time is long enough and the concentration exceeds 20 µg/mL, it exerts a fungicidal effect on yeasts and dermatophytes [264]. In vitro, the effect is limited to proliferating fungi and is greatly dependent on the nutrient medium.

Pharmacokinetics. Ciclopirox penetrates rapidly into and through the skin and nail keratin. After dermal application of a 1% cream formulation, about 1.3% of the dose is absorbed. Serum levels range around 10 µg/L after application of a dose of 37 mg of ciclopirox [268]. After vaginal administration, 15 – 20% of the dose is absorbed [269]. Ciclopirox penetrates well into the skin and nail matrix. Absorbed ciclopirox is mainly eliminated by the kidneys; 1.1 – 1.7% of the topical dose is eliminated in the urine within four days. Only a fraction thereof is unaltered ciclopirox. Approximately 96% of the absorbed drug is bound to human serum proteins [270].

Indications. Ciclopirox is indicated for the topical treatment of dermatophytoses and vaginal infections caused by *Candida* species [271].

Side Effects. Topically applied, ciclopirox is well tolerated. The incidence of local irritations such as itching and burning sensations range between 1 and 4% and are often due to the galenic formulation.

7. Thiocarbamates

Thiocarbamates are inhibitors of fungal squalene epoxidase (see also Chap. 8). By this mechanism, tolciclate and tolnaftate inhibit biosynthesis of fungal ergosterol [272].

7.1. Tolciclate

Tolciclate [*50838-36-3*], methyl (3-methlphenyl)carbamothioic acid *O*-(1,2,3,4-tetrahydro-1,4-methanonaphtalen-6-yl) ester, $C_{20}H_{21}NOS$, M_r 323.46, *mp* 120–121 °C, white crystals from ethanol–diethylether.

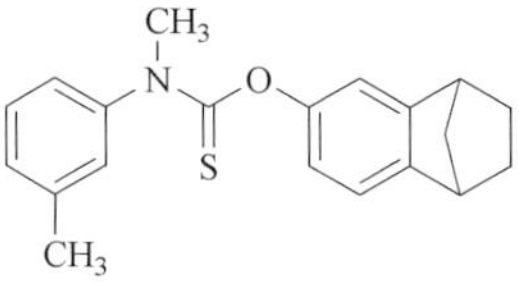

Synonym(s). KC 9147.

Solubility. The substance shows high liposolubility. Soluble in *n*-hexane (15 mg/mL) and *n*-octanol (24 mg/mL).

Stability. The substance is stable under normal storage conditions, but should be protected from light.

Synthesis. [273]:

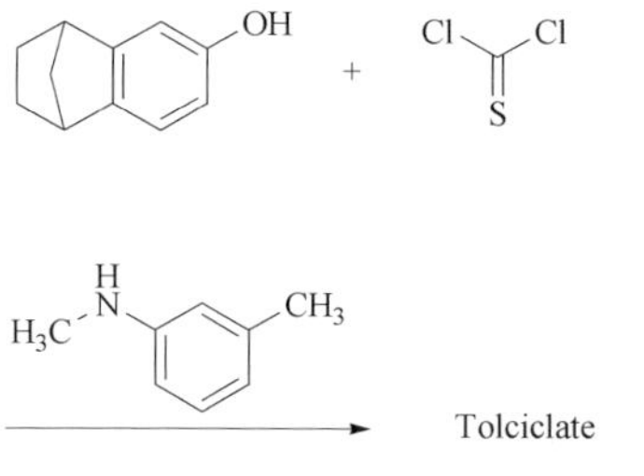

Description. Tolciclate was first synthesized at Carlo Erba Pharmaceuticals [274], [275].

Formulations. 1% solutions, ointments, and cream formulations.

Trade Names: Fungifos (Basotherm), Kilmicen (Carlo Erba), Tolmicen (Carlo Erba, Wing Yee), Tolmicol (Carlo Erba).

Antimycotic Properties. In vitro, tolciclate has a narrow activity spectrum with good efficacy only towards dermatophytes. Isolates of susceptible species showing primary resistance or developing secondary resistance are rarely observed.

Indications. The indications for tolciclate are defined by its narrow spectrum and limited to the treatment of dermatophytoses. Because of inadequate penetration into the nail keratin, it is unlikely to be effective for the treatment of onychomycoses. On average, the efficacy of tolciclate in topical therapy of dermatomycoses is lower than that of most other antimycotics for topical treatment.

Side Effects. Tolciclate is well tolerated on the skin.

7.2. Tolnaftate

Tolnaftate [2398-96-1], (3-methylphenyl)carbamothioic acid O-2-naphthalenylmethylester, $C_{19}H_{17}NOS$, M_r 307.4, mp 110.5 – 111.5 °C, colorless crystals.

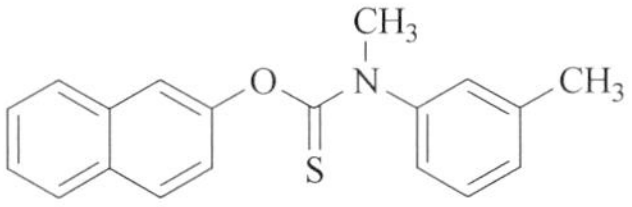

Synonym(s). Naphthiomate-T.

Solubility. The substance shows a good solubility in ethanol, ether, and glacial acetic acid; is quite soluble in DMF, chloroform, dichloromethane, and benzene; and practically insoluble in water.

Stability. Tolnaftate is stable under normal storage conditions, but schould be protected from light.

Synthesis. Tolnaftate is synthesized from β-naphthol, thiophosgene, and N-methl-m-tolulidine by the following pathway [276]. Another route, which does not use thiophosgene, has been described [277].

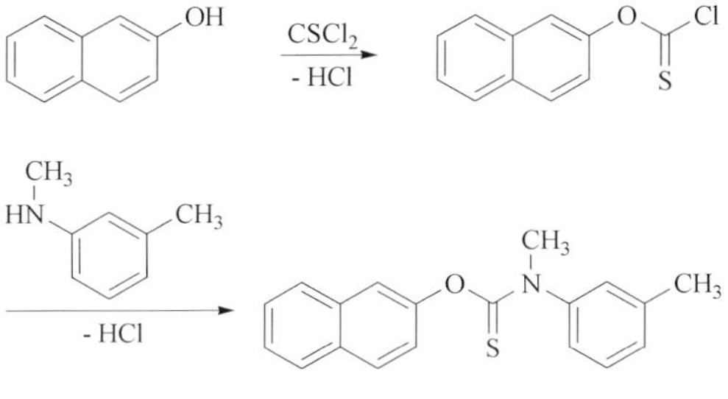

Description. Tolnaftate was first synthesized by Japan Soda and first described as a topical antimycotic in 1962 [278]–[281].

Formulations. 1% solutions, ointments, and cream preparations.

Trade Names: Aftate (JDH, Plough), Alber-T (Hing Yip), Chinofungin (Chinoin), Chlorisept (Chinosol), Ezon-T (Yamanouchi), Focusan (Lundbeck), Footwork (Lederle), Hi-Alarzin (Hing Yip, Yamanouchi), Mikoderm (Akdeniz), NP-27 (Thompson), Pedesal (Krka), Pediderm (Nelson), Pedimycose (Scholl), Pitrex (Taro), Separin (Sumitomo), Sorgoa (Scheurich), Sporiderm (Cétrane), Sporiline (Schering-Plough), Tinacidin (Schering Corp.), Tinactin (Schering-Plough), Tinatox (Brenner), Tinavet (Schering), Tineafax (Wellcome), Ting (Fisons), Tonoftal (Schering), Zeasorb-AF (Stiefel).

Antimycotic Properties. In vitro, tolnaftate has a narrow activity spectrum with fungistatic effect on towards dermatophytes and a few moulds such as some isolates of selected *Aspergillus* species. The MIC values range between 0.1 and 3 µg/mL for susceptible fungi. Isolates of susceptible species showing primary resistance or developing secondary resistance are rarely observed. Tolnaftate is almost ineffective against yeasts and moulds.

Pharmacokinetics. Tolnaftate combines good penetration into the stratus corneum of the skin with poor penetration into nail keratin.

Indications. The indications for tolnaftate are defined by its narrow spectrum and limited to the treatment of dermatophytoses. Tolnaftate is effective when topically administered several times daily over a period of several weeks [231], [232]. Because of its inadequate penetration into nail keratin, tolnaftate is unlikely to be effective for the treatment of onychomycoses. On the average, the efficacy of tolnaftate in topical therapy of dermatomycoses is lower than that of most other antimycotics for topical treatment [282].

Side Effects. Tolnaftate is well tolerated on the skin. Erythemas and allergic reactions are very rare and might also be due to the galenic formulation. No other side effects are known [232], [283].

8. Allylamines

The allylamines are inhibitors of the enzyme squalene epoxidase. This results in an accumulation of squalene, which exhibits lethal effects towards the fungal organism at high concentrations. Thereby, the inhibition of squalene epoxidase blocks ergosterol biosynthesis in the fungal cell [284]. The essential portion of the allylamine molecule is the conjugated enzyme group with a *trans* configuration in the side chain [285]. The enzyme squalene epoxidase is not linked to the cytochrome P450 complex.

8.1. Naftifine

Naftifine [*65472-88-0*], (*E*)-*N*-Methyl-*N*-(3-phenyl-2-propenyl)-1-naphthalenemethenamine, $C_{21}H_{21}N$, M_r 287.40, bp 162–167 °C, colorless, viscous oil.

Naftifine hydrochloride [*65473-14-5*], $C_{21}H_{22}ClN$, M_r 323.9, mp 177 °C, crystals from propanol.

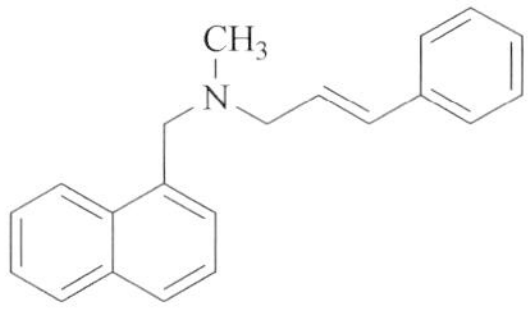

Synthesis. [286], [287]:

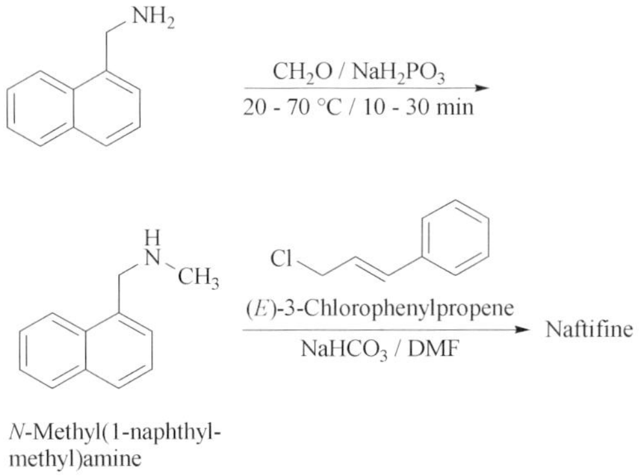

N-Methyl(1-naphthylmethyl)amine

Description. See [288]–[290].

Formulations. 1 % creams, ointments, gels, and solutions.

Trade Names (Naftifine Hydrochloride): Exoderil (Allergan, Novartis, Rentschler, Schering), Fotimin (KRKA), Nafteryl (Schering), Naftin (Allergan).

Antimycotic Properties. Naftifine is fungicidal against dermatophytes such as *Epidermophyton floccosum, Trichophyton* and *Microsporum* species. Against pathogenic yeasts such as *Candida* species and moulds, it shows only an intermediate fungistatic activity in vitro [291], [292].

Pharmacokinetics. Naftifine penetrates well into the stratum corneum of the skin; 2–4 % of the topically administered dose were absorbed after administration of a 1 % gel preperation. After occlusion, an absorption of 6 % of the administered dose was observed [293].

Indications. Naftifine is primarily indicated in the therapy of superficial dermatomycoses [294]–[296]. It also shows some clinical efficacy in superficial infections caused by yeasts and moulds.

Side Effects. Naftifine is well tolerated although rare cases of local skin irritations and contact dermatitis have been found, which might also be due the galenic formulation [293].

8.2. Terbinafine

Terbinafine [*91161-71-6*], (*E*)-*N*-(6,6-dimethyl-2-hepten-4-ynyl)-*N*-methyl-1-naphthalenemethanamine, $C_{21}H_{25}N$, M_r 291.44, colorless crystals.

Terbinafine hydrochloride [*78628-80-4*], $C_{21}H_{26}ClN$, M_r 327.90, *mp* 195–198 °C, colorless crystals from 2-propanol diethyl ether, change in crystal structure begins at 150 °C.

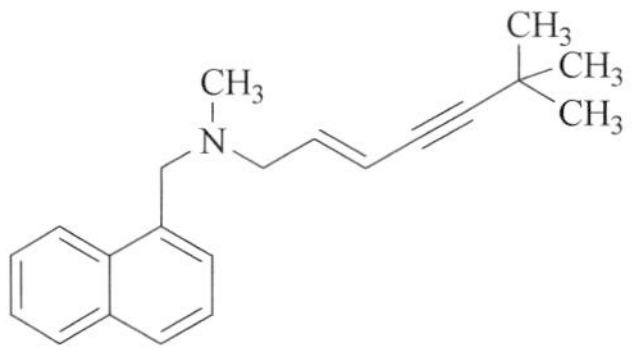

Synonym(s). SF 83627, SF 83-627 (both hydrochloride).

Solubility. Terbinafine is a highly lipophilic substance.

Stability. The substance is stable under normal storage conditions.

Synthesis. [297]:

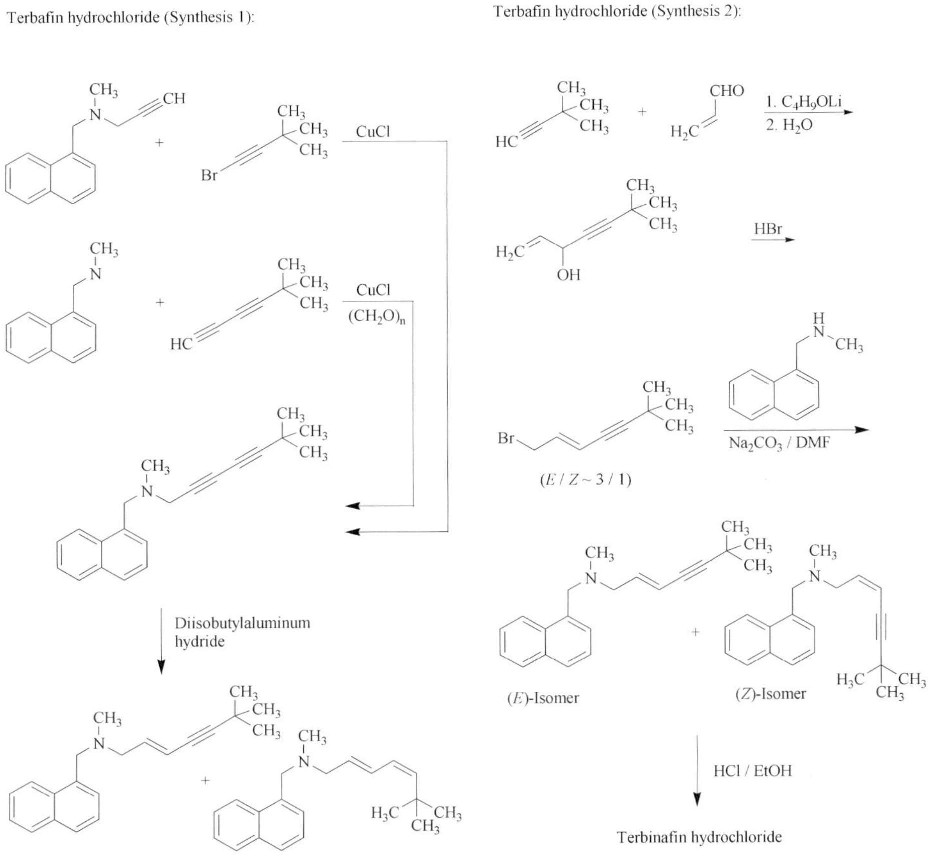

Terbafin hydrochloride (Synthesis 1):

Terbafin hydrochloride (Synthesis 2):

Description. The substance was first synthesized at Sandoz Pharmaceuticals [298].

Formulations. 1% cream and ointment preparations, and tablets à 250 mg.

Trade Names (Terbinafine): Lamisil (Edward Keller).

Trade Names (Terbinafine Hydrochloride): Lamisil (Sanabo, Sandoz-Wander).

Antimcotic Properties. Terbinafine shows a good and mostly fungicidal in vitro antimycotic activity against dermatophytes such as *Epidermophyton floccosum*, *Trichophyton*, and *Microsporum* species. The substance also shows a high but often fungistatic in vitro activity against isolates of various *Aspergillus* species. The activity against pathogenic yeasts is quite variable. Many isolates of *Cryptococcus neoformans* and *Candidia parapsilosis* show an in vitro sensitivity towards terbinafine, although isolates of *Candida albicans*, *Candida (Torulopsis) glabrata*, and *Candida tropicalis* are mostly resistant or have rather high MIC levels [299].

Pharmakocinetics. Terbinafine is well absorbed from the gastrointestinal tract with a bioavailability of 70–80%. Peak plasma concentrations of 0.97 µg/mL were observed following a single oral dose of 250 mg within 2 h of administration. Steady state concentrations are reached within two weeks of therapy [300]. The substance is highly bound to plasma proteins. After oral administration, terbinafine distributes well into tissues, including skin, the nail plate, sweat, sebum, and hair. Terbinafine is metabolized in the liver to inactive metabolites which are excreted in the urine. Elimination occurs biphasically. Plasma elimination half-lives of 11–17 h have been reported, as has a slower elimination half-life of 90–100 h [301]. After topical administration, the lipophilic substance shows a good penetration into the skin and nail matrix and is hardly absorbed (< 5% of the administered dose) [302].

Indications. Oral and topical treatment of dermatophytoses and onychomycoses [303]. The substance also shows activity in the topical treatment of pityriasis versicolor and cutanous candidiasis. 250 mg of terbinafine are given once daily by mouth for 2 to 4 weeks for therapy of tinea cruris and tinea corporis. Especially for tinea pedis, treatment may have to be continued for up to 6 weeks [292].

Side Effects. The following side effects were observed after oral administration:
CNS. Headache.
GIT. Nausea, diarrhoea, anorexia, mild abdominal pain (5%) [301].
Arthralgia, myalgia, and skin reactions such as rush and urticaria rarely occur. Severe skin reactions including erythema multiforme (1:100,000), Stevens–Johnson syndrome and Lyell syndrome have occurred occasionally. Loss or disturbance of taste, which is mostly reversible [304], [305], and liver disfunction have been observed [306]. After topical administration, the substance is well tolerated, although side effects such as skin irritations (e.g., itching and burning sensations) and allergic reaction may occur, which may also often be due to the galenic formulation.

9. Amorolfine

Amorolfine [*78613-35-1*], (±)-*cis*-2,6-dimethyl-4-[2-methyl-3(*p-tert*-pentylphenyl)propyl]morpholine, $C_{21}H_{35}NO$, M_r 317.60, *bp* 134 °C.

Amorolfine hydrochloride [*78613-38-4*], $C_{21}H_{36}ClNO$, M_r 354.06.

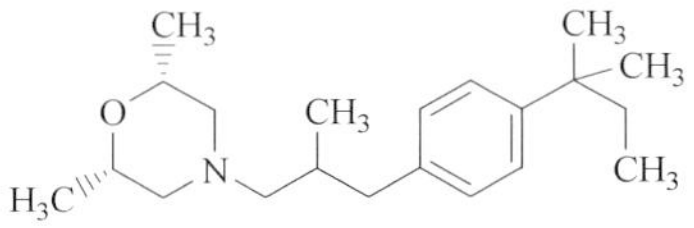

Synonym(s). Ro 14-4767/000 (free base); Ro 14-4767/002 (hydrochloride).

Description. The substance was first synthesized at Roche Pharmaceuticals [307], [308].

Formulations. 0.25 % cream preparations, 5 % nail lacquers, 50 mg vaginal tablets.

Trade Names (Amorolfine Hydrochloride). Loceryl (Roche, Sauter), Locéryl (Roche).

Mechanism of Action. Amorolfine is a morpholine derivative and selectively inhibits Δ^{14}-reduction and Δ^8-Δ^7-isomerase, which are required for Δ^{14}-reductase and Δ^8-Δ^7-isomerization, following C14 demethylation of lanosterol or 24-methlenedihydrolanosterol in the ergosterol biosnthesis pathway of the fungal cell [309].

Antimycotic Properties. Amorolfine is a broad-spectrum antimycotic [310] with in vitro activity against dermatophytes, pathogenic *Candida* species, and a variety of pathogenic dimophic fungi such as *Blastomyces dermatitidis*, *Histoplasma capsulatum*, and *Sporothrix schenckii*. The substance also shows some activity against various isolates of *Aspergillus* species.

Pharmacokinetics. Amorolfine is not absorbed after topical administration on normal skin. After application of a single vaginal tablet containing 50 mg of amorolfine, plasma concentrations between 27 and 83 ng/mL were observed [311].

Indications. For the treatment of skin infections, a 0.25 % cream preparation is applied once daily until clinical cure is achieved and then for a further 3 – 5 d [312]. For the treatment of nail mycoses caused by dermatophytes, yeasts, and moulds, a lacquer preparation containing 5 % amorolfine is painted onto the affected nail once or twice a week until the nail has regenerated. Treatment of onychomycoses normally needs to be continued for up to 1 year. Amorolfine is also active in the treatment of vaginal candidiasis.

Side Effects. Skin irritations, such as erythema, pruritus, or a burning sensation, rarely occur.

10. Unspecific Topical Antimycotics

Other unspecific antimycotics for topical use are listed in Table 12. These somewhat "historical" substances are of minor importance in clinical use nowadays, as highly effective and safe specific substances have become available.

Table 12. Other unspecified topical antimycotics

Generic name	Structural formula	Activity spectrum
Quaternary ammonium compounds Benzalkonium chloride [8001-54-5]		Dermatophytes, yeasts, moulds
Domiphen bromide [13900-14-6]	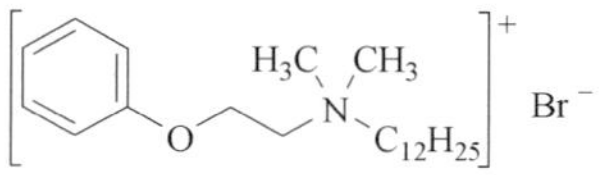	Dermatophytes, yeasts, moulds
Dibromosalane [87-12-7]	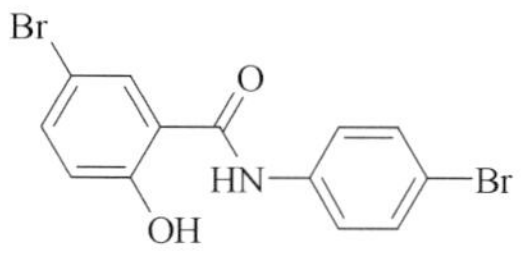	Dermatophytes, yeasts, gram-positive bacteria
Chloroquinaldol [72-80-0]	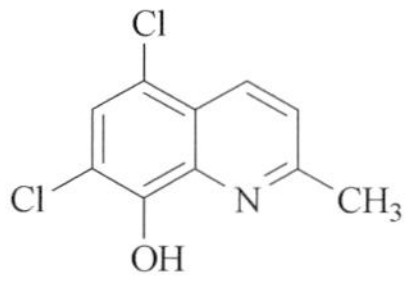	Dermatophytes, yeasts
Dichlorophene [97-23-4]	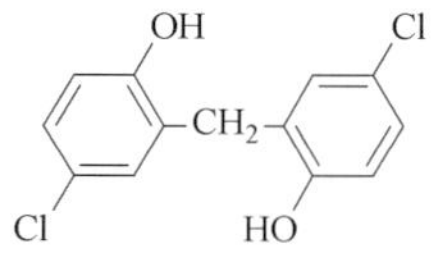	Dermatophytes, yeasts
Hexachlorophene [70-30-4]	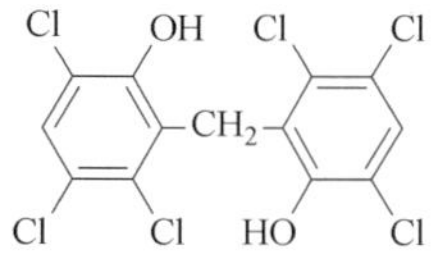	Dermatophytes, yeasts, gram-positive bacteria

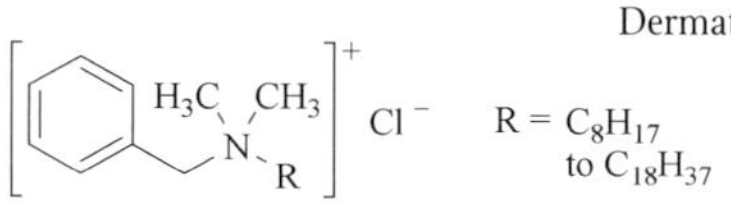

Table 12. (continued)

Generic name	Structural formula	Activity spectrum
Dibenzthione [350-12-9]	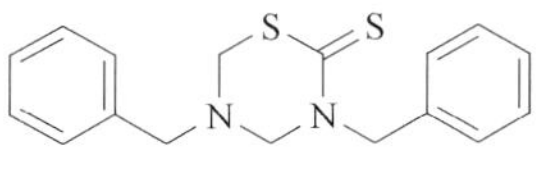	Dermatophytes, yeasts
Chlormidazole [3689-76-7]	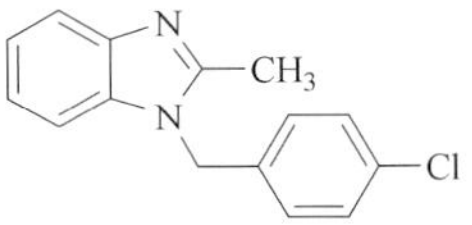	Dermatophytes, yeasts
Mercury compounds Merbromin (Mercurochrome) [129-16-8]	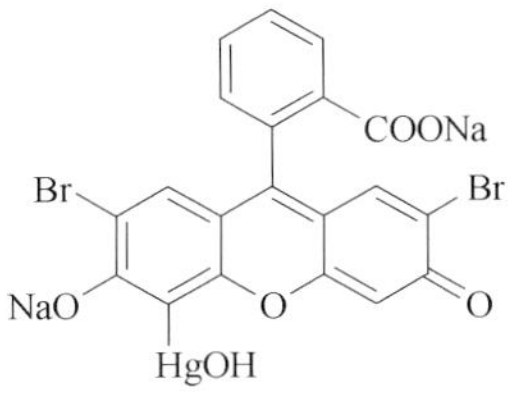	Dermatophytes, yeasts, bacteria
Thimerosal [54-64-8]	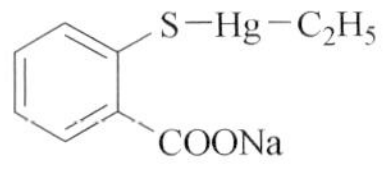	Dermatophytes, yeasts, bacteria
Triphenylmethane dyes Fuchsin (basic violet) [632-99-5]	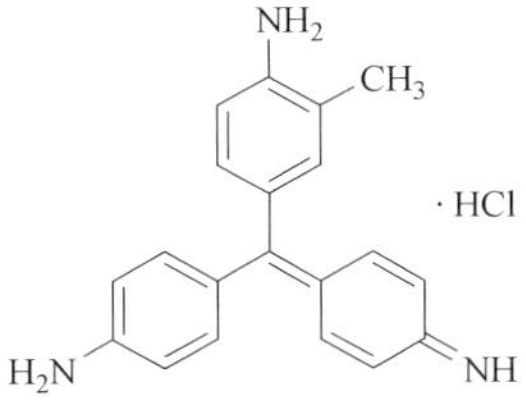	Dermatophytes, yeasts
Undecylenic acid [112-38-9]	$CH_2=CH(CH_2)_8COOH$	Dermatophytes, yeasts

11. Recent Developments and Outlook

Currently, the antimycotic market is clearly dominated by azoles, especially for the treatment of systemic mycoses. The latter market is led by fluconazole and itraconazole. Improved azoles are in clinical development (mostly phase I and II); pradimicins, aureobasidins, echinocandins, and pneumocandins for the treatment of systemic mycoses. The new azoles are reported to have activity against fluconazole-resistant strains and species as well as a broader spectrum of antifungal activity. Except the azoles, all other reported compounds habe a relatively low bioavailability and/or tolerability on the average which may limit their use to therapy of life threatening infections within the hospital segment. Production costs of all these compounds seem to be rather high as compared, e.g., to fluconazole.

Significant increase in research activities in antifungals for systemic application has been observed within the last few years in many pharmaceutical companies and institutes worldwide due to the high medical need and the fast growing market. Therefore, the chances for the development of new antifungal agents with new chemical structures and new mechanisms of action are improving.

12. References

[1]   K. J. Kwon-Chung et al.: *Medical Mycology*, Lea & Febiger, Philadelphia 1992.
[2]   K. H. Büchel et al., in J. S. Bindra et al. (eds.): *Chronicles of Drug Discovery*, vol. **2**, J. Wiley, New York 1984.
[3]   H. Yamaguchi et al., *Sabouraudia* **17** (1979) 311.
[4]   H. Van den Bossche et al., *Biol. Interact.* **21** (1978) 59.
[5]   H. Van den Bossche et al., in W. Siegenthaler et al. (eds.): *Current chemotherapy*, Am. Soc. Microbiol., Washington, D.C. 1978 p. 228.
[6]   J. Haller, *Abstracts XII – Intl. Congr. Microbiol.*, Munich Sept. 1978, no. 38/3.
[7]   H. Van den Bossche et al., *Arch. Int. Physiol. Biochim.* **87** (1979) 849.
[8]   H. Van den Bossche in M. R. McGinnis (ed.): *Current Topics in Medical Mycology*, vol. **1** Springer, New York 1985, p. 313.
[9]   W. R. Ness et al., *J. Biol. Chem.* **253** (1978) 6218.
[10] K. E. Bloch, *CRC Crit. Rev. Biochem.* **7** (1979) 1.
[11] Pfizer, GB 2 099 818, 1982.
[12] Pfizer, US 4 404 216, 1983.
[13] T. E. Rogers et al., *Antimicrob. Agents Chemother.* **30** (1986) 418.
[14] M. R. McGinnis et al. in V. Lorian (ed.): *Antibiotics in Laboratory Medicine*, Williams & Wilkins, Baltimore 1996.
[15] K. W. Brammer et al., *Rev. Infect. Dis.* **12** (1990) Suppl. 1, S 318.
[16] J. E. Thorpe et al., *Antimicrob. Agents Chemother.* **34** (1990) 2032.
[17] M. J. Humphrey et al., *Antimicrob. Agents Chemother.* **28** (1985) 648.

[18] E. T. Houang et al., *Antimicrob. Agents Chemother.* **34** (1990) 909.
[19] R. M. Tucker et al., *Antimicrob. Agents Chemother.* **32** (1988) 369.
[20] K. W. Brammer, *Br. J. Obstet, Gyn.* **96** (1989) 226.
[21] B. Dupont et al., *J. Med. Vet. Mycol.* **26** (1988) 67.
[22] J. A. Como et al., *N. Engl. J. Med.* **33** (1994) 263.
[23] J. E. Mangino et al., *Lancet* **345** (1995) 6.
[24] J. H. Rex et al., *Antimicrob. Agents Chemother.* **39** (1995) 1.
[25] J. E. Bennet et al., *N. Engl. J. Med.* **301** (1979) 126.
[26] W. G. Powderly et al., *N. Engl. J. Med.* **326** (1992) 793.
[27] M. Rozenberg-Arska et al., *J. Antimicrob. Chemother.* **27** (1991) 369.
[28] J. Heeres et al., *J. Med. Chem.* **27** (1984) 894.
[29] F. von Bruchhausen et al. (eds.): *Hagers Handbuch der Pharmazeutischen Praxis,* vol. **8**, Springer, Berlin–Heidelberg–New York 1993, p. 634.
[30] Janssen, EP 6 711, 1980.
[31] Janssen, US 4 267 179, 1981.
[32] D. A. Borelli, *Rev. Infect. Dis.* **9** (1987) 57.
[33] M. Bogers et al., *Rev. Infect. Dis.* **9** (1987) 33.
[34] G. Cauwenbergh et al., *Drug Dev. Res.* **8** (1986) 317.
[35] J. D. Cleary et al., *Ann. Pharmacother.* **26** (1992) 502.
[36] D. W. Denning et al., *Arch. Intern. Med.* **149** (1989) 2301.
[37] T. S. Jennings et al., *Ann. Pharmacother.* **27** (1993) 1206.
[38] R. Negroni et al., *Rev. Infect. Dis.* **9** (1987) 47.
[39] J. W. Van't Wout et al., *J. Infect.* **22** (1991) 45.
[40] J. W. Van't Wout et al., *J. Infect.* **20** (1990) 147.
[41] J. Heykants et al., *Mycoses* **32** (1989) Suppl. 1, 67.
[42] T. C. Hardin et al., *Antimicrob. Chemother.* **32** (1988) 1310.
[43] J. Heykants et al. in R. A. Fromtling (ed.): *Recent Trends in the Discovery, Development and Evaluation of Antifungal Agents,* J. R. Prous, Barcelona 1987, p. 223.
[44] S. M. Grant et al., *Drugs* **37** (1989) 310.
[45] J. Faegermann, *Mycoses* **31** (1988) 377.
[46] E. Van Hecke et al., *Mycoses* **31** (1988) 641.
[47] T. Piepponen et al., *J. Antimicrob. Chemother.* **29** (1992) 195.
[48] G. E. Stein et al., *Antimicrob. Agents Chemother.* **37** (1993) 89.
[49] W. E. Dismukes et al., *Am. J. Med.* **93** (1992) 489.
[50] J. S. Hostetler et al., *Chemotherapy* **38** (1992) Suppl. 1, 12.
[51] J. R. Graybill et al., *Am. J. Med.* **89** (1990) 282.
[52] A. P. Lavrijsen et al., *Lancet* **340** (1992) 251.
[53] R. M. Tucker et al., *J. Antimicrob. Chemother.* **26** (1990) 561.
[54] Janssen Pharmaceutica, DE-OS 2 804 094, 1978 (J. V. Heeres et al.).
[55] J. V. Heeres et al., *J. Med. Chem.* **22** (1979) 1003.
[56] Janssen, DE 2 804 096, 1978.
[57] Janssen, US 4 144 346, 1979.
[58] Janssen, US 4 223 036, 1980.
[59] N. R. Blatchford et al., *Lancet* **320** (1982) 770.
[60] A. Tavihan et al., *Gastroenterology* **90** (1986) 443.
[61] T. K. Daneshmend et al., *Clin. Pharmacokin.* **14** (1988) 13.
[62] G. Lake-Bakaar et al., *Ann. Int. Med.* **109** (1988) 471.

[63] T. K. Daneshmend et al., *Antimicrob. Agents Chemother.* **25** (1984) 1.
[64] M. D. Ene et al., *Br. J. Clin. Pharmacol.* **17** (1984) 173.
[65] A. M. Sugar et al., *Antimicrob. Agents Chemother.* **31** (1987) 1874.
[66] G. Cauwenbergh, *Mycoses* **32** (1989) Suppl. 2, 59.
[67] R. De Felice et al., *Antimicrob. Agents Chemother.* **19** (1981) 1073.
[68] A. Pont et al., *Arch. Int. Med.* **142** (1982) 2137.
[69] J. H. Lewis et al., *Gastroenterology* **86** (1984) 503.
[70] Janssen Pharmaceutica, US 3 717 655, 1969 (E. F. Godefroi et al.).
[71] E. F. Godefroi et al., *J. Med. Chem.* **12** (1969) 784.
[72] Janssen, DE 1 940 388, 1970.
[73] Janssen, US 3 717 655, 1973.
[74] L. Brugmans et al., *Arch. Dermatol.* **102** (1970) 428.
[75] R. Godts et al., *Arzneim. Forsch. / Drug Res.* **21** (1971) 256.
[76] R. C. Heel et al., *Drugs* **19** (1980) 7.
[77] J. R. Graybill et al., *Drugs* **25** (1983) 41.
[78] A. Kucers et al., in A. Kucers et al., (eds.): *The Use of Antibiotics,* Lippincott, Philadelphia 1987, p. 1492.
[79] T. K. Daneshmend et al., *Clin. Pharmacokin.* **8** (1983) 17.
[80] D. A. Stevens, *Drugs* **26** (1983) 347.
[81] V. Feinstein et al., *Ann. Intern. Med.* **93** (1980) 432.
[82] Bayer, DE-OS 2 461 406, 1976 (E. Regel et al.).
[83] Bayer, DE 2 461 406, 1976.
[84] Bayer, FR 2 295 747, 1976.
[85] Bayer, DE 2 643 563, 1978.
[86] Bayer, US 4 118 487, 1978.
[87] W. Ritter, *Proc. VII Int. Congr. ISHAM Palmerston North* (1982) 105.
[88] H. Weber, *Proc. VII Int. Congr. ISHAM Palmerston North* (1982) 105.
[89] S. Stettendorf, *Arzneim. Forsch. / Drug Res.* **33** (1983) 750.
[90] T. Anik et al., *J. Pharm. Sci.* **70** (1981) 8.
[91] F. von Bruchhausen et al. (eds.): *Hagers Handbuch der Pharmazeutischen Praxis,* vol. **7,** Springer, Berlin – Heidelberg – New York 1993, p. 581.
[92] Syntex, US 4 078 071, 1978.
[93] K. A. M. Walker et al., *J. Med. Chem.* **21** (1978) 840.
[94] F. C. Odds et al., *J. Antimicrob. Chemother.* **14** (1984) 105.
[95] W. Droegemueller et al., *Obstst. Gynecol.* **64** (1984) 530.
[96] J. B. Jacobson et al., *Acta. Obstet. Gynecol. Scand.* **64** (1985) 241.
[97] C. S. Brandbeer et al., *Genitourin. Med.* **61** (1985) 270.
[98] W. A. Van Dyk, *J. Reproduct. Med.* **31** (1986) Suppl. 7, 662.
[99] K. H. Büchel et al., *Arzneim. Forsch. / Drug Res.* **22** (1972) 1260.
[100] Bayer, DE-OS 1 617 481, 1976 (K. H. Büchel et al.).
[101] M. Plempel et al., *Dtsch. Med. Wochenschr.* **94** (1969) 1356.
[102] Bayer, DE 1 940 626, DE 1 940 627, 1969.
[103] Bayer, ZA 6 805 392, ZA 6 900 039, 1969.
[104] Bayer, US 3 657 445, US 3 705 172, 1972.
[105] W. Ritter, *Proc. VII Int. Congr. ISHAM Palmerston-North* (1982) 105.
[106] H. Weber, *Proc. VII Int. Congr. ISHAM Palmerston-North* (1982) 105.
[107] S. Stettendorf et al., *Arzneim. Forsch. / Drug Res.* **33** (1983) 750.

[108] Shionogi, BE 883 665, 1980.
[109] Shionogi, US 4 328 348, 1982.
[110] F. Toshiaka et al., *Iyakuhin Kenkyu* **9** (1978) 448.
[111] E. F. Godefroi et al., *J. Med. Chem.* **12** (1969) 784.
[112] Janssen, DE 1 940 388, 1970.
[113] Janssen, US 3 717 655, 1973.
[114] V. Tullio et al., *Mycoses* **33** (1990) 257.
[115] R. C. Heel et al., *Drugs* **16** (1978) 177.
[116] W. Rindt et al., *Arzneim. Forsch. / Drug Res.* **29** (1979) 697.
[117] A. Vukovich et al., *Clin. Pharmacol. Ther.* **21** (1977) 121.
[118] L. E. Millikan et al., *J. Am. Acad. Dermatol.* **18** (1988) 52.
[119] G. Gabriel et al., *Br. J. Vener. Dis.* **59** (1983) 56.
[120] J. S. Bingham et al., *Br. J. Vener. Dis.* **57** (1981) 204.
[121] J. E. Benett in L. S. Goodman et al. (eds.): *The Pharmacological Basis of Therapeutics*, Pergamon Press, New York 1990 p. 1165.
[122] A. Tajana et al., *Arzneim. Forsch. / Drug Res.* **31** (1981) 2127.
[123] Recordati, DE 2 917 244, 1979.
[124] Recordati, US 4 221 803, 1980.
[125] A. Finzi et al., *Mykosen* **29** (1986) 41.
[126] E. M. Kokoschka et al., *Mykosen* **29** (1986) 45.
[127] A. Gastaldi, *Curr. Ther. Res.* **38** (1985) 489.
[128] E. F. Godefroi et al., *J. Med. Chem.* **12** (1969) 784.
[129] Janssen, DE 1 940 388, 1970.
[130] Janssen, US 3 717 655, 1973.
[131] Janssen, US 3 839 574, 1974.
[132] H. J. Kessler, *Arzneim. Forsch. / Drug Res.* **29** (1979) 1344.
[133] H. J. Kessler et al., *Arzneim. Forsch. / Drug Res.* **29** (1979) 1352.
[134] U. Täubler, *Arzneim. Forsch. / Drug Res.* **37** (1987) 461.
[135] H. Wendt et al., *Arzneim. Forsch. / Drug Res.* **29** (1979) 846.
[136] Siegfried, DE 2 839 388, 1980.
[137] Siegfried, US 4 210 657, 1980.
[138] Siegfried, US 4 554 356, 1985.
[139] M. Mosse et al., *Pathol. Biol.* **34** (1986) 684.
[140] G. Mixich et al., *Arzneim. Forsch. / Drug Res.* **29** (1979) 1510.
[141] F. von Bruchhausen et al. (eds.): *Hagers Handbuch der Pharmazeutischen Praxis*, vol. **8**, Springer, Berlin – Heidelberg – New York 1993, p. 1262.
[142] Siegfried, DE 2 657 578, 1977.
[143] Siegfried, US 4 124 767, 1978.
[144] A. Polak, *Arzneim. Forsch. / Drug Res.* **32** (1982) 17.
[145] G. Stüttgen et al., *Mykosen* **28** (1985) 138.
[146] W. H. Beggs, *IRCS Med. Sci.* **11** (1983) 677.
[147] M. Raga et al., *J. Med. Chem. Chim. Ther.* **21** (1986) 329.
[148] Ferrer, EP 151 477, 1985.
[149] Ferrer, US 5 135 943, 1992.
[150] E. Martin-Mezuelos et al., *Chemotherapy* **42** (1996) 112.
[151] E. Drouhet et al., *Arzneim. Forsch. / Drug Res.* **42** (1992) 705.
[152] C. Palacin et al., *Arzneim. Forsch. / Drug Res.* **42** (1992) 699.

[153] C. Palacin et al., *Arzneim. Forsch. / Drug Res.* **42** (1992) 714.
[154] O. Azcona et al., *Curr. Ther. Res.* **49** (1991) 1046.
[155] F. von Bruchhausen et al. (eds.): *Hagers Handbuch der Pharmazeutischen Praxis*, vol. **9**, Springer, Berlin – Heidelberg – New York 1993, p. 691.
[156] Syntex, DE 2 541 833, 1976.
[157] Syntex, FR 2 285 126, 1976.
[158] Syntex, US 4 038 409, US 4 039 677, 1977.
[159] P. Benfield et al., *Drugs* **35** (1988) 143.
[160] A. Yoshida et al., *Chemotherapy* **32** (1984) 477.
[161] F. Iwata et al., *Shinkin to Shinkinsho* **25** (1984) 147.
[162] J. Heeres et al., *J. Med. Chem.* **26** (1983) 611.
[163] F. von Bruchhausen et al. (eds.): *Hagers Handbuch der Pharmazeutischen Praxis*, vol. **9**, Springer, Berlin – Heidelberg – New York 1993, p. 808.
[164] Janssen, DE 2 804 096, 1978.
[165] Janssen, US 4 144 346, 1979.
[166] Janssen, US 4 223 036, 1980.
[167] R. A. Fromtling, *Clin. Microbiol. Rev.* **1** (1988) 187.
[168] A. Kjaeldgaard, *Pharmacotherapeutica* **4** (1986) 525.
[169] U. M. Moebius, *Lancet* **332** (1988) 966.
[170] F. von Bruchhausen et al. (eds.): *Hagers Handbuch der Pharmazeutischen Praxis*, vol. **9**, Springer, Berlin – Heidelberg – New York 1993, p. 944.
[171] Pfizer, BE 841 309, 1976.
[172] Pfizer, US 4 062 966, 1977.
[173] S. P. Clissold et al., *Drugs* **31** (1986) 29.
[174] S. Jevons, *Antimicrob. Agents Chemother.* **15** (1979) 597.
[175] F. C. Odds, *J. Antimicrob. Chemother.* **6** (1980) 749.
[176] Y. M. Clayton et al., *Clin. Exp. Dermatol.* **7** (1982) 543.
[177] D. Brunelli et al., *Contact Dermatitis* **27** (1992) 120.
[178] R. Izu et al., *Contact Dermatitis* **26** (1992) 130.
[179] R. W. Holz in F. E. Hahn (ed.): *Antibiotics*, vol. **5**, Springer, Berlin – Heidelberg – New York 1979, p. 313.
[180] J. Vandeputte et al., *Antibiot. Anm.* (1955/56), 587.
[181] J. M. T. Hamilton-Miller, *J. Pharm. Pharmacol.* **25** (1973) 401.
[182] J. F. Gallelli, *Drug Intell.* **1** (1967) 102.
[183] E. Borowski et al.: *Vth Intl. Congr. Biochem. Moscow 1961*, Pergamon Press, London 1961, p. 3.
[184] F. von Bruchhausen et al. (eds.): *Hagers Handbuch der Pharmazeutischen Praxis*, vol. **7**, Springer, Berlin – Heidelberg – New York 1993, p. 237.
[185] C. P. Schaffner et al., *Antibiot. Chemother.* **11** (1961) 724.
[186] E. Borowski et al., *Tetrahedron Lett.* (1970) 3909.
[187] K. C. Nicolaou et al., *Chem. Commun.* (1986) 413.
[188] O. Mathieson, US 2 908 611, 1959.
[189] R. Janknegt, *Clin. Pharmacokinet.* **23** (1992) 279.
[190] H. Lackner et al., *Pediatrics* **89** (1992) 1259.
[191] A. Zoubek et al., *Pediat. Hematol. Oncol.* **9** (1992) 187.
[192] E. D. Ralph et al., *Antimicrob. Agents Chemother.* **35** (1991) 188.
[193] C. E. Hughes et al., *Antimicrob. Agents Chemother.* **25** (1984) 560.
[194] T. K. Daneshmend et al., *Clin. Pharmacokin.* **8** (1983) 17.

[195] J. Brajtburg et al., *J. Infect. Dis.* **149** (1984) 986.
[196] K. J. Christiansen et al., *J. Infect. Dis.* **152** (1985) 1037.
[197] N. Collette et al., *Antimicrob. Agents Chemother.* **33** (1989) 362.
[198] A. J. Atkinson et al., *Antimicrob. Agents Chemother.* **13** (1978) 271.
[199] A. Kucers et al. in A. Kucers et al. (eds.): *The Use of Antibiotics*, J. B. Lippincott, Philadelphia 1987, p. 1441.
[200] D. J. Drutz, *Drugs* **26** (1983) 337.
[201] G. Medoff et al., *Proc. Natl. Acad. Sci. USA* **69** (1972) 196.
[202] J. E. Bennett et al., *N. Engl. J. Med.* **301** (1979) 126.
[203] H. A. Gallis et al., *Rev. Infect. Dis.* **12** (1990) 308.
[204] F. Gigliotti et al., *J. Infect. Dis.* **156** (1987) 784.
[205] M. S. Maddux et al., *Drug Intell. Clin. Pharm.* **14** (1980) 177.
[206] W. T. Butler et al., *Ann. Intern. Med.* **61** (1964) 175.
[207] G. Lopez-Berenstein et al., *J. Infect. Dis.* **151** (1985) 704.
[208] A. P. Struyk et al., *Antibiot. Ann.* (1957/58) 878.
[209] W. E. Meyer, *Chem. Commun.* (1968) 470.
[210] T. Strittmatter, *Chem. Abstr.* **94** (1981) 214469.
[211] Koninklijke Nederlandsche Gist en Spiritus-Fabriek, GB 844 289, 1960.
[212] American Cyanamid, GB 846 933, 1960.
[213] E. Skytte Christensen et al., *Acta Obstst. Gynecol. Scand.* **61** (1982) 325.
[214] D. Kerridge, *Ad. Microb. Physiol.* **27** (1986) 1.
[215] T. Wegemann, *Internist* **30** (1989) 46.
[216] M. I. Maurad, *Assiut. Vet. Med.* **11** (1984) 236.
[217] D. J. Sutton et al., *Vet. Rec.* **118** (1986) 27.
[218] E. L. Hazen et al., *Proc. Soc. Exp. Biol. Med.* **76** (1951) 93.
[219] C. N. Chong et al., *Tetrahedron Lett.* (1970) 5145.
[220] A. C. Moffat: *Ckarke's Isolation and Identification of Drugs*, Pharmaceutical Press, London 1986.
[221] J. R. Carlson et al., *Antibiot. Chemother.* **9** (1959) 139.
[222] O. Mathieson, US 2 832 719, 1958.
[223] American Canamid, US 3 517 100, 1970.
[224] R. Duschinsky et al., *J. Am. Chem. Soc.* **79** (1957) 4559.
[225] Hoffmann-La Roche, US 2 945 038, 1960 (R. Duschinsky et al.).
[226] Hoffmann-La Roche, US 2 802 005, 1957.
[227] Hoffmann-La Roche, US 2 945 038, 1960.
[228] Hoffmann-La Roche, US 3 040 026, 1962.
[229] Hoffmann-La Roche, BE 628 615, 1963.
[230] Hoffmann-La Roche, US 3 368 938, 1968.
[231] D. C. E. Speller (ed.): *Antifungal Chemotherapy*, J. Wiley & Sons, New York 1980.
[232] H. Otten et al.: *Antibiotika-Fibel*, Thieme, Stuttgart 1975, p. 666.
[233] A. Polak et al., *Eur. J. Biochem.* **32** (1973) 276.
[234] J. Schönbeck et al., *Sabouraudia* **11** (1973) 10.
[235] R. E. Cutler et al., *Clin. Pharmacol. Ther.* **24** (1978) 333.
[236] A. Polak, *Postgrad. Med. J.* **55** (1979) 667.
[237] F. von Bruchhausen et al. (eds.): *Hagers Handbuch der Pharmazeutischen Praxis*, vol. **8**, Springer, Berlin – Heidelberg – New York 1993, p. 226.
[238] T. K. Daneshmend et al., *Clin. Pharmacokin.* **8** (1983) 17.
[239] R. B. Diasio et al., *Antimicrob. Agents Chemother.* **14** (1978) 903.

[240] B. E. Harris et al., *Antimicrob. Agents Chemother.* **29** (1986) 44.
[241] K. F. Wagner in G. L. Mandell et al. (eds.): *Principles and Practice of Infectious Diseases*, Churchill Livingstone, New York 1990, p. 1975
[242] D. Amstrong in J. F. Ryley (ed.): *Chemotherapy of Fungal Diseases*, Springer, Berlin – Heidelberg – New York 1990, p. 439.
[243] A. Kucers et al. in A. Kucers et al. (eds.): *The Use of Antibiotics*, J. B. Lippincott, Philadelphia 1987, p. 1534.
[244] A. Brossi et al., *Helv. Chim. Acta.* **43** (1960) 1444.
[245] Glaxo, US 3 069 328, US 3 069 329, 1962.
[246] P. W. Brian, *Trans. Br. Mycol. Soc.* **43** (1960) 1.
[247] W. Dittmar, *Mykosen* **9** (1966) 104.
[248] R. Brehme et al., *Pharmazie* **34** (1979) 372.
[249] T. Nogichi et al., *Antimicrob. Agents Chemother.* **1962** (1961 – 70) 259.
[250] W. Adam et al., *Ärztl. Forsch.* **14** (1960) 144.
[251] M. Rowland, *J. Pharm. Sci.* **57** (1968) 984.
[252] M. Kraml, *Antibiot. Chemother.* **12** (1962) 239.
[253] V. Shah, *J. Pharm. Sci.* **61** (1972) 634.
[254] C. Lin et al., *Drug Metabol. Rev.* **4** (1975) 75.
[255] B. R. Lecky, *Lancet* **336** (1990) 230.
[256] G. Mion et al., *Lancet* **334** (1989) 1331.
[257] G. Lohhaus et al., *Chem. Ber.* **100** (1967) 658.
[258] Hoechst, DE-AS 2 214 608, 1973 (G. Lohhaus et al.).
[259] Hoechst, ZA 6 906 039, 1970.
[260] Hoechst, DE 2 214 608, 1973.
[261] Hoechst, US 3 883 545, 1975.
[262] Hoechst, DE 2 795 831, 1978.
[263] W. Dittmar, *Proc. III. Intern. Congr. Trop. Der.*, Sao Paulo 1974.
[264] K. Iwata et al., *Arzneim. Forsch. / Drug Res.* **31** (1981) 1323.
[265] K. Sakurai et al., *Chemotherapy* **24** (1978) 68.
[266] K. Iwata et al., *Arzneim. Forsch. / Drug Res.* **31** (1981) 1323.
[267] W. Dittmar et al., *Arzneim. Forsch. / Drug Res.* **31** (1981) 1317.
[268] H. M. Kellner et al., *Arzneim. Forsch. / Drug Res.* **31** (1981) 1337.
[269] S. G. Jue et al., *Drugs* **29** (1985) 330.
[270] H. G. Alpermann et al., *Arzneim. Forsch. / Drug Res.* **31** (1981) 1328.
[271] H. G. Peil, *Arzneim. Forsch. / Drug Res.* **31** (1981) 1366.
[272] N. S. Ryder et al., *Antimicrob. Agents Chemother.* **29** (1986) 858.
[273] I. De Carneri et al., *Arzneim. Forsch. / Drug Res.* **26** (1976) 769.
[274] Carlo Erba, DE 2 313 845, 1973.
[275] Carlo Erba, US 3 855 263, 1974.
[276] Nippon Soda K.K. Tokio, DE-AS 1 468 388, 1962 (M. K. Taoka et al.).
[277] R. Brehme et al., *Pharmazie* **34** (1979) 372.
[278] T. Nogichi et al., *Antimicrob. Agents Chemother.* (1962) 259.
[279] Japan Soda, FR 1 337 797, 1963.
[280] Japan Soda, BE 627 322, 1962.
[281] Japan Soda, US 3 334 126, 1967.
[282] S. C. Harvey in L. S. Goodman et al. (eds.): *The Pharmacological Basis of Therapeutics*, Macmillan Publ., New York 1985 p. 959.

[283]   A. H. Gould, *South. Med. J.* **59** (1966) 176.
[284]   G. Petranyi et al., *Science* **224** (1984) 1239.
[285]   A. Schütz et al., *J. Med. Chem.* **27** (1984) 1539.
[286]   F. von Bruchhausen et al. (eds.): *Hagers Handbuch der Pharmazeutischen Praxis*, vol. **8**, Springer, Berlin – Heidelberg – New York 1993, p. 1068.
[287]   H. Loibner et al., *Tetrahedron Lett.* **25** (1984) 2535.
[288]   Sandoz, BE 853 976, 1977.
[289]   Sandoz, US 4 282 251, 1981.
[290]   Lab. Frumtost-Prem S.A., ES 504 432, 1982.
[291]   A. Georgopoulos et al., *Antimicrob. Agents Chemother.* **19** (1981) 386.
[292]   G. Petranyi et al., *Antimicrob. Agents Chemother.* **19** (1981) 390.
[293]   J. P. Monk et al., *Drugs* **42** (1991) 659.
[294]   S. Nolting et al., *Mykosen* **28** (1985) 69.
[295]   L. E. Millikan et al., *J. Am. Acad. Dermatol.* **18** (1988) 52.
[296]   I. Effendy et al., *Therapiewoche* **36** (1986) 848.
[297]   F. von Bruchhausen et al. (eds.): *Hagers Handbuch der Pharmazeutischen Praxis*, vol. **9**, Springer, Berlin – Heidelberg – New York 1993, p. 802.
[298]   Sandoz, EP 24 587, 1981.
[299]   Y. M. Clayton et al., *Br. J. Dermatol.* **130** (1994) Suppl. 43, 7.
[300]   J. C. Jensen, *Clin. Exp. Dermatol.* **14** (1989) 110.
[301]   J. A. Balfour et al., *Drugs* **43** (1992) 259.
[302]   P. J. Dykes et al., *J. Dermatol. Treat.* **1** (1990) Suppl. 2, 19.
[303]   T. C. Jones, *Br. J. Dermatol.* **132** (1995) 683.
[304]   L. Juhlin, *Lancet* **339** (1992) 1483.
[305]   J. P. Ottervanger et al., *Lancet* **340** (1992) 728.
[306]   G. Lowe et al., *B.M.J.* **306** (1993) 248.
[307]   Hoffmann-La Roche, DE 2 752 096, 1978.
[308]   Hoffmann-La Roche, US 4 202 894, 1980.
[309]   A. Polak in J. F. Ryley (ed.): *Chemotherapy of Fungal Diseases*, Springer, Berlin – Heidelberg – New York 1990, p. 153.
[310]   A. Polak in J. F. Ryley (ed.): *Chemotherapy of Fungal Diseases*, Springer, Berlin – Heidelberg – New York 1990, p. 505.
[311]   E. Rhode et al. in R. A. Fromtling (ed.): *Recent Trends in the Discovery, Development and Evaluation of Antifungal Agents*, J. R. Prous, Barcelona 1987, p. 575.
[312]   M. Haria et al., *Drugs* **49** (1995) 103.

Anthelmintics

Peter Andrews, Bayer AG, Wuppertal, Federal Republic of Germany
Achim Harder, Bayer AG, Wuppertal, Federal Republic of Germany

1.	Introduction	1265	2.3.1.	Roundworm	1272
2.	Parasitic Worms	1266	2.3.2.	Hookworms	1273
2.1.	Trematodes (Flukes)	1266	2.3.3.	Pinworm	1273
2.1.1.	Blood Flukes	1266	2.3.4.	Whipworm	1274
2.1.2.	Lung Flukes	1267	2.3.5.	Threadworm	1274
2.1.3.	Liver Flukes	1268	2.3.6.	Trichina	1275
2.1.4.	Giant Intestinal Fluke	1268	2.3.7.	Larvae of Toxocara canis, Toxocara cati, Ancylostoma brasiliense and Ancylostoma caninum	1275
2.1.5.	Lesser Intestinal Flukes	1269			
2.2.	Cestodes (Tapeworms)	1269			
2.2.1.	Broad or Fish Tapeworm	1269	2.3.8.	Guinea Worm	1276
2.2.2.	Beef Tapeworm	1270	2.3.9.	Lymphatic Filariae	1276
2.2.3.	Pork Tapeworm	1270	2.3.10.	Convoluted Filaria	1277
2.2.4.	Dwarf Tapeworm	1271	2.3.11.	African Eye Worm	1277
2.2.5.	Dog Tapeworm	1271	3.	Anthelmintic Drugs	1277
2.3.	Nematodes (Roundworms)	1272	4.	References	1287

1. Introduction

Parasitism is a special form of intimate relationship between two species. One species, the host, is to some degree injured through the activities of the other species, the parasite. The parasite need not be parasitic through all stages of its existence. Parasites living within the host are distinguished as endoparasites, e.g., helminths, from those found on the surface of the body, the ectoparasites, e.g., arthropods. There are forms of close biologic relationships other than parasitism. Commensalism denotes an association beneficial to one partner and at least not disadvantageous to the other. Mutualism (or symbiosis) denotes an association that is mutually beneficial.

This article describes only the helminth infections of humans. Infections caused by other parasitic organisms, such as protozoa and ectoparasites, are described elsewhere (→ Synthetic Chemotherapeutic Agents, → Drugs Used in Dermatology), as are the veterinary parasitoses.

This article is not intended to be a guide to therapy, and mention of a drug is not to be taken as an endorsement. Chapter 2 describes the helminth parasites of humans and the diseases they cause, and Chapter 3 describes the drugs that may be used to control the parasites.

2. Parasitic Worms

Helminths are the wormlike organisms classified in two phyla, the flatworms (Platyhelminthes), comprising the flukes (or trematodes) and the tapeworms (or cestodes), and the roundworms (Nemathelminthes). The roundworms are also called nematodes. Therefore, "helminth" refers to three morphologically different groups: flukes, tapeworms, and roundworms. Most flatworms are hermaphroditic, having both male and female reproductive organs within the same individual, and are parasites. In contrast, the roundworms have separate sexes and are overwhelmingly free living. Nevertheless, a number of roundworm species are parasitic, affecting humans, animals, and plants.

2.1. Trematodes (Flukes)

2.1.1. Blood Flukes (*Schistosoma* spp.)

Three species of schistosomes cause most human schistosomiasis infections: *Schistosoma mansoni* and *S. japonicum* (locally also *S. intercalatum* and *S. mekongi*) cause the intestinal form of the disease, whereas *S. haematobium* (rarely *S. mattheii*) causes the urinary form.

Distribution. *S. haematobium*: all over Africa, in the Near East (Syria to Iran and Yemen), islands off East Africa. *S. mansoni*: throughout most of Africa southeast of a line from Gambia to Libya, Arabian peninsula, Brazil, Venezuela, Surinam, some Antilles Islands. *S. japonicum*: southern half of continental China, the Philippines; however, virtually extinct in Japan. *S. mekongi*: Laos, Cambodia. *S. intercalatum*: Cameroun to the Congo Basin. *S. mattheii*: restricted to southern Africa. About 270 million people are infected [13]. Annually, about one million deaths are caused by schistosomiasis [14].

Development. The schistosomes differ in many aspects from all other human flukes. They are elongate, wormlike, and live within the blood vessels. They have separate sexes. The adult parasites *S. mansoni* and *S. japonicum* are found as pairs in the mesenteric veins, whereas *S. haematobium* inhabits the veins of the vesical plexus. Adult males are 10–15 mm long and 0.5–1 mm wide. Their lateral body seams are folded downwards, forming a tube, the gynecophoric canal, which holds the female, which can be up to 28 mm long. The females deposit eggs, a few dozen to several thousand per day, depending on the species. About three fourths of the eggs laid are retained in the tissues of the body — mainly the liver and the intestinal and bladder walls — where they die within three weeks. The other eggs pass through the tissues into the lumen of the intestine or the bladder. They contain a larva, the miracidium, which hatches as soon as the eggs are voided into fresh water. The miracidium swims through the water in search of an appropriate snail (its intermediate host), penetrates the snail's tissues, and begins to multiply. After 4–7 weeks another type of larva, the cercaria, is released from the snail. These cercariae locate human skin exposed to water and penetrate it. Within another 6–12 weeks the juvenile schistosomes migrate through the vascular system and lungs to their predilected sites, mature, pair, and start laying eggs. The normal lifespan of schistosomes is 2–7 years. However, some parasites can survive and produce eggs for as long as thirty years.

Clinical Features. Severe itching (cercarial dermatitis) may occur at the site of skin penetration, and fever may occur during the migration phase of the parasite. In established infections, the egg incites an inflammatory reaction around the site where it is deposited in the tissues. This leads to fibrosis and calcification of the bladder, hematuria, uretric and renal involvement, and bladder cancer (*S. haematobium*). In the intestinal form of the disease (*S. mansoni* and *S. japonicum*) symptoms encountered are diarrhea, weakness, abdominal pain, splenomegaly, hepatic and portal fibrosis, portal hypertension, ascites, and carcinoma of the colon. Severe infections with *S. japonicum* can interfere with normal growth and mental development.

Therapy: praziquantel, oxamniquine, metrifonate, niridazole, hycanthone.

2.1.2. Lung Flukes (*Paragonimus* spp.)

Distribution. *P. westermani:* East Asia (Korea, Japan, China, Philippines, Indonesia). *P. kellicotti:* North America. *P. africanus:* Gabon to Zaire; several other species ranging from Mexico to Peru. About 20 million people are infected [15].

Development. The adult parasites measure 8–16 mm by 4–6 mm. They are found in pulmonary cysts that are formed by destroyed tissue and that connect to the respiratory passages of the lung. Eggs are voided with the sputum or the feces. Once eggs reach fresh water, a larva, the miracidium, develops within 2–4 weeks. The miracidium hatches from the egg, actively penetrates into specific aquatic snails, and multiplies. After 8–10 weeks, cercariae begin to be released from the snail. They locate crabs or crayfish, penetrate them, and encyst in the musculature as metacercariae. The crustaceans can also become infected by feeding on infected snails. Humans acquire the infection by ingesting raw or improperly cooked crab meat or juice. The juvenile flukes excyst in the small intestine, penetrate the gut wall, and migrate through the diaphragm into the pleural cavity, where they penetrate the serosal layers of the lungs. Finally they arrive in the vicinity of the bronchioles. Within six weeks they develop into adult worms in tissue capsules that rupture and thus connect with the air ducts. These parasites can live for up to twenty years. Many mammals can act as carriers for human lung flukes. However, their importance as reservoir hosts is largely unknown.

Clinical Features. Mild infections: fever and cough. Severe infections: dry cough, violent blood spitting, weight loss, fever, pleural effusion, chest pain.

Therapy: praziquantel, bithionol.

2.1.3. Liver Flukes

Infections are caused by *Clonorchis sinensis*, *Opisthorchis viverrini*, or *O. felineus*. Infections with the cosmopolitan liver flukes of ruminants (*Fasciola hepatica*, *Dicrocoelium dendriticum*) may be important locally.

Distribution. *C. sinensis:* Korea, China, Southeast Asia, Japan, Taiwan, Philippines, Indonesia. *O. viverrini:* Laos, Thailand, Indian subcontinent. *O. felineus:* USSR and some European foci. About 50 million people are infected [15].

Development. Adult *C. sinensis* and *O. viverrini* (10–20 and 8–12 mm long, respectively) live in the intrahepatic bile ducts. Their eggs contain fully developed larvae, the miracidia, which are voided with the feces and ingested by aquatic snails. The parasitic larva (miracidium) hatches in the intestine of the snail, enters its tissues, and starts multiplying. After some time cercariae are released into the water. They infect cyprinid fish percutaneously, encyst in the fish musculature, and become infective metacercariae in about four weeks. Infection is acquired through the consumption of raw or improperly cooked fish. The young flukes excyst in the small intestine, migrate into the bile ducts, and mature within four weeks. They may then live for several decades. Dogs, cats, and pigs are important reservoir hosts.

Clinical Features. Mild infections: generally asymptomatic. Severe infections: dilation and fibrosis of the bile ducts, hemorrhage, digestive disturbances. Chronic infections: epigastric pain, anorexia, cholangitis, liver necrosis, bile duct and pancreas carcinoma, intrahepatic calculi.

Therapy: praziquantel, bithionol.

2.1.4. Giant Intestinal Fluke (*Fasciolopsis buski*)

Distribution. South China, Southeast Asia, Indonesia, and the Indian subcontinent. Two to ten million people are infected [15].

Development. The eggs of the giant intestinal fluke are voided with the feces. In fresh water a miracidium develops in 3–7 weeks, hatches, locates an aquatic snail, penetrates it, and begins multiplying. After 4–8 weeks cercariae begin to be released, and they encyst as metacercariae on aquatic plants. Infection is acquired when such plants (water caltrop, *Trapa* spp.; water chestnut, *Eleocharis* sp.) are peeled with the teeth. The metacercarial cysts are thus freed, are swallowed, and develop to sexually mature flukes (5–7 cm long) in the small intestine within about three months. Pigs, dogs, and rabbits are important reservoir hosts.

Clinical features. Mild infections are often asymptomatic. In severe infections there are generalized toxic and allergic symptoms, diarrhea, weakness, malabsorption, edema, ascites, abdominal pain, and gastrointestinal hemorrhage.

Therapy: praziquantel, niclosamide.

2.1.5. Lesser Intestinal Flukes

Several species of these small flukes (2–7 mm in length) are found in humans: *Echinostoma ilocanum*, *E. lindoense*, *Heterophyes heterophyes*, *Metagonimus yokogawai*, *Gastrodiscoides hominis*, *Watsonius watsoni*.

Distribution. East and South Asia, West Africa, Mediterranean countries, especially Egypt (*Heterophyes*). About 16 million people are infected [13].

Development. These parasites live in the small intestine. They have typical trematode life cycles, with a single intermediate host (*Gastrodiscoides* and *Watsonius*; metacercariae on aquatic plants) or two intermediate hosts. The second host can be fish (*Heterophyes* and *Metagonimus*) or snails and mussels (*Echinostoma*). Infection is acquired through the consumption of raw or improperly cooked plants, fish, or mollusks. The lesser intestinal flukes live and produce eggs for several months.

Clinical Features. Mild infections are often asymptomatic or show only nonspecific symptoms: diarrhea, abdominal discomfort, and headache.

Therapy: praziquantel, thiabendazole, niclosamide, bephenium.

2.2. Cestodes (Tapeworms)

2.2.1. Broad or Fish Tapeworm (*Diphyllobothrium latum*)

Distribution. Northern temperate zone (Finland, USSR, Alaska, Canada; small foci in northern Italy, Switzerland, and Chile). *D. pacificum*: coastal regions of Peru. There are several related species that infect humans (e.g., *Diplogonoporus* in Japan). About 2–9 million people are infected [6].

Development. *D. latum* is a typical ribbonlike tapeworm of extreme length (up to 10–15 m) living in the small intestine. After the eggs are voided with the feces into fresh water, a larva, the coracidium, is released. Larvae are ingested by small copepod crustaceans. When these are ingested by cyprinid fish, the parasites penetrate into the musculature of the fish and develop into the infective stage, the plerocercoid. Humans acquire the infection by consuming raw or improperly cooked fish.

Clinical Features. Infections are generally asymptomatic. When they do appear, symptoms include abdominal discomfort, dizziness, and fatigue. About 4% of those infected develop clinical symptoms of a vitamin B_{12} deficiency indistinguishable from pernicious anemia.

Therapy: praziquantel, niclosamide.

Several related species that cannot develop to maturity in humans, but do so in other mammals, may infect humans accidentally. They can proliferate and cause painful inflammatory tissue reactions. This rare condition, sparganosis, can be treated only by surgical removal. Praziquantel may prove to be effective.

2.2.2. Beef Tapeworm (*Taenia saginata* =*Taeniarhynchus saginatus*)

Distribution. Cosmopolitan, about 77 million people are infected [13].

Development. Adult beef tapeworms generally measure 3–10 m. Each day about ten proglottids of the ribbonlike body, the strobila, are shed, containing several hundred thousand eggs. When these eggs are ingested by cattle, a larva, the oncosphere, hatches from the egg and reaches the musculature (presumably by the circulatory system), where an infective larva, the cysticercus, develops within 10–12 weeks. Cattle may carry infective cysticerci in their muscles up to three years. Humans acquire the infection by consuming raw or improperly cooked beef. Within 3–4 months the parasite reaches maturity in the intestine. It may live up to twenty years.

Clinical Features. Infections often go unnoticed. In symptomatic cases: vague abdominal pain, nausea, weakness, loss of weight, increased or decreased appetite, headache.

Therapy: niclosamide, praziquantel, mebendazole.

2.2.3. Pork Tapeworm (*Taenia solium*)

Distribution. Almost cosmopolitan, wherever people eat improperly cooked pork. About 3–5 million people are infected with adult *T. solium* [15].

Development, clinical features, and therapy are all similar to those of the beef tapeworm.

Cysticercosis. Although pigs usually are the intermediate hosts in the life cycle of the pork tapeworm, humans can also become infected with the larvae of this parasite by ingesting the eggs, which remain viable in the soil for many weeks. The resulting infection, *cysticercosis,* occurs in Central and South America, South and East Asia, East Africa, and eastern Europe and may afflict more people than are infected with the adult tapeworm [15]. This condition is often accompanied by myositis and high fever. When the brain is involved (neurocysticercosis, developing in 1–5% of the infected persons), this condition is often accompanied by epileptic attacks, meningoencephalitis, and intracranial hypertension. In South America alone, there are about 350000 cases [16].

Formerly, the only *treatment* for cysticercosis was surgical removal of larval cysts.

Therapy: praziquantel, niclosamide, mebendazole.

2.2.4. Dwarf Tapeworm (*Hymenolepis nana*)

Distribution. Cosmopolitan, about 45–50 million people, mainly children, are infected [15].

Development. The dwarf tapeworm is unique in that the adult can develop following ingestion of the eggs by humans. The larva, the oncosphere, that hatches from the egg develops to the cysticercoid stage in the intestinal mucosa within 4–7 days. Alternatively, infections may be acquired through accidental ingestion of grain beetles containing infective cysticercoid larvae. The adult tapeworm, developing from the cysticercoid to maturity in 1–2 weeks, grows up to 40 cm long.

Clinical Features. Mild infections are asymptomatic. Severe infections: abdominal pain, diarrhea, headache, dizziness, anorexia.

Therapy: praziquantel, niclosamide.

2.2.5. Dog Tapeworm (*Echinococcus granulosus, E. multilocularis*)

Distribution. *E. granulosus:* cosmopolitan, especially in sheep-rearing countries. *E. multilocularis:* small foci in North, Central, South, and East Europe, northern United States. About 100000 people are infected [15].

Development. Many mammals (*humans,* sheep, goats, cattle, pigs, dromedaries, horses, rodents) serve as intermediate hosts of these small (3–5 mm long) tapeworms that live as adults in the intestine of dogs or other canines. When humans ingest the eggs, larvae hatch, penetrate the gut wall, and generally lodge in the liver or lungs, although all other tissues can also be affected. The larvae grow extremely quickly, forming unilocular (*E. granulosus*) or multilocular (*E. multilocularis*) cysts. The cysts contain many protoscolices, which mature to adult worms when offal containing cystic material is fed to dogs.

Clinical Features. Cysts are often asymptomatic until they grow large, but then the symptoms become progressively severe. Symptoms depend on the location of the cysts. Liver cysts can cause symptoms resembling a mucoid carcinoma. Lung cysts give rise to coughing and chest pain, whereas cerebral cysts can cause serious neurological damage.

Therapy: surgical removal, in inoperable cases, mebendazole, which does not kill the parasite but arrests the progressive course of the disease.

2.3. Nematodes (Roundworms)

The parasitic nematodes are conveniently divided into two groups: the intestinal nematodes, including the roundworm, hookworm, pinworm, whipworm, threadworm, and trichina, and the extra-intestinal nematodes, including the guinea worm and several species of filariae. The second group live in the tissues of fluids of the body and require an arthropod vector for the completion of their life cycles.

2.3.1. Roundworm (*Ascaris lumbricoides*)

Distribution. Cosmopolitan, common in humid tropical climates. About 1300 million people are infected [13].

Development. This large parasite (20–40 cm long) lives in the small intestine and feeds on gut contents. Each female produces over 200000 eggs per day. The eggs are passed with the feces, become infective in about one month, and can remain infective in the soil for up to fifteen years. After infective eggs have been ingested, e.g., with vegetables, larvae are released, penetrate the duodenal wall, and migrate through the liver and the blood vessels to the lungs. There they break through the alveolae into the air passages, ascend the trachea, are swallowed, and attain maturity in the gut about 8–10 weeks later. The females live about one year.

Clinical Features. The majority of infections are light and asymptomatic. Larvae in the lungs may cause pneumonitis (cough, wheezing, dyspnea). Complications are intestinal obstruction and worm irritation after medication, causing hepatic duct obstruction, appendicitis, and intestinal perforation.

Therapy: levamisole, pyrantel, piperazine, mebendazole, thiabendazole, albendazole. Ivermectin is in clinical development

2.3.2. Hookworms

Two species, *Ancylostoma duodenale* (common or Old World hookworm) and *Necator americanus* (American or New World hookworm), parasitize humans, causing one of the most important tropical diseases.

Distribution. *N. americanus:* the Americas, tropical Africa, South and East Asia. *A. duodenale:* from the Mediterranean through India to China and Southeast Asia, Brazil. *A. ceylanicum* is of local importance. About 930 million people are infected and about 60000 deaths are caused by hookworm infections each year [13], [14].

Development. The mature worms ($\approx$ 10 mm long) attach to the gut mucosa and suck blood (*Necator* 5 – 10 times less than *Ancylostoma*). Each female produces 10000 (*Necator*) or 20000 (*Ancylostoma*) eggs per day. These are voided with the feces and release larvae that develop in the soil within seven days to the infective stage, which is able to survive up to one month. Infection is percutaneous (*Ancylostoma* also oral). The larvae migrate to the heart and lungs. Subsequent development is as described for the roundworm (*Ascaris*).

Clinical Features. Mild, cutaneous reaction to the infective larvae and lung reactions (cough, wheezing) to the migratory larvae. Chronic phase: epigastric pain, tenderness, peptic symptoms, iron-deficiency anemia, protein-loss enteropathy, hypoproteinemia. Severe infections: high-output cardiac-failure dyspnea, edema. The symptoms of hookworm infections are often aggravated by malnutrition and secondary bacterial or other parasitic infections.

Therapy: pyrantel, thiabendazole, mebendazole, bephenium, levamisole, tetrachloroethylene, bitoscanate.

2.3.3. Pinworm (*Enterobius vermicularis*)

Distribution. Cosmopolitan, common in the developed countries of the northern hemisphere. About 350 million people are infected [13].

Development. The pinworm, which is more commensal than parasitic, lives in the lumen of the colon and feeds on gut contents. Gravid females migrate out of the anus to deposit their eggs (5000 – 10000), which are fully embryonated and infective within a few hours. Infections and reinfections are acquired through accidental oral intake of eggs. Larval development occurs in the intestine within about six weeks.

Clinical features. Perianal itching. Complications caused by ectopic migration are rare.

Therapy: pyrantel, piperazine, mebendazole, pyrvinium.

2.3.4. Whipworm (*Trichuris trichiura*)

Distribution. Cosmopolitan, especially common in warm and humid climates. About 690 million people are infected [13].

Development. The parasites (3–5 mm long) live attached to the cecal mucosa. The eggs are voided with the feces and require a period of at least three weeks to become infective. Eggs remain infective in the soil for over one year. The larvae hatch after eggs have been ingested, e.g., with vegetables or dirt, and mature in the intestine within 2–3 months. The adults live 3–10 years.

Clinical Features. Infections are generally asymptomatic; occasionally, abdominal discomfort and, in children, diarrhea occur.

Therapy: oxantel, mebendazole, thiabendazole, albendazole. Ivermectin is in clinical development.

2.3.5. Threadworm (*Strongyloides stercoralis*)

Distribution. Tropical and subtropical climates, extending into southern Europe and southern United States. About 80 million people are infected [15].

Development. The female worms live attached to the mucosa of the small intestine, on which they feed. They produce, usually parthogenetically, larvae that are infective shortly after passage of the feces and infect humans percutaneously. Their further development to maturity takes about 17 days and resembles that of hookworms. Autoinfection is possible when larvae mature in the gut and infections may be perpetuated up to twenty years. When environmental conditions are favorable, threadworms can exist for some time as free-living nematodes, completing several generations in the soil.

Clinical Features. Infections may be asymptomatic, but gastrointestinal symptoms (diarrhea, katarrhal enteritis, central epigastric pain) are common. In cases of severe infection the prognosis is poor, and the mortality rate is high.

Therapy: thiabendazole, mebendazolealbendazole. Ivermectin is in clinical development.

Infections with other nematodes (e.g., *Angiostrongylus, Anisakis, Capillaria, Dipetalonema, Gnathostoma, Oesophagostomum, Trichostrongylus*) are less important and generally respond to thiabendazole and/or mebendazole. Thiabendazole is also effective against migrating larvae of nematodes that normally mature in other mammals (cutaneous and visceral larva migrans). About 5–10 million people are infected [15].

2.3.6. Trichina (*Trichinella spiralis*)

Distribution. Cosmopolitan except Australia. About 46 million people are infected [13].

Development. Infection is initiated by the consumption of raw or improperly cooked meat (especially pork) containing the encysted larvae. These excyst in the gut and mature to adult worms in the gut wall within 3–4 days. Sexes mate in the gut lumen, and each female then produces up to 2000 larvae within two weeks. The larvae penetrate the gut wall, enter the lymphatic vessels, disseminate via the circulatory system throughout the body, develop in the muscle, and finally encyst in muscle fibers.

Clinical Features. Infections are generally asymptomatic during the intestinal phase. The muscle infection causes myositis, eosinophilia, leukocytosis, and occasional fever.

Therapy: thiabendazole, mebendazolealbendazole. Ivermectin is in clinical development.

2.3.7. Larvae of Toxocara canis, Toxocara cati, Ancylostoma brasiliense and Ancylostoma caninum

Distribution. Cosmopolitan.

Development. Humans, especially children are the "wrong hosts" for these larvae. Human infection are rare and occur through oral ingestion of larvae following contact with contaminated faeces of *Toxocara canis* infected dogs or *T. cati* infected cats or with contaminated soil, sand, or playgrounds. The larvae of *Ancylostoma brasiliense* or *Ancylostoma caninum* penetrate the skin ("creeping eruption") following contact with contaminated dog or cat faeces.

Clinical Features. The disease is determined by the site where larvae reside. Larvae migrantes viscerales of *T. canis* and *cati* are mostly observed in liver and lungs, occasionally in eyes or brain. Larvae of *Ancylostoma brasiliense* are called larvae migrantes cutaneae, larvae of *Ancylostoma caninum* are called epidermis larvae.

Therapy: Thiabendazole, diethylcarbamazine.

2.3.8. Guinea Worm (*Dracunculus medinensis*)

Distribution. Focally in West, North, and East Africa, the Middle East, from Iran through Pakistan into Indonesia. About 50–100 million people are infected [15].

Development. The parasites develop in the body cavity and deeper connective tissue for about 12–14 months. The mature females then migrate to the subcutaneous tissues, mainly of the legs. There they cause blisters and ulceration, through which large numbers of larvae are discharged when the skin contacts water. If taken up by copepod crustaceans, the larvae develop in the body cavity and become infective in 2–3 weeks. Humans acquire the infection by drinking water containing infected copepods. Dogs are known reservoir hosts.

Clinical Features. Infections are usually asymptomatic for one year. Specific signs of the disease are painful blisters and ulcerations of the skin where the females penetrate the skin. Secondary bacterial infections often cause additional complications.

Therapy: Careful mechanical extraction, thiabendazole. Both niridazole and metronidazole reduce the inflammatory tissue reactions without affecting the parasites.

2.3.9. Lymphatic Filariae (*Wuchereria bancrofti, Brugia malayi*)

Distribution. *Wuchereria:* throughout the tropics and subtropics. *Brugia malayi:* from India to Korea and Indonesia. *B. timori:* Timor. About 380 million people are infected [13].

Development. The adult parasites, the macrofilariae, live in lymphatic ducts and nodes. The females deposit eggs, from which larvae, the microfilariae, hatch. The larvae are distributed with the blood throughout the body but congregate in the peripheral vasculature during the night. From there they are taken up by mosquitoes and mature to infective larvae in their musculature. They then enter the proboscis of the mosquito, and when it takes another blood meal they enter the human skin through the puncture. In the new host, they mature and mate, and the females start producing the next generation of microfilariae.

Clinical Features. Infections generally are asymptomatic, but chronic lesions, presumably of an allergic nature, can cause lymphangitis and obstruction of lymphatic drainage, leading to elephantiasis and hydrocele.

Therapy: diethylcarbamazine.

2.3.10. Convoluted Filaria (*Onchocerca volvulus*)

Distribution. Africa south of the Sahara, Yemen, Mexico to Venezuela. About 40 million people are infected [14].

Development. The adult worms (macrofilariae) live in subcutaneous nodules. The larvae (microfilariae) collect in the superficial layers of the skin and are transmitted by black flies (*Simulium*), in which they develop to the infective stage. Humans are infected when the fly takes another blood meal.

Clinical Features. Microfilariae may cause chronic cutaneous lesions. Invasion of the eye: conjunctivitis, keratitis, iridocyclitis, chorioretinitis, glaucoma, opacity of the lens, leading to optic atrophy and blindness.

Therapy: diethylcarbamazine (microfilaricidal), suramin (macrofilaricidal), ivermectin (microfilacricidal).

2.3.11. African Eye Worm (*Loa loa*)

Distribution. Rain forests of West and Central Africa. More than 10 million people are infected [15].

Development. The adult parasites (macrofilariae) actively migrate through the subcutaneous tissue. The microfilariae are found in the peripheral blood during the day and are transmitted by flies (*Chrysops*).

Clinical Features. Allergic swelling of the skin.

Therapy: diethylcarbamazine.

3. Anthelmintic Drugs

The anthelmintic drugs are listed alphabetically. The *mechanism of action* of the older anthelmintics can be found in [17]. The numerous *trade names* for the older drugs are given for some twenty countries in [18]. A number of trade names are given in [19]. The national drug directories can also be consulted.

Albendazole [*54965–21–8*], methyl(5-(propylthio)-2-benzimidazol)carbamate, $C_{12}H_{15}N_3O_2S$, M_r 265.3, *mp* 208 – 210 °C, is a colorless crystalline solid. It is insoluble in water, sparingly soluble in dimethylformamide or dimethylsulfoxide, and soluble in formic acid.

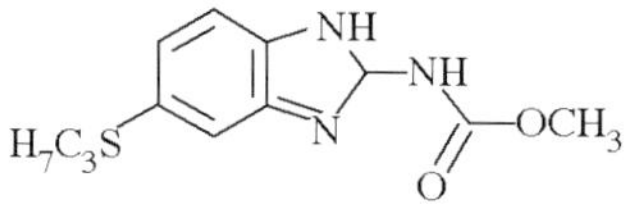

The compound is synthesized in a five step synthesis. 4-chloro-2-nitroaniline is first acetylated with acetic anhydride to give 1-acetamido-4-chloro-2-nitrobenzene. Treatment of this intermediate compound with potassium thiocyanate furnishes the key intermediate 1-acetamido-2-nitro-4-thiocyanatobenzene. Conversion of the thiocyanato group into the required *n*-propylthio analogue, with a simultaneous conversion of the acetamido group to the free amine, is effected by treating the last intermediate with 1-bromopropane in the presence of a base. Further reduction of the nitro group provides the diamine, which is subsequently ring closed to albendazole [20].

Albendazole is used against nematode infections: *Ascaris, Necator, Ancylostoma, Trichuris, Enterobius,* and systemic nematodes such as *Trichinella spiralis, Gnathostoma spinigerum,* and larval *Angiostrongylus cantonensis.* In addition it is used against the larval stages of the cestodes *Echinococcus granulosus* and *E. multilocularis* and for treatment of neurocysticercosis caused by *Taenia solium* [21].

Occasional *adverse effects* include fever, reversible leucopenia, headache, nausea, dizziness, sleeplessness, epigastric pain and diarrhoea. When higher dosages are given over a long period elevated levels of transaminases are observed, seldom hepatitis and occasionally alopecia. Contraindications are pregnancy and liver cirrhosis. Albendazole interferes selectively with the polymerization of β-tubulin to microtubuli. The microtubules play a crucial role in several important cell functions such as material transport within cells and neurotransmission [22].

Trade names: Eskadole (Smith Kline Beecham), Valbazen (Pfizer).

Bephenium [3818-50-6], benzyldimethyl(2-phenoxyethyl)ammonium 3-hydroxy-2-naphthoate, $C_{28}H_{29}NO_4$, M_r 443.5, *mp* 170 – 171 °C, is a bitter tasting, odorless, yellow crystalline solid. It is insoluble in water and slightly soluble in ethanol.

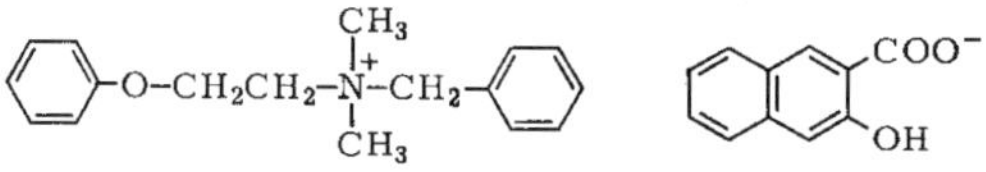

The compound is synthesized by the reaction of dimethyl(2-phenoxyethyl)amine with benzyl halide [23].

Bephenium can be used against *Ascaris, Ancylostoma,* and *Trichostrongylus* infections; it is less effective against *Necator.* Occasional adverse reactions occur: nausea, diarrhea, vomiting, headache, vertigo.

Trade names: Alcopar(a) (Wellcome, Tanabe); Befen (Andromaco); Lecibis (Andromaco, Columbia); Debephenium (ICN Usafarma); Fedal Uncin (Hosbon-Fedal); Hebe (Farmacon); Befeval (Valmorca); Befenium (Laquifa).

Bithionol [*97-18-7*], 2,2′-thiobis(4,6-dichlorophenol), $C_{12}H_6Cl_4O_2S$, M_r 356.1, *mp* 188 °C, is a tasteless, colorless, crystalline solid. It is insoluble in water but soluble in ethanol.

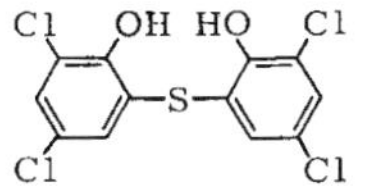

Bithionol is synthesized by the reaction of chlorophenol with sulfur dichloride and sulfuryl chloride [24].

Bithionol can be used against liver, lung, and intestinal flukes. Frequent adverse reactions are abdominal pain, diarrhea, anorexia, nausea, vomiting, dizziness, headache, skin rashes, urticaria. Hepatic and renal involvement are rare complications.

Trade names: Actamer (Monsanto); Bitin (Tanabe); Bitin S (Tanabe), which contains the sulfoxide of bithionol.

Diethylcarbamazine [*1642-54-2*], N,N-diethyl-4-methyl-1-piperazinecarboxamide citrate, $C_{16}H_{29}N_3O_8$, M_r 391.4, *mp* 141 – 143 °C, is a white, odorless, crystalline solid with a bitter acid taste. It is freely soluble in water or hot ethanol but insoluble in acetone, chloroform, ether, or dioxane.

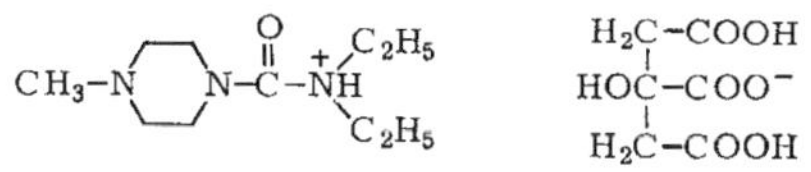

The compound is synthesized by the reaction of diethylcarbamoyl chloride with methylpiperazine [25].

Diethylcarbamazine dihydrogen citrate is used against the microfilarial stage of filariae. It also kills the adult filariae of *Wuchereria, Brugia,* and *Loa.* Occasional adverse reactions are anorexia, nausea, vomiting, headache, and drowsiness. Allergic reactions occur, especially in onchocercosis patients: intense pruritus, edema, lymphadenitis, dermatitis, papular rash, fever, tachycardia.

The most likely mechanism of action is a rapid effect on the surface of the microfilariae. Previously hidden antigenic determinants become accessible to the immune system of the host, which then eliminates the microfilariae from the circulating blood [8].

Trade names: Hetrazan and Hetrazan 1949 (Lederle); Banocide (Wellcome); Notezine (Specia); Supatonin (Tanabe); Carbilazine (Willows Francis); Caricide (Am. Cyanamid).

Ivermectin (synonyms: 5-*O*-Demethyl-22,23-dihydroavermectin A_{1a}, 22,23-Dihydroabamectin, 22,23-Dihydroavermectin B_1). Ivermectin B_{1a} [*71827–03–7*], $C_{48}H_{74}O_{15}$, M_r 891.1, *mp* 155 – 157 °C; ivermectin B_{1b} [*70209–81–3*], $C_{47}H_{72}O_{15}$, M_r 877.1, is a white powder and forms crystals from ethanol – water. It is soluble in methyl ethyl ketone, propylene glycol, and poly(ethylene glycol).

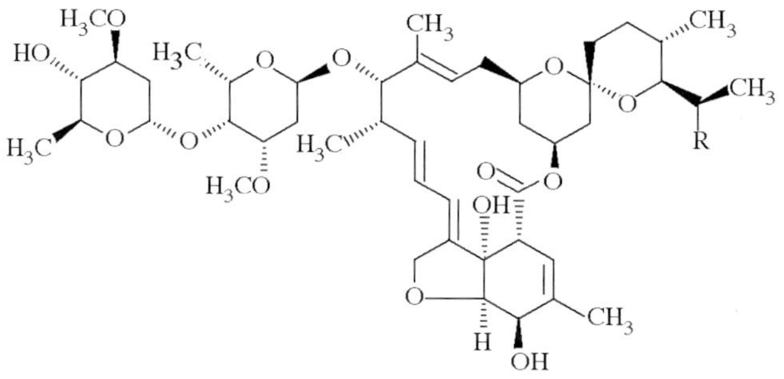

Ivermectin is a mixture of at least 80% ivermectin B_{1a} (R=CH_3) and not more than 20% ivermectin B_{1b} (R=H).

Ivermectin is a semisynthetic compound derived from abamectin (avermectin B_1) a fermentation product of *Streptomyces avermitilis* by selective hydration with chlorotris-(triphenylphosphin)rhodium (I) (Wilkinson catalyst) [26]–[28].

In *human medicine* ivermectin is the drug of choice for onchocerciasis (river blindness) although it only kills *Onchocerca volvulus* microfilariae and not the adults. Although the drug has no curative effect, a single dose of 0.2 mg per kilogram body weight every 6–12 months controls the disease [29]. The drug is under clinical development for treatment of intestinal nematode infections (*Ascaris, Enterobius, Trichuris, Strongyloides*, and *Trichinella spiralis*).

In *veterinary medicine* ivermectin is used in cattle against a wide variety of nematodes and arthropods. It is inactive against cestodes and trematodes.

Occasional *adverse effects* in humans include headache, fever, pruritus, oedema, and arthralgies. However, as ivermectin is used only 1–2 times a year at low dosages, side effects are rare. Veterinary usage has shown that collies are particular sensitive in developing gabergic side effects (CNS depression with dizziness, ataxia, tremor, salivation, mydriasis, and coma with exitus).

Presumably, ivermectin exerts its mode of action by interfering with GABA, glutamate or glycine gated chloride channels, all of which are inhibitory neurotransmitters [30]–[32]. This action leads to an increased permeation of chloride ions, followed by a hyperpolarization of nerve-muscle membranes and a flaccid paralysis of the parasites. The action of ivermectin can be antagonised by bicuculline and picrotoxin [33].

Trade names: Mectizan in France (MSD), Ivomec (MSDAGVET).

Levamisole [*16595-80-5*], L(−)-2,3,5,6-tetrahydro-6-phenylimidazo[2,1-*b*]thiazole hydrochloride, $C_{11}H_{12}N_2S \cdot HCl$, M_r 240.8, *mp* 227–229 °C, is a colorless, odorless, crystalline solid. It is soluble in water and slightly soluble in chloroform.

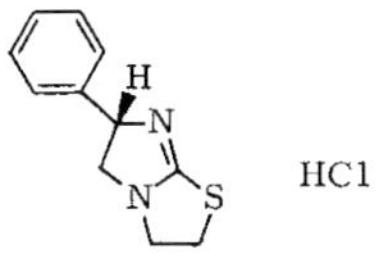

HCl

The compound is synthesized by the reaction of 4-phenyl-2-thioimidazoline with dibromoethane [34].

Levamisole is used against roundworm (*Ascaris*) infections. It is less effective against *Ancylostoma*, *Necator*, or *Strongyloides*. Occasional adverse reactions are nausea, vomiting, abdominal discomfort, headache, dizziness, hypertension.

Levamisole causes rapid spastic contraction in nematodes by persistently depolarizing the muscle-cell membrane. It also acts as a ganglion-stimulating compound [35]. The paralyzed nematodes are then eliminated from the intestine. Levamisole is likely to be hydrolyzed to L(–)-2-oxo-3-(2-mercaptoethyl)-5-phenylimida-zoline under alkaline conditions. This metabolite is a strong and stereospecific inhibitor of the enzyme fumarate reductase through its interaction with –SH groups of the active center.

Trade names: Ascaridil, Decaris, and Stimamizol (Johnson & Johnson, Janssen); Ketrax (ICI); Solaskil (Specia).

Mebendazole [*31431-39-7*], methyl (5-benzoyl-1*H*-benzimidazol-2-yl)carbamate, $C_{16}H_{13}N_3O_3$, M_r 295.3, *mp* 288.5 °C, is an off-white amorphous powder. It is insoluble in water, sparingly soluble in dimethylformamide or dimethyl sulfoxide, and soluble in formic acid.

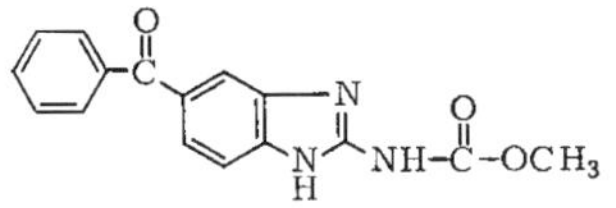

The compound is synthesized by the reaction of 3,4-diaminobenzophenone hydrochloride with *N*-carboxymethyl-*S*-methylisothiourea [36].

Mebendazole is used against nematode infections: *Ascaris, Trichuris, Enterobius, Ancylostoma, Necator, Capillaria*. It is also effective against tapeworms (*Taenia*). Occasionally adverse reactions, abdominal pain and diarrhea, occur. A rare complication is leukopenia. Pregnancy is a contraindication.

Mebendazole interferes with glucose uptake by nematodes and cestodes in vivo and in vitro. This interference occurs by the selective interaction with intracellular tubulin and subsequent inhibition of the assembly of microtubules. The microtubules participate in several important cell functions, e.g., the transport of materials within cells [37].

Trade name [18]: Vermox (Janssen, Ortho).

Metrifonate [*52-68-6*], *O,O*-dimethyl-2,2,2-trichloro-1-hydroxyethylphosphonate, $C_4H_8Cl_3O_4P$, M_r 257.4, *mp* 83–84 °C, is a colorless, crystalline solid. It is soluble in water, chloroform, or ether and very slightly soluble in pentane or hexane.

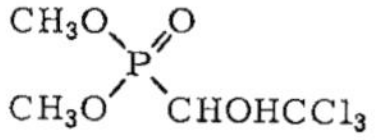

$$\text{CH}_3\text{O}\diagdown\underset{\text{CH}_3\text{O}}{\overset{\displaystyle\text{P}}{\diagup}}\!\!\overset{\displaystyle\text{O}}{\diagdown}\text{CHOHCCl}_3$$

The compound is synthesized by reaction of chloral with dimethyl phosphite [38].

Metrifonate is used against the blood fluke *Schistosoma haematobium*. The occasional adverse reactions are nausea, vomiting, bronchospasm, weakness, diarrhea, and abdominal pain.

In vivo, metrifonate rapidly rearranges to *O,O*-dimethyl *O*-(2,2-dichlorovinyl) phosphate, which is a potent inhibitor of schistosome acetylcholinesterase. This paralyzes the parasites because of the accumulation of acetylcholine, which functions as inhibitory transmitter [39].

Trade name: Bilarcil (Bayer).

Niclosamide [*50-65-7*], 2′,5-dichloro-4′-nitrosalicylanilide, $C_{13}H_8Cl_2N_2O_4$, M_r 327.5, *mp* 225–230 °C, is a tasteless, pale yellow, crystalline solid. It is almost insoluble in water; sparingly soluble in ethanol, chloroform, or ether; and soluble in acetone.

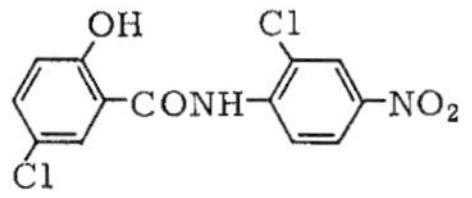

The compound is synthesized by condensation of 5-chlorosalicylic acid with 2-chloro-4-nitroaniline [40].

Niclosamide is used against all kinds of tapeworms. It is also effective against intestinal flukes. Occasional adverse reactions are abdominal pain and nausea.

Niclosamide causes uncoupling of oxidative phosphorylation and respiration. It also activates ATPases, and this disturbs the energy balance of cestodes, which degrade endogenous glycogen, and impairs glucose uptake. Protein synthesis is also affected; e.g., a trypsin inhibitor is no longer formed, which facilitates proteolytic attack of the tapeworm by host digestive enzymes [41].

Trade names: Yomesan, Cestocide, and Nasemo (Bayer); Niclocide (Miles); Trédémine (Roger Bellon); Radeverm (VEB Arzneimittelwerk Dresden); Sulqui (Ella); Atenase (ICN Usafarma); Teniarene (Amsa); Teniamida (Bial); Fedal Telmin (Hosbon-Fedal); Copharten (Cophar). Some preparations contain niclosamide monohydrate.

Niridazole [*61-57-4*], 1-(5-nitro-2-thiazolyl)-2-imidazolidinone, $C_6H_6N_4O_3S$, M_r 214.2, *mp* 260–262 °C, is a yellow, crystalline solid. It is almost insoluble in water and most organic solvents, but it is soluble in dimethylformamide.

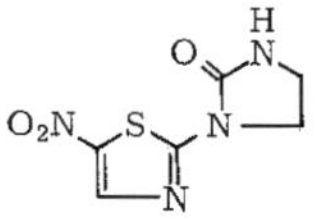

The compound is synthesized by reaction of 2-amino-5-nitrothiazole and β-chloroethylisocyanate [42].

Niridazole can be used against blood flukes, especially *Schistosoma haematobium*. Tolerance and efficacy are reduced in *S. mansoni* and especially in *S. japonicum* infections. It is also used in *Dracunculus* infections. Immunosuppression, vomiting, cramps, dizziness, and headache are among the frequent adverse reactions. Occasional adverse reactions are diarrhea, electrocardiographic changes, rash, insomnia, and paresthesia. Psychosis, hemolytic anemia, and convulsions are rare. Contraindications are impaired liver function, glucose-6-phosphatedehydrogenase deficiency, epilepsy, and severe heart diseases.

Niridazole causes a depletion of glycogen in schistosomes by inducing a reduced rate of conversion of active glycogen phosphorylase to its inactive form. This is achieved through inhibition of the enzyme phosphorylase phosphatase, which normally inactivates glycogen phosphoryl- ase. It is possible that the active moiety is not niridazole, but its 5-imino analog, which can be formed by schistosomes in vitro under anaerobic conditions [43].

Trade names: Ambilhar (Ciba-Geigy); Ambilhar Ciba (Biogalenica Q Farm).

Oxamniquine [*21738-42-1*], 1,2,3,4-tetrahydro-2-[(isopropylamino)methyl]–7-nitro-6-quinolinemethanol, $C_{14}H_{21}N_3O_3$, M_r 279.3, mp 151 – 152 °C, is a light orange, crystalline solid. It is slightly soluble in water and soluble in methanol, acetone, or chloroform.

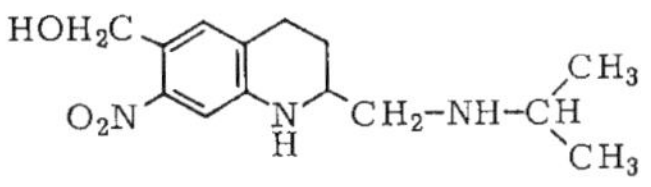

The compound is manufactured by fermentative oxidation of its 6-methyl analog by *Aspergillus scleroticum* [44].

Oxamniquine is used against the blood fluke *Schistosoma mansoni*. The adverse reactions observed are vertigo, vomiting, nausea, abdominal pain, anorexia, dizziness, drowsiness, and somnolence. Hallucinations are occasional, and epileptiform convulsions are rare. Intramuscular injection causes severe pain at the site of injection.

The antischistosomal effect of oxamniquine as a function of time correlates well with the activity of the enzyme ornithine-δ-transaminase. The enzyme may interfere with normal arginine formation and the maintenance of a normal nitrogen balance in the parasite [43].

Trade names: Mansil, Vansil, and Vancil (all Pfizer).

Oxantel [*68813-55-8*], (*E*)–3-[2-(1,4,5,6-tetrahydro-1-methyl-2-pyrimidinyl)ethenyl]-phenol 4,4′-methylenebis(3-hydroxy-2-naphthalenecarboxylate), $C_{36}H_{32}N_2O_7$, M_r 604.7, mp 207 – 208 °C (hydrochloride), is a yellow crystalline solid. It is practically insoluble in water.

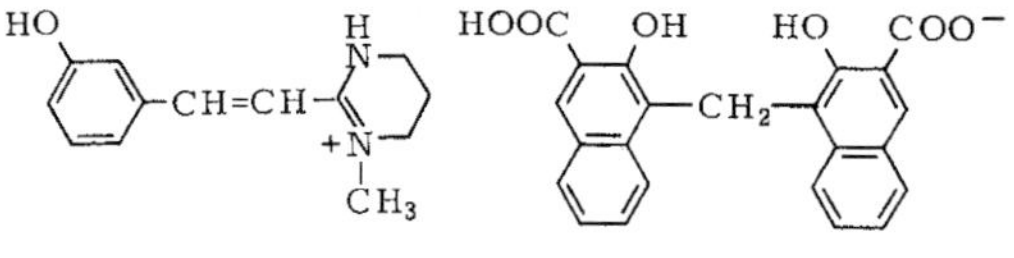

It is synthesized by condensation of 3-hydroxybenzaldehyde and 1,2-dimethyl-1,4,5,6-tetrahydropyrimidine in refluxing ethyl formate [45].

The cation is the active part of the compound. Oxantel is used against whipworm (*Trichuris*). No adverse reactions to this relatively new drug have been found, but it can be expected to behave much like the closely related pyrantel.

Trade name: Telopar (Pfizer).

Piperazine [*110-85-0*], piperazine hexahydrate, $C_4H_{10}N_2 \cdot 6\,H_2O$, M_r 194.2, mp 43–45 °C, is a colorless salty tasting crystalline solid. It is soluble in water and insoluble in ether.

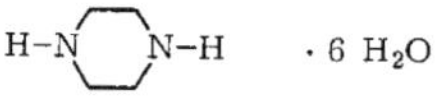

Piperazine is used against roundworms (*Ascaris*) and pinworms (*Enterobius*). Adverse reactions are nausea, vomiting, diarrhea, abdominal pain, headache, and paresthesia. Urticaria may occur occasionally. Contraindications are impaired renal or hepatic function. Various salts are also used: Calcium ethylenediaminetetraacetate (edetate), calcium citrate, citrate, tartrate, maleate, phosphate, adipate, sebacate, sulfate. Piperazine increases the resting potential of the somatic musculature of nematodes, especially in the syncytial region, by increasing the permeability of the membrane to chloride ions. This results in flaccid paralysis of the parasites, which are expelled from the intestine [17].

Trade names [18]: Uvilon (Bayer); Antepar 1953 (Wellcome); Helmazine (Midy); Multifuge (Glaxo); Neox (Rocador).

Praziquantel [*55268-74-1*], 2-cyclohexylcarbonyl-1,2,3,6,7,11b-hexahydro-4H-pyrazino[2,1-*a*]isoquinolin-4-one, $C_{19}H_{24}N_2O_2$, M_r 312.4, mp 136–140 °C, is a bitter tasting, colorless, crystalline solid. It is sparingly soluble in water and soluble in ethanol, chloroform, or dimethyl sulfoxide.

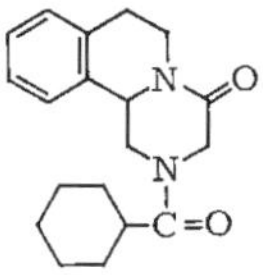

The compound is synthesized by reacting 1,2,3,6,7,11b-hexahydropyrazino[2,1-*a*]isoquinolin-4-one with cyclohexylcarbonyl chloride in chloroform [46].

Praziquantel is used against all infections caused by trematode parasites (blood, liver, lung, and intestinal flukes) or adult tapeworms. It is also effective in cysticercosis. Occasional adverse reactions are epigastric pain, nausea, dizziness, diarrhea, and vomiting. Rare adverse reactions are skin manifestations, headache, and tiredness.

Praziquantel causes a very rapid increase in the permeability of membranes to divalent cations. In the musculature of trematodes and cestodes, this permeability increase results in a calcium concentration increase, which initiates rapid contracture and subsequent spastic contraction and paralysis. In the tegument, praziquantel causes rapid and progressive vacuolization, which leads to partial disintegration of the body surface of the parasites. This, in turn, renders them susceptible to attacks by the host defense system and its digestive enzymes [47].

Trade names: Biltricide (Bayer, Miles); Cestox, Cesol, and Cisticid (all E. Merck).

Pyrantel [*22204-24-6*], (*E*)-1,4,5,6-tetrahydro-1-methyl-2-[2-(2-thienyl)vinyl]pyrimidinium 4,4′-methylenebis(3-hydroxy-2-naphthalenecarboxylate), $C_{34}H_{30}N_2O_6S$, M_r 594.7, *mp* 178 – 179 °C (free base), is a tasteless, odorless, yellow, crystalline solid. It is insoluble in water, slightly soluble in dimethylformamide, and soluble in dimethyl sulfoxide.

The compound is synthesized by reacting 2-thiophenealdehyde with 1,2-dimethyltetrahydropyrimidine [48].

The cation is the active part of the compound. Pyrantel is used against nematodes: *Ascaris*, *Enterobius*, *Ancylostoma*, and *Necator*. Adverse reactions are anorexia, abdominal cramps, diarrhea, nausea, vomiting. Occasional adverse reactions are headache, dizziness, drowsiness, and skin rash. Impaired hepatic function is a contraindication.

Pyrantel is a depolarizing neuromuscular blocking agent. In nematodes it causes a slowly developing contraction and paralysis. The immobilized parasites are then eliminated from the intestine. The neuromuscular junction of *Ascaris* is 100-fold more sensitive to pyrantel than to acetylcholine [49].

Trade names: Lombpiareu (Areu); Aut (Elea); Aguipiran (Aguilar); Tamoa (North Medicamenta); Antiminth (Roerig); Combantrin, Trilombrin, Cobantril, and Helmex (Pfizer); Tricocel and Piranver (ICN Usafarma); Verdal (Columbia); Perverme (Biofarma); Piranver and Piranver F (ICN Farmaceutica).

Pyrvinium [*3546-41-6*], 6-(dimethylamino)-2-[2-(2,5-dimethyl-1-phenyl-1*H*-pyrrol-3-yl)ethenyl]–1-methylquinolinium 4,4′-methylene-bis(3-hydroxy-2-naphthalenecarboxylate), $C_{75}H_{70}N_6O_6$, M_r 1151.4, *mp* 210 – 215 °C, is a tasteless, odorless, bright orange to black, crystalline solid. It is insoluble in water or ether, slightly soluble in chloroform, and soluble in glacial acetic acid.

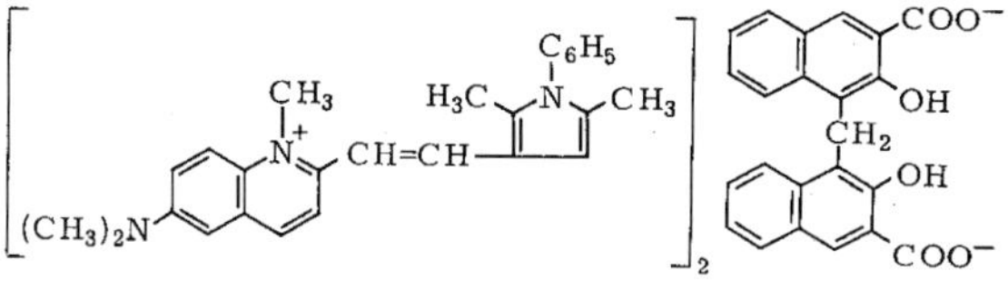

The compound is synthesized by the reaction of 1,2-dimethyl-6-dimethylaminoquinoline iodide with 2,5-dimethyl-1-phenyl-3-pyrrolaldehyde [50].

The cation is the active part of the compound. Pyrvinium is used against pinworms (*Enterobius*). Occasional adverse reactions are blood dyscrasia and folic acid deficiency. Rare adverse reactions are rash, vomiting, convulsion, shock, photosensitivity, and headache.

Trade names: Vanquil, Vanpar, Vanquin, Povan(yl) Molevac, Povan 1959, Paquil, Polyquil, and Povanyl (all Parke-Davis and Substantia).

Suramin [*129-46-4*], hexasodium 8,8'-[carbonylbis[imino-3,1-phenylenecarbonylimino(4-methyl-3,1-phenylene)carbonylimino]]bis-1,3,5-naphthalenetrisulfonate, $C_{51}H_{34}N_6Na_6O_{23}S_6$, M_r 1429.2, is a pinkish white, white, or cream-colored, odorless, hygroscopic powder with a slightly bitter taste. It is soluble in water and insoluble in chloroform or ether.

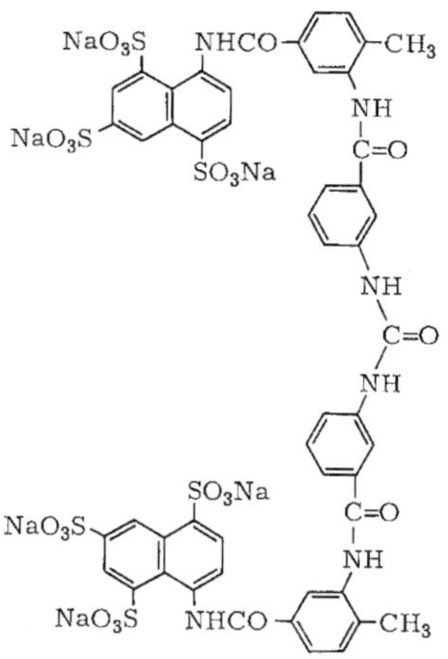

The compound is synthesized by condensation of 1-naphthylamine-4,6,8-trisulfonic acid with 3-nitro-4-methylbenzoyl chloride, reduction of the product, condensation with 3-nitrobenzoyl chloride, renewed reduction, and final condensation with phosgene [51].

Suramin is used as a macrofilaricide against *Onchocerca*. Frequent adverse reactions are nausea, vomiting, colic, and urticaria. Occasionally, kidney damage, blood dyscrasias, shock, and optic atrophy occur. Allergic reactions to filarial proteins (pruritus, rash, fever, edema, hyperesthesia, photophobia, lachrymation) are also encountered.

The mechanism of action is not well understood. Suramin inhibits many different enzyme systems by binding to free cationic amino acid residues in the area of the active center. This appears to lead to a specific, but slight, interference with the DNA-RNA replication mechanism. The inhibition of protein kinase, an enzyme intimately involved in intracellular regulation, also may be involved in the effects of suramin [52].

Trade names: Germanin (Bayer); Antrypol (ICI); Moranyl (Specia).

Thiabendazole [*148-79-8*], 2-(4-thiazolyl)-1*H*-benzimidazole, $C_{10}H_7N_3S$, M_r 201.3, mp 304–305 °C, is a tasteless, colorless, crystalline solid. It is insoluble in water, slightly soluble in ethanol or chloroform, and soluble in acetone, dimethyl sulfoxide, or dimethylformamide.

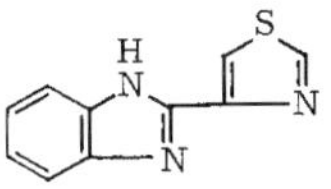

The compound is synthesized by heating 4-thiazolecarboxamide with *o*-phenylenediamine in polyphosphoric acid [53].

Thiabendazole is used against nematodes and nematode larvae: *Enterobius, Ascaris, Strongyloides, Necator, Trichinella, Ancylostoma, Capillaria, Dracunculus,* and *Trichostrongylus*. It is also effective against intestinal flukes. Frequent adverse reactions are dizziness, anorexia, vomiting, and epigastric distress. Diarrhea, fever, headache, skin rashes, hallucinations, and olfactory disturbances occur occasionally. Tinnitus, hypotension, and syncope are rare complications. The mechanism of action is not yet understood. Thiabendazole inhibits fumarate reductase, an enzyme essential for mitochondrial energy production in many nematodes, but this cannot be the primary or sole mechanism. Possibly thiabendazole, like mebendazole, affects tubulin polymerization, but there is no good experimental evidence for this [43].

Trade names: Mintezol and Minzolum (Merck Sharp & Dohme).

4. References

[1] H. Loewe: "Anthelminthica," in G. Ehrhardt, R. Ruschig (eds.): *Arzneimittel, Entwicklung, Wirkung, Darstellung,* vol. 5, Verlag Chemie, Weinheim 1972, pp. 11–110.
[2] R. B. Borrows: "Human and Veterinary Anthelmintics," *Progr. Drug. Res.* **17** (1973) 108–209.
[3] P. J. Islip: "Anthelmintic Agents," in M. E. Wolff (ed.): *Burger's Medicinal Chemistry,* vol. **2**, J. Wiley & Sons, New York 1979, pp. 481–530.
[4] R. Cavier (ed.): "Chemotherapy of Helminthiasis," *Int. Encycl. Pharmacol. Ther.,* section 64, vol. 1, Pergamon, Oxford 1973.
[5] "Intestinal Protozoan and Helminthic Infections," *WHO Tech. Rep. Ser.* 1981, no. 666.
[6] "Parasitic Zoonoses," *WHO Tech. Rep. Ser.* 1979, no. 637.

[7] M. A. Gemmell, P. P. Johnstone: "Cestodes," *Antibiot. Chemother.* (Basel) 30 (1981) 54–114.
[8] F. Hawking: "Chemotherapy of Filariasis," *Antibiot. Chemother.* (Basel) 30 (1981) 135–162.
[9] E. A. Malek (ed.): *Snail-Transmitted Parasitic Diseases,* CRC Press, Boca Raton, Florida, 1980.
[10] A. E. R. Taylor, R. Muller (eds.): "The Relevance of Parasitology to Human Welfare Today," *Symp. Br. Soc. Parasitol.*, vol. 16, Blackwell Scientific Publications, Oxford 1978.
[11] E. K. Markell, M. Voge: *Medical Parasitology,* Saunders, Philadelphia 1981.
[12] P. E. C. Manson-Bahr, F. I. C. Apted: *Manson's Tropical Diseases,* Baillière Findall, London 1982.
[13] W. Peters, in [10], pp. 25–40.
[14] J. Walsh, K. S. Warren, *N. Engl. J. Med.* **301** (1979) 967–974.
[15] D. Stürchler: *Endemiegebiete tropischer Infektionskrankheiten,* Huber, Bern 1981.
[16] H. Schenone in A. Flisser et al. (eds.): *Cysticercosis, Present State of Knowledge and Perspectives,* Academic Press, New York 1982, pp. 25–38.
[17] J. del Castillo in M. Florkin, B. T. Scheer (eds.): *Chemical Zoology,* vol. **3,** Academic Press, New York 1969, pp. 521–554.
[18] Chemindex International 1983, IMS A.G., Zug, Switzerland.
[19] Index Nominum 1984, International Drug Directory, Laboratory of the Swiss Pharmaceutical Society, Zürich, Switzerland.
[20] L. B. Townsend, D. S. Wise: "The synthesis and chemistry of certain anthelmintic benzimidazoles," *Parasitology Today* **6** (1990) 107–112.
[21] G. C. Cook: "Use of benzimidazole chemotherapy in human helminthiases : indications and efficacy," *Parasitology Today* **6** (1990) 133–136.
[22] E. Lacey: "Mode of action of benzimidazoles," *Parasitology Today* **6** (1990) 112–115.
[23] Burroughs Wellcome, US 2918401, 1959 (C. Copp).
[24] I. G. Farbenind., DE 583055, 1933 (F. Muth).
[25] American Cyanamid, US 2467893, 1949 (S. Kushner, L. Brancone).
[26] J. C. Chabala et al., *J. Med. Chem.* **23** (1980) 1134–1136.
[27] H. Mrozik et al., *Tetrahedron Lett.* **23** (1982) 2377.
[28] M .E. Jung, *Tetrahedron Lett.* **28** (1987) 5977.
[29] H. Bradshaw, *Parasitology Today* **5** (1989) 63–64.
[30] D. F. Cully et al., *Nature (London)* **371** (1994) 707–712.
[31] D. F. Cully et al., *J. Biol. Chem.* **271** (1996) 20187–20191.
[32] W. Forth in W. Forth, D. Henschler, W. Rummel, K. Starke, (eds): Anthelmintika, *Pharmakologie und Toxikologie,* Wissenschaftsverlag, Osnabrück, 1993, 711–712.
[33] S. S. Pong, C .C. Wang, L .C. Fritz, *J. Neurochem.* **34** (1980) 351–358.
[34] Janssen, US 3274209, 1966 (A. H. M. Raeymaekers, D. C. J. C. Thienpont, P. J. A. W. Demoen).
[35] P. A. J. Janssen: "The Levamisole Story," *Prog. Drug Res.* **20** (1976) 347–383.
[36] Janssen, US 3657267, 1972 (J. L. H. Van Gelder, L. F. C. Roevens, A. H. M. Raeymaekers).
[37] H. Van den Bossche, F. Rochette, C. Hörig: "Mebendazole and Related Anthelmintics," *Adv. Pharmacol. Chemother.* **19** (1982) 67–128.
[38] Bayer, US 2701225, 1955 (W. Lorenz).
[39] B. Holmgren, I. Nordgren, M. Sandoz, A. Sundwall: "Metrifonate. Summary of Toxicological and Pharmacological Information Available," *Arch. Toxicol.* **41** (1978) 3–29.
[40] Bayer, GB 824345, 1959.
[41] P. Andrews, J. Thyssen, D. Lorke: "The Biology and Toxicology of Molluscicides, Bayluscide," *Pharmacol. Ther.* **19** (1983) 245–295.
[42] Ciba, BE 632989, 1963.

[43] P. Andrews, A. Haberkorn, H. Thomas: "Antiparasitic Drugs: Mechanism of Action, Pharmacokinetics, and in Vitro and in Vivo Assays of Drug Activity," in V. Lorian (ed.): *Antibiotics in Laboratory Medicine*, 2nd ed., Williams and Wilkins, Baltimore 1985, Chap. 9.

[44] H. C. Richards, R. Forster, *Nature (London)* **222** (1969) 581–582.

[45] Pfizer, ZA 6804589, 1967; FR 1584069, 1967 (J. W. McFarland).

[46] E. Merck, US 4113867, 1978.

[47] P. Andrews, H. Thomas, R. Pohlke, J. Seubert: "Praziquantel," *Med. Res. Rev.* **3** (1983) 147–200.

[48] Pfizer, BE 658987, 1965.

[49] M. L. Aubry, P. Cowell, M. J. Davey, S. Shevde: "Aspects of the Pharmacology of a New Anthelmintic: Pyrantel," *Br. J. Pharmacol.* **38** (1970) 332–344.

[50] Eastman Kodak, US 2515912, 1950 (E. Van Lare, L. G. S. Brooker).

[51] I. G. Farbenind., GB 224849, 1924.

[52] F. Hawking: "Suramin: With Special Reference to Onchocerciasis," *Adv. Pharmacol. Chemother.* **15** (1978) 289–322.

[53] Merck & Co., US 3017415, 1962 (L. H. Sarett, H. D. Brown).

HIV and AIDS Therapeutics

PIERRE L. BEAULIEU, Boehringer Ingelheim (Canada) Ltd., Bio-Méga Research Division, Laval (Québec), Canada

JOHN PROUDFOOT, Boehringer Ingelheim Pharmaceuticals Inc., Ridgefield, Connecticut 06877, United States

1.	Introduction — HIV and AIDS	1292
1.1.	HIV Cytopathogenesis and AIDS	1292
1.2.	Virus Life Cycle	1292
1.3.	Opportunities for Drug Intervention	1294
2.	Reverse Transcriptase Inhibitors	1294
2.1.	Introduction	1294
2.2.	Nucleoside Analogs	1295
2.3.	Nonnucleoside Analogs	1301
2.4.	Compounds in Development	1304
3.	Protease Inhibitors	1304
3.1.	Introduction	1304
3.2.	Hydroxyethylamine Isosteres	1305
3.3.	Hydroxyethylene Isosteres	1310
3.4.	Compounds in Development	1314
4.	Emerging Concepts in AIDS Therapy	1315
5.	References	1315

Abbreviations used in this article

Asn	asparagine
Boc	*tert*-butoxycarbonyl
DCC	dicyclohexylcarbodiimide
DDQ	2,3-dichloro-5,6-dicyano-1,4-benzoquinone
DIAD	diisopropyl azodicarboxylate
EDC	1,2-dichloroethane
LHMDS	lithium hexamethyldisilazide
tBDMSOTf	*tert*-butyldimethylsilyl trifluoromethanesulfonate
TFA	trifluoroacetic acid
TMDA	tetramethylethylenediamine
TMSCI	trimethylsilyl chloride
TsOH	*p*-toluenesulfonic acid

1. Introduction — HIV and AIDS

The human immunodeficiency virus (HIV) has been identified as the etiologic agent causing AIDS [1], [2]. In the last twenty years, AIDS has become the world's most deadly infectious disease, ranking fourth after ischemic heart disease, strokes and acute respiratory infections in terms of mortalities (2.28 million deaths worldwide in 1998 according to a recent World Health Organization report) [3]. It is estimated that 33.4 million people worldwide are currently suffering from HIV/AIDS infection [3]. The magnitude of this tragedy has triggered an unprecedented search for a cure for this deadly disease.

1.1. HIV Cytopathogenesis and AIDS

HIV-1 and HIV-2 are members of the lentivirus group of retroviruses. While the HIV-1 strain of the virus predominates in the western world, HIV-2 is most prevalent in Africa. More than 95% of HIV-infected people live in the developing world [3].

The acquired immunodeficiency syndrome (AIDS) results from infection of components of the human immune system by HIV. The consequence of this event is a severe immunosuppression resulting from destruction of CD4-positive T-helper lymphocytes and macrophages (see also → Blood). HIV can remain latent or chronically expressed at a low level for extended periods (averages about 10 years from initial infection to clinical disease). Following this prolonged disease-free period, the CD4 cell-count decreases to a level at which the immune system is no longer effective against opportunistic pathogens, resulting in the emergence of multiple infections, tumors, and, ultimately, death [4], [5].

1.2. Virus Life Cycle

The HIV-1 genome consists of ten open reading frames encoding structural, regulatory and accessory proteins (Table 1, Fig. 1) [6], [7]. Products of the *gag* gene form the major components of the virion core. The *pol* gene encodes all important enzymes required for DNA synthesis and integration, and processing of polyproteins. The *env* gene product is a structural glycoprotein that is important in AIDS pathogenesis. Regulatory proteins (*tat, rev* gene products, as well as others shown in Table 1) are required for control of the viral replication cycle. Finally, accessory gene products (*vpu, vif*) perform important functions related to virion maturation, release and infectivity.

HIV infection is mediated by the interaction of the *gp*120 glycoprotein on the surface of the virus and its receptor, the CD4 molecule on the targeted host cells of the immune system (usually T-cell lymphocytes and macrophages, Fig. 1). Following fusion with the

Table 1. HIV-gene functions

Genes	Functions
Structural	
gag	matrix, virion maturation and stability, capsid, virus particle maturation and release
pol	protease, reverse transcriptase, integrase
env	external envelope, receptor binding, virion infectivity
Regulatory	
tat	transcription transactivator of gene expression
rev	regulator of protein expression
tev/tnv	undefined
nef	negative factor, virus propagation
vpr	early regulatory protein
Accessory	
vif	cell-free virus transmission, *env* processing
vpu	virus maturation/release

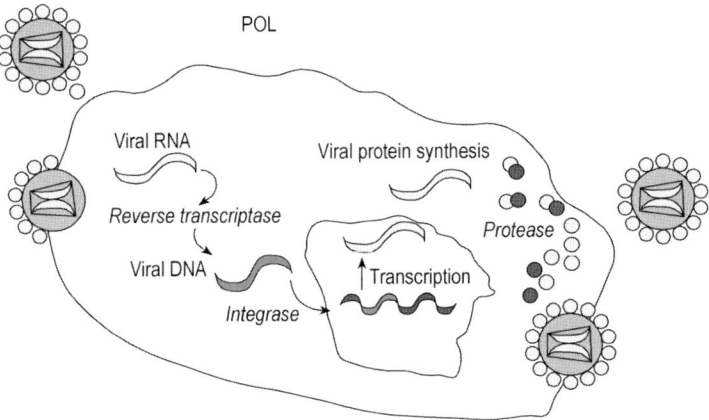

Figure 1. HIV virus life cycle

host cell membrane and uncoating of the viral particle, viral single-stranded RNA is transcribed into double-stranded DNA (dsDNA) by the reverse transcriptase (RT) enzyme. After transport into the cell nucleus, integration of viral dsDNA into the cellular chromosomal DNA generates *provirus*. The virus now permanently infects the cell. Following a latency period, expression of the provirus leads to a virus replication. Viral RNAs are spliced and transported into the cytoplasm where viral protein synthesis occurs. In the final stages of replication, *gag* and *gag-pol* precursor proteins and two molecules of viral RNA assemble with envelope protein to form an immature virion. The autocatalytically released HIV protease then cleaves the remaining protein precursors to form a new, mature and infectious viral particle that is released from the cell.

1.3. Opportunities for Drug Intervention

Studies on the life cycle of the virus have identified several opportunities for therapeutic intervention [8].

The reverse transcriptase is unique to retroviruses and since no analogous enzyme activity is present in normal mammalian cells, inhibitors of this enzyme were viewed as particularly attractive therapeutics, with the potential for a large margin of safety [9]. RT transforms the single-stranded RNA virus genome into dsDNA, prior to integration into the host genome by the integrase enzyme. Firstly, it transcribes a strand of DNA complementary to the viral RNA. Secondly, as this process is occurring, RT degrades the original RNA strand, leaving single stranded DNA. Finally, it synthesizes a DNA strand complimentary to the DNA strand just synthesized, giving dsDNA.

HIV protease, another essential protein in the virus life cycle, processes the *gag* and *gag-pol* gene products. Cleavage of these polyproteins by the protease at specific sites generates the proteins and enzymes required for viral maturation [10], [11]. The discovery that inhibition of the protease resulted in the production of noninfectious, immature virions [12], [13] triggered an extensive search for inhibitors of this enzyme [14]–[16].

2. Reverse Transcriptase Inhibitors

2.1. Introduction

Two classes of reverse transcriptase inhibitor are used in clinical practice. The first identified were the *nucleoside analogs*, which constitute the backbone of first line therapeutic intervention. After conversion to their triphosphates, which are the pharmacologically active species within cells, the nucleoside analogs mimic the natural substrates of the enzyme. However, since the analogs lack the 3'-OH group necessary for DNA chain elongation, they prematurely terminate the transformation of the viral RNA into dsDNA. They may also act as competitive inhibitors with respect to the natural nucleoside substrates. These inhibitors typically still have some affinity for endogenous DNA polymerase enzymes, which can lead to limiting side effects.

The *nonnucleoside inhibitors*, commonly referred to as NNRTIs, constitute the second class of RT inhibitors. In contrast to the nucleoside analogs, these drugs bind to a lipophilic site of the reverse transcriptase close to the active site and reduce the conformational flexibility of the enzyme. Nonnucleoside inhibitors are characterized by extremely high selectivity for RT, and are ineffective even against the closely related HIV-2 RT. Their significant disadvantage is the rapid emergence of resistant virus upon monotherapy with NNRTIs, resistance that typically renders the virus insensitive to other inhibitors of this particular class. These drugs are used in combination with the various nucleoside RT inhibitors in multiple therapy regimens.

2.2. Nucleoside Analogs

Zidovudine [30516-87-1], 3′-azido-3′-deoxythymidine, $C_{10}H_{13}N_5O_4$, M_r 267.2, mp 116–118 °C [17], [18], was the first nucleoside inhibitor approved for use in the treatment of AIDS, and still constitutes the backbone of combination regimens that include other nucleoside RT inhibitors, nonnucleoside RT inhibitors and protease inhibitors.

Synthesis [17], [18]:

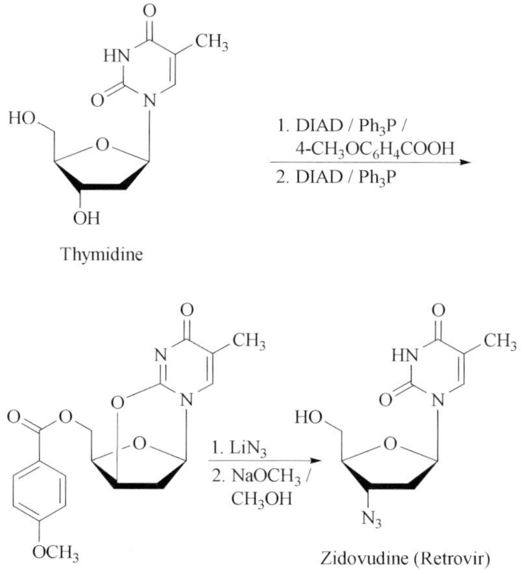

Trade names: Retrovir (Glaxo Wellcome); as a combination with lamivudine (see below), Combivir (Glaxo Wellcome).

Didanosine [*69655-05-6*], 2′,3′-dideoxyinosine, $C_{10}H_{12}N_4O_3$, M_r 236.2, *mp* 177–180 °C (several values in the literature).

Synthesis [19]:

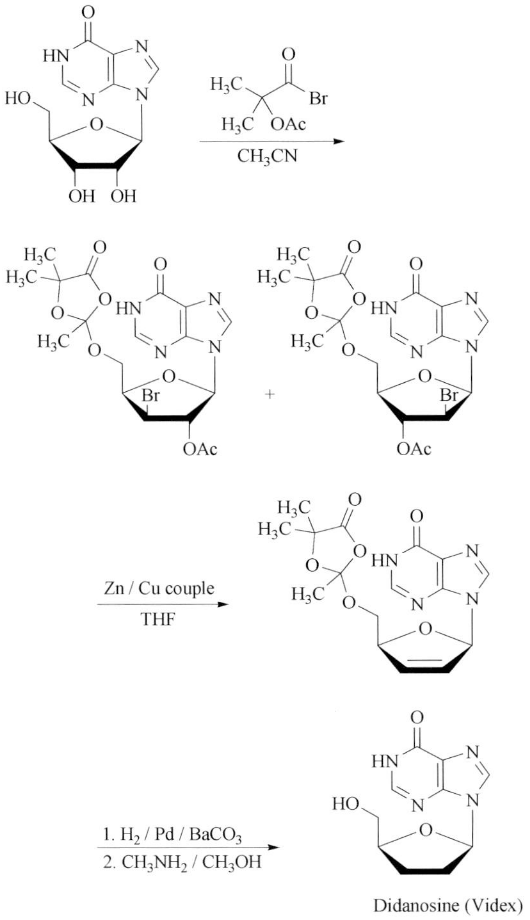

Trade name: Videx (Bristol-Myers Squibb).

Zalcitabine [7481-89-2], 2′,3′-dideoxycytidine, $C_9H_{13}N_3O_3$, M_r 211.2, mp 215–217 °C.

Synthesis [20]:

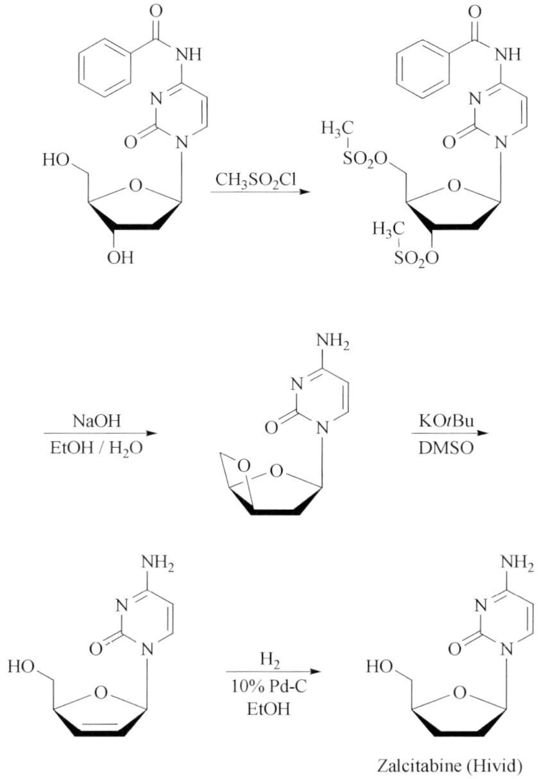

Trade name: Hivid (Hoffmann-LaRoche).

Stavudine [*3056-17-5*], 2′,3′-didehydro-3′-deoxythymidine, $C_{10}H_{12}N_2O_4$, M_r 224.2, mp 165–166 °C.

Synthesis [21]:

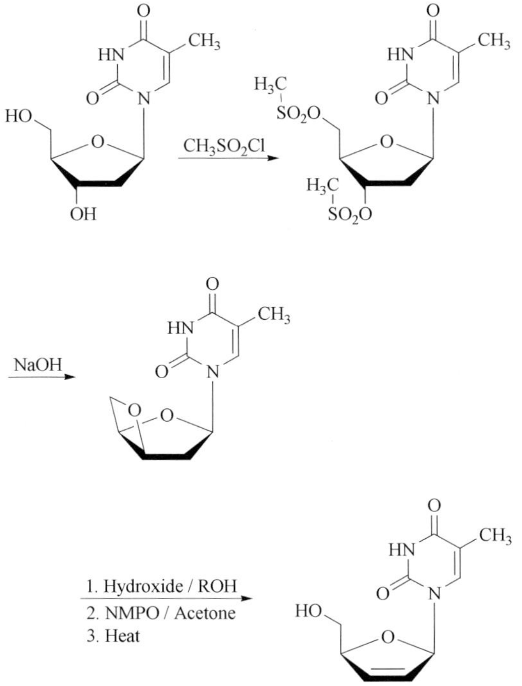

Stavudine (Zerit)

Trade name: Zerit (Bristol-Myers Squibb).

Lamivudine [*134678-17-4*], (2*R*,5*S*)-4-amino-1-[2-(hydroxymethyl)-1,3-oxathiolan-5-yl]-2(1*H*)-pyrimidinone, $C_8H_{11}N_3O_3S$, M_r 229.3, mp 158–160 °C.

Synthesis [22], [23]:

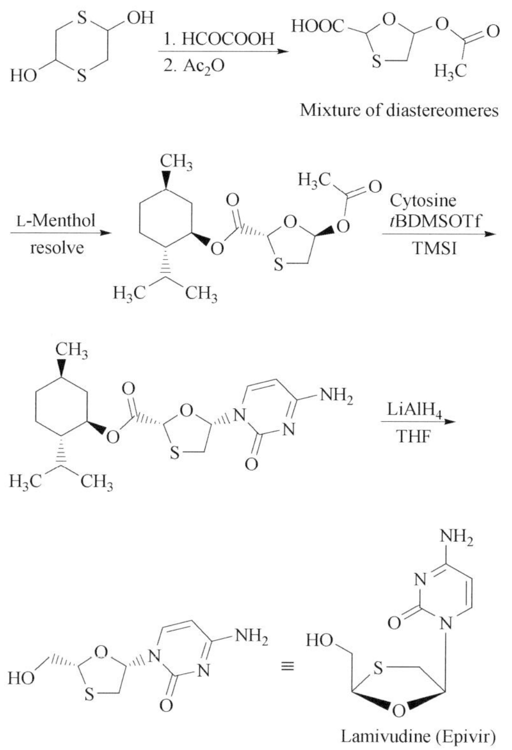

Trade names: Epivir (BioChem Pharma, Glaxo Wellcome), Combivir (lamivudine and zidovudine combination tablets, Glaxo Wellcome).

Abacavir [*136470-78-5*], (1*S*,4*R*)-4-[2-amino-6-(cyclopropylamino)-9*H*-purin-9-yl]-2-cyclopentene-1-methanol, $C_{14}H_{18}N_6O$, M_r 286.3, foam; abacavir sulfate [*188062-50-2*], (1*S*,4*R*)-4-[2-amino-6-(cyclopropylamino)-9*H*-purin-9-yl]-2-cyclopentene-1-methanol, sulfate salt (2:1), $(C_{14}H_{18}N_6O)_2H_2SO_4$, M_r 670.76, mp 224–225 °C.

Synthesis [24], [25]:

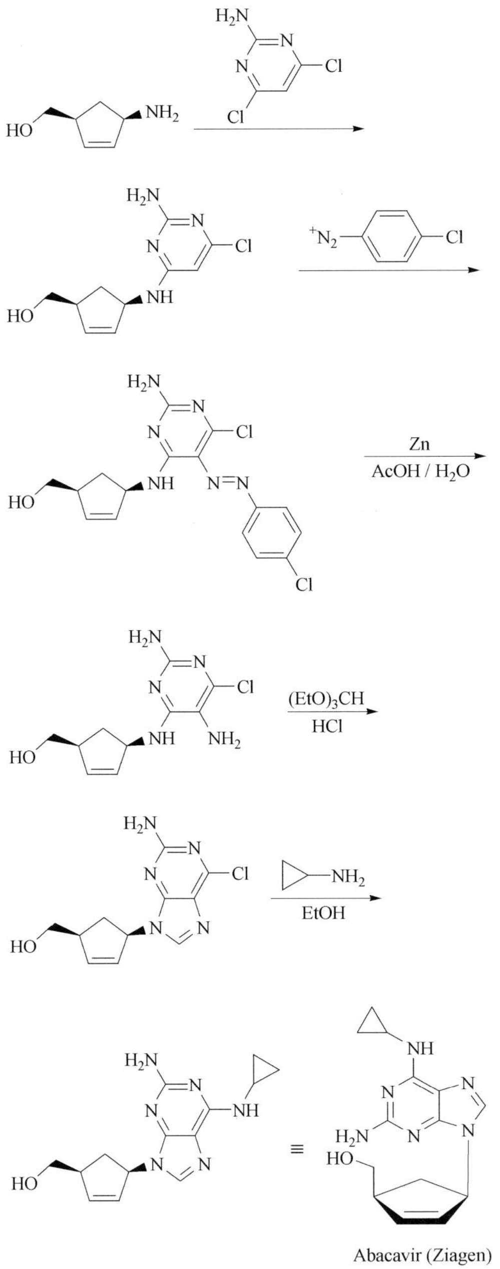

Trade name (abacavir sulfate): Ziagen, Glaxo Wellcome.

2.3. Nonnucleoside Analogs

Nevirapine [*129618-40-2*], 11-cyclopropyl-5,11-dihydro-4-methyl-6*H*-dipyrido[3,2-b:2′,3′-e][1,4]diazepin-6-one, $C_{15}H_{14}N_4O$, M_r 266.3, *mp* 247–249 °C, is a noncompetitive inhibitor of HIV-1 RT, and was the first of the nonnucleoside RT inhibitor class to receive approval for use in the treatment of AIDS.

Synthesis [26]:

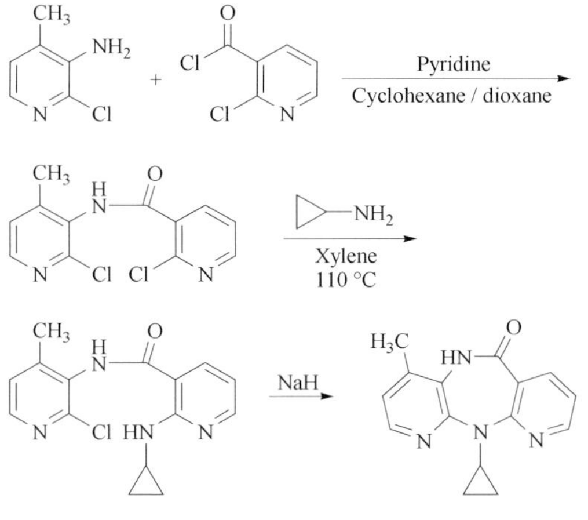

Trade name: Viramune (Boehringer Ingelheim).

Delavirdine [*136817-59-9*], 1-[3-[(1-methylethyl)amino]-2-pyridinyl]-4-[[5-[(methylsulfonyl)amino]-1*H*-indol-2-yl]carbonyl]piperazine, $C_{22}H_{28}N_6O_3S$, M_r 456.56, *mp* 226–228 °C; delavirdine methanesulfonate (mesylate) [*147221-93-0*] $C_{22}H_{28}N_6O_3S \cdot CH_4O_3S$, M_r 552.68, *mp* 220–222 °C.

Synthesis [27], [28]:

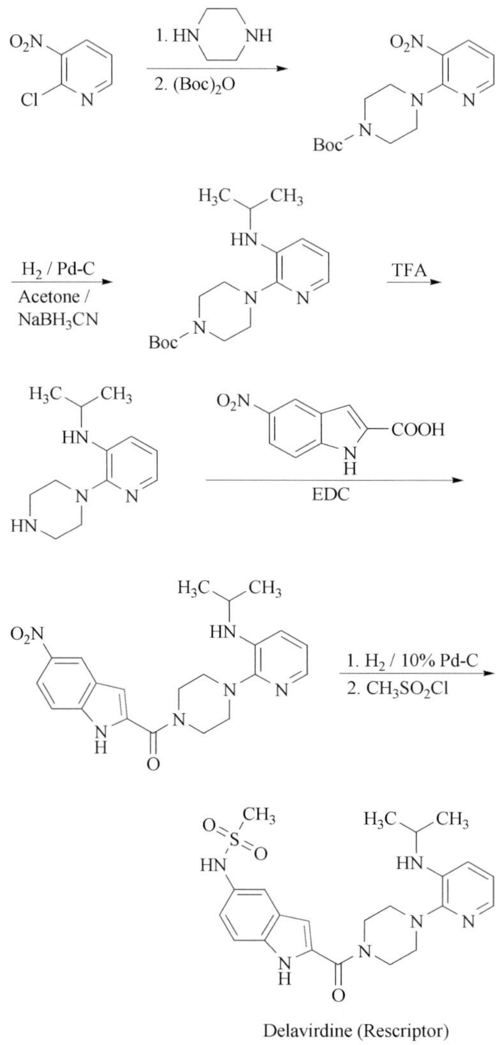

Trade name: Rescriptor (Pharmacia & Upjohn).

Efavirenz [*154598-52-4*], (4*S*)-6-chloro-4-(cyclopropylethynyl)-1,4-dihydro-4-(trifluoromethyl)-2*H*-3,1-benzoxazin-2-one, $C_{14}H_9ClF_3NO_2$, M_r 315.7, *mp* 139–141 °C.

Synthesis [29]:

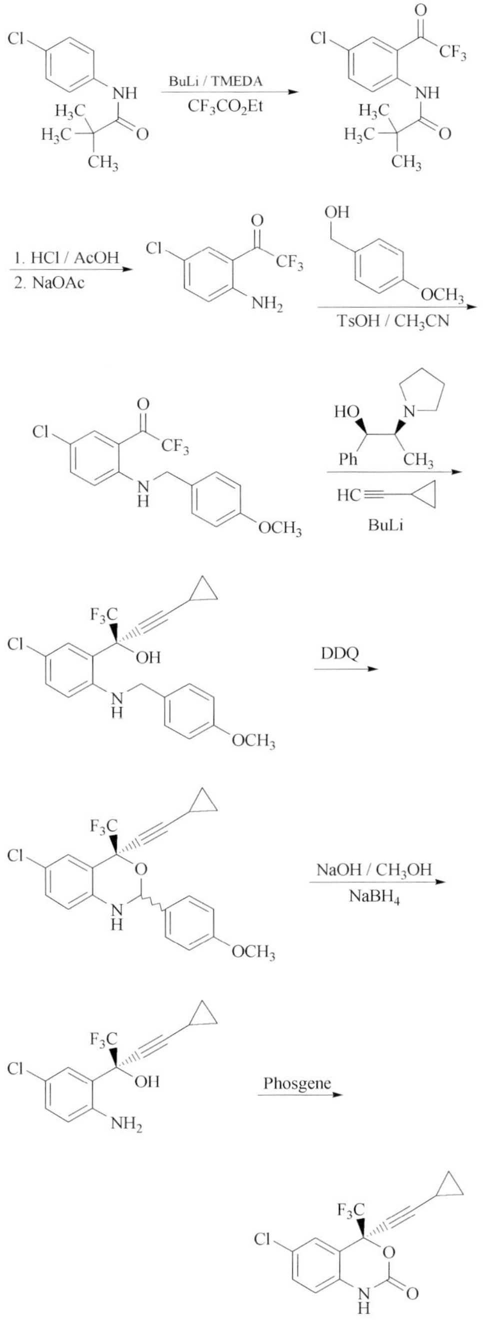

Trade name: Sustiva (DuPont Pharmaceuticals).

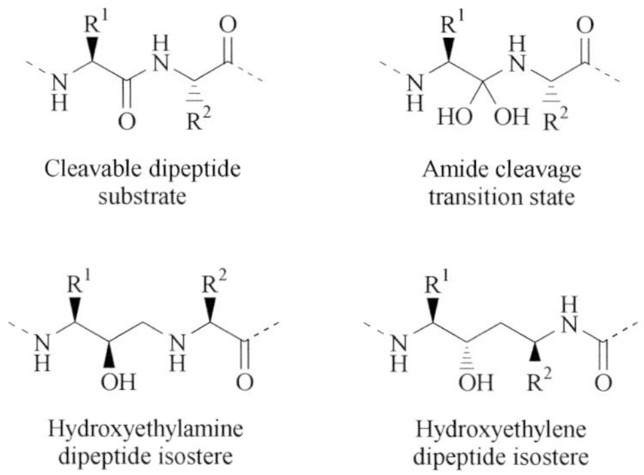

Figure 2. Protease inhibitors mimic the transition state of hydrolysis of an amide bond

2.4. Compounds in Development

Several additional nucleoside and nonnucleoside RT inhibitors are currently in various stages of clinical development. In general, these drug candidates attempt to improve on the various disadvantages associated with the currently used therapies. For nucleoside inhibitors, the emphasis is on improved safety margins, and for NNRT inhibitors, newer compounds attempt to demonstrate better profiles against a variety of resistant mutant reverse tranicriptases.

3. Protease Inhibitors

3.1. Introduction

Being a member of the aspartyl protease class of enzymes [30], HIV protease, a homodimeric enzyme, is susceptible to inhibition by peptide-like structures incorporating various isosteres that mimic the transition state for the hydrolysis of an amide bond (Fig. 2) [14]–[16].

Among them, hydroxyethylamine and hydroxyethylene isosteres have led to extremely potent inhibitors of the enzyme. Despite their high peptidic character (often resulting in low oral bioavailability and rapid excretion), high molecular masses (575–750) and structural complexities, these classes of compounds effectively reduce viral load in infected patients and, when used in combination with reverse transcriptase inhibitors are providing physicians with new tools for combating the deadly disease. To date, five peptidomimetic-based HIV protease inhibitors are on the market, and several others are undergoing evaluation in the clinic.

3.2. Hydroxyethylamine Isosteres

Saquinavir [*127779-20-8*], [3S-[2[1R*(R*),2S*],3α,4aβ,8aβ]]-N^1[3-[3-[[(1,1-dimethylethyl)amino]carbonyl]octahydro-2(1H)-isoquinolinyl]-2-hydroxy-1-(phenylmethyl)propyl]-2-[(2-quinolinylcarbonyl)amino]butanediamide; free base, $C_{38}H_{50}N_6O_5$, M_r 670.85, $[\alpha]_D^{20}$ −55.9 ° (c = 0.5, MeOH); monomethanesulfonate [*149845-06-7*], $C_{38}H_{50}N_6O_5 \cdot CH_4O_3S$, M_r 766.95.

Saquinavir is a potent and specific inhibitor of HIV-1 and HIV-2 proteases ($K_i \leq 0.12$ nM). It is the first peptidomimetic-based HIV protease inhibitor to be approved in the United States (1995) and Europe (1996) [31]–[33]. It is used in combination with RT-inhibitor nucleoside analogs (zidovudine or zalcitabine) for the treatment of advanced HIV infection. A new soft gel formulation (fortovase) was launched in the United States (1997) and provides improved bioavailability.

Synthesis [34]:

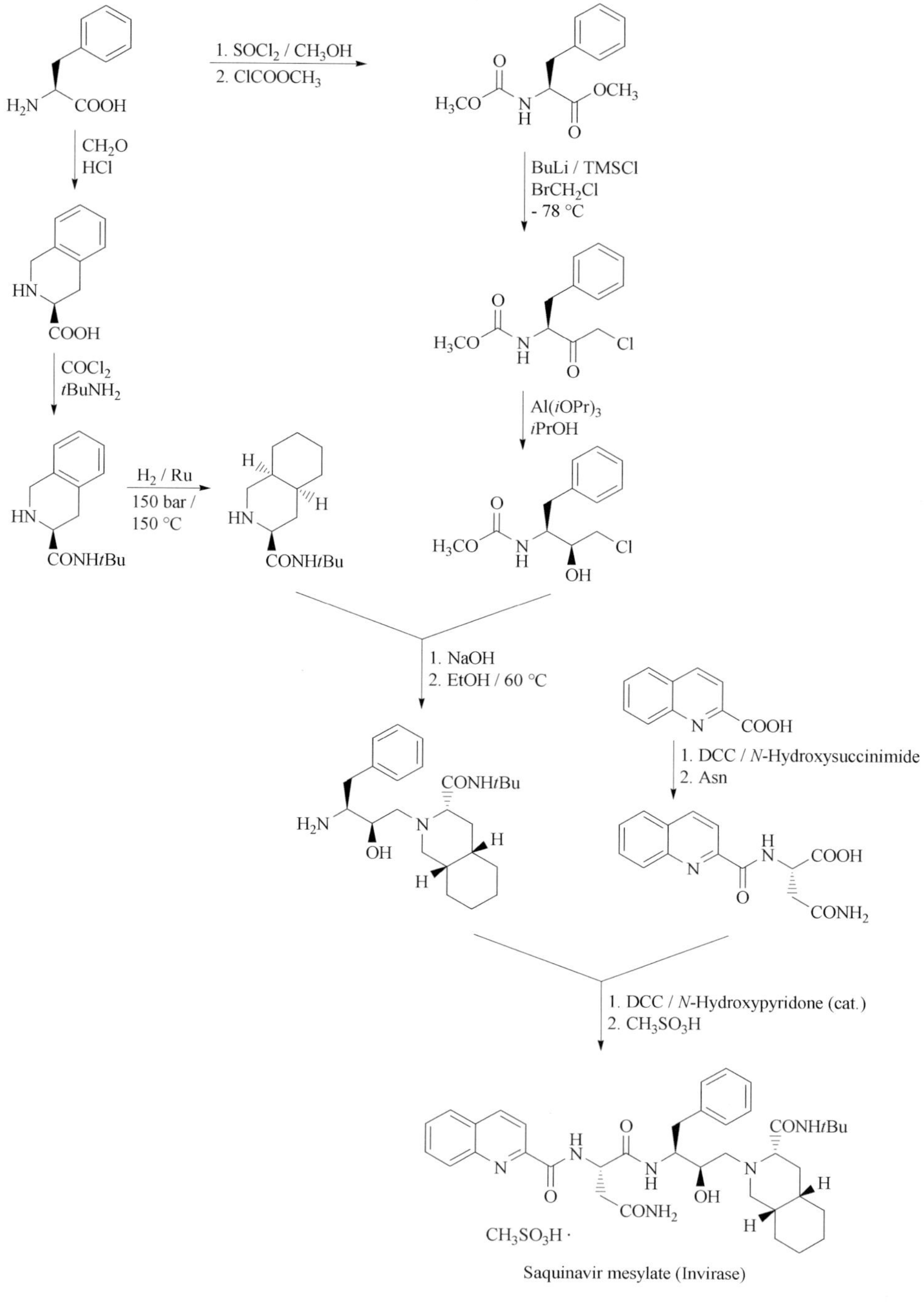

Saquinavir mesylate (Invirase)

Trade names: Invirase (saquinavir mesylate, Ro 318959, Hoffmann-La Roche, Switzerland), Fortovase (soft gel formulation, Ro 318959 003, Hoffmann-La Roche, Switzerland).

Nelfinavir [*159989-64-7*], [3S-[2(2S*,3S*),3α4aβ,8aβ]]-N-(1,1-Dimethylethyl)decahydro-2-[2-hydroxy-3-[(3-hydroxy-2-methylbenzoyl)amino]-4-(phenylthio)butyl]-3-isoquinolinecarboxamide; free base $C_{32}H_{45}N_3O_4S$, M_r 567.80; monomethanesulfonate (mesylate) [*159989-65-8*], $C_{32}H_{45}N_3O_4S \cdot CH_4O_3S$.

Nelfinavir mesylate (Viracept) is a nonpeptidic HIV protease inhibitor ($K_i = 2$ nM) co-developed by Agouron Pharmaceuticals and Japan Tobacco [35], [36]. It was launched in 1997 for use in adults and children, in combination with reverse transcriptase inhibitors (zidovudine and lamivudine or stavudine and lamivudine).

Synthesis [37], [38]:

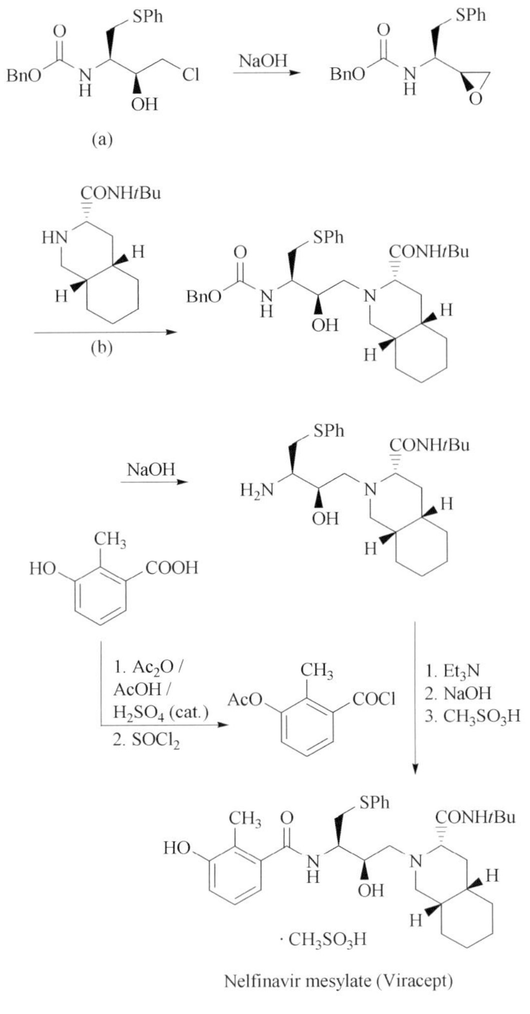

(a: [39], b: [40])

1307

Trade names (nelfinavir mesylate): Viracept (Agouron, USA; Japan Tobacco, Japan; Hoffmann-La Roche, Switzerland).

Amprenavir [*161814-49-9*], [3*S*-3*R**(1*R**,2*S**)]-3-[[(4-aminophenyl)sulfonyl](2-methylpropyl)amino]-2-hydroxy-1-(phenylmethyl)propyl]carbamic acid tetrahydro-3-furanyl ester; free base $C_{25}H_{35}N_3O_6S$, M_r 505.64. Amprenavir is a white to cream-colored solid with a solubility of $\approx$ 0.04 mg/mL in water at 25 °C.

Amprenavir is the latest HIV protease inhibitor to be approved in the United States (1999), in combination with approved RT-inhibitors, for the treatment of HIV-1 infections. Its long half-life allows twice daily dosing thus simplifying combination therapies and patient compliance.

Synthesis [41]:

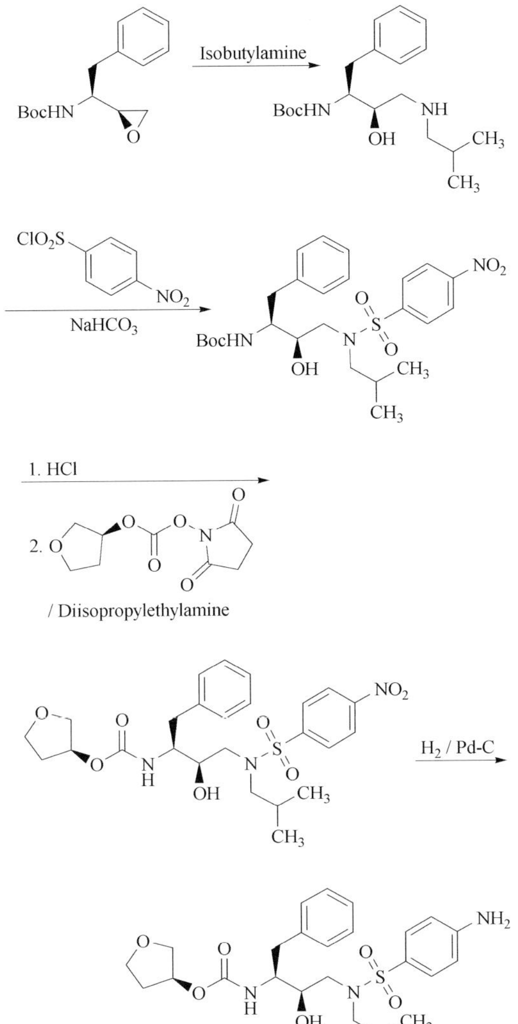

Amprenavir (Agenerase)

Trade name: Agenerase (Vertex Pharmaceuticals, USA; Glaxo Wellcome, UK; Kissei, Japan).

3.3. Hydroxyethylene Isosteres

Ritonavir [*155213-67-5*], [5S-(5R*,8R*,10R*,11R*)]-10-hydroxy-2-methyl-5-(1-methylethyl)-1-[2-(1-methylethyl)4-thiazolyl]-3,6-dioxo-8,11-bis(phenylmethyl)-2,4,7,12-tetraazatridecan-13-oic acid 5-thiazolylmethyl ester; free base $C_{37}H_{48}N_6O_5S_2$, M_r 720.96.

Ritonavir is a peptidomimetic inhibitor of HIV-1 and HIV-2 proteases, that was approved in 1996 for use in combination with RT nucleoside analogs, or as monotherapy for the treatment of HIV-1 infection. It reversibly inhibits cytochrome P450 enzymes, thus inhibiting its own metabolism by these enzymes [42].

Synthesis [43] – [45]:

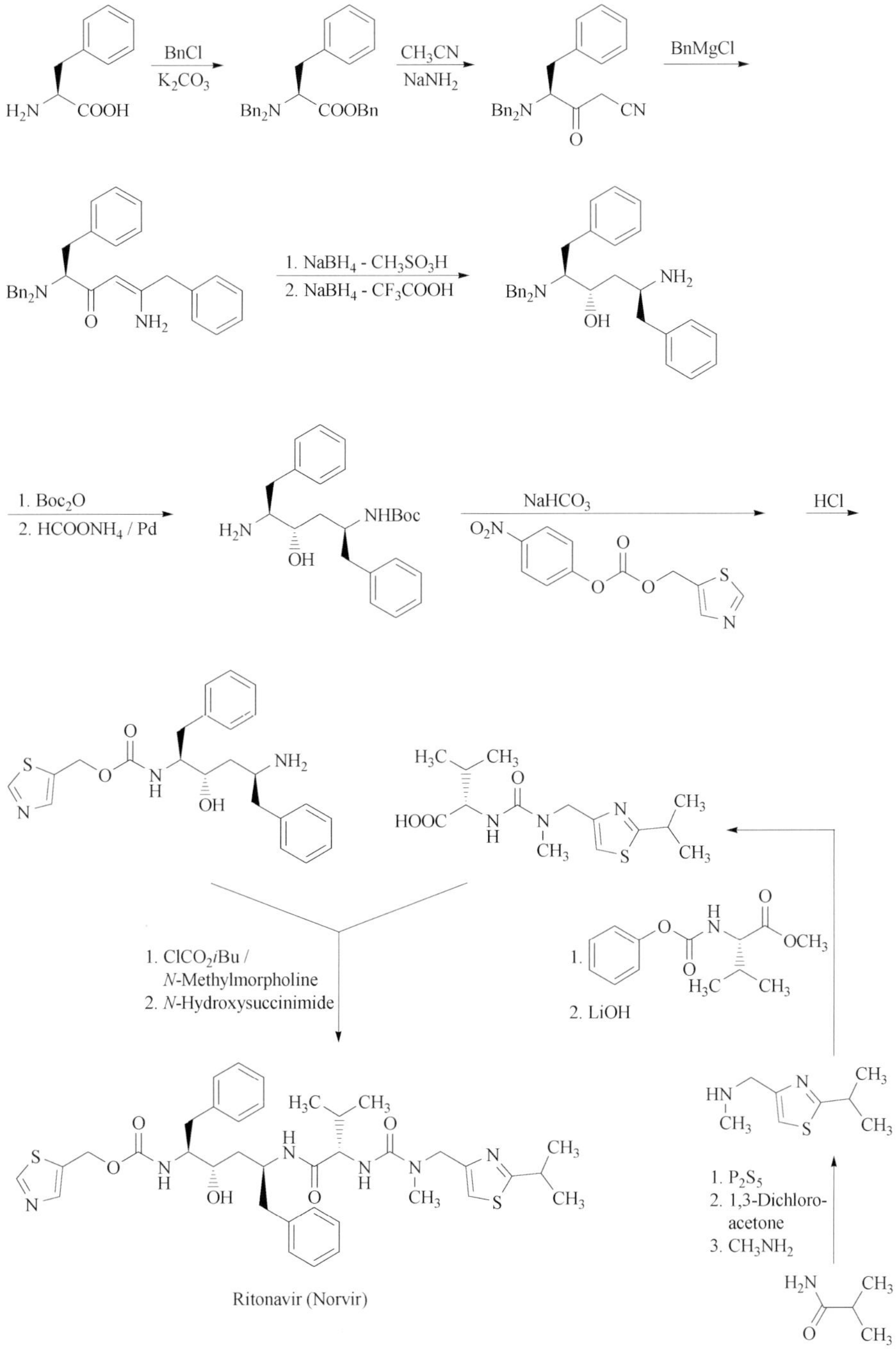

Ritonavir (Norvir)

Trade name: Norvir (Abbott, USA).

Indinavir [*150378-17-9*], [1(1*S*,2*R*),5(*S*)]-2,3,5-trideoxy-*N*-(2,3-dihydro-2-hydroxy-1*H*-inden-1-yl)-5-[2-[[(1,1-dimethylethyl)amino]carbonyl]-4-(3-pyridinylmethyl)-1-piperazinyl]-2-(phenylmethyl)-D-erythropentonamide; free base $C_{36}H_{47}N_5O_4$, M_r 613.80, *mp* 153–154 °C (anhydrous form), $[\alpha]_D^{22}$ +24.1 ° (c = 0.0133, CHCl$_3$); sulfate [*157810-81-6*], $C_{36}H_{47}N_5O_4 \cdot H_2SO_4$, M_r 711.88, crystals from absolute ethanol, softens at 135 °C, *mp* 150–153 °C (decomp.).

Indinavir was launched in 1996 and is now approved for use alone or in combination with zidovudine and other reverse transcriptase inhibitors. It is a potent inhibitor (K_i = 0.36 nM) of HIV-1 protease [46], [47].

Synthesis [48]:

Indinavir sulfate (Crixivan)

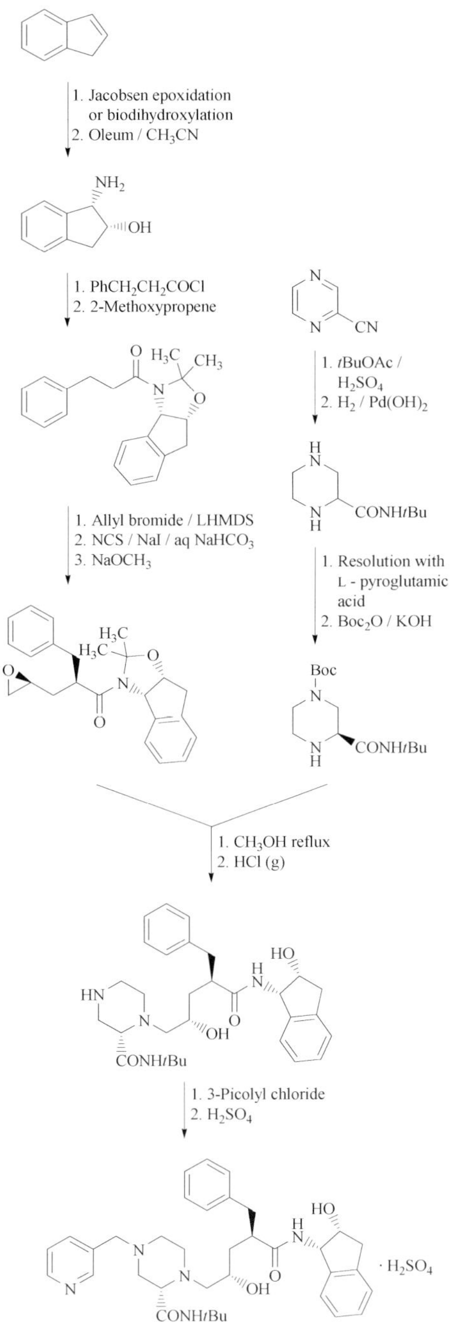

Trade name (indinavir sulfate): Crixivan (Merck & Co. USA).

3.4. Compounds in Development

Several second-generation HIV protease inhibitors are presently being evaluated in the clinic (Fig. 3).

Lopinavir (Abbott) is a hydroxyethylene-based peptidomimetic inhibitor in phase III trials, and is being evaluated for use in combination with ritonavir. BMS is evaluating peptidomimetic inhibitors, including lasinavir (a hydroxyethylene isostere in phase I/II) and CGP-73547 (a hydroxyethylhydrazine isostere in phase I). Japan Energy has KNI-272, an AHPBA peptide isostere in phase II trials and a C_2-symmetric nonpeptide cyclic urea licensed from DuPont Merck is being developed by Triangle Pharmaceuticals (phase II). Finally, tipranavir (PNU 140690) is a third generation, nonpeptide inhibitor of HIV protease based on a 5,6-dihydro-4-hydroxy-2-pyrone scaffold. It is currently in phase II trials (Pharmacia & Upjohn).

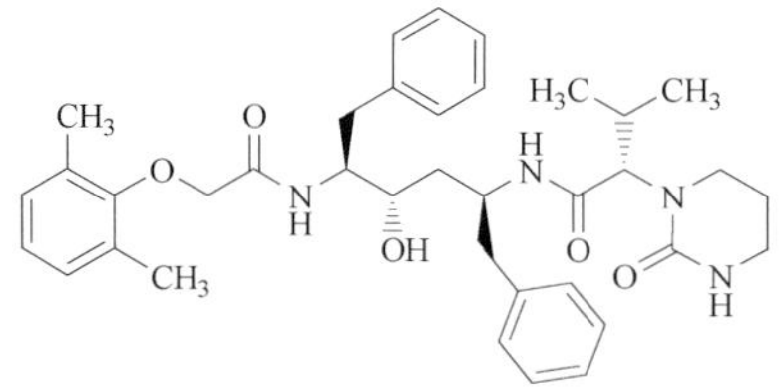

Lopinavir (ABT-378, Abbott, phase III)

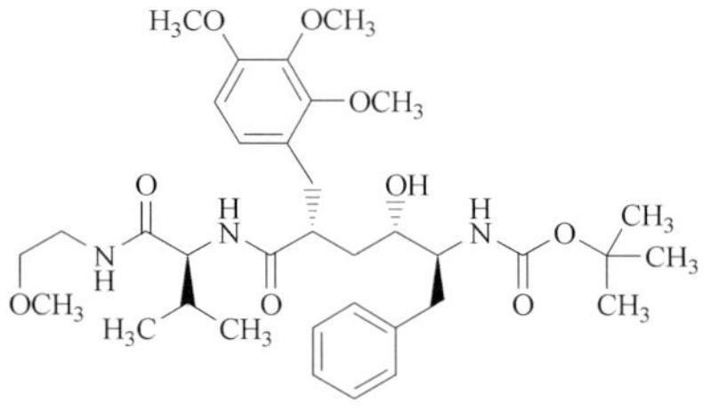

Lasinavir (CGP-61755, BMS, phase I / II)

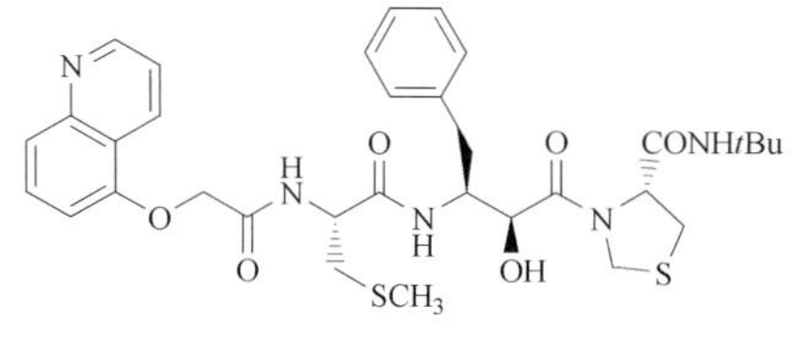

KNI-272 (Japan Energy Corporation, phase II)

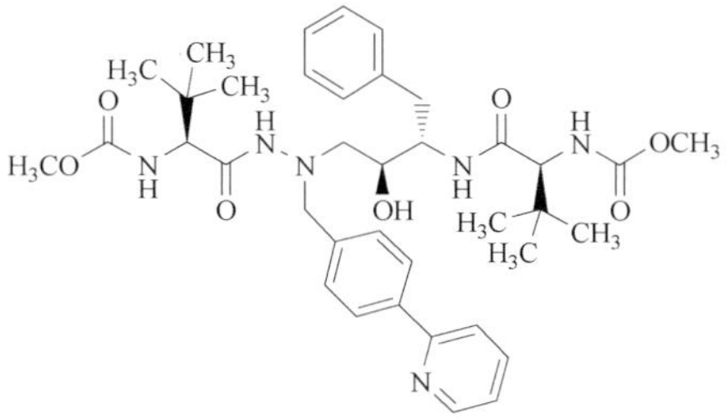

CGP-73547 (BMS, phase I)

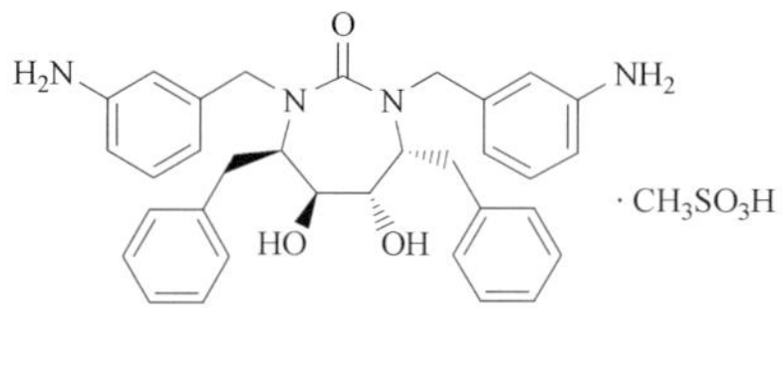

DMP-450 (Triangle Pharmaceuticals, phase II)

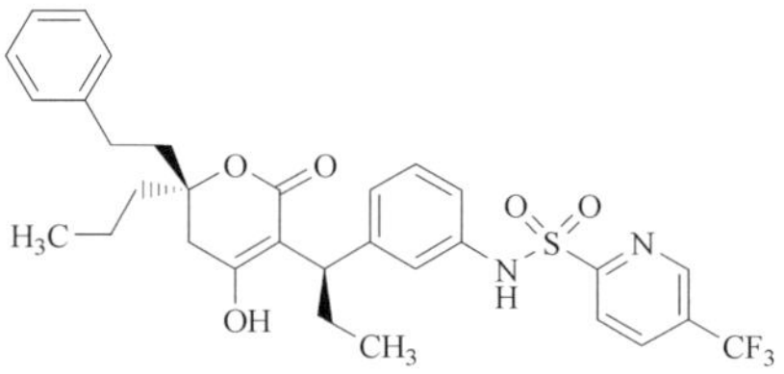

Tipranavir (PNU 140690, Pharmacia & Upjohn, phase II)

Figure 3. HIV therapeutics in clinical development

4. Emerging Concepts in AIDS Therapy

Despite evidence that the immune response can not destroy HIV, it is believed that boosting of the body's defense mechanism in preparation for the viral attack, may succeed in containing the virus. Important efforts are being directed at the development of vaccines that can protect against HIV infection. Current strategies include the search for agents that stimulate humoral and/or cellular immune responses by targeting the *env* surface glycoprotein of HIV. On another front, the use of live-attenuated HIV to elicit an immune response is being hampered by potential reversion to pathogenic strains. Coupled to the high variability, a characteristic of this virus, it is unlikely that suitable vaccines for use in humans will become available in the next five years [49]–[51].

As a consequence, most efforts to combat the spread of the epidemic caused by HIV infection have concentrated on the development of drugs, which arrest the replication cycle of the virus. Most successful at the present time, reverse transcriptase and protease inhibitors have been the major focus of the search for antivirals. Due to the emergence of drug resistance [52], drug failure and side effects associated with the use of such agents, continued efforts are being directed at the discovery of novel therapies, which target other essential features of the virus's life cycle. These include inhibitors of the Tat protein, therapies based on CD4 and the HIV envelope, the viral integrase and regulatory *nef* and *rev* gene products. In addition, immune-based approaches such as monoclonal antibodies, inhibitors of cytokine-mediated virus activation [53], [54] and antisense oligonucleotides [56], [57] are all being actively explored. To date however, most of this research remains at the early exploratory stage.

While greater emphasis is being placed on combination therapies based on the use of RT and protease inhibitors [55], a definitive cure to the disease remains elusive at this time and an eventual vaccine remains the only true hope for arresting the worldwide epidemic.

5. References

Acknowledgement: we are grateful to Dr. Janice Kelland and Georges Medawar for help with literature searches.

[1] F. Barré-Sinoussi et al., *Science* **220** (1983) 868.
[2] R. C. Gallo et al., *Science* **224** (1984) 500.
[3] Press release, April 1999, UNAIDS. For up to date reports on the epidemiology of HIV and AIDS, consult the Joint United Nations Programme on HIV/AIDS (UNAIDS) web site at http://www.unaids.org.
[4] C. A. Andrews, R. A. Koup, *J. Antimicrob. Chemoth.* **37** Suppl. B (1996) 13.

[5] A. S. Fauci, *Ann. Int. Med.* **114** (1991) 678.
[6] M. Stevenson, M. Bukrinsky, S. Haggerty, *AIDS Res. Hum. Retrov.* **8** (1992) 107.
[7] L. Ratner, *Persp. Drug Disc. Des.* **1** (1993) 3.
[8] E. De. Clercq, *J. Med. Chem.* **38** (1995) 2491.
[9] H. Mitsuya, S. Broder, *Nature* **325** (1987) 773.
[10] P. L. Darke et al., *Biochem. Bioph. Res. Commun.* **156** (1988) 297.
[11] P. A. Darke, J. R. Huff, *Adv. Pharmacol.* **25** (1994) 399.
[12] N. E. Kohl et al., *Proc. Natl. Acad. Sci. USA* **85** (1988) 4686.
[13] R. A. Kramer et al., *Science* **231** (1986) 1580.
[14] P. L. Beaulieu et al., *J. Med. Chem.* **40** (1997) 2164.
[15] S. Ren, E. J. Lien, *Prog. Drug. Res.* **51** (1998) 1.
[16] C. Debouck, *AIDS Res. Hum. Retroviruses* **8** (1992) 153.
[17] S. Czernecki, J.-M. Valéry, *Synthesis* (1991) 239.
[18] Burroughs Wellcome Co., US 4 724 232, 1988.
[19] V. Bath, E. Stocker, B. G. Ugarkar, *Synth. Commun.* **22** (1992) 1481.
[20] J. P. Horwitz, J. Chua, M. Noel, J. T. Donatti, *J. Org. Chem.* **32** (1967) 817.
[21] Bristol Myers Squibb Company, US 5 539 099, 1996.
[22] H. Jin et al., *J. Org. Chem.* **60** (1995) 2621.
[23] BioChem Pharma Inc., US 5 696 254, 1997.
[24] Glaxo Wellcome, WO 98/52949, 1998.
[25] Glaxo Wellcome, US 5 641 889, 1997.
[26] K. D. Hargrave et al., *J. Med. Chem.* **34** (1991) 2231.
[27] D. L. Romero, *Drugs of the Future* **19** (1994) 238.
[28] Upjohn Company, US 5 563 142, 1996.
[29] M. E. Pierce et al., *J. Org. Chem.* **63** (1998) 8536.
[30] A. Wlodawer, J. W. Erickson, *Annu. Rev. Biochem.* **62** (1993) 543.
[31] N. A. Roberts, *Int. Antiviral News* **3** (1995) 2.
[32] G. Moyle, *Expert. Opin. Invest. Drugs* **5** (1996) 155.
[33] S. L. Nightingale, *J. Am. Med. Assoc.* **275** (1996) 273.
[34] W. Göhring et al., *Chimia* **50** (1996) 532.
[35] S. W. Kaldor et al., *J. Med. Chem.* **40** (1997) 3979.
[36] X. Rabasseda, A. M. Martel, J. Castañer, *Drugs Future* **22** (1997) 371.
[37] T. Inaba et al., *J. Org. Chem.* **63** (1998) 7582.
[38] Agouron Pharmaceuticals Inc., WO 98/09951, 1998.
[39] Agouron Pharmaceuticals Inc., 5 484 926, 1996, (B. A. Dressmann et al.).
[40] Hoffmann-La Roche Inc., US 5 256 783, 1993, (S. Gokhale, M. Schlageter).
[41] Vertex Pharmaceuticals Inc., US 5 783 701, 1998.
[42] D. J. Kempf et al., *J. Med. Chem.* **41** (1998) 602.
[43] A. R. Haight et al., *Org. Process Res. Dev.* **3** (1999) 94.
[44] Abbott Laboratories, US 5 567 823, 1996.
[45] Abbott Laboratories, WO 94/14436, 1994.
[46] B. D. Dorsey et al., *J. Med. Chem.* **37** (1994) 3443.
[47] J. P. Vacca et al., *Proc. Natl. Acad. Sci. USA* **91** (1994) 4096.
[48] P. J. Reider, *Chimia* **51** (1997) 306.
[49] D. Baltimore, C. Heilman, *Scientific American,* July 1998, 98.
[50] D. R. Burton, J. P. Moore, *Nature Medicine, vaccine supplement* **4** (1998) 495.
[51] C. Heilman, D. Baltimore, *Nature Medicine, vaccine supplement* **4** (1998) 532.

[52] G. J. Moyle, *Immun. Infect. Dis.* **5** (1995) 170.
[53] N. Cammack, *Antivir. Chem. Chemoth.* **10** (1999) 53.
[54] S. W. Hunt, III, G. J. LaRosa, *Ann. Rep. Med. Chem.* **33** (1998) 263.
[55] A.-M. Vandamme, K. Van Vaerenbergh, E. De Clercq, *Antivir. Chem. Chemoth.* **9** (1998) 187.
[56] R. F. Rando, *Curr. Res. Mol. Ther.* **1** (1998) 67.
[57] D. S. Reddy, *Drugs Today* **32** (1996) 113.

Endocrine and Metabolic Drugs

Endocrine related drugs comprise slightly less than 10 % of the worldwide pharmaceutical market, half of which relate to treatments for diabetes and another quarter which relate to sex hormones. Lilly and Novo Nordisk are well established in the insulin market. More importantly for the focus of this review, a number of new products have recently entered the Type II diabetes market. The glitazones are good examples of the latter. Sankyo/Warner Labert/Glaxo Wellcome's troglitazone (Rezulin) was the first but apparently was associated with liver toxicity in some patients. SKB's rosiglitazone (Avandia) and Lilly/Takeda's pioglitazone followed. The market for Type II diabetes continues to grow as the incidence of the disease grows. In the following pages, these drugs and their therapeutic uses, as well as other hormones or hormone derivatives, are reviewed.

Steroids

RENATE MÜLLER, Schering AG, Berlin, Federal Republic of Germany

1.	Nomenclature	1321	3.	Bile Acids 1328
2.	Sterols	1322	4.	Steroid Hormones 1330
2.1.	Cholesterol and Related		5.	Sapogenins. 1332
	Compounds	1322	6.	Steroid Alkaloids 1333
2.2.	Vitamin D	1325	7.	Steroid Lactones 1335
2.3.	Methyl Sterols	1326	8.	References 1336

1. Nomenclature

Steroids are compounds that contain a cyclopenta[α]-phenanthrene ring skeleton which, when fully hydrogenated, is called gonan or steran.

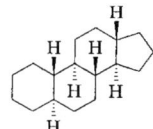

5α-Gonan or 5α-Steran

This parent structure can be modified by opening one or more bonds, by ring enlargement, or by ring contraction. The carbon atoms are numbered and the rings are indicated by letters (Fig. 1). There are angular methyl groups on C-10 and C-13. If, for instance, C-10 does not bear a methyl group, the resulting compounds are called 19-norsteroids.

For chiral centers at carbon atoms 8, 9, 10, 13, and 14, the absolute stereochemistry (Fig. 2) is specified in the parent names. Rings *B*, *C*, and *D* are predominantly *trans* linked. All atoms or groups below the plane of the ring are represented by a dashed line and the letter α, and those above the plane are indicated by a solid line and the letter β. If the configuration is unknown, a wavy line and the letter ξ is used.

Hydrogen atoms at 8, 9, and 14 are not shown in the structural formula as long as there is *trans* linkage. The position of the hydrogen at C-5 should always be indicated. The parent compounds are shown in Figures 3 and 4.

The great variety of natural and synthetic steroid compounds results from the presence of double bonds and various substituents on the ring skeleton and the side-chain. The systematic names are formed according to IUPAC rules [1]. Trivial

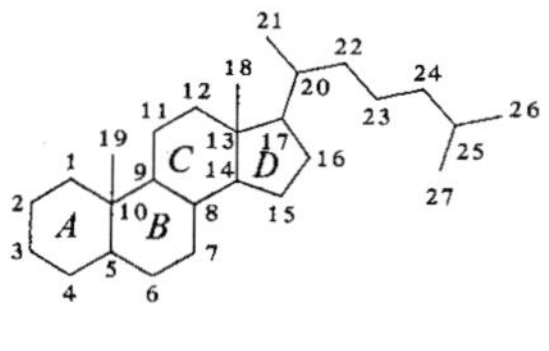

Figure 1. Steroid numbering

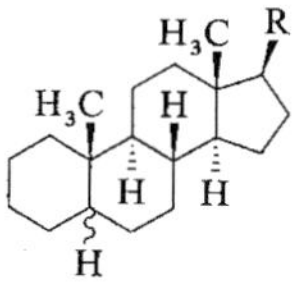

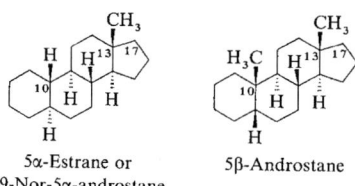

Figure 2. Steroid stereochemistry
R = Oxygen functional group or alkyl group, which may be substituted

Figure 3. Parent compounds with no side-chain at C-17

5α-Estrane or
19-Nor-5α-androstane

5β-Androstane

names are permitted for some steroids, mostly natural steroids with high biological activity.

Owing to the chair conformation of the cyclohexane ring (Fig. 5), substituents on the steroid skeleton are either approximately in the plane of the ring, equatorial (e), or vertical, axial (a). The chemical reactivity depends on this conformation [2].

2. Sterols

The first compounds of this class to be studied in detail were isolated from the unsaponifiable parts of animal and plant oils and fats. They are solid crystalline secondary alcohols with 26–30 carbon atoms, and can occur as such or as fatty acid esters and glycolipids. They were given the group name sterols (Greek, *stereos*, solid).

2.1. Cholesterol and Related Compounds
(Figs. 6, 7)

Cholesterol (**1**) is the most widely distributed sterol. Its name indicates that it is the main constituent of human gallstones. It is found in many animal tissues, in wool fat, and in plants. A survey of the literature on the occurrence, isolation, determination, metabolism, etc., is presented by COOK [3]. In humans and other mammals, cholesterol

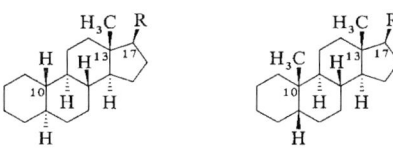

Figure 4. Parent compounds with a side-chain at C-17

R	Configuration	5α-Series	5β-Series
		5α-pregnane	5β-pregnane
	20R	5α-cholane	5β-cholane
	20R	5α-cholestane	5β-cholestane
	20R, 24S	5α-ergostane	5β-ergostane
	20R, 24R	5α-campestane	5β-campestane
	20R, 24S	5α-poriferastane	5β-poriferastane
	20R, 24R	5α-stigmastane	5β-stigmastane
	20S, 22R, 23R, 24R	5α-gorgostane	5β-gorgostane

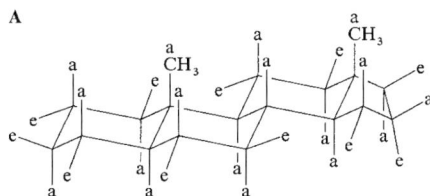

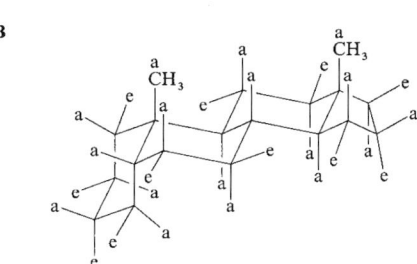

Figure 5. Conformations of steroids
A) 5α-Series; B) 5β-Series

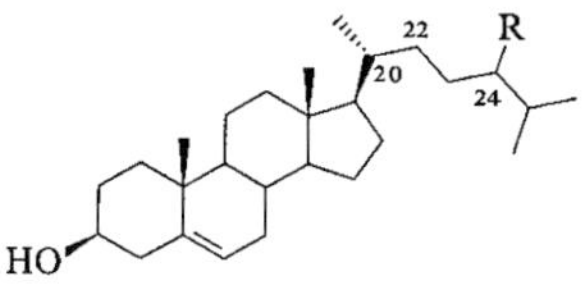

Figure 6. Δ^5-Sterols

Sterol	Δ	R	Configuration	CAS No.
Cholesterol (1)		H	20 R	[57-88-5]
Stigmasterol (2)	22	Et	20 R,24 R	[83-48-7]
β-Sitosterol (3)		Et	20 R,24 R	[83-46-5]
Brassicasterol (4)	22	Me	20 R,24 R	[474-67-9]
Campesterol (5)		Me	20 R,24 R	[474-62-4]
24-Methylene-cholesterol (9)		=CH$_2$	20 R	[474-63-5]
Poriferasterol (11)	22	Et	20 R,24 S	[481-16-3]
Clionasterol (12)		Et	20 R,24 S	[83-47-6]
24-Dehydro-cholesterol (14)	24	H	20 R	[313-04-2]

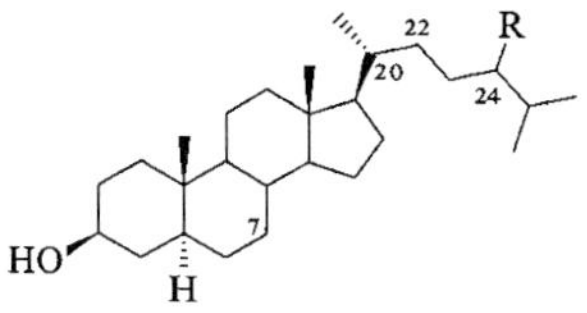

Figure 7. 5α-Sterols

Sterol	Δ	R	Configuration	CAS No.
Spinasterol (6)	7,22	Et	20 R,24 R	[481-18-5]
Zymosterol (7)	8,24	H	20 R	[128-33-6]
Faecosterol (8)	8	=CH$_2$	20 R	[516-86-9]
Spongesterol (10)	22	Me	20 R,24 S	[474-66-8]
Chondrillasterol (13)	7,22	Et	20 R,24 S	[481-17-4]

is a metabolic precursor of gallstones [4] and steroid hormones (→ Hormones). At a low developmental level, diverse, poorly separable sterol mixtures occur, but in the course of evolution, cholesterol becomes increasingly important until, in vertebrates, it is practically the only sterol [5]. In the 1930s, cholesterol served as the starting compound for androgen synthesis [2]. Although it was for some time replaced by sapogenins, it has become important again, owing to progress in microbiological fermentation [6]. Other important starting sterols for the partial chemical synthesis of compounds of the androstane and estrane series are stigmasterol (2) and β-sitosterol (3) (→ Hormones). Both sterols are contained, e.g., in soybean oil. Further phytosterols include brassicasterol (4) and campesterol (5) [7], spinasterol (6) from spinach [8] or alfalfa [9], and the yeast sterols zymosterol (7) [10] and faecosterol (8) [7]. The sterols isolated from invertebrate sea animals exhibit structural changes, especially in the side chain. 24-Methylenecholesterol (9), spongesterol (10), poriferasterol (11), clionasterol

(12), and chondrillasterol (13) occur in various fungi, and 24-dehydrocholesterol (14) occurs in the goose barnacle [2]. A cyclopropane ring in the side-chain, as in gorgosterol (22 S,22 R,23 R,24 R) [29782-65-8] (15), appears to be characteristic of a whole class of marine steroids [11].

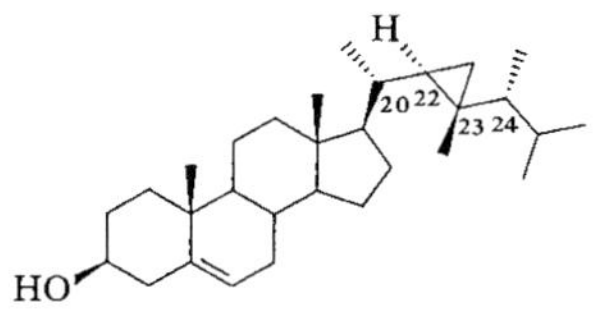

15

Calysterol [57331-04-1] (16) contains a cyclopropene ring [12].

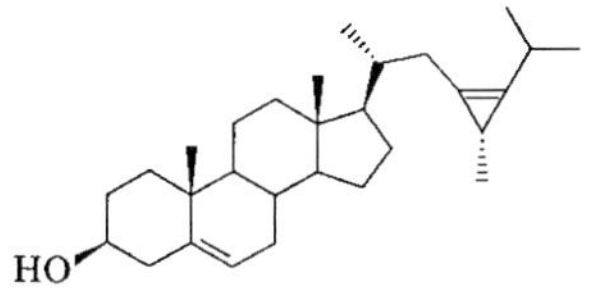

16

A list of natural steroids is given in [7, Table 19]. Some cholesterol derivatives, e.g., cholesteryl benzoate, form liquid crystals.

2.2. Vitamin D

It was recognized around 1920 that rickets could be cured by cod-liver oil or solar irradiation. The antirachitic constituent of cod-liver oil was found to be a fat-soluble compound which was called vitamin D. It was also discovered that plant oils became antirachitic on irradiation [13], and that the potentially active compounds were in the unsaponifiable part [14], which consisted of a mixture of sterols. Whereas cod-liver oil apparently contained preformed vitamin D, the phytosterol mixture contained a provitamin. It was found [2] that ergosterol is the precursor of vitamin D_2 (ergocalciferol) (→ Hormones) and 7-dehydrocholesterol yields vitamin D_3 (cholecalciferol). Vitamin D_3 is metabolized to 1α,25-dihydroxyvitamin D_3 (calcitriol) in the liver and kidney. Calcitriol is the agent responsible for calcium and phosphate transport in the body. Thus, cholecalciferol and its biologically active metabolites must be regarded as vitamins as well as hormones (→ Hormones). Since the only abundant source of vitamin D is fish-liver oil, syntheses are of great importance [2]. Attempts were made to find other antirachitic substances by irradiating sterols that differed from ergosterol only in the side-chain [15]. Methods for the synthesis of calcitriol have been developed more recently, and attempts are being made to find derivatives which are more efficient in separating the systemic effect on calcium metabolism from the antiproliferative activity

(→ Hormones). In the last few years, side-chain analogs have been tested for possible use in some forms of cancer and in psoriasis [16] – [19].

2.3. Methyl Sterols

"Sterols" with additional methyl groups on the ring skeleton are found together with cholesterol in many plants and fungi, and, abundantly, in the wool fat of sheep. The best known are lanosterol [79-63-0] (**17**) and agnosterol [472-29-7] (**18**) from wool fat [20], and kryptosterol from yeast [21].

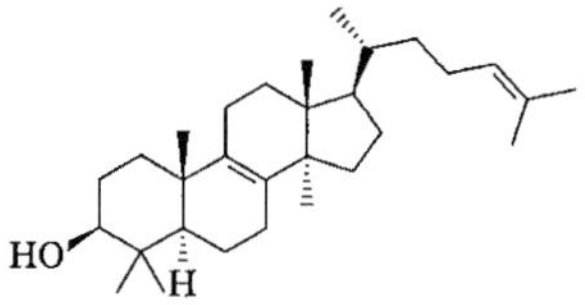

The identity of kryptosterol and lanosterol was established later [22]. Although it is analytically difficult to distinguish these compounds from the real sterols, they are regarded as triterpenes, which are mainly pentacyclic isoprenoid compounds with 30 carbon atoms. Lanosterol plays a key role in the biosynthesis of cholesterol. Acetate is converted to squalene in ten steps with the help of enzymes, and the transformation of squalene to cholesterol involves another 19 enzymatic steps [23], [24]. Squalene gives rise to lanosterol, which is then converted to ergosterol, an important sterol in many fungi. Substances which interfere with steroid biosynthesis are used in plant protection to control pathogenic fungi. Other compounds of the lanostane type are mainly 4,4,14α-trimethyl derivatives of cholestane or ergostane; some are acids (Fig. 8), e.g., eburicoic acid (**19**), polyporenic acid C (**20**), tumulosic acid (**21**), and pinicolic acid (**22**). They all occur in parasitic fungi that live on wood [2]. Others have a cyclopropane ring at position 9, 10 (Fig. 9), e.g., cycloartenol (**23**) from the latex of the fruit of *Artocorpus integrifolia*, cyclolaudenol (**24**) from opium, and cycloeucalenol (**25**), with only one methyl group at C-4, from eucalyptus. The significance of cycloartenol in the biosynthesis of steroids in higher plants is comparable to that of lanosterol [25].

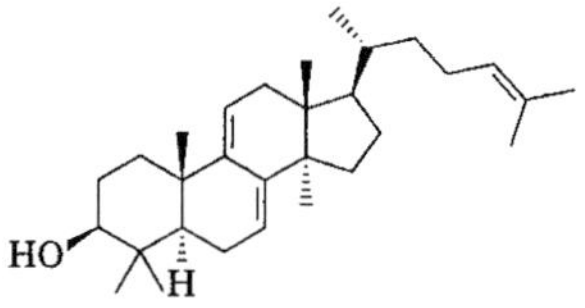

Compounds of the euphane and lanostane types are epimers with respect to carbon atoms 13, 14, and 17. Thus, euphol, obtained from the resin of a species of *Euphorbia*, corresponds to 13α, 14β, 17α-lanosterol. The 20-epimer is tirucallol [26].

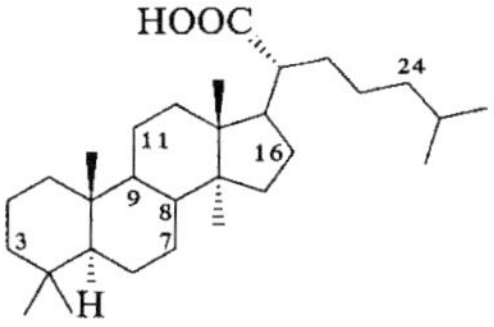

Figure 8. Lanostane acids

	Δ	3	16	24	CAS No.
Eburicoic acid (**19**)	8	β-OH		=CH$_2$	[560-66-7]
Polyporenic acid C (**20**)	7,9(11)	=O	α-OH	=CH$_2$	[465-18-9]
Tumulosic acid (**21**)	8	β-OH	α-OH	=CH$_2$	[508-24-7]
Pinicolic acid (**22**)	8,24	=O			[466-05-7]

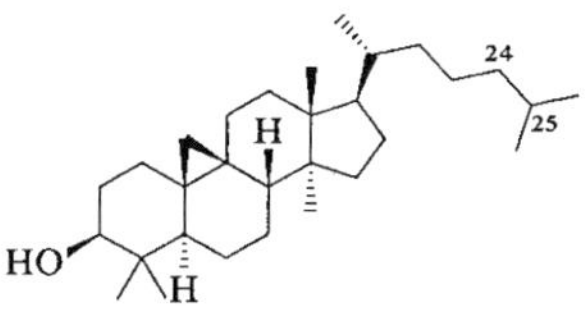

Figure 9. 9,19-Cyclo lanostane compounds

Cycloartenol (**23**) Δ^{24} [469-38-5]
Cyclolaudenol (**24**) Δ^{25}-(24S)-methyl [511-61-5]
Cycloeucalenol (**25**) 24-methylene [469-39-6]

The cucurbitacins exhibit natural bactericidal activity, and have been tested for cytotoxic and antitumor activity [27], and are regarded as kairomones. The fusidane series should also be mentioned here, fungal metabolites which have antibiotic properties and are especially active in staphylococcal infections. These compounds include fusidic acid [6990-06-3] (**26**) (→ Antibiotics) and helvolic acid [29400-42-8] (**27**).

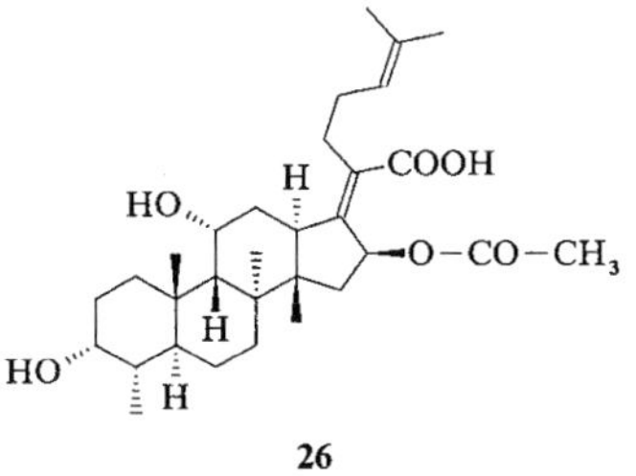

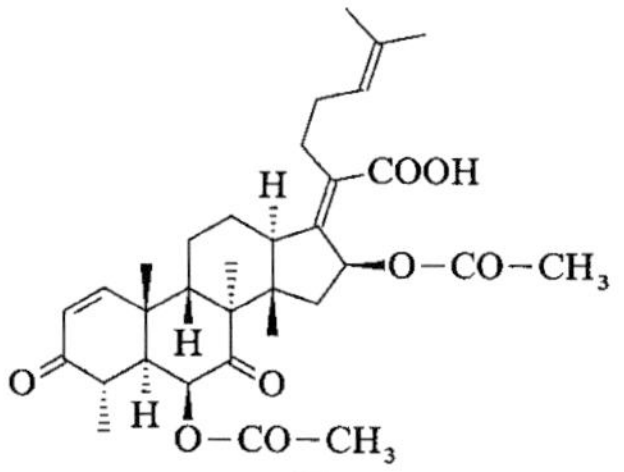

The stereochemical linkage, *trans-syn-trans-anti-trans*, which forces ring *B* into the boat form, is unusual [28].

3. Bile Acids

The bile acids are among the oldest known steroid compounds. In 1828, cholic acid was isolated in the impure state by the saponification of ox bile. The first careful studies were carried out by STRECKER in 1848 [29]. Later, it was primarily WIELAND and coworkers who studied the constitution and configuration of the bile acids. They are mostly mono-, di-, or trihydroxy derivatives of cholanic acid [2], [4]. Choleinic acid, the second bile acid to be discovered, is a molecular compound consisting of 1 mol fatty acid and 8 mol deoxycholic acid [30]. Deoxycholic acid reacts with many acids, esters, alcohols, ethers, phenols, and hydrocarbons to form molecular compounds, now collectively known as cholenic acids. They are very stable, and X-ray analysis showed that they are inclusion compounds [31]. Apocholic acid (8, 14-dehydrodeoxycholic acid) behaves in the same way [32]. Only cholic acid and deoxycholic acid (from ox bile) and hyodeoxycholic acid (from pig bile) can be used as starting compounds for syntheses, because they are available in large quantities from natural sources. Even microbiological methods of cortisone production, starting with plant steroid sapogenins, have not been able to completely replace cholic acid and deoxycholic acid as base materials (→ Hormones). Whereas mammalian bile acids are derived from 5β-cholanic acid, bile acids with side-chains extended by 3–4 carbon atoms have been found in the bile of phylogenetically old species of toads and crocodiles, e.g., 3α-7α-dihydroxy-5β-cholestanic acid in the bile of *Alligator mississipiensis*. The bile alcohol scymnol [6785-34-8] (**28**) was isolated from shark bile, and has also been found in other species of fish and frogs [33].

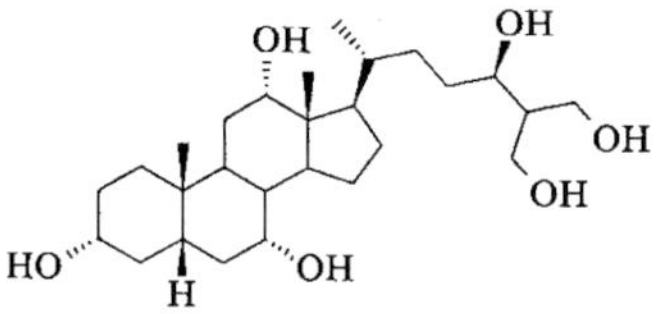

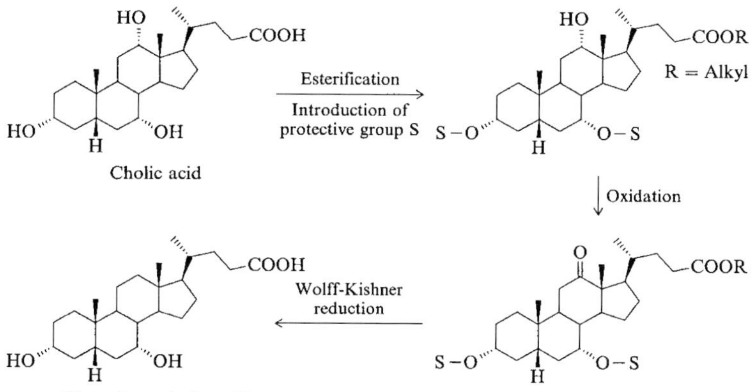

Figure 10. Production of chenodeoxycholic acid

A survey of the bile acids found in fish, amphibia, reptiles, birds, and mammals is presented by HASLEWOOD [34]–[36]. Bile acids with extended side-chains are regarded as the phylogenetic precursors of mammalian bile acids [37].

The bile acids are the end-products of cholesterol metabolism in the liver [38]. First, changes occur at the ring skeleton. The side-chain is then subjected to oxidative degradation [39]. The bile acids possess a hydrophobic and a hydrophilic part. They form water-soluble micelles with lecithins and water-insoluble lipophilic food components; thus, fatty acids, monoglycerides, steroids, fat-soluble vitamins, and cholesterol are kept in solution in the bile [40]. However, if the concentration of bile acids falls below a certain value, gallstones can form. Gallstones differ in chemical composition; ca. 10–20 % are pure cholesterol stones [41], which can be dissolved by oral administration of chenodeoxycholic acid [*474-25-9*] (**29**) or ursodeoxycholic acid [*128-13-2*] (→ Gallbladder and Liver Therapeutics). Ursodeoxycholic acid can also be used in the treatment of liver cirrhosis [42] and to ameliorate the symptoms of biliary reflux disease [43]. Cholic acid [*81-25-4*], deoxycholic acid [*83-44-3*], and dehydrocholic acid [*81-23-2*] are used as choleretics (→ Gallbladder and Liver Therapeutics). The sodium salt of deoxycholic acid is used on a small scale to improve the resorption of drugs, e.g., absorption of insulin through the nasal mucous membrane [40].

Chenodeoxycholic acid (Fig. 10) is produced on an industrial scale from cholic acid according to the method of FIESER [44]. Ursodeoxycholic acid can be obtained from chenodeoxycholic acid (**29**) by oxidation of the 7α-OH group to the 7-ketone, followed by reduction with alkali metal in amyl alcohol or liquid ammonia [45], by the selective microbiological hydroxylation of lithocholic acid [46], or from cholic acid [47].

4. Steroid Hormones

Estrone [53-16-7] was the first sex hormone to be obtained in the pure state [48]. A short time later, estriol [50-27-1] was isolated from the urine of pregnant women [49]. In 1934, four different teams succeeded almost simultaneously in preparing pure progesterone [50]–[53]. The preparation of testosterone [58-22-0] was achieved in 1935 [54]. At the same time, investigation of extracts of the adrenal cortex intensified. The research groups of REICHSTEIN and KENDALL introduced the practice of naming their isolated substances by letters. By 1943, as many as 28 steroids had been identified, including cortisone, cortisol, cortexolone (Reichstein S), 11-dehydrocorticosterone, corticosterone, and deoxycorticosterone [55]. Aldosterone was isolated in 1953 [56]. Industrial production of pregnenolone from diosgenin, which occurs in the root of the Mexican plant *Dioscorea*, began in 1940 [57]. The discovery in 1948 that cortisone had a strong effect on the symptoms of arthritis, initiated an intensive search for further anti-inflammatory steroids and the development of new methods for the commercial production of cortisone. The discovery of the reduction of phenol ethers to dihydro compounds (1950) made possible the production of 19-norsteroids from estrone methylether [58]. Many steroid reactions had been described by 1950, but little was known about the reaction mechanisms. BARTON showed that many rules that had been established empirically could be rationalized by the concept of axial and equatorial bonds [59]; conformation analysis gave steroid research a strong impetus. It was also recognized that UV [60] and IR spectra [61] are valuable in structural elucidation.

Another important physical quantity for the characterization of steroid compounds is optical rotation. Steroids exhibit several asymmetric centers; all naturally occurring steroids are optically active (→ Hormones).

Nuclear magnetic resonance sprectra, mass spectrometry, and X-ray analysis are also used for structure determination [62].

Steroid hormones are divided into three main physiological groups:

1) Sex hormones
   Estrogens (→ Hormones)
   Gestagens (→ Hormones)
   Androgens (→ Hormones)
2) Adrenal steroids
   Glucocorticoids (→ Hormones)
   Mineralocorticoids (→ Hormones)
3) Calcium-regulating sterols (→ Hormones)
   Vitamin D
4) Insect skin-shedding hormones

Cholesterol is the common precursor in the biosynthesis of the first three groups (→ Hormones). Apart from steroid compounds with hormonal activity, some compounds act as hormone antagonists. Depending on the mode of action, distinction is made between substances that occupy specific receptors and those that block hormone production. Thus, not only are antiestrogens known (→ Hormones), but also antiandrogens (→ Hormones), antigestagens (→ Hormones), and antagonists of mineralocorticoids (→ Hormones).

In the synthesis of steroid hormones, four processing modes are distinguished:

1) Partial chemical synthesis from natural products (→ Hormones)
2) Microbiological degradation of sterols (→ Hormones)
3) Total synthesis (→ Hormones)
4) Isolation from natural sources

The starting materials for partial chemical synthesis must contain the steroid nucleus, they must be convertible to steroid hormones via intermediates by simple chemical reactions, and they must be available in sufficient quantities. The most important starting substances are cholesterol, stigmasterol, β-sitosterol, and diosgenin (→ Hormones) as well as bile acids, ergosterol, and hecogenin (→ Hormones). The most important compounds obtained are dehydropregnenolone, progesterone, androstenolone, and estrone (→ Hormones). The androgenic steroids also have a more-or-less strong anabolic activity which can be intensified by chemical modification of the steroid nucleus. However, some androgenic activity always remains (→ Hormones). The glucocorticoids, another group of steroid hormones, have also been subjected to various chemical modifications. Attempts have been made to reduce the mineralocorticoid side effects. The introduction of a double bond at position 1,2 gave compounds with glucocorticoid, anti-inflammatory, and antiallergenic activity 4–5 times that of cortisone (→ Hormones). The introduction of 9α-halogen (→ Hormones), 16α-hydroxyl (→ Hormones), 6α-methyl (→ Hormones), 16α-methyl (→ Hormones), 16-methylene (→ Hormones), 6α-fluoro, 6α-chloro, and 6α,9α-difluoro substituents (→ Hormones) produced further improvements. In 1964, RASPÉ and coworkers found that 16α-methylcorticoids lacking the 17-hydroxyl group are still relatively strong anti-inflammatory substances, with very low mineralocorticoid activity (→ Hormones). The 11β-hydroxyl group has been replaced by a halogen (→ Hormones). Other derivatives of well-known, active corticoids gave a series of compounds with little or no systemic activity, so these compounds can be very successful in treating dermatosis in children (→ Hormones). It has not been possible to find either the exact mechanism of the anti-inflammatory and immunosuppressive activity of the glucocorticoids (→ Drugs Used in Dermatology), or a new class of compounds to replace the glucocorticoids.

The insect skin-shedding hormones are completely different. Each stage of skin shedding from larva to pupa to adult insect is subject to hormonal control. Whereas growth is controlled by the juvenile hormones, the skin-shedding hormones are steroids. In 1954, the first 25 mg of a crystalline skin-shedding hormone was isolated from the silkworm *Bombyx mori* [63]. This substance, α-ecdysone (**30**), was found to be a steroid by chemical and X-ray studies [64]. A very similar compound, β-ecdysone (**31**), also occurs in the silkworm. Other skin-shedding hormones, or zooecdysones (Fig. 11) have been isolated from insects or crustaceans, e.g., inokosterone (**32**) and makisterone

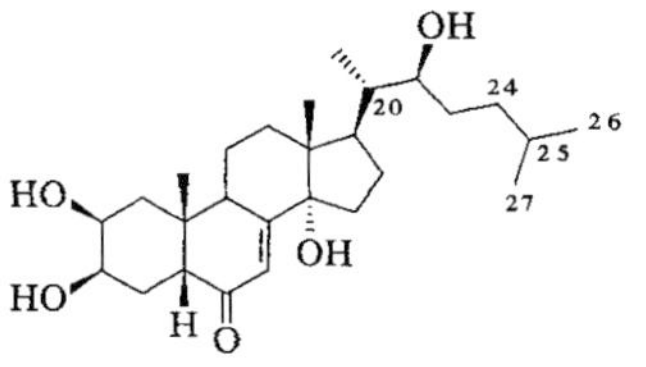

Figure 11. Zooecdysones

Name	20	24	25	26	CAS No.
α-Ecdysone (30)			OH		[3604-87-3]
β-Ecdysone (31)	OH		OH		[5289-7-7]
Inokosterone (32)	OH			OH	[15130-85-5]
Makisterone A (33)	OH	CH$_3$	OH		[20137-14-8]
26-Hydroxy-β-Ecdysone (34)	OH		OH	OH	[52717-49-4]

A (33) from crabs [65] and 26-hydroxy-β-ecdysone (34) from the tobacco hornworm *Manduca sereta (L)* [66]. Shortly after the elucidation of the structure of α-ecdysone (30), a number of steroids with skin-shedding activity, the phytoecdysones, were also found in plants [67]. They are widely distributed, and occur in relatively large quantities [68]. By 1979, a total of 47 different phytoecdysones were known, and most of the zooecdysones had been found in plants as well [69]. The first synthesis was published by Schering AG–Hoffmann LaRoche [70]. This was followed by several improvements and other starting materials were used [71]–[73]. The direct introduction of a 2,7-diene-6-one group from 3β-tosylate-5-ene resulted in a considerable shortening of the synthetic pathway [74].

5. Sapogenins

Plant glycosides that form a soapy foam in water are called sapogenins. They are composed of an aglycon, sapogenin, and one or more sugars. Apart from a few exceptions, the sugar moiety is bound to the steroid through a 3β-OH group. The first saponins were derived from digitalis preparations and were called digitonin and gitonin, and the corresponding aglycons, digitogenin and gitogenin. Even later, the names were often derived from the plants from which the sapogenins were isolated. Raw plant extracts were used as detergents because they are neither alkaline nor precipitable by hard water. Saponin preparations without cardiac effects have been used in fire extinguishers, and as fish poisons. Although fish are numbed or killed by saponins, they are still edible because the oral consumption of saponins is not harmful to humans. The structures of the aglycons are not as varied as those of the cardiac glycosides. These steroids belong predominantly to the spirostanes. The most important is diosgenin, the starting compound for the partial synthesis of steroid hormones (→ Hormones). Hecogenin was also of importance for a time (→ Hormones).

6. Steroid Alkaloids

Steroid alkaloids occur in both plants and animals. Plant steroid alkaloids are widely distributed, and have various structures, depending on the plant species. Funtumine [474-45-3] (**35**), conessine, holarrhimine Δ^5 [468-31-5] (**36**), and holarrhidine 5α-H [82182-51-2] (**37**) have been isolated from certain species of *Holarrhena*.

The first three compounds were considered as starting substances for the commercial synthesis of steroids [75], but without success.

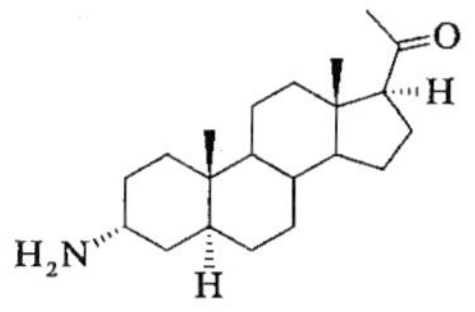

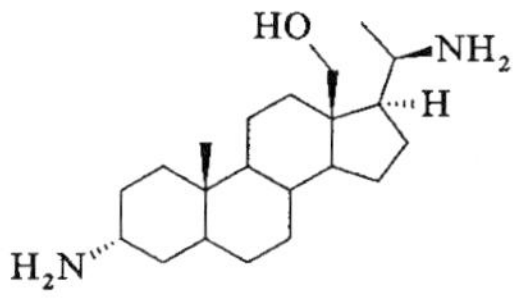

Irehdiamine A is found in *Funtumia elastica*, and cyclobuxine D and buxocine C [76]–[78] in certain species of *Buxus*. The solanum alkaloids solasodine, solanidine [80-78-4] (**38**), and tomatidine [77-59-8] (**39**) should also be mentioned [79].

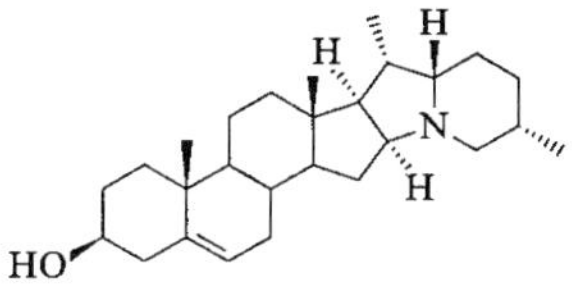

Solasodine and tomatidine have been converted to 16-dehydropregnenolone by the Marker process [80]. The solanum alkaloids occur in plants in the form of glycosides.

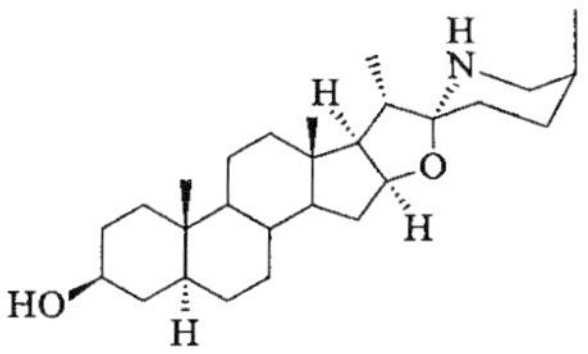

All the alkaloids mentioned above have an essentially unchanged steroid skeleton. Attempts have been made to prepare solanum alkaloids with the help of biotechnology, but without commercial success [81].

In the veratrum and fritillaria alkaloids, ring C is contracted to a five-membered ring and ring D is enlarged to a six-membered ring. Hence, they are C-nor-D-homosteroids [82], [83]. Jervine and verticine, as well as the ester alkaloid protoverine [76-45-9] (**40**) belong to this group.

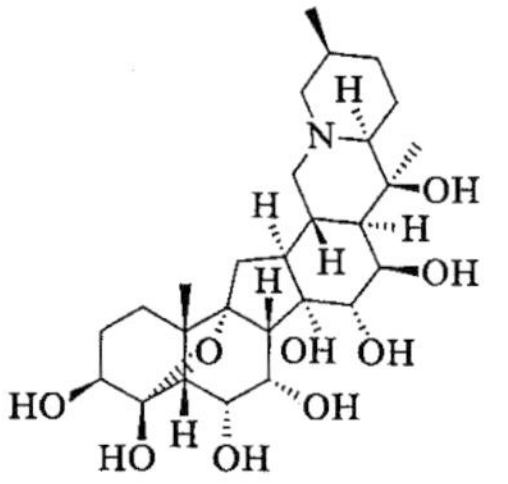

40

The ester alkaloids contain 7–9 oxygen atoms, esterified with organic acids, such as acetic, angelic, veratric, vanillic, or tiglic acid. Protoverine possesses vasodilative properties.

The animal steroidal alkaloids are poisonous, and are excreted though the skin. The main poisons of the salamander are samandarin [467-51-69] (**41**) and cycloneosamandione [631-72-1] (**42**).

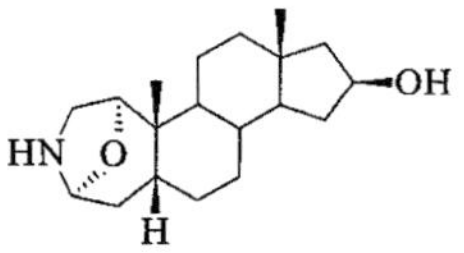

41

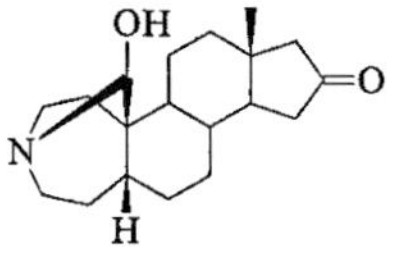

42

Samandarin poisoning leads to cramps and respiratory paralysis. In fact, salamanders can die from their own poison if it enters the circulatory system. The poison of a small, brightly colored frog, *Phyllobates amotaenia*, is used as an arrow poison in Colombia. The active substances are 20-esters of batrachotoxins with various pyrrole carboxylic acids [84]. One of the most poisonous substances known is 2,4-dimethyl-pyrrole-3-carboxylic acid ester. Its subcutaneous LD_{50} is 2 µg/kg in the mouse, compared to 500 µg/kg for strychnine.

7. Steroid Lactones

The steroid lactones include the cardenolides and bufadienolides, aglycons of the cardiac glycosides (→ Cardiac Glycosides and Synthetic Cardiotonic Drugs, → Cardiac Glycosides and Synthetic Cardiotonic Drugs) as well as the holothurigenins, withanolides, and antheridiols. The sea cucumber excretes saponins that are toxic to fish, and exhibit hemolytic and oncolytic properties [85]. Acid hydrolysis gives the holothurigenins, e.g., seychellogenin [24041-68-7] (**43**).

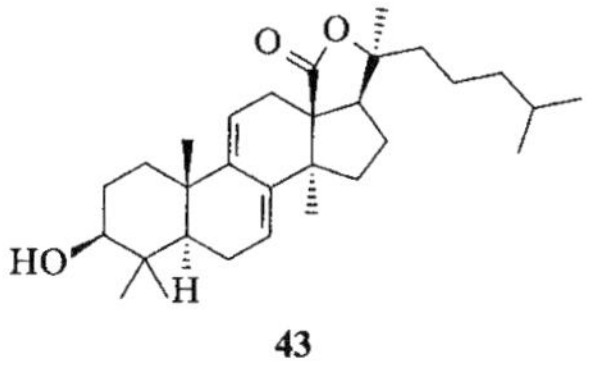

43

The withanolides are found in the Solanacae plant family [86]. The best-known representative is withaferin A [5119-48-2] (**44**).

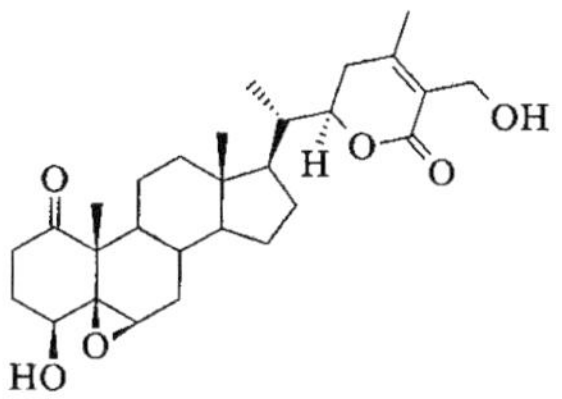

44

Antheridiol [22263-79-2] (**45**) is the first specifically acting steroid sex hormone to be discovered in plants [87].

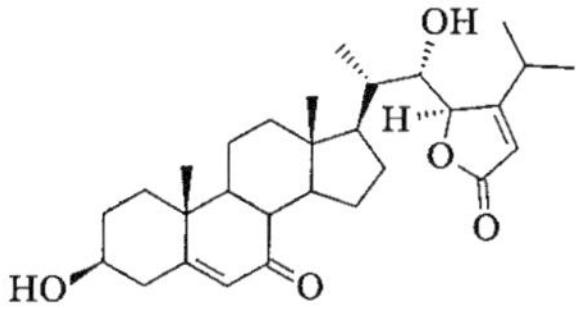

45

Antheridiol is secreted by the female plant and brings about formation of antheridial hyphae on the male plant of *Achyla bisexualis* [88], which lives in water.

8. References

[1] IUPAC-IUB 1989 *Pure Appl. Chem.* **61** (1989) 1783.
[2] L. F. Fieser, M. Fieser: *Steroide,* Verlag Chemie, Weinheim 1961.
[3] R. P. Cook: *Cholesterol,* Academic Press, New York 1958.
[4] A. Aigner, A. Bauer, *Med. Monatsschr. Pharm.* **11** (1988) 369.
[5] W. Bergmann: "Evolutionary Aspects of Sterols," in [3].
[6] M. Nagasawa, M. Bae, G. Tamura, K. Arima, *Agric. Biol. Chem.* **33** (1969) 1636.
[7] C. J. W. Brooks in S. Coffey (ed.): *Rodd's Chem. Carbon Compounds,* 2nd ed., vol. **2 D,** Elsevier, Amsterdam, pp. 150–156.
[8] M. C. Hart, F. W. Heyl, *J. Biol. Chem.* **95** (1932) 311.
[9] E. Fernholz, M. L. Moore, *J. Am. Chem. Soc.* **61** (1939) 2467.
[10] H. Wieland, Y. Kanaoka, *Justus Liebigs Ann. Chem.* **530** (1937) 146.
[11] R. L. Hale et al., *J. Am. Chem. Soc.* **92** (1970) 2179. R. L. Hale, N. C. Ling, C. Djerassi, *J. Am. Chem. Soc.* **92** (1970) 5281.
[12] E. Fattorusso et al., *Tetrahedron* **31** (1975) 1715.
[13] A. F. Hess, M. Weinstock, *J. Biol. Chem.* **62** (1924) 301.
[14] A. F. Hess, M. Weinstock, F. D. Helman, *J. Biol. Chem.* **63** (1925) 305.
[15] A. Windaus, R. Langer, *Justus Liebigs Ann. Chem.* **508** (1933) 105.
[16] T. Eguchi et al., *Chem. Pharm. Bull.* **38** (1990) 1246.
[17] L. Binderup et al., *Biochem. Pharmacol.* **42** (1991) 1569.
[18] N. Kubadera, H. Watanabe, T. Kawanashi, M. Matsumoto, *Chem. Pharm. Bull.* **40** (1992) 1494.
[19] W. H. Okamura, J. A. Palenzuela, J. Plumet, M. M. Midland, *J. Cell. Biochem.* **49** (1992) 10.
[20] H. Windaus, R. Tschesche, *Hoppe-Seyler's Z. Physiol. Chem.* **190** (1930) 51.
[21] H. Wieland, H. Pasedach, A. Ballauf, *Justus Liebigs Ann. Chem.* **529** (1937) 68.
[22] L. Ruzicka, R. Deuss, O. Jeger, *Helv. Chim. Acta* **28** (1945) 759. L. Ruzicka, R. Deuss, O. Jeger, *Helv. Chim. Acta* **29** (1946) 204.
[23] J. R. Sabine: *Cholesterol,* Marcel Dekker, New York 1977, pp. 79–103.
[24] L. J. Goad in H. L. J. Makin (ed.): *Biochemistry of Steroid Hormones,* Blackwell Scientific Publ., Oxford 1975, p. 17.
[25] R. B. Boar, C. R. Roamer, *Phytochemistry* **14** (1975) 1143.
[26] F. L. Warren, K. H. Watling, *Chem. Ind. (London)* 1956, 24.
[27] D. Lavie, E. Glotter, *Fortschr. Chem. Org. Naturst.* **29** (1971) 307.
[28] W. F. Johns (ed.): *MTP International Review of Science,* **8**, Section 1, Butterworths, London 1973, 198; W. F. Johns (ed.): *MTP International Review of Science,* **8**, Section 2, 1976, 195.
[29] A. Strecker, *Justus Liebigs Ann. Chem.* **67** (1848) 1.
[30] H. Wieland, H. Sorge, *Hoppe-Seyler's Z. Physiol. Chem.* **97** (1916) 1.
[31] F. Cramer: *Einschlußverbindungen,* Springer Verlag, Heidelberg 1954.
[32] G. Cilento, *J. Am. Chem. Soc.* **74** (1952) 908.
[33] P. Nair in D. Kritchevsky (ed.): *The Bile Acids,* vol. **1,** Plenum Press, New York 1971; vol. **2,** 1973.
[34] G. A. D. Haslewood, W. M. Wootton, *Biochem. J.* **47** (1950) 584.
[35] G. A. D. Haslewood, *Physiol. Rev.* **35** (1955) 178.
[36] I. G. Anderson, G. A. D. Haslewood, I. D. P. Wootton, *Biochem. J.* **67** (1957) 323.
[37] P. Back, *Klin. Wochenschr.* **58** (1980) 55.
[38] K. Bloch, B. N. Berg, D. Rittenberg, *J. Biol. Chem.* **149** (1943) 511.

[39] H. Danielsson, J. Sjovall, *Annu. Rev. Biochem.* **74** (1975) 233.
[40] A. Aigner, A. Bauer, *Med. Monatsschr. Pharm.* **11** (1988) 369.
[41] A. Sieg et al., *DMW, Dtsch. Med. Wochenschr.* **111** (1986) 1760.
[42] R. Poupon, R. E. Poupon, Y. Calenus, *Lancet* 1987, 834.
[43] A. B. Stefanicosky et al., *Gastroenterology* **89** (1985) 1000.
[44] L. F. Fieser, S. Rajagapolan, *J. Am. Chem. Soc.* **72** (1950) 5530.
[45] Roussel Uclaf, EP 0 072 293, 1983 (J. Bulidon, R. Demuynck, C. Pavan).
[46] Yakult Honsha, EP 119040, 1984 (H. Sawada, M. Watanuki).
[47] Erregierre, EP 0 063 106, 1982 (A. Bonaldi, E. Molinari).
[48] A. Butenandt, U. Westphal, *Hoppe-Seyler's Z. Physiol. Chem.* **191** (1930) 127.
[49] G. F. Marian, *Biochem. J.* **24** (1930) 435; 1021.
[50] A. Butenandt, U. Westphal, W. Hohlweg, *Hoppe-Seyler's Z. Physiol. Chem.* **227** (1934) 84. A. Butenandt, *Wiener Klin. Wochenschr.* **30** (1934) 934.
[51] K. H. Slotta, H. Ruschig, E. Fels, *Ber. Dtsch. Chem. Ges.* **67** (1934) 1270.
[52] W. M. Allen, O. Wintersteiner, *Science (Washington, D.C. 1883)* **80** (1934) 190. O. Wintersteiner, W. M. Allen, *J. Biol. Chem.* **107** (1934) 321.
[53] M. Hartmann, A. Wettstein, *Helv. Chim. Acta* **17** (1934) 1365.
[54] K. David, E. Dingemanse, J. Freud, E. Laqueur, *Hoppe-Seyler's Z. Physiol. Chem.* **233** (1935) 281.
[55] T. Reichstein, C. W. Shoppee, *Vitam. Horm. (N.Y.)* **1** (1943) 345.
[56] S. A. Simpson et al., *Experientia* **9** (1953) 333. S. A. Simpson et al., *Helv. Chim. Acta* **37** (1954) 1163.
[57] P. A. Lehmann, A. Bolvar, R. Quintero, *J. Chem. Educ.* **50** (1973) 195.
[58] A. J. Birch, *J. Chem. Soc.* 1950, 367.
[59] D. H. R. Barton, *Experientia* **6** (1950) 316. D. H. R. Barton, *J. Chem. Soc.* 1953, 1027.
[60] L. Dorfman, *Chem. Rev.* **53** (1953) 47.
[61] W. Neudert, H. Röpke: *Steroid-Spektrenatlas*, Springer Verlag, Berlin 1965.
[62] E. Heftmann (ed.): *Modern Methods of Steroid Analysis*, Academic Press, New York 1973.
[63] A. Butenandt, P. Karlson, *Z. Naturforsch. B Anorg. Chem. Org. Chem.* **9 B** (1954) 389.
[64] R. Huber, W. Hoppe, *Chem. Ber.* **98** (1965) 2403.
[65] K. Nakanishi in K. Nakanishi et al. (eds.): *Natural Products Chemistry*, vol. **1**, Academic Press, New York 1974, p. 527.
[66] J. N. Kaplanis, W. E. Robbins, M. J. Thompson, S. R. Dutky, *Steroids* **27** (1976) 675.
[67] K. Nakanishi et al., *Chem. Commun.* 1966, 915.
[68] D. A. Shooley, G. Weiss, K. Nakanishi, *Steroids* **19** (1972) 377.
[69] A. Prakash, S. Ghosal, *J. Sci. Ind. Res.* **38** (1979) 632.
[70] U. Kerb et al., *Helv. Chim. Acta* **49** (1966) 1601.
[71] A. Furlenmeier et al., *Helv. Chim. Acta* **50** (1967) 2387.
[72] H. Mori, K. Shibata, K. Tsuneda, M. Sawai, *Chem. Pharm. Bull.* **16** (1968) 2416.
[73] E. Lee, Y.-T. Liu, P. H. Solomon, K. Nakanishi, *J. Am. Chem. Soc.* **98** (1976) 1634.
[74] D. H. R. Barton, P. G. Feakins, J. P. Poyser, P. G. Sammes, *J. Chem. Soc. C* 1970, 1584.
[75] R. Pappo, *J. Am. Chem. Soc.* **81** (1959) 1010.
[76] R. Goutarel, *Alkaloids (London)* **1** (1971) 382.
[77] F. Kuony-Hu, R. Goutarel, *Alkaloids (London)* **3** (1973) 258.
[78] F. Kuony-Hu, R. Goutarel, *Alkaloids (London)* **4** (1974) 246.
[79] R. B. Herbert, *Alkaloids (London)* **3** (1973) 280.
[80] K. Schreiber, *Pure Appl. Chem.* **21** (1970) 131.

[81] R. Verpoorte, R. van der Heuden, W. M. van Gulik, H. J. G. ten Hoopen, *Alkaloids (N.Y.)* **40** (1991) 157.
[82] R. B. Herbert, *Alkaloids (London)* **4** (1974) 383.
[83] J. V. Greenhill, *Alkaloids (N.Y.)* **41** (1992) 177.
[84] T. Tokuyama, J. Daly, B. Witkop, *J. Am. Chem. Soc.* **91** (1969) 3931.
[85] J. S. Grossert, *Chem. Soc. Rev.* **1** (1972) 1.
[86] A. B. Kundu, A. Mukherjee, A. K. Dey, *J. Sci. Ind. Res.* **35** (1976) 616.
[87] G. P. Arsenault, K. Biemann, A. W. Barksdale, T. C. Mc. Morris, *J. Am. Chem. Soc.* **90** (1968) 5635.
[88] T. C. Mc. Morris, A. W. Barksdale, *Nature* **215** (1967) 320.

Peptides and Protein Hormones

Separate keyword: → *Interferons*

WOLFGANG KÖNIG, Hoechst Aktiengesellschaft, Frankfurt am Main, Federal Republic of Germany

1.	Introduction	1340
1.1.	Synthesis of Peptide and Protein Hormones	1341
1.2.	Mechanism of Action of Peptides and Protein Hormones	1341
1.3.	Nomenclature and Abbreviations	1342
2.	Gonadoliberin, Thyroliberin, Gonadotropins, Thyrotropin, Inhibin, and Related Hormones	1348
2.1.	Gonadoliberin	1350
2.2.	Thyroliberin	1353
2.3.	Gonadotropins	1355
2.4.	Thyrotropin	1357
2.5.	Inhibins, Activins, and Related Hormones	1358
2.5.1.	Inhibins	1358
2.5.2.	Activins	1359
2.5.3.	Transforming Growth Factor-β	1359
2.5.4.	Bone Morphogenetic Proteins and Osteogenic Proteins	1361
2.5.5.	Muellerian Inhibiting Substance	1361
2.6.	Follistatin	1362
3.	Parathyroid Hormone and the Calcitonin Family	1362
3.1.	Parathyroid Hormone and PTH-Related Protein	1363
3.2.	Calcitonin	1365
3.3.	Calcitonin Gene Related Peptide	1367
3.4.	Amylin	1369
4.	Corticoliberin – Proopiomelanocortin Cascade	1369
4.1.	Corticoliberin	1371
4.2.	Corticotropin	1373
4.3.	Melanotropins	1376
4.4.	Opioid Peptides	1379
4.4.1.	β-Endorphin	1380
4.4.2.	Preproenkephalins A and B	1382
4.4.2.1.	Peptides Derived from Preproenkephalin A	1382
4.4.2.2.	Peptides Derived from Preproenkephalin B	1384
4.4.3.	Dermorphin and Deltorphins	1385
4.4.4.	Exorphins	1386
4.4.5.	Femarfarmamide and Related Structures	1387
5.	Blood Pressure Regulating Peptides	1388
5.1.	The Angiotensin – Kinin System	1388
5.1.1.	Angiotensins	1389
5.1.2.	Kinins	1392
5.2.	Substance P, Neurokinins, and Tachykinins	1394
5.2.1.	Introduction	1394
5.2.2.	Amyloid A4 Protein	1397
5.3.	Vasopressin, Oxytocin, and Melanostatin	1398
5.3.1.	Vasopressin and Oxytocin	1398
5.3.2.	Melanostatin	1403
5.4.	Endothelin	1403
5.5.	Atrial Natriuretic Factor	1405
5.6.	Pancreatic Peptide Family	1408
5.6.1.	Pancreatic Peptide	1408

5.6.2.	Neuropeptide Y	1409	7.4.2.	Glicentin and Enteroglucagon	1428
5.6.3.	Peptide with N-terminal Tyrosine and C-terminal Tyrosinamide	1411	7.4.3.	Glucagon-Like Peptide-1-(7–36)amide	1428
5.6.4.	Neuromedin U	1411	**7.5.**	**Gastrin Inhibiting Peptide**	1429
5.6.5.	Delta-Sleep-Inducing Peptide	1412	**7.6.**	**Growth Hormone Releasing Hormone and the Growth Hormone Cascade**	1430
6.	**Cholecystokinin and Gastrin**	1413			
6.1.	**Trypsin-Sensitive Cholecystokinin Releasing Peptide**	1413	7.6.1.	Galanine	1431
			7.6.2.	Growth Hormone Releasing Hormone	1432
6.2.	**Cholecystokinin and Cerulein**	1414	7.6.3.	Growth Hormone, Placental Lactogen, and Prolactin	1434
6.3.	**Gastrin Releasing Peptide and Bombesin**	1416	7.6.3.1.	Growth Hormone	1435
6.4.	**Gastrin**	1418	7.6.3.2.	Prolactin	1437
7.	**Secretin Family**	1421	7.6.3.3.	Placental Lactogen	1439
7.1.	**Secretin**	1421	7.6.4.	Somatomedins, Insulin, and Relaxin	1439
7.2.	**Vasoactive Intestinal Peptide**	1423	7.6.4.1.	Somatomedins	1439
7.3.	**Pituitary Adenylate Cyclase Activating Polypeptide (PACAP)**	1425	7.6.4.2.	Insulin	1442
			7.6.4.3.	Relaxin	1445
			7.6.5.	Somatostatin	1446
7.4.	**Glucagon and Related Peptides**	1426	7.6.6.	Pancreastatin	1449
7.4.1.	Glucagon	1426	**8.**	**References**	1449

1. Introduction

In the past, peptide hormones were mainly classified according to their site of synthesis or release, e.g., peptide hormones of the hypothalamus, pituitary gland, gastrointestinal tract, or thyroid gland. This classification is no longer expedient because peptide hormones are not just produced at a single site in the body. Classification according to clinical indications is also unsuitable because these hormones can develop different activities depending on the site of action and on the dose. Peptides that are formed in endocrine cells can have an endocrine or a paracrine function. An *endocrine peptide* is secreted directly into the bloodstream and produces an effect at a distant receptor. A peptide exerts a *paracrine effect* if, after being released, it affects the metabolism of either its own cell or a neighboring cell. Peptides that exert a paracrine effect are also referred to as peptide regulatory factors (PRFs) [1].

An attempt will be made to classify structurally similar peptides into families and, at the same time, to integrate the activity cascades [2].

1.1. Synthesis of Peptide and Protein Hormones [3]

Peptide and protein hormones are synthesized in the cell in the form of large prepropeptides which are then cleaved and modified by enzymes to give active products [4]. Many biologically active peptides have an amide group at the C-terminus. This amide group is formed from the C-terminal glycine by peptidylglycine-α-amidating monooxygenase.

Naturally occurring peptides can be isolated by conventional separation techniques (precipitation, chromatography etc.). However, the demand for increased purity and economic considerations have resulted in the development of methods for synthesizing peptides and proteins. R. GEIGER calculated that peptides up to a chain length of about 20 amino acids could be produced economically using peptide chemistry [3]. For peptides and proteins with a chain length exceeding 20 amino acids, genetic engineering methods are preferred.

Peptides are usually synthesized by condensation of amino acids starting at the C-terminus. Synthesis may be performed in solution or using a solid phase. In solid-phase synthesis the C-terminus of the amino acid is bound to a polymeric support. Various groups are used to protect side-chain functional groups. The hydroxyl groups of tyrosine, serine, and threonine are mostly protected by the formation of ethers. Amino groups are blocked by urethane protecting groups and carboxyl groups are esterified with alcohols. Peptide synthesis using proteases is becoming increasingly important alongside the purely chemical methods.

Foreign peptides and proteins can also be produced in genetically altered microorganisms (e.g., *Escherichia coli*) [5], [6]. Plasmids or cosmids are used to introduce synthetic, semisynthetic, or natural genetic material (DNA) into the microorganisms. The foreign DNA is then transcribed into messenger RNA which is finally translated into peptides and proteins by the protein synthesis "machinery" of the microorganism.

DNA recombination using genetic engineering is also often used for cloning DNA sequences whose structure is to be elucidated. The structures of many proteins and peptides have been deduced from the base sequences of the DNA used to direct their synthesis.

1.2. Mechanism of Action of Peptides and Protein Hormones

Peptide and protein hormones bind to specific receptors where they stimulate the synthesis and activation of enzymes via second messengers such as intracellular calcium, cyclic 3′,5′-adenosine monophosphate (cAMP), cyclic 3′,5′-guanosine monophosphate (cGMP), inositol triphosphate, diacylglycerol, or lysophosphatidylinositol.

The hormones also stimulate the release of prostaglandins, steroid hormones, thyroid hormones, peptides, and glycoproteins. Occasionally, internalization of the receptor may occur in which excess peptide hormones causes disappearance of the receptors (receptor down regulation). This results in temporary stimulation and subsequent inhibition of the target cells.

Peptide *agonists* are peptide hormone analogues which stimulate the cell. *Antagonists* are competitive analogues which bind to the peptide receptors, and thus compete with but do not stimulate the cell. They compete with the agonists for the peptide receptor and thus inhibit the activation of the cell stimulated by agonists without receptor down regulation.

1.3. Nomenclature and Abbreviations

According to IUPAC, amino acids can be abbreviated with three-letter [5] or one-letter symbols [6]. The one-letter nomenclature is becoming increasingly popular. Both systems are used here.

The regulatory peptides [7] and their analogues [8] will also be designated in accordance with IUPAC, if possible. Small letters in front of the name of the peptide refer to the origin of the peptide and provide information on the species. The first letters of the latin name are used for this purpose. The position of an amino acid in a peptide chain is denoted by a superscript, after the three-letter symbol (e.g., His^5). Regions within peptide chains are indicated in parentheses, e.g., β-EP-(1 – 27) denotes amino acids in positions 1 to 27 of β-endorphin.

One-letter symbols for L-amino acids [a]

A	alanine
B	aspartic acid or asparagine[b]
C	cysteine
D	aspartic acid
E	glutamic acid
F	phenylalanine
G	glycine
H	histidine
I	isoleucine
K	lysine
L	leucine
M	methionine
N	asparagine
Ng	asparagine with *N*-glycosidic carbohydrate residue
P	proline
Ph	4-hydroxyproline
Q	glutamine

*Q	pyroglutamic acid
R	arginine
S	serine
Sg	serine with *O*-glycosidic carbohydrate residue
T	threonine
V	valine
W	tryptophan
X	unknown amino acid
Xa	amino acid amide
X-ol	amino acid with carboxyl group reduced to alcohol function
Y	tyrosine
Ys	*O*-sulfated tyrosine
Z	glutamic acid or glutamine[b]
/	amino acid deletion

[a] D-amino acids are denoted by the corresponding small letters
[b] uncertainty in structural elucidation

Three-letter and other symbols for natural and unnatural amino acids and protecting groups

Abu	α-aminobutyric acid
Ac	acetyl
Aca	ε-aminocaproic acid
Ace	1-aminocyclopropane-1-carboxylic acid
Acm	acetamidomethyl
Ada	adamantylalanine
Agm	agmatin
Ahx	6-aminohexanoic acid
Aib	α-aminoisobutyric acid
Ala	alanine
Ams	unknown amino acid
Ams-NH$_2$	amino acid amide
Ams-ol	amino acid with carboxyl group reduced to alcohol function
Aoa	8-aminooctanoic acid
Aoc	azabicyclo[3.3.0]octane-3-carboxylic acid
Arg	arginine
Asn	asparagine
Asp	aspartic acid
Asu	α-aminosuberic acid
Asx	aspartic acid or asparagine[b]
Aza	denotes amino acid in which the α-CH group is replaced by N
Bip	biphenylalanine
Boc	*tert*-butoxycarbonyl
Bzl	benzyl
Car	carboranylalanine
Chg	cyclohexylglycine

	Cit	citrulline
	4-ClPhe	4-chlorophenylalanine
	Cys	cysteine
	de	denotes a deletion
	Dip	4-β-diphenylalanine
	Et	ethyl
	4-FPhe	4-fluorophenylalanine
	Gln	glutamine
	Glu	glutamic acid
	Glx	glutamic acid or glutamine[b]
	Gly	glycine
	Har	homoarginine
	His	histidine
	Hmp	L-2-hydroxy-3-mercaptopropionic acid
	Hyp	4-hydroxyproline
	Ile	isoleucine
	imBzl	imidobenzyl
	Lac	L-lactic acid
	Leu	leucine
	Lys	lysine
	Me	methyl
	Met	methionine
	Met (O)	methionine sulfoxide
	Met (O_2)	methionine sulfone
	Mpr	3-mercaptopropionic acid
	Nal	3-(2-naphthyl)alanine
	Nic	nicotinoyl
	Nle	norleucine
	Oic	octahydroindolecarboxylic acid
	Orn	ornithine
	Pal	3-(3-pyridyl)alanine
	Peg	phenethylglycine
	Pen	penicillinamine
	Pgl	phenylglycine
	pGlu	pyroglutamic acid
	Phe	phenylalanine
	Phe(Cl_2)	3,4-dichlorophenylalanine
	Phe(NO_2)	4-nitrophenylalanine
	Phg	phenylglycine
	Pic	picolyl
	Pmp	β,β-cyclopentamethylene-3-mercaptopropionic acid
	Pro	proline
	Rha	α-L-rhamnopyranosyl
	Sar	sarcosine (N-methylglycine)
	Ser	serine
	Sta	statine
	Thi	β-(2-thienyl)alanine

Thr	threonine
Tic	tetrahydroisoquinolinecarboxylic acid
Trp	tryptophan
Tyr	tyrosine
Tyr (Me)	*O*-methyltyrosine
Z	benzyloxycarbonyl
Δ	dehydro
Ψ	denotes a substitution of the peptide bond

Prefixes indicating species origin

anq-	eel (*Anquilla*)
b-	cow (*Bos*)
ca-	dog (*Canis*)
cap-	goat (*Capra*)
cav-	guinea pig (*Caviidae*)
cu-	rabbit (*Cuniculus*)
g-	chicken (*Gallina*)
go-	gudgeon (*Gobius*)
h-	man (*Homo*)
m-	mouse (*Mus*)
o-	sheep (*Ovis*)
p-	pig (*Porcus*)
pl-	flounder (*Platichtys flesus*)
r-	rat (*Rattus*)
ra	European common frog (*Rana temporaria*)
sa-	salmon (*Salmo*)
xe-	xenopus (*Xenopus laevis*)

Abbreviations for peptides and other compounds

A4	amyloid A4 protein
ACE	angiotensin converting enzyme
ACTH	corticotropin (adenocorticotropic hormone)
AMP	adenosine 5′-monophosphate
ANF	atrial natriuretic factor
ANP	atrial natriuretic peptide
APP	amyloid precursor protein
AT	angiotensin
ATG	angiotensinogen
ATP	adenosine 5′-triphosphate
ATV	activin
BB	bombesin
BK	bradykinin
BMP	bone morphogenic protein
BNP	brain natriuretic peptide
cAMP	cyclic 3′,5′-adenosine monophosphate

Peptides and Protein Hormones

CCK	cholecystokinin
CCKRP	cholecystokinin releasing peptide
CG	chorionic gonadotropin
cGMP	cyclic 3′,5′-guanosine monophosphate
CGRP	calcitonin gene related peptide
CLIP	corticotropin-like intermediate lobe peptide
CM	casomorphin
CNP	C-type natriuretic peptide
CPON	C flanking peptide of neuropeptide Y
CRH	corticoliberin (corticotropin releasing hormone)
CRL	cerulein
CT	calcitonin
CTP	celltropin
DFT	dentinal fluid transport stimulating peptide
DM	dermorphin
DP	dynorphin
DSIP	delta-sleep-inducing peptide
DT	deltorphin
EDRF	epidermis derived relaxing factor (nitrogen monoxide)
EGF	epidermal growth factor
EK	enkephalin
EP	endorphin
ET	endothelin
FMRF	femarfarmamide
FSH	follitropin (follicle stimulating hormone)
GABA	γ-aminobutyric acid
GAL	galanine
GAP	gonadotropin releasing hormone associated peptide
GG	glucagon
GH	growth hormone (also known as somatotropin)
GHRH	growth hormone releasing hormone (also known as somatoliberin)
GIP	gastric inhibitory peptide
GLI	glicentin
GLP	glucagon-like peptide
GMP	guanosine 5′-monophosphate
GnRH	gonadoliberin (gonadotropin releasing hormone)
GnTH	gonadotropin (gonadotropic hormone)
GRP	gastrin releasing peptide
GT	gastrin
IGF	insulin-like growth factor
IGFBP	insulin-like growth factor binding protein
IHB	inhibin
IL	interleukin
KD	kallidin
KG	kininogen
Lac	lactic acid
LH	lutropin (luteinizing hormone)

LHRH	lutropin releasing hormone (luteinizing hormone releasing hormone)
LPH	lipotropin (lipolytic hormone)
MG	menopausal gonadotropin (urogonadotropin)
MIF	melanostatin (melanotropin inhibiting factor)
MIS	muellerian inhibiting substance
MPF	melanotropin potentiating factor
MRF	melanoliberin (melanotropin releasing factor)
MSH	melanotropin (melanocytes stimulating hormone)
MT	motilin
NK	neurokinin
NMU	neuromedin U
NPK	neuropeptide K
NPY	neuropeptide Y
NT	neurotensin
OP	osteopoetin
OT	oxytocin
PACAP	pituitary adenylate cyclase activating polypeptide
PDGF	platelet derived growth factor
PG	prostaglandin
PH	parotid hormone
PHI	peptide with N-terminal histidine and C-terminal isoleucinamide
PHM	peptide with N-terminal histidine and C-terminal methioninamide
PIF	prolactin inhibiting factor
PL	placental lactogen
PLA_2	phospholipase A_2
PLC	prolactin related cDNA
PLF	proliferin
PLP	prolactin-like proteins
PMSG	pregnant mare serum gonadotropin
POMC	proopiomelanocortin
PP	pancreatic peptide
PRC	prolactin related cDNA
PRL	prolactin
PRP	proliferin related protein
PSP	pancreatic spasmolytic peptide
PST	pancreastatin
PTH	parathyroid hormone
PTHrP	parathyroid hormone related protein
PTSI	pancreatic secretory inhibitor
PYY	peptide with N-terminal tyrosine and C-terminal tyrosinamide
RLX	relaxin
SEC	secretin
SL	somatolactin
SM	somatomedin
SP	substance P
SRH	somatoliberin (somatotropin releasing hormone; growth hormone releasing hormone, GHRH)

SRIH	somatostatin (somatotropin release inhibiting hormone)	
STH	somatotropin (somatotropic hormone; growth hormone, GH)	
T_3	triiodothyronine	
T_4	thyroxine	
TGF	transforming growth factor	
TNF	tumor necrosis factor	
TRH	thyroliberin (thyrotropin releasing hormone)	
TSH	thyrotropin (thyroid stimulating hormone)	
VIP	vasoactive intestinal peptide	
VP	vasopressin	
VT	vasotocin	
vWF	von Willebrand factor	

2. Gonadoliberin, Thyroliberin, Gonadotropins, Thyrotropin, Inhibin, and Related Hormones

Thyroliberin (thyrotropin releasing hormone, TRH), and gonadoliberin (gonadotropin releasing hormone, GnRH) are formed in the hypothalamus and release thyrotropin (thyroid stimulating hormone, TSH) and the gonadotropins (follitropin, follicle stimulating hormone, FSH) or lutropin (luteinizing hormone, LH) respectively from the anterior pituitary gland (adenohypophysis).

TSH stimulates the formation of the thyroid hormones, triiodothyronine (T_3) and thyroxine (T_4) in the thyroid gland; T_3 and T_4, in turn inhibit the release of TSH but stimulate the release of TRH (Fig. 1).

The gonadotropins LH and FSH stimulate the synthesis of the steroid sex hormones in the gonads (Fig. 2). These steroid hormones, in turn, inhibit the release of gonadotropins, but stimulate the release of GnRH. In the gonads, FSH stimulates the release of the inhibins (IHB), which exert a negative feedback to inhibit further secretion of FSH from the hypophysis. The hormone follistatin formed in the gonads also inhibits the release of FSH. The activins (homo- and heterodimers of the IHB β-chains), like the transforming growth factor β (TGF-β), stimulate the formation of FSH.

The tripeptide TRH and the decapeptide GnRH have structural similarities:

GnRH:	*Q	H	W	S	Y	G	L	R	P	Ga
TRH:	*Q	H	/	/	/	/	/	/	Pa

The gonadotropins (LH and FSH) and TSH also belong to one family. They consist of two noncovalently bound glycoproteins termed the α- and the β-chain; the α-chains of TSH and gonadotropins are identical.

TGF-β, the inhibins, the muellerian inhibiting substance, and the bone morphogenic proteins are structurally related. The inhibins consist of two glycoprotein chains that

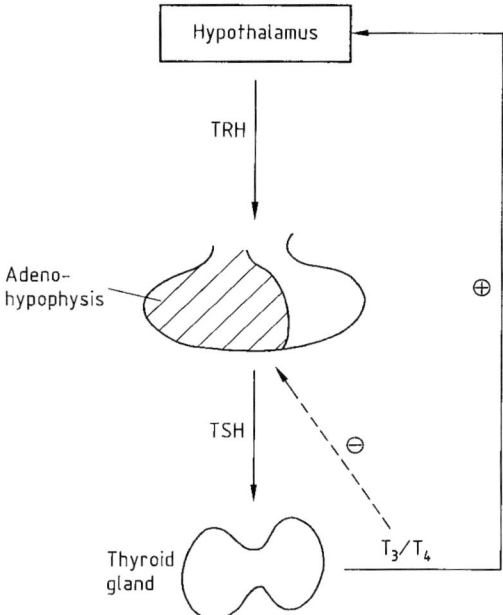

Figure 1. Control of thyroid hormone release

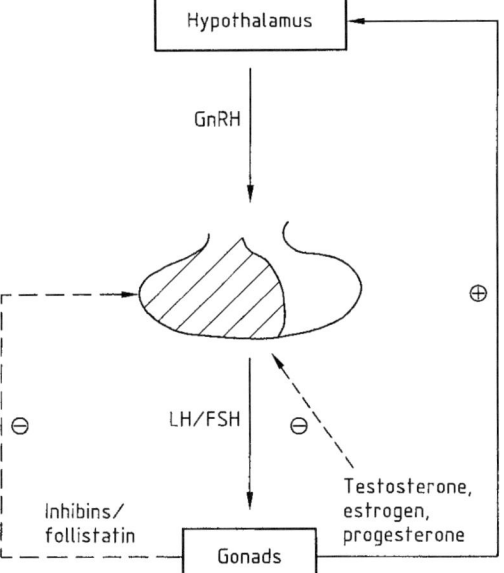

Figure 2. Control of steroid sex hormone release

are linked together by disulfide bridges; their α-chains are identical. The activins are homo- and heterodimers of the inhibin β-chains. TGF-β is a homodimeric compound.

Follistatin is a single-chain glycoprotein which has no structural similarity to the TGF-β/inhibin family.

2.1. Gonadoliberin

Occurrence. The decapeptide gonadoliberin [9034-38-2], [9034-40-6], also known as gonadotropin releasing hormone (GnRH), M_r 1182, is formed as a precursor peptide consisting of 92 amino acids (prepro-GnRH) [9]. The C-terminal peptide of prepro-GnRH (gonadotropin releasing hormone associated peptide, GAP) consists of 56 amino acids and inhibits prolactin synthesis [10]. GnRH is found not only in the hypothalamus, but also in the brain, heart, liver, pancreas, kidneys, adrenals, small intestine, gonads, and in the milk of lactating rats [11].

No differences are observed in the GnRHs of different mammalian species. Two GnRHs (g-GnRH I and II) are found in chicken, fish [12], and amphibians [13]. α-Factors that are similar to GnRH have been isolated from microorganisms (e.g., brewer's yeast) and cause cell aggregation and the exchange of chromosomes.

h-,p-,o-GnRH	*Q	H	W	S	Y	G	L	R	P	Ga		
g-GnRH I	*Q	H	W	S	Y	G	L	Q	P	Ga		
g-GnRH II	*Q	H	W	S	H	G	W	Y	P	Ga		
α-Factor α1	W	H	W	/	L	Q	L	K	P	G	Q	P	M	Y

Release [14]. The synthesis of GnRH from its precursor requires a peptidylglycine α-amidating monooxygenase that is dependent on Cu^{2+} and ascorbic acid [15]. GnRH is secreted from the hypothalamus in short pulses at intervals of 60–120 min and reaches the anterior pituitary via the hypophyseal portal system where it induces the release of the gonadotropins LH and FSH.

Although gonadotropins should be expected to inhibit release of GnRH by a negative feedback mechanism, they stimulate the release of GnRH from GnRH-containing neurons of male rats [16].

The steroid hormones of the gonads (testosterone, estrogen, and progesterone) stimulate the release of GnRH [17]. The post-mortem hypothalamic level of GnRH is lowered in women after the menopause [18]. Vasoactive intestinal peptide (VIP) [19], neuropeptide Y (NPY) [20], thymosin-$β_4$, and compounds of the arachidonic acid cascade [21] also stimulate the release of GnRH.

Prolactin [14], stress, and the stress hormones (CRH, ACTH, β-endorphin, oxytocin, and glucocorticoids) inhibit the formation of GnRH [14], [22]–[25]. The effect of steroid hormones depends on the endogenous opioids. Testosterone stimulates the release of GnRH when endogenous opioids are blocked and inhibits GnRH release when the endogenous opioids are stimulated [26].

Receptors [27]. Receptors for GnRH are found not only in the pituitary, but also in the gonads, placenta [28], corpus luteum [29], and on the oocytes [30]. In human breast cancer tissue, two classes of [D-Trp^6]GnRH receptors were detected: one with a high affinity and low capacity and the other with a low affinity and a high capacity. These correlate positively with estrogen and progesterone binding [31].

In female rats, physiological doses of GnRH stimulate the formation of GnRH receptors in the hypophysis [32], high doses have the opposite effect [33]. Highly active GnRH agonists reduce the number of GnRH receptors, while the GnRH antagonists have no effect [33]. GnRH agonists are rapidly taken up into the cell by endocytosis, while the antagonists remain on the membrane [34].

Biological Effects. Under physiological conditions (pulsatile application every 60–120 min), GnRH stimulates the calcium-dependent hypophyseal synthesis and release [35] of LH and FSH in the pituitary. However, high nonphysiological doses suppress LH and FSH secretion in rats and humans by a down regulation of GnRH receptors, this process does not require calcium [33].

Estrogen stimulates the GnRH-induced secretion of LH when low doses of GnRH are infused, but inhibits secretion at high GnRH doses [36]. Estradiol, insulin [33], adrenaline, noradrenaline [37], and neuropeptide Y (NPY) [38] enhance the LH-releasing effect of GnRH. Progesterone and dihydrotestosterone, as well as stress and glucocorticoids in female rats [39]–[41] inhibit the GnRH-induced release of LH.

GnRH directly affects steroid synthesis in the gonads of experimental animals. Small doses or short treatment have a stimulating [42] and high doses or long-term treatment an inhibitory effect [43].

In chondrasarcoma and other tumors [44], high doses of GnRH agonists inhibit tumor growth. Similar to physical castration, chemical castration with high doses of GnRH agonists regenerates the thymus in old male rats [45].

Structure–Activity Relationships [46], [47]. *Gonadoliberin Agonists.* In mammals substitution of His^5 for Tyr^5 and Trp^7 for Leu^7 gives peptides with full GnRH activity: [His^5]-GnRH is twice as effective as GnRH in releasing FSH [48].

The activity of GnRH is increased 100–200-fold by substituting an ethylamine residue for Gly^{10}-amide and by replacing Gly^6 by D-amino acids. It is important that the D-amino acid should contain an α-CH and a β-CH$_2$ group at position 6. Still higher activities can be achieved with synthetic aromatic D-amino acids, e.g., D-3-(2,4,6-(Me)$_3$Phe) [49]. A combination of the hydrophilic AzaGly-amide in position 10 with hydrophobic aromatic amino acids led to [D-Nal^6, AzaGly^{10}]GnRH, one of the most active compounds now available (230 times more active than GnRH) [50]. Hydrophilic D-Ser-*O*-glycosides [51] or β-Asp α-esters [52] in position 6 also give very active compounds.

Gonadoliberin Antagonists. Modification at positions 1, 2, 3, 6, and 10 lead to GnRH antagonists. Highly active compounds are:

1) Ac-[D-Nal^1,D-4-ClPhe^2,D-Trp^3,D-Har(Et)$_2^6$,D-Ala^{10}] GnRH (detirelix) [53], [54]
2) Ac-[D-Nal^1,D-4-ClPhe^2,D-Pal^3,D-Har(Et)$_2^6$,Har(Et)$_2^8$, D-Ala^{10}]GnRH (RS-26306) [47]
3) Ac-[D-Nal^1,D-4-ClPhe^2,D-Trp^3,D-Ser(Rha)^6,AzaGly^{10}] GnRH (HOE 013) [55]
4) Ac-[D-Nal^1,D-4-ClPhe^2,D-Pal^3,Arg^5,D-4-methoxybenzoyl-Abu^6,D-Ala^{10}]GnRH (ORF 21 243 = "Nal-Glu") [56], [57]
5) Ac-[D-Nal^1,D-4-ClPhe^2,D-Pal^3,Lys(Pic)^5,D-Lys(Pic)^6, *N*-isopropyl-Lys^8,D-Ala^{10}]GnRH (ORF 23 541, "Nal-Lys" = antide) [58]
6) Ac-[D-Nal^1,D-4-ClPhe^2,D-Pal^3,D-Cit^6,D-Ala^{10}]GnRH (SB-75, D 20 453, Cetrorelix) [59]

A problem associated with GnRH antagonists is the anaphylactic, histamine-releasing effect, mainly caused by the additional basic charges at position 6 and by the strongly lipophilic N-terminal amino acids [60]. This effect is reduced if the arginine in position 6 is moved to position 5 (e.g., compound 4), D-Trp^3 is replaced by D-Pal^3 (e.g., 2, 4–6) and the arginine in position 6 or 8 is modified (e.g., 2) or suitably substituted (e.g., 3, 4, 6) [59], [61]–[64].

These antagonists protect germ cells from X-ray damage by reversibly blocking the GnRH receptors [65].

Uses [47]. Cryptorchism is treated by the intranasal application of GnRH (gonadorelin; Kryptocur, Hoechst) [66].

Delayed puberty, hypogonadism, anorexia nervosa, Kallmann syndrome, and secondary amenorrhea can be treated with a pulsatile GnRH infusion (Cyclomat, Ferring Arzneimittel). GnRH is also used in the differential diagnosis of fertility disturbances.

In anorexia nervosa, the parenteral application of 5 µg of triptorelin every second day increased the release of LH and initiated the menstrual cycle [67]. A case of acute intermittent porphyria could be controlled by the subcutaneous administration of 5 µg/day of histrelin [68].

Analogues that have a longer and stronger effect are used predominantly in high doses to suppress the gonads. The following GnRH analogues are either on the market or being clinically tested.

1) [*tert*-Butyl-D-Ser^6]GnRH-(1–9)nonapeptide-ethylamide (buserelin [*57982-77-1*])
2) [*tert*-Butyl-D-Ser^6,AzaGly^{10}]GnRH (goserelin [*65807-02-5*]) [69]
3) [D-Trp^6]GnRH (triptorelin [*57773-63-4*]) [70]
4) [D-Trp^6]GnRH-(1–9)nonapeptide-ethylamide (deslorelin)
5) [D-Leu^6]GnRH-(1–9)nonapeptide-ethylamide (leuprolid acetate, leuprorelin [*74381-53-6*]) [71]
6) [D-Trp^6,N-MeLeu^7]GnRH-(1–9)nonapeptide-ethylamide (lutrelin [72])
7) [D-Nal^6]GnRH (nafarelin [*76932-56-4*]) [73]
8) [D-His(imBzl)^6]GnRH-(1–9)nonapeptide-ethylamide (histrelin) [74], [75])

GnRH agonists can be applied parenterally, nasally, or as a biodegradable implant. They are being clinically tested or applied for the inhibition of testosterone- and estrogen-dependent tumors (carcinoma of the prostate and the breast), in endometriosis, as contraceptives [76], in pubertas precox [77], and for in vitro fertilization. Other indications are regression of benevolent uterus myoma and leiomyoma [78], hirsutism [79], premenstrual intermittent porphyria (disturbance of heme synthesis) [80], and normal prostate hyperplasia [81].

In women, suppressive therapy with GnRH agonists lowers the mineral content of the bones [82].

2.2. Thyroliberin [2]

Occurrence. Thyroliberin [*24305-27-9*], also known as thyrotropin releasing hormone or factor (TRH or TRF), M_r 362, is found primarily in the brain, gastrointestinal tract, pancreas, and prostate. It is formed as prepro-TRH; r-prepro-TRH consists of 225 amino acids and contains 5 copies of the sequence:
– K R Q H P G K R –
which contains the pro-TRH sequence QHPG. Prepro-TRH is cleaved by endopeptidases to release pro-TRH. The N-terminal glutamine of pro-TRH is cyclized to pyroglutamic acid either spontaneously or enzymatically and the C-terminal glycine is converted to the amide function with a peptidylglycine-α-amidating monooxygenase. The peptides located between the TRH sequences are called the prepro-TRH connecting peptides.

The anorexigenic peptide, pGlu-His-Gly-OH, isolated from the urine of women suffering from anorexia nervosa, and a metabolite of TRH, cyclo-(His-Pro), have a central appetite-inhibiting effect [83]. Another structurally similar peptide, pGlu-Pro-amide, was found in rabbit prostate and human sperm [84].

Release. TRH is released centrally by electric shocks [85], 6-hydroxydopamine [86], dopamine [87], cold [88], and thyroid hormones [89]. It is inhibited by opioid peptides [90]. In the periphery, TRH release is increased by histamine [91], testosterone [92], and serotonin [93], and decreased by T_3 [94], carbachol [93], streptozotocin [95], and dopamine [96].

An enzyme (thyroliberinase, pyroglutamyl-peptidase II, E.C.3.4.19) that specifically cleaves the pyroglutamyl–histidine bond of TRH is found in brain homogenates [97]. It deactivates TRH and thus decreases TSH and PRL synthesis.

Receptors. The receptors responsible for TSH secretion are down regulated by chronic pulsatile TRH application, the receptors for PRL secretion are not as sensitive [98].

Thyroidectomy [99] and thyroxine treatment [100] increase the number of TRH binding sites in the rat pituitary. Both κ- and δ-opioids inhibit the binding of radioactively labeled [N^τ-MeHis^2]TRH to brain membranes, but μ-opioids are inactive [101]. The blood pressure of spontaneously hypertensive rats correlates positively with the number of TRH receptors in the brain, which could be responsible for the pathophysiology of high blood pressure [102].

Chlorodiazepoxide is a competitive antagonist of the TRH receptor in GH3 cells [103] and the rat amygdala [104]. Other TRH receptor antagonists are diazepam, midazolam, and (–)-4-methylmidazolam [105].

Biological Effects [106]. *Biochemical and metabolic functions* comprise the stimulation of 1,2-diacylglycerol and inositol-1,4,5-triphosphate levels, the increase of intracellular calcium, and the decrease of serum calcium.

Endocrine functions. TRH secreted in the hypothalamus stimulates the release of thyrotropin (TSH) and prolactin (PRL) in the adenohypophysis of the pituitary. The TRH-stimulated TSH secretion can be inhibited by somatostatin, dopamine [107], and calcitonin [108]. The TRH-stimulated release of PRL is enhanced by estrogen and inhibited by testosterone [109] and dopamine. TRH increases the release of T_3, T_4, and thyroglobulins from the thyroid gland either via TSH [110] or directly [111]. In the pancreas it inhibits the release of glucagon.

Vascular effects are species specific [112]. Vasodilatation occurs via activation of the vagus and vascular contraction occurs via activation of the sympathetic nervous system [113]. Blood pressure is increased [114].

Effects on the nervous system [115] include the potentiation of K^+-stimulated acetylcholine release [116], reversal of natural or narcotic sedation, anticonvulsant activity, inhibition of cortical neuron activity (antinociceptive action) [117], muscle tone improvement, as well as increased blood pressure, heart rate, cerebral blood flow, renal renin release [118], and catecholamine release from the adrenal medulla.

TRH induces hypo- or hyperthermia (depending on species) and reverses induced hypothermia. It also increases the respiratory rate.

Gastrointestinal effects comprise decreased food and water intake, increased gastric secretion (gastric ulcer) and gastric emptying, increased or decreased motility (depending on species), vagus-mediated contraction of the stomach [119], and increased pancreatic exocrine secretion.

Structure–Activity Relationships. Only four synthetic analogues are more effective than TRH in releasing TSH. [2-(L-β-(pyrazolyl-1)Ala]TRH and [1-(3-carboxy-5-oxothiamorpholine)]TRH are 1.5 times more active than TRH. [2-(N^τ-MeHis]TRH and the corresponding N-amylamide [120] have eight times and approximately ten times the potency of TRH, respectively.

Many TRH analogues exhibit the CNS activities of TRH, without releasing TSH to the same extent [121]–[130].

1) L-Pyro-2-aminoadipyl-His-thiaPro-amide (MK 771)
2) 6-Methyl-5-oxothiamorpholinyl-3-carbonyl-His-Pro-amide (montirelin, CG 3703) [122]
3) [(S)-4-Oxo-2-azetidinyl)-carbonyl-His-Pro-amide (YM-14673, azetirelin dihydrate) [124]
4) γ-Butyrolactone-γ-carbonyl-His-Pro-amide (DN-1417) [125]

MK-771 and CG-3703 are stable to the TRH-specific pyroglutamylpeptidase II, but DN-1417 is degraded faster than TRH. All three are competitive inhibitors of pyroglutamylpeptidase II [130].

Uses. TRH (generic name Protirelin) is used as a diagnostic reagent for thyroid and hypophyseal functions (Relefact TRH, Hoechst; Antepan, Henning; Thyroliberin/TRH, Merck). TRH preparations for parenteral (0.2–0.4 mg), oral (40 mg), and intranasal (1 mg, Relefact TRH nasal) application are available.

Possible therapeutic applications are listed below [106]:

1) Motor neuron disease, amyotrophic lateral sclerosis (ALS)
2) Spinal injury
3) Cerebral ischaemia (stroke)
4) Circulatory shock
5) CNS trauma
6) Alzheimer's disease, senile dementia

7) Down's syndrome
8) Schizophrenia
9) Depression
10) Antinociception/analgesia (with morphine)
11) Respiration
12) Reversal of anesthesia and narcotic overdose
13) Spinocerebellar degeneration
14) Epilepsy

2.3. Gonadotropins [131]

Occurrence. The gonadotropins (gonadotropic hormones, GnTH) are glycoproteins and comprise lutropin [*9002-67-9*] (luteinizing hormone, LH), M_r (protein part) 23 113 (h-LH); follitropin [*9002-68-0*] (follicle stimulating hormone, FSH), M_r (protein part) 22 395 (h-FSH); and chorionic gonadotropin [*9002-61-3*] (CG), M_r (protein part) 25 726 (h-CG). LH and FSH are formed in the adenohypophysis and CG is produced in the chorion. They consist of two (α and β) protein chains, each chain contains disulfide bridges. Disulfide bridges between the two chains do not exist. The α-chains of all three hormones are identical but species dependent (89 amino acids in h-LH and h-FSH, 92 amino acids in h-CG) [132]. The β-chains (145 amino acids in h-CG, 121 in h-LH, and 111 in h-FSH) are responsible for activity [133]. The chains are synthesized separately as procompounds.

The structures of GnTHs from humans, horse, sheep, cow, and pig as well as LH sequences from rat, rabbit, salmon, and pike eel have been elucidated [134].

Thyrotropin (TSH) is structurally related to the gonadotropins because it has an identical α-chain.

Release. The action of pulsatile GnRH on the gonadotropic cells of the adenohypophysis controls the release of the gonadotropins, LH, and FSH (see Section 2.1). FSH is preferentially released at a slow pulse frequency and LH is preferentially released at a faster GnRH frequency. Centrally applied GnRH lowers LH secretion [135]. Endothelin is as effective as GnRH in releasing the gonadotropins from anterior pituitary cells but this process does not involve GnRH receptors [136].

Low doses of estrogen and testosterone stimulate the release of gonadotropins by stimulating GnRH formation [137]–[139]. However, the long-term administration of higher doses directly inhibits gonadotropin release (negative feedback) [137], [140]. Calcitriol also potentiates GnRH-induced gonadotropin release [141]. Phospholipase A_2 [142], prostaglandin E_2 [143], and diacylglycerol [144] stimulate release, whereas cyclooxygenase inhibitors inhibit release [145].

The frequency of LH secretion in women increases from the early follicular phase to the mid and the late follicular phases of the menstrual cycle. The LH amplitude decreases in the midfollicular phase (negative feedback by estrogen). The LH pulse frequency and the amplitude sink during the luteal phase (effect of progesterone). Levels of FSH correlates with those of LH [146]. The inhibitory influence of the steroid hormones is lacking in the menopause and the level of LH and FSH rises. Treatment with progesterone [147] or ethinylestradiol [148] lowers the increased plasma gonadotropin levels. Testosterone is responsible for this effect in men [149]. FSH release is inhibited by the inhibins and stimulated by the activins and TGF-β.

Gonadotropin release is lowered by several factors, usually via the inhibition of GnRH release: deprivation of food [150], stress [151], morphine [152], [153], endogenous opioids [154], marihuana [155], glucocorticoids [156], dopamine agonists [157], alcohol [158], and lithium salts [159]. Central noradrenergic mechanisms inhibit the release of LH via the activation of β-adrenergic receptors [160].

Receptors. The h-LH/CG [161] and r-LH/CG [162] receptors have been characterized. The former consists of 696 amino acids with a signal peptide containing 27 amino acids.

FSH stimulates the formation of LH receptors in rat granulosa cells. Estrogen or testosterone is important for the FSH-stimulated synthesis of LH receptors [163]. Androgens [164] and interleukin-1 (IL-1) [165] inhibit the FSH-stimulated synthesis of LH receptors.

Biological Effects. In the ovaries, LH stimulates the synthesis of cAMP and inositol triphosphate [166] and reduces levels of cGMP by inhibiting guanylate cyclase [167].

The activities of LH and FSH complement each other. In the male organism, spermatogenesis can occur only if both hormones are present. FSH is responsible for sperm development, and LH stimulates the testosterone secretion in the interstitial cells, a process necessary for sperm maturation.

Although FSH activates follicular growth in women, ovulation and formation of the corpus luteum do not occur. Ovulation, corpus luteum formation, and estrogen secretion require the simultaneous presence of LH.

Structure – Activity Relationships. The intercystine loops in positions (38–57) of the β-chains of h-CG and h-LH slightly inhibit the binding of ^{125}I-h-CG to rat ovarian membranes and stimulate testosterone synthesis [168].

The tetrapeptides h-FSH-β-(34–37) and h-FSH-β-(49–52) inhibit the receptor binding of ^{125}I-h-FSH. The peptide h-FSH-(33–53) binds strongly and stimulates estradiol synthesis [169]. The peptides h-FSH-β-(81–95)amide and h-FSH-β-(31–45)amide bind tightly to the FSH receptor, the former stimulates estradiol synthesis in Sertoli cell cultures [170]. Partial sequences of the α-chain also displace h-CG from the receptor [171].

Antibodies against h-FSH-β-(76–118) or h-FSH-β-(1–33) bind to h-FSH, LH, and TSH [172]. A monoclonal antibody directed against h-CG binds in the vicinity of Lys^{45} of the α-chain and Asp^{112} of the β-chain. A peptide in which these two regions are combined stimulates the production of a specific antibody against h-CG and could be used for the development of a synthetic vaccine for castration [173].

The sugar residues on the peptide chains are of great importance for biological activity. Although deglycosylated h-CG binds more strongly to the receptors than intact h-CG, it does not activate cAMP synthesis in vitro and it inhibits the activity of glycosylated h-CG. However, deglycosylated or desialylated h-CH stimulates the synthesis of testosterone in vivo (monkeys) [174].

Uses [175]. h-CG or h-MG (menopausal gonadotropin, urogonadotropin, a mixture of h-CG and h-FSH), and h-FSH (Metrodin, Serono) are used predominantly in human medicine. h-CG (Choragon, Ferring; Predalon, Organon; Pregnesin, Serono; Primogonyl, Schering) and h-MG (Humegon, Organon; Pergonal, Serono) are used to treat gonadal disorders, e.g., primary and secondary amenorrhea, sterility, spermatogenesis, delayed puberty, cryptorchism, oligospermia, and menorrhagia. The main indication is polycystic ovary disease and in vitro fertilization [175]. Pulsatile therapy with Metrodin [176] or Pergonal [177] has advantages over daily injections.

2.4. Thyrotropin [178]

Occurrence. Thyrotropin [9002-71-5], also known as thyroid stimulating hormone (TSH), M_r (protein part) 22 813 (h-TSH), is formed in the adenohypophysis.

Release [179]. The release of TSH from the hypophysis is stimulated by TRH. High doses of the thyroid hormones, triiodothyronine (T_3) and thyroxine (T_4) inhibit TSH synthesis in the hypophysis by a negative feedback mechanism. Small doses of T_3 and T_4 stimulate the release of TSH via the activation of TRH. TSH release is also stimulated by calcitriol [180], prostaglandins [181], and Arg-vasopressin [182].

TSH release is inhibited by neuromedin B [183], glucocorticoids [184], estrogen, testosterone, α-endorphin [185], and somatostatin [186].

Receptors. The structures of the TSH receptor in the dog [187] and humans [188] have been established via the elucidation of the cDNA. The human receptor has 744 amino acids. There are sequence similarities between these receptors and the LH/h-CG receptors.

Biological Effects. TSH stimulates the synthesis of thyroglobulin and the synthesis and secretion of T_3 and T_4 in the thyroid gland through cAMP [189]. It enhances DNA synthesis and the proliferation of epithelial follicular cells from human thyroid tissue [190], a process that can be inhibited by somatostatin [191]. Lithium salts inhibit the TSH-stimulated synthesis of T_3 and T_4 and produce hypothyreosis in women [192].

Structure–Activity Relationships. The sequences (26–46), (31–45), and (21–35) of the α-chain [193] and (101–112), (71–85), (31–45), (41–55), and (1–5) of the h-TSH-β-chain [194] inhibit the binding of TSH to human thyroid membrane preparations.

2.5. Inhibins, Activins, and Related Hormones

The inhibins, activins, and the transforming growth factor-β (TGF-β) consist of two protein chains linked by disulfide bridges. The inhibins (IHB) are heterodimers consisting of an α- and a β-chain. The α-chains of IHB-A and IHB-B are identical and there are structural similarities between the α- and β-chains. Activin A and TGF-β are homodimeric compounds consisting of two identical chains of the β-type. Related proteins include muellerian inhibiting substance, bone morphogenetic proteins, and osteogenic proteins. They are high molecular mass glycoproteins consisting of 400–500 amino acids. Their C-terminal regions are very similar to those of TGF-β and the IHB β-chains.

2.5.1. Inhibins [195], [196]

Occurence. Inhibins occur mainly in the gonads. Other sites of synthesis are the placenta (in humans) [197] and the brain [198]. The structure of IHB-A from humans [M_r (protein part) 27 667], pigs, cows, and rats and that of IHB-B from humans [M_r (protein part) 27 508], pigs, and rats have been elucidated. The α-chain (134 amino acids in humans) and the β-chain (h-IHB-βA 115 amino acids, h-IHB-βB 116 amino acids) are formed from different messenger RNAs in the form of prepropeptides.

Release. Inhibin is released primarily by FSH, presumably via an increase in cAMP [199]. h-CG, LH, testosterone [200], [201], somatomedin C [202] and, in the case of hypophysectomized female rats, estradiol stimulate in vitro inhibin release. GnRH agonists inhibit both basal as well as PMSG-stimulated IHB release, and a GnRH antagonist potentiates the stimulating effect of PMSG on IHB release [203].

The plasma levels of both estradiol and inhibin are low during the early and mid follicular phase of menstruation and the FSH level is slightly elevated. In the late follicular phase, inhibin increases parallel to estradiol, while the levels of FSH and LH decrease and increase, respectively. Inhibin reaches a first maximum when LH and FSH exhibit a maximum [204]. The activity of IHB is reduced in the early luteal phase of the menstrual cycle (parallels that of estrogen) [204]. In the mid luteal phase, inhibin increases parallel to progesterone, while plasma levels of FSH and LH fall.

The plasma inhibin level in men suffering from idiopathic hypogonadotropic hypogonadism is low [205]. The level falls by 70% during treatment with h-CG, which is associated with a complete loss of endogenous FSH [206].

Female patients with ovarian granulosa cell tumors [207], polycystic ovarian diseases [208], and duodenal ulcers [209] have high serum inhibin levels.

Biological Effects. IHB-A and IHB-B inhibit the synthesis and release of FSH in the hypophysis by specifically inhibiting messenger RNA synthesis [210]. Inhibin suppresses the GnRH-induced formation of GnRH receptors in cell cultures of rat pituitary [211]. In vitro, it inhibits progesterone synthesis [212] and aromatase activity [213] in the gonads.

2.5.2. Activins [195]

Occurrence. Activins (ATV) are found in the ovarian follicular fluid and in leukemic cells. h-ATV-A [*104625-48-1*], M_r (protein part) 25 934, is a homodimer of the β-chain of IHB-A consisting of 115 amino acids each, [214], [215] and h-ATV-A-B [*114949-23-4*], M_r (protein part) 25 775 is a heterodimer consisting of the β-chains of IHB-A and IHB-B with 115 and 116 amino acids, respectively [216].

Follistatin [*117628-82-7*] is an activin-binding protein that has been isolated from rat ovaries [217].

Biological Effects. The activins stimulate the release of FSH from the hypophysis. Unlike GnRH, which releases FSH and LH very rapidly, the activins require >4 h to bring about the secretion of FSH. Activins also potentiate the FSH-induced aromatase activity which is important for estrogen synthesis [216], but inhibit progesterone synthesis [218].

Activin produced by leukemic cells also acts like erythropoietin (stimulates blood formation) [219].

2.5.3. Transforming Growth Factor-β [219], [220]

Occurrence. Transforming growth factor-β (TGF-β) [*98726-64-8*], M_r (protein part) 25 572 (h-TGF-β), consists of two identical peptide chains of 112 amino acids each, linked by disulfide bridges [221]. It is stored in platelets in the form of a high molecular mass, inactive complex [222]. TGF-β occurs widely; messenger RNA coding for TGF-β occurs in the liver cells of adult animals, but also in all cell lines studied so far [221].

Release. TGF-β is activated and released from its inactive storage complex in platelets by thrombin [223], acid, or urea (i.e., by blood clotting) [222]. Heparin cleaves the complex between TGF-β and α_2-macroglobulin formed in the blood plasma and, thus, potentiates the effect of TGF-β [224]. Bone-degrading factors (e.g., parathyroid hormone and calcitriol), release TGF-β from the skulls of rat fetuses and of newborn rats. Calcitonin decreases TGF-β activity [225]. Activated osteoclasts release TGF-β from the TGF-β complex formed by bone organ cultures [226]. Osteoclasts are cells responsible for bone degradation.

Receptors. Three different types [219] of receptor for TGF-β have been found in practically all cells that have been studied, they appear to be insensitive to down regulation. In osteoblasts (bone-forming cells) the binding of TGF-β to the receptors is reduced by parathyroid hormone [227].

Biological Effects. h-TGF-β stimulates both the secretion of FSH from rat anterior pituitary cells [228] and the aromatase activity in ovarian granulosa cells [213], [218], [229]. These effects are similar to those produced by the activins, and are inhibited by the epidermal growth factor-like tumor growth factor-α (TGF-α) [229].

In adrenal cells TGF-β inhibits cholesterol synthesis and thus basal and ACTH-stimulated steroidogenesis [230]. In rat pancreas cells, TGF-β_1 and TGF-β_2 stimulate insulin release, this process is glucose-dependent [231].

TGF-β is a general regulator of cell growth. Both TGF-α and TGF-β stimulate the growth of mesenchymal cell cultures via the induction of platelet derived growth factor (PDGF) [232]. However, TGF-β inhibits the growth of epithelial cells [233] and many different human cancer cell cultures [221], [234], [235]. It potentiates the PDGF-stimulated growth of fibroblasts, but inhibits growth of these cells stimulated by the epidermal growth factor (EGF) [235]. TGF-β inhibits or stimulates cell differentiation; it inhibits the differentiation of B-lymphocytes and muscle myoblasts [236], but stimulates the conversion of 3T3 fibroblasts to adipocytes.

Differentiation and proliferation of epithelial cells are enhanced by TGF-β and these cells exhibit increased production of fibronectin, collagen, and other cell-adhesive proteins [237]. Formation of receptors for these cell-binding proteins is also enhanced by TGF-β [219]. In smooth muscle cells from the porcine aorta, TGF-β stimulates the production of elastin [238]. In the extracellular matrix, TGF-β increases the expression of chondroitin and dermatan sulfate proteoglycan [239], it also inhibits protease activity induced by interleukin 1 (IL-1) and proteoglycan degradation [240].

TGF-β also acts as a regulator of the immune system. For example, it inhibits the basal and the TNF-α-stimulated (tumor necrosis factor α) adhesion of neutrophil blood cells to endothelial cells [241], inactivates macrophages [242], and inhibits the IL-1-stimulated degradation of cartilage [243].

TGF-β appears to be of importance for early bone development and bone repair. In rat fetal skulls it increases the synthesis of DNA and collagen [244], stimulates the formation of osteopontin in osteosarcoma cells [245], and inhibits the production of osteoclast-like cells in cell cultures [246]. The effect of TGF-β is inhibited by parathyroid hormone in osteoblasts [227].

Uses. TGF-β stimulates collagen matrix contraction by fibroblasts, a process that should accelerate the healing of wounds [247].

2.5.4. Bone Morphogenetic Proteins and Osteogenic Proteins

Occurrence. Decalcified bone matrix stimulates the formation of new bone [248]. Bone-growth-inducing substances isolated from decalcified bones include osteogenin [249], osteopoetin (OP) [250], and bone morphogenetic proteins (BMP) [251]. h-BMP-2A, h-BMP-2B, and h-BMP-3 contain 396, 408, and 427 amino acids, respectively [251]. The structures of h-BMP-2A and h-BMP-2B are very similar. The C-terminal ends of all three glycoproteins and the TGF-β/IHB family are also similar. Sequenced fragments of b-osteogenin have almost the same amino acid sequence as BMP-3 [252]. The glycoprotein, h-OP-1 contains 431 amino acids and a signal sequence of 29 amino acids [253].

Release. Vitamin D stimulates BMP synthesis [254]. BMP production decreases with age [255]. Extracts of rat chondrasarcoma and osteosarcoma show no BMP activity, and inhibit the bone-inducing activity of the BMPs [256].

Biological Effects. The BMPs initiate the formation of cartilage, resulting in the production of new bone. Recombinant h-BMP-2A [257] and isolated b-osteogenin [252] induce the formation of cartilage (chondrogenesis) and bone (osteogenesis) in rats. The activity of demineralized bone matrix is reduced in rats deficient in vitamin D [258]. Aluminum salts [259] and vitamin A [260] have an adverse effect on chondrogenesis and osteogenesis induced by demineralized bone matrix. In older rats, PDGF potentiates the cartilage and bone formation induced by demineralized bone matrix [261].

Uses. Gelatin that contains osteogenin corrects diaphyseal defects in sheep [262]. Bone defects in patients with cholesteatomas could be rectified with osteopoetin implants [250].

2.5.5. Muellerian Inhibiting Substance

Occurrence. The muellerian inhibiting substance (MIS) [*80497-65-0*] is a glycoprotein found in large amounts in the Sertoli cells of the testes of newborn calfs. Native MIS is probably a glycosilated disulfide-linked dimer (M_r ca. 140 000). h-Mis, M_r (protein chain) 58 000, consists of 535 amino acids, its C-terminal sequence shows similarities to the TGF-β/inhibin family [263].

Biological Effects. MIS inhibits the development of the muellerian duct during gonadal differentiation in the male embryo. In the female embryo the muellerian duct develops into fallopian tubes, uterus, and vagina. Complete virilization requires the presence of MIS. Preparations of b-MIS inhibit the in vitro and in vivo growth of tumor cells derived from the ovaries and the endometrium [263]. The growthinhibiting effect of MIS on the muellerian duct can be reversed in vitro by epidermal growth factor [264].

Structure–Activity Relationships. A C-terminal fragment (M_r 25 000) that is similar to TGF-β forms a complex with the N-terminal region of MIS. This complex can be made by limited digestion of MIS with pepsin and has full biological activity [265].

2.6. Follistatin [266]

Occurrence. Follistatins [*117628-82-7*] are single-chain glycoproteins that were originally isolated from ovarian follicular fluid. Two human follistatin precursors containing 344 and 317 amino acids give rise to two follistatins consisting of 315 (M_r ca. 34 800) and 288 amino acids (M_r ca. 31 600), respectively [267].

Follistatin has three homologous domains: amino acid sequences (66–135), (139–210), and (216–287). These domains have a ca. 50% sequence similarity both among themselves and with the human pancreatic secretory inhibitor (h-PTSI). The 309–314 region (ISSILE) is identical to a tyrosine kinase domain of the EGF receptor.

Biological Effects. In hypophyseal cell cultures, both follistatins inhibit the release of FSH. Like the inhibins, the follistatins also inhibit the FSH-stimulated synthesis of estrogen. Follistatin is a binding protein for the activins [268].

3. Parathyroid Hormone and the Calcitonin Family

Parathyroid hormone (PTH) and calcitonin (CT) are secreted from the parathyroid and thyroid glands, respectively. They regulate the plasma calcium level and bone formation. The plasma calcium level is increased by PTH and decreased by CT; PTH stimulates the turnover of bone and CT inhibits it. PTH also stimulates synthesis and release of calcitriol (1α,25-dihydroxyvitamin D_3) in the kidneys. Like high serum calcium levels, this substance inhibits release of PTH and stimulates the synthesis of calcitonin.

The reproductive hormones described in Chapter 2 also affect bone formation. TGF-β not only stimulates the release of FSH, but also initiates chondrogenesis and osteogenesis (i.e., synthesis of cartilage and bone) [269]. Lack of steroidal sex hormones (estrogen, testosterone, and progesterone), which are important for the release of gonadoliberin, leads to osteoporosis. The chaotic release pattern of PTH appears to be dependent on estrogen levels in women [270].

PTH-related peptide, PTHrP, is formed primarily in tumors and has properties similar to those of PTH. A CT-gene related peptide (CGRP) has also been found.

Amylin is a peptide produced by the β-cells of the pancreas that has a structural similarity to CGRP.

3.1. Parathyroid Hormone and PTH-Related Protein [271]–[273]

Occurrence. Parathyroid hormone (PTH) [9002-64-6] (IUPAC name parathyrin) is formed in the parathyroid gland but is also found in the brain and the pituitary gland. A PTH-related protein (PTHrP) with 141 amino acids (M_r 16 005, h-PTHrP) is formed in tumors that are accompanied by hypercalcemia [274] and in lectin-stimulated lymphocytes [275].

h-, b-, p-, and r-PTH consist of 84 amino acids (M_r 9425, h-PTH); g-PTH has 88 amino acids [276]. The precursor of PTH is prepro-PTH, a peptide containing 31 additional N-terminal amino acids [277].

Release of PTH. Low plasma calcium increases the synthesis and release of PTH, while high plasma calcium, an infusion of calcium, or a calcium diet [278] lowers the level of PTH. A circadian rhythm is observed with two daily PTH maxima followed by calcium peaks with a delay of 2 h [279]. PTH secretion is also stimulated by insulin-induced hypoglycemia [280], a phosphate-rich diet [281], calcium antagonists [282], a NaCl-rich diet [283], a diet low in vitamin D_3 and calcium [284], lithium [285], stress (induced hypocalcemia) [286], noradrenaline [287], histamine, 17-β-estradiol, and progesterone [288]. In osteoporotic women, the frequency and amplitude of the chaotic pulsatile microsecretion of PTH are greatly reduced after the menopause and estrogen therapy partly regenerates this pulsatile secretion [289].

PTH stimulates the synthesis of calcitriol in the kidneys. Calcitriol inhibits the synthesis and release of PTH [290]. See also → Hormones.

Magnesium slightly inhibits the release of PTH [291]. Dietary magnesium correlates positively with the bone mineral density. The release of PTH is also inhibited by calcium canal activators [282], cholinergic agonists [292], H_2-receptor blockers [293], and β-receptor blockers [294].

Occurrence of high PTH levels increases with age [295]. Raised levels of PTH are also observed in nephrocalcinosis, osteitis fibrosa cystica, chronic kidney failure [296], liver disease [297], myotonic dystrophy [298], obesity [299], and coronary arterial disease [300].

The release of PTHrP is reduced by glucocorticoids, calcitriol [301], and octreotide [302].

Receptors. PTH–receptor complexes are internalized by bone cells, thus lowering receptor density on the cell surface.

Biological Effects. The physiological role of PTH is to maintain the extracellular concentration of calcium (1.1 – 1.3 mmol/L) and to prevent hypocalcemia. It is also the principal regulator of bone turnover. PTH stimulates intracellular formation of cAMP and the phosphorylation of intracellular proteins, it also increases intracellular calcium levels.

Low doses of PTH stimulate bone synthesis and high doses result in bone resorption [303]. Hyperparathyroidism causes osteoporosis, while hypoparathyroidism results in increased bone formation [304].

In cultured, osteoblast-like cells, PTH stimulates DNA synthesis [305], alkaline phosphatase activity, amino acid incorporation [306], the release of mediators (e.g., interleukin 6) [307], which stimulate the bone-resorbing activity of the osteoclasts [308]. In osteoclasts, PTH stimulates carbonic anhydrase [309] and acid phosphatase [310].

In vivo, PTH also potentiates bone formation induced by insulin-like growth factor I [311] and the activity of TGF-β [312]. In the kidney, PTH stimulates the hydroxylation of 25-hydroxyvitamin D_3 to calcitriol which increases enteral absorption of calcium. High plasma calcium levels lower the formation of calcitriol [313] and low calcitriol levels are therefore observed in hyperparathyroid patients with high calcium levels [314]. In the kidneys, PTH causes calcium derived from bone to be returned to the blood and phosphate is excreted in the urine. After prolonged infusion of PTH (> 12 d), calcium and magnesium excretion is increased [315].

PTH lowers blood pressure and causes vasodilatation [316]; cAMP plays a role in the lowering of blood pressure [317]. PTH inhibits the entry of calcium into the cell which may also play a part in its vascular action [318]. PTH-(1 – 34) causes a marked reduction of the blood pressure of spontaneously hypertensive rats [319]. However, about 40% of patients suffering from hyperparathyroidism are hypertensive because PTH potentiates the hypertensive effect of hypercalcemia [320].

At high calcium concentrations PTH (unlike CT) enhances the arginine-induced secretion of glucagon, which can lead to hyperglycemia. Low PTH doses stimulate the glucose- and calcium-dependent release of insulin in isolated rat pancreas cells, high doses have an inhibitory effect.

Parenteral administration of PTH promotes the development of gastric ulcers, which can be prevented by CT [321]. In contrast, intracerebroventricular application of r-PTH-(1 – 34) inhibits the secretion of gastric acid and the development of gastric ulcers in rats [322].

The properties of PTHrP are similar to those of PTH [323]. It increases the level of calcium and stimulates the formation of calcitriol. The calcium-mobilizing activity of PTHrP-(1 – 34) amide is comparable with that of b-PTH-(1 – 34) [324].

Structure – Activity Relationships. The (1 – 34) N-terminal sequence of PTH contains all the structural requirements for full biological activity. Extension and shortening at the C-terminus of PTH-(1 – 34) results in loss of biological activity. The region (25 – 34) is required for receptor binding, while the first two amino acids at the amino end are of great importance for biological activity. g-PTH-(1 –

34), mainly differs structurally from b-PTH between positions 15 and 27 and has only one-tenth the biological activity [325].

Oxidation of Met^8 and Met^{18} in PTH-(1–34) abolishes its hypotensive and vasodilating effects without affecting its hypercalcemic effect. Oxidation of PTH-(1–34) reduces adenylate cyclase activation in renal blood capillaries, but not in the tubuli [326].

The activity of b-PTH-(1–34) is increased by the substitution of Tyr or D-Tyr for Phe^{34} and the replacement of the two Met residues by Nle. The blocking of the terminal carboxyl group by an amide function also enhances biological activity. Thus, [Nle^{8,18},D-Tyr^{34}]b-PTH-(1–34)amide is about four times more effective than b-PTH-(1–34).

PTH Antagonists. The peptides PTH-(3–34) and [Nle^{8,18},Tyr^{34}]b-PTH-(3–34)amide are weak in vitro inhibitors of PTH. [Tyr^{34}]b-PTH-(7–34)amide inhibits the PTH-induced excretion of phosphate in urine, cAMP formation [327], and the increase in plasma calcium levels [328]. Cleavage of PTH-(1–34) with cathepsin D yields PTH-(8–34), which also acts as an antagonist in vitro [329]. Subsitution of Gly^{12} by D-Trp leads to antagonists that are 10–30 times more effective [330].

Uses. PTH-(1–34) has been used in the treatment of osteoporosis. Alternating therapy with calcitriol [331] or sa-calcitonin [332] appears to be especially favorable. PTH-(1–34) can also serve as a diagnostic aid to differentiate between pseudohypoparathyroidism and hypoparathyroidism [333].

3.2. Calcitonin [334]

Occurence. Calcitonin (CT) [9007-12-9] is mainly synthesized in the C-cells of the thyroid gland, CT-like material is also found in the brain, hypothalamus, pituitary, lungs, thymus, liver, gastrointestinal tract, adrenals, muscle, parathyroid gland, cerebrospinal fluid, seminal fluid [335], and in breast milk [336]. The calcitonins are peptide amides containing 32 amino acids, M_r 3418 (h-CT). Marked structural differences are observed between species. Calcitonin is formed as prepro-CT which contains 109 amino acids. The structures of h-prepro-CT [337] and of g-prepro-CT [338] were elucidated via the cDNA.

Release of CT. CT is released from the C-cells in response to calcium, strontium, barium, cholinergic peptides (e.g., cholecystokinin and cerulein), secretin, and glucagon. Chronic hypercalcemia has an inhibitory effect on the acute calcium-stimulated release of CT [339]. 17-β-Estradiol and progesterone stimulate the in vitro secretion of CT from the C-cells of the thyroid glands of eight day old rats [340]. Somatostatin and low plasma levels of calcium inhibit the release of CT. Increased intracellular calcium and cAMP appear to be the second messengers for CT release [341].

Men have a higher plasma CT level than women [342]. A raised plasma level of CT is found in cancer of the thyroid gland [343], heroin addicts [344], and urticaria pigmentosa [345]. A reduced plasma CT level is found in hypothyroidism [346], Cushing's syndrome [347], Turner's syndrome [348], postmenopausal women [349], and hypogonadal osteoporotic men [350].

Receptors. Receptors for CT in osteoblast-like cells are increased by calcitriol [351] and decreased by CT (via internalization) [352].

Biological Effects. Low doses of CT stimulate cGMP formation. High doses activate adenylate cyclase (cAMP formation) and inhibit the release of cGMP [353], [354]. In mammals, CT lowers plasma calcium and phosphate levels by inhibiting bone resorption and promoting renal excretion [355].

In women, the ovarian steroid hormones potentiate the hypocalcemic effect of CT [356]. CT exerts antidiuretic and natriuretic effects [357], promoting the excretion of uric acid [358]. It also stimulates 25-hydroxyvitamin D_3-1α-hydroxylase (i.e., calcitriol synthesis in the kidneys) [359].

The mobility of osteoclasts (giant cells in the bone marrow that are responsible for bone resorption) is inhibited by CT [360] via increased uptake of calcium [361] and phosphate [362]. CT stimulates the proliferation of osteoblasts [363] and activates cartilage synthesis [364].

In humans, CT lowers the insulin level [365], reduces the suppressive effect of glucose on glucagon release, and thus increases blood sugar levels. CT inhibits the release of glucagon induced by arginine or hypoglycemia, as well as glucagon-stimulated glycogenolysis. This may explain why CT lowers plasma glucagon levels and the blood sugar concentration in insulin-dependent diabetics [366]. Intracerebroventricular administration of sa-CT increases the glucose-stimulated release of insulin in rats [367].

Fatty acid synthesis in the liver is increased by CT via calcium-dependent activation of fatty acid synthetase [368] and ATP citrate lyase [369].

CT has both stimulating and inhibitory effects on the hypophyseal hormones. It stimulates the release of the stress hormones ACTH [365] and β-endorphin [365] (analgesic effect) and thus release of cortisol from the adrenals [365]. CT inhibits the release of growth hormone [370], prolactin [371], lutropin, follitropin [372], and thyrotropin [373].

Subcutaneous application of CT reduces the stress-induced secretion of gastric acid and the formation of ulcers [374]. In the gastrointestinal tract, CT releases somatostatin, inhibits the release of gastrin [375], and increases motility [376]. Secretion of pancreatic enzymes and the contraction of the gallbladder are also reduced [377]. These effects may be due to action on the CNS because they are also obtained after intracerebroventricular application. Furthermore, central application also causes hypocalcemia in rats [334], delays the emptying of the stomach [378], and reduces absorption of food and water [379]. Intracerebroventricular application of CT increases blood pressure, heart rate [380], and body temperature [381], but intravenously applied CT has no effect [380].

Structure – Activity Relationships. In mammals sa- and anq-CTs produce a 10 – 30 times stronger hypocalcemic effect than that exerted by the mammalian peptides. The C-terminal prolinamide is especially important for biological activity. Substitution of the C-terminal carboxamido group or elimination of the C-terminal proline reduces activity by > 99%. Acylation of the amino group of the N-terminal cysteine or its substitution by hydrogen slightly increases activity. The N-terminus has a ring structure due to disulfide bond formation between Cys^1 and Cys^8. This is important for the activity of h-CT, but not for sa-CT [382]. The disulfide structure or the size of the ring is not essential for the activity of anq-CT [383]. The amino acids Ser^{29} and Thr^{31} are partly responsible for the high activity of sa-CT. In sa-CT, Val^8 can be replaced by Gly without loss of activity. However, this does not apply to Met^8 of h-CT [383].

de-Leu^{16}-sa-CT and de-Phe^{16}-h-CT exhibit only one- tenth of the corresponding CT activity in binding tests [384], whereas Tyr^{22} [383], [385], Leu^{19}, Ser^{13} [386], Gln^{20}, and Thr^{21} [385] of sa-CT can be deleted without loss of activity.

Active sa-CT analogues with a stable α-helical structure or with a low α-helicity exhibit high biological activity [387], [388]. The α-helical structure seems to be important for action on the kidney, but not for that on the brain [389].

Uses. CT can be administered by the subcutaneous, intramuscular, rectal, or intranasal route. Intranasal administration has the fewest side effects [390]. Nasal administration of only 50 IU of sa-CT per day inhibits bone resorption without having a hypoglycemic effect or increasing plasma cAMP levels [391].

sa-CT (Calcitonin L and Karil, Sandoz; Calsynar, Rover), h-CT (Cibacalcin, Ciba-Geigy), p-CT (Calcitonin S, Rover) and, in Japan, [1,7-L-aminosuberic acid]anq-CT (Elcatonin, Carbocalcitonin, and Turbocalcitonin, Toyo, Ioza) are used to treat Paget's disease, hypercalcemia, hyperparathyroidism, osteoporosis [392], osteolysis [393], the Sudeck syndrome, acute pancreatitis, chronic polyarthritis [394], and tumorosteolysis. CT is also used as a centrally acting analgesic [395].

3.3. Calcitonin Gene Related Peptide [396]

Occurrence. The messenger RNA for calcitonin gene related peptide (CGRP) [*83652-28-2*] is found primarily in the central and peripheral nervous systems [397], [398]. Circulating CGRP is formed primarily by the perivascular and cardiac nerves.

CGRP consists of 37 amino acids, M_r 3790 (h-α-CGRP). Two human and rat CGRPs have been observed: α-CGRP is coded in the α-gene and β-CGRP in the β-gene [399]. The α-gene is also responsible for coding CT.

Release. CGRP is released by capsaicin [400] and glucocorticoids [401]. Testosterone [402] and δ- and μ-opioid receptor agonists [403] inhibit the release of CGRP.

Biological Effects. CGRP is involved in sensory, motor, and autonomous nervous functions. It stimulates the formation of cAMP.

Intravenous injections result in a rapid decrease of blood pressure and an increase in the heart rate. CGRP is probably of physiological importance for the prevention of cerebral ischemia caused by excessive vasoconstriction and for the control of the peripheral resistance [404]. CGRP stimulates proliferation of human endothelial cells and may be important for the formation of new blood vessels, e.g., in the healing of wounds [405]. Noradrenaline release in mice is stimulated by intracerebroventricular application resulting in increased arterial blood pressure and tachycardia. In the lungs, CGRP exerts a higher bronchoconstricting effect than substance P [406]. Intradermal administration of CGRP causes a more pronounced reddening of the skin than substance P [407]. CGRP may potentiate the release of substance P from primary afferent terminals and promote the transmission of nociceptive information induced by mechanical noxious stimuli [408]. Administration of CGRP to rats inhibits secretion of growth hormone stimulated by a variety of agents [409]. Stress-induced release of prolactin (PRL) is reduced by intraperitoneal or subcutaneous application of CGRP [410].

In the adrenals, CGRP increases the secretion of cortisol and aldosterone and inhibits the release of Met-enkephalin [411].

CGRP appears to be a central mediator of ingestive behavior because central application suppresses food intake. Similar to CT, both the intravenous and the central application of CGRP inhibits gastric acid secretion [412]; this may be a direct effect or/ and mediated through the release of somatostatin [413]. CGRP relaxes the intestinal muscle [414] and inhibits the emptying of the stomach in rats [415]. Low doses of α- and β-CGRP have an antisecretory effect on the epithelium of the colon, but high doses stimulate secretion [416].

CGRP inhibits insulin secretion [417] and insulin-stimulated glycogen synthesis [418], but promotes glycogenolysis [418]. CGRP (and/or amylin) may be responsible for insulin resistance in type II diabetes mellitus.

CGRP inhibits proliferation of cells from the lymph nodes and the spleen of mice [419]. The effect of plasma calcium level and on bone metabolism is similar, but less potent, than that of CT.

Structure–Activity Relationships. The complete structure is required for full biological activity. Acetylation of Lys^{24}, and Lys^{35}, or the N-terminal amino group significantly reduces biological activity [420]. The N-terminal region appears to be important for activity and the C-terminal region for receptor binding.

The N-terminal fragments (1–12), (1–15), and (1–22) of CGRP reduce the blood pressure of rats anesthetized with urethane in a dose-dependent process [421]. h-α-CGRP-(12–37) has a weak antagonistic effect in guinea pig heart preparations [422].

3.4. Amylin [423]

Occurrence. Amylin [*112938-42-8*] (islet amyloid polypeptide), a peptide amide containing 37 amino acids, M_r 3903 (h-amylin), was isolated from amyloid deposits in the pancreatic islet cells of patients suffering from type II diabetes (non-insulin-dependent) [424]. There is a 46% structural similarity between amylin and CGRP. The disulfide ring and the C-terminal amide are important for full biological activity [425].

Amylin also occurs in normal β-cells of various mammals. Not only patients suffering from type II diabetes, but also healthy persons have amylin in their islet β-cells. However, amyloid deposits are only found in type II diabetics and not in healthy persons, indicating a pathogenic effect of amylin [426].

Release. Amylin, like insulin, is released from the rat pancreas by glucose or arginine stimulation [427]. The human plasma level of amylin is low in insulin-dependent diabetes [428].

Biological Effects. Amylin, like CGRP, inhibits the basal and insulin-stimulated glucose uptake and glycogen synthesis in skeletal muscle [429]. Amylin (and/or CGRP) may be responsible for insulin resistance in type II diabetes mellitus.

Amylin exerts similar effects to CT: it inhibits the osteoclastic resorption of bone at low levels of plasma calcium in rats and rabbits and lowers the plasma calcium level in rats [430].

4. Corticoliberin – Proopiomelanocortin Cascade

Corticoliberin (corticotropin releasing hormone or factor, CRH or CRF) and the peptides derived from proopiomelanocortin (POMC) are called the stress hormones because they are released in response to stress. The hypothalamic factor, CRH, stimulates cAMP synthesis and a cAMP-dependent protein kinase [431], resulting in secretion of POMC from the corticotrophic cells of the adenohypophysis, the endocrine cells of the pars intermedia, and central neurons that originate in the hypothalamus. The prohormone, POMC (h-POMC has 241 amino acids, M_r 26 678) is then enzymatically cleaved to form POMC-(1–108)-NH$_2$ (HP-N-108), corticotropin (ACTH), and β-lipotropin (β-LPH), see Figure 3. Further degradation of HP-N-108 gives POMC-(1–76) and HP-N-30 [POMC-(79–108)-amide] [432]. The former is degraded to POMC-(1–28) [433], δ-melanotropin (δ-MSH), and γ-MSH [434]. ACTH can be cleaved to form α-MSH and the corticotropin-like intermediate lobe peptide (CLIP). Cleavage of the N-terminal tetrapeptide of CLIP gives β-cell tropin (β-CTP), and β-LPH yields γ-LPH, β-MSH, β-endorphin (β-EP), and the melanotropin potentiating factor (MPF). Acetyla-

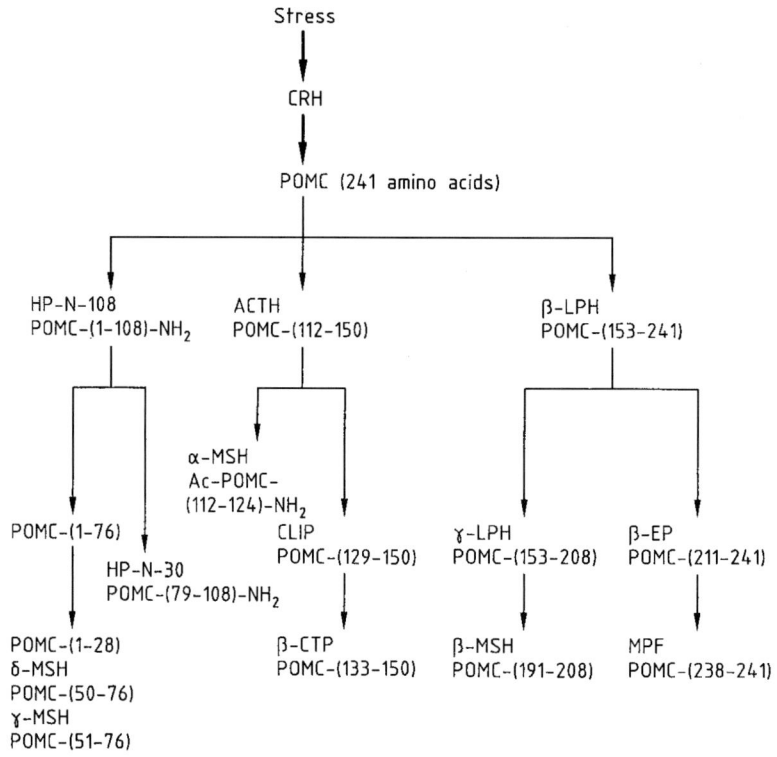

Figure 3. Degradation of proopiomelanocortin (POMC)

tion by opiomelanotropin acetyltransferase increases the biological activity of α-MSH, decreases the activity of ACTH, and abolishes the opiate activity of β-EP.

In the adrenal cortex, ACTH stimulates the formation of the glucocorticoids and α-MSH releases aldosterone in the presence of angiotensin II (AT II). The gluco- and mineralocorticoids play an important role in stress situations. The glucocorticoids enhance liver gluconeogenesis and exhibit immunosuppressive and anti-inflammatory activity. For further details, see → Hormones. The release of ACTH and β-EP is inhibited by the glucocorticoids via a negative feedback mechanism. The mineralocorticoid aldosterone exerts antidiuretic and antinatriuretic effects which are important during blood loss.

Both CRH and POMC-derived stress hormones usually inhibit release of the reproductive hormones [435] and enhance release of the atrial natriuretic factor (ANF). ANF then down regulates the stress hormones and stimulates release of the reproductive hormones; it is presumably important as a regulator in coping with stress. The androgens and gonadotropins stimulate POMC synthesis [436], whereas estrogen and testosterone inhibit the formation of the stress hormones [437]. Expression of the POMC gene in the testes correlates closely with the maturation of the Leydig cells. In contrast to the adenohypophysis, the POMC gene in the testes is not regulated by glucocorticoids [436].

Permanent stress probably desensitizes the immunosuppressive ACTH – glucocorticoid system. This condition is characterized by chronic fatigue and increased formation of antibodies against the body's own tissues, a preliminary stage of autoimmune diseases [438].

The structures of h- [439], r- [440], b- [441], sa-, and xe-POMC [442] have been elucidated via their DNA sequences.

4.1. Corticoliberin

Occurrence. Corticoliberin [9015-71-8], also known as corticotropin releasing hormone or factor (CRH or CRF), is found in nerve fibers throughout the brain, in the pancreas [443], adrenal medulla [444], placenta [445], stomach, and testes [446].

Although h-CRH is identical to r-CRH (peptide amide with 41 amino acids, M_r 4757), the precursor proteins of these two species have different sequences [447]. p-CRH [448] is very similar to h- and r-CRH.

Release. There is a circadian rhythm in the secretion of CRH. Similar to ACTH, the plasma level of CRH in humans peaks at 6 a.m. and has its lowest values at 6 and 10 p.m. [449]. CRH is released in response to stress (e.g., hunger [450], loss of blood [444], peripheral increase in noradrenaline [451]). Very small doses of catecholamines stimulate and large doses inhibit the central release of CRH [452].

The pulse frequency of CRH in the rat is greatly increased by the chronic administration of alcohol [453]. The release of CRH from the hypothalamus is also stimulated by morphine [454], acetylcholine [455], and interleukins [456]. Endogenous glucocorticoids appear to be important for the synthesis and release of CRH.

The release of CRH from the hypothalamus is inhibited by ACTH [457] and endogenous opioids [458] via a negative feedback mechanism. Very small doses (0.1 – 1 nmol/L) of β-endorphin stimulate the release of CRH in vitro, but higher doses (> 0.1 µmol/L) inhibit both basal and acetylcholine-stimulated release [459].

CRH is involved in the paracrine formation of ACTH in the placenta. Plasma CRH levels increase continually during pregnancy from the second trimester up to birth and fall rapidly after delivery [460].

In Alzheimer's disease, a reduced CRH level is observed in the brain [461]. Patients suffering from anorexia nervosa [462] and depression [463] exhibit enhanced release of CRH.

Receptors. CRH receptors are found in the adenohypophysis and in regions of the brain that are related to the limbic and autonomic nervous systems [464].

In spite of the increased plasma ACTH level in adrenal-ectomized rats, the number of CRH receptors in the hypophysis falls. This down regulation is due to enhanced secretion of vasopressin [465]. In contrast, the CRH receptors on membranes of the liver, spleen, testes, prostate, and pancreas [466] increase in number after adrenalectomy.

Biological Effects. CRH stimulates the synthesis and release of proopiomelanocortin as well as its degradation in the adenohypophysis and placenta [467] to form ACTH, α-melanotropin (α-MSH) [468], γ_3-MSH [469], and β-endorphin (β-EP). It also increases plasma levels of corticosterone, aldosterone [470], and 18-hydroxycorticosterone [471] via release of ACTH. CRH also directly potentiates ACTH-stimulated synthesis of corticosterone [472]. The physiological importance of CRH is emphasized by the fact that CRH antigens cause atrophy of the adrenal glands and lower plasma levels of ACTH and β-EP.

Extracellular Ca^{2+}, the entry of Ca^{2+} into the cell, calmodulin [473], and cAMP-dependent protein kinase [474] all play a role in the CRH-stimulated release of ACTH.

Intravenously applied CRH raises the plasma level of oxytocin and vasopressin, which potentiate the activity of CRH [475]. Angiotensin II [476] and gastrin-releasing peptide-(14–27) [477] also potentiate the ACTH-releasing effect of CRH. Somatostatin, melanophore concentrating hormone [478], delta-sleep-inducing peptide [479], glucocorticoids, morphine [480], and endogenous opiates inhibit the ACTH-secreting effect of CRH.

In vitro, CRH stimulates the release of endorphin and dynorphin from the rat pituitary [481]. Intravenous dynorphin A-(1–13) and intravenous or intradermal CRH inhibit the inflammatory response to scalding in rats independent of the pituitary and adrenal functions [482]. The anti-inflammatory effect is not dependent on the hypophysis, but on adrenal function [483].

CRH inhibits the reproductive functions of both sexes [484], [485]. The release of GnRH to the hypophysis from the hypothalamus is prevented by CRH [486]. The prolactin-dependent inhibition of GnRH release is partly due to the activation of CRH-containing neurons [487]. CRH also centrally inhibits the release of growth hormone by stimulating somatostatin release in the hypothalamus [488]. Intravenous CRH raises plasma levels of prolactin [489].

High intravenous doses of CRH selectively dilate the blood vessels of the mesenterium; this is responsible for a decline in peripheral resistance and systemic arterial blood pressure. CRH activates prostaglandin synthesis. The prostaglandins, in turn, inhibit the hypotensive effect of CRH by a negative feedback mechanism [490]. The central application of CRH raises blood pressure and pulse rate by increasing adrenaline and noradrenaline levels. In vivo, CRH potentiates the cholecystokinin-induced pancreatic secretion of protein and the secretin-induced secretion of HCO_3^- [491].

Intravenous CRH results in a strong and persistent lowering of the level of pentagastrin-stimulated gastric acid, while the plasma gastrin level increases [492]. Intraperitoneal CRH prevents formation of stress-induced gastric ulcers in rats [493]. CRH inhibits the emptying of the stomach [494], reduces the transit in the small intestine, increases the transit in the large intestine, and causes diarrhea [495].

In humans, intravenous CRH increases the breath-time volume, breath frequency [496], and attentiveness; it also improves the mood and well-being of depressive patients [497].

CRH exerts a central antipyretic effect [498]. In experimental animals (mainly rats) central application of CRH increases locomotoric activity, emotionalism [499], and cleaning behavior; reduces food uptake [500] and aggressiveness; and enhances defensive behavior [501].

Structure – Activity Relationships. The three amino acids at the N-terminus of CRH are not important for biological activity. However, vasodilating and ACTH-releasing activity fall to below 0.1% on further deletion of N-terminal amino acids.

The C-terminal region is of extreme importance for CRH activity. Both 41-deamido-o-CRH-(1 – 41) with a free C-terminal carboxyl group and o-CRH-(1 – 39)amide have < 0.1% of the activity of CRH. Met^{21} is of some significance for activity because [Met(O)^{21}]o-CRH exhibits only 10% of the biological activity. However, norleucine and norvaline can be substituted for this amino acid with practically no loss of activity.

Uses. Subcutaneous injection is advisable and is equally as effective as intravenous administration. Activity after intranasal administration is only about 1% of that obtained on parenteral administration [502].

A pulsatile injection of CRH normalizes ACTH and cortisol secretion after glucocorticoid therapy (secondary adrenal insufficiency) [503] and apparently does not desensitize the pituitary [504]. Pulsatile application (6.25 µg every 30 min) is preferable to long-term infusion (50 or 100 µg/4 h) [505].

CRH can be used as a diagnostic aid for hypophyseal function [506]. After intravenous administration of 100 µg of CRH, the plasma ACTH and cortisol levels in patients suffering from Cushing's syndrome increase to a higher extent than in healthy persons [507]. In depressive patients with a high level of cortisol, CRH produces a lower ACTH response than in normal persons [508]. An abnormal increase in growth hormone after application of CRH is observed in patients suffering from congenital thyrotropin deficiency [509].

4.2. Corticotropin [510]

Occurrence. Corticotropin [*9002-60-2*], also known as adrenocorticotropic hormone (ACTH), has 39 amino acids, M_r 4541 (h-ACTH), and is formed in the adenohypophysis from the precursor POMC. It is also found in the interstitial fluid of the testes [511] and in the pancreas [512]. Enzymatic degradation and acetylation of ACTH give rise to the N-terminal derivative α-melanotropin (α-MSH) and two C-terminal derivatives: the corticotropin-like intermediate lobe peptide (CLIP) and β-celltropin (β-CTP). ACTH sequences from various species are given in [513].

Release. ACTH is formed from the precursor POMC in response to stress, CRH, and stimuli that release CRH or potentiate its activity. The formation of ACTH is accompanied by an increase in cAMP (see Section 4.1) [510]. ACTH is secreted in micropulses lasting a few minutes. Pulse frequency remains the same throughout the day, but the amplitude is higher in the morning than in the afternoon and night [514], it is also higher in men than in women [515]. Serotonin increases the formation of ACTH and cortisol by stimulating release of CRH. It appears to be partly responsible for the circadian increase in cortisol [516].

In vivo, corticosteroids inhibit the secretion of ACTH [510] in the adenohypophysis and cerebrospinal fluid. Glucocorticoids do not affect the in vitro CRH-stimulated release of ACTH from the pituitary [517]. γ-Aminobutyric acid (GABA) appears to reduce the release of ACTH via inhibition of CRH release [516]. Increased salt intake depresses the formation of ACTH in response to stress or CRH [518].

Inhibitors of angiotensin converting enzyme inhibit the release of ACTH via decreased synthesis of angiotensin II.

ACTH secretion is increased in Cushing's syndrome, but lowered in Sheehan's syndrome. The circadian ACTH rhythm is disturbed in patients suffering from depression; hypersecretion of ACTH and cortisol is observed [519].

Receptors. An ACTH receptor (M_r 225 000) from mouse adrenal cells consists of four different subunits [520]. In humans, ACTH receptors have also been found on splenocytes and peripheral mononuclear leukocytes [521].

IGF-I potentiates the ACTH-stimulated formation of ACTH receptors [522] and TGF-β reduces the number of ACTH receptors on adrenocortical cells [523].

Biological Effects. In the adrenal cortex, ACTH stimulates the synthesis of glucocorticoids (cortisol, cortisone, corticosterone) and mineralocorticoids (cortexone and aldosterone) in a calcium-dependent process [524] that involves activation of adenylate cyclase. ACTH must be administered in the morning for maximum stimulation of glucocorticoid synthesis. The sensitivity of adrenal cells to ACTH is reduced when they are exposed to ACTH for longer periods [525].

The ACTH-stimulated secretion of aldosterone probably proceeds via angiotensin II [526]. ACTH also enhances the release of adrenaline and noradrenaline in the adrenal gland [527]. It exerts a trophic effect on the adrenocorticotropic cells of the adrenal gland. Adrenal hyperplasia is observed as a result of the hypersecretion of ACTH (Cushing's disease). The adrenal glands shrink as a result of ACTH insufficiency of the pituitary (Sheehan's syndrome).

Corticostatin from granulocytes is a peptide containing 34 amino acids that inhibits the ACTH-stimulated formation of corticosterone [528].

ACTH and N-terminal fragments of ACTH, e.g., ACTH-(4–10) [529] also have a therapeutic effect in cases of shock [530]. Cholinergic mechanisms in the central nervous system play an important role here [531].

ACTH-(1–24) stimulates the pancreatic secretion of $NaHCO_3$ and protein via cholinergic mechanisms [532]. The ACTH molecule also has insulinotropic and hypoglycemic activity [512]. However, it also potentiates the diabetogenic activity of growth hormone induced by hypoglycemia [533].

ACTH binds specifically to splenocytes, inhibiting the production of antibodies and interferon (direct immunosuppressive activity) [521].

ACTH-(4–10) is responsible for the influence of ACTH on behavior and on learning processes (see Section 4.3): it increases attentiveness, sexual activity, and the memory capacity of rats, at the same time reducing anxiety. Cholinergic neurons and muscarinic receptors play a role in ACTH-dependent behavior patterns [534].

The central application of ACTH, ACTH-(1–24), and α-MSH decreases food uptake [535]. Centrally applied ACTH increases blood pressure and heart rate, it also prevents persons from falling asleep, while de-Ac-α-MSH and CLIP intensify slow-wave sleep and paradoxical sleep [536].

Structure–Activity Relationships. A fragment as small as ACTH-(4–7) (10^{-4} mol/L) stimulates in vitro corticosterone synthesis. A dramatic increase in activity occurs with the inclusion of Met^3. However, the full activity of the N-terminal sequence is attained with a free amino group at the N-terminal Ser^1. Thus, ACTH-(1–10) is 100 times more active than ACTH-(5–10), and ACTH-(1–18) has the full activity of native ACTH.

ACTH-(1–24) (tetracosactid) is more active in releasing corticosteroids than ACTH in vitro. ACTH can be divided into at least four segments. Sequence (11–18) is very important for receptor binding. Sequence (4–10) is responsible for the corticotropic effect, and the N-terminal tripeptide (1–3) is the "amplifier". The C-terminal fragment, ACTH-(25–39), is responsible for antigenicity and safe transport. ACTH-(11–24) is a competitive antagonist for the corticotropic activity of ACTH. [Phe^9]ACTH-(1–24) also inhibits glycolysis and steroidogenesis.

ACTH from guinea pigs in which Pro^{24} is replaced by Ala has a higher aldosterone-releasing activity than hACTH or ACTH-(1–24)amide [513]. [Cys(carboxamidomethyl)^{25}]ACTH-(1–26) has three to four times the activity of ACTH in stimulating aldosterone secretion, but a lower corticosterone-stimulating activity. The substitution of D-Ser or β-Ala for Ser^1, of Lys for Arg^{17}, and of Lysin-amide or 1,4-diaminobutylamine for Arg^{18} gives long-acting analogues that have about up to eight times the activity of ACTH.

[β-Ala^1, Lys^{17}]ACTH-(1–17)-4-amino-n-butylamide (ACTH-(1–17), alsactide) is stable to aminopeptidases and carboxypeptidases and has a long duration of action.

CLIP exhibits insulin-releasing activity which decreases after shortening or acetylation of the molecule at the N-terminus [537]. CLIP, like α-MSH and de-Ac-α-MSH, lowers the β-endorphin-stimulated secretion of prolactin in rats [538].

Uses. Natural ACTH from the pig is marketed under the name Acethropan (Hoechst). The synthetic ACTH derivatives, tetracosactide (Synacthen, CIBA) and alsactide (Synchrodyn 1–17, Hoechst), are also used. ACTH preparations are used as diagnostic aids for the functioning of the adrenal cortex and as a therapeutic agent for insufficient functioning of the adrenal cortex, in multiple sclerosis, inflammatory rheumatic diseases, collagen diseases, acute gout, radicular pain syndrome, severe allergic skin diseases, and collitis ulcerosa. ACTH also has an antiemetic effect and is administered to cancer patients treated with *cis*-platinum [539].

4.3. Melanotropins

Occurrence. α-Melanotropin [37213-49-3], also known as α-melanocytes stimulating hormone (α-MSH), M_r 1665, is derived from POMC (POMC-112–124). In mammals its amino acid sequence is
Ac–S Y S M E H F R W G K P Va
It is found primarily in the hypophyseal pars intermedia and the hypothalamus.

γ- and δ-MSH are derived from POMC-(1–28) and β-MSH from γ-LPH (see Fig. 3). In many species, two different β-MSH peptides are found.

Release. Release of α-MSH is probably regulated through the enzymatic degradation of oxytocin as the prohormone for melanostatin (MIF) and melanoliberin (MRF). Stress and corticoliberin increase brain and plasma α-MSH levels. β-Endorphin can also release α-MSH in the central nervous system. The α-MSH is released in the septum in response to fever induced by interleukin-1, especially when the temperature increases (shivering phase) [540].

Increased levels of γ-MSH are observed in physiological stress, cardiovascular distress, blood loss [541], and cardiac arrest [542].

Biological Effects. Melanin is a black-brown skin pigment that consists mainly of polymerized dihydroxyphenylalanine. In melanocytes, α-MSH stimulates tyrosinase and thus the synthesis of melanin (melanogenesis); it also promotes pigment transport of pigment granules (melanosomes). Dispersion of the melanosomes through the numerous dendrites of the melanocytes causes the skin to darken, the skin is lightened by aggregation of melanosomes. Human skin becomes darker within 24 h after administration of α-MSH. Darkening of the skin is observed in kidney damage and diseases in which the plasma α-MSH and ACTH levels are increased (Cushing's, Addison's, Nelson's disease, and ACTH-secreting tumors). The skin becomes lighter in hypophyseal insufficiency. The C-terminal tetrapeptide of β-endorphin, the melanotropin potentiating factor (MPF), enhances the melanotropic activity of α-MSH.

The second messenger of α-MSH appears to be cAMP. Inhibitors of α-MSH (e.g., melatonin) inhibit the α-MSH-stimulated formation of cAMP but increase the concentration of cGMP. Melatonin, a hormone found in the pineal body, stimulates the aggregation of melanosomes.

α-MSH has many other activities in mammals [543]:

Increase in testicular Sertoli cell cAMP
Secretion of estradiol and plasminogen activator
Increase in lipogenesis and sebum production in the skin
Increase in lipolysis in adipose tissue
Increase in plasma free fatty acid levels
Increase in adrenal steroidogenesis related to fetal growth and development

Increase in aldosterone synthesis and secretion by the adrenal zona glomerulosa (angiotensin II seems to be important [544])
Increase in pineal serotonin levels
Decrease in pineal melatonin levels
Increase in growth hormone secretion from the pituitary
Inhibits stress and β-endorphin-stimulated prolactin release [545]
Increase in plasma luteinizing hormone [546]
Increase in plasma glucagon and insulin
Decrease in blood pressure
Decrease in plasma calcium levels
Decrease in bone resorption
Decrease in immunomodulatory and inflammatory activities of IL-1 [547]
Reduction of body temperature following intracerebroventricular or parenteral administration [548]
Behavior (increased arousal, attention, learning, memory retention, sexuality)
Improved nerve regeneration

h-β-MSH has the same activity as α-MSH in the Anolis skin test (darkening of the skin). β-MSH sequences from other species also show high melanotropic activity, which is usually slightly lower than that of α-MSH. Like α-MSH, β-MSH also raises the plasma levels of glucose, glucagon, insulin, and free fatty acids in the rabbit.

Of the γ-MSH peptides, Ac-γ$_1$-MSH exerts the highest melanotropic effect but still has < 0.1% of the α-MSH activity. γ$_2$-MSH inhibits the release of α-MSH from the hypophysis and reduces the β-endorphin-induced analgesic and hypothermic effect. The direct infusion of γ-MSH into the renal artery results in prompt excretion of sodium and potassium [549]. The ACTH-stimulated synthesis of glucocorticoids and aldosterone is potentiated by γ$_3$-MSH [550]. Intraventricular application of γ-MSH leads to a longterm increase in blood pressure [551]. The functions of POMC-(1–108) and its fragments are probably based on its growth-promoting effect on the adrenal cortex and its hypertensive effect.

Structure–Activity Relationships. The N-terminal acetyl group is important for the melanotropic activity of α-MSH. α-MSH can be subdivided into three regions:

1) The classical messenger sequence, His-Phe-Arg-Trp (α-MSH-(6–9))
2) The C-terminal tetrapeptide Gly-Lys-Pro-Val-amide which has seven times the melanotropic activity of α-MSH-(6–9)
3) The N-terminal sequence Ac-Ser-Tyr-Ser-Met-Glu which acts as a potentiator

Within the messenger sequence, -Phe-Arg- appears to be crucial for melanotropic activity. The minimum effective sequence is Ac-His-Phe-Arg-Trp-amide [552]. The lysine in the C-terminal tetrapeptide is important for activity. Met^4, Gly^{10}, and Pro^{12} are important for MSH activity [552]. If Met^4 and Gly^{10} are replaced by a cysteine disulfide bridge ([Cys^4,Cys^{10}]α-MSH), a cyclopeptide is obtained which contains the two important messenger sequences. This peptide is a superagonist and is 10 000 times more active than α-MSH in the frog skin test. The cyclic analogue Ac-[Cys^4,Cys^{10}]α-MSH-(4–10)amide is less active than α-MSH, whereas Ac-[Cys^4,Cys^{10}]α-MSH-(4–13)-amide is again superactive and has the same activity as [Cys^4,Cys^{10}]α-MSH in the frog skin test. Thus, the C-terminal sequence is of great significance for MSH activity. The disulfide bridge is important for biological activity. Reduction of the disulfide bridge results in a 1000–10 000-fold decrease in biological activity [553].

Met^4 is sensitive to oxidation, and can be replaced by norleucine without loss of activity. Treatment of α-MSH with hot alkali prolongs the duration of action. [Nle^4,D-Phe^7]-α-MSH is 60 times as active as α-MSH in the frog skin test. In mice, [Nle^4,D-Phe^7](4–11) or -(4–10)α-MSH has 100 times the activity of α-MSH [554]. Even Ac-[Nle^4,D-Phe^7]α-MSH-(4–9)amide is still ten times as active as α-MSH in the melanoma tyrosinase and lizard skin tests, but it is ten times less active in the frog skin test [555]. The central application of [Nle^4,D-Phe^7]α-MSH has an antipyretic effect that is ten times greater than that of α-MSH. The antipyretic effect after intravenous application is, however, not pronounced. The C-terminal tripeptide appears to be very important for antipyretic activity [556].

Cyclic compounds containing D-Phe^7, an acidic amino acid in position 5, and a basic amino acid in position 10 are more active than α-MSH [557].

The activity of [Cys^4,D-Phe^7,Cys^{10}]α-MSH on melanocytes is about as high as that of [Cys^4,Cys^{10}]α-MSH, but it acts considerably longer. [Cys^4,D-Phe^7,Cys^{10}]α-MSH-(1–12)amide has the same potency as [Cys^4,D-Phe^7, Cys^{10}]α-MSH, indicating that Val^{13} is not required for melanotropic activity [558].

An analogue of α-MSH-(4–9), Met(O$_2$)-Glu-His-Phe-D-Lys-Phe-OH (ORG-2766) [559], [560] is about 100–1000 times more active than α-MSH-(4–9) in promoting learning. However, high doses have the opposite effect. The subcutaneously or orally applied peptide has an antiamnesic effect, antagonizes pentobarbital anesthesia, and reduces morphine uptake in the brain. Met(O$_2$)-Glu-His-Phe-D-Lys-Phe-NH-(CH$_2$)$_8$-NH$_2$ (HOE 427, Ebiratide [561]) and Met(O)-Glu-His-Phe-D-Lys-Phe-NH-(CH$_2$)$_8$-NH$_2$ have 100 times the ORG-2766 activity in this test. In learning tests with rats, HOE 427 is about 500 times more active than ORG-2766 [562]; the sequence Phe-D-Lys-Phe appears to be especially important [563].

Antagonists [543]. Ac-α-MSH-(7–10)amide is a weak, selective α-MSH antagonist in the lizard skin bioassay. Ac-[D-Trp^7,D-Phe^{10}]α-MSH-(7–10)amide is a competitive inhibitor of α-MSH in the frog and lizard skin assays [564]. A strong antagonist for frog skins is the uncyclized peptide Ac-Nle-Asp-Trp-D-Phe-Nle-Trp-Lys-amide; the cyclic lactam of this compound is a full agonist. Other antagonists are the growth hormone releasing peptide, His-D-Trp-Ala-Trp-D-Phe-Lys-amide and its analogues.

Uses. [Nle^4,D-Phe^7]α-MSH (intermedin alpha) can be applied topically and is absorbed through the skin causing increased pigmentation in the yellow mouse [565]. It is being tested as a suntanning agent.

α-MSH-(6–9) exerted a positive transdermal effect (better mood, as well as less anxiety, pain, spasticity, and muscular weakness) in patients suffering from multiple sclerosis [566].

ORG-2766 improved the mood and level of performance of patients without influencing their sleep [554]. Even in elderly, mentally weak patients, ORG-2766 increased attentiveness and induced social behavior [567]. Ebiratide is being tested on patients suffering from Alzheimer's disease.

4.4. Opioid Peptides

Many peptides that have an effect on opioid receptors are formed from prohormones. β-*Endorphin* (β-EP, Section 4.4.1) is derived from proopiomelanocortin (Fig. 3), its activity is based on the N-terminal pentapeptide sequence Tyr-Gly-Gly-Phe-Met. This sequence is also found in *Met-enkephalin* (Met-EK), which is produced from prepro-EK A of the adrenal medulla. A series of *extended Met-EKs* that are structurally related to prepro-EK A have been isolated from the adrenal medulla (Section 4.4.2.1). The *extended Leu-EKs* isolated from the pituitary and hypothalamus (e.g., dynorphins, neoendorphins, and rimorphin) are derived from a common precursor, prepro-EK B (Section 4.4.2.2). Prepro-EK A and B are structurally similar.

The C-terminal tetrapeptide sequence of prepro-EK A is, apart from the missing amide group, identical to that of femarfarmamide (FMRF-amide), a peptide isolated from molluscs. FMRF-amide is not an opiate. It is not derived from prepro-EK A because the C-terminal glycine important for the formation of the amide group is missing. In molluscs, it is formed from a prepropeptide which contains 21 copies of the precursor FMRFG sequence.

Kyotorpin, a Tyr-Arg dipeptide with analgesic activity, was isolated from the bovine brain. It is probably derived from the propeptide neokyotorpin (Thr-Ser-Lys-Tyr-Arg).

The *dermorphins* and *deltorphins* (Section 4.4.3) are heptapeptide amides from the skin of the South American frogs *Phyllomedusa sauvagii*, *P. rhodei*, and *P. bicolor*. They contain D-alanine or D-methionine.

The *exorphins* (Section 4.4.4) are opioid peptides which are formed during food digestion. Examples are α- and β-casomorphins.

Modified morphines (e.g., naloxone or certain EK analogues) are antagonists for endogenous opioids and exogenous opiates. In a certain dosage range, cholecystokinin (CCK-8) also acts as an endogenous selective antagonist for the analgesic activity of opioids.

Opioid Receptors. Prior to the discovery of endogenous opioids, three opioid receptors were postulated: the μ-receptor (for morphine), κ-receptor (for ketocyclazocin), and σ-receptor (for SKF 10 047). The discovery of the enkephalins (EK) led to the introduction of the δ-receptors. β-EP prompted the postulation of the ε-receptors, and the naloxone-sensitive effect on thermoregulation gave rise to γ-receptors. Iota-receptors have been postulated in the dog and rabbit intestine, and λ-receptors with a high affinity for 4,5-epoxymorphinane have been described. The μ-receptors have been subdivided into $μ_1$- and $μ_2$-receptors. There are three types of κ-receptor [568].

All endogenous opioids are ligands for the μ-, δ-, κ-, and ε-receptors. Met-EK, Leu-EK, and β-EP are the main ligands for the μ- and δ- receptors. The dermorphins are specific for the μ-receptors, the deltorphins for the δ-receptors, and the dynorphins for the κ-receptors. β-EP binds primarily to the ε-receptors. Opiates (e.g., morphine) bind to the $μ_2$-receptors.

In vitro bioassays for the receptors are described in [569]. Selective ligands for the µ- and δ-receptors are described in [570]. Peptide ligands for the κ- and ε-receptors are given in [571]–[574].

Biological Activity. The most important effects mediated by the opioid receptors follow [575]:

$µ_1$-*mediated*: supraspinal analgesia, inhibition of gastrointestinal transit and motility [576], suppression of experimentally induced diarrhoea [576], and development of anorexia.

δ-*mediated*: spinal analgesia, inhibition of acetylcholine release from rat corpus striatum [577], impairment of avoidance learning in rats [578], and inhibition of SP-stimulated plasma extravasation and vasodilation [579].

κ-*mediated*: analgesia, increased food intake [580], ACTH release in rats [581], and diuresis (mediated by the adrenal medulla) [582].

ε-*mediated*: supraspinal analgesia.

σ-*mediated*: psychomimetic effects and behavioral changes.

4.4.1. β-Endorphin

Occurrence [583]. β-Endorphin (β-EP) [*60617-12-1*], M_r 3465 (h-β-EP), contains 31 amino acids and is derived from proopiomelanocortin (see Fig. 3). It has been isolated primarily from the pituitary, but is also found in human placenta cell cultures, ovaries [584], sperm [585], endometrium [586], gallbladder [587], pancreas, and small intestine.

β-EP derivatives such as β-EP-(1–27) (C fragment or δ-EP), Ac-β-EP, Ac-β-EP-(1–27) [588], and β-EP-(1–18) are found in the rat adenohypophysis. Acetyl-β-EP-(1–18) mainly occurs in the rat neurohypophysis [589]. In rats, the degree of acetylation of the hypothalamic endorphins increases with age [590].

The C-terminal tetrapeptide of β-EP (Lys-Lys-Gly-Glu in humans) is the melanotropin potentiating factor (MPF).

Release. β-EP is secreted in the hypophysis in response to stress, corticoliberin, angiotensin II, lipoxygenase or epoxygenase products [591], insulin-induced hypoglycemia (via cholinergic mechanisms [592]), adrenaline, food uptake, and chronic alcohol consumption [593]. β-EP is released from the mucosa of the small intestine by gastric acid or bile acid and from the gallbladder mucosa by CCK-8 [588]. Estrogen and testosterone increase the plasma level of β-EP [594]. Both h-CG and PMSG stimulate formation of β-EP in the ovaries [595] and in the testicular interstitial fluid [596].

In humans, somatostatin and oxytocin have no effect on the basal plasma β-EP level, but lower the increased levels of β-EP, β-LPH, and cortisol caused by insulin-induced hypoglycemia [597].

In humans, release of β-EP exhibits a circadian rhythm similar to that of cortisol: a high level in the morning and a low level at night [598]. The basal β-EP level is lower in people who bear a high risk of becoming alcoholics compared to those who do not. The plasma level of β-EP increases with alcohol consumption in the high risk group, but not in the low risk group [599]. Plasma levels of β-EP are significantly raised in depressive patients [600] and in obesity [601].

Biological Effects. β-EP has analgesic and lipolytic activity [602]. The intracerebroventricular application of CCK-8 [603] and α-MSH inhibits the analgesic activity of β-EP. β-EP lowers phosphodiesterase activity, somatostatin secretion in the isolated pancreas of the dog, the release of GnRH in the mesencephal central grey substance, plasma LH, the oxytocin release induced by sucking, and, in high doses, the formation of CRH, ACTH, LPH, and cortisol. Very low doses release CRH in rat hypothalami [604].

Intraventricular and intravenous application of β-EP causes release of growth hormone and prolactin. The β-EP-stimulated release of prolactin can be inhibited by α-MSH or CLIP [605]. β-EP increases plasma insulin, glucagon, and glucose, especially in obese humans [606].

Pharmacological doses of β-EP reduce the left ventricular systolic and diastolic pressure in the isolated rat heart [607]. Depending on the dose and conditions, the intracerebroventricular application of β-EP results in hyper- or hypothermia. Other central effects are the release of catecholamines, Met-EK [608], thyrotropin, vasopressin, and α-MSH as well as increased food uptake.

Structure – Activity Relationships. β-EP has a highly specific opiate recognition sequence at the N-terminus (positions 1–5) which is linked to an amphiphilic helix (positions 13–31) by a hydrophilic region (positions 6–12) [609]. Since the N-terminal Tyr-Gly bond of β-EP is more resistant to aminopeptidases than that of the EKs, β-EP has a longer duration of action. Acetylated sa-β-EPs have no activity in an opiate receptor assay.

Incorporation of Gln or Arg at position 8 doubles the analgesic effect, Trp in position 27 quadruples the analgesic effect. [Gln^8,Trp^{27}]h-β-EP has almost eight times the receptor binding ability of h-β-EP, but its analgesic effect corresponds to that of h-β-EP. The replacement of Glu^8 appears to be of great importance for receptor binding and may find application for the design of β-EP antagonists [610]. [Gln^8,Gly^{31}]h-β-EP-Gly-Gly-amide is a strong antagonist for β-EP-induced analgesia, it is 200 times more active than naloxone [611].

The substitutions D-Ala^2 and MePhe^4 increase the binding of the β-EPs to the δ-receptors. Increasing hydrophobicity at position 5 correlates with decreasing analgesic activity [612].

Substitution of the dermorphin sequence for the seven amino acids at the N-terminus gives a highly analgesic peptide which is 4.4 times more active than h-β-EP and about as active as dermorphin [612].

Analogues which do not show homology with β-EP in the twelve amino acids at the C-terminus but have the helical structure of the C-terminus are as active or more active than β-EP. Cysteine bridges between positions 14 and 26, 15 and 26, 16 and 26, as well as 17 and 26 are tolerated. These derivatives exhibit stronger receptor binding than h-β-EP [613].

β-EP-(1–27) has only 0.2% of the β-EP activity in the opiate receptor displacement assay. β-EP-(1–27) inhibits β-EP-induced analgesia and release of growth hormone [614], and the β-EP-induced hypothermia in the mouse [615], but not the β-EP-induced release of prolactin [614].

The N-terminal tyrosine is very important for the analgesic activity of β-EP: de-Tyr^1-β-EP does not bind to opiate receptors. h-β-EP-(6–31) inhibits β-EP-induced analgesia, the β-EP stimulated release of prolactin, but has no effect on the release of TSH [616]. h-β-EP-(6–31), h-β-EP-(28–31), and h-β-EP-(30–31) inhibits the α-MSH-induced grooming, stretching, and yawning syndrome, as well as β-EP-induced grooming and catatonia [617]. The C-terminus is of prime importance for the lipolytic activity of β-EP. Derivatives in which the two C-terminal amino acids are deleted have no lipolytic activity [618].

4.4.2. Preproenkephalins A and B

There is considerable structural similarity between the two preproenkephalins (pre-pro-EKs) isolated from endocrine and nerve tissue. The structures of h-, b-, r- [619], xe- [620] prepro-EK A and h- and b-prepro-EK B have been elucidated. Prepro-EK A is mainly cleaved to give Met-EK and extended Met-EK peptides (e.g., peptide E and adrenorphin). Prepro-EK B is cleaved to give Leu-EK, the neoendorphins, and the dynorphins.

Prepro-EKs are widely distributed in the body and are processed differently according to their location [621].

4.4.2.1. Peptides Derived from Preproenkephalin A

Occurrence. Preproenkephalin A (prepro-EK A) [*88895-24-3*], M_r 30 781 (h-prepro-EK A), contains 267 amino acids. Numerous peptides derived from prepro-EK A have been isolated from the adrenal medulla [622]; these peptides are in turn precursors of Met-EK (Tyr-Gly-Gly-Phe-Met, M_r 574) and Leu-EK (Tyr-Gly-Gly-Phe-Leu, M_r 556). Plasma proteins treated with pepsin also generate peptides related to Met-EK [623].

Release. The release of the enkephalins is stimulated by GABAnergic mechanisms, insulin-induced hypoglycemia [624], endotoxin shock [625], electroshock [626], intraventricular application of β-endorphin [627], and glucocorticoids [628]. Dopamine inhibits the formation of Met-EK [629]. Synthesis of messenger RNA for prepro-EK A is stimulated by stress [630], angiotensin II [631], FSH, and cAMP [632], but inhibited by chronic administration of glucocorticoids [630]. The plasma level of Met-EK increases considerably in marathon runners [633].

Met- and Leu-EK are rapidly degraded by aminopeptidases and enkephalinase, which cleaves the C-terminal dipeptide.

Biological Effects [634]. For a description of the activity of the enkephalins, see Section 4.4 (μ_1-, δ-, and κ-mediated opioid actions). Enkephalins inhibit transmitter release from nerve terminals in the central and peripheral nervous systems by blocking calcium channels [635].

Met- and Leu-EK mainly bind to δ-receptors and only have an analgesic effect if applied centrally. Analogues that preferentially bind to the μ- and δ-receptors exhibit an analgesic effect, even after peripheral application. The tachyphylaxis and substance dependency produced by morphine are also observed after chronic application of the highly active EK analogues.

Endocrine and exocrine pancreatic and gastrointestinal secretion are modulated by the enkephalins [636]. Enkephalins contract the lower esophagus and the pyloric sphincter [637], delay gastrointestinal transit [638], trigger gallbladder contraction [639], relax the Oddi sphincter [640], cause vasoconstriction in the lungs [641], and inhibit bladder motility [642].

Depending on the dosage and species, enkephalins exert a hypotensive [643] or hypertensive effect [644]. The intracerebroventricular application of μ- and δ-agonists to rats inhibits the release of oxytocin [645] and vasopressin (diuresis).

The enkephalins modulate the release of ACTH and cortisol via extrahypophyseal mechanisms [646]. They potentiate the ACTH-stimulated release of corticosterone [647] and inhibit the CRH-induced increase in the plasma levels of ACTH, β-EP, and cortisol [648].

The enkephalins increase the plasma level of prolactin by inhibiting dopamine release. They stimulate the release of growth hormone via the growth hormone releasing hormone.

The enkephalins inhibit ovulation and the release of lutropin by inhibiting GnRH synthesis. Endogenous opioids inhibit gonadotropin secretion; this is dependent on the gonadal steroids. The FSH-stimulated formation of progesterone is promoted by the enkephalins [649], while the secretion of testosterone from the testes of immature rats is suppressed [650]. In experimental animals, endogenous opioids inhibit the release of TRH and, thus, the formation of TSH [651].

The endogenous opioids can also act as immunostimulants [652] and immunosuppressants [653].

Structure–Activity Relationships. Tyr^1 is essential for opiate activity. However, its N-terminal amino group can be methylated, guanylated, or extended by amino acids without significantly affecting the activity. The *N*-allyl-EKs (particularly *N*-allyl-Met-EK) are, like naloxone, antagonists of morphine and EK.

Gly^2 can be replaced by α-aminoisobutyric acid or D-amino acids to give compounds that are considerably more active and more resistant to enzymatic degradation. The more lipophilic the substitution, the stronger the binding to the μ-receptors and, therefore, the analgesic activity. Substitution with hydrophilic groups increases affinity for the δ-receptors.

Gly^3 is important for the biological activity of the EKs. Replacement of Phe^4 by other amino acids leads to loss of activity. However, N-methylation or substitution by AzaPhe increases analgesic activity.

The more lipophilic the amino acid in position 5, the stronger the analgesic effect after parenteral administration. For example, the intracerebroventricular application of the *O*-galactosyl derivative of [D-Met^2,Hyp^5]EK-amide produces an analgesic effect that is 50 000 times that of morphine [654].

Position 5 does not necessarily have to be occupied by an amino acid: methioninol sulfoxide, the thiolactone of homocysteine, and substituted hydrazides are also suitable.

Shortening of the chain at the C-terminus leads to the tripeptide-N-methylphenethylamides or tripeptide-2-amino-4-methylpentane amides, with a still higher analgesic activity. H-Tyr-NH(Me)-$(CH_2)_4$-CO-NH(Me)-CH_2-CH_2-C_6H_5 also shows high analgesic activity when applied intravenously [655]. An EK analogue that is cyclized via a disulfide bridge is [D-Cys^2-D-Cys^5]EK-amide also has a powerful analgesic effect. All of the above highly active compounds are more stable to enzymatic degradation than the rapidly degradable natural EKs.

The hydrophilic EK analogues, [D-Arg^2,Phe(NO$_2$)^4, Pro^5]EK-amide (BW 942C) [656] and [D-Met(O)^2, Phe(NO$_2$)^4,Pro^5]EK-amide (nifaltide) [657] are effective against diarrhoea.

4.4.2.2. Peptides Derived from Preproenkephalin B

Occurrence. Preproenkephalin B (prepro-EK B) [88895-25-4], M_r 28 422 (h-prepro-EK B), contains 254 amino acids. The most important peptides derived from prepro-EK B are the pentapeptide Leu-enkephalin (Leu-EK), M_r 556 [658]; the longer dynorphins (DPs); and the neoendorphins (neo-EP). The DPs and neo-EPs were first isolated from the hypothalamus, hypophysis, and duodenum of the pig. Later, DPs were found in the bovine adrenal gland [659], guinea pig heart [660], and rat duodenum [661].

Leu-EK	Y G G F L
DP-A-8	Y G G F L R R I
DP-A-9	Y G G F L R R I R
DP-A-11	Y G G F L R R I R P K
DP-A-13	Y G G F L R R I R P K L K

Release. During dehydration, immunoreactive DP is elevated in the hypothalamus but lowered in the hypophysis.

In ovariectomized rats, the level of immunoreactive DP increases in the adenohypophysis, this increase is prevented by estrogen [662]. Accumulation of immunoreactive DP-8 in the hippocampus and frontal cortex of rats appears to be accompanied by reduced learning ability [663]. After traumatic injuries, immunoreactive dynorphin increases in the spinal cord of the rat. This appears to be responsible for hind limb paralysis, an attendant symptom of traumatic injuries [664]. Immunoreactive DP levels are low in the spinal cord of schizophrenic patients [665].

Biological Effects. See Section 4.4 (κ-receptor-mediated opioid actions). DPs potentiate the glucose- or amino-acid-stimulated release of insulin [666], but inhibit the release of somatostatin [666], TRH [667], oxytocin [668], vasopressin [669], and counteract morphine tolerance [670].

The central application of DPs in rats decreases body temperature, increases food [671] and water intake [672], suppresses motor activity [673], reduces the response to acoustic signals [673], and leads to hindlimb paralysis similar to that observed after spinal cord injury [674].

Structure – Activity Relationships. DPs are specific ligands for the κ-receptors. The affinity for the δ-receptors increases with decreasing chain length [675]. Short-chain DPs are rapidly degraded by peptidases and therefore have a considerably shorter biological activity. DP-A-8 and DP-A-9 are assumed to have a neural transmitter or modulator function at the κ-receptors, and the more stable DP-A-13 and DP-A-17 to have a more hormonal function. As with other opioids, the N-terminal tyrosine is essential for opiate activity.

The basic amino acids, particularly Arg^6 and Arg^7, are important for the biological activity of DP-A-13 [676]; substitution of Ala for Ile^8 gives a more active (2 – 9 times) compound [676]. [D-Cys^2,Cys^5-N-MeArg^7,D-Leu^8]DP-(1 – 8)ethylamide is highly analgesic, it binds more strongly to the μ- and δ-receptors than to the κ-receptors [677].

The substitution of D-Trp in DP-A-11 gives rise to weak, nonselective antagonists [678]. N-diallyl derivatives of DP-A-11 are also opioid antagonists with weak selectivity for κ-receptors [679]. De-Tyr^1-rimorphin inhibits morphine-induced effects [680].

Uses. The intrathecal application of 15 μg of DP-A-13 to cancer patients produces a nociceptive effect that lasts for > 4 h [681]. [D-Ala^2]DP-A-6 [*081733-79-1*] (Dalargin) is used to treat duodenal peptic ulcers [682].

4.4.3. Dermorphin and Deltorphins

Occurrence. Dermorphin (DM) [*77614-16-5*], M_r 803, and deltorphin (DT) [*119975-64-3*], M_r 955, each contain a D-amino acid:

Dermorphin	Y	a	F	G	Y	P	Sa
Deltorphin	Y	m	F	H	L	M	Da

They were isolated from the skin of the South American frogs, *Phyllomedusa sauvagii*, *P. rhodei*, and *P. bicolor*. One precursor of DM contains five copies of a sequence of 35 amino acids, the C-terminus of this sequence contains the DM sequence [683]. Another precursor contains four copies of DM and one copy of DT [684]. The precursors contain alanine and methionine in the L form, which are converted to the D conformation in a posttranslational reaction.

Immunoreactive DM has also been found in the brain, adrenal glands, and the gastrointestinal tract of the rat [685].

Biological Effects. DM binds preferentially to the μ-receptors. When given intravenously, it has a powerful analgesic effect (10 times that of morphine).

DM inhibits the secretion of gastric acid, the emptying of the stomach, secretion from the pancreas, and intestinal motility. It increases the plasma levels of prolactin [686], growth hormone [687], thyrotropin [688], somatostatin, gastrin, glucagon [689], blood pressure, and heart rate. It inhibits the release of ACTH, β-LPH, β-EP [690], LH [691], secretin, and pancreatic peptide [692].

DT binds to the δ-receptors. When given intracerebroventricularly it improves the memory of mice [693].

Structure–Activity Relationships. The C-terminal amide group is important for the biological activity of DM [694]. In the guinea pig ileum, [Tyr^7]DM has twice the activity of DM and 1.4 times the analgesic activity [695]. The substitution of Hyp for Pro^6 gives a compound with the same analgesic and gastrointestinal activity, but a higher prolactin-releasing activity [694]. [Tyr(OMe)^5]DM has a higher affinity for the μ-receptors and [Phe^5]DM binds more strongly than DM to the δ-receptors [696]. [D-Arg^2]DM has about the same analgesic activity as DM [697]. Shortening the [D-Arg^2]DMs at the C-terminus enhances activity [697]. DM guanylated at the N-terminus has a higher analgesic effect and inhibits the gastrointestinal transit [698].

In the case of the tetrapeptide analogues, amides with a bulky side chain, N-terminal guanylated compounds, and the [Sar^4], [D-Arg^2], and [D-Met(O)^2] substitutions all increase activity. Tetrapeptide analogues are described in [699]–[706].

The selectivity of DM and DT for the opioid receptors depends on charge effects and the hydrophobicity of the C-terminus [707]. DT-(1–4)amide binds almost exclusively to the μ-receptors. Met^6, Asp^7, and Leu^5 are important for binding to the δ-receptors.

4.4.4. Exorphins

The term exorphins refers to peptides with opiate activity that are produced from food during digestion.

Occurrence. β-Casomorphin-7 (β-CM-7) [79805-24-6], M_r 790, was isolated from casein peptone and corresponds to b- or o-β-casein-(60–66) [708]. The more active N-terminal β-CM tetrapeptide amide (morphiceptin) has been synthesized and isolated from enzymatically digested milk proteins [709].

α-Casomorphins were isolated from α-casein after treatment with pepsin (amino acid sequences 90–95 and 90–96). An exorphin-like sequence was found between positions 43 and 49 of α-gliadin [710].

α-CM-7	R Y L G Y L E
α-CM-6	R Y L G Y L
β-CM-7	Y P F P G P I
β-CM-5	Y P F P G
β-CM-4	Y P F P
Morphiceptin	Y P F P a

Biological Effects. The β-CMs have only slight opiate activity. Intragastric β-CMs cause the release of somatostatin. In a process that can be reversed by naloxone, digested gluten and β-CM (oral) stimulate the release of insulin and glucagon in dogs after a test meal [711].

The β-CMs exert an opioid effect on intestinal electrolyte transport [712]. When applied parenterally, they stimulate the postprandial release of pancreatic peptide and the amino-acid- or glucose-stimulated release of insulin [713], increase the plasma level of prolactin [714], and inhibit the release of somatostatin, thyroliberin, and thyrotropin [715].

Structure – Activity Relationships. The N-terminal tripeptide of β-CM has no activity. Morphiceptin, the N-terminal tetrapeptide amide, has the same analgesic activity as Met-EK in the guinea pig ileum assay and is a specific ligand for the μ-receptors. The peptides, h-β-CM-(1 – 4) amide (valmuceptin) and h-[D-Val^4]β-CM-(1 – 4)amide (devalmuceptin) bind more tightly to the μ-opiate receptors than morphiceptin [716].

Substitution of D-Pro or D-pipecolic acid for Pro^4 gives [D-Pro^4]β-CM-5 (deprolorphin), [D-Pro^4]-morphiceptin (deproceptin, Wellcome PL 017) [717], and [D-pipecolic acid^4]β-CM-5 (depilorphin), which are more active than morphine both in vivo and in vitro. These compounds have a high analgesic activity in the rat [718]. Substitution of D-Phe for Phe^3 increases antinociceptive activity [718]. Replacement of D-Pro^2 leads to loss of opioid activity, whereas [D-pipecolic acid^2]β-CM-5 has a higher long-lasting analgesic effect [718]. Substitution of D-Pro for Pro^4 and D-Phe for Phe^3 promotes binding to the μ-receptors.

4.4.5. Femarfarmamide and Related Structures

Occurrence. The tetrapeptide femarfarmamide (FMRF-amide), M_r 599, was isolated from molluscs. It is not derived from prepro-EK A (which also contains the FMRF sequence at its C-terminus) but is coded in a gene that contains 21 copies of the FMRF sequence [719].

Immunocytochemical methods have shown that FMRF-like material occurs in the pancreas of chicken, ileum of the dog, and in brain neurons of the frog and rat. The first FMRF-like peptide from vertebrates, LPLRF-amide, was isolated from chick brain.

Other structurally related compounds are found in many animals (e.g., cockroach [720] and hawk moth [721]).

Biological Effects [722]. In molluscs, FMRF-amide exerts both a stimulating and an inhibitory effect on the heart. It hyperpolarizes neurons in the snail. In rats, both FMRF-amide and LPLRF-amide [723] (intravenously and centrally applied) increase arterial blood pressure.

FMRF-amide inhibits the spontaneous or acetylcholine-induced contraction of the anterior gizzard of *Aplysia california* and the stomach of *Navanax* [724]. It also inhibits colon motility in mammals [722].

FMRF-amide has no opiate activity, but acts as an opiate antagonist, e.g., [725], [726]. However, it only binds weakly to κ- or μ-receptors [727].

On intracerebroventricular application in rats, FMRF-amide exhibits amnesic activity [728] and increases the plasma level of growth hormone [729].

Structure – Activity Relationships. The C-terminal amide group and the full length of the molecule are important for the biological activity of FMRF-amide. Acetylation and benzoylation increase its contractile effect. A hydrophobic amino acid is required in position 2, while short-chain amino acids decrease activity [730].

5. Blood Pressure Regulating Peptides

This chapter deals with the angiotensin–kinin system, the neurokinins and tachykinins, vasopressin and the structurally related peptides, oxytocin and vasotocin, the endothelins, the atrial natriuretic factor (ANF), and neuropeptide Y (NPY). These peptide hormones influence blood pressure and function as modulators of the reproductive hormones (Chap. 2) and of the stress-induced proopiomelanocortin (POMC) cascade (Chap. 4). AT II and, above all, the kinins appear to be involved in ovulation and the formation of semen. AT II stimulates the POMC cascade, presumably via CRH, and the formation of gluco- and mineralocorticoids; it also raises the level of luteotropin (LH) when applied centrally. Vasopressin has a peripheral stimulating effect and inhibits the POMC cascade when administered centrally. Oxytocin plays a role during mating as well as during and after birth. Endothelin releases LH from the pituitary; ANF, which can be released by endothelin, inhibits the formation of the stress hormone cascade and the synthesis of aldosterone. NPY also exerts an effect on the secretion of LH through the release of LHRH from the hypothalamus.

5.1. The Angiotensin–Kinin System

The kinins and angiotensins are formed in biological fluids by the enzymatic cleavage of protein precursors. Kallikreins cleave kininogens (KG) to yield kinins, which are blood pressure lowering peptides. The species-specific enzyme renin can also be generated by kallikrein from inactive prorenin and is responsible for the formation of the decapeptide angiotensin I (AT I) from the N-terminus of the α-globulin angiotensinogen (ATG). The structures of r-prepro-ATG and h-prepro-ATG have been elucidated [731]. Glucocorticoids stimulate the formation of ATG [732], [733], IL-6 potentiates the glucocorticoid-stimulated synthesis of ATG [733], and AT II or [Sar^1]AT II inhibits the production of ATG in hepatocytes [734].

Although AT I has no effect on blood pressure, it is acted upon by the membrane-bound angiotensin converting enzyme (ACE or kininase II), a zinc-containing carboxydipeptidase, to form the octapeptide angiotensin II (AT II), a vasopressor (i.e., blood pressure increasing agent). The hypotensive kinins are, however, degraded by ACE. Therefore, the inhibition of ACE has a double hypotensive effect because both the formation of the blood pressure increasing AT II as well as the degradation of blood pressure lowering kinins are inhibited. The first inhibitors of ACE were isolated from snake venom and called bradykinin potentiating peptides. Examples of orally active ACE inhibitors are captopril, enalapril, and ramipril. Inhibitors of renin also exert a hypotensive effect. Potent inhibitors of renin are ATG analogues in which the peptide

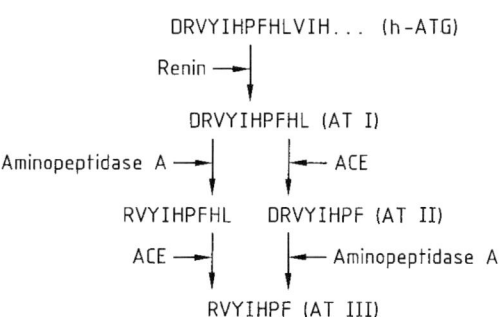

Figure 4. Formation of angiotensins (AT) from angiotensinogen (ATG)

bond between positions 10 and 11 is reduced or replaced by statin, an amino acid occurring in pepstatin, or by similar compounds.

5.1.1. Angiotensins

Occurrence. Angiotensins (ATs) are found both in the periphery and in the brain. Plasma AT II is formed primarily in the lungs. AT II and AT III are also synthesized in the juxtaglomerula cells of the kidney [735], the adrenal cortex [736], the gonadotropic cells of the pituitary [737], the hypothalamus [738], the ovarian follicle [739], and the Leydig cells of the testes. Roughly equimolar amounts of AT II-(1–7), AT II, and AT I have been found in the rat hypothalamus [738]. The sequences of human and rat angiotensins are as follows:

```
D R V Y I H P F H L
←——————  AT I  ——————→
←—————— AT II →
    ←—— AT III →
```

Release. See Figure 4. AT I [*9041-90-1*], M_r 1296.5, is formed by the action of renin on α-globulin angiotensinogen (ATG). AT I is in turn cleaved by the angiotensin converting enzyme (ACE), also known as kininase II, or carboxypeptidases to form AT II [*11128-99-7*], M_r 1046.2 [740]. AT III [*12687-51-3*], M_r 931.1, is formed from AT II (aminopeptidase A) or de-Asp^1-AT I (ACE or carboxypeptidases). Large amounts of aminopeptidase A are found in the adrenal cortex.

Estradiol or a combination of estradiol and progesterone increases the plasma level of AT I and II in rats, whereas progesterone alone exerts no such effect [741]. A low-NaCl diet raises and a high-NaCl diet lowers the serum level of AT II [742]. The intracerebroventricular application of sodium chloride appears to result in a pressor effect through secretion of AT II [743]. The plasma AT II level is increased in thirst [744], dehydration, loss of blood [745], and endotoxin shock [746].

Receptors. Two AT II receptors occur in the adrenal gland: The *AT 1 receptor* (adrenal cortex) is sensitive to DuP-753 (Du Pont) [747], the *AT 2 receptor* (adrenal medulla) is sensitive to PD 123 319 (Parke Davis), Exp-655 (Merck MSD/Du Pont) [747], and WL-19 (Werner Lambert) [748].

AT III has its own subclass of receptors in the vascular smooth muscles: [Sar^1,Ile^7]-AT III is a selective antagonist for AT III [749].

The density of the AT II receptors can be increased by dehydration [750], stress [751], or IGF-I [752]. Treatment with AT II leads to the down regulation of the receptors in rat hepatocytes [753]. The density of the AT II receptors on the glomeruli in the kidney is reduced by a low intake of sodium chloride [744], diabetes mellitus [754], and by ACTH-(1–24) [755]. Chronic treatment with estrogen reduces the AT II receptor density in the hypophysis, adrenal cortex [756], and the rat placenta [741].

Biological Effects. AT II and AT III increase intracellular levels of calcium, phosphatidic acid, phosphoinositides [757], and cGMP [758], and increase the formation of PGE$_2$ and PGI$_2$ by activating phospholipase A$_2$ [759].

In the kidney, AT II inhibits the activity of renin [760] and the synthesis of messenger RNA for renin by a negative feedback effect through a lipoxygenase product [761].

AT II causes contraction of the vascular smooth muscles of the uterus, intestine, aorta, myocardium, and kidney, resulting in a potent vasoconstricting and blood pressor activity. Prostaglandins counteract the activity of AT II in the kidney [762].

AT II has many other effects. It increases plasma vasopressin levels [763] and exerts a positive inotropic effect on the myocardium [764]; in the kidney it reduces the glomerular filtration rate and, in physiological doses, the excretion of sodium ions and water [765], but it exerts a natriuretic and diuretic effect when given in high doses [766]. It reduces water absorption by the jejunum via the synthesis of prostaglandins [767] and stimulates the formation of aldosterone and corticosterone in the adrenal glands after short-term administration [768]. It may play a role in ovulation [769]: it stimulates the secretion of androgen and estrogen from rat ovaries [739] and reduces the LH-stimulated synthesis of progesterone in bovine luteal cells [770]. Like AT II, AT III stimulates the release of aldosterone and prostaglandin.

Structure–Activity Relationships. The C-terminal carboxyl group and the aromatic amino acids are important for the activity of AT II [771]. The N-terminal sequence is responsible for the specificity, intensity, and duration of its biological activity. However AT II-(1–7), like AT II, releases vasopressin and probably plays a physiological role in its formation [772].

h-[Sar^1]AT II, h-[Aib^1]AT II, and [D-Ser^1]AT II [773] are more stable to aminopeptidases and thus more potent than AT II. The activity is lowered if Asn occupies position 1 [774].

Although the N-terminal amino group is not important for the pressor activity of AT II, acylation decreases the myotropic and aldosterone-stimulating effects.

Tyr^4 is very important for biological activity: fragments without Tyr^4 are devoid of activity [771]. The activity depends on the electronegativity of the aromatic ring in position 4. Increasingly electronegative substituents lower activity. [Sar^1,3-NH$_2$-Tyr^4]AT II [775] and monoiodo-AT II [776] are, however, more active than [Sar^1]AT II and AT II. A β-branched amino acid in position 5 is important for agonistic activity. *O*-Methylthreonine is more effective than Ile in position 5 [777].

If Pro^7 in AT I or AT II is replaced by Ala, cardioselective substances are obtained which have a positive inotropic activity and lower pressor effect [764]. [Cys^3,Cys^5]AT II is a cyclic analogue with about the same activity as AT II [778].

AT II Antagonists. Positions 4 and 8 are responsible for the agonistic or antagonistic action of AT II. Competitive antagonists result when the side chain of Phe^8 is extended or shortened by a methylene group or when aliphatic or branched aromatic [779] amino acids are substituted for Phe^8. One of the most active antagonists of this series is [Sar^1,Bip^8]AT II [779] (Bip = biphenylalanine). The presence of D-amino acids, such as D-phenylglycine (D-Phg) or D-phenethylglycine (D-Peg), in position 8 gives active antagonists, which are not degraded by carboxypeptidases [780]. [Sar^1,Val^5,Ala^8]AT II (saralasin) and [Sar^1,Ile^5,Thr^8]AT II (DE-3489, sarthran) have been examined more closely. [Sar^1,Thr(Me)^{5,8}]AT II and [Sar^1,Chg^5,Lac^8]AT II [781] (Chg = L-cyclohexylglycine, Lac = L-lactic acid) are more active than saralasin. Unlike the agonists, the antagonists do not require a β-branched amino acid in position 5. For instance, [Sar^1,Tyr^5,Ile^8]AT II has three times the antagonistic activity of [Sar^1,Ile^{5,8}]AT II [777].

Alkylation of the phenolic hydroxyl group of Tyr^4 [782] or the substitution of Phe [783] or 4-ClPhe [784] for Tyr^4 also gives AT II antagonists. However, [Sar^1,Tyr(Me)^4]AT II (sarmesin) is less active than saralasin.

Shortening the molecule at the C-terminus also yields AT II antagonists. Saralasin and sarmesin are about six times as active as [Sar^1] AT II-(1 – 7)amide [785]. The latter compound has no agonistic properties and does not inhibit the central dipsogenic activity of AT II [786].

Nonpeptide AT II receptor antagonists are imidazole derivatives and are described in [747], [787], [788].

AT III Analogues. Substitution of D-N-methylalanine for Arg^1 in AT III produces a more potent analogue. [D-N-MeAla^1,Ile^7]AT III is a selective AT III antagonist [789]. [Sar^1,Ile^7]AT III is a selective vascular AT III antagonist. [Ile^7]AT III inhibits both AT III and AT II [749].

Uses. [Asn^1,Val^5]AT II (angiotensin amide [*53-73-6*], Hypertensin CIBA, Ciba), an AT II agonist, is administered in states of shock and collapse when normal blood pressure should be restored as quickly as possible.

The AT II antagonist saralasin [*34273-10-4*] (Sarenin, Röhm Pharma) is used in the diagnosis of AT II-dependent forms of hypertonia and in the preliminary treatment of donor kidneys before transplantation to minimize loss of functions of the transplanted kidney caused by ischemia.

5.1.2. Kinins [790]

Occurrence. Kininogens (KG) are cleaved by kallikrein to give kinins. This system is widely distributed, e.g., in the liver, in acinar cells of the rat submandibular gland, in the human kidney, in platelets, endothelial cells, rat vascular smooth muscle cells (important for the regulation of the local vascular tone) [791], in the brain [792], Leydig cells, ovaries, and in milk [793]. The amino acid sequences of mammalian kinins follow:

Bradykinin (BK) [58-82-2],
M_r 1060				R	P	P	G	F	S	P	F	R
[Hyp³]BK				R	P	Ph	G	F	S	P	F	R

Kallidin [342-10-9],
M_r 1188			K	R	P	P	G	F	S	P	F	R
[Hyp³]Lys-BK			K	R	P	Ph	G	F	S	P	F	R
Met-Lys-BK		M	K	R	P	P	G	F	S	P	F	R

T-Kinin [86030-63-9],
M_r 1260	I	S	R	P	P	G	F	S	P	F	R	
T-Kinin-Leu	I	S	R	P	P	G	F	S	P	F	R	L

Large amounts of BK, kallidin, [Hyp³]Lys-BK, and [Hyp³]BK were found in human urine [794], plasma [795], and in ascites of patients suffering from stomach cancer [796].

T-Kininogens have also been found in the rat [797]. These precursors are cleaved by T-kininogenase [798] (endopeptidase K [799]) or high concentrations of trypsin to give T-kinin (Ile-Ser-BK).

Structural similarities have been observed between kinins and various hormones of insects and crustaceans [800], [801].

Release. Injury lowers tissue respiration and thus the pH [802] which activates the Hagemann factor (the blood clotting factor XII) resulting in the release of bradykinin (BK) [803]. Kallekrein in the plasma cleaves kininogen to produce BK, whereas tissue kallikrein cleaves kininogen to form kallidin (KD = Lys-BK), which can in turn be converted to BK by aminopeptidases.

Kinins are released during muscular work [804], PMSG/h-CG induced ovulation [805], and inflammation.

Prostaglandins inhibit the formation of kinins. Inhibitors of prostaglandin synthesis therefore potentiate BK synthesis [806]. Aspirin, for instance, potentiates the antihypertensive effect of Captopril in spontaneously hypertensive rats [807].

Inhibitors of kallikrein retard the release of kinins. Endogeneous kallikrein inhibitors include α-macroglobulin and aprotinin (Trasylol, Bayer; Antagosan, Behring Werke). Synthetic inhibitors are substrate analogues of the kininogens, an example is D-Chg-L-Chg-L-Arg-4-nitroanilide, a structural analogue of the C-terminal cleavage sequence, Pro-Phe-Arg-Ser [808]. The bradykinin B_2-receptor antagonists also inhibit kallikrein [809].

BK is deactivated primarily in the lungs by the dipeptidase kininase II, which is identical to ACE. Inhibitors of ACE (e.g., Captopril, Squibb; Enalapril, Merck; Ramipril, Hoechst) potentiate BK [810]–[812]. Other endopeptidases, carboxypeptidases (kininase I activity), and aminopeptidase degrade BK [813]–[815].

In patients suffering from hypertension, the concentrations of kallikrein, prekallikrein, and kinin in the urine are significantly lower than in persons with normal blood pressure [816]. Captopril decreases blood pressure in hypertensive patients with a normal urine kallikrein content, but not in patients who have a low urine kallikrein content [817]. Kallikrein activity decreases in old age.

An increased level of kinin or kallikrein is found in diabetics with orthostatic hypotension [818], allergic persons [819], and patients suffering from inflammatory stomach diseases, ulcerative colitis, pancreatitis, rheumatic inflammation, or myocardial infarction.

Only in rats, do prepro-T-kininogen messenger RNA [820] and T-kininogen [821] appear to be formed primarily in response to inflammatory stimulants. Expression of the T-kininogen gene increases with age [822].

Receptors. The B_1- and B_2-receptors for BK have been studied intensively. The B_1-receptors bind de-Arg^9-BK more strongly than BK, while the B_2-receptors have a greater affinity for BK. Studies with agonists and antagonists on different tissues have led to the assumption that there are multiple B_2-receptors for kinins [823]. The vasorelaxing effects of BK exerted via B_2-receptors in isolated large vessels require intact vascular endothelial cells. The vasoconstricting effects of BK do not require an intact endothelium. The BK-receptor complex is internalized, causing temporary down regulation of the BK-receptor [824].

Biological Effects [825]. *B_1-receptor-mediated effects.* The most important B_1-receptor-mediated effects [826] are contraction of rat duodenum, relaxation of rabbit coeliac artery [827], capsaicin-induced inflammation in mice [828], and release of IL-1 and TNF-α in macrophages.

B_2-receptor-mediated effects [826] include release of EDRF and calcium from intracellular stores, leading to activation of PLA_2 and increase in PGE_2 production; the PGE_2-dependent activation of adenylate cyclase in arterial smooth muscle cells [829]; hypotensive activity: reduces blood pressure increased by AT II [830], adrenaline [831], and vasopressin [790]; increase of vascular permeability and pain [832] in rat skin; improvement of heart function [833]; and increase in human sperm motility.

Other effects of the kinins include stimulation of ovulation [805], [834], rhinitis [790], release of SP- and CGRP-like materials in neurons sensitive to capsaicin [835], and glucose uptake in working skeletal muscle via prostaglandins and phosphofructokinase [836].

Structure–Activity Relationships. De-Arg^9-BK and de-Arg^{10}-Met-kallidin (stronger) are ligands for B_1-receptors. The C-terminal arginine residue is very important for binding to the B_2-receptors. De-Arg^9-[D-Phe^8]BK has four times and Sar-[D-Phe^8]de-Arg^9-BK ten times the effect of de-Arg^9-BK on B_1-receptors [837].

1393

De-Arg^9-[Leu^8]BK, the corresponding KD derivative, de-Arg^9-[Gly^7]BK, and de-Arg^9-[D-Ala^7]BK [838] are powerful B$_1$-receptor antagonists.

The action of KD on B$_2$-receptors is greater than that of BK. The guanido groups of Arg^1 and Arg^9 and the proline residue in positions 2 and 3 of BK are important for biological activity at the B$_2$-receptors. Strong agonists are obtained by the substitution of β-(2-thienyl)-Ala (Thi) or dehydro-Phe (Δ-Phe) for Phe^5 and Phe^8, Hyp for Pro^3, α-aminoisobutyric acid for Pro^7, Tyr(Me) for Phe^8, or by reduction of the peptide bonds between positions 8 and 9 [839] and 6 and 7 [840]. The effect of [Hyp^3,Tyr(OMe)^8]BK on B$_2$-receptors is about three times that of BK [837]. [Phe^8-ψ-(CH$_2$-NH)-Arg^9]BK is a selective B$_2$-receptor agonist, which is five times as active as BK [839]. The effect of [Δ-Phe^5]BK on the uterus and ileum is twice that of BK and the effect on blood pressure is 23 times that of BK.

The replacement of Pro^7 in BK by D-Phe, D-Nal, D-Pal, or D-tetrahydroisoquinolinecarboxylic acid (D-Tic) [841], and the substitution of octahydroindolecarboxylic acid (Oic) [841] for the Phe^8 gives strong B$_2$-receptor antagonists. Depending on the test, D-Arg-[Hyp^3,Thi^5, D-Tic^7,Oic^8]BK (Hoe 140) [841] is a 100–1000 times stronger BK antagonist than the standard peptide, D-Arg-[Hyp^3,Thi^{5,8},D-Phe^7]BK [842].

Modification of the BK antagonists at positions 1, 2, 3, 8, and 9, alters tissue selectivity; extension with D-Arg at the N-terminus improves affinity and inhibits enzymatic degradation [843].

BK-receptor antagonists with Phe^8 or Thi^8 also act as B$_1$-receptor blockers [844] because carboxypeptidases convert these peptides to biologically active B$_1$-receptor antagonists by cleaving the C-terminal arginine [845]. The histamine-releasing effect of BK antagonists can be lowered by acetylating the N-terminal amino group [846].

Uses. BK antagonists may find application for treating inflammation, pain, rheumatoid arthritis, osteoarthritis, inflammatory stomach diseases, rhinitis, asthma, and gout [790].

5.2. Substance P, Neurokinins, and Tachykinins [847]

5.2.1. Introduction

Occurrence. The neurokinins include substance P (SP) [*516-47-2*], neurokinin A (NKA) [*86933-74-6*], and neurokinin B (NKB) [*86933-75-7*]. Similar compounds which occur in cold-blooded animals are called tachykinins.

h-,b-Substance P	R P K P Q Q F F G L Ma, M_r 1348
h-,g-,p-Neurokinin A	H K T D S F V G L Ma, M_r 1133
p-Neurokinin B	D M H D F F V G L Ma, M_r 1210
Scyliorhinin II	S P S N S K C P D G P D C F V G L Ma, M_r 1851

Substance P was the first of a large number of peptides that were found in the gastrointestinal tract and the brain. Substance P and the neurokinins are produced in the form of preproneurokinins. Three preproneurokinins (α, β, and γ) have been characterized in the cow and the rat [848]. There is a very great similarity between h-β-prepro neurokinin and b-β-prepro neurokinin [849]. Neuropeptide K (NPK) is neurokinin A that is extended by 26 amino acids at the N-terminus [850]. NPK, NKA [851], NKA-(3–10), and NKA-(4–10) [852] occur in the plasma and tumor tissue of carcinoid patients and in the bronchi. Apart from NKA, NKB (neuromedin K) has also been isolated from the spinal cord of the pig.

Substance P occurs in the spinal cord, the sensory nuclei of the brain stem, and the sensory nerve terminals in the gastrointestinal tract, urogenital tract, bile duct, bronchi, skin [853], and thymus [854]. Neurokinin A has been isolated from the rabbit iris [855].

Tachykinins are found in the salivary glands of the octopus, the skin of amphibians [856], and the locust [857]. Scyliorhinin I and II are the tachykinins of the mud fish, *Scyliorhinus caniculus* [856]. There are indications that tachykinin-like substances also occur in mammals [852]. There is a structural similarity between the tachykinins and the neurotoxic β-amyloid protein-(25–35) [858].

Release. Substance P is released after stimulation of the sensory neurons. It is assumed to be a pain transmitter, especially since the level of SP in the spinal cord is increased by mechanical pressure or hot water applied to the extremities [859], and analgesics inhibit the release of SP from sensory nerve ends [859].

Substance P and NKA are released by capsaicin, bombesin [860], CCK-8 [861], and by cholinergic [862], serotoninergic [863], and dopaminergic [864] agents.

A high NKA- and SP-like immunoreactivity is found in tumor tissue and plasma NKA immunoreactivity can be used as tumor marker. "Flushing" episodes and diarrhoea correlate positively with plasma NKA [865]. The neurokinin level is also elevated in the joint fluid of patients with rheumatic inflammatory diseases [866].

In the brain, SP is preferentially cleaved by metalloendopeptidase E.C.3.4.24.11 (also called enkephalinase or neutral endopeptidase) to form the N-terminal heptapeptide SP-(1–7). The C-terminal heptapeptide SP-(5–11) is, however, also formed in the brain after cleavage with the post-proline-cleaving enzyme [867].

Digestion in the plasma occurs preferentially through a post-proline-cleaving enzyme to give dipeptides [868]. SP is also a substrate for the angiotensin converting enzyme (ACE). Levels of SP are increased by smoking (inhibition of the neutral endopeptidase) [869] and by ACE inhibitors [870].

Receptors. Three neurokinin receptors have been found: NK1 for SP, NK2 for NKA, and NK3 for NKB [871]. The r-NK1 receptor, the r-NK2 receptor [872], and the b-NK2 receptor [873] have been characterized via their cDNA. They occur in a wide variety of tissues [874].

Selective ligands of neurokinin receptors are described in the literature [874]:

NK1 receptor agonists	[871], [877]–[879]
NK2 receptor agonists	[878], [880]–[882]
NK3 receptor agonists	[871], [876], [883]
NK1 receptor antagonists	[884]–[886]
NK2 receptor antagonists	[875], [887]–[889]
NK3 receptor antagonists	[859], [890]

Biological Effects [874]. *NK1-receptor-mediated effects* The most important NK1-receptor-mediated effects are hyperalgesia, hypotension, salivary secretion, and increase of capillary permeability.

NK2-receptor-mediated effects include bronchoconstriction [850], [891], [892], activation of guinea pig alveolar macrophages [893], protection against gastric lesions in rats [894], and tachycardia.

NK3-receptor-mediated effects are analgesia [895], bradycardia, and increase in capillary permeability.

Other effects of SP include activation of phospholipase C and D [896], stimulation of the proliferation of synoviocytes and the release of PGE_2 and collagenase (rheumatoid arthritis) [897], modulation of gastrointestinal motility [898], [899], and immunostimulating activity [900], [901].

Structure–Activity Relationships. In the brain, SP-(5–11) is a pain neurotransmitter. It causes aggressive, stress-induced behavior, impairs memory, and increases blood pressure when applied centrally. SP-(1–7) exerts an analgesic effect, increases learning capacity, improves memory, and lowers blood pressure [867], [902], [903]. Based on the opposing effects of the N- and C-terminal peptides of SP, STEWART postulated the existence of SP-N and SP-C receptors [867]. The regeneration of nerve fibers is stimulated by N-terminal fragments of SP [904].

SP Agonists. The shortest active SP analogues are acylated SP-(7–11) derivatives [905], some are more active than SP. The most effective compound in the guinea pig ileum test is SP-(7–11) with a 4-hydroxyphenylacetic acid residue at its N-terminus.

The C-terminal amide group, a neutral lipophilic amino acid in position 11 [906], Leu^{10}, and Gly^9 are important for biological activity. Phe^8 can be replaced by Tyr(OMe), cyclohexylalanine, or Ile without loss of activity, whereas substitution of these residues for Phe^7 produces only slightly active compounds. The remaining N-terminal amino acids are not essential for biological activity.

pGlu-Gln-Phe-MePhe-Sar-Leu-Met-amide is an SP analogue with a long-lasting effect. It has only 10% of the spasmogenic activity, but all of the aggressive CNS activity of SP and is resistant to the proteolytic enzymes of the hypothalamus.

A derivative that is dimerized with succinic acid via the N-terminal amino groups of SP-(3–11) has 2.4 times the receptor affinity and 75 times the saliva-producing effect of SP [907].

NKA-(4–10) is equally as active as NKA. Asp^4, Phe^6, and Val^7 are important for the biological activity of NKA and NKB [908].

SP Antagonists. The standard antagonist is spantide I which binds primarily to the NK1 and NK2 receptors.

Spantide I	[D-Arg^1,D-Trp^{7,9},Leu^{11}]SP
Spantide II	[D-Lys(Nic)^1,Pal^3,D-Phe(Cl$_2$)^5,Asn^6,D-Trp^{7,9}, Nle^{11}]SP

D-Phe(Cl$_2$) = 3,4-D-dichlorophenylalanine
D-Lys(Nic) = *N*-nicotinoyl-D-lysine

The activity of SP antagonists is enhanced by the replacement of Gln^6 by Asn^6 [909]. As a result of the substitution of D-Lys(Nic) for D-Arg^1, spantide II, unlike spantide I, has no neurotoxic properties [910] and a reduced histamine-releasing activity [911]. Spantide I acts as a bombesin antagonist but spantide II has no effect on bombesin-induced contractions. Spantide II binds to all three neurokinin receptors [911] and exerts a short antinociceptive effect when applied intrathecally [912].

Reduction of peptide bonds also gives SP antagonists which have been evaluated by the displacement of SP from SP receptors in guinea pig acinar cells [913].

SP antagonists with basic N-terminal amino acids release histamine, whereas shortened antagonists (e.g., [D-Pro^4,D-Trp^{7,9,10}]SP-(4–11) can inhibit the SP-induced release of histamine in vitro.

[D-Arg^1,D-Phe^5,D-Trp^{7,9},Leu^{11}]SP was the most suitable peptide for inhibiting the bombesin- and vasopressin induced mitogenesis in small cell lung cancer cells. It has the same affinity for the bombesin and SP receptors and is ten times more active than spantide I on the bombesin receptor [914]. Under comparable conditions, spantide I and other N-terminal-shortened SP antagonists did not inhibit the bombesin- or vasopressin-induced growth of Swiss 3T3 cells.

5.2.2. Amyloid A4 Protein [915]

Occurrence. In Alzheimer's disease and in Down's syndrome (Mongolism), fibrillary amyloid is found within the cortical neurons as neurofibrillary tangles and extracellulary in meningeal and intracortical blood vessels as amyloid plaques. These deposits also occur in the skin, subcutaneous tissue, and in the intestine of Alzheimer patients and healthy elderly persons [916].

The plaques mainly consist of an insoluble peptide called the amyloid A4 protein (A4) [*12627-51-9*], M_r 4514 [917], or amyloid-β protein [918]. It contains 42 amino acids [919]. The gene that codes for A4 produces at least three messenger RNAs for amyloid precursor protein, they form the APP$_{695}$, APP$_{751}$, and APP$_{770}$ proteins [920].

APP$_{695}$ consists of 695 amino acids and contains A4 in positions 597–638. APP$_{751}$ is identical to APP$_{695}$ apart from an insert of 56 amino acids (HL 124i) C-terminal from Arg^{288} and the substitution of Ile for Val^{289}. Apart from an insert of 19 amino acids C-terminal from HL 124i and the substitution of Leu for Ile^{289}, APP$_{770}$ is identical to APP$_{751}$. The HL 124i insert has the structure of a Kunitz inhibitor, which specifically inhibits serine proteases, such as trypsin, chymotrypsin, elastase, plasmin, and cathepsin G. In Alzheimer's disease, the total increase in the messenger RNA for APP in neurons from the locus ceruleus and nucleus basalis is due solely to an increase in APP$_{695}$ that lacks this inhibitor domain. Therefore, it is assumed that A4 is readily formed from this precursor by the action of cerebral proteases [920]. These A4 precursors are transmembrane proteins and receptors for neuronal adherons [921]. Adherons are extracellular, adhesion-mediating particles (mostly glycoproteins) found in the extracellular matrix.

The gene for APP is expressed in certain neurons, some glia cells, and in brain macrophages. The A4 formed in the glia cells and macrophages is responsible for the extracellular formation of the amyloid plaques [919]. Human platelets contain APP_{751} in high concentrations [922].

The A4 of patients suffering from Alzheimer's disease and Down's syndrome is identical to the intraneural amyloid of the Parkinson dementia of Guam and the vascular amyloid of sporadic cerebral amyloid angiopathy. The amyloid protein of patients with leptomeningiosis haemorrhagica interna is shorter by three amino acids at the C-terminus [923].

Release of APP and A 4. The APP gene is localized on chromosome 21 and amplified in Down's syndrome but not in Alzheimer's disease. It is assumed that the A4 protein is formed by increased proteolytic activity in Alzheimer's disease [924]. Proteolysis of APP in membranes at the C-terminus of A4 does not change the aggregation properties. However, enzymatic cleavage at the extracellular N-terminus of A4 is of great importance for plaque formation. Proteinase K cleaves a pro-A4 fragment to peptides of the size of A4 [925]. Platelets also release APP during degranulation [926].

Biological Effects. APP is widely distributed in the neurons of the rat brain and appears to play a role in cell–cell contact, which is required for memory [917]. APP released from cells has an autocrine effect on growth regulation [927].

Low concentrations of A4 and A4-(25–35), which is structurally similar to the tachykinins, have a neurotrophic and at higher concentrations a neurotoxic effect on mature neurons. This neurotoxic activity is also exhibited by some tachykinin antagonists and can be abolished by tachykinin agonists [928].

5.3. Vasopressin, Oxytocin, and Melanostatin

5.3.1. Vasopressin and Oxytocin

Occurrence. The nonapeptides oxytocin (OT) [*50-56-6*], M_r 1007, and vasopressin (VP) [*11000-17-2*], M_r 1084, are formed as prohormones. OT and VP are located at the N-termini of these prohormones, the residual C-termini are the neurophysins (neurophysin I for OT and neurophysin II for VP). OT and VP are transported together with the neurophysins as transport molecules in neurosecretory vesicles from the pituitary down the axon to the neurohypophysis. OT and VP are also synthesized in the pineal body, the retina of mammals, the thymus [929], corpus luteum [930], and testes [931].

	1	2	3	4	5	6	7	8	9
Oxytocin	C	Y	I	S	N	C	P	Q	Ga
Vasopressin	C	Y	F	Q	N	C	P	R	Ga
Vasotocin	C	Y	I	Q	N	C	P	R	Ga

Vasotocin (VT) [*9034-50-8*], M_r 1050, occurs in the pineal body of mammals and in the cerebrospinal fluid of humans. Nα-Ac-OT has been found in the pineal body of the cow [932] and rat [933], and, along with Nα-Ac-VP, also in the pars intermedia and the brain of the rat [933].

Naturally occurring peptides with OT and VP activity are described in [934].

Release. Pneumadin, a decapeptide amide from the lungs, releases VP from the pituitary and has antidiuretic activity [935]:

h-Pneumadin: A G E P K L D A G Va

The plasma VP level increases in response to potassium uptake [936], dehydration and thirst [937], blood loss, intravenous infusion of a hypertonic NaCl solution or hydrochloric acid [938], smoking [939], hypoxia (lack of oxygen), hypercapnia (increase in the CO_2 concentration in arterial blood), and to sexual arousal in men [940]. The plasma level of OT also increases on infusion of sodium chloride [941]. VP release is stimulated by acetylcholine (mediated by the hypothalamus) [942], AT II, AT II-(1–7) [943], histamine [944], and dopamine [945]. The activation of central $α_1$-adrenoceptors increases the release of VP [946], whereas the stimulation of the $α_2$-adrenoceptors decreases release [947].

A low-NaCl diet [948], water uptake [949], chronic alcohol consumption [950], and GABA [951] decrease the release of VP.

The vasoactive intestinal peptide (VIP) [952], angiotensin II (AT II), corticoliberin [953], fever [954], and endotoxin [955] release both VP and OT. β-Endorphin and other endogenous opiates inhibit the release of OT and VP.

The plasma level of OT is increased by manual stimulation before milking [956], during labor [957], by CCK, after the intake of food [958], during orgasm in the woman, and ejaculation in the man [959].

Receptors. There are two VP receptors, the V_1-receptor for pressor activity and the V_2-receptor for antidiuretic activity. They are down regulated by VP agonists [960].

The density of the OT receptors in the brain is increased by estrogen or testosterone [961]. Progesterone increases the binding of OT in the posterior part of the ventromedial nuclei of the hypothalamus, the site at which feminine mating behavior is induced by infusion with OT [962].

Biological Effects of Vasopressin. *V_1-receptor-mediated effects.* The most important V_1-receptor-mediated effects [963] are stimulation of phosphatidylinositol turnover and aldosterone synthesis in rat adrenal glomerulosa cells [964], vasoconstriction and hypertension [965], stimulation of prostaglandin synthesis in the rabbit kidney

[966], promotion of platelet aggregation via serotonin release from platelets [967], and stimulation of the clotting factors vWF and factor VIII [968].

V_2-receptor-mediated effects [963] are stimulation of adenylate cyclase, antidiuretic effect, and induction of hypoosmolality and cerebral edema in rats [969].

Other effects include direct release of ACTH [970], potentiation (VP and OT) of the effect of corticoliberin on ACTH release [970]–[972], improvement of memory and learning capacity of older persons [973], inhibition of the diuretic [974] and potentiation of the natriuretic effect of ANF [975].

Biological Effects of Oxytocin. Oxytocin stimulates adenylate cyclase and the release of arachidonic acid and $PGF_2\alpha$ from human decidual cells [976]; induces contraction of the uterus and lactation; prevents the fall in blood pressure associated with hemorrhage [977]; releases relaxin; stimulates the release of insulin and glucagon [978]; increases plasma glucose levels [979]; and releases POMC peptides, α-MSH, β-endorphin, and ACTH from human placenta [980].

Biological Effects of Vasotocin [981]. Vasotocin inhibits the stimulating action of pregnant mare's serum on mice uteri and ovaries; lowers emotional arousal in rats [982]; and increases rapid eye movement (REM) sleep in children via serotonin release.

Structure–Activity Relationships [983]. The ring structure is of special importance for the activity of OT and VP agonists. However, a disulfide bridge is not essential. 1-carba-OT, where the sulfur from Cys^1 is substituted by carbon, is more potent than OT. Extension of the N-terminus (e.g., with triglycine) causes a depot effect. For instance, Gly_3-[Lys^8]VP (terlipressin; Gycylpressin, Ferring) has clear advantages over VP in the treatment of bleeding of esophageal varices. De-9-Gly-amide-VP (Org 5667, Organon) has no cardiovascular or antidiuretic activity, but improves the learning and memory capacity of rats and also has the VP-stimulating effect on penis erection [984]. The memory-enhancing effect of VP is possibly caused by a metabolite.

Substitution of Phe for Ile^3 increases the vasopressor activity at the expense of the OT activity (uterine contraction), the replacement of Arg^8 or Lys^8 by Leu has the opposite effect. Other examples for the alteration of activity as a result of amino acid substitution are given in Table 1. Highly specific analogues can be produced by combining these substitutions. Compounds in which OT activity predominates are deamino^1-dicarba-[Gly^7]OT (cargutocin), deamino^1-1-monocarba-[$Tyr(Me)^2$]OT (carbetocin, weaker but acts longer than OT) [992], [Thr^4,Gly^7]OT, and [Thr^4,Sar^7]OT [993]. Deamino^1-[4-EtPhe^2]6-carba-OT (narcartocin) has a higher natriuretic and diuretic activity, but is less effective in initiating uterine contraction than OT [994]. Deamino^1-[D-4-EtPhe^2]6-carba-OT is an OT antagonist, which inhibits the uterotonic and galactogogic activity of OT [995].

A powerful antidiuretic preparation with slight pressor activity but with vasodilating activity [996] is deamino^1-[D-Arg^8]VP (dVDAVP, desmopressin, trade name Minirin, Ferring). Deamino^1-[Val^4,D-Arg^8]VP (dKDAVP) lacks pressor activity but has antidiuretic activity. Deamino^1-[Tyr(Me)^2,Val^4,D-Arg^8]VP is another antidiuretic compound, but also a strong OT antagonist [997]. Deamino^1-[MeAla^7,D-Arg^8]VP is another antidiuretic with only slight vasopressor activity [986].

Pressor activity is especially pronounced in [Phe^2, Lys^8]VP (felypressin, octapressin) and [Phe^2,Ile^3,Orn^8]VP.

Table 1. Alteration of the activity of oxytocin (OT) and vasopressin (VP) by amino acid substitution

Amino acid substitution	Effect
Substitution in [Arg^8]VP	
Deamino^1 Tyr(Me)^2, Phe^2, Val^4, Pro^4 [938], Arg^4 [985], Δ^3-Pro^7, 4-Hyp^7, Sar^7, MeAla^7 [986], D-Arg^8, Ala^9, D-Ala^9 [987], Acc^9 [988]	increases ratio of antidiuretic: pressor activity
4-Hyp^7 [981]	increases ratio of contraction of uterus: pressor activity
Ala° [989]	increases ratio of pressor: antidiuretic activity
Substitution in [Lys^8]VP	
Phe^2, Ile^3, Orn^8	increases ratio of pressor: antidiuretic activity
Substitution in OT	
4FPhe^2, Thr^4, Gly^7, thiazolidine carboxylic acid^7, Sar^7,Δ^3-Pro^7 [990]	increases ratio of contraction of uterus: antidiuretic activity
Deamino^1, carba-1, Tyr(Me)^2, carba-6, Gln^8	increases ratio of antidiuretic: contraction of uterus
Asn^4, Val^4, Glu(OMe)^4, β-acetylamino-L-α-aminopropionic acid^4, carba-1, MeAla^7, Gly^7, Acc^7 [990], α-tert-butylGly^8	increases ratio of milk ejection: contraction of uterus
Deamino^1, Hmp^1, carba-1 [964], Thr^4 [991]	increases ratio of contraction of uterus: milk ejection activity
Leu^4	diuretic effect, VP antagonist
3-ITyr^2	oxytocin inhibitor

* The superscript ° denotes N-terminal extension; carba denotes that the sulfur atom of cysteine is replaced by carbon.

Antagonists [998]. Strong inhibitors with antioxytocin and antivasopressor activity and slight anti-antidiuretic (aquaretic) activity are variants of peptides with β,β-dialkylated or β,β-cyclopentamethylene-3-mercaptopropionic acid (Pmp) in position 1, O-alkylated tyrosine in position 2, and ornithine in position 8, e.g., the OT antagonists, [β-mercapto-β,β-diethylpropionic acid^1, Tyr(Me)^2,Orn^8] VT or deamino^1-[Tyr(Me)^2,Orn^8]VT. Relative to [Pen^1]OT and [Pen^1,Phe^2,Thr^4]OT (Pen = penicillinamine), the prototypes for OT antagonists, additional Orn^8 and alkylated Tyr^2 increase duration of action but not potency; D-amino acids in position 2 increase activity but not duration of action. Thus, although [Pen^1,D-Phe(Me)^2,Thr^4,Orn^8]VT is somewhat more potent than [Pen^1,Phe(Me)^2,Thr^4,Orn^8]VT, it has a shorter duration of action [999]. Very strong and selective OT antagonists are [Pmp^1,Tyr(Me)^2,-Thr^4,Orn^8,Leu^9]VT with a low antivasopressor and antidiuretic activity and [Pmp^1,Tyr(Me)^2,Thr^4, Orn^8]VT-(1–8) with a weak antivasopressor and aquaretic activity [1000].

Other powerful OT antagonists are deamino^1-[D-Tyr(Et)^2,Thr^4,Orn^8]VT (ORF 22 164) [1001] and [MePmp^1]OT [1002]. ORF 22 164 strongly inhibits contraction of the human myometrium, and has a very slight vasopressor and antidiuretic activity [1003].

Deamino^1-[Pen^1,Tyr(Me)^2]VP and [Pmp^1,Tyr(Me)^2]VP (SK&F-100 273) have high antivasopressor activity (V$_1$ antagonists) [1004]. Highly selective V$_1$ antagonists with low antidiuretic activity are produced by introducing a 4-methyl or 4-phenyl group in Pmp^1 [1005], [1006].

In these analogues, the aquaretic activity can be increased by substituting D-Phe, D-Tyr, O-alkylated D-Tyr, or D-Ile for Tyr^2. The aquaretic–antivasopressor ratio is further improved by Ile^4, Val^4, Abu^4, or Ala^4 [1007], [1008]. Introduction of a *cis*-4-methyl group in Pmp^1 can increase antagonistic activity and greatly reduce the agonistic activity of antagonists of the type [Pmp^1,D-Tyr(Et)^2,Val^4]VP [1009]. The deletion of Gly-amide and Pro^7 does not significantly affect the aquaretic activity of the antagonists

[1010]. Replacement of the C-terminal tripeptide by, for example, Arg-NH$_2$, Arg-D-Arg-amide [1011], or diaminoalkylamides [1012] gives potent V$_2$ antagonists with aquaretic activity. One of the strongest aquaretic VP antagonists is dicarba-[Pmp^1,D-Tyr(Et)^2,Val^4,Arg^7,D-Arg^8]VP-(1–8)amide (SK&F 105 494) [1013]. A very selective V$_2$ antagonist with an effective dose ratio (V$_2$:V$_1$) of 440 is [Pmp^1,D-Ile^2,Ile^4]VP-(1–8) [1014]. Nonpeptide structures are also tolerated at the C-terminus [1015].

The cyclic structure is not required for antagonistic activity. Thus, a reduced antagonist with free sulfhydryl groups, [Pmp^1,D-Tyr(Et)^2,Val^4]VP, still has a high aquaretic activity. Adamantylacetyl-[D-Tyr(Et)^2,Val^4,Abu^6,Arg^9]VP-(2–9)amide has a higher aquaretic activity than [Pmp^1,D-Tyr(Et)^2,Val^4]VP [1016].

Substitution of sarcosine or *N*-methylalanine for Pro^7 in VP antagonists yields OT antagonists. Thus, [Pmp^1,D-Phe^2,Sar^7]VP or [Pmp^1,D-Phe^2,MeAla^7]VP binds to OT receptors and V$_1$-receptors, but has only slight antidiuretic activity. The OT and V$_1$ antagonist, [Pmp^1,D-Phe^2,MeAla^7]VP is a very strong antiglycogenolytic compound with antigalactogenic activity [1017].

Novel OT antagonists are based on a natural product of *Streptomyces silvestris*. L-156,373 [cyclo-(Pro-D-Phe-*N*-OH-Ile-D-piperazyl-L-piperazyl-*N*-Me-D-Phe)] binds to OT receptors with an approximately 20-fold selectivity compared with V$_1$- and V$_2$- receptors. Derivatives of this compound are described in [1018]–[1020].

Uses. Oxytocin (Orasthin, Orasthin "stark", Hoechst; Oxytocin 3 "horm", Oxytocin forte "horm", Hormonchemie; Partocon, Ferring; Pitocin, Parke Davis; Syntocinon, Sandoz) is used to induce birth in primary and secondary uterine inertia, in placenta retention, postpartum hemorrhage, uterine atonia, lactation disturbances, Cesarean section, and mastitis prophylaxis.

Vasopressin (Pitressin, Pitressin Tannat, Parke Davis) is used to treat the bleeding of esophageal varices and intestinal atonia, to stop bleeding in gynecological operations, and in diabetes insipidus. [Lys^8]VP (lypressin) [*50-57-7*] is available as Postacton (Ferring) or Vasopressin-Sandoz (Sandoz) for pituitary function tests, paralytic ileus, and diabetes insipidus. [Orn^8]VP (ornipressin) [*3397-23-7*] is available as Por 8 Sandoz (Sandoz) and triglycyl-[Lys^8]VP (glypressin, terlipressin [*14636-12-5*]) as Glycylpressin (Ferring). These compounds are used primarily to treat bleeding of esophageal varices, acute uterine bleeding, and in operations. Deamino^1-[D-Arg^8]VP (desmopressin [*16679-58-6*]) is administered as Minirin (Ferring) in diabetes insipidus, after hypophysectomy, for increasing Factor VIII activity in hemophiliacs and in von Willebrand's disease [1021]. Since desmopressin reduces blood loss during operations, it is recommended by Jehovah's Witnesses [1022].

In clinical studies, the V$_1$ antagonist SK&F-100 273, [Pmp^1,Tyr(Me)^2]VP inhibits the lypressin-induced increase in blood pressure [1023].

The OT antagonist deamino^1-[Tyr(Et)^2,Thr^4,Orn^8]VT can be administered in premature labor [1024].

5.3.2. Melanostatin

Occurrence. Melanostatin (MIF) [*9083-38-9*], M_r 284, has been isolated from extracts of the median eminence and has the structure Pro-Leu-Gly-amide. It is identical to the C-terminal tripeptide of oxytocin (OT).

Biological Effects. MIF inhibits the release of α-MSH from the pituitary. As a weak μ-receptor antagonist, it impedes the antinociceptive activity of morphine, β-endorphin, and morphiceptin [1025]. It reduces aggression [1026], impedes morphine- [1027] or alcohol- [1028] induced tolerance, and has an antiamnesic [1029] and inhibitory effect on tremor induced by oxotremorin in mice. MIF increases the affinity of dopamine agonists for dopamine receptors and reduces dopamine D_2-receptor supersensitivity [1030]. Some of its CNS effects also appear to proceed via dopamine receptors [1031].

Structure – Activity Relationships. Structure–activity studies using the oxotremorin tremor test have shown that Pro can be replaced by a pGlu residue and the amide group can be alkylated. For instance, pGlu-Leu-Gly-N(Me)$_2$ has 32 times the potency of MIF. [N-Me-D-Leu^2]MIF (pareptide) potentiates L-dopa-induced behavior patterns and has antiparkinsonian and antidepressant activity. [D-Pip^1]MIF has a stronger antiamnesic activity than MIF [1029]. [3-(+)-thiazolidine-2-carboxamide]MIF, [3-(–)-thiazolidine-2-carboxamide]MIF and [3-L-3,4-dehydroprolinamide]MIF [1032] cause a two- to threefold stimulation of the affinity of dopamine agonists for dopamine receptors and 3(R)-(N-L-prolylamino)-2-oxo-1-pyrrolidine acetamide is 10^4 times more potent than MIF [1033].

Uses [1034]. MIF and its analogues are reported to exert an antidepressive effect, potentiate the learning capacity, and serve in the treatment of Parkinson's disease (→ Parkinsonism Treatment).

5.4. Endothelin [1035]

Occurrence. Endothelin (ET) [*116243-73-3*], M_r 2492 (h-ET 1), is formed in the endothelial cells and occurs in the kidneys [1036], hypothalamus, pituitary [1037], and human milk [1038].

Three different endothelins have been found in humans (h-ET 1, h-ET 2, h-ET 3) and all contain 21 amino acids with two disufide bridges, one between Cys^1 and Cys^{15} and the other between Cys^3 and Cys^{11} [1039]. p-ET [1040] is identical to h-ET 1 and r-ET is identical to h-ET 3. There is great structural similarity between the endothelins and the sarafotoxins from the venom of the snake, *Atractaspis engaddensis* [1041]. The structure of the prepro-compounds of p-ET 1, h-ET 2 [1042], and h-ET 3 [1043] have been elucidated via their cDNA.

Release. Endothelin is released in response to shear stress [1044], thrombin, angiotensin II, vasopressin [1045], TGF-β, TNF-α, and IL-1β [1046]. The release of ET is reduced by EDRF [1047], dehydration [1037], and lack of oxygen (hypoxia) [1048].

The plasma level of ET is low in genetically hypertensive rats [1049].

Plasma ET levels are elevated in patients suffering from essential hypertension [1050], cardiogenic shock, pulmonary hypertension, and in dialysis patients [1051].

Receptors. Two different ET receptors are found in the rat lung with molecular masses of 44 000 and 32 000, respectively. ET 1 and ET 2 bind tightly to the former and ET 3 binds preferentially to the latter [1052].

Treatment with ET [1053] or AT II [1054] lowers the number of ET receptors.

Biological Effects. The endothelins are regulators of blood pressure and circulation [1029]. After a quick drop in blood pressure (via EDRF release [1055]), a strong pressor effect is observed [1056]. They contract vascular and nonvascular smooth muscles [1057]. They stimulate the release of aldosterone and corticosterone in dispersed zona glomerulosa cells [1058], modulate the synthesis of noradrenaline and have a very strong stimulating effect on the release of α-ANP [1059], [1060]. Independent of gonadoliberin receptors, they also stimulate the release of gonadotropins from anterior hypophyseal cells [1061].

Possible pathological conditions arising from an increased formation of endothelin include coronary and cerebral vasospasm, cardiovascular diseases, bronchospasm, atherosclerosis and hypertension, stomach and duodenal ulcers, and acute renal failure [1062].

Structure – Activity Relationships. ET 1 and ET 2 have a higher muscle-contracting activity than ET 3 [1063].

Procompounds of ET 1 that are extended at the C-terminus [p-ET-(1 – 39) and h-ET-(1 – 38)] are 100 times less active than ET in vitro, but almost equally as active in vivo [1064].

The N-terminal amino group, the C-terminal carboxyl group, the carboxyl groups of Asp^8 and Glu^{10}, the aromatic ring in position 14, and Trp^{21} are important for activity [1065]. Loss of Trp^{21} results in a thousand-fold decrease in activity, further shortening of the molecule at the C-terminus reduces the activity still further. The N-terminal 1 – 16 fragment has no vascular activity. ET-(16 – 21) causes contraction of the isolated guinea pig bronchus (ED_{50} = 0.228 μmol/L) but not, of the rat aorta [1066].

Both disulfide bridges (Cys^1-Cys^{15}, Cys^3-Cys^{11}) are important, but not necessary, for activity. Thus, a cyclic derivative with a disulfide bridge between Cys^1 and Cys^{15}, acetamidomethyl protection in positions 3 and 11, and a C-terminal amide have no effect on the arterial blood pressure [1056]. Derivatives with incorrect disulfide bridges are also less active. [$Ala^{1,15}$]ET 1 and [$Ala^{3,11}$]ET 1 have vasoconstrictive activity, but [$Ala^{1,3,11,15}$]ET 1 is devoid of activity [1067].

5.5. Atrial Natriuretic Factor [1068]

Occurrence. A group of peptides (atrial natriuretic factor, ANF, and atrial natriuretic peptides, ANP) that exert natriuretic (i.e., increased renal sodium excretion), diuretic, and vasodilating effects have been isolated from the atrium of the mammalian heart. They are all derived from a protein precursor with 151 amino acids (prepro-ANF). Prepro-ANF is enzymatically degraded or processed and shortened at its N- or C-terminus. Depending on the species, the first 23–25 amino acids at the N-terminus of prepro-ANF belong to the precursor sequence. Asn^1 is the first amino acid in the 126 amino acids of ANF, M_r 13 679 (h-ANF). ANF (also called γ-ANP or cardionatrin IV) was isolated from atrial tissue. In the rat, ANF is stored in the granulocytes of the atrium and not metabolized further [1069]. α-ANP, M_r 3080.5 (h-α-ANP), [ANF-(99–126)], and α-ANP-(3–28) are found in the circulating blood [1070]. In humans, α-ANP and β-ANP (an antiparallel dimer of h-α-ANP) are found in the plasma [1071]; h-β-ANP can be converted to h-α-ANP [1072]. In the rat, atriopeptin III [r-ANF-(103–126)] also occurs as a circulating peptide [1073]. ANF-(1–98) and α-ANP are secreted from the cardiocytes of the atrium into the circulation, the former accumulates to a greater extent than α-ANP because of its longer half life [1074]. ANF-(1–30) and ANF-(31–67) are also found in human serum and show ANF activity [1075]. The structures of ANF from various species are described in [1076].

Immunoreactive ANF has been detected in a wide variety of tissues (e.g., brain [1077], pituitary [1078], and gastrointestinal tract [1079]). ANP-like peptides have also been isolated from the hearts of chicken [1080], eels [1081], and frogs [1082].

An ANF-like peptide, the brain natriuretic peptide (p-BNP) has been found in porcine brain [1083]. The precursors of the BNPs have been characterized through the cDNAs. The BNP structure is located at the C-terminus. b-prepro-BNP is cleaved to b-γ-BNP with 106 amino acids and to smaller forms, such as b-BNP-26 and b-BNP-32 [1084].

Another ANF-like peptide, the C-type natriuretic peptide (CNP), has been found in the porcine brain. The pharmacological spectrum of CNP is similar to that of ANP and BNP.

Release. The calcium-dependent formation of α-ANP appears to proceed via the phosphoinositide and cAMP system. The plasma levels of ANF correlate positively with the pressure of the pulmonary artery and the pressure in the right ventricle of the heart [1087].

The release of α-ANP is stimulated by an increase in volume of the blood, in hypertension, by vasopressin [1088], oxytocin, angiotensin II [1089], adrenaline [1090], by the stimulation of $α_1$-adrenergic [1091] and $α_2$-adrenergic receptors [1092], by substance P [1093], endothelin [1094], thyroliberin [1095], thyroid hormones [1096], by the stimulation of muscarinic and cholinergic receptors, by stress [1097], glucocorticoids [1098], mineralocorticoids [1099], NaCl infusion, a high-NaCl diet [1100], and anesthesia [1097].

Plasma ANP levels are especially high in newborn children and premature babies [1101]. In humans the plasma level of α-ANP increases with age [1102], in congestive cardiac defects [1103], attacks of supraventricular tachycardia and atrial bradycardia, Cushing's syndrome [1104], primary aldosteronism [1105], liver cirrhosis [1106], hyperthyroidism [1107], anorexia nervosa [1108], during exercise [1109], in pregnant women, and in chronic kidney failure [1110].

ANF secretion is lowered by hypophysectomy [1111], hypothyroidism [1112], after operations [1113], after a decrease in the plasma volume, after dehydration [1114], and by a low- NaCl diet.

The endopeptidase E.C.3.4.24.11 cleaves h-α-ANP at four sites [1115]. UK 69 578, an inhibitor of endopeptidase E.C.3.4.24.11, increases the plasma level of ANF in healthy persons [1116] and in patients with chronic cardiac defects; it also reduces blood pressure, and increases natriuresis and diuresis [1117].

Receptors. Three different receptors (ANF A, B, and C) have been isolated by molecular cloning [1118]. ANF A is a membrane-bound guanylate cyclase, it binds ANF and BNP peptides. ANF B is also a membrane-bound guanylate cyclase but binds BNP peptides preferentially. ANF C is coupled to the adenylate cyclase–cAMP system [1119]. The three receptors have a high sequence similarity.

Specific receptors exist in smooth muscle for ANF-(1–30) and ANF-(31–67), these receptors differ from those for ANF-(99–126) [1120].

Glucocorticoids [1121] and dehydration increase the number of ANP binding sites [1122]. The receptor density on the blood platelets decreases with a high-NaCl diet and increases with a low-NaCl diet [1123]. The number of ANF receptors in the rat aorta is reduced by pretreatment with α-ANP (receptor down regulation) [1124].

Biological Effects [1125]. Peripheral application of ANF peptides results in vasodilation, reduces the plasma volume [1126], lowers blood pressure, stimulates natriuresis and diuresis, and inhibits the renin–angiotensin system and the stress hormone cascade.

The vasodilating activity of ANF peptides is based on inhibition of the intracellular increase in calcium [1127].

In models that are not dependent on renin, α-ANP reduces the cardiac output and increases the total peripheral resistance. In hypertensive models in which the renin–angiotensin system is stimulated, α-ANP lowers the blood pressure by reducing the total peripheral resistance [1128]. The decreased blood pressure and cardiac output are due to the lower amount of venous blood reaching the heart and the lower blood volume [1129].

In humans, the diuretic activity of α-ANP is highest on infusion with 0.075 µg kg^{-1} min^{-1}. The natriuretic activity is highest at 0.015 µg kg^{-1} min^{-1} (2 h). A hypotensive effect is only observed at higher doses. The renal effects disappeared at doses of 0.3 µg kg^{-1} min^{-1} [1130]. In healthy subjects, vasopressin inhibits the diuretic effect of ANF and potentiates its natriuretic effect [1131]. In numerous animal models for kidney failure, ANF peptides improve kidney functions, e.g., [1132].

Other effects of ANF peptides include relaxation of the pulmonary artery [1133] and blood vessels of the lung in patients with chronic obstructive pulmonary disease [1134]; inhibition of adenylate cyclase activity [1135] and stimulation of guanylate cyclase and release of cGMP [1136]; inhibition of the release of CRH and proopiomelanocortin [1137]; inhibition of the release and activity [1138], [1139] of vasopressin (the vasorelaxing [1140] and the diuretic [1141], but not the natriuretic activity of ANF are inhibited by vasopressin); inhibition of the AT II-stimulated formation of aldosterone [1142]; inhibition of the synthesis of pregnenolone, deoxycorticosterone, progesterone, cortisol, and corticosterone; and stimulation of the formation of testosterone in interstitial cells of the mouse [1143].

Structure–Activity Relationships [1144]. The α-ANP nomenclature is used for small C-terminal molecules and the ANF-(1–126) for larger molecules:

h-α-ANP

```
S L R R S S C F G G R M D R I G A Q S G L G C N S F R Y
1         10              20            28
```

[= h-ANF-(99–126)]

α-ANP peptides that are cyclized via a disulfide bridge or by means of α-aminosuberic acid [1145] exhibit good activity. The C-terminal Tyr can be replaced by an amide function. N-terminal extension of α-ANP leads to more active derivatives [1146]: urodilatin, ANF-(95–126), has a stronger renal effect than α-ANP [1147]. The aromatic ring of Phe^8, but not that of Phe^{26} is essential for the activity of α-ANP [1148]. Deletion of Gly^9 in h-α-ANP produces an inactive product, whereas full biological activity is retained on substitution of D-Ala for Gly^9 [1145]. Position 12 of α-ANP must be lipophilic (Met or Ile). Asp^{13} is not essential for activity [1145]. α-ANP activity is increased by the phosphorylation of Ser^6 [1149].

The sequence Arg^{11}–Gly^{16} is important for receptor binding and Phe^8–Gly^{16} for the vasodilating activity [1150].

Superactive analogues are obtained by deleting the N-terminal peptide chain up to Cys^7 and by replacing Cys^7 by 3-mercaptopropionic acid (Mpr), 2-mercaptoacetic acid, or 4-mercaptobutyric acid [1151]. [Mpr^7,D-Ala^{9,16}]α-ANP-(7–27)amide has about two to three times the biological activity (receptor binding, drop in blood pressure, and natriuresis in rats) of h-α-ANP [1152]. [Mpr^7,D-Ala^9]α-ANP-(7–28) has twice the hypotensive effect and five times the cGMP-stimulating effect of α-ANP-(5–28) [1153]. r-[D-Tyr^6]α-ANP-(6–28) (dextronatrin) has the same diuretic activity as α-ANP but a lower hypotensive effect [1154]. [Aoc^{6,28},D-Ala^9]α-ANP-(6–28) [1155] exhibits strong receptor binding (IC$_{50}$ = 10^{-11}–10^{-12} M).

A cyclopeptide from [Phe^6,Pro^7,Glu^{13}]r-α-ANP-(6–22), cyclo-(F P F G G R I E R I G A Q S G L G), still produces the full biological effect, but with a low activity [1156].

Inhibition of the CRH-stimulated release of ACTH requires an intact amino terminal of α-ANP-(1–28): α-ANP-(5–28) is inactive but α-ANP-(1–11) still exerts a weak inhibitory effect [1157].

Surprisingly ANF-(1–30), ANF-(31–67), and ANF-(79–98) have a vasodilating and a cGMP-stimulating activity similar to that of ANF [1158]. ANF-(1–30), ANF-(31–67), ANF-(79–98), and ANF-(99–126) increase sodium excretion by 231, 973, 167, and 1405%, respectively. All these peptides except ANF-(79–98) have diuretic activity; ANF-(79–98) exhibits kaliuretic activity (i.e., increases renal potassium excretion) [1159].

h-α-ANP-(7–23)amide and substituted [Cys(R)^{7,23}]h-α-ANP derivatives inhibit h-α-ANP-stimulated synthesis of cGMP [1160]. In rats, cyclo-(Phe-Gly-Gly-Arg-Leu-Asp) and cyclo-(Phe-Gly-Gly-Arg-Met-Asp-Arg-Ile-Gly) have hypertensive activity [1148].

5.6. Pancreatic Peptide Family

The pancreatic peptide family includes three peptide amides containing 36 amino acids: pancreatic peptide (PP), neuropeptide Y (NPY), and the peptide with N- and C-terminal tyrosine (PYY).

There is a certain structural similarity between these peptides and neuromedin U (NMU) and the delta-sleep-inducing peptide (DSIP).

	1																	18
h-PP	A	P	L	E	P	V	Y	P	G	D	N	A	T	P	E	Q	M	A-
h-,r-NPY [1161]	Y	P	S	K	P	D	N	P	G	E	D	A	P	A	E	D	M	A-
h-PYY [1162]	Y	P	I	K	P	E	A	P	G	E	D	A	S	P	E	E	L	N-
cu-DSIP					W	A	G	G	D	A	S	G	E					

	19																	36
h-PP	Q	Y	A	A	D	L	R	R	Y	I	N	M	L	T	R	P	R	Ya
h-,r-NPY	R	Y	Y	S	A	L	R	H	Y	I	N	L	I	T	R	Q	R	Ya
h-PYY	R	Y	Y	A	S	L	R	H	Y	L	N	L	V	T	R	Q	R	Ya
p-NMU-8									Y	F	L	F	R	P	R	Na		

5.6.1. Pancreatic Peptide

Occurrence. Pancreatic peptide (PP) [59763-91-6], M_r 4182 (h-PP), is produced primarily in the pancreas. In squirrels, it was found in the same cells that produce glucagon [1163]. The precursor of h-PP has been characterized [1164]. Sequences of PP from various species are given in [1165].

Release. The pulsatile release of PP in the pancreas is coupled to an insulin pulse [1166]. The plasma level of PP is increased by electrical stimulation of the vagus, food intake [1167], stimulation of β-adrenergic receptors, the μ-receptor ligands morphine and morphiceptin, hypoglycemia induced by insulin or fasting, acetylcholine, cholecystokinin, cerulein, secretin, neurotensin [1168], bombesin [1169], gastrin, GIP, and VIP. The PP level oscillates in a 90-min rhythm in test persons with an empty stomach. The plasma level of PP increases with age [1170]. In general, patients with a low exocrine secretion from the pancreas have an excessively high plasma PP level. In half of patients suffering from pancreatic endocrine tumors, plasma PP levels are raised and cannot be lowered by atropine [1171].

The plasma PP level is lowered by stimulating the α-adrenergic receptors, by δ-receptor agonists ([D-Ala^2,D-Leu^5]enkephalin and [D-Ala^2]-enkephalin), by the intravenous administration of glucose or fatty acids, by glucocorticoids, p-PYY, atropine, and somatostatin [1172].

Biological Effects. PP inhibits the exocrine secretion of enzymes from the pancreas which is stimulated by, for example, secretin, cerulein, cholecystokinin, bethanechol, HCl, or sodium oleate [1173], and cerulein-induced pancreatic hypertrophy [1171].

Other effects of PP include antagonization of VIP-induced vasodilatation of vascular smooth muscle [1174], decrease in plasma insulin levels (rat), relaxation of the Oddi sphincter in small doses and contraction in high doses [1175], contraction of the lower esophageal sphincter [1176], release of NaHCO$_3$ from the gastric and duodenal mucosa [1177], and reduction of food intake.

Structure – Activity Relationships. The C-terminal tyrosinamide residue is very important for biological activity. Since the C-terminal hexapeptide of PP simulates the inhibitory effect of PP on exocrine pancreatic secretion, the active center appears to be located in this region. In contrast to PP, however, both the C-terminal hexa- and decapeptides are unable to lower the insulin level in rats.

5.6.2. Neuropeptide Y [1178]

Occurrence. Neuropeptide Y (NPY) [82785-45-3], M_r 4272 (h-NPY), is widely distributed in the nervous system, usually in the vicinity of arteries. In the brain, the highest concentrations are found in the hypothalamus. Immunoreactive NPY is associated with somatostatin in many neurons of the cerebral cortex, it also occurs in some cells of the diencephalon. In neurons of the arcuate nucleus, NPY is associated with growth hormone releasing hormone (GHRH) [1179]. In the adrenal medulla, NPY is found in the majority of noradrenaline-releasing granules; Met- and Leu-enkephalin have also been observed in some of these cells. NPY has been found together with VIP and PHI in the nonadrenergic nerve fibers of the small intestine [1180].

h- [1181] and r-NPY [1182] are synthesized in the form of a preprocompound with 97 amino acids. The 30 amino acid C-terminal peptide of prepro-NPY (C flanking peptide of neuropeptide Y, CPON) is associated with NPY in the brain and the peripheral nervous system [1183]. p-NPY-(12 – 36)amide (p-NPY-25) has been isolated from the porcine brain [1184]. Sequences of NPY from various species are described in [1163].

Release. NPY is released along with noradrenaline on activation of the sympathetic nerves. Release from the perfused guinea pig heart depends on extracellular calcium, calcium influx, and activation of protein kinase C [1185]. NPY is released by tyramine [1186], insulin-stimulated hypoglycemia [1187], loss of blood [1188], stress [1189], and deprivation of food [1190].

An increased level of NPY is found in platelets of spontaneously hypertensive rats [1191]. The peptide is assumed to play a role in the pathophysiology of congestive heart failure [1192]. A high plasma NPY level is observed in patients with pheochromocytoma (tumor of the adrenal medulla) [1193] and septicemia [1194].

The release of NPY is reduced by sympathectomy, reserpine, or CCK-8 [1195]. The NPY concentration in the hypothalamus decreases in old rats [1196].

Receptors. There are two receptors for NPY. The Y_1-receptor is a glycoprotein (M_r 70 000) which binds NPY, but not NPY-(13–36). The Y_2-receptor is a glycoprotein (M_r 50 000) which binds both NPY and NPY-(13–36). Both receptors also bind PYY [1197], [1198]. The vascular effects of NPY are presumably mediated by the Y_1-receptors and the cardiac effects by the Y_2-receptors [1199]. NPY inhibits the release of noradrenaline in the heart and reduces heart rate [1200]. [Leu^{31},Pro^{34}]NPY is a selective Y_1-receptor agonist [1201].

Biological Effects. NPY increases intracellular calcium in vascular smooth muscle cells [1202], inhibits cAMP formation which is stimulated by, e.g., forskolin [1203], and the calmodulin-stimulated phosphodiesterase [1204]. NPY exerts a strong, calcium-dependent vasoconstricting effect on cerebral arteries in vitro. It inhibits the VIP-induced relaxation of arteries [1205] and causes a persisting increase in blood pressure when applied parenterally.

NPY has many other effects:

Stimulation of diuresis and natriuresis through release of ANF and inhibition of vasopressin [1206].

Inhibition of the secretin- and cholecystokinin-stimulated exocrine pancreatic secretion and reduction of the glucose or β-adrenergic stimulated release of insulin [1207].

Inhibition of the pentagastrin-stimulated secretion of gastric acid [1208].

Increase in the serum LH level in male rats [1209] and in ovariectomized, estrogen-treated rats via the release of gonadoliberin [1210] and the potentiation of gonadoliberin activity [1211].

When applied centrally, NPY stimulates the release of ACTH and cortisol via the potentiation of CRH [1212], increases food and water intake [1213] in the rat, delays the emptying of the stomach [1214], and reduces the blood pressure [1215] and body temperature [1216]. NPY may play a role in bulimia nervosa.

Structure–Activity Relationships. Both the N- and C-termini of NPY are important for binding to the Y_1-receptors, while C-terminal fragments suffice for the binding to the Y_2-receptors. The sequence 5–24 is not necessary for binding to Y_1-receptors. Thus, short sequences such as

[D-Cys^7,Aoa$^{8-17}$, Cys^{20}]p-NPY

[Cys^5,Aoa$^{7-20}$,D-Cys^{24}]p-NPY

bind to mouse brain membranes [1217] (Aoa = 8-amino-octanoic acid). [Aca$^{5-24}$]NPY (Aca = ε-amino-caproic acid) still has about one-third the receptor binding capacity of NPY [1218] and [Ahx$^{5-24}$]NPY (Ahx = 6-aminohexanoic acid) has about one-tenth the activity of NPY [1219].

[Pro^{34}]p-NPY [1220], [His^{34}]p-NPY, and [Leu^{31},Pro^{34}]NPY are selective agonists for Y$_1$-receptors [1221]. [Leu^{31},Pro^{34}]NPY has a higher blood pressure increasing activity than NPY [1221].

Arg^{33} and Arg^{35}, but not Arg^{25} are important for biological activity [1219]. The twelve C-terminal amino acids and the amide function are required to activate the Y$_2$-receptor when tested on the electrically stimulated rat vas deferens. NPY-(26–36)amide and PYY-(26–36)amide have no activity, NPY-(16–36)amide is most active in this test [1222]. NPY-(12–36)amide, NPY-(22–36)amide, and NPY-(22–33)amide inhibit calmodulin-stimulated phosphodiesterase more than NPY. N-terminal fragments are devoid of activity [1204].

NPY-(18–36) is an NPY antagonist for binding to membranes of the heart [1192].

5.6.3. Peptide with N-terminal Tyrosine and C-terminal Tyrosinamide

Occurrence. Peptide with N-terminal tyrosine and C-terminal tyrosinamide (PYY), [*4985-46-0*], M_r 4310 (h-PYY), occurs primarily in the small intestine. In humans, PYY-forming cells are found primarily in the duodenum; in the dog and the pig, PYY is mainly synthesized in the mucous membranes of the ileum and the colon [1223]. Sequences of PYY in various species are described in [1163].

Release. The plasma level of PYY increases after a meal and after the application of p-gastrin releasing peptide [1224].

A raised basal and postprandial level of PYY is observed in steatorrhea (fat diarrhoea) caused by atrophy of the small intestine and in chronic pancreatitis [1225]. In patients suffering from Dumping syndrome (sweating, weakness and diarrhoea caused by stomach operations), plasma levels of PYY increase 6–7-fold on oral glucose administration; somatostatin inhibits this increase [1225].

Biological Effects. PYY reduces the plasma CCK level [1226], the postprandial [1227], or secretin- or CCK-8-stimulated [1228] exocrine secretion from the pancreas and the postprandial secretion of gastric acid [1227]. The vasoconstricting and antisecretory effects on the pancreas are partially abolished by adrenalectomy [1229].

Other effects include increase in blood pressure, reduction of plasma PP and motilin levels [1230], prolongation of transit in the small intestine [1231], and strong emetic activity (PYY >NPY >> PP) [1232].

5.6.4. Neuromedin U

Occurrence. Neuromedin U (NMU) [*111745-44-9*] was first isolated from the spinal cord of the pig (p-NMU-25, M_r 3143; and p-NMU-8). r- [1233], ra- [1234], and cav-NMU [1235] were recovered from the gastrointestinal tract.

NMU-like immunoreactivity is widely distributed, however. In the rat, it is found primarily in the pituitary and the gastrointestinal tract, but also occurs in the brain, spinal cord, and genital tract [1236].

Biological Effects. NMU increases blood pressure. It also contracts the rat uterus [1233], the isolated human ileum [1237], and muscle strips from the human bladder [1237].

Structure–Activity Relationships. r-NMU-25 is about twice as active as p-NMU-25 on the rat uterus. The C-terminal heptapeptide amide is essential for activity [1233].

5.6.5. Delta-Sleep-Inducing Peptide [1238]

Occurrence [1239]. Delta-sleep-inducing peptide (DSIP)[*69431-45-4*], M_r 849, was first isolated from a dialysate of venous blood in rabbits which were kept awake by electrically stimulating the thalamic region. Although the nonapeptide is widely distributed in the brain and other tissues, it is most concentrated in the thalamus and pineal gland. DSIP has been found in the pituitary [1240], urine, cerebrospinal fluid, and human milk. The immunoreactive DSIP in plasma, urine, and cerebrospinal fluid is bound to proteins which protect it from proteolysis.

Release. The concentration of DSIP in the brain increases during hibernation. In rats addicted to alcohol, however, the DSIP concentration in the brain falls. Corticoliberin and vasopressin inhibit the release of DSIP from the adenohypophysis [1240].

Biological Effects [1239]. DSIP can pass through the blood–brain barrier. Infusion into the ventricular system of the rabbit brain stimulates the δ waves of the electroencephalogram which are characteristic of natural sleep. Intracerebroventricular injection of DSIP in mice produces an analgesic effect [1241]. Although naloxone abolishes the sleep-inducing [1242] and analgesic effects [1241], DSIP does not bind to opioid receptors [1243]. DSIP increases resistance to acute emotional stress in rats [1244], inhibits the release of prolactin [1245], and the CRH-stimulated increase in ACTH [1246] and glucocorticoids [1247].

DSIP directly stimulates the release of growth hormone [1248] and inhibits the release of growth hormone releasing hormone [1249] via dopaminergic mechanisms. Similar to NPY and PP, intracerebroventricular application of DSIP stimulates the release of LH in estrogen- and progesterone-treated ovariectomized rats. On the other hand, DSIP has no LH-releasing activity in vitro (hypophysis) [1250].

Structure–Activity Relationships. Shortening the chain at the N- or C-terminus or replacement of α-Asp by β-Asp produces inactive peptides. Replacement of Trp^1 by D-Phe or 4-nitrophenylalanine is without effect, but [D-Trp^1]- and [Phe^1]DSIP increase the motor activity. Phosphorylation of the hydroxyl group of Ser^7 leads to a more potent and longer-acting product (DSIP-P) that is more stable to enzymatic degradation than DSIP.

Uses. In patients with sleep disturbances, the duration of sleep has been significantly prolonged by the administration of DSIP. It increases the power of concentration and tolerance to stress in the waking phase. DSIP exerts an analgesic and antidepressive effect in patients with migraine, headaches, psychogenic pain attacks, and tinnitus [1251]. Treatment with DSIP produces a favorable effect in opiate or alcohol withdrawal, where it abolishes increased blood pressure, tachycardia, tremor, and sweating [1238].

6. Cholecystokinin and Gastrin

Cholecystokinin (CCK), which stimulates contraction of the gallbladder, is identical to pancreozymin, which was made responsible for enzyme release from the pancreas. The counterparts of CCK in amphibians are cerulein (CRL) and related peptides, their common C-terminal pentapeptide is identical to that of gastrin.

Gastrin stimulates the release of gastric acid and pepsin in the stomach. Its activity spectrum is dictated mainly by its C-terminal pentapeptide, whereas that of CCK and CRL depends on the C-terminal heptapeptide, which should contain a sulfated tyrosine residue.

Release of CCK and GT are initiated by special peptides. A trypsin inhibitor with a structure similar to that of the epidermal growth factor has been characterized as the CCK-releasing peptide. The gastrin-releasing peptide is structurally related to bombesin from amphibians.

6.1. Trypsin-Sensitive Cholecystokinin Releasing Peptide

A peptide with 61 amino acids has been isolated from the pancreatic juice of rats and stimulates pancreatic enzyme secretion by inducing the release of CCK. This cholecystokinin releasing peptide (CCKRP) is released in response to a diet rich in protein [1252]. It is structurally similar to the pancreatic trypsin inhibitors of the Kazal type and is classified as such [1252]. Structural similarities are found between CCKRP and the epidermal growth factor (EGF) [1253]. CCKRP also binds to EGF receptors and exerts a growthpromoting effect [1253].

6.2. Cholecystokinin and Cerulein

Occurrence. Cholecystokinin (CKK) [*9011-97-6*] is formed as prepro-CCK. The structures of h- [1254], p- [1255], and r-prepro-CCK [1256] have been elucidated via their cDNA.

CCK has been isolated in the form of peptides containing 58 (CCK-58) [1257], 39 (CKK-39), 33 (CCK-33), 22(CCK-22) [1258], 12 (CCK-12), 8 (CCK-8) [1258], 5 (CCK-5) [1259], and 4 (CCK-4 and GT-4) C-terminal amino acids. These peptides are synthesized in the brain, small intestine, and the pancreas.

The r-prepro-(24–32)nonapeptide (V-9-M) is found in the rat brain. It has a persisting memory-improving effect when injected into the lateral ventricle [1260].

CRL has a similar sequence to CCK-8 and has been isolated from the Australian frog, *Litoria caerulea*. The prepro-CRL from the skin of *Xenopus laevis* codes for five heterogeneous pro-CRL peptides [1261].

h-CCK-8 M_r 1143   D Ys M G W M D F a
CRL M_r 1352   *Q Q D Ys T G W M D F a

Release. Ingestion of unsaturated fats and fatty acids causes a greater release of CCK than ingestion of the corresponding saturated compounds. CCK is also released in response to the acidification of the duodenum [1262], trypsinsensitive cholecystokinin releasing peptide, bombesin [1263], depolarizing K^+ concentrations, and veratridin (a Na^+ channel activator) [1264].

Bile acids [1265], somatostatin, and atropine [1266] inhibit the release of CCK. Stress modulates the CCK content in the rat hypothalamus [1267].

The postprandial level of CCK in the plasma of patients with pancreatic insufficiency is lower than that of healthy persons [1268]. The plasma CCK level in patients with non-insulin-dependent diabetes [1269], hepatic cirrhosis [1270], and chronic pancreatitis [1271] is elevated.

Receptors [1272]. There are two main types of CCK receptor: CCK-A and CCK-B [1261]. The CCK-A receptors are found on pancreatic acinar cells, in the gallbladder, on adenohypophysis cells, and on inhibitory neurons of the lower esophageal sphincter. Only sulfated CCK derivatives bind to these high-affinity receptors. Gastrin exhibits a 500 to 1000 times weaker binding. CCK-B receptors are found in the cerebral cortex and in the central nervous system. They have a tenfold higher preference for sulfated than for nonsulfated CCK derivatives. A third class of receptors termed the gastrin receptors are found on the parietal cells and the gastrointestinal smooth muscles. They have about the same affinities for sulfated and nonsulfated CCK derivatives [1273].

Binding of CCK to pancreatic receptors leads to an up regulation of the CCK receptors [1274]. In newborn rats, the number of CCK receptors is increased by hydrocortisone [1275]. The CCK receptors on rat pancreatic acini are increased fourfold in diabetic rats [1276].

Selective ligands for the CCK-A receptor are described in [1277]. Selective ligands for the CCK-B receptor are described in [1278]–[1282].

Biological Effects. The activity spectrum of CCK was mainly studied with synthetic CCK-8 (sincalide) and CRL (ceruletide).

Receptor activation leads to the breakdown of phosphoinositides, mobilization of cellular calcium, activation of protein kinase C, and the phosphorylation of intracellular proteins.

The most important physiological effects of CCK are the CCK-A-receptor-mediated stimulation of bile flow [1283] (contraction of the gallbladder, relaxation of the bile duct and the Oddi sphincter), the exocrine pancreatic secretion [1284], [1285], and potentiation of the secretin-induced exocrine pancreatic secretion. The CCK-8-induced contraction of the gallbladder is increased in patients suffering from constipation and lowered in patients with diarrhoea [1286] and celiac disease [1287]. VIP and secretin potentiate the effect of CCK on the gallbladder and pancreas. Somatostatin [1288] and atropine [1289] inhibit the CCK-induced contraction of the gallbladder.

cAMP is activated in CCK-induced relaxation. In contractions, cGMP is activated and intracellular calcium and acetylcholine are released [1290].

Other CCK-A-receptor-mediated effects include decreased food intake in rats after peripheral injection of CCK-8 [1291], trophic effects on the pancreas [1292], increased insulin and glucagon secretion in mice [1293], inhibition of pentagastrin-stimulated gastric acid secretion [1294], and increased dopamine efflux in rat brain [1295].

CCK-B-receptor-mediated effects are CCK/opiate interactions [1296] and anxiety [1279]. CCK has an analgesic effect through δ-opioid receptors via the release of endogenous opioids [1297] but it reduces morphine- or β-endorphin-induced analgesia [1298] via μ-receptors [1299].

Other effects of CCK include relaxation of the corpus of the stomach and the lower esophageal sphincter, contraction of the pylorus and prolongation of the time required to empty the stomach, increase in motility from the distal part of the duodenum up to the rectum [1300], stimulation of gastric acid secretion, and inhibition of the gastrin-stimulated secretion of gastric acid.

Structure – Activity Relationships. The C-terminal octapeptide of CCK (CCK-8) has the full activity spectrum of CCK. The C-terminal heptapeptide CCK-7 and CRL-(4–10) [1301] are required for CCK activity. In vivo, deamino^1-CCK-7 has the same protein secretory activity as CCK-7, but is more effective in contracting the gallbladder than CCK-7. CCK-6 has no activity.

The sulfated tyrosine residue is important for binding to CCK-A receptors and the activity spectrum of CCK and CRL. The activity of desulfated CCK-8 resembles that of gastrin [1302]. The substitution of Phe(CH$_2$SO$_3$H) for Tyr(SO$_3$H) gives more stable, highly active derivatives [1303]. If the N-terminal Tyr(SO$_3$H) in CCK-7 is replaced by a 3-(4-sulfoxyphenyl)-2-methylpropanoyl group, the analgesic activity is about five times that of CCK-7 [1259].

Met^3 in CCK-8 is not essential for activity and can be replaced by Leu or Thr; Met^6 can only be replaced by Nle without loss of activity. Thus, Arg-[Thr^3,Nle^6]CCK-8 [1304] and Boc-[Nle^{3,6}]CCK-8-(2–8) [1305] are stable to oxidation and fully active.

Gly^4 can be replaced by D-amino acids. [D-Ala^4]CCK-8 and [D-Trp^4]CCK-8 are almost as active as CCK-8, but they are stable to brain proteases and act longer [1306].

The effect of Boc-[Nle^3-ψ-(COCH$_2$)Gly^4]CCK-8-(2 – 8) on the release of amylase from dispersed rat pancreatic acini is about the same as that of CCK-8, but it also has a central antagonistic effect [1307].

[β-Asp^7]CCK-8 loses its cholecystokinin properties but retains its pancreozymin activity. The substitution of Ser(SO$_3$H), Thr(SO$_3$H), or Hyp(SO$_3$H) for Asp^7 in Ac-CCK-8-(2 – 8) gives preparations that have 2 – 3 times the activity of CCK-8 on isolated strips of gallbladder [1308].

Phe^8 can be replaced by 3-(2-naphthyl)alanine or by 3-cyclohexylalanine without loss of activity [1309].

The C-terminal amide function is very important for the activity of CCK peptides [1310]. CCK receptor antagonists are obtained by deleting C-terminal amino acids. The shortest active fragment is CCK-8-(1 – 5) [1311]. Stronger antagonists are succinyl–CCK-8-(2 – 7)2-phenethylamide [1312] and Boc-[D-Trp^5,Nle^6]CCK-8-(2 – 7)2-phenethyl ester [1313]. Surprisingly strong antagonists are also obtained by substituting D-Orn(Z) for Met^6 in CCK-8 [1282].

Relatively simple derivatives of glutamic acid and tryptophan act as CCK antagonists [1314]. Examples are the inhibitor of acid secretion proglumide (*N*-benzoyl-L-glutamic acid-α-dipropylamide; Milid, Offermann), lorglumide (CR-1409, 3,4-dichlorobenzoyl-Glu-dipentylamide), loxiglumide (CR-1505, 3,4-dichlorobenzoyl-Glu-pentyl-3-methoxypropylamide), and 4-chlorobenzoyltryptophan (benzotript) [1315].

Strong CCK antagonists are asperlicin, a nonpeptide component of *Aspergillus alliaceus*, and its benzodiazepine variants, devazeptides (MK-329 = L-364,718), L-365,260, trifluadom, and L-156,440. L-365,260 preferentially binds to CCK-B receptors, the others bind to CCK-A receptors [1314].

Uses. CCK-8 (sincalide, Squibb – Heyden) [*25126-32-3*], and CRL (ceruletide; Takus, Farmitalia) [*17650-98-5*] [1281] are used in X-ray diagnostics and in the diagnosis of pancreatic function. Their administration halves the time required to pass the small intestine and thus permits quick examination of the intestine with an oral contrast medium. Ceruletide is used therapeutically in postoperative intestinal atonia and paralytic ileus because of its stimulating effect on the small intestine.

Ceruletide can be employed for the expulsion of gallstones [1316] and in biliary colic [1317]. Sincalide and ceruletide (nasal application three times daily, 100 µg) alleviate chronic pancreatitis [1318]. Clinical reports confirm the analgesic properties of CCK-8 [1319].

6.3. Gastrin Releasing Peptide and Bombesin

Occurrence. Gastrin releasing peptide (GRP) [*80043-53-4*], M_r 2859 (h-GRP), is a peptide amide consisting of 27 amino acids and has been isolated from the gastric tissue of pigs, chicken, and dogs and from human lung tumors [1320]. Its C-terminal heptapeptide is identical to that of bombesin. Bombesin (BB) [*31362-50-2*], M_r 1620, 14 amino acids, is a peptide amide from the skin of the frog *Bombina bombina*. Alytesin, litorin, and ranatensin are peptides with related structure and activity that have been isolated from amphibians (frog) [1321].

The neuromedins B, B-32, B-30, and C have been isolated from the brain and spinal cord of the pig; p-neuromedin C corresponds to the C-terminal decapeptide of p-GRP.

Immunoreactive GRP has been found in the central nervous system and peripheral tissue (e.g., gastrointestinal tract). Material similar to GRP-(14–27) has been found in endocrine tumors [1322] and cow's milk [1323].

Release. Electrical stimulation of the vagus increases formation of GRP in the pancreas [1324]. The β-adrenergic stimulation of gastrin and somatostatin appears to be mediated by GRP [1325].

Receptors. At least two different types of receptor are assumed for GRP, BB, and the neuromedins. BB and GRP bind to pancreatic acinar cells and murine 3T3 cells with high affinity, while neuromedin B binds with low affinity. The esophageal mucosa possesses receptors which have a high affinity for neuromedin B, but a low affinity for GRP. The GRP antagonist [D-Phe^6]BB-(6–13)OEt binds preferably to the pancreas, while [Tyr^4,D-Phe^{12}]BB and the SP antagonists [D-Pro^4,D-Trp^{7,9,10}]SP-(4–11) and [D-Arg^1,D-Trp^{7,9},Leu^{11}]SP bind better to the esophagus than to the pancreas [1326].

Biological Effects. BB and GRP have similar effects. In healthy persons, a BB infusion increases the plasma levels of gastrin, CCK, motilin, pancreatic peptide, vasoactive intestinal peptide, gastrin inhibiting peptide, glucagon, insulin, and trypsin [1327], as well as the plasma level of gonadoliberin-stimulated lutropin and follitropin [1328]. The blood glucose level is lowered and plasma Ca^{2+} levels are raised. Somatostatin inhibits BB-induced hormone secretion [1329]. In healthy test persons, the food-induced release of insulin [1324] and the thyroliberin-stimulated levels of thyrotropin and prolactin in the plasma are lowered by an infusion with BB [1328]. These effects are due to increased plasma somatostatin levels. Other effects of GRP/BB include stimulation of gastric acid secretion and exocrine pancreatic secretion, increase in bile flow [1330], trophic action on the gastrin-producing cells of the stomach [1331] and on the pancreas [1332], as well as autocrine growth in cancer cells [1333].

Central application of BB induces hypothermia and adrenaline release which, in turn, inhibits insulin release, and stimulates glucagon release. Other central effects are the reduction of the plasma growth hormone level [1334], stimulation of dopamine synthesis in the hypothalamus [1335], suppression of gastric acid and pancreatic secretion, and inhibition of food and liquid intake [1334].

Structure–Activity Relationships. GRP-(14–27) acts like BB [1336]. The C-terminal nonapeptide of BB and Ac-GRP-(20–27) have minimum length and maximum effect. GRP-(23–27) is the smallest active compound. Shortening the molecule at the C-terminus yields preparations with low affinity and activity. Acetylated peptides shortened at the N-terminus are biologically more active than the corresponding compound with a free amino group. Trp^8 and His^{12} are very important for GRP activity.

Antagonists. [D-Phe^{12},Leu^{14}]BB is a competitive BB antagonist. It inhibits in vitro BB-induced amylase secretion from guinea pig acini cells. [D-Phe^{12}]BB and [Tyr^4,D-Phe^{12}]BB have similar activity [1337]. Introduction of D-Phe^6 enhances the antagonistic activity, while D-Phe^5 has no effect [1338].

BB antagonists are also formed by deleting the C-terminal methionine [1339]. Me₃-C-CO-His-Trp-Ala-Val-D-Ala-His-Leu-OMe exhibits good receptor binding and inhibits mitogenesis of mice Swiss 3T3 cells.

The best (100%) in vivo inhibition of BB-stimulated amylase secretion was produced by Me₂-CH-CO-His-Trp-Ala-Val-D-Ala-His-Leu-NH-Me (ICI 216 140) [1339].

Other good de-Met-antagonists in the receptor binding assay are [D-Phe⁶]BB-(6–13)ethylamide, [D-Phe⁶]BB-(6–13)propylamide [1340], Ac-GRP-(20–26)OEt [1341], and [D-Phe⁶]BB-(6–13)OEt [1342].

Reduction of the peptide bond between positions 13 and 14 yields potent BB antagonists, e.g., [Leu¹³-ψ-CH₂NH-Leu¹⁴]BB [1343]. Shortening of this antagonist at the N-terminus reduces antagonistic activity. Introduction of D-Phe at position 6 and N-terminal shortening produce a compound with ten times the activity [1344].

Uses. BB is used as a diagnostic aid in the gastrin stimulation test. For instance, patients suffering from antral gastritis have an abnormally low gastrin level after BB stimulation.

In severe chronic pancreatitis [1345] and in patients with duodenal ulcers [1346], BB increases the gastrin level to a greater extent than in healthy persons. BB increases the plasma trypsin level in healthy persons, but not in patients with pancreatic insufficiency [1347].

6.4. Gastrin

Occurrence. Gastrin (GT) [9002-76-0], M_r 2098 (h-GT-17), was the first gastro-intestinal hormone to be structurally characterized. It is synthesized primarily in the antrum and the duodenum as prepro-GT. Immunoreactive GT has also been found in the pituitary, pancreas, and nerves. In humans, tyrosine sulfation depends on the site of synthesis and the degree of development [1348].

GT occurs in several molecular sizes: GT-34 (big gastrin), GT-17 (little gastrin), GT-14 (mini gastrin), GT-6, and GT-4. It is structurally related to the transformation protein of polyoma virus (TPPV) and the chicken antral peptide (g-AP) which stimulates gastric acid production in chickens [1349]. The main physiologically active component is GT-17. Although GT-14 has full biological activity, it only occurs in very small amounts.

A sulfated myotropic neuropeptide, leucosulfakinin, has been isolated from the head of the cockroach [1350]; its intestinal myotropic activities are similar to those of GT. Related peptides are found in other insects.

h-GT-34	*Q L G P Q G P P H L V A D P S K K Q G P W L E E E E E A Y G W M D Fa
h-GT-17	*Q G P W L E E E E E A Y G W M D Fa
h-GT-14	W L E E E E E A Y G W M D Fa
GT-4	W M D Fa

Release. The atropine-resistant secretion of GT-17 and GT-34 occurs postprandially, especially in response to protein or on stimulation of the vagus. GT is also released by the gastrin releasing peptide, bombesin, carbachol [1351], Ca^{2+}, vitamin D_3 [1352], tolbutamide [1353], PGE_2 [1354], hypoglycemia [1355], amino acids and the corresponding amines [1356], inhibitors of monoamine oxidase [1357], and testosterone [1358].

The release of GT is reduced by Mg^{2+}, acidification of the food in the antrum to pH 1.2 – 1.4 [1359], inhibitors of amino acid decarboxylases, and estrogen [1358].

In humans, the GT level rises during the day and reaches a maximum at about 6 p.m. [1360]. The plasma level of gastrin is increased in the Zollinger – Ellison syndrome (a GT-producing tumor), pernicious anemia (some patients develop autoantibodies against gastrin receptors [1361]), gastric ulcers [1360], and hyperthyroidism [1362].

Biological Effects. GT increases the formation of cAMP and of inositol phosphates [1363] and decreases the plasma level of calcium.

In the stomach, GT increases the formation of gastric acid and pepsin and stimulates the blood circulation in the mucosa [1364]. GT-17 stimulates the secretion of gastric acid in patients with duodenal ulcers more than in healthy persons [1365]. The gastrin inhibitory peptide (GIP), somatostatin, secretin, neurotensin, glucagon, glicentin, thyroliberin, the anorexigenic peptide, cerulein, and cholecystokinin inhibit the GT-stimulated secretion of gastric acid.

The synthesis and marketing of Boc-β-Ala-GT-4 [*5534-95-2*] (pentagastrin) a GT analogue that is fully active in humans, stimulated intensive investigation of the activity of GT. An infusion with pentagastrin, like a protein meal, increases endogenous GT, insulin, GHRH, and calcitonin, but decreases the plasma level of somatostatin (SRIH). The primary effect of pentagastrin appears to be the inhibition of SRIH release; gastrin, insulin, growth hormone, and calcitonin can then be released.

Other effects of gastrin include trophic effects on the oxyntic mucose, enterochromaffin-like cells, A-cells [1366], pancreas (potentiated by secretin [1367]), tumors of the human stomach and colon [1368], [1369], and rat pancreatic adenocarcinoma [1370]; stimulation of histamine secretion; increase in the pressure of the lower esophageal sphincter; and inhibition of the CCK-induced contraction of the pylorus.

The common C-terminal tetrapeptide amide of gastrin and CCK (GT-4, trymafan, tetrin) stimulates the endocrine secretion of insulin, glucagon, somatostatin, and pancreatic peptide. Exocrine secretion from the pancreas is primarily stimulated by the higher molecular forms of gastrin and CCK rather than by GT-4.

In healthy subjects, intravenous GT-4 causes short attacks of panic or symptoms of fear [1371]. In rats, intracerebroventricular application causes deterioration of memory [1372].

Structure – Activity Relationships. GT-34 is more potent and acts longer than GT-17 when tested in humans [1373]. The five glutamic acid residues at positions 6 – 10 in h-GT-17 are of importance for biological activity [1374].

GT-4 has the structure required for gastrin activity. The C-terminal amide is important for biological activity. Substitution of other amino acids for Asp results in substantial loss of activity.

Replacement of the GT-4 amino acids by the corresponding D-amino acids decreases activity. Agonists and antagonists can be obtained by reducing the peptide bonds. Boc-[Trp^1-ψ-(CH$_2$NH)-Leu^2]GT-4 is an agonist which binds more tightly to gastric mucosa cells than Boc-GT-4 [1375]. Trp^1 is not required for the insulin-releasing activity of GT-4, it can be replaced by L- or D-Orn [1376]. [Pro^2]GT-4 is more effective than GT-4 in releasing insulin from isolated islet cells of the rat pancreas [1377].

Very potent analogues are obtained by N-terminal extension of GT-4; pentagastrin in which GT-4 is extended by Boc-β-alanine is the best known. Glutaroyl-[Leu^5]GT-7 (deglugastrin) is more active by a factor of about two. Peptides having a GT-4 or GT-6 structure with a glycosylated N-terminal amino group have a higher water solubility combined with excellent biological activity.

Oxidation of Met causes loss of gastrin activity. However, both Met residues of GT-7 can be replaced (e.g., by Leu) without loss of activity.

Antagonists. C-terminal GT sequences that lack the C-terminal Phe-amide inhibit GT activity (e.g., BocGT-8-(1–7)amide). The smallest antagonistic sequence is Boc-Trp-Met-Asp-amide [1378]. Somewhat stronger antagonists are obtained by substituting β-homoaspartic acid for Asp, e.g., Boc-Trp-Leu-β-homo-Asp-Phe-amide [1379].

Antagonists are also obtained by reducing the peptide bond between Leu and Asp [1375].

Uses. Pentagastrin (Gastrodiagnost, Merck) is used as a diagnostic aid to determine the maximal stimulation of gastric acid secretion.

7. Secretin Family

The first peptide hormone of the gastrointestinal tract, secretin (SEC), was discovered in 1902 by BAYLISS and STARLING. The structure of secretin was established in the 1960s by V. MUTT and coworkers. There is considerable structural similarity between SEC and the vasoactive intestinal peptide (VIP), p-PHI-27, h-PHM-27, pituitary adenylate cyclase activating polypeptide (PACAP), glucagon (GG), two glucagon-like peptides (GLP-1 and GLP-2), glucose-dependent insulinotropic and gastrin inhibiting peptide (GIP), and growth hormone releasing hormone (GHRH).

GHRH activity triggers the growth hormone cascade: it releases growth hormone in the pituitary which in turn stimulates the synthesis of the somatomedins. Hormones that are structurally related to growth hormone are placental lactogen, prolactin, proliferin, and prolactin-like and prolactin-related proteins. The somatomedins are structurally related to insulin and relaxin. On the basis of their activity, somatostatin, galanine, and pancreastatin also belong to this group. Galanine triggers the growth hormone cascade by releasing GHRH and, like pancreastatin, inhibits the glucose-induced release of insulin. Somatostatin inhibits the release and activity of GHRH, insulin, glucagon, and many other peptide hormones.

7.1. Secretin

Occurrence. Secretin (SEC) [*1393-25-5*], M_r 3039.5 (h-SEC), contains 27 amino acids and is secreted primarily by the mucosa of the duodenum and upper jejunum. It is also found in the antrum, brain [1380], hypothalamus, and pituitary. Propeptides of p-secretin have been isolated and characterized [1381]. The structures of the prepro-hormones of r- and p-SEC have been established via the corresponding cDNA [1382].

h-SEC
H S D G T F T S E L S R L R E G A R L Q R L L Q G L Va

There is some similarity between secretin and the dentinal fluid transport stimulating peptide (DFT), also called the parotid hormone (PH). DFT has been isolated from the parotid gland [1383] and increases the transport of fluid in the teeth [1384]. The C-terminal regions of SEC show structural similarity with allatostatin (a peptide from the brain of the cockroach) [1385] and the bombilitins (mast cell degranulating peptides from the poison of the bumble bee) [1386].

Release. In the small intestine, hydrochloric acid (pH < 4) stimulates the secretion of SEC into the blood stream. Acid-stimulated release of SEC is mediated by a SEC-releasing peptide [1387]. Stress produces a 3–6-fold increase in the plasma levels of secretin. Secretin secretion from the intestinal mucosa is stimulated by high extracellular concentrations of K$^+$ and inhibited by the absence of extracellular Na$^+$ and Ca^{2+} [1388].

SRIH inhibits the release of SEC and exocrine pancreatic secretion. Patients with chronic kidney damage and Zollinger–Ellison syndrome (a gastrin-producing tumor) have a significantly raised level of SEC.

Receptors. SEC binds to the perfused pancreas and the gastric mucosa of rats. In contrast to SEC bound to the pancreas, SEC bound to the gastric mucosa can be displaced by tetragastrin [1389].

Biological Effects. The activity of SEC is generally expressed in clinical units (CU) which are determined in the dog (pancreatic secretion of aqueous NaHCO$_3$): 1 mg of SEC has 4000–5000 CU. The physiological level of about 20 pg/mL, observed in the human jejunum after H$^+$ stimulus, should be achieved after an infusion of 0.03 CU kg^{-1} h^{-1}.

SEC stimulates the release of an aqueous NaHCO$_3$-containing secretion from the panreas which neutralizes any hydrochloric acid entering the small intestine [1390]. Cholecystokinin, neurotensin [1391], and adenosine [1392] potentiate the SEC-stimulated pancreatic secretion. SEC-induced protein secretion from the pancreas is described frequently but is due to impurities, such as cholecystokinin [1393].

Other effects of SEC include:

1) Activation of adenylate cyclase and cAMP synthesis
2) Proliferative effect in the pancreas which is potentiated by CCK [1394]
3) Reduction of the plasma gastrin level and of the formation of gastric acid (via SRIH and prostaglandins [1395], potentiated by neurotensin [1391])
4) Reduction of stress-induced hemorrhage from the stomach [1396]
5) Vasodilative and motility inhibiting effect (dilation of the pancreatic ducts [1397], reduction of the lower esophageal sphincter pressure [1398], increase in the circulation to the pancreas and duodenum)
6) Contraction of the pyloric sphincter (delay in the emptying of the stomach [1399])
7) Potentiation of the CCK-stimulated flow of bile [1400]
8) Release of insulin in vitro

Structure–Activity Relationships. Analogues of secretin that are more active than the native peptide have not been found. Shortening the molecule at the N- or the C-terminus produces less active products. For instance, SEC-(2–27)amide has only 1% of the activity of secretin. The shortest known active sequence is SEC-(1–13): 0.1 mg/kg of this substance still exerts a significant effect on the pancreas. However, 1 mg/kg of SEC-(1–11)amide shows no activity in the rat. Similar to His^1, Asp^3, and Phe^6, Arg^{12} is important for secretin activity.

Secretin extended at the C-terminus by Gly-Lys-Arg (natural pro-SEC) still has about 80% of the SEC activity [1401].

The Asp^3–Gly^4 bond of secretin is responsible for its frequently reported instability. In an acid medium the β-carboxyl group of aspartic acid acylates the adjacent C-terminal peptide bond. This aspartoyl peptide then rearranges to give a peptide in which the peptide bond is formed with the β-carboxyl group of aspartic acid (and not the α-carboxyl group). The Asp-Gly-(3–4) and the Asp-Ser-(15–16) bonds in p-SEC are cleaved in aqueous solution at pH 5 [1402]. The main decomposition product in neutral phosphate buffer is [β-Asp³]SEC [1403]. In acidic acetate buffer (pH 4), SEC-(16–27), SEC-(4–27), [β-Asp³]SEC, [3-aspartoyl]SEC, and [βAsp¹⁵]SEC are formed [1403]. These decomposition products have little or no biological activity.

Uses. Secretin is generally administered parenterally, but the intranasal application of a dry powder has also been described. Secretin (Sekretolin, Hoechst) is applied parenterally in doses of 75–100 CU to test pancreatic function and to diagnose the Zollinger–Ellison syndrome. In contrast to healthy subjects, patients with the Zollinger–Ellison syndrome react to a diagnostic intravenous dose of secretin (75 CU) with increased plasma levels of gastrin and GIP and not a decreased level of gastrin [1402]. SEC normally increases the diameter of the pancreatic duct, but this is not the case in patients with chronic pancreatitis [1404]. Infusion of sekretolin (0.25–0.5 CU kg^{-1} h^{-1}) stops bleeding of the upper gastrointestinal tract [1405]. SEC improves the condition of patients suffering from peptic ulcers [1406]. The relaxing effect on the lower esophageal sphincter can be exploited in patients with achalasia [1398].

The insulin-releasing effect of SEC (2 CU/kg) can be used to differentiate between single insulinomas or nesidioblastosis and multiple B-cell adenomas or hyperplasia [1407].

7.2. Vasoactive Intestinal Peptide [1408]

Occurrence. Vasoactive intestinal peptide (VIP) [37221-79-7], M_r 3326, was first isolated along with secretin from the intestine of the pig. It is distributed throughout the body, primarily in the nervous system.

VIP is a peptide amide containing 28 amino acids. It is formed from the same prepropeptide as PHI-27 (pig) and PHM-27 (humans) that are peptide amides containing 27 amino acids. PHI-27 stands for peptide with N-terminal histidine and C-terminal isoleucinamide, PHM-27 for peptide with N-terminal histidine and C-terminal methioninamide. An extended version of VIP with Gly-Lys-Arg at the C-terminus has also been found [1409].

p-,b-,r-,ca-,cu-,cap-,o-, and h-VIP are identical [1410]:

H S D A V F T D N Y T R L R K Q M A V K K Y L N S I L Na

There is little difference in the sequences of h-PHM-27, p-, b- [1411], and r-PHI-27 [1412]; cu-PHI-27 is identical to p-PHI-27 [1413].

VIP and PHI-27 are found in the neurons of the central nervous system, in the digestive tract, lungs, urogenital tract, gallbladder [1414], exocrine glands, thyroid gland, adenohypophysis [1415], and the adrenal gland. It occurs associated with acetylcholine and galanine [1416].

Release. VIP is released by electrical stimulation of nerve fibers, from the gallbladder wall by stimulation of the vagus, and from the hypothalamus by prostaglandin E_1. Plasma levels are increased by oxytocin, sexual stimulation in women, the sucking stimulus [1417], physical stress [1418], operation stress [1419], and corticosterone [1420]. VIP is also released by Met-enkephalin and β-endorphin [1421] in the hypothalamus, by insulin in the intestine [1422], and by estrogen in the adenohypophysis [1423].

The release of VIP is inhibited by glucose infusions [1418], peripheral application of endogenous opiates [1418], pancreatic peptide, and somatostatin.

Patients with multiple sclerosis have a significantly lower level of VIP in their cerebrospinal fluid than healthy persons. Impotent men have a low plasma VIP level. In patients suffering from cystic fibrosis, fewer VIP-immunoreactive nerves are found in the eccrine sweat glands [1424]. Intravenous glucagon or secretin raises the plasma levels of gastrin and VIP in patients with Zollinger–Ellison syndrome [1425]. The treatment of patients with congestive heart failure with ACE inhibitors increases plasma VIP levels [1426].

Receptors [1427]. At least two high-affinity VIP receptors and one low-affinity VIP receptor may exist. The antagonist [4-Cl-D-Phe^6,Leu^{17}]VIP differentiates between two high-affinity VIP receptors in guinea pig acinar cells [1428]; carbachol binds to the low-affinity VIP receptors [1429]. The number of VIP receptors on human mononuclear leucocytes increases drastically in response to physical activity and an energy deficit [1430].

Biological Effects. In the cell, VIP stimulates the synthesis of cAMP and catecholamines (activation of tyrosine hydroxylase) [1431]. It also potentiates the muscarinic stimulation of the turnover of phosphoinositides [1432]. VIP inhibits the activation of the calmodulin-sensitive phosphodiesterase [1433]. VIP and acetylcholine act synergistically in secretion and circulation. PHI exhibits VIP-like activity. VIP has many biological effects [1408]. The most important are listed below:

1) Cardiovascular system: vasodilation, hypotension, positive chronotropic and inotropic effects [1434].
2) Respiratory system: bronchodilation, augmented ventilation, pulmonary vasodilation, stimulation of bronchial secretion (VIP and asthma [1435]).
3) Gastrointestinal tract: relaxation of lower esophageal sphincter; relaxation of smooth muscle in the stomach and suppression of acid secretion; stimulation of pancreatic water and HCO_3 secretion; increased bile flow; inhibition of intestinal absorption; stimulation of intestinal water and chloride secretion; relaxation of circular muscles and contraction of longitudinal muscles in the colon [1436], [1437].
4) Urogenital tract: increase of blood circulation in the vagina and erection of the penis.
5) Metabolic effects: stimulation of glycogenolysis and lipolysis, bone resorption, hair growth [1438]; regulation of mitosis, differentiation and survival of cultured sym-

pathetic neuroblasts [1439]; stimulation of in vitro meiosis in follicle-enclosed rat oocytes [1440]; and regulation of neuronal survival.

6) Endocrine functions: increased pituitary release of ACTH, prolactin [1441], growth hormone, and LH; suppression of somatostatin release; an increased release of gonadoliberin in the hypothalamus and of iodine-containing hormones in the thyroid; ACTH-like action in the adrenals (stimulation of aldosterone and corticosterone synthesis [1442]); increase of serum testosterone and dihydrotestosterone in women [1443]; increased aromatase activity in the neonatal rat ovary [1444]; increase of serum estradiol in women [1443]; stimulation of estradiol and progesterone release from rat ovaries [1445]; and stimulation of testosterone, progesterone, and pregnenolone release in cultured neonatal testicular cells [1446].

Structure – Activity Relationships. The C-terminal amide is not required for the biological activity of VIP; deamido^{28}-VIP, VIP-Gly, and [Leu^{17}]VIP-Gly-Lys have similar activities [1447]. Shortening the molecule at either the C- and N-terminus decreases biological activity. VIP-(10 – 28) exerts an antagonistic effect on the VIP-stimulated synthesis of cAMP and secretion of pepsinogen [1448]. Met^{17} can be replaced by norleucine without loss of activity.

The hydrophobic amino acids of the helical domain (6 – 28) are important for receptor binding, the hydrophilic amino acids in this region appear to be less important. Substitutions of the amino acids in positions 6, 10, 12, 17, 19, 22, 23, and 27 produce less active derivatives [1449]. Iodination leads to at least two VIP derivatives with full biological activity [1450].

Displacement of [^{125}I]VIP bound to receptors on rat liver membranes by various analogues occurs in the following order: VIP > [D-Ala^4]VIP > [D-Asp^3]VIP > [D-Ser^2]VIP > [D-His^1]VIP > [D-Phe^2]-VIP > [D-Arg^2]VIP. [D-Phe^2]VIP is a competitive VIP antagonist, the other analogues have agonistic activity [1451]. Still better VIP antagonists are [D-4-ClPhe^6,Leu^{17}]VIP [1452] and Ac-[Tyr^1,D-Phe^2]GHRH-(1 – 29)amide [1453]. [D-Phe^4]PHI-27 is a highly selective VIP agonist [1454].

7.3. Pituitary Adenylate Cyclase Activating Polypeptide (PACAP)

Occurrence. There is considerable homology between VIP and o-PACAP-38 [1455] and o-PACAP-27 [1456], two peptide amides from the sheep hypothalamus. The prepropeptide of h-, o- [1457], and r-PACAP [1458] has been characterized via the cDNA. Highly specific receptors have been found primarily in the hypothalamus, brain stem, cerebellum, and lung. PACAP-27 binds more tightly than PACAP-38.

In the rat hypophysis, o-PACAP-38 and o-PACAP-27 stimulate adenylate cyclase 1000 times more effectively than VIP [1455], [1456]. Like VIP, o-PACAP-27 exhibits vasodepressor activity [1456]. Concentrations of 10^{-10} mol/L release growth hormone, prolactin, and ACTH from rat hypophyseal cells, 10^{-9} mol/L releases LH as well [1459].

```
1                                                    30
R S L Q D T E E K S R S F S A S Q A D P L S D P D Q M N E D K R-

33                              53            61
H S Q G T F T S D Y S K Y L D S R R A Q D F V Q W L M N T-
           ←————————— Glucagon —————————→

62  64       69  71
K R N R N N I A K R
   ← SpP–I →

72       78                                              108
H D E F E R H A E G T F T S D V S S Y L E G Q A A K E F I A W L V K G R G-
   ←——————————————————— GLP I ———————————————————→

109 111              123 125
R R D F P E E V A I V E E L G R R-
       ←———— SpP–II ————→

126                                                    159
H A D G S F S D E M N T I L D N L A A R D F I N W L I Q T K I T D R
   ←——————————————— GLP II ———————————————→
```

Figure 5. Amino acid sequences of h-prepro-GG and its cleavage products

7.4. Glucagon and Related Peptides

Glucagon (GG) is formed together with the two glucagon-like peptides, GLP-1 and GLP-2, as prepro-GG (r-[1460], h-[1461], b-[1462], p-[1463]). Glucagon is prepro-GG-(33–61), GLP-1 is prepro-GG-(72–108), and GLP-2 is prepro-GG-(126–159). In the pancreatic A-cells, prepro-GG is cleaved to GG and longer C-terminal peptides that contain GLP-1 and GLP-2. In the small intestine, however, prepro-GG is cleaved to form glicentin [prepro-GG-(1–69)], enteroglucagon [prepro-GG-(33–69)], GLP-1, and GLP-2 [1464]. In the small intestine the six N-terminal amino acids of GLP-1 are removed to give the C-terminal amide GLP-1-(7–36)amide [prepro-GG-(78–107)amide] [1465].

The sequences of h-prepro-GG [1461] and its cleavage products are shown in Figure 5.

7.4.1. Glucagon

Glucagon (GG) [642-83-1], M_r 3483 (h-GG), is formed in the endocrine A-cells of the pancreas. It is, however, also found in the rat pituitary.

Release. Pulsatile GG release is observed in humans at 13–20 min intervals [1466]. The Ca^+-dependent secretion of GG is stimulated by amino acids, stress [1467], β-endorphin [1468], morphine [1469], dopamine, carbachol, cholecystokinin [1470], bombesin, xenopsin, substance P, and high doses of oxytocin and vasopressin. Parathyroid hormone potentiates the arginine-stimulated release of GG.

GG secretion is inhibited by glucose [1471], Mg^{2+}, thyroliberin, calcitonin, SRIH, glicentin-(1–16), glicentin-(62–69) [1472], and sulfonyl urea derivatives, e.g., glibenclamide [1473]. The plasma level of GG is elevated in overweight persons [1474] and is reduced in type I diabetics [1475].

Biological Effects [1476]. Glucagon plays a role in carbohydrate, fat, and protein metabolism. In the liver, its most important target organ, GG stimulates the formation of glucose from glycogen (glycogenolysis) and of ketone bodies from amino acids. Glycogen synthesis is simultaneously inactivated via a phosphorylation reaction [1477] and the binding of insulin in the liver is reduced [1478]. Intracerebroventricular administration of GG increases blood glucose levels via cholinergic and α-adrenergic mechanisms much more than intravenous administration [1479].

In hepatocytes, GG inhibits fatty acid synthesis via activation of a cAMP-dependent protein kinase and subsequent phosphorylation of an acetyl coenzyme A (CoA) carboxylase [1480]. It stimulates lipolysis in brown adipose tissue [1481]. It reduces the activity of β-hydroxy-β-methylglutaryl CoA reductase [1482] and lowers the plasma level of apolipoprotein E, cholesterol, triacyclglycerol, and phospholipids [1483].

Other effects of GG include stimulation of the plasma level of cAMP, reduction of the blood levels of Ca^{2+} and PO_4^{3-}, spasmolytic activity in high doses, relaxation of the esophagus, contraction of the pyloric sphincter [1484], relaxation of the Oddi sphincter [1485], stimulation of bile flow [1486], bronchodilatory effect in asthmatic patients [1487], and inhibition of anaphylactic histamine release.

Structure–Activity Relationships. The shortening of GG and the replacement of its N-terminal histidine reduces glycogenolytic activity [1488]. GG-(1–21) is fully active as regards spasmolytic activity but only about 0.1% as active as regards cAMP-releasing activity. Thus, GG-(1–21) exerts a relaxing effect on the Oddi sphincter [1489] and an analgesic effect in biliary colic [1490]. Further shortening of the compound, e.g., GG-(1–20) and GG-(2–21), reduces the spasmolytic activity to ≤2% of the GG activity. GG-(19–29) does not activate adenylate cyclase, but is 1000 times more active than GG in inhibiting the Ca^{2+}-activated and Mg^{2+}-dependent ATPase [1491] and the liver plasma membrane calcium pump [1492].

Substitutions in the N-terminal region of GG usually lower biological activity. However, [D-Ser^2]GG [1493] and [D-Phe^4]GG had twice and six times the glycogenolytic activity in the rat, respectively [1494].

The amino acid region 10–13 is not essential for biological activity, but is important for receptor binding [1495].

S-Methylation of Met^{27} reduces glucagon activity to 1%; [S-carboxymethyl-Met^{27}]GG [1496], [2-thiol-Trp^{25}]GG, [oxindolyl-Ala^{25}]GG, [Met(O$_2$)^{27}]GG, monoiodo-GG, and [N-ε-alkylamidino-Lys^{12}]GG are fully active.

N-α-Trinitrophenyl-GG and N-α-trinitrophenyl-[12-homoarginine]GG are weak antagonists that exert a partially agonistic effect. Antagonists include de-His^1-[Glu^9]GG-amide [1497], de-His^1-

[Glu⁹,Orn¹²]GGamide, de-His¹-[Glu⁹,Glu¹⁶,Lys²⁹]GG-amide, de-His¹-[Glu⁹,Lys²⁹]GG-amide [1498], [Asp³,D-Phe⁴,Ser⁵, Lys¹⁷,¹⁸,Glu²¹]GG, and [D-Phe⁴,Tyr⁵,I₂-Tyr¹⁰,Arg¹², Lys¹⁷,¹⁸,Glu²¹]GG [1499].

Uses. Glucagon (Glucagon Novo, Novo Ind.; Glukagon Lilly, Eli Lilly) is used to treat hypoglycemia. In McArdle's disease, subcutaneous injection of glucagon increases the blood sugar level and thus improves muscle function.

The spasmolytic activity of GG facilitates endoscopic and radiological investigations of the gastrointestinal tract and computer tomography of the liver, pancreas, and kidney. GG can also bring relief to persons suffering from esophageal cramps [1500] and hiccups [1501]. Its spasmolytic activity in combination with the enhanced flow of bile has been useful in the treatment of biliary colic.

As a result of its effect on the plasma Ca^{2+} level, glucagon has been used in combination with calcitonin and mithramycin in the treatment of Paget's disease.

GG can be applied parenterally (also with a portable pump [1502] or as a depot preparation [1503]), nasally [1504], or as eye drops [1505].

7.4.2. Glicentin and Enteroglucagon

Glicentin [71567-776], M_r 8101, [h-prepro-GG-(1–69)] and enteroglucagon [62340-29-8], M_r 4450, also called oxyntomodulin [h-prepro-GG-(33–69)], are synthesized in the small intestine from prepro-GG [1464]. The plasma level of glicentin rises during hypoglycemia [1506] and physiological doses of both peptides decrease the secretion of gastric acid [1507]. Glicentin has only 10% of the cAMP-stimulating activity of GG. Thus, glicentin and enteroglucagon assume the function of a hypothetical enterogastrone in the intestine (an enterogastrone is a substance formed in the intestine which decreases gastric acid release). The common C-terminal octapeptide appears to be especially important for activity and can inhibit the pentagastrin-stimulated secretion of gastric acid (although 100 times weaker) [1508].

Enteroglucagon is also 10 times more active than GG in inhibiting cerulein-stimulated exocrine pancreatic secretion in rats [1509].

7.4.3. Glucagon-Like Peptide-I-(7–36)amide

The peptide GLP-1 is processed primarily in the small intestine to give the C-terminal amide GLP-1-(7–36)amide, M_r 3298 (h-GLP-1-(7–36)-amide) [1465].

Release. Intestinal release of GLP-1-(7–36)amide is stimulated by orally applied glucose, D-galactose, D-glucuronic acid, 3-O-methyl-D-glucose, maltose, sucrose, or maltitol [1510]. Fructose, D-fucose, D-mannose, D-xylose, and lactose have no effect [1510]. The postprandial increase in the plasma level of GLP-1-(7–36)amide is enhanced in the postgastrectomy dumping syndrome [1511].

Receptors. GLP-1-(7–36)amide binds specifically to a cloned, insulin-producing cell line from the rat. Even micromolar concentrations of GLP-1-(1–36)amide or GIP do not displace GLP-1-(7–36)amide from the receptor [1512]. Dexamethasone reduces the number of GLP-1-(7–36)amide receptors on the B-cells [1513].

Biological Effects. GLP-1-(7–36)amide is more effective in stimulating cAMP synthesis in rat preparations than GLP-1 or glucagon [1514], [1515]. GLP-1-(7–36)amide is more active than GLP-1 in stimulating the glucose-dependent release of insulin from pancreatic islet cells of the rat [1515]. Tests on humans have shown that GLP-1-(7–36)amide has the properties of inkretin (a hypothetical substance that stimulates the glucose-dependent endocrine secretion from the pancreas): its formation in the gastrointestinal tract is enhanced after an oral glucose meal, it releases insulin in a glucose-dependent process, and lowers the plasma levels of glucose and glucagon [1511]. GLP-1-(7–36)amide and GIP have a synergistic effect on the release of insulin [1516] but not in the stomach. In the pancreas GLP-1-(7–36)amide inhibits the secretion of glucagon by post-transcriptional mechanisms and decreases the secretion of somatostatin [1517], [1518].

Structure–Activity Relationships. The amide structure of GLP-1-(7–36)amide is apparently not essential for its insulin-releasing activity. GLP-1-(7–37) is as active as GLP-1-(7–36)amide [1519].

Uses. In type II diabetes (non-insulin-dependent), GLP-1-(7–36)amide stimulates the release of insulin and inhibits the secretion of glucagon and somatostatin. Both in type I (insulin-dependent) and type II diabetes, the isoglycemic-meal-related insulin requirement is lowered by GLP-1-(7–36)amide [1520].

7.5. Gastrin Inhibiting Peptide

Occurrence. The gastrin inhibiting peptide (GIP) [59392-49-3] also known as the glucose-dependent insulinotropic peptide, M_r 4984, contains 42 amino acids and is found in the duodenum, jejunum, and together with glucagon in the A-cells of the pancreas.

Release. GIP is secreted in the duodenum and jejunum in response to a diet rich in fat and carbohydrate. The pancreatic enzymes are also involved in stimulating this release [1521]. During physical exercise, the secretion of GIP in response to orally administered glucose is greatly reduced [1522]. The effect of GIP is enhanced by bombesin or low doses of motilin. High doses of motilin, somatostatin, and dexamethasone inhibit the release of GIP [1523].

The raised level of GIP in insulin-dependent (type I) diabetes can be lowered with insulin [1524]. This is, however, not the case with non-insulin-dependent diabetes mellitus [1525]. The postprandial plasma level of GIP is low in obese, non-insulin-dependent diabetics [1526].

Biological Effects. GIP inhibits the gastrin-stimulated secretion of gastric acid, as well as the secretion of pepsin, and motor activity in the stomach. It stimulates the intestinal secretion and increases the glucose-dependent release of insulin, glucagon, pancreatic peptide, and somatostatin in the pancreas. cAMP is activated in this process. The GIP-mediated release of insulin is dependent on phospholipase A_2 [1527] and is linked to a cholinergic mechanism [1528].

In adipocytes, GIP increases the receptor affinity for insulin [1529] and potentiates the inhibitory effect of insulin on hepatic glycogenolysis [1530].

The physiological importance of GIP is probably based on its insulinotropic activity rather than on the reduction of the gastric acid concentration. Like GLP-1-(7–36)amide, GIP is an important candidate for the role of the postulated inkretin.

Structure–Activity Relationships. Shortening the molecule at the N-terminus by only two amino acids reduces the insulin-releasing effect to about 2%. Endo-29a-Gln-GIP-(1–38) and b-GIP-(1–39) [1531] still possess full insulinotropic activity. GIP-(1–28) and GIP-(3–24) do not release insulin [1532].

7.6. Growth Hormone Releasing Hormone and the Growth Hormone Cascade

The growth hormone cascade is shown in Figure 6. The growth hormone releasing hormone or somatoliberin (GHRH or SRH) and galanine (GAL) stimulate the release of the growth hormone (also known as somatotropin, STH) from the pituitary which, in turn, releases the somatomedins (SMs). Somatostatin (SRIH), which inhibits the release of GHRH and GH, is of great importance in the regulation of growth hormone. GAL inhibits the release of SRIH, which promotes the release of GHRH and GH. GHRH and GH stimulate the synthesis of SRIH and thus permit a negative feedback. A synthetic hexapeptide (SK&F 110 679) also releases GH from the pituitary, independently of GHRH receptors.

There is a high degree of structural similarity between GH and placental lactogen, somatolactin, prolactin, proliferin, and prolactin-related or -like proteins. The insulin-like growth factors, IGF-I (somatomedin C) and IGF-II, as well as somatomedin B belong to the somatomedins. Insulin and relaxin are structurally related to IGF-I and IGF-II.

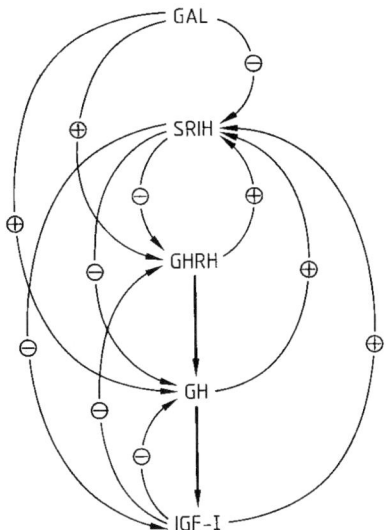

Figure 6. The growth hormone cascade
GAL = galanine; GH = growth hormone (somatotropin); GHRH = growth hormone relesing hormone (somatoliberin); IGF-I = insulin-like growth factor I; SRIH = somatotropin release inhibiting hormone (somatotropin)

7.6.1. Galanine [1533]

Occurrence. Galanine (GAL) [*119418-04-1*], M_r 3211 (p-GAL), was first isolated from the pig intestine [1534]. Its name is derived from its N- and C-terminal amino acids (glycine and alanine). It is a 29 amino acid peptide amide with a C-terminal that is similar to the neurokinins, e.g., neurokinin A or substance P. Both p- and b-GAL are processed in the form of a prepropeptide containing 123 amino acids [1535].

In humans, GAL is not only found in the nerve cells of the small intestine, but also in the brain, spinal cord, pancreas, bladder, pituitary [1536], adrenal gland [1537], and the genital tract [1538]. In the central nervous system, the highest concentrations are found in the hypothalamus [1539] and the median eminence. Unusually high concentrations of GAL are found in pheochromocytomas [1540].

GAL is localized with gonadoliberin in neurons of the hypothalamus [1541].

Release. The release of GAL in the adenohypophysis and hypothalamus is stimulated by estrogen [1536] and the thyroid hormones, e.g., T_4 [1542]. The formation of GAL decreases after the withdrawal of food [1543] and treatment with capsaicin.

Biological Effects. GAL contracts preparations from the rat fundus, ileum, colon, bladder [1534], and jejunum [1544]. It potentiates the contraction of the rat vas deferens that is induced by electrical stimulation or noradrenaline [1545]. GAL inhibits, however, contraction induced by electrical stimulation in the guinea pig colon. It also inhibits the motility of the small intestine in the rabbit. GAL inhibits the release of acetylcholine and substance P from motor nerves [1545] and transmission at the cholinergic nicotinic synapses in the myenteric plexus [1546].

In humans, GAL inhibits gastrointestinal motiliy, delays gastric emptying, and increases the mouth to cecum transit time. It also suppresses the postprandial increase in glucose, PYY, insulin, neurotensin, enteroglucagon, glucagon, pancreatic peptide (PP), and somatostatin (SRIH) levels [1547]. GAL directly inhibits the exocytosis of insulin [1548]. Glucagon is stimulated in vitro. In mice, GAL inhibits the basal and stimulated insulin level, increases the plasma glucagon level temporarily and causes hyperglycemia [1549]. In the perfused porcine pancreas, however, GAL increases the release of insulin and inhibits the secretion of SRIH [1550].

GAL inhibits the release of gastrin and SRIH [1551] and the pentagastrin-stimulated secretion of gastric acid from the rat stomach [1552].

The intracerebroventricular application of GAL to rats or the intravenous infusion in humans increases the plasma levels of growth hormone and prolactin [1539] via the release of adrenaline [1553] and GHRH [1554], or via the inhibition of SRIH. It acts synergistically with GHRH [1555].

In rats, injections of GAL into the paraventricular nucleus cause an increase in the food uptake. It protects hippocampal neurons from the functional effects of anoxia [1556].

Structure–Activity Relationships. The molecule can be shortened at the C-terminus without loss of activity; in contrast, deletions at the N-terminus result in complete loss of activity.

GAL-(1–10) strongly contracts the rat jejunum, but it does not inhibit the electrically stimulated contraction of strips of guinea pig colon [1544]. The action of GAL-(1–20) is species and muscle dependent, and is similar to or weaker than that of GAL [1557]. While GAL-(1–16) has an agonistic effect on the rat hippocampal GAL receptor, GAL-(3–29) [1558] and GAL-(17–29) [1559] are inactive. The effect on the guinea pig ileum is as follows: GAL > GAL-(2–9) > GAL-(1–15) [1560]. In rats, the pentagastrin-stimulated secretion of gastric acid is inhibited by GAL-(9–29) with 5% of the GAL activity, while GAL-(15–29) is devoid of activity [1552].

7.6.2. Growth Hormone Releasing Hormone
[1561]

Occurrence. In humans, growth hormone releasing hormone (GHRH) [*9034-39-3*], also called somatotropin releasing hormone (SRH) or somatoliberin, M_r 5040 (h-GHRH), is a peptide amide containing 44 amino acids. Although GHRH is mainly synthesized in the hypothalamus, immunoreactive GHRH has also been found in the gastrointestinal tract, placenta, human and rat milk, and spinal fluid [1562].

In the neurons of the arcuate nucleus, GHRH is associated with NPY [1563], dopamine [1564], and tyrosine hydroxylase [1565].

During the processing of prepro-h-GHRH, prepro-h-GHRH-(78–107)amide with a pyroglutamyl residue at the N-terminus (CTPG or anorectin) is formed. This peptide reduces the intake of food when injected into the third ventricle [1566].

Release. The release of GHRH is stimulated by central α_2-adrenergic stimuli, high potassium concentrations, GABA, opioids, galanine [1567], dopa [1568], thyroid hormones, and by a glucose meal [1569]. More GHRH is found in the hypothalamus of male rats than in female rats [1570]. The plasma level of immunoreactive GHRH increases greatly during puberty (more so in girls than in boys) [1571].

Somatostatin (SRIH) and pentobarbital anesthesia inhibit the secretion of GHRH. Growth hormone [1572], insulin-like growth factors IGF-I and IGF-II [1573], GHRH itself [1574], and dopamine also inhibit the release of GHRH by increasing SRIH release. The GHRH-stimulated secretion of SRIH is presumed to be regulated by β-endorphin-containing neurons [1575].

The plasma level of GHRH is lower in children suffering from dwarfism and in acromegaly [1576]. Less GHRH is found in the hypothalamus of old rats than in young rats [1577].

Receptors. Chronic infusion (in vivo) of GHRH [1578] and adrenalectomy [1579] lead to the down regulation of GHRH receptors in the rat.

Biological Effects. GHRH stimulates transcription of the GH gene and the release of GH [1580] in the pituitary via an increase in cAMP, arachidonic acid, and PGE_2. Chronic administration does not totally inhibit the release of GH [1581]. It is assumed that melatonin [1582], cholinergic agonists [1583], β_1-adrenergic receptor blockers [1584], and hypoglycemia [1561] increase the GHRH-stimulated GH level via the inhibition of SRIH release. The GHRH-stimulated release of GH is inhibited by SRIH, cholinergic antagonists [1583], α-receptor blockers [1585], dopamine [1586], hyperglycemia [1587], and by IGF-I and II [1573]. The GHRH-induced release of GH is reduced in children with thalassemia major [1588], in patients suffering from non-insulin-dependent (type II) diabetes [1589], hypothyroidism [1590], Cushing's syndrome [1591], depression [1592], and obesity [1561]. The GHRH-stimulated secretion of GH is raised in diabetics with retinopathy [1561], in acromegaly, and in young Alzheimer patients [1593]. In patients with hypothyroidism, T_4 potentiates the GHRH-induced release of GH [1590].

Glucocorticoids exert a biphasic effect on the release of GH. They potentiate GHRH activity by stabilizing the GHRH receptors and enhancing the transcription rate of the GH gene. In cases of chronic misuse of glucocorticoids, however, an inhibition of the release of GH is observed in humans and experimental animals [1594].

Other effects of GHRH include release of calcitonin and neurotensin from rat C-cells [1595], proliferation of endothelial cells in the mucosa of the fundus and antrum [1596], increase in the production of milk in cows and sheep, and increase in REM sleep [1597].

Structure–Activity Relationships. GHRH can be shortened at the C-terminus without loss of activity. In rats, h-GHRH-(1–29)amide has 50% of the activity of h-GHRH-(1–44)amide. However, h-GHRH-(1–29)amide has the same activity as h-GHRH-(1–44)amide in children [1598]. Further shortening at the C-terminus significantly reduces activity [1599]. The N-terminal amino group is of no importance for biological activity because its acetylation gives more potent compounds.

Position 1 should be occupied by an aromatic amino acid [1600]. Nle^{27} and D-Ala^2 increase activity. Although [D-Ala^2]GHRH-(1–29)amide had 50 times the activity in some tests and Ac-[D-Tyr^1,D-Ala^2]-GHRH-(1–29)amide had 38 times the GHRH-(1–29)amide activity in the rat, these compounds do not exhibit a higher activity in humans [1601]. The replacement of Arg^{29} by agmatin (Agm) increases activity. Potent long-acting analogues are [D-Ala^2,Nle^{27}]h-GHRH-(1–28)Agm [1602] and [deamino-Tyr^1,Ala^{15},Nle^{27}]GHRH-(1–28)Agm [1603].

[D-Arg^2]h-GHRH-(1–29)amide exhibits GHRH antagonistic activity in concentrations of 10^{-7} mol/L [1604]. More potent antagonists are [D-Arg^{2,29},Arg^{30}]h-GHRH-(1–30)amide and Ac-[His^1,D-Arg^2,Ala^{15}]r-GHRH-(1–29)amide [1605].

An antagonist that acts in the 10 µmol/L range is [Ser^9-ψ-CH$_2$NH-Tyr^{10}]GHRH-(1–29)amide [1606].

The activity of h-GHRH-(1–27)amide can also be increased: [D-Asp^3,D-Asn^8,Leu^{27}]h-GHRH-(1–27)amide has twice the activity of h-GHRH-(1–29)amide [1607].

Uses. GHRH can be used as a diagnostic aid to differentiate between hypothalamic- and hypophyseal-dependent growth hormone deficiency diseases. Only patients with a defective hypothalamus usually react to GHRH by producing GH. The older the patient (i.e., the longer the endogenous GHRH deficiency disease exists) the poorer the reaction of the hypophysis.

GHRH is used in the treatment of children with GH deficiency. The linear growth of children was improved by subcutaneous administration of 250 µg of GHRH-(1–29)amide twice a day [1608] or after pulsatile treatment (subcutaneous application of 1–3 µg/kg every 3 h) [1609]. Children suffering from dwarfism responded better to h-GH than to h-GHRH-(1–29)amide [1610]. In postmenopausal women GHRH increases bone formation and plasma osteocalcin [1611].

Of the large number of smaller synthetic peptides [1612], one of the most active is the hexapeptide amide SK&F 110 679 (sequence: H w A W f Ka), which has about 10% of the GHRH activity [1613]. There is no structural similarity between this peptide and GHRH. It releases GH from hypophyseal cell cultures in vitro and acts synergistically with GHRH. Thus, it evidently does not bind to the same receptor as GHRH [1614]. In humans, doses as low as 0.25–1 µg kg^{-1} 30 min^{-1} stimulate the release of GH. The compound appears to be more effective in this experiment than GHRH [1615].

7.6.3. Growth Hormone, Placental Lactogen, and Prolactin

The pituitary hormones prolactin (PRL) and growth hormone (GH) are structurally related to placental lactogen (PL) and two GH variants (GH-V and GH-V2) from the placenta. PL has lactogenic properties and some somatotropic activity (tibia test). PRL has no somatotropic activity, but GH has weak PRL activity.

Other substances related to PRL are somatolactin (pl-SL) [1616], proliferin (m-PLF), prolactin related cDNA (b-PRC I, II, and III), proliferin related protein (m-PRP), and prolactinlike proteins (r-PLP A and B, b-PLP I, II, and III) [1617].

There is great similarity in the structures of secretin-(5–20), h-GH-(1–16), and h-PL-(1–16). There is also amino acid homology between h-GH-(30–38), gastrin, and the C-terminus of the B-chain of insulin.

7.6.3.1. Growth Hormone

Occurrence. Growth hormone (GH) [9002-72-6], also known as somatotropin (STH), M_r 22 125 (h-STH), has 191 amino acids and is formed in the pituitary as pro-GH. Circulating GH occurs in several forms: monomeric (little GH), dimeric (big GH) and oligomeric (big-big GH, $M_r > 60\,000$) [1618]. In addition, 20Kh-GH, a de-(32–46)GH, and the two GH variants from the placenta, GH-V and GH-V2 [1619] have also been found. The amino acid sequences of GH in various species are described in [1620], [1621].

Release [1622]. GH is regulated primarily by the hypothalamic hormones galanine (GAL, see Section 7.6.1), growth hormone releasing hormone (GHRH, see Section 7.6.2), and somatostatin (SRIH, see Section 7.6.5). GAL and GHRH stimulate and SRIH inhibits the pituitary release of GH. Treatment with SRIH enhances the GHRH-stimulated release of GH [1623]. GH is also released in response to stimulation of α_1-adrenergic receptors [1624] and cholinergic muscarinic receptors [1625], inhibition of α_2-adrenergic receptors [1626], SKF-110 679, insulin, hypoglycemia, glucagon, arginine, physical stress [1627], α-melanotropin, opioids [1628], angiotensin II [1629], pentagastrin [1630], GIP, neurotensin, combined application of thyroliberin with ACTH [1631], slow wave sleep [1632], intracerebroventricular application of delta-sleep-inducing peptide, interleukin-1β [1633], serotonin [1634], dopamine and dopamine agonists [1635], and indomethacin [1636].

In women, the release of GH is dependent on estrogen. Thus, the plasma level of GH falls during menopause [1637] and in suppression therapy with high doses of GnRH agonists [1638]. In young hypogonadal males, testosterone stimulates the GH formation and growth [1639]. Glucocorticoids also regulate the synthesis of GH. Hypoadrenalism [1640] and high doses of glucocorticoids [1641] lower the release of GH. However, glucocorticoids and triiodothyronine (T_3) enhance the release of GH in vitro through GHRH.

GH release is inhibited by recombinant Met-GH [1642], centrally applied gastrin releasing peptide [1643], corticoliberin [1644], calcitonin [1645], calcitonin gene related peptide [1646], and by TRH [1647], mainly via the secretion of SRIH.

Plasma GH levels in persons with acromegaly and type I diabetes [1648] are higher than in healthy persons, but are reduced in depressive patients with an increased level of somatomedin C [1649]. TRH has no influence on the release of GH in normal subjects, but it causes a paradoxical increase in the plasma level of GH in patients suffering from acromegaly [1650], depression [1651], or metastatic testicular cancer [1652]. GnRH also raises the plasma GH level in type I and type II diabetics [1653].

Receptors. The amino acid sequences of the GH receptors from humans, rabbits [1654], and rats [1655] have been elucidated via their cDNA.

GH binding proteins are mainly the extracellular fragments of GH receptor [1656]. In patients with Laron-type dwarfism (high plasma GH level and low IGF-I level), the GH receptor is defective [1657] and the high-affinity GH binding protein is missing [1658]. African pygmies have a low plasma level of GH-binding protein [1659].

Biological Effects. GHRH is species specific and promotes the longitudinal growth and regeneration of bone. It has a strong mitogenic effect on osteoblasts [1660] and pancreatic β-cells [1661], increases the osteocalcin level [1662], and enhances DNA, RNA, and protein synthesis. It possesses lipolytic, diabetogenic, and insulinotropic properties; it also has an antidiuretic and antinatriuretic effect because it activates renin and increases the plasma level of aldosterone [1663].

In a negative feedback mechanism, GH inhibits the release of GHRH in the hypothalamus [1664] via the release of SRIH [1665].

Pulsatile intravenous GH stimulates the formation of somatomedin C [1666] in the liver and skeletal tissue. This substance is identical to the insulin-like growth factor I (IGF-I) and is largely responsible for the growth effect of GH [1667].

In vitro, GH stimulates the release of insulin from pancreatic islet cells [1668] and the insulin-dependent uptake and oxidation of glucose in adipose and muscle tissues [1669]. The insulin-like effect of GH is only found in vivo in GH deficiency models or children suffering from GH deficiency. Chronic therapy with GH or very high doses of GH have a diabetogenic effect in hypophysectomized animals. After prolonged GH treatment the fasting plasma glucose level in humans is raised [1670] and in patients with acromegaly, glucose metabolism is reduced in spite of the increased release of insulin [1671].

Physiological doses of GH also stimulate the production of peroxide anions in macrophages. It increases the activity of cytolytic T-cells and of natural killer cells, stimulates the formation of antibodies and thymulin synthesis [1672], and regenerates the thymus in dogs [1673].

Structure–Activity Relationships. Similar to h-GH, recombinant Met-h-GH, which acts like natural h-GH in children with GH deficiency [1674], has diabetogenic and insulin-like activity [1675]. Met-h-GH has a somewhat higher tendency to initiate antibody formation [1676]. Other acidic GH derivatives (deamidated h-GH) are as active as h-GH [1620].

The epiphyseal plate of the long bones of the rat declines following hypophysectomy and increases after treatment with GH. This increase in the width of the plate of the proximal end of the tibia has been used for the quantitative measurement of GH. Various GH fragments exhibit biological activity in the tibia test in hypophysectomized animals, e.g., GH-(88–124), GH-(125–156), b-GH-(96–133), and GH-(151–191). h-GH, but not 20K-h-GH, reduces the concentration of free fatty acids in children deficient in GH [1677]. The binding of 20K-h-GH to lactogenic receptors is about 500 times less than that of GH [1678]. The lactogenic activity of h-GH and o-PRL is competitively inhibited by [Met^{14}]h-GH-(14–191) [1621].

h-GH-(4–15) and h-GH-(6–13) stimulate glucose oxidation, glucose uptake in adipose tissue, and the synthesis of glycogen [1679]. The minimal GH sequence with this insulin-potentiating effect is h-GH-(8–13). An artifact, the aspartoylimide variant of these peptides, appears to be responsible for this effect [1680]. Other insulinotropic GH fragments are h-GH-(31–44), h-GH-(32–38) [1681], and h-GH-(32–46) [1682]. Both h-GH-(177–191) and h-GH-(178–191) have diabetogenic activity.

Uses. Met-h-GH and h-GH are produced by genetic engineering and are now used instead of natural h-GH in the therapy of hypophyseal dwarfism. Trade names include Genotropin (Pfrimmer Kabi), Grorm, Saizen (Serono), Humatrop (Lilly), and Norditropin (Nordisk).

h-GH has also been used to activate growth in persons with Turner's syndrome where growth is reduced in spite of normal GH formation [1683]. h-GH also finds application for the healing of wounds [1684] and for increasing the mass [1685] and density of bones [1686]. In older patients with low plasma GH levels, GH increases bone density, lean body mass, and skin thickness, and reduces the amount of adipose tissue [1687].

7.6.3.2. Prolactin

Occurrence. Prolactin (PRL) [9002-62-4], M_r 22 806 (h-PRL, protein part), has 198 amino acids and is formed in the pituitary where it is glycosylated at Asn^{31} [1688] or dimerized via disulfide bridges [1688]. Other sites of synthesis are the placenta, myometrium, ovaries [1689], and hypothalamus. The sequences of PRL from various species are listed in [1690].

Release [1691]. The pulsatile secretion of PRL from the pituitary is regulated by hypothalamic factors. It is stimulated by TRH and VIP and inhibited by dopamine. The C-terminal glycopeptide of vasopressin-neurophysin, which contains 39 amino acids, is a specific PRL-releasing factor [1692]. PRL is also released in response to K$^+$ [1693], hypocalcemia [1694], calcitriol [1695], cholecystokinin, insulin, arginine [1696], epidermal growth factor [1697], the sucking stimulus through endogenous opioids [1698] and VIP [1699], stress and physical exercise [1700] through serotoninergic mechanisms [1701], β-endorphin [1702], AT II [1703], gonadoliberin [1704] through the formation of AT II [1705], neurotensin [1706], and to glutamate [1707]. It is also released by the stimulation of α$_1$-adrenergic receptors [1708] or histamine H$_1$-receptors [1709].

Dopamine from the hypothalamus and pituitary [1710] which is released by PRL is probably one of the most important inhibitors of PRL release.

The secretion of PRL is also inhibited by ascorbic acid [1711], δ opiates [1712], hypercalcemia [1713], the prolactin inhibiting factor (PIF), PRL itself [1714], stimulation of α_2-adrenergic [1708], GABA-A- [1715], or histamine H_2-receptors [1709], chronic nicotine intake [1716], somatostatin and its analogues [1717], calcitonin [1718], IL-1 [1719], T_3 [1720], melatonin [1721], glucocorticoids [1722], vasopressin (presumably via a dopaminergic pathway [1723]), and central neurotensin. A competitive antagonist of PRL is a recombinant h-GH, which lacks the 13 amino acids at the N-terminus [1724].

The steroid sex hormones modulate the release of PRL. Estrogen potentiates the secretion of PRL in experimental animals [1725], but can also potentiate the dopamine-stimulated inhibition of PRL release in humans. 17-β-estradiol potentiates the somatostatin-stimulated inhibition of PRL in pituitary cells [1726]. Testosterone appears to inhibit the secretion of PRL because the TRH-stimulated release of PRL is greatly enhanced in delayed puberty (males) [1727].

A period of breast feeding is followed by a reduced PRL level for at least 12 – 13 years [1728], which could be related to the low risk of breast cancer after pregnancies. Women who are not lactating have a pronounced PRL deficiency [1729]. In humans, the PRL response to TRH falls with age [1730]. Patients with chronic kidney damage and rats with experimentally induced kidney damage develop hyperprolactinemia. The plasma level of PRL is raised in spontaneously hypertensive rats and in patients with hypertension.

Receptors. Receptors for PRL are widely distributed: in the mammary gland, liver, kidneys, brain, prostate, testes, ovaries, and lymphocytes [1731]. The receptors have a similar structure to that of the GH receptors [1732], [1733].

In the rat, the number of PRL receptors in the mammary gland increases on the day of estrus. Together with estradiol, LH stimulates the formation of PRL receptors [1734]. The number of testicular PRL receptors in the golden hamster is increased by PRL, FSH, and LH; ovariectomy and hypophysectomy reduce the synthesis of PRL receptors [1735].

PRL is displaced from the receptors on lymphocytes by cyclosporin [1731].

Biological Effects. In female mammals, PRL increases the production of milk and initiates maternal behavior. The PRL-stimulated biosynthesis of casein and lipid is reduced by inhibitors of phospholipase A_2 [1736].

PRL stimulates the synthesis of dopamine and increases the density of dopamine receptors [1737]. Thus, the dopamine level in the central nervous system and the dopaminergic inhibition of aldosterone are increased in patients suffering from hyperprolactinism [1738].

Similar to placental lactogen, PRL also has luteotropic properties and stimulates the synthesis of progesterone [1739], but inhibits the formation of estradiol and testosterone [1739]. In men with chronically elevated PRL levels, the plasma level of testosterone is low and the excretion of LH in the urine is reduced [1740]. In women, hyperprolactinemia leads to irregular menstruation or even amenorrhea [1741].

PRL increases the blood pressure of experimental animals which is dependent on the adrenal glands and presumably occurs via the potentiation of noradrenaline activity. As a stress hormone, PRL potentiates the antidiuretic activity of Arg-vasopressin [1742] and the AT II induced uptake of water and retention of fluids [1743].

Hyperprolactinemia increases the production of androgens [1744] in the adrenal gland, DNA synthesis in the liver [1745], the susceptibility to chemically induced tumors [1746] and spontaneous mammacarcinoma, and the number of estrogen receptors in the mammary gland [1747]. It also prevents stress-induced stomach ulcers [1748], induces hypercalcemia in several species [1749], and reduces the mineral content of the bone [1750].

PRL is also immunostimulatory. Thus, the reactivity of the lymphocytes is reduced by blocking pituitary PRL release with bromocryptin. The immunosuppressive effect of cyclosporin can be abolished by PRL [1731].

7.6.3.3. Placental Lactogen

Placental lactogen (PL) [9035-54-5], M_r 22 212 (h-PL), is formed in the placenta in the first trimester of pregnancy and can be detected in the blood and urine of pregnant women. Secretion of PL decreases after birth.

h-PL first stimulates the development of the breast tissue. When the level of h-PL decreases after birth, pituitary h-PRL stimulates the secretion of milk.

h-PL has luteotropic properties: it stimulates the release of progesterone and estrogen from the corpus luteum. This effect is potentiated by h-choriogonadotropin (h-CG). The growth-promoting effect of h-PL is somewhat weaker than that of h-STH, but appears to play a more important role in the development of the fetus.

7.6.4. Somatomedins, Insulin, and Relaxin

Somatomedins of the IGF type (insulin-like growth factors I and II) are structurally related to insulin and relaxin. They are synthesized as single-chain prepropeptides. In the case of insulin and relaxin, three disulfide bridges are formed and the C-peptide (connecting peptide) between the N-terminal B-chain and the C-terminal A-chain is enzymatically removed to give two chains joined together by two disulfide bridges. In the case of IGF-I (M_r = 7648.8), and IGF-II (M_r = 7469.5), the C-peptide is not cleaved. Relaxin has no influence on the level of glucose but IGF-I, IGF-II, and insulin are functionally related.

7.6.4.1. Somatomedins [1751]

Occurrence. The somatomedins C [67763-96-6] and A [62046-94-0] (SM-C and SM-A) were previously called the sulfation factor, thymidine factor, nonsuppressible insulin-like activity in serum (NSILA-S), or the multiplication stimulating activity (MSA). The insulin-like growth factor I (IGF-I) is identical to SM-C, SM-A appears to be a deamidated IGF-I [1752]. Somatomedin B [*63774-77-6*] (SM-B) has a completely different structure which is similar to the N-terminal 44 amino acids of vitronectin.

The somatomedins are found in the liver, lungs, prostate, testes, seminal and follicular fluid [1753], thymus, thyroid gland [1754], skeleton, brain, heart, kidneys [1755], gastrointestinal tract, saliva [1756], milk [1757], fibroblasts, chondrocytes [1758], skeletal muscles, and in adipose deposits. A large number of tumors produce the messenger RNA for IGF-I and/or IGF-II [1759].

h-IGF-I, M_r 7649, and h-IGF-II, M_r 7470, each consists of a peptide chain linked with three disulfide bridges. Analogous to proinsulin, the peptide chain is divided into the N-terminal B-chain, the connecting C-peptide, the A-chain, and the C-terminal D-chain. The propeptides also contain the C-terminal E-peptide and the prepropeptides, the N-terminal prepeptide. The structures of IGF-I and IGF-II in various species are described in [1760]–[1764].

Release. The release of SM is stimulated by growth hormone, prolactin (in the liver [1765]), insulin, injuries [1766], exercise [1767], increased vascular strain [1768], gonadoliberin, follitropin, lutropin, h-choriogonadotropin [1769], estradiol [1770], [1771], ACTH, androgens, parathyroid hormone [1759] and by cAMP (in osteoblast cultures [1772]). Prior to birth, neither IGF-I nor IGF-II appears to be dependent on GH [1773]. Here, placental lactogen, but not GH, insulin, or other growth factors stimulates the synthesis of IGF-II.

SM secretion is reduced by insufficient food intake [1774], low-protein diet [1773], chronic alcohol consumption [1775], and suppression treatment with GnRH agonists [1776].

Serum SM levels fall in patients with hypophyseal insufficiency, diabetes mellitus, and after the age of sixty. Women taking contraceptives and patients with hyperthyroidism have elevated SM levels [1777]. In Cushing's syndrome, the SM level in the spinal fluid is raised [1777].

Receptors [1759]. IGF-I and IGF-II bind with high affinity to specific receptors [1778].

Like the insulin receptor, the IGF-I receptor has two α- and two β-subunits. The α-subunit binds the hormones in the cysteine-rich domain between His^{223} and Met^{274} [1779]. The β-subunit contains tyrosine kinase which phosphorylates intracellular proteins after ligand binding.

The IGF-II receptor is very similar to the mannose 6-phosphate receptor and also binds mannose 6-phosphate [1780]. It consists of a single polypeptide chain, has no kinase activity, does not bind insulin, and binds only poorly to IGF-I.

An IGF binding component found in the serum is the extracellular part of the IGF-II receptor [1781]. In addition, four structurally related IGF binding proteins (IGFBP 1, 2, 3, and 4) have been characterized [1782], [1783]. IGFBP-1 binds IGF-I and IGF-II with similar affinity. IGFBP-2 preferentially binds IGF-II. The majority of serum IGF is bound in a complex consisting of IGF-I or IGF-II (γ-subunit), IGFBP-3 (β-subunit), and an α-subunit that does not bind IGF.

The serum IGF bioactivity is inhibited by h-IGFBP-1 and h-IGFBP-2 [1784], [1785]. In contrast, IGFBP-3 has a stimulating effect on the activity of IGF-I [1759].

Biological Effects. IGF-I stimulates intracellular synthesis of 1,2-diacylglycerol and induces the degradation of phosphatidyl choline, a process which involves protein kinase [1786].

In hypophysectomized animals, administration of IGF-I, like GH, increases body weight and bone growth. GH stimulates the proliferation of osteoblasts via the local synthesis of IGF-I [1787]. The GH-dependent increase in calcitriol after a low-phosphate diet is mediated by IGF-I [1788].

IGF-I and IGF-II bind 100–200 times less tightly to insulin receptors than insulin but, like insulin, they increase glucose uptake and oxidation in adipose and muscle tissue. In humans, they have about 6% of the hypoglycemic activity of insulin [1789] and inhibit endogenous insulin release [1790].

The basal and GHRH-stimulated release of GH is inhibited by a negative feedback mechanism by IGF-I. It acts directly on the pituitary, where it inhibits transcription of the GH gene [1791]. It also releases somatostatin in the hypothalamus [1773]. IGF-II does not have a negative feedback effect on the basal or GHRH-stimulated GH [1751]. IGF-I inhibits the secretion of PRL in the pituitary [1792], but stimulates the release of PRL in cells of the human decidua [1793].

In Leydig cells, IGF-I stimulates the growth of h-CG receptors [1794] as well as the basal and h-CG-stimulated secretion of testosterone [1795]. IGF-I stimulates progesterone synthesis and pregnenolone in ovarian cells [1796].

IGF-I and IGF-II also act as autocrine growth factors in muscle development [1797] and cause proliferation of tumor cells [1798]. They increase the motility of melanoma cells and, therefore, could also increase the metastatic potential [1799].

IGF-I has a modulating effect on the immune system: it stimulates the growth of the thymus [1800], but inhibits the IL-2-induced proliferation and antibody formation of splenocytes in vitro [1801].

Structure–Activity Relationships. The A-chain domains of IGF-I are of importance for receptor binding and the growth effect [1802], the B-chain sequences play a role in the interaction with IGFBPs [1803].

Tyr^{24}, Tyr^{31}, and Tyr^{60} are important for binding of IGF-I to the IGF-I receptor and Tyr^{60} for binding to the IGF-II receptor [1804].

IGF-I-(6–70) has only one hundredth of the activity of IGF-I (in stimulating protein synthesis in rat myoblasts), IGF-I-(4–70) and IGF-I-(5–70) are ten times and twice as active as IGF-I, respectively [1805].

Uses. IGF-I reduces the blood sugar and insulin levels in Mendenhall's syndrome with severe insulin resistance [1806] and in Laron-type dwarfism with a defective GH receptor [1807].

7.6.4.2. Insulin

Occurrence. Insulin [*9004-10-8*], M_r 5808 (h-insulin), was first used in 1921 as a therapeutic agent in diabetes mellitus. It is synthesized in the β-cells of the pancreatic islets of Langerhans. However, immunoreactive insulin has also been found in other tissues (e.g., brain, testes, and liver).

Insulin consists of two peptide chains which are joined together by two disulfide bridges. It is formed from single-chain proinsulin by enzymatic removal of the C-peptide (connecting peptide) [1808] between the C-terminal A-chain and the N-terminal B-chain. In most species the A-chain (acid chain) contains 21 amino acids, while the B-chain (basic chain) contains 30 amino acids. The C-chain with its N- and C-terminal basic amino acids contains 35 amino acids. In vivo, the C-chain is found without the N- and C-terminal basic amino acids. In physiological concentrations, insulin occurs in human plasma as a monomer which does not bind to serum proteins [1809]. Insulin sequences of various species are given in [1810].

Insulin-like molecules are also observed in the silkworm *Bombys mori* [1811], in the mollusc *Lymnaea stagnalis* [1812], and in the brain of the insect *Locusta migratoria* [1813].

h-Insulin

```
      B-chain
F V N Q H L C G S H L V E A L Y L V C G E R G F F Y T P K T
              |                     |
G I V E Q C C T S I C S L Y Q L E N Y C N
      A-chain
```

Release [1814]. Insulin is released by an increased plasma level of glucose or amino acids, cAMP, and Ca^{2+} (Mg^{2+}). The plasma insulin level increases after ingestion of food. Incretins are postulated to stimulate the secretion of insulin from the pancreas in a glucose-dependent process. Incretin release from the intestine is increased by oral glucose and inhibited by insulin. The criteria for the postulated incretin are best fulfilled by the glucagon-like peptide 1-(7–36)-amide [1815], the glucose-dependent insulinotropic peptide, and oxyntomodulin [1816]. Glucose-dependent secretion of insulin is also stimulated by the gastrin releasing peptide, neurotensin, substance P, gastrin, pentagastrin, CCK-8 (sulfated or unsulfated), the C-terminal tetrapeptide of CCK, cholinergic substances, $α_2$-adrenoreceptor antagonists [1817], TGF-$β_1$, TGF-$β_2$ [1818], oxytocin [1819], ACTH, glucagon, secretin [1820], calcitriol [1821], estradiol [1822], thyroid hormones [1823], prostaglandin D_2, lipoxygenase products [1824], and phospholipase A_2 products [1825].

Low doses of endogenous opioids and exorphins stimulate insulin secretion while high doses have a more inhibitory effect.

The glucose-stimulated release of insulin is inhibited by adrenaline [1826], prostaglandin E_2 [1827], somatostatin, pancreastatin, pancreatic peptide, galanine [1828], calcitonin gene related peptide [1829], amylin, calcitonin, exogenous insulin [1830], growth hormone, interferon [1831], interleukin 1β, and interleukin 6 [1832].

The cell-specific destruction of the pancreatic β-cells, which precedes insulin-dependent diabetes mellitus occurs by an autoimmune mechanism. Autoantibodies have been found in more than 80% of newly diagnosed patients. The target of the autoantibodies is a β-cell autoantigen which has been identified as glutamic acid decarboxylase [1833] or as heat shock protein 65 [1834].

Receptors [1835]. The insulin receptor consists of two α- and two β- subunits joined together by disulfide bridges to give the configuration: (β-chain-S-S-α-chain)-S-S-(α-chain-S-S-β-chain). The cysteine-rich domain between Asn^{230} and Ile^{285} of the α-chain is important for insulin binding [1836]. The β-subunit is a transmembrane protein and contains a tyrosine-specific protein kinase. Insulin stimulates phosphorylation of the β-subunit. Autophosphorylation of the insulin receptor is restricted by triiodothyronine [1837].

The receptor is synthesized in the form of a prepropeptide.

Diabetes, resistance to insulin, can result from receptor defects (e.g., a deletion in the tyrosine kinase domain [1838] or the substitution of valine for Gly^{996} [1839]). Trp^{1200} is also important for the normal function of insulin receptor kinase [1840]. In Hodgkin's disease, antibodies against the insulin receptors can cause hypoglycemia because of their insulin-like activity [1841].

The formation of insulin receptors is stimulated by glucose [1842] and down regulated by insulin and proinsulin [1843]. In cancerous breast tissue, the concentration of insulin receptors is more than six times that in normal tissue and correlates with the size of the tumor and with the estrogen receptor concentration [1844].

The following positions in insulin appear to be important for receptor binding: A-chain: Gly^1, Gln^5, Tyr^{19}, and Asn^{21}; B-chain: Val^{12}, Tyr^{16}, Arg^{22}, Gly^{23}, Phe^{24}, Phe^{25}, and Tyr^{26}.

Estrogen, progesterone [1845], h-GH, h-GH-(7–13) [1846], and relaxin [1847] increase the binding of insulin to the receptor; prolactin [1847], heparin [1848], or contraceptive therapy in women [1849] reduces binding.

Biological Effects. Insulin lowers blood glucose by promoting glucose uptake in skeletal muscle, heart, and adipose tissue, by activating glycogen synthesis [1850], and by inhibiting glycogenolysis in the liver [1851]. Glucose transport proteins are involved in the uptake of glucose [1852]. Taurine [1853], the basic peptides Arg-Lys and Arg-Arg [1854], and somatostatin potentiate the insulin effect on the blood glucose level.

Apart from stimulating glucose and amino acid transport, insulin also increases the secretion of somatomedins, the affinity of the IGF-II receptors [1855], the proliferation and differentiation of cells (healing of wounds), ion transport, renal Na^+/K^+ ATPase [1856], lipogenesis, potentiates the FSH-stimulated synthesis of estrogen and progestin [1857] and the h-CG-stimulated synthesis of testosterone [1858], increases food intake [1859], stimulates the secretion of gastric acid [1860] and bile flow, and is important for postprandial cholecystokinin- or secretin-stimulated exocrine pancreatic secretion [1861].

Hypoglycemia induced by insulin is a stress reaction which leads to the release of ACTH and cortisol [1862]. Insulin also increases the K^+-stimulated production of aldosterone [1863]. In type I diabetics, insulin lowers the level of cortisol raised by fasting [1864].

Insulin inhibits the synthesis and release of somatostatins and enteroglucagon [1865], reduces the basal and PTH-stimulated excretion of phosphate in the urine [1866] as well as renal Ca^{2+} reabsorption, stimulates the synthesis of collagen [1867], and potentiates the PTH-stimulated synthesis of calcitriol in the kidney [1868].

Insulin inhibits the noradrenaline- or AT II-stimulated contraction of arteries and veins (hypotensive effect [1869]), increases the heart rate [1870], and is positively inotropic [1871].

The inability of a given concentration of insulin to produce an expected biological effect is called insulin resistance. This resistance can be caused by circulating antagonists of insulin activity (antibodies, prostaglandins, and glucocorticoids [1872]), down regulation of the receptors, or receptor defects.

Structure – Activity Relationships. The biological activity of insulin is expressed in international units (IU). The activity is measured in the mouse cramp test or by determining the decrease in the blood sugar level in the rabbit. Crystalline insulin contains 26 – 28 IU/mg.

Proinsulin is only about one-third as active as insulin in lowering the blood sugar level, but its action is significantly prolonged [1873]. The antilipolytic activity of h-proinsulin is higher than that of h-insulin [1874].

Hydrophobic interactions are responsible for the formation of insulin dimers. The hexamer of insulin is formed from three dimers with two or four zinc atoms [1875]. Monomeric insulin is assumed to be the biologically active form, i.e., binds to the receptor. The regions around A1, A2, A19, A21, and B22 – B25 are important for biological activity.

Fully active insulin is obtained when the B-chain is shortened at the N-terminus by two amino acids and at the C-terminus by three amino acids.

[$AspB^{10}$]h-insulin [1876] and de-(B^{26}–B^{30})-[$AspB^{10}$,Tyr-NH_2B^{25}]insulin [1877] are more active than h-insulin. $GlyB^{23}$ can be replaced by D-Ala without loss of activity, but [$AlaB^{23}$]insulin has a greatly reduced biological activity [1878]. Both [D-$PheB^{24}$]b-insulin and [D-$AlaB^{24}$]b-insulin still possess activity [1879]. Substitution of Glu for $TyrB^{26}$ or $ThrB^{27}$ or of Asp for $ProB^{28}$ leads to monomeric analogues having a slightly higher activity [1880]. $ThrB^{29}$ can be replaced by many different amino acids and still retain activity [1881].

The N-terminal Gly of the A-chain can only be replaced by D-amino acids without loss of activity. [D-$LeuA^1$]insulin possesses full insulin activity and [D-$TrpA^1$]insulin is even slightly more potent than native insulin. The hydroxyl group of $TyrA^{19}$ forms a hydrogen bond with the carboxyl group of $GlyA^1$ and stabilizes the molecule. Modifications of $AsnA^{21}$ lead to less active compounds. The cysteines are exremely important for the tertiary structure.

Less soluble analogues with prolonged activity are obtained by introducing Arg into the molecule [1882] – [1884], amidating the Glu-γ-carboxyl groups and the C-terminus of the B-chain [1885], and by the palmitoylation of the A-chain [1886]. Long-acting analogues include [$GlyA^{21}$,$ArgB^{27}$,$ThrB^{30}$-amide]h-insulin (NovoSol-Basal) [1882]. [$AspB^9$,$GluB^{27}$]h-insulin [1887] and [$AspB^{10}$]h-insulin [1887] are monomeric derivatives that act especially quickly.

Uses [1888], [1889]. Many insulin preparations are available for the parenteral treatment of diabetes mellitus. Bovine, porcine, and human insulin are generally used. Commercially available h-insulin preparations are either enzymatically modified p-insulin [H-Insulin, Depot-H-Insulin, Komb-H-Insulin (Hoechst); Insulin Monotard

HM, Insulin Actrapid HM (Novo); Insulin Velasulin Human, Insulin Mixtard Human (Nordisk)] or chain recombinant insulin produced by genetic engineering in bacteria (Huminsulin, Eli Lilly). In enzymatically modified products the C-terminal alanine of p-insulin is substituted by a threonine ester with the aid of trypsin. The ester function is then hydrolyzed to give h-insulin. In chain recombinant h-insulin, the A- and B-chains are synthesized separately and subsequently linked by oxidation.

De-PheB^1-insulin derivatives (Insulin Defalan, Hoechst) have full insulin activity, but are more soluble and less antigenic than insulin. In patients with insulin allergy, de-PheB^1-p-insulin has been used successfully because it has a low antigenicity [1890]. However, de-Phe-insulins are no longer on the market.

Insulin pumps have been developed for better insulin adjustment, pulsatile infusion appears to be more efficient than continuous infusion [1891].

Treatment with insulin is also recommended in total parenteral feeding [1892], in cancer-induced anorexia, and loss of weight [1893]. Topically applied insulin promotes the healing of wounds [1894].

7.6.4.3. Relaxin [1895]

Occurrence. Relaxin (RLX) [9002-69-1], M_r 6333 (h-RLX 2), is found primarily in the serum and tissues of pregnant mammals. It occurs in especially high concentrations in the corpus luteum of pregnant animals, and is isolated from this source. In addition, RLX is found in the endometrium during pregnancy, breast [1896], testes, prostate, and seminal fluid.

RLX, like insulin, consists of an A- (24 amino acids in humans) and a B-chain (32 amino acids in humans) linked by sulfide bridges and is synthesized as a single-chain prohormone. h-RLX occurs in two forms, h-RLX-2 is formed in the ovaries during pregnancy [1897]. p-RLX occurs in three forms. RLX sequences from various species are described in [1897].

h-Relaxin-2
```
*QLYSALANKCCHVGCTKRSLARFC
         |     /
  SWMEEVIKLCGRELVRAQIAICGMSTWSKRSL
```

Release. The plasma level of RLX increases greatly at the end of pregnancy and falls rapidly after birth. In hypophysectomized animals, endogenous RLX can be activated by estrogens and progesterone. The release of RLX is increased by h-CG [1898], oxytocin, and prostaglandin F_2. Indomethacin prevents the increase in the RLX level and delays birth.

In trophoblastic disease, which is accompanied by an increase in the plasma h-CG level, RLX is detectable in the plasma [1899].

Biological Effects. Purified RLX (NIH standard, RXN-P1) has about 3000 U/mg. One unit is the threshold dose which causes a softening of the interpubic disc in two-thirds of the guinea pigs tested.

In advanced pregnancy, RLX reduces the contraction of the uterus. This effect is supported by progesterone and is important for the maintenance of pregnancy. RLX has a growth-promoting and relaxing effect on the uterus and cervix and a growth-promoting and lactation-inhibiting effect on the mammary gland. RLX alone inhibits progesterone synthesis, but increases estrogen synthesis [1900]. RLX also appears to prepare the myometrium for parturition because without it oxytocin cannot induce birth. RLX inhibits prostaglandin synthesis during pregnancy, but stimulates it during birth [1901].

The relaxing effect of RLX on the interpubic disc (pelvic symphysis) occurs because the compact cartilaginous tissue between the two pubic bones is converted to a more flexible structure. The water content of the tissue is increased and the secretion of proteolytic enzymes is stimulated, loosening the network of collagen and glycosaminoglycan [1902]. Estrogen and hypophyseal hormones (e.g., growth hormone) are also involved in this deaggregation of collagen. Intravaginal RLX (2 mg/patient) causes cervical maturation in humans and induces birth.

RLX increases sperm motility and exerts a hypotensive effect [1903]. It inhibits the pressor activity of angiotensin II, vasopressin, and noradrenaline [1904].

Structure–Activity Relationships. The shortening of the A-chain of p-RLX by two amino acids at the N-terminus does not affect biological activity, whereas deletion of five or six amino acids halves the biological activity [1905].

Oxidation of $TrpB^{27}$ in p-RLX with N-bromosuccinimide has no influence on biological activity. However, p-RLX becomes inactive when $TrpB^{18}$ is oxidized.

7.6.5. Somatostatin [1906]

Occurrence. Somatostatin [38916-34-6], also called somatotropin release inhibiting hormone or factor (SRIH or SRIF), M_r 1638, is formed as prepro-SRIH. The main product of gene expression in the stomach and pancreas of mammals is pro-SRIH-(1–64), which is cleaved at the C-terminus to form SRIH-28 with 28 amino acids and SRIH-14 with 14 amino acids [1907].

SRIH-14 and SRIH-like substances have been found in the hypothalamus, nervous system, pancreatic islet D-cells, gastrointestinal mucosa, thymus, extrahepatic bile ducts, plasma [1908], in the C-cells of the thyroid gland, ovaries, and in plants (e.g., spinach [1909]). Somatostatin sequences of various species are described in [1910].

Release. SRIH is released in the antrum (lower part of the stomach) in response to fasting [1911], acid in the antrum, intraduodenal infusion of fat and HCl [1908], stress

[1912], cholinergic and β-adrenergic mechanisms, dopamine D_2 agonists [1913], estrogen and androgens [1913], and numerous peptide hormones including growth hormone releasing hormone and somatomedins [1908], [1914]–[1917].

The secretion of SRIH is reduced shortly after meals [1911], by electrical stimulation of the vagus, sham feeding, α-adrenergic stimuli, serotonin [1918], and various peptides (e.g., endogenous opioids, galanine, pancreatic peptide, and insulin).

The plasma level of SRIH is lower at night [1919]. The plasma SRIH level in children below the age of two is raised. Above two years its level is normal and increases continuously with age [1920].

The plasma SRIH level is increased in patients with chronic renal failure, cirrhosis [1921], and primary hypothyroidism [1922]. The level of SRIH is raised in the cerebrospinal fluid of patients with brain tumors, metabolic encephalopathy, spinal cord compression, and intracranial hypertension [1923].

The plasma level of SRIH is reduced in patients with pernicious anemia, achlorhydria [1920], cluster headache [1924], psoriasis [1925], obesity [1926], and in women who are taking estrogen-containing contraceptives [1927]. The level of SRIH in the cerebrospinal fluid is reduced in patients with Alzheimer's disease [1928], unipolar depression, and active multiple sclerosis [1912]. The level of SRIH is reduced in the cortex and hippocampus of patients suffering from Parkinson's disease [1929].

Receptors. Several SRIH types of receptors are found in the brain.

SRIH receptors are detected in most GH- and TSH-producing pituitary adenomas, carcinoids, islet cell carcinomas, EGF-receptor-negative glial tumors, and small cell lung carcinomas [1930]. The number of SRIH receptors is reduced in the cerebral cortex of patients with Alzheimer's disease [1931].

Biological Effects. SRIH-14 [1932] stimulates phosphodiesterase, inhibits cAMP synthesis, blocks calcium influx, and the turnover of phosphoinositides in the cytosol [1933]. It also inhibits the release and action of hormones (e.g., growth hormone (GH), prolactin, thyrotropin, stress-induced corticotropin [1934]), calcitonin, insulin, glucagon, GIP, VIP, secretin, pancreatic polypeptide, gastrin releasing peptide [1908], gastrin, cholecystokinin, and motilin. GH is the most sensitive to SRIH inhibition, followed by thyrotropin [1935].

SRIH inhibits exocrine secretion of gastric acid, pepsin (anti-ulcer effect), pancreatic enzymes, pancreatic bicarbonate, colonic fluid (reduction of secretory diarrhoea [1936]), bile flow (patients with somatostatinoma usually have gall-stones [1937]). It also potentiates human platelet aggregation stimulated by collagen and arachidonic acid [1938] (control of upper gastrointestinal hemorrhage).

SRIH enhances gastric emptying [1939] and intestinal transit [1940]. It may be useful in the management of the dumping syndrome [1941]. SRIH inhibits cell proliferation in the exocrine pancreas [1942], rabbit jejunum [1943], liver after hepatectomy [1944], and tumor growth. The antitumor effect of SRIH is mediated by the stimulation of a tyrosine phosphatase [1945].

Immunomodulatory actions include increase of leukocyte migration inhibitory factor [1946], enhancement of human lymphocyte natural killer activity [1947], and mouse spleen lymphocyte proliferation [1948].

Structure–Activity Relationships. The substitution of D-Trp for Trp^8 produces an analogue that has 8 times the activity of SRIH in inhibiting GH release.

Potent, long-acting SRIH analogues of the urotensin II type are of tremendous interest [1949]–[1952]. For instance, octreotide [*49474-41-2*], [*79517-01-4*] is 70 times more effective than SRIH-14 in inhibiting GH. It has a long duration of action when given intramuscularly [1949].

SRIH-14   A G C̄ K N F F W K T F T S C̄

go-Urotensin II A G T A D C̄ F W K Y C̄ V

Octreotide   f C̄ F W K T C̄ T-ol

Cyclic peptides without a disulfide structure are also potent, long-acting compounds. Examples are cyclo-(*N*-Me-Ala-Tyr-D-Trp-Lys-Val-Phe) (seglitide) and cyclo-(Pro-Phe-D-Trp-Lys-Thr-Phe).

Cyclic retropeptides of the latter, e.g., cyclo-(Phe-Thr-Lys-Trp-Phe-D-Pro), inhibit the uptake of cholate and phallotoxin in liver cells [1953] and protect against taurocholate- and ceruletide-induced pancreatitis [1954], and against ethanol-induced stomach lesions [1955], without possessing true SRIH activity.

Analogues of octreotide which usually contain penicillamine instead of Cys^7 are poor ligands for SRIH receptors, but exhibit stronger binding to opioid receptors [1956]. They act as selective antagonists for μ-opioid receptors, i.e., they inhibit morphine induced analgesia [1957].

Uses. SRIH-14 infusions (Stilamin, Serono; Aminopon, UCB; Somatostatin Ferring, Ferring; Somatostatin Labaz, Sanofi) have proved useful for the treatment of bleeding peptic ulcers and gastrointestinal lesions, for preventive treatment against stress-induced ulcers, and in the healing of fistulae of the small intestine and gallbladder.

Octreotide (Sandostatin, Sandoz) [1958] is used in the treatment of GH- and thyrotropinsecreting pituitary tumors, carcinoid tumors, glucagonomas, insulinomas, and gastrinomas. The use of octreotide in acute esophageal variceal bleeding, pancreatic pseudocysts, external gastrointestinal and pancreatic fistulae, short bowel syndrome, dumping syndrome, and refractory hypersecretory diarrhoea related to the acquired immunodeficiency syndrome has provided encouraging results. Preliminary reports indicate the efficiency of octreotide in the treatment of psoriasis, autonomic neuropathy [1959] and of carcinoid flush [1960], and its ability to reduce growth in tall adolescents. Other potential applications include the control of pain [1961], rheumatoid arthritis, and diabetic microangiopathy as well as the prevention of postoperative pancreatic complications (fistulae, pancreatitis, and abscess).

7.6.6. Pancreastatin

Occurrence. Pancreastatin (PST) [*117148-67-1*] from the pancreas of the pig contains 49 amino acids [1962]. Sequences from other species are given in [1963]. h-PST contains 52 amino acids, M_r 5509.

The precursor of pancreastatin is chromogranin A, an acidic glycoprotein which is widely distributed in the neuroendocrine system and released in large amounts from the chromaffin cells of the adrenal medulla [1963]. Pancreastatin contains the penta-Glu sequence of gastrin (GT) and has a C-terminus that is similar to that of gonadoliberin (GnRH).

In patients with non-insulin-dependent diabetes mellitus, the plasma pancreastatin level increases after the digestion of glucose, which is not the case in healthy persons [1964].

Biological Effects. Pancreastatin inhibits the first phase of glucose- or arginine-induced release of insulin and the arginine-induced release of somatostatin, but potentiates the arginine-induced secretion of glucagon [1965] from the rat pancreas. These changes are reminiscent of non-insulin-dependent diabetes.

In the rat, pancreastatin lowers the plasma insulin level and increases the plasma level of glucose in response to intragastric glucose [1966] and raises the plasma glucagon level in response to intravenous arginine [1967]. Like glucagon, it stimulates hepatic glycogenolysis [1968].

Pancreastatin also inhibits the CCK-stimulated secretion of protein and fluid from the pancreas [1967] and the CCK-stimulated growth of human pancreatic adenocarcinoma [1969].

Structure–Activity Relationships. The inhibitory effect on the glucose-stimulated release of insulin is restricted to the C-terminal region of the molecule [1962]. The shortest, active C-terminal fragment is p-PST (35–49) [1970]. The C-terminal amide function is required for biological activity [1970].

8. References

[1] A. R. Green, *Lancet I* (1989) 705.
[2] W. König in A. Kleemann et al. (eds.): *Arzneimittel, Fortschritte 1972 bis 1985*, VCH Verlagsgesellschaft, Weinheim 1987, pp. 572–772.
[3] R. Geiger, *Coll. Soc. Fr. Etudes Fertil.* **26** (1988) 63.
[4] E. Bürger, *Arzneim.-Forsch./Drug Res.* **38** (1988) 754.
[5] IUPAC-IUB Commission on Biochemical Nomenclature, *J. Biol. Chem.* **247** (1972) 977.
[6] IUPAC-IUB Commission on Biochemical Nomenclature, *Eur. J. Biochem.* **5** (1968) 151.
[7] IUPAC-IUB Commission on Biochemical Nomenclature, *J. Biol. Chem.* **250** (1975) 3215.

[8] IUPAC-IUB Commission on Biochemical Nomenclature, *Hoppe-Seyler's Z. Physiol. Chem.* **348** (1967) 262.
[9] J. P. Adelman et al., *Proc. Natl. Acad. Sci. USA* **83** (1986) 179.
[10] K. Nikolics et al., *Nature (London)* **316** (1985) 511.
[11] S. S. Smith, S. R. Ojeda, *Endocrinology (Baltimore)* **115** (1984) 1973.
[12] J. A. King, R. P. Millar, *Peptides (N.Y.)* **6** (1985) 689.
[13] J. A. King, R. P. Millar, *Peptides (N.Y.)* **7** (1986) 827–834.
[14] D. D. Rasmussen, *J. Endocrinol. Invest.* **9** (1986) 427.
[15] B. A. Eipper et al., *Endocrinology (Baltimore)* **116** (1985) 2497.
[16] P. A. Melrose, *Endocrinology (Baltimore)* **121** (1987) 200.
[17] M. J. Kelly et al., *Neuroendocrinology* **49** (1989) 88.
[18] C. R. Parker, Jr., J. C. Porter, *J. Clin. Endocrinol. Metab.* **58** (1984) 488.
[19] S. Ohtsuka et al., *Acta Endocrinol. (Copenhagen)* **117** (1988) 399.
[20] L. G. Allen et al., *Endocrinology (Baltimore)* **121** (1987) 1953.
[21] K. Gerozissis et al., *Neuroendocrinology* **40** (1985) 272.
[22] A. K. Dubey, T. M. Plant, *Biol. Reprod.* **33** (1985) 423.
[23] T. J. Cicero et al., *Life Sci.* **37** (1985) 467–474.
[24] F. Petraglia et al., *Endocrinology (Baltimore)* **120** (1987) 1083.
[25] H. Rosen et al., *Endocrinology (Baltimore)* **122** (1988) 2873.
[26] K. E. Nikolarakis et al., *Neuroendocrinology* **44** (1986) 314.
[27] Z. Naor, *Endocr. Rev.* **11** (1990) 326.
[28] A. Jagannadha Rao, N. R. Moudgal, *IRCS Med. Sci.* **12** (1984) 1105.
[29] T. A. Bramley et al., *J. Endocrinol.* **113** (1987) 317–327.
[30] N. Dekel et al., *Endocrinology (Baltimore)* **123** (1988) 1205.
[31] M. Fekete et al., *J. Clin. Lab. Analysis* (1988).
[32] D. R. Pieper et al., *Endocrinology (Baltimore)* **115** (1984) 1190.
[33] P. M. Conn et al., *Fed. Proc. Fed. Am. Soc. Exp. Biol.* **43** (1984) 2351.
[34] P. C. Wynn et al., *Endocrinology (Baltimore)* **119** (1986) 1852.
[35] A. Starzec et al., *Endocrinology (Baltimore)* **119** (1986) 561.
[36] W.-S. A. Wun, I. H. Thorneycroft, *Mol. Cell. Endocrinol.* **54** (1987) 165.
[37] S. R. Swartz, G. P. Moberg, *Endocrinology (Baltimore)* **118** (1986) 2425.
[38] W. R. Crowley et al., *Endocrinology (Baltimore)* **120** (1987) 941.
[39] H. Dobson, *Acta Endocrinol. (Copenhagen)* **115** (1987) 63.
[40] D. E. Suter, N. B. Schwartz, *Endocrinology (Baltimore)* **117** (1985) 849.
[41] P. S. Li, *Life Sci.* **41** (1987) 2493.
[42] A. P. N. Themmen et al., *J. Endocrinol.* **108** (1986) 431.
[43] S. Y. Bin et al., *Contraception* **35** (1987) 79.
[44] J. A. Foekens et al., *Biochem. Biophys. Res. Commun.* **140** (1986) 550.
[45] B. D. Greenstein et al., *J. Endocrinol.* **112** (1987) 345.
[46] M. J. Karten, J. E. Rivier, *Endocr. Rev.* **7** (1986) 44.
[47] J. J. Nestor, Jr., B. H. Vickery, *Annu. Rep. Med. Chem.* **23** (1988) 211.
[48] K. Folkers et al., *Proc. Natl. Acad. Sci. USA* **82** (1985) 1070.
[49] J. J. Nestor, Jr., et al., *J. Med. Chem.* **27** (1984) 320.
[50] T. L. Ho et al., *Int. J. Pept. Protein Res.* **24** (1984) 79.
[51] W. König et al. in T. Shiba, S. Sakakibara (eds.): *Peptide Chemistry 1987*, Protein Research Found., Osaka 1988, p. 591.
[52] J. Seprodi et al.,*Biochem. Biophys. Res. Commun.* **144** (1987) 1214.

[53] J. J. Nestor, Jr., et al. in F. Labrie et al. (eds.): *LH-RH and its Analogues,* Elsevier, Excerpta Medica, Amsterdam – New York – Oxford 1984, p. 24.
[54] J. J. Nestor, Jr., et al., *J. Med. Chem.* **31** (1988) 65.
[55] W. König et al. in G. Jung, E. Bayer (eds.): *Peptides 1988,* Walter de Gruyter, Berlin – New York 1989, p. 334.
[56] J. E. Rivier et al., *J. Med. Chem.* **29** (1986) 1846.
[57] The Salk Institute for Biological Studies, EP 0 102 492, 1985.
[58] A. Ljungqvist et al., *Biochem. Biophys. Res. Commun.* **148** (1987) 849.
[59] S. Bajusz et al., *Int. J. Pept. Protein Res.* **32** (1988) 425.
[60] M. V. Nekola et al., *Int. Arch. Allergy Appl. Immunol.* **84** (1987) 3161.
[61] K. Folkers et al., *Z. Naturforsch. B. Anorg. Chem. Org. Chem.* **42B** (1987) 101.
[62] S. Bajusz et al., *Proc. Natl. Acad. Sci. USA* **85** (1988) 1637–1641.
[63] A. Phillips et al., *Life Sci.* **41** (1987) 2017.
[64] A. Phillips et al., *Life Sci.* **43** (1988) 883.
[65] A. V. Schally et al., *Proc. Natl. Acad. Sci. USA* **84** (1987) 851–855.
[66] A. Bertelsen et al., *Eur. J. Pediatr.* **146** (1987) Suppl. 2, 40.
[67] S. Zgliczynski et al., *Acta Endocrinol. (Copenhagen)* **105** (1984) 161.
[68] K. E. Anderson et al., *N. Engl. J. Med.* **311** (1984) 643.
[69] *Drugs Future* **9** (1984) 295.
[70] *Drugs Future* **3** (1978) 645; **11** (1986) 812.
[71] *Drugs Future* **8** (1983) 1038; **9** (1984) 930.
[72] *Drugs Future* **10** (1985) 355; **12** (1987) 404.
[73] *Drugs Future* **11** (1986) 623–624; **12** (1987) 720–722.
[74] G. C. Doelle et al., *Horm. Metab. Res.* **18** (1986) 201.
[75] *Drugs of the Future* **11** (1986) 745.
[76] A. Lemay et al., *Fertil. Steril.* **47** (1987) 60.
[77] M. W. van Maarschalkerweerd, M. Gons, *Eur. J. Pediatr.* **146** (1987) 272.
[78] V. Perl et al., *Fertil. Steril.* **48** (1987) 383.
[79] J. L. Andreyko et al., *J. Clin. Endocrinol. Metab.* **63** (1986) 854.
[80] K. E. Anderson et al., *Clin. Pharmacol. Ther. (St. Louis)* **41** (1987) 180.
[81] J. L. Gabrilove et al., *N. Engl. J. Med.* **318** (1988) 580.
[82] W. H. Matta et al., *Br. Med. J.* **294** (1987) 1523.
[83] J. F. Wilber et al., *Clin. Res.* **34** (1986) 272A.
[84] S. M. Cockle et al., *FEBS Lett.* **252** (1989) 113–117.
[85] M. J. Kubek et al., *Life Sci.* **36** (1985) 315.
[86] T. M. Engber et al., *Regul. Pept.* **12** (1985) 51.
[87] B. M. Lewis et al., *J. Endocrinol.* **115** (1987) 419.
[88] M. T. Lin et al., *Neuroendocrinology* **50** (1989) 177.
[89] M. Mori, M. Yamada, *J. Endocrinol.* **114** (1987) 443.
[90] T. Mitsuma et al., *Horm. Metab. Res.* **21** (1989) 301.
[91] T. Mitsuma et al., *Peptides (N.Y.)* **8** (1987) 473.
[92] S. Bhasin et al., *Endocrinology (Baltimore)* **114** (1984) 946.
[93] P. Lamberton et al., *Endocrinology (Baltimore)* **117** (1985) 1834.
[94] S. Aratan-Spire et al., *Acta Endocrinol. (Copenhagen)* **106** (1984) 209.
[95] A. Dutour et al., *Peptides (N.Y.)* **10** (1989) 523.
[96] T. Mitsuma et al., *Horm. Res.* **25** (1987) 223.
[97] S. Wilk, *Life Sci.* **39** (1986) 1487.

[98] L. D. Keith et al., *Neuroendocrinology* **43** (1986) 445.
[99] M. Mori et al., *Neuroendocrinology* **48** (1988) 153.
[100] H. N. Bhargava et al., *Arch. Int. Pharmacaodyn. Ther.* **297** (1989) 247.
[101] H. N. Bhargava, *Ann. N. Y. Acad. Sci* **553** (1989) 526.
[102] H. N. Bhargava, *Ann. N. Y. Acad. Sci* **553** (1989) 528.
[103] A. H. Drummond, *Biochem. Biophys. Res. Commun.* **127** (1985) 63.
[104] R. K. Rinehart et al., *J. Pharmacol. Exp. Ther.* **238** (1986) 178.
[105] A. H. Drummond, P. J. Hughes, F. Ruiz Larrea, L. A. Joels, *Ann. N. Y. Acad. Sci.* **553** (1989) 197.
[106] E. C. Grtiffiths, *Clin. Science* **73** (1987) 449.
[107] A. Sanchez, E. Montoya, *Horm. Metab. Res.* **19** (1987) 604.
[108] I. Zofkova, J. Bednar, *Exp. Clin. Endocrinol.* **83** (1984) 263.
[109] H. Watanobe et al., *J. Endocrinol. Invest.* **8** (1985) 459.
[110] W. Schäfgen et al., *Horm. Metab. Res.* **16** (1984) 615.
[111] J.-R. Attali et al., *Endocrinology (Baltimore)* **116** (1985) 561.
[112] L. O. D. Koskinen, *Ann. N. Y. Acad. Sci.* **553** (1989) 353.
[113] L. O. D. Koskinen, *Ann. N. Y. Acad. Sci.* **553** (1989) 608.
[114] M. Burnier et al., *Clin. Res.* **37** (1989) 249A.
[115] S. Reichlin, *Acta Endocrinol. (Copenhagen) Suppl.* **276** (1986) 21.
[116] T. Suzuki et al., *Gen. Pharmacol.* **20** (1989) 239–242.
[117] T. Mitsuma et al., *Exp. Clin. Endocrinol.* **89** (1987) 55.
[118] J. Mattila, R. D. Bunag, *J. Pharmacol. Exp. Ther.* **238** (1986) 232.
[119] T. Garrik et al., *Life Sci.* **40** (1987) 649.
[120] J.-P. Roussel et al., *Acta Endocrinol. (Copenhagen)* **114** (1987) 314.
[121] *Drugs Future* **9** (1984) 807–808.
[122] *Drugs Future* **8** (1983) 1007; **11** (1986) 1059; **12** (1987) 1153; **14** (1989) 1224.
[123] H. Maeda et al., *Chem. Pharm. Bull.* **36** (1988) 190.
[124] *Drugs Future* **14** (1989) 203; **15** (1990) 285.
[125] *Drugs Future* **7** (1982) 21; **8** (1983) 56; **9** (1984) 62; **10** (1985) 75; **11** (1986) 6; **12** (1987) 70; **13** (1988) 73; **14** (1989) 78; **15** (1990) 92.
[126] H. Maeda et al., *Int. J. Pept. Protein Res.* **33** (1989) 403–411.
[127] *Drugs Future* **7** (1982) 167; **8** (1983) 282; **14** (1989) 295; **15** (1990) 319.
[128] *Drugs Future* **13** (1988) 420–423; **14** (1989) 48.
[129] M. Oka et al., *Arzneim. Forsch.* **39** (1989) 297.
[130] R. Lanzara et al., *Ann. N. Y. Acad. Sci.* **553** (1989) 559.
[131] J. G. Pierce, T. F. Parsons, *Ann. Rev. Biochem.* **50** (1981) 465.
[132] D. N. Ward et al., *J. Protein Chem.* **1** (1982) 263–280.
[133] F. S. Esch et al., *Proc. Natl. Acad. Sci. USA* **83** (1986) 6618.
[134] C.-S. Liu et al., *J. Biochem.* **186** (1989) 105.
[135] A. M. Naylor et al., *Neuroendocrinology* **49** (1989) 531.
[136] S. S. Stojilkovic et al., *Science (Washington, D.C.)* **248** (1990) 1663.
[137] G. Saade et al., *J. Endocrinol.* **114** (1987) 95.
[138] G. Emons et al., *Acta Endocrinol. (Copenhagen)* **113** (1986) 219.
[139] G. A. Schuiling et al., *Acta Endocrinol. (Copenhagen)* **110** (1985) 329.
[140] L. N. Sidneva et al., *Probl. Endocrinol.* **33** (1987) 67–70.
[141] I. Porsova et al., *Horm. Res.* **28** (1987) 288.
[142] L. Kiesel et al., *Mol. Cell. Endocrinol.* **51** (1987) 1.

[143] L. Wogensen, J. Warberg, *Acta Endocrinol. (Copenhagen)* **106** (1984) 30.
[144] P. M. Conn et al., *Biochem. Biophys. Res. Commun.* **126** (1985) 532.
[145] C. G. Brown, N. L. Poyser, *J. Endocrinol.* **103** (1984) 155.
[146] M. Filicori et al., *J. Clin. Endocrinol. Metab.* **62** (1986) 1136.
[147] P. A. Steele, S. J. Judd, *Clin. Endocrinol.* **29** (1988) 1.
[148] T. Nishi et al., *Acta Obstet. Gynecol. Scand.* **66** (1987) 309.
[149] A. Tsatsoulis et al., *Clin. Endocrinol.* **32** (1990) 73.
[150] F. R. A. Cagampang et al., *Horm. Metab. Res.* **22** (1990) 269.
[151] E. M. McColl et al., *Clin. Endocrinol.* **31** (1989) 617.
[152] F. Petraglia et al., *Life Sci.* **38** (1986) 2103.
[153] S. E. Lenahan et al., *Neuroendocrinology* **44** (1986) 89.
[154] M. S. Smith et al., *Neuroendocrinology* **50** (1989) 308.
[155] J. H. Mendelson et al., *J. Pharmacol. Exp. Ther.* **237** (1986) 862.
[156] I. R. Reid, *Ann. Intern. Med.* **106** (1987) 639.
[157] K. Seki, I. Nagata, *Acta Endocrinol. (Copenhagen)* **122** (1990) 211.
[158] C. R. Pohl et al., *Endocrinology (Baltimore)* **120** (1987) 849.
[159] S. H. Sheika et al., *Life Sci.* **44** (1989) 1363–1369.
[160] S. Taleisnik, C. H. Sawyer, *Neuroendocrinology* **44** (1986) 265.
[161] H. Loosfelt et al., *Science (Washington, D.C.)* **245** (1989) 525.
[162] K. C. McFarland et al., *Science (Washington, D.C.)* **245** (1989) 494.
[163] R. Farookhi, J. Desjardins, *Mol. Cell. Endocrinol.* **47** (1986) 13.
[164] X.-C. Jia et al., *Endocrinology (Baltimore)* **117** (1985) 13.
[165] P. E. Gottschall et al., *Biochem. Biophys. Res. Commun.* **149** (1987) 502.
[166] J. S. Davis et al., *Biochem. J.* **238** (1986) 597.
[167] V. V. Patwardhan, A. Lanthier, *J. Endocrinol.* **101** (1984) 305.
[168] H. T. Keutmann et al., *Proc. Natl. Acad. Sci. USA* **84** (1987) 2038.
[169] T. A. Santa Coloma et al., *Biochemistry* **29** (1990) 1194.
[170] T. A. Santa Coloma, L. E. Reichert, Jr., *J. Biol. Chem.* **265** (1990) 5037.
[171] M. C. Charlesworth et al., *J. Biol. Chem.* **262** (1987) 13409.
[172] B. B. Saxena, P. Rathnam, *Biochemistry* **24** (1985) 813.
[173] J.-M. Bidart et al., *Science (Washington, D.C.)* **248** (1990) 736.
[174] L. Liu et al., *Endocrinology (Baltimore)* **124** (1989) 175.
[175] A. Traub, Jr., *J. Med. Sci.* **155** (1986) Suppl. 30.
[176] A. Lanzone et al., *Fertil. Steril.* **48** (1987) 1058.
[177] W. G. Rossmanith et al., *Int. J. Fertil.* **32** (1987) 460.
[178] J. A. Magner, *Endocrine Reviews* **11** (1990) 354.
[179] J. A. Franklyn, M. C. Sheppard, *J. Endocrinol.* **117** (1988) 161.
[180] K. Törnquist, C. Lamberg-Allardt, *Horm. Metab. Res.* **18** (1986) 69.
[181] N. W. Kasting, J. B. Martin, *Neuroendocrinology* **39** (1984) 201.
[182] M. D. Lumpkin et al., *Science (Washington, D.C.)* **235** (1987) 1070.
[183] V. Rettori et al., *Proc. Natl. Acad. Sci. USA* **86** (1989) 4789.
[184] S. W. J. Lamberts et al., *J. Endocrinol. Invest.* **7** (1984) 313.
[185] F. Sanchez-Franco, L. Cacicedo, *Hormone Res.* **24** (1986) 55.
[186] R. J. Comi et al., *N. Engl. J. Med.* **317** (1987) 12.
[187] M. Parmentier et al., *Science (Washington, D.C.)* **246** (1989) 1620.
[188] M. Misrahi et al., *Biochem. Biophys. Res. Commun.* **166** (1990) 394.
[189] E. Bone et al., *Biochem. Biophys. Res. Commun.* **141** (1986) 1261.

[190] P. P. Roger, J. E. Dumont, *Biochem. Biophys. Res. Commun.* **149** (1987) 707.
[191] G. Zerek Melen et al., *Res. Exp. Med.* **187** (1987) 415.
[192] A. Nilsson, R. Axelsson, *Curr. Ther. Res.* **46** (1989) 85.
[193] J. C. Morris III et al., *Endocrinology (Baltimore)* **123** (1988) 456.
[194] J. C. Morris et al., *J. Biol. Chem.* **265** (1990) 1881.
[195] N. Ling et al., *Vitam. Horm. (N.Y.)* **44** (1988) 1.
[196] S.-Y. Ying, *Endocr. Rev.* **9** (1988) 267.
[197] R. I. McLachlan et al., *Biochem. Biophys. Res. Commun.* **140** (1986) 485.
[198] P. E. Sawchenko et al., *Nature (London)* **334** (1988) 615, 2406–2410.
[199] T. Suzuki et al., *Mol. Cell. Endocrinol.* **54** (1987) 185–195.
[200] C. G. Tsonis et al., *J. Endocrinol.* **112** (1987) R11.
[201] F. Petraglia et al., *Science (Washington, D.C.)* **237** (1987) 187–189.
[202] Z. Zhiwen et al., *Endocrinology (Baltimore)* **120** (1987) 1633.
[203] C. Rivier, W. Vale, *Endocrinology (Baltimore)* **124** (1989) 195.
[204] R. I. McLachlan et al., *J. Clin. Endocrinol. Metab.* **65** (1987) 954–961.
[205] C. B. Sheckter et al., *J. Clin. Endocrinol. Metab.* **67** (1988) 1221.
[206] R. I. McLachlan et al., *J. Clin. Endocrinol. Metab.* **67** (1988) 1305.
[207] R. E. Lappöhn et al., *N. Engl. J. Med.* **321** (1989) 790–793.
[208] C. P. Channing et al., *Proc. Soc. Exp. Biol. Med.* **178** (1985) 339.
[209] S. A. Shanbhag et al., *J. Endocrinol.* **103** (1984) 389–393.
[210] J. E. Mercer et al., *Mol. Cell. Endocrinol.* **53** (1987) 251–254.
[211] Q. F. Wang et al., *Endocrinology (Baltimore)* **124** (1989) 363.
[212] B. V. Bapat et al., *Int. J. Fertil.* **31** (1986) 71–76.
[213] S.-Y. Ying et al., *Biochem. Biophys. Res. Commun.* **136** (1986) 969.
[214] W. Vale et al., *Nature (London)* **321** (1986) 776.
[215] N. Ling et al., *Biochem. Biophys. Res. Commun.* **138** (1986) 1129.
[216] N. Ling et al., *Nature (London)* **321** (1986) 779–782.
[217] T. Nakamura et al., *Science (Washington, D.C.)* **247** (1990) 836.
[218] L. A. Hutchinson et al., *Biochem. Biophys. Res. Commun.* **146** (1987) 1405.
[219] J. Massague, *Cell* **49** (1987) 437.
[220] M. B. Sporn et al., *Science (Washington, D.C.)* **233** (1986) 532.
[221] R. Derynck et al., *Nature (London)* **316** (1985) 701.
[222] R. Pircher et al., *Biochem. Biophys. Res. Commun.* **136** (1986) 30.
[223] L. M. Wakefield et al., *J. Biol. Chem.* **263** (1988) 7646.
[224] T. A. McCaffrey et al., *J. Cell. Biol.* **109** (1989) 441.
[225] J. Pfeilschifter, G. R. Mundy, *Proc. Natl. Acad. Sci. USA* **84** (1987) 2024–2028.
[226] R. O. C. Oreffo et al., *Biochem. Biophys. Res. Commun.* **158** (1989) 817.
[227] M. Centrella et al., *Proc. Natl. Acad. Sci. USA* **85** (1988) 5889.
[228] S. Y. Ying et al., *Biochem. Biophys. Res. Commun.* **135** (1986) 950.
[229] E. Y. Adashi, C. E. Resnick, *Endocrinology (Baltimore)* **119** (1986) 1879.
[230] M. Hotta, A. Baird, *Proc. Natl. Acad. Sci. USA* **83** (1986) 7795.
[231] Y. Totsuka et al., *Biochem. Biophys. Res. Commun.* **158** (1989) 1060.
[232] E. B. Leof et al., *Proc. Natl. Acad. Sci. USA* **83** (1986) 2453.
[233] H. L. Moses et al., *Cancer Cells* **3** (1985) 65.
[234] T. Nakamura et al., *Biochem. Biophys. Res. Commun.* **133** (1985) 1042.
[235] A. B. Roberts et al., *Proc. Natl. Acad. Sci. USA* **82** (1985) 119.
[236] J. Massague et al., *Proc. Natl. Acad. Sci. USA* **83** (1986) 8206.

[237]   E. Balza et al., *FEBS Lett.* **228** (1988) 42.
[238]   J. Liu, J. M. Davidson, *Biochem. Biophys. Res. Commun.* **154** (1988) 895.
[239]   A. Bassols, J. Massague, *J. Biol. Chem.* **263** (1988) 3039.
[240]   S. Chandrasekhar, A. K. Harvey, *Biochem. Biophys. Res. Commun.* **157** (1988) 1352.
[241]   J. R. Gamble, M. A. Vadas, *Science (Washington, D.C.)* **242** (1988) 97.
[242]   S. Tsunawaki et al., *Nature (London)* **334** (1988) 260.
[243]   H. J. Andrews et al., *Biochem. Biophys. Res. Commun.* **162** (1989) 144.
[244]   M. Centrella et al., *Endocrinology (Baltimore)* **119** (1986) 2306.
[245]   M. Noda et al., *J. Biol. Chem.* **263** (1988) 13916.
[246]   C. Chenu et al., *Proc. Natl. Acad. Sci. USA* **85** (1988) 5683.
[247]   R. Montesano, L. Orci, *Proc. Natl. Acad. Sci. USA* **85** (1988) 4894.
[248]   M. R. Urist, *Science (Washington, D.C.)* **150** (1965) 893.
[249]   R. W. Katz, A. H. Reddi, *Biochem. Biophys. Res. Commun.* **157** (1988) 1253.
[250]   H. Heumann, K. Schmidt, *Bild Wiss.* (1989) no. 5, 46.
[251]   J. M. Wozney et al., *Science (Washington, D.C.)* **242** (1988) 1528.
[252]   F. P. Luyten et al., *J. Biol. Chem.* **264** (1989) 13377.
[253]   E. özkaynak et al., *EMBO J.* **9** (1990) 2085.
[254]   T. K. Sampath et al., *Biochem. Biophys. Res. Commun.* **124** (1984) 829.
[255]   A. H. Reddi, *Isr. J. Med. Sci.* **21** (1985) 312.
[256]   N. Muthukumaran, A. H. Reddi, *Clin. Orthop. Relat. Res.* **200** (1985) 159.
[257]   E. A. Wang et al., *Proc. Natl. Acad. Sci. USA* **87** (1990) 2220.
[258]   J. J. Vandersteenhoven et al., *Calcif. Tissue Int.* **42** (1988) 39.
[259]   H. S. Talwar et al., *Kidney Int.* **29** (1986) 1038.
[260]   D. P. Desimone, A. H. Reddi, *Calcif. Tissue Int.* **35** (1983) 732.
[261]   R. Howes et al., *Calcif. Tissue Int.* **42** (1988) 34.
[262]   F. W. Thielemann et al., *Z. Orthop. Ihre Grenzgeb.* **122** (1984) 843.
[263]   R. L. Cate et al., *Cell* **45** (1986) 685.
[264]   J. P. Coughlin et al., *Mol. Cell. Endocrinol.* **49** (1987) 75.
[265]   R. B. Pepinsky et al., *J. Biol. Chem.* **263** (1988) 18961.
[266]   S.-Y. Ying, *Endocr. Rev.* **9** (1988) 267.
[267]   S. Shimasaki et al., *Proc. Natl. Acad. Sci. USA* **85** (1988) 4218.
[268]   T. Nakamura et al., *Science (Washington, D.C.)* **247** (1990) 836.
[269]   M. E. Joyce et al., *J. Cell. Biol.* **110** (1990) 2195.
[270]   H. M. Harms et al., *J. Clin. Endocrinol. Metab.* **69** (1989) 843.
[271]   J. F. Habener et al., *Physiol. Rev.* **64** (1984) 985.
[272]   T. J. Martin, L. J. Suva, *Clin. Endocrinol.* **31** (1989) 631.
[273]   J. J. Orloff et al., *Endocr. Rev.* **10** (1989) 476.
[274]   L. J. Suva et al., *Science (Washington, D.C.)* **237** (1987) 893.
[275]   E. af Ekenstam et al., *Clin. Endocrinol.* **32** (1990) 323.
[276]   J. Russell, L. M. Sherwood, *Mol. Endocrinol.* **3** (1989) 325.
[277]   L. J. Suava et al., *Science (Washington, D.C.)* **237** (1987) 893.
[278]   K. L. Insogna et al., *N. Engl. J. Med.* **313** (1985) 1126.
[279]   M. E. Markowitz et al., *J. Clin. Endocrinol. Metab.* **67** (1988) 1068.
[280]   S. Ljunghall et al., *Exp. Clin. Endocrinol.* **84** (1984) 319.
[281]   M. S. Calvo et al., *J. Clin. Endocrinol. Metab.* **66** (1988) 823.
[282]   L. A. Fitzpatrick et al., *Biochem. Biophys. Res. Commun.* **138** (1986) 960.
[283]   A. Goulding, E. Gold, *Horm. Metab. Res.* **20** (1988) 743.

[284] M. S. Seshadri et al., *Endocrinology (Baltimore)* **117** (1985) 2417.
[285] E. W. Seely et al., *Acta Endocrinol. (Copenhagen)* **121** (1989) 174.
[286] S. Balabanova et al., *Horm. Res.* **24** (1986) 302.
[287] S. Ljunghall et al., *Exp. Clin. Endocrinol.* **84** (1984) 313.
[288] B. Duarte et al., *J. Clin. Endocrinol. Metab.* **66** (1988) 584.
[289] H.-M. Harms et al., *J. Clin. Endocrinol. Metab.* **69** (1989) 843.
[290] J. Russell et al., *Endocrinology (Baltimore)* **119** (1986) 2864.
[291] O. Fermet et al., *J. Endocrinol.* **113** (1987) 117.
[292] G. A. Williams et al., *Metab. Clin. Exp.* **34** (1985) 612.
[293] S. A. Rabbani et al., *J. Biol. Chem.* **259** (1984) 2949.
[294] J. Lefebvre, *Horm. Res.* **32** (1989) 104.
[295] L. E. Mallette, *Am. J. Med. Sci.* **293** (1987) 239.
[296] S. Hirschel-Scholz et al., *Calcif. Tissue Int.* **40** (1987) 103.
[297] E. F. Rittinghaus et al., *Acta Endocrinol. (Copenhagen)* **111** (1986) 62.
[298] H. Yoshida et al., *J. Clin. Endocrinol. Metab.* **67** (1988) 488.
[299] T. Andersen et al., *Metab. Clin. Exp.* **35** (1986) 147.
[300] S. Ljunghall et al., *Exp. Clin. Endocrinol.* **88** (1986) 95.
[301] K. Ikeda et al., *J. Biol. Chem.* **264** (1989) 15743.
[302] D. Wynick et al., *J. Endocrinol.* **123** (1989) Suppl., 36.
[303] A. K. Hall, I. R. Dickson, *Acta Endocrinol. (Copenhagen)* **108** (1985) 217.
[304] T. Okazaki et al., *Metab. Clin. Exp.* **33** (1984) 710.
[305] A. van der Plas et al., *Biochem. Biophys. Res. Commun.* **129** (1985) 918.
[306] J. A. Yee, *J. Bone Mineral. Res.* **3** (1988) 211.
[307] C. W. G. M. Löwik et al., *Biochem. Biophys. Res. Commun.* **162** (1989) 1546.
[308] P. M. J. McSheehy, T. J. Chambers, *Endocrinology (Baltimore)* **119** (1986) 1654.
[309] S. F. Silverton et al., *Am. J. Physiol.* **253** (1987) E670.
[310] G. E. Hall, A. D. Kenny, *J. Pharmacol. Exp. Ther.* **238** (1986) 778.
[311] E. M. Spencer et al., *Acta Endocrinol. (Copenhagen)* **121** (1989) 435.
[312] M. Centrella et al., *Proc. Natl. Acad. Sci. USA* **85** (1988) 5889.
[313] K. Hove et al., *Endocrinology (Baltimore)* **114** (1984) 897.
[314] J. Wortsman et al., *J. Clin. Endocrin. Metab.* **62** (1986) 1305.
[315] H. N. Hulter, J. C. Peterson, *Metab. Clin. Exp.* **33** (1984) 662.
[316] L. L. S. Mok et al., *Endocr. Rev.* **10** (1989) 420.
[317] P. K. T. Pang et al., *Can. J. Physiol. Pharmacol.* **64** (1986) 1543.
[318] M. C. M. Yang et al., *J. Pharmacol. Exp. Ther.* **252** (1990) 840.
[319] S. Anderson et al., *Hypertension (Dallas)* **5** (1983) Suppl. I, 59.
[320] K. Iseki et al., *Am. J. Physiol.* **250** (1986) F924.
[321] M. A. Gorozhankina et al., *Patol. Fiziol. Eksp. Ter.* **2** (1986) 71.
[322] G. Clementi et al., *Eur. J. Pharmacol.* **166** (1989) 549.
[323] N. Horiuchi et al., *Science (Washington D.C.)* **238** (1987) 1566.
[324] D. D. Thompson et al., *Proc. Natl. Acad. Sci. USA* **85** (1988) 5673.
[325] M. P. Caulfield et al., *Endocrinology (Baltimore)* **123** (1988) 2949.
[326] A. L. Frelingerl, III., J. E. Zull, *J. Biol. Chem.* **259** (1984) 5507.
[327] R. L. Shew, P. K. T. Pang, *Peptides (Fayetteville, N.Y.)* **5** (1984) 485.
[328] S. H. Doppelt et al., *Proc. Natl. Acad. Sci. USA* **83** (1986) 7557.
[329] J. W. Hamilton et al., *Mol. Cell. Endocrinol.* **44** (1986) 179.
[330] M. E. Goldman et al., *Endocrinology (Baltimore)* **123** (1988) 2597.

[331] J. Reeve et al., *Eur. J. Clin. Invest.* **17** (1987) 421.
[332] E. F. Rittinghaus et al., *Acta Endocrinol. (Copenhagen)* Suppl. **117** (1988) no. 287, 167.
[333] L. E. Mallette, *Ann. Intern. Med.* **109** (1988) 800.
[334] D. Goltzman, G. S. Tannenbaum, *Brain. Res.* **416** (1987) 1.
[335] E. Bucht et al., *Acta Physiol. Scand.* **126** (1986) 289.
[336] E. Bucht et al., *Acta Endocrinol. (Copenhagen)* **113** (1986) 529.
[337] J. M. Le Moullec et al., *FEBS Lett.* **167** (1984) 93.
[338] J. Glowacki et al., *Endocrinology (Baltimore)* **116** (1985) 827.
[339] F. Raue et al., *Endocrinology (Baltimore)* **115** (1984) 2362.
[340] C. Greenberg et al., *Endocrinology (Baltimore)* **118** (1986) 2594.
[341] H. Scheruebl et al., *Acta Endocrinol. (Copenhagen)* Suppl. **117** (1988) no. 287, 66.
[342] O. Torring et al., *Horm. Metab. Res.* **17** (1985) 536.
[343] B. Rasmusson, *Acta Endocrinol. (Copenhagen)* **106** (1984) 112.
[344] F. Tagliaro et al., *J. Endocrinol. Invest.* **7** (1984) 331.
[345] A. Lundin, K. öberg, *Acta Med. Scand.* **215** (1984) 281.
[346] J. J. Body et al., *J. Clin. Endocrinol. Metab.* **62** (1986) 700.
[347] G. Luisetto et al., *J. Endocrinol. Invest.* **9** (1986) 239.
[348] J. Zseli et al., *Calcif. Tissue Int.* **39** (1986) 297.
[349] J.-Y. Reginster et al., *Gynecol. Endocrinol.* **2** (1988) 195.
[350] C. Foresta et al., *Horm. Metab. Res.* **19** (1987) 275.
[351] N. Takahashi et al., *Endocrinology (Baltimore)* **123** (1988) 1504.
[352] H.-G. Schneider et al., *Mol. Cell. Endocrinol.* **58** (1988) 9.
[353] C. M. Rotella et al., *Eur. J. Pharmacol.* **107** (1985) 347.
[354] J. Barsony, S. J. Marx, *Endocrinology (Baltimore)* **122** (1988) 1218.
[355] A. N. K. Yusufi et al., *Am. J. Physiol.* **252** (1987) F598.
[356] F. Pansini et al., *J. Endocrinol.* **116** (1988) 155.
[357] D. R. Roy, *Can. J. Physiol. Pharmacol.* **63** (1985) 89.
[358] J. Puig et al., *Acta Pharmacol. Toxicol. Suppl.* **59** (1986) no. 5, part 2, 126.
[359] M. Yamada et al., *Endocrinology (Baltimore)* **116** (1985) 693.
[360] T. J. Chambers et al., *J. Clin. Endocrinol. Metab.* **63** (1986) 1080.
[361] A. Malgaroli et al., *J. Biol. Chem.* **264** (1989) 14342.
[362] H. I. Khouja et al., *J. Endocrinol.* **119** Suppl. (1988) 152.
[363] J. R. Farley et al., *Calcif. Tissue Int.* **45** (1989) 214.
[364] W. M. Burch, *Endocrinology (Baltimore)* **114** (1984) 1196.
[365] G. P. Ceda et al., *Acta Endocrinol. (Copenhagen)* **120** (1989) 416.
[366] C. Beglinger et al., *Gut* **29** (1988) 243.
[367] G. H. Greeley, Jr., et al., *Regul. Pept.* **24** (1989) 259.
[368] M. Yamaguchi, M. Toyoizumi, *Horm. Metab. Res.* **18** (1986) 378.
[369] M. Yamaguchi, K. Momose, *Horm. Metab. Res.* **18** (1986) 22.
[370] B. J. Looij, Jr., et al., *Ned. Tijdschr. Geneeskd.* **132** (1988) 1558.
[371] I. Zofkova, J. Nedvidkova, *Exp. Clin. Endocrinol.* **92** (1989) 262.
[372] T. S. Mahrous, A. M. Nakhla, *Acta Endocrinol. (Copenhagen)* **119** (1988) 525.
[373] I. Zofkova, J. Bednar, *Exp. Clin. Endocrinol.* **92** (1988) 268.
[374] Y. Goto et al., *Can. J. Physiol. Pharmacol.* (1986) July, Suppl., 110.
[375] W. Woloszczuk et al., *Horm. Metab. Res.* **18** (1986) 197.
[376] P. Demol et al., *Gastroenterology* **86** (1984) part 2, 1060.
[377] D. V. Kleist et al., *Gastroenterology* **86** (1984) part 2, 1290.

[378] H. J. Lenz, *Am. J. Physiol.* **254** (1988) G920.
[379] Y. Tache et al., *Can. J. Physiol. Pharmacol.* (1986) July, Suppl., 53.
[380] P. Bauerfeind et al., *Dig. Dis. Sci.* **30** (1985) 368.
[381] M. J. Fargeas et al., *Regul. Pept.* **11** (1985) 95.
[382] R. C. Orlowski et al., *Eur. J. Biochem.* **162** (1987) 399.
[383] D. M. Findlay et al., *Endocrinology (Baltimore)* **117** (1985) 801.
[384] D. M. Findlay et al., *Endocrinology (Baltimore)* **112** (1983) 1288.
[385] R. M. Epand et al., *J. Med. Chem.* **31** (1988) 1595.
[386] R. M. Epand et al., *Eur. J. Biochem.* **159** (1986) 125.
[387] G. R. Moe et al., *J. Am. Chem. Soc.* **105** (1983) 4100.
[388] R. M. Epand et al., *Biochem. Biophys. Res. Commun.* **152** (1988) 203.
[389] M. J. Twery et al., *Eur. J. Pharmacol.* **155** (1988) 285.
[390] T. Buclin et al., *Calcif. Tissue Int.* **41** (1987) 252.
[391] G. Thamsborg et al., *Lancet I* (1988) 413.
[392] M. T. McDermott, G. S. Kidd, *Endocr. Rev.* **8** (1987) 377.
[393] L. Nogarin, *Clin. Trials J.* **23** Suppl. 1 (1986) 87.
[394] K. Krueger, M. Schattenkirchner, *Z. Rheumatol.* **47** (1988) 287.
[395] G. Schiraldi et al., *Eur. J. Respir. Dis.* **69** Suppl.146 (1986) A150.
[396] L. H. Breimer et al., *Biochem. J.* **255** (1988) 377.
[397] T. Inui et al., *Life Sci.* **45** (1989) 1199.
[398] A. Bjurhom et al., *Peptides (N.Y.)* **9** (1988) 165.
[399] P. H. Steenbergh et al., *FEBS Lett.* **209** (1986) 97.
[400] X.-Y. Hua, J. M. Lundberg, *Acta Physiol. Scand.* **128** (1986) 453.
[401] A. F. Russo et al., *J. Biol. Chem.* **263** (1988) 5.
[402] P. Popper, P. E. Micevych, *Neuroendocrinology* **50** (1989) 338.
[403] M. Pohl et al., *Neuropeptides (Edinburgh)* **14** (1989) 151.
[404] H. Kawasaki et al., *Nature (London)* **335** (1988) 164.
[405] A. Haegerstrand et al., *Proc. Natl. Acad. Sci. USA* **87** (1990) 3299.
[406] J. B. D. Palmer et al., *Br. J. Pharmacol.* **91** (1987) 95.
[407] J. Wallengren, R. Hakanson, *Eur. J. Pharmacol.* **143** (1987) 267.
[408] R. Oku et al., *Brain Res.* **403** (1987) 350.
[409] C. Netti et al., *Neuroendocrinology* **49** (1989) 242.
[410] C. Elie et al., *Neuropeptides (Edinburgh)* **16** (1990) 109.
[411] J. P. Hinson, G. P. Vinson, *Neuropeptides (Edinburgh)* **16** (1990) 129.
[412] H. J. Lenz et al., *Gut* **26** (1985) 550.
[413] W. S. Helton et al., *Am. J. Physiol.* **256** (1989) G715–G720.
[414] L. Bartho et al., *Eur. J. Pharmacol.* **135** (1987) 449.
[415] H. J. Lenz, *Am. J. Physiol.* **254** (1988) G920.
[416] H. M. Cox et al., *Br. J. Pharmacol.* **97** (1989) 996.
[417] B. Ahren et al., *Diabetologia* **30** (1987) 354.
[418] B. Leighton et al., *FEBS Lett.* **249** (1989) 357.
[419] Y. Umeda et al., *Biochem. Biophys. Res. Commun.* **154** (1988) 227.
[420] J. R. Tippins et al., *Biochem. Biophys. Res. Commun.* **134** (1986) 1306.
[421] C. A. Maggi et al., *Eur. J. Pharmacol.* **179** (1990) 217.
[422] T. Dennis et al., *J. Pharmacol. Exp. Ther.* **251** (1989) 718.
[423] M. Nishi et al., *J. Biol. Chem.* **265** (1990) 4173.
[424] S. Mosselman et al., *FEBS Lett.* **239** (1988) 227.

[425] A. N. Roberts et al., *Proc. Natl. Acad. Sci. USA* **86** (1989) 9662.
[426] A. Clark et al., *Lancet II* (1990) 231.
[427] A. Ogawa et al., *J. Clin. Invest.* **85** (1990) 973.
[428] E. Hartter et al., *Lancet I* (1990) 854.
[429] G. J. S. Cooper et al., *Proc. Natl. Acad. Sci. USA* **85** (1988) 7763.
[430] H. K. Datta et al., *Biochem. Biophys. Res. Commun.* **163** (1989) 876.
[431] J. R. Dave et al., *Endocrinology (Baltimore)* **120** (1987) 305.
[432] M. Fenger, A. J. Johnsen, *Biochem. J.* **250** (1988) 781.
[433] F. E. Estivariz et al., *J. Endocrinol.* **116** (1988) 207.
[434] H. P. J. Bennett, *Peptides (N.Y.)* **7** (1986) 615.
[435] D. C. Cumming et al., *J. Clin. Endocrinol. Metab.* **60** (1985) 810.
[436] C.-L. C. Chen, M. B. Madigan, *Endocrinology (Baltimore)* **121** (1987) 590.
[437] J. N. Wilcox, J. L. Roberts, *Endocrinology (Baltimore)* **117** (1985) 2392–2396.
[438] A. Poteliakhoff, *Lancet II* (1985) 326.
[439] A. C. Y. Chang et al., *Proc. Natl. Acad. Sci. USA* **77** (1980) 4890.
[440] J. Drouin et al., *FEBS Lett.* **193** (1985) 54.
[441] S. Nakanishi et al., *Nature (London)* **278** (1979) 423.
[442] G. J. M. Martens et al., *J. Biol. Chem.* **260** (1985) 13685.
[443] P. Petrusz et al., *Proc. Natl. Acad. Sci. USA* **80** (1983) 1721.
[444] T. O. Bruhn et al., *Endocrinology (Baltimore)* **120** (1987) 25.
[445] A. Sasaki et al., *J. Clin. Endocrinol. Metab.* **67** (1988) 768.
[446] D. J. Yoon et al., *Endocrinology (Baltimore)* **122** (1988) 759.
[447] H. Jingami et al., *FEBS Lett.* **191** (1985) 63.
[448] M. Patthy et al., *Proc. Natl. Acad. Sci. USA* **82** (1985) 8762.
[449] T. Watabe et al., *Life Sci.* **40** (1987) 1651.
[450] S. Suemaru et al., *Life Sci.* **39** (1986) 1161.
[451] C. Rivier, W. Vale, *Endocrinology (Baltimore)* **114** (1984) 2409.
[452] P. M. Plotsky, *Endocrinology (Baltimore)* **121** (1987) 924.
[453] E. Redel et al., *Endocrinology (Baltimore)* **123** (1988) 2736.
[454] S. Suemaru et al., *Endocrinol. Jpn.* **33** (1986) 441.
[455] T. Suda et al., *Life Sci.* **40** (1987) 673.
[456] B. Bateman et al., *Endocr. Rev.* **10** (1989) 92.
[457] T. Suda et al., *J. Clin. Endocrinol. Metab.* **64** (1987) 909.
[458] F. Yajima et al., *Life Sci.* **39** (1986) 181.
[459] J. C. Buckingham, T. A. Cooper, *Br. J. Pharmacol.* **90** Suppl. (1987) 167P.
[460] T. Laatikainen et al., *Neuropeptides (Edinburgh)* **10** (1987) 343.
[461] E. B. De Souza et al., *Nature (London)* **319** (1986) 593.
[462] T. Shibasaki et al., *Life Sci.* **43** (1988) 1103.
[463] A. E. Calogero et al., *Clin. Res.* **35** (1987) 393A.
[464] G. Aguilera et al., *Neuroendocrinology* **43** (1986) 79.
[465] M. C. Holmes et al., *Endocrinology (Baltimore)* **121** (1987) 2093.
[466] J. R. Dave, R. L. Eskay, *Biochem. Biophys. Res. Commun.* **134** (1986) 255.
[467] A. N. Margioris et al., *J. Clin. Endocrinol. Metab.* **66** (1988) 922.
[468] J. Kraicer et al., *Neuroendocrinology* **41** (1985) 363.
[469] J. H. Meador-Woodruff et al., *Neuropeptides (Edinburgh)* **9** (1987) 269.
[470] J. V. Conaglen et al., *J. Clin. Endocrinol. Metab.* **58** (1984) 463.
[471] C. Maser-Gluth et al., *Life Sci.* **35** (1984) 879.

[472] E. De Souza, G. R. Van Loon, *Experientia* **40** (1984) 1004.
[473] D. O. Sobel, *Peptides (N.Y.)* **7** (1986) 443.
[474] S. Guild, T. Reisine, *J. Pharmacol. Exp. Ther.* **241** (1987) 125.
[475] D. M. Gibbs, *Endocrinology (Baltimore)* **116** (1985) 723.
[476] M. Keller-Wood et al., *Am. J. Physiol.* **250** (1986) R396.
[477] A. C. Hale et al., *J. Endocr.* **102** (1984) R1.
[478] B. I. Baker et al., *J. Endocr.* **106** (1985) R5.
[479] M. V. Graf et al., *Neuroendocrinology* **41** (1985) 353.
[480] R. S. Rittmaster et al., *J. Clin. Endocrinol. Metab.* **60** (1985) 891.
[481] K. E. Nikolarakis et al.], *Brain Res.* **399** (1986) 152.
[482] J. G. Kiang, E. T. Wei, *J. Pharmacol. Exp. Therap.* **243** (1987) 517. E. T. Wei et al., *J. Pharmacol. Exp. Therap.* **247** (1988) 1082.
[483] K. M. Hargreaves et al., *Eur. J. Pharmacol.* **170** (1989) 275.
[484] C. Rivier, W. Vale, *Endocrinology (Baltimore)* **114** (1984) 914.
[485] B. Miskowiak et al., *Exp. Clin. Endocrinol.* **88** (1986) 25.
[486] F. Petraglia et al., *Endocrinology (Baltimore)* **120** (1987) 1083.
[487] A. Kooy et al., *Neuroendocrinology* **51** (1990) 261.
[488] H. Katakami et al., *Neuroendocrinology* **41** (1985) 390.
[489] A. R. M. M. Hermus et al., *Clin. Endocrinol. (Oxford)* **21** (1984) 589.
[490] M. Vlaskovska et al., *Endocrinology (Baltimore)* **115** (1984) 895.
[491] S. J. Konturek et al., *Peptides (N.Y.)* **8** (1987) 575.
[492] Y. Tache et al., *Gastroenterology* **86** (1984) 281.
[493] R. Murison et al., *Psychoneuroendocrinology (Oxford)* **14** (1989) 331.
[494] M. Burlage et al., *Klin. Wochenschr.* **66** (1988) Suppl. 13, 207.
[495] C. L. Williams, T. F. Burks, *Gastroenterology* **92** (1987) part 2, 1808.
[496] D. Oppermann et al., *Klin. Wochenschr.* **63** (1985) 231.
[497] G. Laux, K. P. Lesch, M. Schwab, *Neuropsychobiology* **19** (1988) 40.
[498] G. L. Bernardini et al., *Peptides (N.Y.)* **5** (1984) 57.
[499] G. F. Koob, F. E. Bloom, *Proc. Fed. Am. Soc. Exp. Biol.* **44** (1985) 259.
[500] D. R. Britton et al., *Life Sci.* **38** (1985) 211.
[501] A. Mele et al., *Peptides (N.Y.)* **8** (1987) 935.
[502] C. R. DeBold et al., *J. Clin. Endocrinol. Metab.* **60** (1985) 836.
[503] P. C. Avgerinos et al., *J. Clin. Endocrinol. Metab.* **62** (1986) 816.
[504] D. Desir et al., *J. Clin. Endocrinol. Metab.* **63** (1986) 1292.
[505] H. Vierhapper, W. Waldhäusl, *Exp. Clin. Endocrinol.* **88** (1986) 355.
[506] O. A. Müller et al., *Horm. Res.* **25** (1987) 185.
[507] D. N. Orth et al., *Proc. Fed. Am. Soc. Exp. Biol.* **44** (1985) 197.
[508] P. W. Gold et al., *N. Engl. J. Med.* **314** (1986) 1329.
[509] Y. Hayashizaki et al., *Horm. Metabol. Res.* **18** (1986) 849.
[510] L. Proulx et al., *J. Endocrinol. Invest.* **7** (1984) 257.
[511] M. M. Valenca, A. Negro-Vilar, *Endocrinology (Baltimore)* **118** (1986) 32.
[512] J. Knudtzon, *Horm. Metabol. Res.* **18** (1986) 579.
[513] A. I. Smith et al., *J. Endocrinol.* **115** (1987) R5.
[514] M. Carnes et al., *Peptides (N.Y.)* **9** (1988) 325.
[515] P. M. Horrocks et al., *Clin. Endocrinol. (Oxford)* **32** (1990) 127.
[516] J. C. Buckingham, *Br. Med. Bull.* **41** (1985) 203.
[517] M. Familari, J. W. Funder, *Aust. N. Z. J. Med.* **17** Suppl. 1 (1987) 164.

[518] J. Dohanics et al., *Peptides (N.Y.)* **11** (1990) 59.
[519] J. F. Mortola et al., *J. Clin. Endocrinol. Metab.* **65** (1987) 962.
[520] K. L. Bost, J. E. Blalock, *Mol. Cell. Endocrinol.* **44** (1986) 1.
[521] E. M. Smith et al., *N. Engl. J. Med.* **317** (1987) 1266.
[522] A. Penhoat et al., *Biochem. Biophys. Res. Commun.* **165** (1989) 355.
[523] W. E. Rainey et al., *J. Biol. Chem.* **264** (1989) 21474.
[524] R. Cheitlin et al., *J. Biol. Chem.* **260** (1985) 5323.
[525] S. W. J. Lamberts et al., *Mol. Cell. Endocrinol.* **52** (1987) 243.
[526] G. Ramirez et al., *J. Clin. Endocrinol. Metab.* **66** (1988) 46.
[527] L. J. Valenta et al., *Horm. Res.* **23** (1986) 16.
[528] A. Singh et al., *Biochem. Biophys. Res. Commun.* **155** (1988) 524.
[529] A. Bertolini et al., *Eur. J. Pharmacol.* **130** (1986) 19.
[530] A. Bertolini et al., *Eur. J. Pharmacol.* **122** (1986) 387.
[531] S. Guarini et al., *Pharmacol. Res. Commun.* **19** (1987) 511.
[532] J. Chariot et al., *Seances Acad. Sci. Ser. 3,* **307** (1988) 235.
[533] M. Kollind et al., *Diabetologia* **27** (1984) 298A.
[534] A. J. Dunn, G. Vigle, *Neuropharmacology* **24** (1985) 329.
[535] R. Poggioli et al., *Peptides (N.Y.)* **7** (1986) 843.
[536] N. Chastrette et al., *Neuropeptides (Edinburgh)* **15** (1990) 61.
[537] S. J. Dunmore et al., *Biochem. J.* **244** (1987) 797.
[538] S. L. Wardlaw, C. E. Markowitz, *Clin. Res.* **34** (1986) 435A.
[539] N. Colbert et al., *Presse Med.* **12** (1983) 1077.
[540] R. C. Bell, J. M. Lipton, *Am. J. Physiol.* **252** (1987) R1152.
[541] K. A. Gruber, M. F. Callahan, *Am. J. Physiol.* **257** (1989) R681.
[542] J. Wortsman et al., *J. Clin. Endocrinol. Metab.* **61** (1985) 355.
[543] A. M. de Lauro Castrucci et al., *Drugs Future* **15** (1990) 41–55.
[544] G. G. Nussdorfer et al., *Biochem. Biophys. Res. Commun.* **141** (1986) 1279.
[545] C. B. Newman et al., *Life Sci.* **36** (1985) 1661.
[546] R. L. Reid et al., *J. Clin. Endocrinol. Metab.* **58** (1984) 773.
[547] R. A. Daynes et al., *J. Invest. Dermatol.* **88** (1987) 483.
[548] F. Di Giovine et al., *Br. J. Rheumatol.* **26** (1987) Suppl. 2, 62.
[549] S. Y. Lin et al., *Hypertension (Dallas)* **10** (1987) 619.
[550] M.-T. Pham-Huu-Trung et al., *J. Clin. Endocrinol. Metab.* **61** (1985) 467.
[551] K. A. Gruber, M. F. Callahan, *Am. J. Physiol.* **257** (1989) R681.
[552] V. J. Hruby et al., *J. Med. Chem.* **30** (1987) 2126.
[553] M. Lebl et al., *Collect. Czech. Chem. Commun.* **49** (1984) 2680.
[554] N. Levine et al., *J. Invest. Dermatol.* **87** (1986) 416.
[555] D. G. Klemes et al., *Biochem. Biophys. Res. Commun.* **137** (1986) 722.
[556] D. B. Richards, J. M. Lipton, *Peptides (N.Y.)* **5** (1984) 815.
[557] F. Al-Obeidi et al., *J. Med. Chem.* **32** (1989) 2555.
[558] W. L. Cody et al., *J. Med. Chem.* **28** (1985) 583.
[559] *Drugs Future* **7** (1982) 319; **8** (1983) 468; **9** (1984) 381; **10** (1985) 431.
[560] *Drugs Future* **11** (1986) 432–433; **12** (1987) 505–506; **13** (1988) 489–490; **14** (1989) 480.
[561] *Drugs Future* **14** (1989) 514–516; **15** (1990) 621.
[562] G. Wiemer et al., *Arch. Pharmacol.* **335** Suppl. (1987) R94.
[563] G. Wolterink, J. M. van Ree, *Brain Res.* **421** (1987) 41.
[564] T. K. Sawyer et al., *Peptides (N.Y.)* **11** (1990) 351.

[565] M. E. Hadley et al., *Life Sci.* **40** (1987) 1889.
[566] W. R. Kiessling, *Arch. Neurol. (Chicago)* **44** (1987) 995.
[567] P. K. Sorensen et al., *Prog. Neuropsychopharmacol. Biol. Psychiatry* **10** (1986) 479.
[568] I. Kitchen, *Gen. Pharmac.* **16** (1985) 79.
[569] A. T. McKnight et al., *Neuropharmacology* **24** (1985) 1011.
[570] R. S. Rapka, *Life Sci.* **39** (1986) 1825.
[571] A. H. Mulder et al., *Neuropeptides (Edinburgh)* **14** (1989) 99.
[572] J. Gairin et al., *J. Pharmacol. Exp. Ther.* **245** (1988) 995.
[573] R. Quirion, A. S. Weiss, *Peptides (N.Y.)* **4** (1983) 445.
[574] P. Sanchez-Blazquez et al., *Neuropeptides (Edinburgh)* **5** (1984) 181.
[575] G. W. Pasternak, *Biochem. Pharmacol.* **35** (1986) 361.
[576] T. F. Burks et al., *Life Sci.* **43** (1988) 2177.
[577] A. H. Mulder et al., *Nature (London)* **308** (1984) 278.
[578] S. B. Weinberger et al., *Regul. Pept.* **26** (1989) 323.
[579] Z. Khalil, R. D. Helme, *Neuropeptides (Edinburgh)* **17** (1990) 45.
[580] A. Jackson, S. J. Cooper, *Neuropharmacology* **25** (1986) 653.
[581] A. Pfeiffer et al., *Endocrinology (Baltimore)* **116** (1985) 2688.
[582] T. P. Blackburn et al., *Br. J. Pharmacol.* **89** (1986) 593.
[583] I. Kitchen, *Gen. Pharmacol.* **16** (1985) 79.
[584] F. A. Aleem et al., *Fertil. Steril.* **45** (1986) 507.
[585] R. Singer et al., *Experientia* **41** (1985) 64.
[586] T. Wahlström et al., *Life Sci.* **36** (1985) 987.
[587] I. Shimizu et al., *Regul. Pept.* **16** (1986) 331.
[588] M. Dennis et al., *J. Chromatogr.* **266** (1983) 163.
[589] O. Vuolteenaho, *Acta Physiol. Scand. Suppl.* **531** (1984) 1.
[590] C. W. Wilkinson, D. M. Dorsa, *Neuroendocrinology* **43** (1986) 124.
[591] W. Knepel, G. Meyen, *Neuroendocrinology* **43** (1986) 44.
[592] D. Copolov et al., *Clin. Endocrinol. (Oxford)* **19** (1983) 575.
[593] C. Gianoulakis et al., *Endocrinology (Baltimore)* **122** (1988) 817.
[594] F. R. Caspari, S. Alapin-Rubillovitz, *J. Clin. Endocrinol. Metab.* **60** (1985) 34.
[595] C. Shaba et al., *Endocrinology (Baltimore)* **115** (1984) 378.
[596] A. Fabbri et al., *Endocrinology (Baltimore)* **122** (1988) 749.
[597] F. Petraglia et al., *Clin. Endocrinol. (Oxford)* **24** (1986) 609.
[598] T. Barreca et al., *Life Sci.* **38** (1986) 2263.
[599] C. Gianoulakis et al., *Life Sci.* **45** (1989) 1097.
[600] F. Facchinetti et al., *J. Endocrinol. Invest.* **9** (1986) 11.
[601] A. R. Genazzani et al., *J. Clin. Endocrinol. Metab.* **62** (1986) 36.
[602] W. O. Richter et al., *Endocrinology (Baltimore)* **120** (1987) 1472.
[603] S. Itow et al., *Can. J. Physiol. Pharmacol.* **63** (1985) 81.
[604] J. C. Buckingham, *Neuroendocrinology* **42** (1986) 148.
[605] S. L. Wardlaw et al., *Endocrinology (Baltimore)* **119** (1986) 112.
[606] D. Giugliano et al., *Metab. Clin. Exp.* **36** (1987) 974.
[607] A. Y. S. Lee et al., *Int. J. Pept. Protein Res.* **24** (1984) 525.
[608] L.-F. Tseng, *J. Pharmacol. Exp. Ter.* **239** (1986) 160.
[609] J. P. Blanc, E. T. Kaiser, *J. Biol. Chem.* **259** (1984) 9549.
[610] D. Yamashiro et al., *Int. J. Pept. Protein Res.* **24** (1984) 520.
[611] P. Nicolas et al., *Proc. Natl. Acad. Sci. USA* **81** (1984) 3074.

[612] D. Yamashiro et al., *Int. J. Pept. Protein Res.* **24** (1984) 516.
[613] J. Blake et al., *Int. J. Pept. Protein Res.* **25** (1985) 575.
[614] D. Collado-Escobar et al., *Eur. J. Pharmacol.* **129** (1986) 385.
[615] H. H. Suh et al., *Peptides (N.Y.)* **8** (1987) 123.
[616] A. E. Panerai et al., *Eur. J. Pharmacol.* **99** (1984) 341.
[617] M. D. Hirsch, T. L. O'Donohue, *J. Pharmacol. Exp. Ther.* **237** (1986) 378.
[618] W. O. Richter et al., *Endocrinology (Baltimore)* **120** (1987) 1472.
[619] K. Yoshikawa et al., *J. Biol. Chem.* **259** (1984) 14301.
[620] G. J. M. Martens, E. Herbert, *Nature (London)* **310** (1984) 251.
[621] N. Zamir et al., *Proc. Natl. Acad. Sci. USA* **81** (1984) 6886.
[622] I. Kitchen, *Gen. Pharmac.* **16** (1985) 79.
[623] E. A. Singer et al., *Endocrinology (Baltimore)* **119** (1986) 1527–1533.
[624] T. Kanamatsu et al., *Proc. Natl. Acad. Sci. USA* **83** (1986) 9245.
[625] J. D. Watson et al., *J. Endocrinol.* **111** (1986) 329.
[626] J. S. Hong et al., *Neuropeptides (Edinburgh)* **5** (1985) 557.
[627] L.-F. Tseng et al., *Regul. Pept.* **14** (1986) 181.
[628] K. Yoshikawa, S. L. Sabol, *Biochem. Biophys. Res. Commun.* **139** (1986) 1.
[629] J. A. Angulo et al., *Eur. J. Pharmacol.* **130** (1986) 341.
[630] S. L. Lightman, W. S. Young III, *Proc. Natl. Acad. Sci. USA* **86** (1989) 4306.
[631] D. C.-C. Wan et al., *Neuropeptides (Edinburgh)* **16** (1990) 141.
[632] D. Kew, D. L. Kilpatrick, *Mol. Endocrinol.* **3** (1989) 179.
[633] D. K. Sommers et al., *Eur. J. Clin. Pharmacol.* **38** (1990) 391.
[634] A. Pfeiffer, A. Herz, *Horm. Metab. Res.* **16** (1984) 386.
[635] A. Tsunoo et al., *Proc. Natl. Acad. Sci. USA* **83** (1986) 9832.
[636] M. Bickl, *Dig. Dis. Sci.* **29** Suppl. (1984) 10S.
[637] J. C. Reynolds et al., *Am. J. Physiol.* **246** (1984) part 1, G130.
[638] A. Dray et al., *Pain (Amsterdam)* **20** Suppl. 2 (1984) S230.
[639] R. F. Crochelt et al., *Gastroenterology* **84** (1983) part 2, 1403.
[640] J. Behar, P. Biancani, *Gastroenterology* **86** (1984) 134.
[641] P. A. Crooks et al., *Biochem. Pharmacol.* **33** (1984) 4095.
[642] A. Dray, R. Metsch, *Eur. J. Pharmacol.* **98** (1984) 155.
[643] M. D. Owen et al., *Peptides (N.Y.)* **5** (1984) 737.
[644] G. E. Sander et al., *Peptides (N.Y.)* **6** (1985) 133.
[645] D. M. Wright, G. Clarke, *Neuropeptides (Edinburgh)* **5** (1984) 273.
[646] K. Hashimoto et al., *Endocrinol. Jpn.* **33** (1986) 813.
[647] P. G. Andreis et al., *Neuropeptides (Edinburgh)* **12** (1988) 165.
[648] B. Allilio et al., *J. Clin. Endocrinol. Metab.* **63** (1986) 1427.
[649] F. Facchinetti et al., *J. Clin. Endocrinol. Metab.* **63** (1986) 1222.
[650] I. Gerendai et al., *Regul. Pept.* **27** (1990) 107.
[651] T. Mitsuma et al., *Horm. Metab. Res.* **21** (1989) 301.
[652] N. F. Plotnikoff et al., *Fed. Proc. Fed. Am. Soc. Exp.* **44** (1985) 118.
[653] Y. Shavit et al., *Science (Washington, D.C.)* **223** (1984) 188.
[654] J. L. Torres et al., *Experientia* **45** (1989) 574.
[655] M. Maeda et al., *Chem. Pharm. Bull.* **32** (1984) 4157.
[656] G. W. Hardy et al., *J. Med. Chem.* **31** (1988) 960.
[657] *Drugs Future* **11** (1986) 657.
[658] N. Zamir et al., *Nature (London)* **307** (1984) 643.

[659] S. Lemaire et al., *Can. J. Physiol. Pharmacol.* **62** (1984) 484.
[660] E. Weihe et al., *Neuropeptides (Edinburgh)* **5** (1985) 453.
[661] H. J. Wolter, *Peptides (N.Y.)* **7** (1986) 389.
[662] S. Spampinato et al., *Life Sci.* **38** (1986) 403.
[663] H.-K. Jiang et al., *Proc. Natl. Acad. Sci. USA* **86** (1989) 2948.
[664] A. I. Faden et al., *Regul. Pept.* **11** (1985) 35.
[665] A. Z. Zhang et al., *Neuropeptides (Edinburgh)* **5** (1985) 553.
[666] R. Schick et al., *Dig. Dis. Sci.* **29** Suppl. (1984) 74S.
[667] T. Mitsuma et al., *Horm. Metab. Res.* **21** (1989) 301.
[668] N. Falke, *Neuropeptides (Edinburgh)* **11** (1988) 163.
[669] J. A. T. Haaf et al., *Acta Endocrinol. (Copenhagen)* **114** (1987) 96.
[670] A. Rezvani, E. L. Way, *Eur. J. Pharmacol.* **102** (1984) 475.
[671] M. A. Della-Fera et al., *Am. J. Physiol.* **258** (1990) R946.
[672] H. Shimizu et al., *Life Sci.* **45** (1989) 25.
[673] H. Tilson et al., *Neuropeptides (Edinburgh)* **8** (1986) 193.
[674] A. I. Faden et al., *Peptides (N.Y.)* **4** (1983) 631.
[675] P. Sanchez-Blazquez et al., *Neuropeptides (Edinburgh)* **4** (1984) 369.
[676] A. Turcotte et al., *Int. J. Pept. Protein Res.* **23** (1984) 361.
[677] H. Yoshino et al., *J. Med. Chem.* **33** (1990) 206.
[678] J. E. Gairin et al., *J. Med. Chem.* **29** (1986) 1913.
[679] J. E. Gairin et al., *Br. J. Pharmacol.* **95** (1988) 1023.
[680] Y. Kiso et al., *J. Pharmacobio Dyn.* **7** (1984) S-91.
[681] H. L. Wen et al., *Peptides (N.Y.)* **8** (1987) 191.
[682] *Drugs Future* **16** (1991) 203.
[683] K. Richter et al., *Science (Washington, D.C.)* **238** (1987) 200.
[684] A. Mor et al., *FEBS Lett.* **255** (1989) 2690.
[685] A. Mor et al., *Neuropeptides (Edinburgh)* **13** (1989) 51.
[686] D. Giudici et al., *Neuroendocrinology* **39** (1984) 236.
[687] E. C. degli Uberti et al., *Horm. Res.* **24** (1986) 251.
[688] E. Roti et al., *J. Endocrinol. Invest.* **7** (1984) 211.
[689] H. G. Gullner et al., *Arch. Int. Pharmacodyn. Ther.* **266** (1983) 155.
[690] E. C. Degli Uberti et al., *J. Clin. Endocrinol. Metab.* **61** (1985) 1018.
[691] F. Petraglia et al., *Peptides (N.Y.)* **6** (1985) 869.
[692] S. J. Konturek et al., *Gastroenterology* **84** (1983) part 2, 1213.
[693] F. Pavone et al., *Peptides (N.Y.)* **11** (1990) 591.
[694] A. Rossi et al., *Peptides (N.Y.)* **4** (1983) 577.
[695] K. M. Sivanandaiah et al., *Int. J. Pept. Protein Res.* **33** (1989) 463.
[696] K. Darlak et al., *Peptides (N.Y.)* **5** (1984) 687.
[697] K. Kisara et al., *Br. J. Pharmacol.* **87** (1986) 183.
[698] M. A. Cervini et al., *Peptides (N.Y.)* **6** (1985) 433.
[699] S. Salvadori et al., *Eur. J. Med. Chem. Chim. Ther. (Paris)* **18** (1983) 489.
[700] S. Salvadori et al., *Hoppe Seyler's Z. Physiol. Chem.* **365** (1984) 1199.
[701] K. Suzuki et al., *Chem. Pharm. Bull.* **33** (1985) 4865.
[702] K. Chaki et al., *Peptides (N.Y.)* **11** (1990) 139.
[703] Y. Sasaki et al., *Biochem. Biophys. Res. Commun.* **120** (1984) 214.
[704] K. Suzuki et al., *Chem. Pharm. Bull.* **36** (1988) 4834.
[705] S. Salvadori et al., *J. Med. Chem.* **29** (1986) 889.

[706] M. Marastoni et al., *J. Med. Chem.* **30** (1987) 1538.
[707] G. Balboni et al., *Biochem. Biophys. Res. Commun.* **169** (1990) 617.
[708] R. Greenberg et al., *J. Biol. Chem.* **259** (1984) 5132.
[709] K. J. Chang et al., *J. Biol. Chem.* **260** (1985) 9706.
[710] L. Graf et al., *Neuropeptides (Edinburgh)* **9** (1987) 113.
[711] V. Schusdziarra et al., *Diabetologia* **24** (1983) 113.
[712] D. Tome et al., *Am. J. Physiol.* **253** (1987) G737.
[713] V. Schusdziarra et al., *Endocrinology (Baltimore)* **112** (1983) 885.
[714] J. Nedvidkova et al., *Exp. Clin. Endocrinol.* **85** (1985) 249.
[715] T. Mitsuma et al., *Exp. Clin. Endocrinol.* **84** (1984) 324.
[716] M. Yoshikawa et al., *Agric. Biol. Chem.* **48** (1984) 3185.
[717] K.-J. Chang et al., *J. Pharmacol. Exp. Ther.* **227** (1983) 403.
[718] H. Matthies et al., *Peptides (N.Y.)* **5** (1984) 463.
[719] M. Schaefer et al., *Cell (Cambridge Mass.)* **41** (1985) 457.
[720] R. H. M. Ebberink et al., *Peptides (N.Y.)* **8** (1987) 515.
[721] D. A. Price et al., *Peptides (N.Y.)* **8** (1987) 533.
[722] R. B. Raffa, *Peptides (N.Y.)* **9** (1988) 915.
[723] C. S. Barnard, G. J. Dockray, *Regul. Pept.* **8** (1984) 209.
[724] T. Austin et al., *Can. J. Physiol. Pharamcol.* **61** (1983) 949.
[725] H. Y. T. Yang et al., *Proc. Natl. Acad. Sci. USA* **82** (1985) 7757.
[726] M. Kavaliers, M. Hirst, *Peptides (N.Y.)* **6** (1985) 847.
[727] X. Z. Zhu, R. B. Raffa, *Neuropeptides (Edinburgh)* **8** (1986) 55.
[728] G. Telegdy, I. Bollok, *Neuropeptides (Edinburgh)* **10** (1987) 157.
[729] A. Ottlecz, G. Telegdy, *Neuropeptides (Edinburgh)* **9** (1987) 161.
[730] K. Kouge et al., *Bull. Chem. Soc. Jpn.* **60** (1987) 4343.
[731] A. Fukamizi et al., *J. Biol. Chem.* **265** (1990) 7576.
[732] E. T. Ben-Ari et al., *J. Biol. Chem.* **264** (1989) 13074.
[733] N. Itoh et al., *FEBS Lett.* **244** (1989) 6.
[734] E. Coezy et al., *Am. J. Physiol.* **257** (1989) C888.
[735] T. Inagami et al., *Fed. Proc. Fed. Am. Soc. Exp. Biol.* **45** (1986) 1414.
[736] M. Naruse et al., *J. Clin. Endocrinol. Metab.* **61** (1985) 480.
[737] C. F. Deschepper et al., *Endocrinology (Baltimore)* **119** (1986) 36.
[738] M. C. Chappell et al., *J. Biol. Chem.* **264** (1989) 16518.
[739] M. Bumpus et al., *Am. J. Med. Sci.* **295** (1988) 406.
[740] D. G. Changaris et al., *Biochem. Biophys. Res. Commun.* **138** (1986) 573.
[741] O. Kurauchi et al., *Horm. Metab. Res.* **21** (1989) 558.
[742] D. D. Smyth, H. Y. M. Fung, *Eur. J. Pharmacol.* **102** (1984) 55.
[743] J. C. Lee et al., *Am. J. Physiol.* **251** (1986) R258.
[744] J. Mann, *Acta Endocrinol. (Copenhagen)* **102,** Suppl. 253 (1983) 10.
[745] N. R. Levens, *Am. J. Physiol.* **246** (1984) Part 1, G634.
[746] H. F. Janssen et al., *Circ. Shock* **23** (1987) 197.
[747] A. T. Chiu et al., *Biochem. Biophys. Res. Commun.* **165** (1989) 196.
[748] R. S. L. Chang et al., *Biochem. Biophys. Res. Commun.* **171** (1990) 813.
[749] R. Tabrizchi et al., *Life Sci.* **43** (1988) 537.
[750] A. Israel et al., *Neuroendocrinology* **42** (1986) 57.
[751] E. Castren, J. M. Saavedra, *Endocrinology (Baltimore)* **122** (1988) 370.
[752] I. Luoveau et al., *Biochem. Biophys. Res. Commun.* **163** (1989) 32.

[753] B. Bouscarel et al., *J. Biol. Chem.* **263** (1988) 14920.
[754] B. J. Ballermann et al., *Am. J. Physiol.* **247** (1984) F110.
[755] G. Andoka et al., *Biochem. Biophys. Res. Commun.* **121** (1984) 441.
[756] P. D. Carriere et al., *Neuroendocrinology* **43** (1986) 49.
[757] J. Bingham Smith et al., *Proc. Natl. Acad. Sci. USA* **81** (1984) 7812.
[758] J. A. Gilbert et al., *Biochem. Pharmacol.* **33** (1984) 2527.
[759] H. Satoh et al., *Biochem. Biophys. Res. Commun.* **126** (1985) 464.
[760] I. Antonipillai et al., *Clin. Res.* **35** (1987) 135A.
[761] A. Nakamura et al., *Am. J. Physiol.* **258** (1990) E1.
[762] M. E. Olson et al., *Clin. Sci.* **72** (1987) 429. K. Yamaguchi et al., *Acta Endocrinol. (Copenhagen)* **118** (1988) 82.
[763] V. L. Brooks et al., *Circ. Res.* **58** (1986) 829.
[764] M. C. Khosla et al., *Hypertension (Dallas)* **11** Suppl. I (1988) 1–38.
[765] J. E. Hall, *Fed. Proc. Fed. Am. Soc. Exp. Biol.* **45** (1986) 1431.
[766] V. L. Schuster, *Federation Proc. Fed. Am. Soc. Exp. Biol.* **45** (1986) 1444.
[767] N. R. Levens, *Am. J. Physiol.* **245** (1983) G511.
[768] G. G. Nussdorfer et al., *Exp. Clin. Endocrinol.* **88** (1986) 158.
[769] L. L. Espey et al., *Proc. Soc. Exp. Biol. Med.* **193** (1990) 249.
[770] D. Stirling et al., *J. Biol. Chem.* **265** (1990) 5.
[771] T. Kono et al., *Endocrinol. Jpn.* **34** (1987) 737.
[772] M. T. Schiavone et al., *Neurobiolog.* **85** (1988) 4095.
[773] K. Y. Hui et al., *Int. J. Pept. Protein Res.* **34** (1989) 177.
[774] T. Kono et al., *Life Sci.* **37** (1985) 365.
[775] G. Guillemette et al., *J. Med. Chem.* **27** (1984) 315.
[776] A. Husain et al., *J. Pharmacol. Exp. Ther.* **239** (1986) 71.
[777] J. Samanen et al., *J. Med. Chem.* **32** (1989) 466.
[778] K. L. Spear et al., *J. Med. Chem.* **33** (1990) 1935.
[779] K.-H. Hsieh et al., *J. Med. Chem.* **32** (1989) 898.
[780] J. Samanen et al., *J. Med. Chem.* **31** (1988) 510.
[781] O. Nyeki et al., *J. Med. Chem.* **30** (1987) 1719.
[782] M. N. Scanlon et al., *Life Sci.* **34** (1984) 317.
[783] M. H. Goghari et al., *J. Med. Chem.* **29** (1986) 1121.
[784] J. Samanen et al., *J. Med. Chem.* **32** (1989) 1366.
[785] P. R. Bovy et al., *J. Med. Chem.* **32** (1989) 520.
[786] P. R. Bovy et al., *J. Med. Chem.* **33** (1990) 1477.
[787] P. C. Wong et al., *Hypertension (Dallas)* **12** (1988) 340.
[788] J. V. Duncia et al., *J. Med. Chem.* **33** (1990) 1312.
[789] J. Samanen et al., *J. Med. Chem.* **31** (1988) 737.
[790] J. E. Taylor et al., *Drug. Dev. Res.* **16** (1989) 1.
[791] N. B. Oza et al., *J. Clin. Invest.* **85** (1990) 597.
[792] D. I. Diz, *Peptides (Fayetteville, N.Y.)* **6** (1985) 57.
[793] W. E. Wilson et al., *J. Biol. Biochem.* **264** (1989) 17777.
[794] H. Kato et al., *FEBS Lett.* **232** (1988) 252.
[795] K.-I. Enjyoji, H. Kato, *FEBS Lett.* **238** (1988) 1.
[796] H. Maeda et al., *J. Biol. Chem.* **263** (1988) 1605.
[797] K.-I. Enjyoji et al., *J. Biol. Chem.* **263** (1988) 973.
[798] A. Barlas et al., *FEBS Lett.* **218** (1987) 266.

[799] N. Gutman et al., *Eur. J. Biochem.* **171** (1988) 577.
[800] J. A. Veenstra, *FEBS Lett.* **250** (1989) 231–234.
[801] G. Gäde, R. Kellner, *Peptides (Fayetteville, N.Y.)* **10** (1989) 1287. Y. Kuroki et al., *Biochem. Biophys. Res. Commun.* **167** (1990) 273. L. Schoofs et al., *Peptides (Fayetteville, N.Y.)* **11** (1990) 427.
[802] H. Wilkens et al. in A. Bertelli, N. Back (eds.): *Shock, Biochemical, Pharmacological and Clinical Aspects*, Plenum Press, New York 1970, pp. 201–214.
[803] D. Proud, A. P. Kaplan, *Ann. Rev. Immunol. (Palo Alto)* **6** (1988) 49.
[804] M. Wicklmayr et al., *Horm. Metab. Res.* **20** (1988) 535.
[805] L. L. Espey et al., *Am. J. Physiol.* **251** (1986) E362.
[806] G. M. Patrassi et al., *Thromb. Haemostasis* **50** (1983) 228.
[807] C. P. Quilley et al., *Hypertension (Dallas)* **10** (1987) 294.
[808] H. Okunishi et al., *Hypertension* **8** Suppl. 1 (1986) 114.
[809] J. Spragg et al., *Peptides (Fayetteville, N.Y.)* **9** (1988) 203.
[810] G. Caspritz et al., *Arzneim. Forsch.* **36** (1986) 1605.
[811] S. Kumakura et al., *J. Pharmacol. Exp. Ther.* **243** (1987) 1067.
[812] R. E. Ferner et al., *Br. J. Clin. Pharmacol.* **25** (1988) 105P.
[813] E. G. Erdös, *J. Cardiovasc. Pharmacol.* **15** Suppl. 6 (1990) S20.
[814] J. M. Stewart, R. J. Vavrek, *J. Cardiovasc. Pharmacol.* **15** (1990) S69.
[815] I. A. Sheikh, A. P. Kaplan, *Biochem. Pharmacol.* **38** (1989) 993.
[816] K. Shimamoto et al., *J. Cardiovasc. Pharmacol.* **15** (1990) S83.
[817] P. Maddedu et al., *J. Hypertens.* **5** (1987) 645.
[818] I. Miyamori et al., *Exp. Clin. Endocrinol.* **87** (1986) 169.
[819] D. Proud et al., paper presented at the congress *Kinin 84*, Savannah, Ga., 1984.
[820] R. Kayegama et al., *J. Biol. Chem.* **260** (1985) 12060.
[821] A. Barlas et al., paper presented at the congress *Kinin 84*, Savannah, Ga., 1984.
[822] F. Sierra et al., *Mol. Cell. Biol.* **9** (1989) 5610.
[823] K. M. Braas et al., *Br. J. Pharmacol.* **94** (1988) 3.
[824] A. A. Roscher et al., *J. Cardiovasc. Pharmacol.* **15** Suppl. 6 (1990) S39.
[825] O. A. Carretero, A. G. Scicli in J. H. Laragh et al. (eds.): *Endocrine Mechanisms in Hypertension*, Raven Press, New York 1989, p. 219.
[826] D. Regoli et al., *J. Cardiovasc. Pharmacol.* **15** Suppl. 6 (1990) S30.
[827] J. M. Ritter et al., *Br. J. Pharmacol.* **96** (1989) 23.
[828] C. R. Mantione, R. Rodriguez, *Br. J. Pharmacol.* **99** (1990) 516.
[829] B. S. Dixon et al., *Am. J. Physiol.* **258** (1990) C299.
[830] B. Waeber et al., *J. Cardiovasc. Pharmacol.* **15** Suppl. 6 (1990) S78.
[831] K. Inokuchi, K. U. Malik, *Am. J. Physiol.* **246** (1984) F387.
[832] E. T. Whalley et al., *Naunym-Schmiedeberg's Arch. Pharmacol.* **336** (1987) 652.
[833] B. A. Schölkens et al., *J. Hypertens.* **5** Suppl. 5 (1987) S7.
[834] Y. Yoshimura et al., *Endocrinology (Baltimore)* **122** (1988) 2540.
[835] P. Geppetti et al., *Br. J. Pharmacol.* **94** (1988) 288.
[836] G. Dietze et al., Paper presented at the congress *Kinin 84*, Savannah, Ga., 1984.
[837] N.-E. Rhaleb et al., *Br. J. Pharmacol.* **99** (1990) 445.
[838] J. Barabe et al., *Can. J. Physiol. Pharmacol.* **62** (1984) 627.
[839] G. Drapeau et al., *Eur. J. Pharmacol.* **155** (1988) 193.
[840] R. J. Vavrek et al. in E. Giralt, D. Andersen (eds.): *Peptides 1990*, ESCOM Sci. Publ., Leiden 1991, p. 642.

[841] F. J. Hook et al., *Br. J. Pharmacol.* **102** (1991) 769.
[842] R. J. Vavrek, J. M. Stewart in C. M. Deber et al. (eds.): *Peptides. Structure and Function*, Pierce Chem. Comp., Rockford, Ill., 1985, p. 655.
[843] J. M. Stewart, R. J. Vavrek, *J. Cardiovasc. Pharmacol.* **15** Suppl. 6 (1990) S69.
[844] D. Regoli et al., *Eur. J. Pharmacol.* **123** (1986) 61.
[845] D. Regoli et al., *Eur. J. Pharmacol.* **127** (1986) 219.
[846] P. Devillier et al., *Eur. J. Pharmacol.* **149** (1988) 137.
[847] A. S. Dutta, *Drugs Future* **12** (1987) 781–792.
[848] Y. Kawaguchi et al., *Biochem. Biophys. Res. Commun.* **139** (1986) 1040.
[849] A. J. Harmar et al., *FEBS Lett.* **208** (1986) 67.
[850] C.-R. Martling et al., *Life Sci.* **40** (1987) 1633.
[851] J. M. Conlon et al., *Peptides (Fayetteville, N.Y.)* **9** (1988) 859.
[852] E. Theodorsson-Norheim et al., *Eur. J. Biochem.* **166** (1987) 693.
[853] A. Bucsics et al., *Peptides (Fayetteville, N.Y.)* **4** (1983) 451.
[854] A. Ericsson et al., *Mol. Endocrinol.* **4** (1990) 1211.
[855] B. Beding-Barnekow, E. Brodin, *Regul. Pept.* **25** (1989) 199.
[856] J. M. Conlon et al., *FEBS Lett.* **200** (1986) 111.
[857] L. Schoofs et al., *FEBS Lett.* **261** (1990) 397.
[858] B. A. Yankner et al., *Science (Washington D.C.)* **250** (1990) 279.
[859] J. L. Vaught, *Life Sci.* **43** (1988) 1419.
[860] F. Angel et al., *Gastroenterology* **84** (1983) Part 2, 1092.
[861] J. Wiley, C. Owyang, *Am. J. Physiol.* **252** (1987) G431.
[862] A. Mandahl, *Eur. J. Pharmacol.* **114** (1985) 121.
[863] P. Tonnesen, O. B. Schaffalitzky de Muckadell, *Allergy (Copenhagen)* **42** (1987) 146.
[864] M. J. Bannon et al., *J. Biol. Chem.* **261** (1986) 6640.
[865] I. Norheim et al., *J. Clin. Endocrinol. Metab.* **63** (1986) 605.
[866] P. Devillier et al., *N. Engl. J. Med.* **314** (1986) 1323.
[867] J. M. Stewart, M. E. Hall in T. Shiba, S. Sakakibara (eds.): *Peptide Chemistry 1987*, Protein Research Foundation, Osaka 1988, p. 449.
[868] J. M. Conlon, L. Sheehan, *Regul. Pept.* **7** (1983) 335.
[869] D. J. Dusser et al., *J. Clin. Invest.* **84** (1989) 900.
[870] W. E. Siems et al., *Pharmazie* **42** (1987) 153.
[871] S. Lavielle et al., *Biochem. Pharmacol.* **37** (1988) 41.
[872] Y. Yokota et al., *J. Biol. Chem.* **264** (1989) 17649.
[873] Y. Masu et al., *Nature (London)* **329** (1987) 836.
[874] D. Regoli et al., *Trends Pharmacol. Sci. TIPS* **9** (1988) 290.
[875] S. H. Buck, S. A. Shatzer, *Life Sci.* **42** (1988) 2701.
[876] S. H. Buck, J. L. Krstenansky, *Eur. J. Pharmacol.* **144** (1987) 109.
[877] I. Haro et al., *J. Pharm. Sci.* **79** (1990) 74.
[878] R. M. Hagan et al., *Br. J. Pharmacol.* **98** (1989) Suppl., 717P.
[879] P. Rovero, V. Pestellini, *Neuropeptides (Edinburgh)* **10** (1987) 355.
[880] P. Rovero et al., *Peptides (Fayetteville, N.Y.)* **10** (1989) 593.
[881] S. H. Buck et al., *Life Sci.* **47** (1990) PL-37.
[882] M. Chorev et al., *Eur. J. Pharmacol.* **127** (1986) 187.
[883] C. J. Mussap, E. Burcher, *Peptides (Fayetteville, N.Y.)* **11** (1990) 827.
[884] J.-M. Qian et al., *J. Biol. Chem.* **264** (1989) 16667.
[885] T. Sakurada et al., *Eur. J. Pharmacol.* **174** (1989) 153.

[886] P. Ward et al., *J. Med. Chem.* **33** (1990) 1848.
[887] J. Mizrahi et al., *Eur. J. Pharmacol.* **118** (1985) 25.
[888] D. Regoli et al., *Eur. J. Pharmacol.* **109** (1985) 121.
[889] P. Rovero et al., *Eur. J. Pharmacol.* **175** (1990) 113.
[890] G. Drapeau et al., *Regul. Pept.* **31** (1990) 125.
[891] C. A. Maggi et al., *Eur. J. Pharmacol.* **177** (1990) 81.
[892] E. Naline et al., *Am. Rev. Respir. Dis.* **140** (1989) 679.
[893] S. Brunelleschi et al., *Br. J. Pharmacol.* **100** (1990) 417.
[894] S. Evangelista et al., *Peptides (Fayetteville, N.Y.)* **10** (1989) 79.
[895] C. A. Maggi et al., *Br. J. Pharmacol.* **100** (1990) 588.
[896] I. Rollandy et al., *Neuropeptides (Edinburgh)* **13** (1989) 175. M. Arisawa et al., *Proc. Natl. Acad. Sci. USA* **86** (1989) 7290.
[897] M. Lotz et al., *Science (Washington D.C.)* **235** (1987) 893.
[898] J. E. T. Fox, E. E. Daniel, *Am. J. Physiol.* **250** (1986) G21.
[899] L. D. Hirning, T. F. Burks, *Gastroenterology* **84** (1983) part 2, 1188.
[900] J. P. McGillis et al., *Fed. Proc. Fed. Am. Soc. Exp.* **46** (1987) 196.
[901] I. Paegelow et al., *Pharmazie* **44** (1989) 145.
[902] M. E. Hall, J. M. Stewart, *Peptides (Fayetteville, N.Y.)* **5** (1984) 85.
[903] T. Sakurada et al., *Jpn. J. Pharmacol.* **43** (1987) Suppl., 272P.
[904] K. D. Yench et al., *Patol. Fiziol. Eksp. Ter.* 1984, no. 3, 75.
[905] R. Michelot et al., *Eur. J. Med. Chem. Chim. Ther.* **23** (1988) 243.
[906] C. Poulos et al., *J. Med. Chem.* **30** (1987) 1512.
[907] Y. Higuchi et al., *Eur. J. Pharmacol.* **160** (1989) 413.
[908] E. Munekata et al., *Peptides (Fayetteville, N.Y.)* **8** (1987) 169.
[909] A. Ljungquist et al., *Regul. Pept.* **24** (1989) 283.
[910] K. Folkers et al., *Proc. Natl. Acad. Sci. USA* **87** (1990) 4833.
[911] R. Hakanson et al., *Regul. Pept.* **31** (1990) 75.
[912] Z. Wiesenfeld-Hallin et al., *Regul. Pept.* **29** (1990) 1.
[913] J.-M. Qian et al., *J. Biol. Chem.* **264** (1989) 16667.
[914] P. J. Woll, E. Rozengurt, *Proc. Natl. Acad. Sci. USA* **85** (1988) 1859.
[915] G. G. Glenner, *Cell* **52** (1988) 307.
[916] C. L. Joachim et al., *Nature (London)* **341** (1989) 226.
[917] J. Kang et al., *Nature (London)* **325** (1987) 733.
[918] R. E. Tanzi et al., *Science (Washington D.C.)* **235** (1987) 880.
[919] D. Goldgaber et al., *Science (Washington D.C.)* **235** (1987) 877.
[920] M. R. Palmert et al., *Science (Washington D.C.)* **241** (1988) 1080.
[921] D. Schubert et al., *Science (Washington D.C.)* **241** (1988) 223.
[922] W. E. van Nostrand et al., *Science (Washington D.C.)* **248** (1990) 745.
[923] F. Prelli et al., *Biochem. Biophys. Res. Commun.* **151** (1988) 1150.
[924] D. Allsop et al., *Proc. Natl. Acad. Sci. USA* **85** (1988) 2790.
[925] T. Dyrks et al., *EMBO J.* **7** (1988) 949.
[926] A. I. Bush et al., *J. Biol. Chem.* **265** (1990) 15977.
[927] T. Saitoh et al., *Cell* **58** (1989) 615.
[928] B. A. Yankner et al., *Science (Washington D.C.)* **250** (1990) 279–282.
[929] V. Geenen et al., *Science (Washington D.C.)* **232** (1986) 508.
[930] R. Ivell, D. Richter, *EMBO J.* **3** (1984) 2351.
[931] S. Lundin, G. Toresson, *Horm. Metab. Res.* **19** (1987) 629.

[932] B. Liu et al., *J. Biol. Chem.* **263** (1988) 72.
[933] B. Liu, J. P. H. Burbach, *J. Neuroendocrinology* **1** (1989) no. 1.
[934] M. Mühlethaler et al., *Experientia* **40** (1984) 777.
[935] V. A. Batra et al., *Regul. Peptides* **30** (1990) 7.
[936] D. P. Brooks et al., *Experientia* **42** (1986) 1012.
[937] P. Soelberg Sorensen, M. Hammer, *Am. J. Physiol.* **248** (1985) R78.
[938] B. C. Wang, L. Share, K. L. Goetz, *Fed. Proc. Fed. Am. Soc. Exp. Biol.* **44** (1985) 72.
[939] B. Waeber et al., *Am. J. Physiol.* **247** (1984) H895.
[940] M. R. Murphy et al., *J. Clin. Endocrinol. Metab.* **65** (1987) 738.
[941] S. W. T. Cheng, W. G. North, *Neuroendocrinology* **42** (1986) 174.
[942] C. M. Gregg, *Neuroendocrinology* **40** (1985) 423–429.
[943] M. T. Shiavone et al., *Proc. Natl. Acad. Sci. USA* **85** (1988) 4095.
[944] R. Cacabelos et al., *Neuroendocrinology* **45** (1987) 368.
[945] T. Ivanyi et al., *Exp. Clin. Endocrinol.* **88** (1986) 303.
[946] O. Willoughby et al., *Neuroendocrinology* **45** (1987) 219.
[947] T. Kimura et al., *Endocrinology (Baltimore)* **114** (1984) 1426.
[948] I. Os et al., *Acta Med. Scand.* **220** (1986) 195.
[949] G. Geelen et al., *Am. J. Physiol.* **247** (1984) R968.
[950] H. Ishizawa et al., *Eur. J. Pharmacol.* **189** (1990) 119.
[951] J. R. Sladek, Jr., C. D. Sladek, *Fed. Proc. Fed. Am. Soc. Exp. Biol.* **44** (1985) 66.
[952] B. Bardrum et al., *Endocrinology (Baltimore)* **123** (1988).
[953] T. O. Bruhn et al., *Endocrinology (Baltimore)* **119** (1986) 1558.
[954] R. Landgraf et al., *Am. J. Physiol.* **259** (1990) R1056.
[955] N. W. Kasting, *Can. J. Physiol. Pharmacol.* **64** (1986) 1575.
[956] D. Schams et al., *J. Endocrinol.* **102** (1984) 337.
[957] S. Thornton et al., *Br. Med. J.* **297** (1988) 167.
[958] K. Uvnäs-Moberg et al., *Acta Physiol. Scand.* **124** (1985) 391.
[959] M. S. Carmichael et al., *J. Clin. Endocrinol. Metab.* **64** (1987) 27.
[960] P. Chang, K. Kimura, *Jpn. J. Pharmacol.* **43** (1987) Suppl., 184P.
[961] A. E. Johnson et al., *Neuroendocrinology* **50** (1989) 199.
[962] M. Shumacher et al., *Science (Washington D.C.)* **250** (1990) 691.
[963] S. Chin Mah, K. G. Hofbauer, *Drugs Future* **12** (1987) 1055.
[964] E. A. Woodcock et al., *Endocrinology (Baltimore)* **118** (1986) 2432.
[965] L. Share, J. T. Crofton, *Fed. Proc. Fed. Am. Soc. Exp. Biol.* **43** (1984) 103.
[966] M. J. S. Miller et al., *Clin. Exp. Pharmacol. Physiol.* **13** (1986) 577.
[967] J. Maslyszko et al., *Thromb. Haemostasis* **61** (1989) 537.
[968] P. A. Flordal et al., *Thromb. Haemostasis* **61** (1989) 541.
[969] F. A. Laszlo et al., *Acta Endocrinol. (Copenhagen)* **106** (1984) 56.
[970] L. Hary et al., *Fundam. Clin. Pharmacol.* **3** (1989) 452.
[971] J. C. Buckingham, *J. Endocrinol.* **113** (1987) 389.
[972] P. Mormede et al., *Regul. Pept.* **12** (1985) 175.
[973] J. J. Legros et al., *Prog. Neuropsychopharmcol. Biol. Psychiatry (Oxford)* **12** Suppl. (1988) S71.
[974] J. McMurray, A. D. Struthers, *Br. Heart J.* **59** (1988) 627.
[975] L. M. Graczak et al., *Am. J. Physiol.* **256** (1989) H925.
[976] T. Wilson et al., *Prostaglandins* **35** (1988) 771.
[977] D. P. Brooks et al., *Neuroendocrinology* **38** (1984) 382.
[978] S. Stock et al., *Regul. Pept.* **30** (1990) 1.

[979] N. Altszuler et al., *Proc. Soc. Exp. Biol. Med.* **182** (1986) 79.
[980] A. N. Margioris et al., *J. Clin. Endocrinol. Metab.* **66** (1988) 922.
[981] A. Buku et al., *J. Med. Chem.* **30** (1987) 1509.
[982] R. Brown, M. G. King, *Peptides (Fayetteville, N.Y.)* **5** (1984) 1135.
[983] S. Chin Mah, K. G. Hofbauer, *Drugs Future* **12** (1987) 1055.
[984] M. R. Melis et al., *Pharmacol. Res. Commun.* **20** (1988) 1117.
[985] P. Rekowski et al., *Acta Chem. Scand. Ser. B* **B39** (1985) 453.
[986] Z. Grzonka et al., *J. Med. Chem.* **29** (1986) 96.
[987] A. Buku et al., *Int. J. Peptide Res.* **23** (1984) 551.
[988] Z. Prochazka et al., *Collect. Czech. Chem. Commun.* **53** (1988) 2604.
[989] B. Lammek et al., *Experientia* **43** (1987) 1211.
[990] Z. Prochazka et al., *Collect. Czech. Chem. Commun.* **49** (1984) 642.
[991] M. Lebl et al., *Collect. Czech. Chem. Commun.* **50** (1985) 418.
[992] G. Sweeney et al., *Curr. Ther. Res.* **47** (1990) 528.
[993] Z. Grzonka et al., *J. Med. Chem.* **26** (1983) 1786.
[994] *Drugs Future* **9** (1984) 830.
[995] M. Lebl et al., *Collect. Czech. Chem. Commun.* **50** (1985) 132.
[996] R. A. Johns, *Anesthesiology* **71** (1989) A487.
[997] B. Lammek et al., *J. Med. Chem.* **32** (1989) 244.
[998] L. B. Kinter et al., *Am. J. Physiol.* **254** (1988) F165.
[999] W. Y. Chan et al., *Proc. Soc. Exp. Biol. Med.* **185** (1987) 187.
[1000] M. Manning et al., *J. Med. Chem.* **32** (1989) 382.
[1001] D. W. Hahn et al., *Am. J. Obstet. Gynecol.* **157** (1987) 977.
[1002] P. Rekowski et al., *Pol. J. Pharmacol. Pharm.* **39** (1987) 303.
[1003] P. Melin et al., *J. Endocr.* **111** (1986) 125.
[1004] *Drugs Future* **13** (1988) 25.
[1005] B. Lammek et al., *Pol. J. Pharmacol. Pharm.* **40** (1988) 423.
[1006] B. Lammek et al., *J. Med. Chem.* **31** (1988) 603.
[1007] M. Manning et al., *J. Med. Chem.* **27** (1984) 423.
[1008] M. Manning et al., *J. Med. Chem.* **26** (1983) 1607.
[1009] B. Lammek et al., *Peptides (Fayetteville, N.Y.)* **10** (1989) 1109.
[1010] M. Manning et al., *Nature (London)* **308** (1984) 652.
[1011] F. El-Fehail Ali et al., *J. Med. Chem.* **30** (1987) 2291.
[1012] W. F. Huffman et al., *J. Med. Chem.* **28** (1985) 1759.
[1013] M. L. Moore et al., *J. Med. Chem.* **31** (1988) 1487.
[1014] M. Manning et al., *J. Med. Chem.* **30** (1987) 2245.
[1015] W. H. Sawyer et al., *Peptides (Fayetteville, N.Y.)* **9** (1988) 157.
[1016] M. Manning et al., *Nature (London)* **329** (1987) 839.
[1017] F. Kasprzykowski et al., *Collect. Czech. Chem. Commun.* **53** (1988).
[1018] D. J. Pettibone et al., *Endocrinology (Baltimore)* **125** (1989) 217.
[1019] R. M. Freidinger et al., *J. Med. Chem.* **33** (1990) 1845.
[1020] M. G. Bock et al., *J. Med. Chem.* **33** (1990) 2321.
[1021] P. Mannucci et al., *Ann. N.Y. Acad. Sci.* **509** (1987) 71.
[1022] P. R. Martens, *Lancet I* (1989) 1322.
[1023] B. Waeber et al., *J. Cardiovasc. Pharmacol.* **8** Suppl. 7 (1986) S111.
[1024] M. Akerlund et al., *Br. J. Obstet. Gynaecol.* **94** (1987) 1040.
[1025] J. E. Zadina et al., *Brain Res.* **409** (1987) 10.

[1026]   G. Campbell Teskey, M. Kavaliers, *Peptides (Fayetteville. N.Y.)* **6** (1985) 165.
[1027]   H. N. Bhargava, *Neuropharmacology* **25** (1986) 737.
[1028]   G. Szabo et al., *Acta Physiol. Hung.* **69** (1987) 115.
[1029]   G. L. Kovacs et al., *Pharmacol. Biochem. Behav.* **31** (1988) 833.
[1030]   R. L. Johnson et al., *J. Med. Chem.* **33** (1990) 1828.
[1031]   L. Pulvirenti et al., A. J. Kastin, *Eur. J. Pharmacol.* **151** (1988) 289.
[1032]   R. L. Johnson et al., *J. Med. Chem.* **29** (1986) 2100.
[1033]   K.-L. Yu et al., *J. Med. Chem.* **31** (1988) 1430.
[1034]   R. K. Mishra et al., *Drugs Future* **11** (1986) 303.
[1035]   M. Yanagisawa et al., T. Masaki, *Biochem. Pharmacol.* **38** (1989) 1877.
[1036]   T. Kosaka et al., *FEBS Lett.* **249** (1989) 42.
[1037]   T. Yoshizawa et al., *Science (Washington D.C.)* **247** (1990) 462.
[1038]   H. C. Lam et al., *FEBS Lett.* **261** (1990) 184.
[1039]   A. Inoue et al., *Proc. Natl. Acad. Sci. USA* **86** (1989) 2863.
[1040]   M. Yanagisawa et al., *Nature (London)* **332** (1988) 411.
[1041]   Y. Kloog et al., *Science (Washington D.C.)* **242** (1988) 268.
[1042]   S. Ohkubo et al., *FEBS Lett.* **274** (1990) 136.
[1043]   H. Onda et al., *FEBS Lett.* **261** (1990) 327.
[1044]   M. Yoshizumi et al., *Biochem. Biophys. Res. Commun.* **161** (1989) 859.
[1045]   T. Emori et al., *Biochem. Biophys. Res. Commun.* **160** (1989) 93.
[1046]   K. Ohta et al., *Biochem. Biophys. Res. Commun.* **169** (1990) 578.
[1047]   C. Boulanger, T. F. Lüscher, *J. Clin. Invest.* **85** (1990) 587.
[1048]   H. Rakugi et al., *Biochem. Biophys. Res. Commun.* **169** (1990) 973.
[1049]   N. Suzuki et al., *Biochem. Biophys. Res. Commun.* **167** (1990) 941.
[1050]   Y. Saito et al., *N. Engl. J. Med.* **322** (1990) 205.
[1051]   P. Cernacek, D. J. Stewart, *Biochem. Biophys. Res. Commun.* **161** (1989) 562.
[1052]   Y. Masuda et al., *FEBS Lett.* **257** (1989) 208.
[1053]   Y. Hirata et al., *FEBS Lett.* **239** (1988) 13.
[1054]   P. Roubert et al., *Biochem. Biophys. Res. Commun.* **164** (1989) 809.
[1055]   N. Fukuda et al., *Biochem. Biophys. Res. Commun.* **167** (1990) 739.
[1056]   R. K. Minkes et al., *Eur. J. Pharmacol.* **164** (1989) 571.
[1057]   R. M. Eglen et al., *Br. J. Pharmacol.* **97** (1989) 1297. Y. Matsumura et al., *Life Sci.* **44** (1989) 149.
[1058]   G. Mazzocchi et al., *Peptides (Fayetteville, N.Y.)* **11** (1990) 763.
[1059]   J. R. Hu et al., *Eur. J. Pharmacol.* **158** (1988) 177.
[1060]   Y. Fukuda et al., *Biochem. Biophys. Res. Commun.* **155** (1988) 167.
[1061]   S. S. Stojilkovic et al., *Science (Washington D.C.)* **248** (1990) 1633.
[1062]   Y. Shibouta et al., *Life Sci.* **46** (1990) 1611.
[1063]   R. K. Minkes et al., *Am. J. Physiol.* **259** (1990) H1152.
[1064]   T. Kashiwabara et al., *FEBS Lett.* **247** (1989) 73.
[1065]   K. Nakajima et al., *Biochem. Biophys. Res. Commun.* **163** (1989) 424.
[1066]   A. Locei et al., *Neuropeptides (Edinburgh)* **16** (1990) 21.
[1067]   M. D. Randall et al., *Br. J. Pharmacol.* **98** (1989) 685.
[1068]   J. Gutowska, M. Nemer, *Endocr. Rev.* **10** (1989) 519.
[1069]   G. Thibault et al., *Biochem. J.* **241** (1987) 265.
[1070]   P. Needleman, *Fed. Proc. Fed. Am. Soc. Exp. Biol.* **45** (1986) 2096.
[1071]   A. Miyata et al., *Biochem. Biophys. Res. Commun.* **142** (1987) 461.

[1072] H. Ioth et al., *Biochem. Biophys. Res. Commun.* **143** (1987) 560.
[1073] R. Eskay et al., *Science (Washington D.C.)* **232** (1986) 636.
[1074] L. Meleagros et al., *Peptides (Fayetteville, N.Y.)* **10** (1989) 545.
[1075] C. J. Winters et al., *Biochem. Biophys. Res. Commun.* **150** (1988) 231.
[1076] G. P. Vlasuk et al., *Biochem. Biophys. Res. Commun.* **136** (1986) 396. R. Takayanagi et al., *Biochem. Biophys. Res. Commun.* **142** (1987) 483.
[1077] N. Zamir et al., *Brain Res.* **365** (1986) 105.
[1078] J. Gutkowska et al., *Peptides (Fayetteville, N.Y.)* **8** (1987) 461. S. H. Kim et al., *Life Sci.* **45** (1989) 1581. A. M. Vollmar et al., *Peptides (Fayetteville, N.Y.)* **9** (1988) 965.
[1079] O. Vuolteenaho et al., *FEBS Lett.* **233** (1988) 79.
[1080] A. Miyata et al., *Biochem. Biophys. Res. Commun.* **155** (1988) 1330.
[1081] Y. Takei et al., *Biochem. Biophys. Res. Commun.* **164** (1989) 537.
[1082] J.-I. Sakata et al., *Biochem. Biophys. Res. Commun.* **155** (1988) 1338.
[1083] T. Sudoh et al., *Nature (London)* **332** (1988) 78.
[1084] K. Maekawa et al., *Biochem. Biophys. Res. Commun.* **157** (1988) 410.
[1085] T. Sudoh et al., *Biochem. Biophys. Res. Commun.* **168** (1990) 863.
[1086] H. Ruskoaho et al., *Biochem. Biophys. Res. Commun.* **139** (1986) 266.
[1087] K. Naruse et al., *Peptides (Fayetteville, N.Y.)* **8** (1987) 285.
[1088] Y. Shenker et al., *J. Cardiovasc. Pharmacol.* **8** (1986) 1300.
[1089] D. Lachance et al., *Hypertension (Dallas)* **9** (1987) 524.
[1090] J. A. Sanfield et al., *Am. J. Physiol.* **252** (1987) Part 1, E740.
[1091] M. G. Currie, W. H. Newman, *Biochem. Biophys. Res. Commun.* **137** (1986) 94.
[1092] B. Baranowska et al., *Biochem. Biophys. Res. Commun.* **143** (1987) 159.
[1093] B. Baranowska et al., *Hypertension (Dallas)* **9** (1987) 545.
[1094] J. R. Hu et al., *Eur. J. Pharmacol.* **158** (1988) 177.
[1095] O. Sergev et al., *J. Endocrinol. Invest.* **13** (1990) 649.
[1096] M. Weissel et al., *Klin. Wochenschr.* **64** Suppl. VI, (1986) 93.
[1097] K. Horky et al., *Biochem. Biophys. Res. Commun.* **129** (1985) 651.
[1098] H. Matsubara et al., *Biochem. Biophys. Res. Commun.* **145** (1987) 336.
[1099] G. Wambach et al., *Klin. Wochenschr.* **64** Suppl. VI (1986) 53.
[1100] K. M. Verburg et al., *Am. J. Physiol.* **251** (1986) R947.
[1101] W. Rascher et al., *Horm. Res.* **28** (1987) 58.
[1102] D. R. J. Singer et al., *Lancet II* (1987) 1394.
[1103] M. R. Arendt et al., *Klin. Wochenschr.* **64** Suppl. VI, (1986) 97.
[1104] T. Yamaji et al., *J. Clin. Endocrinol. Metab.* **67** (1988) 348.
[1105] T. Yamaji et al., *J. Clin. Endocrinol. Metab.* **63** (1986) 815.
[1106] A. L. Gerbes et al., *N. Engl. J. Med.* **313** (1985) 1609.
[1107] M. Kohno et al., *Am. J. Med.* **83** (1987) 648.
[1108] M. Ohashi et al., *Horm. Metab. Res.* **20** (1988) 705.
[1109] A. E. G. Raine et al., *N. Engl. J. Med.* **315** (1986) 533.
[1110] Y. Shenker et al., *Life Sci.* **41** (1987) 1635.
[1111] N. Zamir et al., *Proc. Natl. Acad. Sci. USA* **84** (1987) 541.
[1112] R. S. Zimmerman et al., *J. Clin. Endocrinol. Metab.* **64** (1987) 353.
[1113] R. Garcia et al., *Biochem. Biophys. Res. Commun.* **136** (1986) 510.
[1114] M. Sakamoto et al., *Biochem. Biophys. Res. Commun.* **135** (1986) 515.
[1115] J. L. Sonnenberg et al., *Peptides (Fayetteville, N.Y.)* **9** (1988) 173.
[1116] A. Jardine et al., *J. Am. Coll. Cardiol.* **13** Suppl. A, (1989) 76A.

[1117] D. B. Northridge et al., *J. Am. Coll. Cardiol.* **13** Suppl. A (1989) 76A.
[1118] M.-S. Chang et al., *Nature (London)* **341** (1989) 68.
[1119] M. B. Anand-Srivastava et al., *J. Biol. Chem.* **265** (1990) 8566.
[1120] D. L. Vesely et al., *Peptides (Fayetteville, N.Y.)* **11** (1990) 193.
[1121] K. L. Lanier-Smith, M. G. Currie, *Eur. J. Pharmacol.* **178** (1990) 105.
[1122] J. M. Saavedra et al., *Endocrinology (Baltimore)* **120** (1987) 426.
[1123] E. L. Schiffrin et al., *Circulation* **72** (1985) part 2, III–293.
[1124] Y. Hirata et al., *Biochem. Biophys. Res. Commun.* **138** (1986) 405.
[1125] A. G. Jardine et al., *Klin. Wochenschr.* **67** (1989) 902.
[1126] N. C. Trippodo, R. W. Barbee, *Am. J. Physiol.* **252** (1987) R915.
[1127] P J. S. Chiu et al., *Eur. J. Pharmacol.* **124** (1986) 277.
[1128] B. L. Pegram et al., *Fed. Proc. Am. Soc. Exp. Biol.* **45** (1986) 2382.
[1129] R. J. Schiebinger, J. Linden, *Am. J. Physiol.* **251** (1986) H1095.
[1130] A. Petrillo et al., *Gastroenterology* **92** (1987) part 2, 1764.
[1131] J. McMurray et al., *Eur. J. Clin. Pharmacol.* **35** (1988) 409.
[1132] K. Schafferhans et al., *Klin. Wochenschr.* **66** Suppl. 13, (1988) 223.
[1133] A. H. Morice et al., *Am. Rev. Respir. Dis.* **135** (1987) Part 2, A300.
[1134] S. Adnot et al., *Am. Rev. Respir. Dis.* **137** (1988) Part 2, 108.
[1135] M. B. Anand-Srivastava, M. Cantin, *Biochem. Biophys. Res. Commun.* **138** (1986) 427.
[1136] K. Iitake et al., *Endocrinology (Baltimore)* **119** (1986) 438.
[1137] T. Takao et al., *Life Sci.* **42** (1988) 1199.
[1138] M. A. Dillingham, R. J. Anderson, *Science (Washington D.C.)* **231** (1986) 1572.
[1139] Z. Zukowska-Grojec et al., *J. Pharmacol. Exp. Ther.* **239** (1986) 480.
[1140] P. Nambi et al., *Proc. Natl. Acad. Sci. USA* **83** (1986) 8492.
[1141] J. P. Briggs, J. Schnermann, *J. Cardiovasc. Pharmacol.* **8** (1986) 1296.
[1142] H. Vierhapper et al., *Hypertension (Dallas)* **8** (1986) 1040.
[1143] F. Bex, A. Corbin, *Eur. J. Pharmacol.* **115** (1985) 125.
[1144] P. R. Bovy, *Med. Res. Rev.* **10** (1990) 115.
[1145] T. Kimura, S. Sakakibara, *Jpn. J. Pharmacol.* **40** Suppl., (1986) 77P.
[1146] J. G. De Mey et al., *J. Pharmacol. Exp. Ther.* **240** (1987) 937.
[1147] H. Saxenhofer et al., *Am. J. Physiol.* **259** (1990) F832.
[1148] Y. Kiso et al., *J. Pharm. Sci.* **76** (1987) S172.
[1149] J. Rittenhouse et al., *J. Biol. Chem.* **261** (1986) 7607.
[1150] C. F. Hassman et al., *Biochem. Biophys. Res. Commun.* **152** (1988) 1070.
[1151] P. W. Schiller et al., *Biochem. Biophys. Res. Commun.* **143** (1987) 499.
[1152] D. L. Song et al., *Eur. J. Pharmacol.* **160** (1989) 141.
[1153] J. D. Mogannam et al., *J. Cardiovasc. Pharmacol.* **8** (1986) 1317.
[1154] R. Deghenghi et al., *Drug. Dev. Res. (New York)* **15** (1988) 87.
[1155] J. Knolle et al., lecture held at Hoechst AG 1990.
[1156] R. F. Nutt et al., *Abstr. Pap. Am. Chem. Soc.* **195** (1988) BIOL1.
[1157] M. S. King, A. J. Baertschi, *Endocrinology (Baltimore)* **124** (1989) 286.
[1158] D. L. Vesely et al., *Biochem. Biophys. Res. Commun.* **148** (1987) 1540.
[1159] D. R. Martin et al., *Am. J. Physiol.* **258** (1990) F1401.
[1160] Y. Kitajima et al., *Biochem. Biophys. Res. Commun.* **164** (1989) 1295.
[1161] R. Corder et al., *Biochem. J.* **219** (1984) 699.
[1162] K. Tatemoto et al., *Biochem. Biophys. Res. Commun.* **157** (1988) 713.
[1163] W. B. Rhoten, *Experientia* **43** (1987) 428.

[1164] A. B. Leiter et al., *J. Biol. Chem.* **259** (1984) 14702.
[1165] D. M. McKay et al., *Regul. Pept.* **31** (1990) 187.
[1166] J. B. Jaspan et al., *Am. J. Physiol.* **251** (1986) E215.
[1167] S. J. Konturek et al., *Gastroenterology* **86** (1984) Part 2, 1140.
[1168] R. D. Beauchman et al., *Gastroenterology* **90** (1986) Part 2, 1341.
[1169] G. R. Greenberg et al., *Dig. Dis. Sci.* **30** (1985) 946.
[1170] L. R. Gingerich et al., *Metab. Clin. Exp.* **34** (1985) 25.
[1171] E. Hidvegi et al., *Acta Physiol. Hung.* **63** (1984) 237.
[1172] A. J. L. de Jong et al., *Regul. Pept.* **17** (1987) 285.
[1173] C. Beglinger et al., *Am. J. Physiol.* **246** (1984) G286.
[1174] S. Ito et al., *J. Pharmacol.* **33** Suppl. (1983) 129P.
[1175] R. L. Conter et al., *Clin. Res.* **32** (1984) 24A.
[1176] W. H. Coltharp et al., *Gastroenterology* **90** (1986) Part 2, 1379.
[1177] S. J. Konturek et al., *Am. J. Physiol.* **248** (1985) G687.
[1178] T. S. Gray, J. E. Morley, *Life Sci.* **38** (1986) 389.
[1179] P. Ciofi et al., *Neuroendocrinology* **45** (1987) 425.
[1180] E. Ekblad et al., *Regul. Pept.* **10** (1984) 47.
[1181] C. D. Minth et al., *J. Biol. Chem.* **261** (1986) 11974.
[1182] J. Allen et al., *Proc. Natl. Acad. Sci. USA* **84** (1987) 2532.
[1183] J. M. Allen et al., *Neuropeptides (Edinburgh)* **6** (1985) 95.
[1184] K. Kitamura et al., *Biochem. Biophys. Res. Commun.* **169** (1990) 1164.
[1185] M. Haass et al., *Am. J. Physiol.* **259** (1990) R925.
[1186] J.-T. Cheng, C. L. Shen, *Eur. J. Pharmacol.* **123** (1986) 303.
[1187] K. Takahashi et al., *Peptides (Fayetteville, N.Y.)* **9** (1988) 433.
[1188] A. Rudehill et al., *Acta Physiol. Scand.* **131** (1987) 517.
[1189] V. Castagne et al., *Regul. Pept.* **19** (1987) 55.
[1190] L. S. Brady et al., *Neuroendocrinology* **52** (1990) 441.
[1191] T. Ogawa et al., *Biochem. Biophys. Res. Commun.* **165** (1989) 1399.
[1192] A. Balasubramaniam, S. Sheriff, *J. Biol. Chem.* **265** (1990) 14724.
[1193] J. M. Lundberg et al., *Regul Pept.* **13** (1986) 169.
[1194] J. D. Watson et al., *J. Endocrinol.* **116** (1988) 421.
[1195] N. Pages et al., *Neuropeptides (Edinburgh)* **17** (1990) 141.
[1196] A. Sahu et al., *Endocrinology (Baltimore)* **122** (1988) 2199.
[1197] S. P. Sheikh et al., *FEBS Lett.* **245** (1989) 209.
[1198] S. P. Sheikh, J. A. Williams, *J. Biol. Chem.* **265** (1990) 8304.
[1199] N. A. Scott et al., *Am. J. Physiol.* **259** (1990) H174.
[1200] L. Edvinsson et al., *Trends Pharmacol. Sci. TIPS* **8** (1987) 231.
[1201] J. Fuhlendorff et al., *Proc. Natl. Acad. Sci. USA* **87** (1990) 182.
[1202] S.-I. Mihara et al., *FEBS Lett.* **259** (1989) 79.
[1203] B. B. Fredholm et al., *Acta Physiol. Scand.* **124** (1985) 467.
[1204] T. Ishiguro et al., *Chem. Pharm. Bull.* **36** (1988) 2720.
[1205] J. L. Morris, *Peptides (Fayetteville, N.Y.)* **11** (1990) 381.
[1206] J. F. Aubert et al., *J. Pharmacol. Exp. Ther.* **244** (1988) 1109.
[1207] G. H. Greeley, Jr. et al., *Am. J. Physiol.* **254** (1988) E513.
[1208] Y.-S. Guo et al., *Am. J. Physiol.* **253** (1987) G298.
[1209] J. F. Rodriguez-Sierra et al., *Peptides (Fayetteville, N.Y.)* **8** (1987) 539.
[1210] W. B. Wehrenberg et al., *Neuroendocrinology* **49** (1989) 680.

[1211] W. R. Crowley et al., *Endocrinology (Baltimore)* **120** (1987) 941.
[1212] H. E. Albers et al., *Am. J. Physiol.* **258** (1990) R376.
[1213] B. G. Stanley et al., *Peptides (Fayetteville, N.Y.)* **6** (1985) 1205.
[1214] T. N. Pappas et al., *Gastroenterology* **90** (1986) Part 2, 1578.
[1215] H. Härfstrand, *Acta Physiol. Scand.* **128** (1986) 121.
[1216] J. Esteban et al., *Life Sci.* **45** (1989) 2395.
[1217] J. L. Krstenansky et al., *Proc. Natl. Acad. Sci. USA* **86** (1989) 4377.
[1218] A. Beck et al., *FEBS Lett.* **244** (1989) 119.
[1219] A. G. Beck-Sickinger et al., *Eur. J. Biochem.* **194** (1990) 449.
[1220] J. L. Krstenansky et al., *Neuropeptides (Edinburgh)* **17** (1990) 117.
[1221] B. Baranowska et al., *Biochem. Biophys. Res. Commun.* **145** (1987) 680.
[1222] L. Grundemar, R. Hakanson, *Br. J. Pharmacol.* **100** (1990) 190.
[1223] G. W. Aponte et al., *Am. J. Physiol.* **249** (1985) G745.
[1224] T. E. Adrian et al., *Gastroenterology* **92** (1987) Part 2, 1289.
[1225] T. E. Adrian et al., *Regul. Pept.* **9** (1984) 320.
[1226] F. Lluis et al., *Gastroenterology* **90** (1986) Part 2, 1525.
[1227] T. N. Pappas et al., *Am. J. Physiol.* **248** (1985) G118.
[1228] T. N. Pappas et al., *Gastroenterology* **89** (1985) 1387.
[1229] W. W. Pawlik et al., *Gastroenterology* **90** (1986) Part 2, 1580.
[1230] T. E. Adrian et al., *J. Clin. Endocrinol. Metab.* **63** (1986) 803.
[1231] R. J. Playford et al., *Lancet* **335** (1990) 1555.
[1232] R. K. Harding, T. J. McDonald, *Peptides (Fayetteville, N.Y.)* **10** (1989) 21.
[1233] N. Minamino et al., *Biochem. Biophys. Res. Commun.* **156** (1988) 355.
[1234] J. Domin et al., *J. Biol. Chem.* **264** (1989) 20881.
[1235] R. Murphy et al., *Peptides (Fayetteville, N.Y.)* **11** (1990) 613.
[1236] J. Domin et al., *Peptides (Fayetteville, N.Y.)* **8** (1987) 779.
[1237] C. A. Maggi et al., *Br. J. Pharmacol.* **99** (1990) 186.
[1238] M. V. Graf, A. J. Kastin, *Peptides (Fayetteville, N.Y.)* **7** (1986) 1165.
[1239] G. A. Schoenenberger, *Eur. Neurol.* **23** (1984) 321.
[1240] A. Bjartell et al., *Neuroendocrinology* **50** (1989) 564.
[1241] A. Nakamura et al., *Eur. J. Pharmacol.* **121** (1986) 157.
[1242] A. M. J. Young, B. J. Key, *Neuropharmacology* **23** (1984) 1347.
[1243] J. E. Taylor, *IRCS Med. Sci.* **14** (1986) 1122.
[1244] K. V. Sudakov et al., *J. Biol. Sci.* **18** (1983) 1.
[1245] M. V. Graf, A. J. Kastin, *Neurosci. Biobehav. Rev.* **8** (1984) 83.
[1246] T. Okajima, G. Hertting, *Horm. Metab. Res.* **18** (1986) 497.
[1247] M. V. Graf et al., *Neuroendocrinology* **41** (1985) 353.
[1248] K. S. Iyer, S. M. McCann, *Neuroendocrinology* **46** (1987) 93.
[1249] K. S. Iyer, S. M. McCann, *Neuroendocrinology* **46** (1987) 93.
[1250] A. Sahu, S. P. Kalra, *Life Sci.* **40** (1987) 1201.
[1251] W. Larbig et al., *Eur. Neurol.* **23** (1984) 372.
[1252] K. Iwai et al., *J. Biol. Chem.* **262** (1987) 8956.
[1253] S.-I. Fukuoka et al., *Biochem. Biophys. Res. Commun.* **145** (1987) 646.
[1254] Y. Takahashi et al., *Proc. Natl. Acad. Sci. USA* **82** (1985) 1931.
[1255] U. Gubler et al., *Proc. Natl. Acad. Sci. USA* **81** (1984) 4307.
[1256] R. J. Deschenes et al., *Proc. Natl. Acad. Sci. USA* **81** (1984) 726.
[1257] V. E. Eysselein et al., *Am. J. Physiol.* **258** (1990) G253.

[1258] Z.-Z. Zhou et al., *Peptides* **6** (1985) 337.
[1259] J. Shively et al., *Am. J. Physiol.* **252** (1987) G272.
[1260] A. Takashima, S. Itoh, *Can. J. Physiol. Pharmacol.* **67** (1989) 228.
[1261] T. Wakabayashi et al., *Gene* **31** (1984) 295.
[1262] Y. F. Chen et al., *Am. J. Physiol.* **249** (1985) G29.
[1263] J. B. J. M. Jansen, C. B. H. W. Lamers, *Life Sci.* **33** (1983) 2197.
[1264] J. J. Benoliel et al., *Fundam. Clin. Pharmacol.* **3** (1989) 141.
[1265] I. Koop, *Eur. J. Clin. Invest.* **20** Suppl. 1 (1990) S51.
[1266] I. C. Forgacs et al., *Clin. Sci.* **66** (1984) 61P.
[1267] R. A. Siegel et al., *Neuroendocrinology* **46** (1987) 75.
[1268] J. B. M. J. Jansen, C. B. H. W. Lamers, *Gut* **27** (1986) A1260.
[1269] I. Nakano et al., *Regul. Pept.* **14** (1986) 229.
[1270] S. Kanayama et al., *Life Sci.* **41** (1987) 1915.
[1271] A. Schafmayer et al., *Digestion* **30** (1984) 95.
[1272] D.-H. Yu et al., *Am. J. Physiol.* **258** (1990) G86.
[1273] M. Praissman, M. Walden, *Gastroenterology* **92** (1987) part 2, 1579.
[1274] M. Otsuki, J. A. Williams, *Gastroenterology* **84** (1983) part 2, 1266.
[1275] Y. K. Leung et al., *Dig. Dis. Sci.* **30** (1985) 981.
[1276] M. Otsuki et al., *Gastroenterology* **87** (1984) 882.
[1277] B. E. Evans, *Drugs Future* **14** (1989) 971.
[1278] L. L. Iversen et al., *J. Endocrinol.* **124** Suppl. (1990) 18.
[1279] J. Hughes et al., *Proc. Natl. Acad. Sci. USA* **87** (1990) 6728.
[1280] B. Charpentier et al., *Peptides* **9** (1988) 835.
[1281] B. Charpentier et al., *J. Med. Chem.* **32** (1989) 1148.
[1282] I. Maraseigne et al., *J. Med. Chem.* **31** (1988) 966.
[1283] R. A. Liddle et al., *J. Clin. Invest.* **84** (1989) 1270.
[1284] M. F. O'Rourke et al., *Am. J. Physiol.* **258** (1990) G179.
[1285] S. J. Konturek et al., *Am. J. Physiol.* **248** (1985) G687.
[1286] J. E. Kellow et al., *Am. J. Physiol.* **253** (1987) G650.
[1287] A. Brown et al., *Gut* **26** (1985) A580.
[1288] L. Gullo et al., *Dig. Dis. Sci.* **31** (1986) 1345.
[1289] L. Marzio et al., *Am. J. Gastroenterol.* **80** (1985) 1.
[1290] G. Schmidt et al., *Klin. Wochenschr.* **64** Suppl. 5 (1986) 56.
[1291] F. Makovec et al., *Regul. Pept.* **16** (1986) 281.
[1292] K. A. Zucker et al., *Am. J. Physiol.* **257** (1989) part 1, G511.
[1293] S. Karlsson, B. Ahren, *Acta Physiol. Scand.* **135** (1989) 271.
[1294] K. C. K. Lloyd et al., *Gastroenterology* **96** (1989) part 2, A305.
[1295] T. W. Vickroy et al., *Eur. J. Pharmacol.* **152** (1988) 371.
[1296] C. T. Dourish et al., *Eur. J. Pharmacol.* **176** (1990) 35.
[1297] E. K. Hong, A. E. Takemori, *J. Pharmacol. Exp. Ther.* **251** (1989) 594.
[1298] M. Rattray et al., *Neuropeptides* **14** (1989) 263.
[1299] D. S. K. Magnuson et al., *Neuropeptides* **16** (1990) 213.
[1300] J. E. Kellow et al., *Am. J. Physiol.* **252** (1987) G345–G356.
[1301] G. Zetler, *Peptides* **5** (1984) 729.
[1302] S. Rattan, R. K. Goyal, *Gastroenterology* **90** (1986) 94.
[1303] D. Pelaprat, M. Reibaud, *J. Med. Chem.* **32** (1989) 445.
[1304] C. Beglinger et al., *Regul Pept.* **8** (1984) 291.

[1305] M. Ruiz-Gayo et al., *Peptides* **6** (1985) 415.
[1306] M. Knight et al., *Peptides* **6** (1985) 631.
[1307] C. Mendre et al., *J. Biol. Chem.* **263** (1988) 10641.
[1308] B. Penke et al., *J. Med. Chem.* **27** (1984) 845.
[1309] I. Marseigne et al., *Int. J. Peptide Protein Res.* **33** (1989) 230.
[1310] M.-C. Galas et al., *Am. J. Physiol.* **254** (1988) G176.
[1311] J. D. Gardner et al., *Am. J. Physiol.* **248** (1985) G98.
[1312] *Drugs Future* **13** (1988) 34.
[1313] M.-F. Lignon et al., *J. Biol. Chem.* **262** (1987) 7226.
[1314] B. E. Evans, *Drugs Future* **14** (1989) 971.
[1315] R. T. Jensen et al., *Biochim. Biophys. Acta* **761** (1983) 269.
[1316] R. Tritapepe, C. di Padova, *Br. Med. J.* **293** (1986) 1102.
[1317] N. Basso et al., *Gastroenterology* **89** (1985) 605.
[1318] A. Pap, V. Varro, *Digesetion* **30** (1984) 118.
[1319] B. S. Barbaz et al., *Neuropharmacology* **25** (1986) 823.
[1320] E. R. Spindel et al., *Proc. Natl. Acad. Sci. USA* **83** (1986) 19.
[1321] D. Barra et al., *FEBS Lett.* **182** (1985) 53.
[1322] J. Price et al., *J. Clin. Endocrinol. Metab.* **60** (1985) 1097.
[1323] G. D. Jahnke, L. H. Lazarus, *Proc. Natl. Acad. Sci. USA* **81** (1984) 578.
[1324] C. Scarpignato, B. Micali, *Gut* **27** (1986) 499.
[1325] F. Porecca et al., *Eur. J. Pharmacol.* **114** (1985) 167.
[1326] T. von Schrenck et al., *Am. J. Physiol.* **259** (1990) G468.
[1327] C. B. H. W. Lamers et al., *Dig. Dis. Sci.* **31** Suppl. (1986) 428S.
[1328] A. E. Pontiroli et al., *Hormone Res.* **23** (1986) 129.
[1329] A. J. L. de Jong et al., *Regul. Pept.* **17** (1987) 285.
[1330] C. E. Sievert, Jr. et al., *Am. J. Physiol.* **254** (1988) G361.
[1331] T. Lehy et al., *Gastroenterology* **90** (1986) 1942.
[1332] C. Stock Damge et al., *Gut* **28** Suppl. (1987) 1.
[1333] F. Cuttitta et al., *Nature (London)* **316** (1985) 823.
[1334] G. de Caro et al., *Peptides* **5** (1984) 607.
[1335] E. Widerlöv et al., *Peptides* **5** (1984) 523.
[1336] F. Girard et al., *Eur. J. Pharmacol.* **102** (1984) 489.
[1337] P. Heinz-Erian et al., *Am. J. Physiol.* **252** (1987) G439.
[1338] Z. A. Saeed et al., *Peptides* **10** (1989) 597.
[1339] R. Camble et al., *Life Sci.* **45** (1989) 1521.
[1340] L.-H. Wang et al., *Biochemistry* **29** (1990) 616.
[1341] D. C. Heimbrook et al., *J. Biol. Chem.* **264** (1989) 11258.
[1342] T. von Schrenck et al., *Am. J. Physiol.* **259** (1990) G468.
[1343] D. H. Coy et al., *J. Biol. Chem.* **263** (1988) 5056.
[1344] D. H. Coy et al., *J. Biol. Chem.* **264** (1989) 14691.
[1345] G. Bonora et al., *Gastroenterology* **84** (1983) Part 2, 1111.
[1346] C. A. Helman, B. I. Hirschowitz, *Gastroenterology* **92** (1987) 1926.
[1347] J. C. M. Hafkenscheid et al., *Clin. Chim. Acta* **136** (1984) 235.
[1348] B. N. Andersen et al., *Regul. Pept.* **10** (1985) 329.
[1349] R. Dimaline et al., *FEBS Lett.* **205** (1986) 318.
[1350] R. J. Nachman et al., *Science (Washington D.C.)* **234** (1986) 71.
[1351] M. M. Wolfe et al., *Clin. Res.* **31** (1983) 477.

[1352] T. Kurose et al., *Life Sci.* **42** (1988) 1995.
[1353] T. Chiba et al., *Horm. Metab. Res.* **15** (1983) 516.
[1354] S. F. Alino et al., *Acta Physiol. Scand.* **126** (1986) 1.
[1355] K. C. Christensen, *Scand. J. Gastroenterol.* **19** (1984) 339.
[1356] J. DelValle, T. Yamada, *J. Clin. Invest.* **85** (1990) 139.
[1357] E. J. Dial et al., *Gastroenterology* **90** (1986) 1018.
[1358] A. Rentis et al., *Horm. Metabol. Res.* **18** (1986) 423.
[1359] B. I. Hirschowitz, E. Molina, *Peptides* **5** (1984) 35.
[1360] C. H. Cho et al., *IRCS Med. Sci.* **13** (1985) 629.
[1361] H. J. de Aizpurua et al., *N. Engl. J. Med.* **313** (1985) 479.
[1362] M. K. Müller et al., *Horm. Metabol. Res.* **18** (1986) 675.
[1363] J. Puurunen, U. Schwabe, *Br. J. Pharmacol.* **90** (1987) 479.
[1364] L. Holm Rutili, T. Berglindh, *Am. J. Physiol.* **250** (1986) part 1, G575.
[1365] A. J. Blair III et al., *J. Clin. Invest.* **78** (1986) 779.
[1366] R. Hakanson et al., *Scand. J. Gastroenterol.* **21** Suppl. 118 (1986) 18.
[1367] T. E. Solomon et al., *Gastroenterology* **92** (1987) 429.
[1368] A. Imdahl et al., *J. Cancer Res. Clin. Oncol.* **115** (1989) 388.
[1369] C. J. Kusyk et al., *Am. J. Physiol.* **251** (1986) G597.
[1370] D. L. Morris et al., *Gut* **30** (1989) A739.
[1371] C. de Montigny, *Arch. Gen. Psychiatry* **46** (1989) 511.
[1372] G. Katsuura, S. Itoh, *Peptides* **7** (1986) 809.
[1373] V. E. Eysselein et al., *J. Clin. Invest.* **73** (1984) 1284.
[1374] W. Göhring et al., *Hoppe-Seyler's Z. Physiol. Chem.* **365** (1984) 83.
[1375] J. Martinez et al., *J. Med. Chem.* **28** (1985) 1874.
[1376] N. V. Sadovnikova et al., *Khim. Farm. Zh.* **21** (1987) 1424.
[1377] D. G. Patel et al., *Diabetes* **38** Suppl. 2 (1989) 206A.
[1378] J. Martinez et al., *J. Med. Chem.* **27** (1984) 1597.
[1379] M. Rodriguez et al., *J. Med. Chem.* **32** (1989) 522.
[1380] T. M. Chang et al., *Peptides (Fayetteville, N.Y.)* **6** (1985) 193.
[1381] G. Gafvelin et al., *Proc. Natl. Acad. Sci. USA* **87** (1990) 6781.
[1382] A. S. Kopin et al., *Proc. Natl. Acad. Sci. USA* **87** (1990) 2299. D. Gossen et al., *Biochem. Biophys. Res. Commun.* **160** (1989) 862. Y. Shinomura et al., *Life Sci.* **41** (1987) 1243. D. Gossen et al., *Peptides (Fayetteville, N.Y.)* **11** (1990) 123.
[1383] T. Yamamoto et al., *Chem. Pharm. Bull.* **34** (1986) 3803.
[1384] J. Leonora et al., *Am. J. Physiol.* **252** (1987) E477.
[1385] G. E. Pratt et al., *Biochem. Biophys. Res. Commun.* **163** (1989) 1243.
[1386] A. Argiolas, J. J. Pisano, *J. Biol. Chem.* **260** (1985) 1437.
[1387] P. Li et al., *J. Clin. Invest.* **86** (1990) 1474.
[1388] S. N. S. Murthy, A. Lavy, *Biochem. Pharmacol.* **37** (1988) 1027.
[1389] S. Iwakawa et al., *J. Pharmacobio-Dyn.* **8** (1985) S136.
[1390] M. V. Singer et al., *Am. J. Physiol.* **248** (1985) G532.
[1391] D. R. Fletcher et al., *Regul. Pept.* **11** (1985) 217.
[1392] L. Bolondi et al., *Dig. Dis. Sci.* **29** (1984) 802.
[1393] G. E. Feurle et al., *Pancreas* **2** (1987) 422.
[1394] H. Haarstad et al., *Scand. J. Gastroenterol.* **21** (1986) 589.
[1395] I. S. Chung et al., *Clin. Res.* **36** (1988) 394A.
[1396] M. Murakami et al., *Dig. Dis. Sci.* **30** (1985) 346.

[1397] F. Yamagishi et al., *Eur. J. Pharmacol.* **118** (1985) 203.
[1398] T. Hashimoto et al., *Can. J. Physiol. Pharmacol. Suppl.* 1986, 40.
[1399] J. H. Kleibeuker et al., *Gastroenterology* **94** (1988) 122.
[1400] D. L. Kaminski, Y. G. Deshpande, *Gastroenterology* **85** (1983) 1239.
[1401] H. Olson et al., *Peptides (Fayetteville, N.Y.)* **9** (1988) 301.
[1402] S. Kiyama et al., *Chem. Pharm. Bull.* **33** (1985) 3205.
[1403] T. Tsuda et al., *J. Pharm. Sci.* **79** (1990) 53.
[1404] J. Glaser et al., *Scand. J. Gastroenterol.* **24** (1989) 179.
[1405] B. Krakamp et al., *MMW Münch. Med. Wochenschr.* **130** (1988) 47.
[1406] Y. S. Malov, *Ter. Arkh.* **57** (1985) 6.
[1407] B. Glaser et al., *J. Clin. Endocrinol. Metab.* **66** (1988) 1138.
[1408] S. I. Said, *J. Endocrinol. Invest.* **9** (1986) 191.
[1409] G. Gafvelin et al., *Peptides (Fayetteville, N.Y.)* **9** (1988) 469.
[1410] G. Gafvelin, *Peptides (Fayetteville, N.Y.)* **11** (1990) 703.
[1411] K. Tatemoto et al., *FEBS Lett.* **174** (1984) 258.
[1412] M. Nishizawa et al., *FEBS Lett.* **183** (1985) 55.
[1413] D. Gossen et al., *Peptides (Fayetteville, N.Y.)* **11** (1990) 123.
[1414] S. Björck et al., *Acta Physiol. Scand.* **128** (1986) 639.
[1415] M. A. Arnaout et al., *Endocrinology (Baltimore)* **119** (1986) 2052.
[1416] D. E. Burleigh, J. B. Furness, *Neuropeptides (Edinburgh)* **16** (1990) 77.
[1417] M. Eriksson et al., *Peptides (Fayetteville, N.Y.)* **8** (1987) 411.
[1418] P. K. Opstad, *Peptides (Fayetteville, N.Y.)* **8** (1987) 175.
[1419] T. A. Crozier et al., *Horm. Metab. Res.* **20** (1988) 352.
[1420] E. Giladi et al., *Isr. J. Med. Sci.* **23** (1987) 924.
[1421] G. Burns et al., *Life Sci.* **40** (1987) 951.
[1422] T. V. Nowak et al., *Gastroenterology* **96** (1989) part 2, A368.
[1423] K. S. L. Lam et al., *Neuroendocrinology* **52** (1990) 417.
[1424] P. Heinz-Erian et al., *Science (Washington D.C.)* **229** (1985) 1407.
[1425] F. E. Bauer et al., *MMW Münch. Med. Wochenschr.* **126** (1984) 1097.
[1426] L. Woie et al., *Gen. Pharmacol.* **18** (1987) 577.
[1427] M. S. O'Dorisio, *Fed. Proc. Fed. Am. Soc. Exp. Biol.* **46** (1987) 192.
[1428] K. McArthur et al., *Gastroenterology* **92** (1987) part 2, 1524.
[1429] M. Murakami et al., *Gastroenterology* **92** (1987) part 2, 1544.
[1430] P. Wiik et al., *Peptides (Fayetteville, N.Y.)* **9** (1988) 181.
[1431] H. Houchi et al., *Biochem. Pharmacol.* **36** (1987) 1551.
[1432] M. Raiteri et al., *Eur. J. Pharmacol.* **133** (1987) 127.
[1433] M. S. Barnette, B. Weiss, *J. Neurochem.* **45** (1985) 640.
[1434] Y. Karasawa et al., *Eur. J. Pharmacol.* **187** (1990) 9.
[1435] S. I. Said, *N. Eng. J. Med.* **320** (1989) 1271.
[1436] A. Bennett et al., *J. Pharm. Pharmacol.* **36** (1984) 787.
[1437] J. Fontaine et al., *Br. J. Pharmacol.* **89** (1986) 599.
[1438] N. Yanaihara, EP 0 225 639, 1985.
[1439] D. W. Pincus et al., *Nature (London)* **343** (1990) 564.
[1440] B. Carlsson et al., *Acta Physiol. Scand.* **129** (1987) 437.
[1441] H. Abe et al., *Endocrinology (Baltimore)* **116** (1985) 1383.
[1442] L. A. Cunningham, M. A. Holzwarth, *Endocrinology (Baltimore)* **122** (1988) 2090.
[1443] B. Ottesen et al., *Regul. Pept.* **16** (1986) 299.

[1444] F. W. George, S. R. Ojeda, *Proc. Natl. Acad. Sci. USA* **84** (1987) 5803.
[1445] C. E. Ahmed et al., *Endocrinology (Baltimore)* **118** (1986) 1682.
[1446] B. G. Kasson et al., *Mol. Cell. Endocrinol.* **48** (1986) 21.
[1447] M. Hamada et al., *Am. Rev. Respir. Dis.* **137** (1988) part 2, 35.
[1448] J. T. Turner et al., *Peptides (Fayetteville, N.Y.)* **7** (1986) 849.
[1449] G. F. Musso et al., *Biochemistry* **27** (1988) 8174.
[1450] G. Hallden et al., *Regul. Pept.* **16** (1986) 183.
[1451] P. Robberecht et al., *Eur. J. Biochem.* **159** (1986) 45.
[1452] S. J. Pandol et al., *Am. J. Physiol.* **250** (1986) G553.
[1453] D. C. Thompson et al., *Peptides (Fayetteville, N.Y.)* **9** (1988) 443.
[1454] P. Robberecht et al., *Eur. J. Biochem.* **165** (1987) 243.
[1455] A. Miyata et al., *Biochem. Biophys. Res. Commun.* **164** (1989) 567.
[1456] A. Miyata et al., *Biochem. Biophys. Res. Commun.* **170** (1990) 643.
[1457] C. Kimura et al., *Biochem. Biophys. Res. Commun.* **166** (1990) 81.
[1458] K. Ogi et al., *Biochem. Biophys. Res. Commun.* **173** (1990) 1271.
[1459] H.-C. Lam et al., *Eur. J. Biochem.* **193** (1990) 725.
[1460] G. Heinrich et al., *Endocrinology (Baltimore)* **115** (1984) 2176.
[1461] G. I. Bell et al., *Nature (London)* **304** (1983) 368.
[1462] L. C. Lopez et al., *Proc. Natl. Acad. Sci. USA* **80** (1983) 5485.
[1463] T. Buhl et al., *J. Biol. Chem.* **263** (1988) 8621.
[1464] L. Thim, A. J. Moody, *Regul. Pept.* **2** (1981) 139.
[1465] C. Orskov et al., *J. Biol. Chem.* **264** (1989) 12826.
[1466] D. S. Weigle, C. J. Goodner, *Endocrinology (Baltimore)* **118** (1986) 1606.
[1467] G. Skoglund et al., *Eur. J. Pharmacol.* **143** (1987) 83.
[1468] G. Paolisso et al., *Acta Endocrinol. (Copenhagen)* **115** (1987) 161.
[1469] L. J. Klaff, G. J. Taborsky, Jr., *Clin. Res.* **34** (1986) 60A.
[1470] D. L. Kaminski et al., *Am. J. Physiol.* **254** (1988) G864.
[1471] D. G. Pipeleers et al., *Endocrinology (Baltimore)* **117** (1985) 817.
[1472] A. Ohneda et al., *Diabetologia* **29** (1986) 397.
[1473] C. G. Ostenson et al., *Diabetologia* **29** (1986) 861.
[1474] A. A. R. Starke et al., *Diabetes* **33** (1984) 277.
[1475] K. F. Hanssen et al., *Diabetologia* **27** (1984) 285A.
[1476] A. E. Farah, *Pharmacol. Rev.* **35** (1983) 181.
[1477] A. Akatsuka et al., *J. Biol. Chem.* **260** (1985) 3239.
[1478] K. Yamauchi, K. Hashizume, *Endocrinology (Baltimore)* **119** (1986) 218.
[1479] S. Marubashi et al., *Acta Endocrinol. (Copenhagen)* **108** (1985) 6.
[1480] R. Holland et al., *Eur. J. Biochem.* **140** (1984) 325.
[1481] R. J. Howland, A. D. Benning, *FEBS Lett.* **208** (1986) 128.
[1482] O. G. Bjornsson et al., *Eur. J. Clin. Invest.* **14** (1984) Part 2, 39.
[1483] C. Guettet et al., *Biochim. Biophys. Acta L* **1105** (1989) 233.
[1484] B. G. Weick, S. Ritter, *Am. J. Physiol.* **250** (1986) R676.
[1485] D. L. Carr Locke et al., *Dig. Dis. Sci.* **28** (1983) 312.
[1486] D. L. Kaminski et al., *Am. J. Physiol.* **254** (1988) G864.
[1487] M. S. Sherman et al., *J. Allergy Clin. Immunol.* **81** (1988) 908.
[1488] K. Corey Flanders et al., *J. Biol. Chem.* **259** (1984) 7031.
[1489] J. F. Rey et al., *Dig. Dis. Sci.* **31** (1986) 355.
[1490] G. Jacobson et al., *Lancet II* (1984) 1149.

[1491] A. Mallat et al., *Nature (London)* **325** (1987) 620.
[1492] P. Blacke et al., *J. Biol. Chem.* **265** (1990) 21514.
[1493] P. Robberecht et al., *Regul. Pept.* **21** (1988) 117.
[1494] J. Sueiras-Diaz et al., *J. Med. Chem.* **27** (1984) 310.
[1495] J. L. Krstenansky et al., *Biochemistry* **25** (1986) 3833.
[1496] K. A. Cornely et al., *Arch. Biochem. Biophys.* **240** (1985) 698.
[1497] C. G. Unson et al., *Proc. Natl. Acad. Sci. USA* **84** (1987) 4083.
[1498] C. G. Unson et al., *J. Biol. Chem.* **264** (1989) 789.
[1499] B. Gysin, V. J. Hruby, *Experientia* **42** (1986) 680.
[1500] S. W. Trenkner et al., *Radiology (Easton, Pa.)* **149** (1983) 401.
[1501] A. M. N. Gardner, *Br. Med. J.* **290** (1985) 822.
[1502] L. van Gaal et al., *Acta Clin. Belg.* **40** (1985) 266.
[1503] J. Buch, A. Buch, *Acta Pharmacol. Toxicol.* **53** (1983) 188.
[1504] A. E. Pontiroli et al., *Eur. J. Clin. Pharmacol.* **37** (1989) 427.
[1505] L. M. Chuang et al., *Pharmacologist* **31** (1989) 125.
[1506] A. Ohneda et al., *Horm. Metab. Res.* **19** (1987) 85.
[1507] B. T. G. Schjoldager et al., *Eur. J. Clin. Invest.* **18** (1988) 499.
[1508] C. Jarrousse et al., *FEBS Lett.* **188** (1985) 81.
[1509] T. M. Biedzinski et al., *Peptides (Fayetteville, N.Y.)* **8** (1987) 967.
[1510] K. Shima et al., *Acta Endocrinol. (Copenhagen)* **123** (1990) 464.
[1511] B. Kreymann et al., *Lancet II* (1987) 1300.
[1512] C. Orskov, J. H. Nielsen, *FEBS Lett.* **229** (1988) 175.
[1513] G. Richter et al., *Acta Endocrinol. (Copenhagen)* **120** Suppl. 1 (1989) 191.
[1514] A. B. Hansen et al., *FEBS Lett.* **236** (1988) 119.
[1515] D. J. Drucker et al., *Proc. Natl. Acad. Sci. USA* **84** (1987) 3434.
[1516] H.-C. Fehmann et al., *FEBS Lett.* **252** (1989) 109.
[1517] E. Yamato et al., *Biochem. Biophys. Res. Commun.* **167** (1990) 431.
[1518] C. Orskov et al., *Endocrinology (Baltimore)* **123** (1988) 2009.
[1519] G. K. Hendrick et al., *Diabetes* **37** Suppl. 1 (1988) 49A.
[1520] M. Gutniak et al., *Diabetologia* **33** Suppl. (1990) A73.
[1521] W. A. Rogers et al., *Dig. Dis. Sci.* **28** (1983) 345.
[1522] P. C. S. Blom et al., *Acta Physiol. Scand.* **123** (1985) 367.
[1523] L. Groop et al., *Diabetes* **33** Suppl. 1 (1984) 166A.
[1524] T. Krarup et al., *Diabetologia* **25** (1983) 173.
[1525] C. K. Lardinois et al., *Diabetes* **33** (1984) 110A.
[1526] F. J. Service et al., *J. Clin. Endocrinol. Metab.* **58** (1984) 1133.
[1527] R. B. Richeson et al., *Clin.Res.* **33** (1985) 66A.
[1528] A. J. McCullough et al., *Am. J. Physiol.* **248** (1985) E299.
[1529] G. Hansen Starich et al., *Am. J. Physiol.* **249** (1985) E603.
[1530] H. Hartmann et al., *Diabetologia* **29** (1986) 112.
[1531] E. Sandberg et al., *Acta Physiol. Scand.* **127** (1986) 323.
[1532] W. E. Schmidt et al., *Eur. J. Clin. Invest.* **16** (1986) part 2, A9.
[1533] A. Rökaeus, *Trends Pharmacol. Sci.* **10** (1987) 158.
[1534] K. Tatemoto et al., *FEBS Lett.* **164** (1983) 124.
[1535] A. Rökaeus, M. Carlquist, *FEBS Lett.* **234** (1988) 400.
[1536] L. M. Kaplan et al., *Proc. Natl. Acad. Sci. USA* **85** (1988) 7408.
[1537] F. E. Bauer et al., *FEBS Lett.* **201** (1986) 327.

[1538] F. E. Bauer et al., *Peptides (Fayetteville, N.Y.)* **7** (1986) 5.
[1539] F. E. Bauer et al., *Lancet II* (1986) 192.
[1540] F. E. Bauer et al., *J. Clin. Endocrinol. Metabol.* **63** (1986) 1372.
[1541] I. Merchenthaler et al., *Proc. Natl. Acad. Sci. USA* **87** (1990) 6326.
[1542] S. C. Hooi et al., *Neuroendocrinology* **51** (1990) 351.
[1543] L. S. Brady et al., *Neuroendocrinology* **52** (1990) 441.
[1544] E. Ekblad et al., *Br. J. Pharmacol.* **86** (1985) 241.
[1545] L. W. Haynes, *Trends Pharmacol. Sci.* **7** (1986) 214.
[1546] K. Tamura et al., *Eur. J. Pharmacol.* **136** (1987) 445.
[1547] F. E. Bauer et al., *Gastroenterology* **97** (1989) 260.
[1548] G. W. G. Sharp et al., *J. Biol. Chem.* **264** (1989) 7302.
[1549] S. Lindskog, B. Ahren, *Acta Physiol. Scand.* **129** (1987) 305.
[1550] T. Messell et al., *Regul. Pept.* **28** (1990) 161.
[1551] Y. N. Kwok et al., *Eur. J. Pharmacol.* **145** (1988) 49.
[1552] W. J. Rossowski, D. H. Coy, *Life Sci.* **44** (1989) 1807.
[1553] S. G. Cella et al., *Endocrinology (Baltimore)* **122** (1988) 855.
[1554] Y. Murakami et al., *Eur. J. Pharmacol.* **136** (1987) 415.
[1555] S. M. Gabriel et al., *Life Sci.* **42** (1988) 1981.
[1556] Y. Ben Ari, M. Lazdunski, *Eur. J. Pharmacol.* **165** (1989) 331.
[1557] J. E. T. Fox et al., *Peptides (Fayetteville, N.Y.)* **9** (1988) 1183.
[1558] B. Gallwitz et al., *Biochem. Biophys. Res. Commun.* **172** (1990) 268.
[1559] X.-J. Xu et al., *Eur. J. Pharmacol.* **182** (1990) 137.
[1560] A. Kuwahara et al., *Regul. Pept.* **29** (1990) 23.
[1561] L. A. Frohman, J.-O. Jansson, *Endocr. Rev.* **7** (1986) 223.
[1562] K. Chihara et al., *J. Clin. Endocrinol. Metab.* **60** (1985) 269.
[1563] P. Ciofi et al., *Neuroendocrinology* **45** (1987) 425.
[1564] B. Meister et al., *Acta Physiol. Scand.* **124** (1985) 133.
[1565] H. Okamura et al., *Neuroendocrinology* **41** (1985) 177.
[1566] K. Arase et al., *Endocrinology (Baltimore)* **121** (1987) 1960.
[1567] Y. Murakami et al., *Eur. J. Pharmacol.* **136** (1987) 415.
[1568] K. Chihara et al., *J. Clin. Endocrinol. Metab.* **62** (1986) 466.
[1569] Y. Kashio et al., *J. Clin. Endocrinol. Metab.* **64** (1987) 92.
[1570] S. M. Gabriel et al., *Neuroendocrinology* **50** (1989) 299.
[1571] J. Argente et al., *J. Clin. Endocrinol. Metab.* **63** (1986) 680.
[1572] I. Ganzetti et al., *Peptides (Fayetteville, N.Y.)* **7** (1986) 1011.
[1573] G. P. Ceda et al., *Endocrinology (Baltimore)* **120** (1987) 1658.
[1574] H. Katakami et al., *Endocrinology (Baltimore)* **118** (1986) 1872.
[1575] M. C. Aguila, S. M. McCann, *Endocrinology (Baltimore)* **120** (1987) 341.
[1576] I. Ganzetti et al., *J. Endocrinol. Invest.* **10** (1987) 241.
[1577] V. de Gennaro Colonna et al., *Peptides (Fayetteville, N.Y.)* **10** (1989) 705.
[1578] W. B. Wehrenberg et al., *Neuroendocrinology* **43** (1986) 266.
[1579] H. Seifert et al., *Endocrinology (Baltimore)* **117** (1985) 424.
[1580] W. B. Wehrenberg, *Endocrinology (Baltimore)* **118** (1986) 489.
[1581] C. E. Brain et al., *J. Endocrinol.* **117** Suppl. (1988) 82.
[1582] R. Valcavi et al., *J. Endocrinol.* **112** Suppl. (1987) 154.
[1583] V. Locatelli et al., *J. Endocrinol.* **111** (1986) 271.
[1584] N. Mauras et al., *Metab. Clin. Exp.* **36** (1987) 369.

[1585] K. Chihara et al., *Endocrinology (Baltimore)* **114** (1984) 1402.
[1586] M. C. White et al., *J. Endocrinol.* **105** (1985) 269.
[1587] M. Press et al., *Diabetes* **33** Suppl. 1 (1984)13A.
[1588] C. Pintor et al., *J. Clin. Endocrinol. Metab.* **62** (1986) 263.
[1589] N. T. Richards et al., *Diabetologia* **27** (1984) 529.
[1590] R. Valcavi et al., *Clin. Endocrinol.* **24** (1986) 693.
[1591] M. Hotta et al., *Life Sci.* **42** (1988) 979.
[1592] K.-P. Lesch et al., *J. Clin. Endocrinol. Metab.* **65** (1987) 1278.
[1593] R. Cacabelos et al., *Acta Endocrinol. (Copenhagen)* **117** (1988) 295.
[1594] F. L. Culler et al., *Clin. Res.* **35** (1987) 221A.
[1595] F. Zeytin, P. Braceau, *Biochem. Biophys. Res. Commun.* **123** (1984) 497.
[1596] T. Lehy et al., *Gastroenterology* **100** (1986) 646.
[1597] F. Obal, Jr. et al., *Am. J. Physiol.* **255** (1988) R310.
[1598] Z. Josefsberg et al., *Isr. J. Med. Sci.* **23** (1987) 837.
[1599] N. Ling et al., *Biochem. Biophys. Res. Commun.* **123** (1984) 854.
[1600] N. Ling et al., *Biochem. Biophys. Res. Commun.* **122** (1984) 304.
[1601] T. J. Aitman et al., *Peptides (Fayetteville, N.Y.)* **10** (1989) 1.
[1602] M. Kovacs et al., *Life Sci.* **42** (1988) 27.
[1603] L. Bokser et al., *Life Sci.* **46** (1990) 999.
[1604] K. Sato et al., *Biochem. Biophys. Res. Commun.* **149** (1987) 531.
[1605] K. Sato et al., *Biochem. Biophys. Res. Commun.* **167** (1990) 360.
[1606] S. J. Hocart et al., *J. Med. Chem.* **33** (1990) 1954.
[1607] D. H. Coy et al., *J. Med. Chem.* **30** (1987) 219.
[1608] R. J. M. Ross et al., *Lancet I* (1987) 5.
[1609] M. O. Thorner et al., *N. Engl. J. Med.* **312** (1985) 4.
[1610] O. Butenandt, B. Staudlt, *Eur. J. Pediatr.* **148** (1989) 393.
[1611] P. Franchimont et al., *Acta Clin. Belg.* **42** (1987) 143.
[1612] Beckmann Instruments, Inc., EP 0018072, 1979.
[1613] C. Y. Bowers et al., *Endocrinology (Baltimore)* **114** (1984) 1573.
[1614] O. Sartor et al., *Endocrinology (Baltimore)* **116** (1985) 952.
[1615] B. E. Ilson et al., *J. Clin. Endocrinol. Metab.* **69** (1989) 212.
[1616] M. Ono et al., *Proc. Natl. Acad. Sci. USA* **87** (1990) 4330.
[1617] M. Yamakawa et al., *J. Biol. Chem.* **265** (1990) 8915.
[1618] I. A. MacFarlane et al., *Acta Endocrinol. (Copenhagen)* **112** (1986) 547.
[1619] N. E. Cooke, J. Ray, J. G. Emergy, S. A. Liebhaber, *J. Biol. Chem.* **263** (1988) 9001.
[1620] A. Skottner et al., *Acta Endocrinol. (Copenhagen)* **118** (1988) 14.
[1621] A. Gertler et al., *Acta Endocrinol. (Copenhagen)* **118** (1988) 720.
[1622] J.-O. Jansson et al., *Endocr. Rev.* **6** (1985) 128.
[1623] H. Soya, M. Suzuki, *Endocrinology (Baltimore)* **122** (1988) 2492.
[1624] P. Maertens, C. Denef, *Mol. Cell. Endocrinol.* **54** (1987) 203.
[1625] H. Yagi et al., *Horm. Metab. Res.* **18** (1986) 723.
[1626] A. D. Struthers et al., *Neuroendocrinology* **44** (1986) 22.
[1627] F. E. Chang et al., *J. Clin. Endocrinol. Metab.* **62** (1986) 551.
[1628] M. Marastoni et al., *Arzneim. Forsch.* **39** (1989) 639.
[1629] E. C. degli Uberti et al., *Metabolism* **39** (1990) 1063.
[1630] L. Altomonte et al., *Exp. Clin. Endocrinol.* **88** (1986) 334.
[1631] P. A. Graham et al., *Clin. Res.* **35** (1987) 396A.

[1632] J. R. Peters et al., *Clin. Endocrinol.* **25** (1986) 213.
[1633] E. W. Bernton et al., *Science (Washington, D.C.)* **238** (1987) 519.
[1634] F. Lopez et al., *Acta Endocrinol. (Copenhagen)* **113** (1986) 317.
[1635] M. L. Vance et al., *J. Clin. Endocrinol. Metab.* **64** (1987) 1136.
[1636] J. K. Schmitt, *Clin. Res.* **37** (1989) 9A.
[1637] K. Y. Ho et al., *J. Clin. Endocrinol. Metab.* **64** (1987) 51.
[1638] M. J. Mansfield et al., *J. Clin. Endocrinol. Metab.* **66** (1988) 3.
[1639] N. Mauras et al., *Metabolism* **38** (1989) 286.
[1640] A. Giustina et al., *Horm. Metab. Res.* **21** (1989) 693.
[1641] A. Giustina et al., *Acta Endocrinol. (Copenhagen)* **122** (1990) 206.
[1642] R. J. M. Ross et al., *Clin. Endocrinol.* **26** (1987) 117.
[1643] S. Kentroti et al., *Endocrinology (Baltimore)* **122** (1988) 2407.
[1644] N. Ono et al., *Proc. Natl. Acad. Sci. USA* **82** (1985) 7787.
[1645] G. P. Ceda et al., *Acta Endocrinol. (Copenhagen)* **120** (1989) 416.
[1646] C. Netti et al., *Neuroendocrinology* **49** (1989) 242.
[1647] H. Katakami et al., *Endocrinology (Baltimore)* **117** (1985) 1139.
[1648] N. C. Schaper, *Acta Endocrinol. (Copenhagen)* **122** (1990) 7.
[1649] K.-P. Lesch et al., *J. Clin. Endocrinol. Metab.* **65** (1987) 1278.
[1650] A. Zarate et al., *Horm. Metab. Res.* **18** (1986) 400.
[1651] C. Marchesi et al., *Pharmacopsychiatry* **20** (1987) 64.
[1652] P. Pietschmann et al., *Horm. Metab. Res.* **22** (1990) 109.
[1653] O. Giampietro et al., *Metabolism* **36** (1987) 1149.
[1654] D. W. Leung et al., *Nature (London)* **330** (1987) 537.
[1655] L. S. Mathews et al., *J. Biol. Chem.* **264** (1989) 9905.
[1656] R. L. Hintz, *Horm. Res.* **33** (1990) 105.
[1657] P. J. Godowski et al., *Proc. Natl. Acad. Sci. USA* **86** (1989) 8083.
[1658] A. Aguirre et al., *Horm. Res.* **34** (1990) 4.
[1659] G. Baumann et al., *N. Engl. J. Med.* **320** (1989) 1705.
[1660] M. C. Slootweg et al., *J. Endocrinol.* **116** (1988) R11–R13.
[1661] J. H. Nielsen et al., *Mol. Endocrinol.* **13** (1989) 165.
[1662] M. E. Markowitz et al., *J. Clin. Endocrinol. Metab.* **69** (1989) 420.
[1663] J. K. Damm, *Acta Endocrinol. (Copenhagen)* **114** (1987) 124.
[1664] I. Ganzetti et al., *J. Endocrinol. Invest.* **10** (1987) 241.
[1665] D. M. Maiter et al., *Neuroendocrinology* **51** (1990) 174.
[1666] J. Isgaard et al., *Endocrinology (Baltimore)* **123** (1988) 2605.
[1667] C. T. Roberts, Jr. et al., *J. Biol. Chem.* **22** (1986) 10025.
[1668] J. H. Nielsen et al., *Acta Endocrinol. (Copenhagen)* **110** Suppl. 273 (1985) 74.
[1669] L. G. Frigert et al., *Horm. Metab. Res.* **19** (1987) 464.
[1670] F. Salomon et al., *Diabetologia* **33** (1990) Suppl., A218.
[1671] M. B. Davidson, *Endocr. Rev.* **8** (1987) 115.
[1672] C. K. Edwards III et al., *Science (Washington, D.C.)* **239** (1988) 769.
[1673] B. Lawier Goff et al., *Clin. Exp. Immunol.* **68** (1987) 580.
[1674] S. L. Kaplan et al., *Lancet I* (1986) 697.
[1675] J. L. Kostyo et al., *Proc. Natl. Acad. Sci. USA* **82** (1985) 4250.
[1676] S. L. Kaplan et al., *Lancet I* (1986) 697.
[1677] F. L. Culler et al., *Horm. Metab. Res.* **20** (1988) 107.
[1678] N. Emoto et al., *Acta Endocrinol. (Copenhagen)* **114** (1987) 283.

[1679] F. M. Ng, J. A. Harcourt, *Diabetologia* **29** (1986) 882.
[1680] V. M. J. Robson et al., *Hoppe-Seyler's Z. Physiol. Chem.* **371** (1990) 423.
[1681] R. W. Stevenson et al., *Acta Endocrinol. (Copenhagen)* **117** (1988) 457.
[1682] C. E. Mondon et al., *Endocrinology (Baltimore)* **123** (1988) 827.
[1683] C. Singer-Granick et al., *Horm. Res.* **24** (1986) 246.
[1684] H. Waago, *Lancet I* (1987) 1485.
[1685] E. A. van der Veen, J. C. Netelenbos, *Horm. Res.* **33** (1990) 65.
[1686] F. Greig et al., *Horm. Res.* **31** Suppl. 1(1989) 20.
[1687] D. Rudman et al., *N. Engl. J. Med.* **323** (1990) 1.
[1688] Y. A. Pankov, V. Y. Butnev, *J. Pept. Prot. Res.* **28** (1986) 113.
[1689] R. Einspanier et al., *FEBS Lett.* **204** (1986) 37.
[1690] M. Watahiki et al., *J. Biol. Chem.* **264** (1989) 5535.
[1691] S. H. Shin et al., *Can. J. Physiol. Pharmacol.* **65** (1987) 2036.
[1692] G. Nagy et al., *Biochem. Biophys. Res. Commun.* **151** (1988) 524.
[1693] J. E. Merritt, B. L. Brown, *Life Sci.* **35** (1984) 707.
[1694] F. Bigi et al., *Exp. Clin. Endocrinol.* **95** (1990) 224.
[1695] K. Törnquist, *Acta Endocrinol. (Copenhagen)* **116** (1987) 459.
[1696] I. S. Kampa et al., *Horm. Metab. Res.* **18** (1986) 419.
[1697] J. S. Ramsdell, A. H. Tashjian, Jr., *Endocrinology (Baltimore)* **117** (1985) 2050.
[1698] P. G. Knight et al., *Neuroendocrinology* **44** (1986) 29.
[1699] H. Abe et al., *Endocrinology (Baltimore)* **116** (1985) 1383.
[1700] E. Rolandi et al., *Horm. Res.* **21** (1985) 209.
[1701] K. D. Meirleir et al., *Horm. Metab. Res.* **17** (1985) 380.
[1702] P. Buydens et al., *Horm. Metab. Res.* **18** (1986) 575.
[1703] P. L. Canonico, R. M. MacLeod, *Endocrinology (Baltimore)* **118** (1986) 233.
[1704] H. Minakami et al., *Endocrinol. Jpn.* **33** (1986) 511.
[1705] T. H. Jones et al., *J. Endocrinol.* **111** Suppl. (1986) 134.
[1706] M. Memo et al., *J. Neurochem.* **47** (1986) 1689.
[1707] I. S. Login, *Life Sci.* **47** (1990) 2269.
[1708] P. Tatar, M. Vigas, *Neuroendocrinology* **39** (1984) 275.
[1709] U. Knigge et al., *J. Clin. Endocrinol. Metab.* **62** (1986) 491.
[1710] I. Murai, N. Ben-Jonathan, *Neuroendocrinology* **43** (1986) 453.
[1711] S. H. Shin, R. Stirling, *J. Endocrinol.* **118** (1988) 287.
[1712] N. Yonehara, D. H. Clouet, *J. Pharm. Exp. Ther.* **231** (1984) 38.
[1713] S. Röjdmark et al., *J. Endocrinol. Invest.* **7** (1984) 635.
[1714] L. S. Frawley, C. L. Clark, *Endocrinology (Baltimore)* **119** (1986) 1462.
[1715] J.-P. Loeffler et al., *Neuroendocrinology* **43** (1986) 504.
[1716] K. Fuxe et al., *Acta Physiol. Scand.* **125** (1985) 437.
[1717] T. Karashima, A. V. Schally, *Proc. Soc. Exp. Biol. Med.* **185** (1987) 69.
[1718] I. Zofkova, J. Nedvidkova, *Exp. Clin. Endocrinol.* **92** (1988) 262.
[1719] E. W. Bernton et al., *Science (Washington, D.C.)* **238** (1987) 519.
[1720] D. F. Wood et al., *J. Endocrinol.* **115** (1987) 497.
[1721] A. I. Esquifino et al., *J. Endocrinol. Invest.* **12** (1989) 171.
[1722] J. D. Wark, V. Gurtler, *Biochem. J.* **241** (1987) 397.
[1723] L. V. dePaolo et al., *Peptides (Fayetteville, N.Y.)* **7** (1986) 541.
[1724] A. Gertler et al., *Endocrinology (Baltimore)* **118** (1986) 720.
[1725] C. L. Bethea, E. Yuzuriha, *Endocrinology (Baltimore)* **119** (1986) 771.

[1726] N. Kimura et al., *Endocrinology (Baltimore)* **119** (1986) 1028.
[1727] W. D. Hetzel et al., *Horm. Metab. Res.* **22** (1990) 648.
[1728] V. C. Musey et al., *N. Engl. J. Med.* **316** (1987) 229.
[1729] A. Kauppila et al., *J. Clin. Endocrinol. Metab.* **64** (1987) 309.
[1730] B. B. Arnetz et al., *Life Sci.* **39** (1986) 135.
[1731] P. C. Hiestand et al., *Proc. Natl. Acad. Sci. USA* **83** (1986) 2599.
[1732] J.-M. Boutin et al., *Cell* **53** (1988) 69.
[1733] M. Edery et al., *Proc. Natl. Acad. Sci. USA* **86** (1989) 2112.
[1734] P. Guillaumot et al., *Biochem. Biophys. Res. Commun.* **135** (1986) 1076.
[1735] H. G. Klemcke et al., *Endocrinology (Baltimore)* **118** (1986) 773.
[1736] J. A. Rillema et al., *Horm. Metab. Res.* **18** (1986) 672.
[1737] R. E. Hruska, *J. Neurochem.* **47** (1986) 1908.
[1738] E. Jungmann et al., *Horm. Metab. Res.* **18** (1986) 704.
[1739] T. H. Welsh, Jr. et al., *Biol. Reprod.* **34** (1986) 796.
[1740] F. T. Murray et al., *J. Clin. Endocrinol. Metab.* **59** (1984) 79.
[1741] A. Canfriez, *Horm. Res.* **22** (1985) 209.
[1742] W. G. North et al., *Horm. Res.* **15** (1981) 55.
[1743] S. Kaufmann, *J. Physiol. (London)* **310** (1981) 435.
[1744] G. J. Pepe, E. D. Albrecht, *Endocrinology (Baltimore)* **117** (1985) 1968.
[1745] A. R. Buckley et al., *Biochem. Biophys. Res. Commun.* **138** (1986) 1138.
[1746] A. R. Buckley et al., *Life Sci.* **37** (1985) 2569.
[1747] T. G. Muldoon, *Endocrinology (Baltimore)* **121** (1987) 141.
[1748] F. Drago et al., *Life Sci.* **36** (1985) 191.
[1749] J. P. Barlet, *J. Endocrinol.* **107** (1985) 171.
[1750] M. P. Caraceni et al., *Calcif. Tissue Int.* **37** (1985) 687.
[1751] R. E. Humbel, *Eur. J. Biochem.* **190** (1990) 445.
[1752] W. H. Daughaday et al., *Endocrinology (Baltimore)* **121** (1987) 1911.
[1753] K. Ramasharma et al., *Biochem. Biophys. Res. Commun.* **140** (1986) 536.
[1754] L. K. Bachrach et al., *Biochem. Biophys. Res. Commun.* **154** (1988) 861.
[1755] M. R. Hammerman, *Am. J. Physiol.* **257** (1989) F503.
[1756] D. C. Costigan et al., *J. Clin. Endocrinol. Metab.* **66** (1988) 1014.
[1757] A. N. Corps et al., *J. Clin. Endocrinol. Metab.* **67** (1988) 25.
[1758] J. Wroblewski et al., *Acta Endocrinol. (Copenhagen)* **115** (1987) 37.
[1759] E.-M. Rutanen, F. Pekonen, *Acta Endocrinol. (Copenhagen)* **123** (1990) 7.
[1760] A. Shimatsu, P. Rotwein, *J. Biol. Chem.* **262** (1987) 7894.
[1761] A. Honegger, R. E. Humbel, *J. Biol. Chem.* **261** (1986) 569.
[1762] P. P. Zumstein et al., *Biochemistry* **82** (1985) 3169.
[1763] P. Rotwein, *Proc. Natl. Acad. Sci. USA* **83** (1986) 77.
[1764] V. R. Sara et al., *Proc. Natl. Acad. Sci. USA* **83** (1986) 4904.
[1765] L. J. Murphy et al., *Endocrinology (Baltimore)* **122** (1988) 2027.
[1766] E. Jennische et al., *Acta Physiol. Scand.* **129** (1987) 9.
[1767] P. Bang et al., *Eur. J. Clin. Invest.* **20** (1990) 285.
[1768] H.-A. Hansson et al., *Acta Physiol. Scand.* **129** (1987) 165.
[1769] K. Ramasharma, C. H. Li, *Proc. Natl. Acad. Sci. USA* **84** (1987) 2643.
[1770] L. J. Murphy, A. Ghahary, *Endocr. Rev.* **11** (1990) 443.
[1771] T. K. Gray et al., *Biochem. Biophys. Res. Commun.* **158** (1989) 407.
[1772] T. L. McCarthy et al., *J. Biol. Chem.* **265** (1990) 15353.

[1773] C. G. Scanes et al., *IRCS Med. Sci.* **14** (1986) 515.
[1774] S. Goldstein, L. S. Phillips, *Metabolism* **38** (1989) 745.
[1775] W. E. Sonntag, R. L. Boyd, *Life Sci.* **43** (1988) 1325.
[1776] D. A. Harris et al., *J. Clin. Endocrinol. Metab.* **61** (1985) 152.
[1777] A. Tham et al., *Horm. Metab. Res.* **18** (1986) 706.
[1778] R. G. Rosenfeld et al., *Biochem. Biophys. Res. Commun.* **143** (1987) 199.
[1779] T. A. Gustafson, W. J. Rutter, *J. Biol. Chem.* **265** (1990) 18663.
[1780] R. G. MacDonald et al., *Science (Washington, D.C.)* **239** (1988) 1134.
[1781] R. L. Hintz, *Horm. Res.* **33** (1990) 105.
[1782] A. L. Albiston, A. C. Herington, *Biochem. Biophys. Res. Commun.* **166** (1990) 892.
[1783] S. Shimasaki et al., *Mol. Endocrinol.* **4** (1990) 1451.
[1784] A. M. Taylor et al., *Clin. Endocrinol.* **32** (1990) 229.
[1785] M. Ross et al., *Biochem. J.* **258** (1989) 267.
[1786] I. Kojima et al., *J. Biol. Chem.* **265** (1990) 16846.
[1787] M. Ernst, E. R. Froesch, *Biochem. Biophys. Res. Commun.* **151** (1988) 142.
[1788] B. P. Halloran, E. M. Spencer, *Endocrinology (Baltimore)* **123** (1988) 1225.
[1789] H.-P. Guler et al., *N. Engl. J. Med.* **317** (1987) 137.
[1790] H.-P. Guler et al., *Proc. Natl. Acad. Sci. USA* **86** (1989) 2868.
[1791] S. Yamashita et al., *J. Clin. Endocrinol. Metab.* **43** (1986) 730.
[1792] C. G. Goodyer et al., *Endocrinology (Baltimore)* **115** (1984) 1568.
[1793] K. M. Thrailkill et al., *Endocrinology (Baltimore)* **123** (1988) 2930.
[1794] M.-H. Perrard-Sapori et al., *Eur. J. Biochem.* **165** (1987) 209.
[1795] T. Lin et al., *Biochem. Biophys. Res. Commun.* **137** (1986) 950.
[1796] J. D. Veldhuis et al., *J. Biol. Chem.* **261** (1986) 2499.
[1797] S. E. Tollefsen et al., *J. Biol. Chem.* **264** (1989) 13810.
[1798] A. J. Stewart et al., *J. Biol. Chem.* **265** (1990) 21172.
[1799] M. L. Stracke et al., *J. Biol. Chem.* **264** (1989) 21544.
[1800] K. Binz et al., *Proc. Natl. Acad. Sci. USA* **87** (1990) 3690.
[1801] P. Hunt, D. D. Eardley, *J. Immunol.* **136** (1986) 3994.
[1802] L. Y.-H. Tseng et al., *Biochem. Biophys. Res. Commun.* **149** (1987) 672.
[1803] M. L. Bayne et al., *J. Biol. Chem.* **263** (1988) 6233.
[1804] M. L. Bayne et al., *J. Biol. Chem.* **265** (1990) 15648.
[1805] C. J. Bagley et al., *Biochem. J.* **259** (1989) 665.
[1806] J. D. Quin et al., *Diabetologia* **32** (1989) 531A.
[1807] Z. Laron et al., *Lancet II* (1988) 1170.
[1808] D. J. Gross et al., *J. Biol. Chem.* **264** (1989) 21486.
[1809] E. Helmerhorst, G. B. Stokes, *Diabetes* **36** (1987) 261.
[1810] W. König in A. Kleemann et al., (eds.): *Arzneimittel, Fortschritte 1972 bis 1985*, VCH Verlagsgesellschaft, Weinheim 1987, p. 728.
[1811] T. Adachi et al., *J. Biol. Chem.* **264** (1989) 7681.
[1812] A. B. Smit et al., *Nature (London)* **331** (1988) 535.
[1813] M. Lagueux et al., *Eur. J. Biochem.* **187** (1990) 249.
[1814] J. N. Fain, *Metabolism* **33** (1984) 672.
[1815] B. Kreymann et al., *Lancet II* (1987) 1300.
[1816] T. Matsuyama et al., *Diabetes* **36** Suppl. 1 (1987) 164A.
[1817] W. H. Hsu et al., *Proc. Soc. Exp. Biol. Med.* **184** (1987) 345.
[1818] Y. Totsuka et al., *Biochem. Biophys. Res. Commun.* **158** (1989) 1060.

[1819] S. B. Richardson et al., *Diabetes* **37** (1988) 103A.
[1820] H. Kofod et al., *Am. J. Physiol.* **154** (1988) E454.
[1821] M. C. d'Emden et al., *Biochem. Biophys. Res. Commun.* **164** (1989) 413.
[1822] A. Faure et al., *Horm. Metab. Res.* **17** (1985) 378.
[1823] T. Ikeda et al., *Biochem. Pharmacol.* **40** (1990) 1769.
[1824] S. B. Pek, M. F. Walsh, *Proc. Natl. Acad. Sci. USA* **81** (1984) 2199.
[1825] N. G. Morgan et al., *Biochim. Biophys. Acta C* **845** (1985) 526.
[1826] G. Holm, *Acta Med. Scand. Suppl.* **672** (1983) 21.
[1827] I. C. Green et al., *Diabetologia* **27** (1984) 282A.
[1828] B. Ahren et al., *Biochem. Biophys. Res. Commun.* **140** (1986) 1054.
[1829] K. Hermansen, B. Ahren, *Regul. Pept.* **27** (1990) 149.
[1830] S. Raptis et al., *Diabetes* **32** Suppl. 1 (1983) 94A.
[1831] F. Shimizu et al., *Endocrinology (Baltimore)* **117** (1985) 2081.
[1832] C. Southern et al., *Biochem. J.* **272** (1990) 243.
[1833] S. Baekkeskov et al., *Nature (London)* **347** (1990) 151.
[1834] D. B. Jones et al., *Lancet* **336** (1900) 583.
[1835] S. Zeuzem, *Aktuel. Endokrinol. Stoffwechsel* **8** (1987) 132.
[1836] T. A. Gustafson, W. J. Rutter, *J. Biol. Chem.* **265** (1990) 18663.
[1837] C. Correze et al., *Biochem. Biophys. Res. Commun.* **126** (1985) 1061.
[1838] M. Taira et al., *Science (Washington, D.C.)* **245** (1989) 63.
[1839] M. Odawara et al., *Science (Washington,D.C.)* **245** (1989) 66.
[1840] D. E. Moller et al., *Mol. Endocrinol.* **4** (1990) 1183.
[1841] E. G. Walters et al., *Lancet I* (1987) 241.
[1842] P. Briata et al., *Biochem. Biophys. Res. Commun.* **169** (1990) 397.
[1843] M. L. Standaert, R. J. Pollet, *J. Biol. Chem.* **259** (1984) 2346.
[1844] V. Papa et al., *J. Clin. Invest.* **86** (1990) 1503.
[1845] G. B. Willars, T. W. Atkins, *J. Endocrinol.* **104** (1985) Suppl., 100.
[1846] J. D. McArmstrong et al., *Acta Endocrinol. (Copenhagen)* **102** (1983) 492.
[1847] J. C. Jarrett II et al., *Am. J. Obstet. Gynecol.* **149** (1984) 250.
[1848] K. Kriaucunas, C. R. Kahn, *Clin. Res.* **32** (1984) 401A.
[1849] G. Kasdorf, R. K. Kalkhoff, *Clin. Res.* **34** (1986) 965A.
[1850] P. Dent et al., *Nature (London)* **348** (1990) 302.
[1851] J. S. Marks, L. H. Parker Botelho, *J. Biol. Chem.* **261** (1986) 2781.
[1852] D. E. James et al., *Nature (London)* **338** (1989) 83.
[1853] W. G. Lampson et al., *Can. J. Physiol. Pharmacol.* **61** (1983) 457.
[1854] H. S. Glauber et al., *N. Engl. J. Med.* **316** (1987) 443.
[1855] C. L. Oppenheimer et al., *J. Biol. Chem.* **258** (1983) 4824.
[1856] M. P. Cohen et al., *Biochim. Biophys. Acta M* **856** (1986) 182.
[1857] J. Ben Davoren, A. J. W. Hsueh, *Mol. Cell. Endocrinol.* **35** (1984) 97.
[1858] T. Lin et al., *Endocrinology (Baltimore)* **119** (1986) 1641.
[1859] J. Rodin et al., *Metab. Clin. Exp.* **34** (1985) 826.
[1860] C. J. H. Ingoldby et al., *Am. J. Gastroenterol.* **79** (1984) 16.
[1861] K. Y. Lee et al., *Am. J. Physiol.* **258** (1990) G268.
[1862] E. Mezey et al., *Science (Washington, D.C.)* **226** (1984) 1085.
[1863] J. H. Pratt et al., *Clin. Res.* **31** (1983) 763A.
[1864] A. J. Spijker et al., *Eur. J. Clin. Invest.* **15** (1985) part 2, A16.
[1865] V. Schusdziarra et al., *Neuropeptides* **7** (1986) 51.

[1866] J. Guntupalli et al., *Am. J. Physiol.* **249** (1985) part 2, F610.
[1867] B. E. Kream et al., *Endocrinology (Baltimore)* **116** (1985) 296.
[1868] N. Wongsurawat, H. J. Armbrecht, *Acta Endocrinol. (Copenhagen)* **109** (1985) 243.
[1869] S. Yagi et al., *Diabetes* **37** (1988) 1064.
[1870] A. Siani et al., *Eur. J. Clin. Pharmacol.* **38** (1990) 393.
[1871] R. Sethi et al., *J. Mol. Cell. Cardiol.* **22** (1990) Suppl. 1, S33.
[1872] C. Grunfeld, D. S. Jones, *Diabetes* **32** (1983) 128A.
[1873] H. Schatz, S. Ammermann, *Horm. Metab. Res., Suppl. Ser.* **18** (1988) 1.
[1874] C. Rosak et al., *Horm. Metab. Res., Suppl. Ser.* **18** (1988) 16.
[1875] G. D. Smith et al., *Proc. Natl. Acad. Sci. USA* **81** (1984) 7093.
[1876] G. P. Schwartz et al., *Proc. Natl. Acad. Sci. USA* **84** (1987) 6408.
[1877] G. T. Burke et al., *Biochem. Biophys. Res. Commun.* **173** (1990) 982.
[1878] S.-C. Chu et al., *Sci. Sin. (Engl. Ed.)* **16** (1973) 71.
[1879] R. G. Mirmira, H. S. Tager, *J. Biol. Chem.* **264** (1989) 6349.
[1880] J. Brange et al., *Nature (London)* **333** (1988) 679.
[1881] H. P. Neubauer et al., *Diabetologia* **27** Suppl. (1984) 129.
[1882] S. Jorgensen et al., *Diabetologia* **32** (1989) 500A.
[1883] P. Balschmidt et al., *Diabetologia* **33** Suppl. (1990) A117.
[1884] L. D. Monti et al., *Diabetologia* **33** Suppl. (1990) A60.
[1885] P. Balschmidt et al., *Diabetes* **38** Suppl. 2 (1989) 157A.
[1886] M. Hashimoto et al., *Pharm. Res.* **6** (1989) 171.
[1887] S. Kang et al., *Diabetologia* **32** (1989) 502A.
[1888] J. N. MacPherson, J. Feely, *Br. Med. J.* **300** (1990) 731–736.
[1889] M. Rodier, *Presse Med.* **19** (1990) 959–964.
[1890] W. Foertsch, *Aktuel. Endokrinol. Stoffwechsel* **7** (1986) 33.
[1891] G. Paolisso et al., *J. Clin. Endocrinol. Metab.* **66** (1988) 1220.
[1892] A. Roza et al., *Clin. Res.* **31** (1983) 622A.
[1893] W. T. Chance et al., *JNCI J. Natl. Cancer Inst.* **77** (1986) 497.
[1894] S. J. Whang, F. L. Greenway, *Clin. Res.* **33** (1985) 67A.
[1895] B. E. Kemp, H. D. Niall, *Vitam. Horm. (N.Y.)* **41** (1984) 79.
[1896] G. Mazoujian, G. D. Bryant-Greenwood, *Lancet* **335** (1990) 299.
[1897] P. Hudson et al., *EMBO J.* **3** (1984) 2333.
[1898] J. S. Ottobre et al., *Biol. Reprod.* **31** (1984) 1000.
[1899] K. Seki et al., *Endocrinol. Jpn.* **33** (1986) 727.
[1900] A. I. Musah et al., *Endocrinology (Baltimore)* **120** (1987) 317.
[1901] A. L. Bernal et al., *Br. J. Obstet. Gynaecol.* **94** (1987) 1045.
[1902] B. Viell, H. Struck, *Horm. Metab. Res.* **19** (1987) 415.
[1903] J. St-Louis, G. Massicotte, *Life Sci.* **37** (1985) 1351.
[1904] G. Massicotte et al., *Clin. Exp. Hypertens. Part B* **B6** (1987) 255.
[1905] E. E. Büllesbach, C. Schwabe, *Biochemistry* **25** (1986) 5998.
[1906] C. H. S. McIntosh, *Life Sci.* **17** (1985) 2043.
[1907] M. Bersani et al., *J. Biol. Chem.* **264** (1989) 10633.
[1908] P. G. Burhol et al., *Acta Physiol. Scand.* **121** (1984) 223.
[1909] D. LeRoith et al., *Endocrinology (Baltimore)* **117** (1985) 2093.
[1910] S. M. Cutfield et al., *FEBS Lett.* **214** (1987) 57.
[1911] S. V. Wu et al., *Metabolism* **39** (1990) 125.
[1912] R. M. Post, D. R. Rubinow, *Lancet II* (1986) 810.

[1913] R. Zorrilla et al., *Neuroendocrinology* **52** (1990) 527.
[1914] T. Chiba et al., *Am. J. Physiol.* **253** (1987) G62.
[1915] M. C. Aguila et al., *Neuroendocrinology* **52** (1990) 238.
[1916] M. Sato et al., *Neuroendocrinology* **50** (1989) 139.
[1917] D. E. Scarborough, *Metabolism* **39** (1990) 108.
[1918] H. Koop, R. Arnold, *Gastroenterology* **84** (1983) part 2, 1214.
[1919] K. Uvnäs-Moberg, L. Wetterberg, *Acta Physiol. Scand.* **120** (1984) 517.
[1920] S. M. Webb et al., *Acta Endocrinol. (Copenhagen)* **110** (1985) 145.
[1921] S. E. Shoelson et al., *Am. J. Physiol.* **250** (1986) E428.
[1922] G. Strazzulla et al., *Horm. Metab. Res.* **20** (1988) 126.
[1923] H. Stepien et al., *Horm. Metab. Res.* **18** (1986) 555.
[1924] F. Sicuteri et al., *IRCS Med. Sci.* **13** (1985) 308.
[1925] D. Neugebauer, G. Weber, *Z. Hautkrankh.* **63** (1988) 585.
[1926] V. Schusdziarra et al., *Diabetes* **34** (1985) 595.
[1927] K. Uvnäs-Moberg et al., *Acta Obstet. Gynecol. Scand.* **68** (1989) 165.
[1928] S. Gomez et al., *Life Sci.* **39** (1986) 623.
[1929] J. M. Radke et al., *Eur. J. Pharmacol.* **134** (1987) 105.
[1930] J. C. Reubi et al., *Metabolism* **39** (1990) 78.
[1931] M. Flint Beal et al., *Science (Washington, D.C.)* **229** (1985) 289.
[1932] S. Reichlin, *N. Engl. J. Med.* **309** (1983) 1556.
[1933] M. J. Toro et al., *Horm. Res.* **29** (1988) 59.
[1934] M. R. Brown et al., *Endocrinology (Baltimore)* **114** (1984) 1546.
[1935] A. Skamene, Y. C. Patel, *Clin. Endocrinol.* **20** (1984) 555.
[1936] S. Mulvihill et al., *N. Engl. J. Med.* **310** (1984) 467.
[1937] S. Bjorck, J. Svanvik, *Scand. J. Gastroenterol.* **19** (1984) 173.
[1938] P. P. Gazzaniga et al., *Experientia* **44** (1988) 892.
[1939] M. Mogard et al., *Gastroenterology* **86** (1984) Part 2, 1186.
[1940] G. J. Krejs, *Diabetes* **33** (1984) 548.
[1941] R. G. Long et al., *Br. Med. J.* **290** (1985) 886.
[1942] J. Morisset, *Regul. Pept.* **10** (1984) 11.
[1943] E. F. Stange et al., *Horm. Metab. Res.* **16** (1984) 74.
[1944] N. Altorki et al., *Gastroenterology* **86** (1984) part 2, 1014.
[1945] C. Liebow et al., *Metabolism* **39** (1990) 163.
[1946] M. Pawlikowski et al., *Peptides (Fayetteville, N.Y.)* **8** (1987) 951.
[1947] M. Pawlikowski et al., *Neuropeptides* **13** (1989) 75.
[1948] M. Pawlikowski et al., *Biochem. Biophys. Res. Commun.* **129** (1985) 52.
[1949] *Drugs Future* **9** (1984) 342; **10** (1985) 435; **11** (1986) 430; **12** (1987) 500; **13** (1988) 482; **14** (1989) 479.
[1950] R.-Z. Cai et al., *Proc. Natl. Acad. Sci. USA* **83** (1986) 1896. *Drugs Future* **14** (1989) 1052; **15** (1990) 1144.
[1951] R.-Z. Cai et al., *Proc. Natl. Acad. Sci. USA* **84** (1987) 2502.
[1952] W. A. Murphy et al., *Life Sci.* **40** (1987) 2515.
[1953] K. Ziegler et al., *Biochim. Biophys. Acta C* **845** (1985) 86.
[1954] G. Rohr et al., *Klin. Wochenschr.* **64** Suppl. 7 (1986) 90.
[1955] S. Szabo et al., *Can. J. Physiol. Pharmacol.* Suppl. (1986) 8.
[1956] J. T. Pelton et al., *Proc. Natl. Acad. Sci. USA* **82** (1985) 236.
[1957] J. M. Walker et al., *Peptides (Fayetteville, N.Y.)* **8** (1987) 869.

[1958]   P. E. Battershill, S. P. Clissold, *Drugs* **38** (1989) 658.
[1959]   J. L. Abelson et al., *J. Clin. Psychopharmacol.* **10** (1990) 128.
[1960]   H. J. Balks et al., *Eur. J. Clin. Pharmacol.* **36** (1989) 133.
[1961]   R. D. Penn et al., *Lancet* **335** (1990) 738.
[1962]   K. Tatemoto et al., *Nature (London)* **324** (1986) 476.
[1963]   A. L. Iacangelo et al., *Endocrinology (Baltimore)* **122** (1988) 2339.
[1964]   A. Funakoshi et al., *Regul. Pept.* **30** (1990) 159.
[1965]   S. Efendic et al., *Proc. Natl. Acad. Sci. USA* **84** (1987) 7257.
[1966]   A. Funakoshi et al., *Regul. Pept.* **24** (1989) 225.
[1967]   A. Funakoshi et al., *Regul. Pept.* **25** (1989) 157.
[1968]   V. Sanchez et al., *Biosci. Rep.* **10** (1990) 87.
[1969]   J. P. Smith, S. T. Kramer, *Gastroenterology* **96** (1989) part 2, A479.
[1970]   T. Zhang et al., *Biochem. Biophys. Res. Commun.* **173** (1990) 1157.

Chemical Contraception

Do Won Hahn, R. W. Johnson Pharmaceutical Research Institute, Raritan, New Jersey 08869, United States

John L. McGuire, Johnson & Johnson, New Brunswick, New Jersey 08933, United States

1.	Introduction	1493	4.3.	Long-Acting Contraceptives .	1504
2.	Natural Methods	1493	5.	Intrauterine Devices (IUDs) .	1506
3.	Barrier Methods	1495	6.	Sterilization	1507
4.	Hormonal Methods	1497	7.	Contragestational Drugs . . .	1508
4.1.	Oral Contraception	1497	8.	New Approaches	1510
4.2.	Progestogen-Only Oral Contraceptives	1503	9.	References	1514

1. Introduction

Oral contraceptives, vaginal contraceptives, injectable/implantable contraceptives, and contragestational agents, are used by large numbers of women today. Research in male contraception is a priority of many reproduction research centers. All of these methods are based on a form of chemical contraception. The large number of methods available reflect the fact that it is generally accepted that none of these methods, any more than other contraceptive methods, such as condoms, induced abortion, surgical sterilization or natural family planning, represent an ideal method of fertility regulation [1]–[3] for everyone. Each has its own advantages and disadvantages.

This chapter will review various contraceptive methods, with particular emphasis on those whose contraceptive action is based on chemical interruption of fertility. A comprehensive listing of contraceptive products can be found in the Directory of Contraceptives [4].

2. Natural Methods

There are a number of natural contraceptive methods. Among them are periodic abstinence and breast feeding.

Periodic Abstinence. Abstinence from sexual intercourse during the fertile period of the menstrual cycle with the intent of avoiding conception is reviewed in [5]. Knaus first found, in 1929, that the fertile period in women occurs before menstruation [6]. A year later Ogino in Japan independently confirmed this finding [7]. From their studies, Knaus and Ogino developed formulas to determine the fertile and infertile days of a woman's menstrual cycle. At about the same time, van de Velde linked ovulation to a cyclic shift in basal body temperature [8], which can be used respectively to determine the time of ovulation. These studies led to a woman's ability to calculate the time she ovulates and is infertile and to abstain from intercourse during her infertile period.

The primary difficulty with utilizing periodic abstinence (rhythm method) to prevent conception lies in the month-to-month variation in the time of ovulation. Whereas the ovum can only be fertilized during the first 12–24 h after its release from the ovary, sperm remain viable in the female reproductive tract, able to fertilize an ovum, for 5–7 d and perhaps longer. Thus, intercourse several days prior to ovulation can result in pregnancy.

More recent findings may better identify the fertile period. Changes in the quality and quantity of cervical mucus occurring 5–6 d prior to the mid-cycle surge of luteinizing hormone (LH), which initiates ovulation, can be used to predict the fertile period. However, it is sometimes difficult to recognize the changes in mucus. In addition, colorimetric enzyme immunoassays have recently been developed for the measurement of LH in urine. Since LH is rapidly excreted, the increase in urine LH levels can be used as a marker to predict impending ovulation. Recent studies suggest that urine LH testing every evening is a more reliable method of predicting ovulation within the ensuing 48 h [9]. However, despite these advances and others which will be made in the future, one must always remember the life span of sperm in the female reproductive tract and be cautious in predicting how useful the method will be in prospectively identifying the onset of the fertile period. Research on other methods of consistently and accurately predicting ovulation are ongoing.

Breast-Feeding. Another natural method is breast feeding. In many societies, it is commonly believed that women who are breast-feeding are incapable of becoming pregnant. Suckling does, in fact, lead to a release of the hormone prolactin, which in turn suppresses the release of estrogen from the ovary and luteinizing hormone (LH) from the pituitary gland. The suppression of these hormones can inhibit ovulation.

However, it is generally accepted that the duration of lactational amenorrhea and infertility is variable [10] and that maximum birth spacing is achieved only when a mother is "fully" or nearly fully breast-feeding. The consensus of researchers in the area of lactational infertility appears to be that, if used in this way, full breast-feeding can provide significant protection from pregnancy during the first six months after giving birth [11].

3. Barrier Methods

Chemical Vaginal Contraceptives. Vaginal contraception dates back to as far as 1850 B.C., when written instructions for vaginal contraceptives appeared in Egyptian papyri. Vaginal contraceptives are among the simplest and most widely utilized forms of birth control available today. Major advancements have been made in improving vaginal contraceptives by conducting research into sperm motility and by the development of spermicidal assays [12] which could be used to search for spermicidal agents for inclusion in vaginal contraceptives.

In 1855, KOLLIKER reported on the effects of chemical agents on sperm motility [13], prompting the first studies in modern-day vaginal contraceptive research. In 1880, RENDELL produced the first commercial vaginal contraceptive: a suppository of cocoa butter containing the spermicide quinine sulfate. Vaginal preparations containing other agents toxic to sperm, such as boric acid, tannic acid, or mercuric dichloride rapidly proliferated. Only a few of these are still in use, due to their lower spermicidal potency or potential toxicity.

In 1932, BAKER [14] provided the basis for in vitro spermicidal assays used to evaluate new spermicidal agents. In the late 1930s Johnson & Johnson's Ortho Research Laboratories developed the Sander Cramer test [15], which has been widely used since that time find new spermicidal agents and to compare spermicidal formulations. In the 1940s, the first surface active spermicidal agents, octoxynol [9002-93-1], 4-diisobutylphenoxy-polyethoxy-ethanol, trade name Triton X-100, and nonoxynol-9 [26027-38-3], 7-nonyl-phenoxypolyethoxyethanol, trade name Triton N, were discovered and developed at the Ortho Research Laboratories in the United States [16]. These two agents rapidly became the principal active ingredients utilized in vaginal contraceptives worldwide. A third surfactant spermicidal agent, menfegol [57821-32-6], 4-menthanylphenylpolyoxyethylene(8,8)ether, was discovered and developed by Eisai in Japan, in the late 1960s [17].

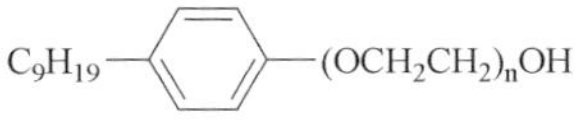

Nonoxynol-9

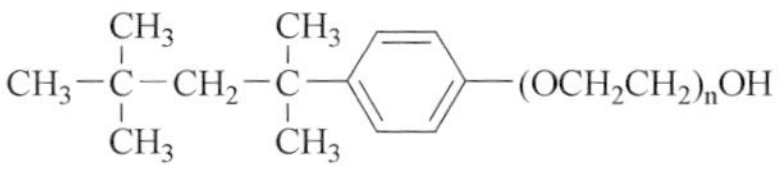

Triton X-100

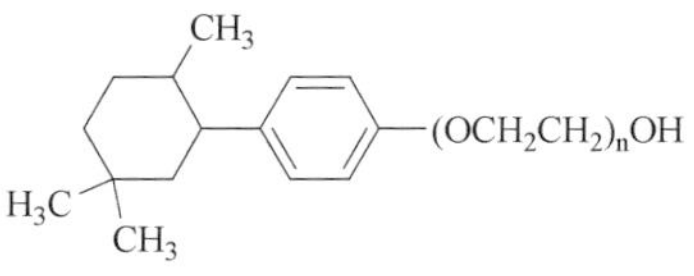

Menfegol

In recent years the prevalence of HIV infection and other sexually transmitted diseases (STD) have led to focused research efforts to develop new spermicidal agents which also have antimicrobial activity [18].

A comment should be made about the formulations, or inactive ingredients, used in vaginal contraceptives. They consist of a relatively inert base material which both blocks to some extent the passage of sperm and which also serves as a carrier for the chemically active spermicide. In the last several decades, it has been recognized that the inert formulation that contains and delivers the spermicide can affect the efficacy of vaginal contraceptives [19], [20]. As a result, physical properties of vaginal contraceptive formulations have been improved to deliver the spermicide more effectively and/or to enhance consumer compliance. Formulations currently available include jellies, creams, suppositories, films, aerosol foams, and foaming tablets.

The major advantages of vaginal contraceptives are simplicity and safety. In addition, they do not require medical supervision. Previously considered to be relatively ineffective, it has since been shown that vaginal contraceptives, if consistently used properly, can indeed be effective [20].

Physical Barriers. Various physical barrier devices are manufactured for use by men and women in regulating fertility. These include condoms, diaphragms and cervical caps and may be used in conjunction with a vaginal contraceptive.

Diaphragms are shallow rubber cups with a flexible metal rim, which are placed over the female cervix. They act both as a mechanical barrier to sperm and as a receptacle for a spermicidal agent [21].

Cervical caps are available in Europe, although they have not yet been approved for use in the United States or Japan. Cervical caps are deeper and smaller in diameter than diaphragms, and they are often made of rigid plastics, such as lucite; others are made of rubber. They cover only the cervix and are held in place by suction. Cervical caps are believed to have about the same degree of effectiveness as the diaphragm.

Another barrier contraceptive is the "female condom", Reality [22]. It is a thin, soft, loose-fitting polyurethane sheath with two flexible rings.

Another type of barrier contraceptive device is the vaginal contraceptive sponge. Studies show that the failure rate for the contraceptive sponge in parous women is higher than the failure rate for the diaphragm. A U.S. study of the contraceptive sponge found a first-year failure rate of 13.9% in nulliparous sponge users and a 28.3% first-year failure rate in parous sponge users. Other disadvantages of the method include allergic reactions in a small percentage of women, and a slightly increased risk of nonmenstrual toxic shock syndrome (TSS). When compared with other over-the-counter methods, the sponge can be expensive [23].

The male barrier contraceptive device is known as the condom, or rubber, and is widely available in most countries. The condom is a sheath, sometimes packaged with a lubricant and spermicide, which serves as a cover for the penis and a receptacle for semen. The method is very effective if the condom is properly used. Usage appears to be increasing due to adjunctive use with other methods of contraception for prevention of

HIV or other sexually transmitted diseases. By rough estimate, condoms may have been used in more than 13 billion acts of sexual intercourse that risked unwanted pregnancy, HIV, and/or other sexually transmitted diseases [24].

4. Hormonal Methods

4.1. Oral Contraception

Combination oral contraceptives, first introduced in the early 1960s, contain a synthetic progestogen and a synthetic estrogen. There are over 60 million users of this method in the world today, and almost 150 million women have used oral contraceptives sometime during their reproductive lives [25], [26]. According to ACOG, in addition to being extremely effective as a contraceptive agent, the use of oral contraceptives is associated with a variety of health and noncontraceptive benefits, including quality of life benefits and protection from certain gynecologic malignancies.

The idea that hormonal control of ovulation might be practical first appeared in 1921 in the work of HABERLANDT, who showed that extracts of the corpus luteum could make mice and rabbits infertile [27]. During the next 20 years, female steroidal sex hormones (progesterone and estradiol-17β) were isolated, identified, and synthesized. Unfortunately, evaluation of their clinical efficacy was severely restricted by insufficient quantities of compounds for testing. By 1943, MARKER had developed a method for synthesizing progesterone and other steroid derivatives from diosgenin, a naturally occurring steroid present in a species of Dioscorea. MARKER's studies led to the synthesis of larger quantities of these and other steroids which could be evaluated clinically for their effects on fertility regulation.

Progesterone [57-83-0] is a natural female hormone which affects ovulation; it is largely deactivated when given orally. However, an elimination at C-19 of progesterone results in a compound that is orally active. Similarly, structural modification of steroids from other hormone classes, such as the elimination at C-19 from testosterone, results in compounds (progestogens) which have progesterone-like effects (they induce secretory changes in estrogen-primed endometrium; induce the formation of thick, viscous cervical mucus; and suppress the release of luteinizing hormone and ultimately ovulation) [28]. In 1951, both The Syntex Laboratories and Searle R & D in the United States produced the substituted synthetic progestogens, norethisterone [68-22-4] (called norethindrone in the United States) and norethynordrel [68-23-5], which are still utilized in oral contraceptives today.

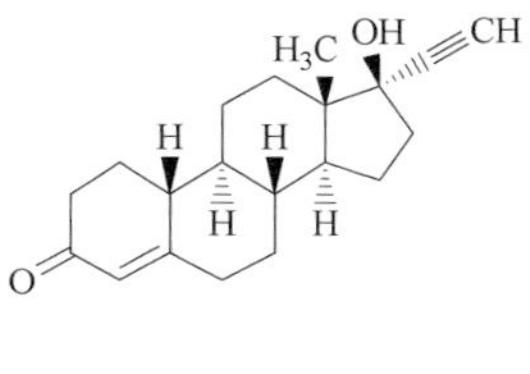

Norethindrone

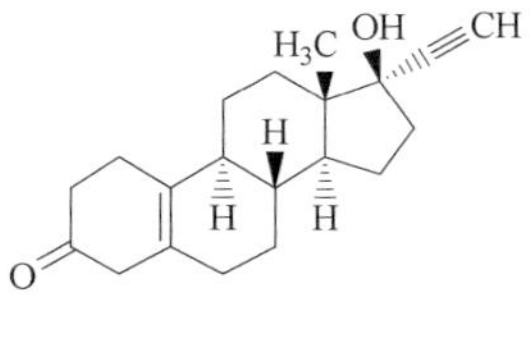

Norethynodrel

Other progestogens which were subsequently synthesized include lynestrenol [52-76-7], chlormadinone acetate [302-22-7], medroxy-progesterone acetate [71-58-9], ethynodiol diacetate [287-76-7], levonorgestrel [6533-00-2], desogestrel [54024-22-5], norgestimate [35189-28-7], and gestodene [60282-87-3].

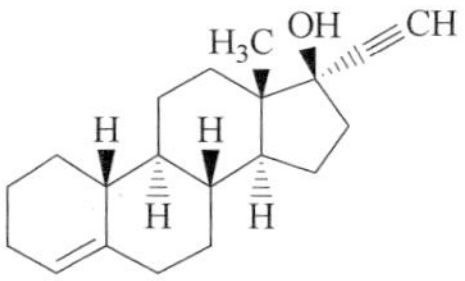

Lynetrenol

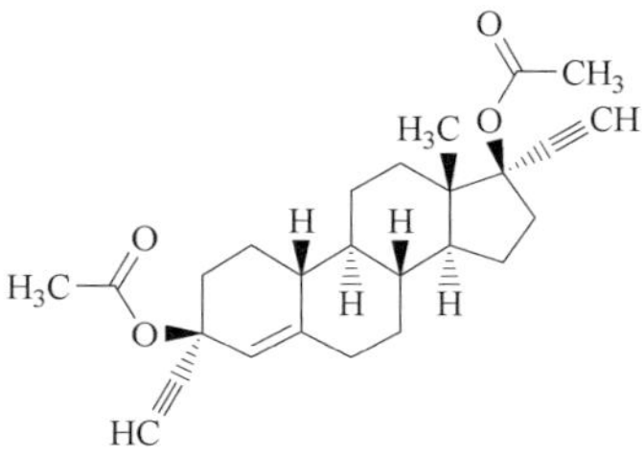

Ethynodiol diacetate

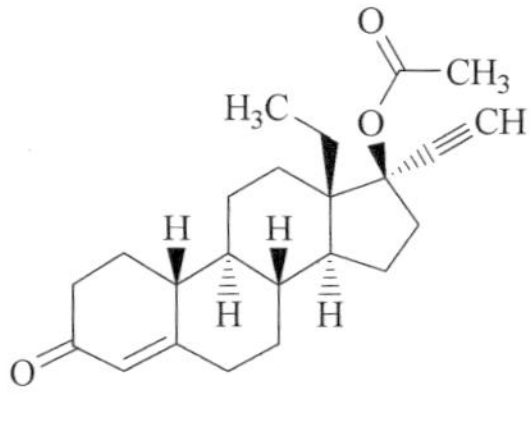

Norgestrel (levonogrestrel)

The first clinical study of an oral contraceptive began with a pill which investigators believed contained only the progestogen norethynodrel. Early clinical data demonstrated that this preparation inhibited ovulation. In subsequent studies, utilizing newly synthesized batches of the progestogen, ovulation was inhibited, but significant intermenstrual vaginal bleeding also occurred. It was then discovered that the original batch of norethynodrel, which gave better menstrual cycle control, was contaminated with an estrogen, mestranol. This led to the subsequent discovery that addition of estrogen to pure progestogen not only controlled intermenstrual bleeding but also increased anti-ovulatory efficacy. This led to the controlled inclusion of estrogen in future oral contraceptives [29].

Most oral contraceptives today are combinations of a synthetic estrogen and a synthetic progestational agent (progestogen). Combination estrogen/progestogen contraceptives depend primarily upon inhibition of ovulation, and secondarily on the mucus and endometrial effects of the progestogen which provide ancillary contraceptive mechanisms by interfering with fertilization [25]. The role of the progestogen is mainly to ensure suppression of luteinising hormone (LH), and it is the principal agent responsible for suppression of ovulation mechanisms. The estrogenic component has a dual function; further contribution to the suppression of ovulation is achieved via estrogen-induced inhibition of follicle-stimulating hormone (FSH) secretion, and stabilization of the pseudo-decidualised endometrium is achieved by the estrogen, thereby preventing much of the undesirable and unpredictable breakthrough bleeding and spotting which characterizes progestogen-only oral contraceptives.

Oral contraceptives available today are actually second or third generation contraceptive products and are much safer than the original products. The original combination products, introduced into clinical use in the 1960's, contained relatively high doses of estrogen and progestogen, in order to ensure maximum contraceptive effectiveness in the absence of dose ranging studies. It must be remember that the development of oral contraceptives preceded the introduction of liberal abortion laws in many countries. The knowledge that a pregnancy could not be terminated if woman became pregnant in a clinical study prevented clinicians in those early studies from titrating the dose of hormone needed to prevent pregnancy down to its lowest effective dose. Epidemiological studies during the 1960s and 1970s indicated that use of these high dose oral contraceptive products at that time was associated with an increased risk of thromboembolic disease. These cardiovascular complications were associated with the relatively high doses of estrogen used in original oral contraceptives, especially in women who smoke. Those findings led to the development of new, low estrogen dose, combination oral contraceptives which are used today (see Table 1).

For example, early oral contraception formulations contained up to 100–150 µg of estrogen. By the early 1990s, the majority of oral contraceptives contained only 30 or 35 µg of estrogen. These low-dose products are commonly referred to as second-generation oral contraceptives, i.e., new products which have almost totally replaced the original high-estrogen dose oral contraceptives. Numerous review articles and reports on cohort studies describing the use of oral contraceptives and the incidence of side effects are available [30]–[44].

As pharmaceutical companies continued to conduct studies to find the lowest effective doses of estrogen and progestogen effective as contraceptives and to market those products, women utilizing the products began to experience increased levels of intermenstrual bleeding. This observation led to the development of multiphasic oral contraceptives, a new approach to low-dose contraception.

Table 1. Representative low-dose monophasic oral contraceptives

Trade name	Progestogen/dose, mg/d	Estrogen* dose, µg/d	Manufacturer
Ortho-Novum 1/35	norethindrone/1.0	35	Ortho McNeil/Janssen Cilag
Ortho Cyclen/Cilest	norgestimate/0.25	35	Ortho McNeil/Janssen Cilag
Desogen	desogestrel/0.15	30	Organon
Brevicon	norethindrone/0.5	35	Syntex
Lo-Ovral	norgestrel/0.3	30	Wyeth
Nordette	levonorgestrel/0.15	30	Wyeth
Demulen 1/35	ethynodiol diacetate/1.0	35	Searl
Loestrin 1/20	norethindrone acetate/1.0	20	Parke-Davis
Loestrin 1.5/30	norethindrone acetate/1.5	30	Parke-Davis
Microgynon 30	levonorgestrel/0.15	30	Schering
Marvelon	desogestrel/0.15	30	Organon

* Ethinyl estradiol in all cases.

Multiphasic oral contraceptives vary the dose of active ingredients or the ratio of progestogen to estrogen throughout the cycle, instead of remaining constant dose as in conventional fixed combination oral contraceptives. Attempts are made to utilize the lowest effective dose of active ingredients, and by varying the ratio, attempt to better control intermenstrual bleeding. Some, but not all, of the low-dose multiphasic oral regimens studied, significantly reduce intermenstrual bleeding and spotting [45]. The success of this approach was seen in the marketplace when the first triphasic oral contraceptive was introduced in the United States (Johnson & Johnson's Ortho-Novum 7/7/7) in 1984. It rapidly became one of the most commonly prescribed oral contraceptives.

The next advance in the development of oral contraceptives was the discovery of new, pharmacologically more selective progestogens. The goal of synthesizing these more selective progestogens was to eliminate undesirable pharmacological activity but retain the needed progestional/antiovulatory activity. Studies published in the 1970s and 1980s demonstrated that androgenicity, associated with the progestional component of some of the original progestogens, is associated with increased weight gain, an increased incidence of acne, (affecting compliance) and changes in lipid metabolism [46]–[48]. These latter changes may impact cardiovascular morbidity. Hence, oral contraceptives that do not disturb the various blood lipid fractions and do not lower HDL may be preferable to the ones that shift the lipid profile in an undesirable direction [49], [50]. Three companies, Johnson & Johnson's Ortho-McNeil division, Organon, and Schering AG, worked in this area, attempting to dissociate the unwanted androgenicity and desirable progestational activities of steroidal progestogens using medicinal chemistry approaches. Norgestimate, desogestrel, and gestodene emerged from this research [51]–[56]. As has been demonstrated in many therapeutic areas, relatively minor differences in molecular structure can be associated with major differences in biochemical activity and that was no less true with this research. These new progestins had unique characteristics [48]. In the case of norgestimate, clinical studies were subsequently done to demonstrate that this progestogen, in combination with ethinyl estradiol, actually improved acne as well as providing contraceptive efficacy. This product, Ortho Tricyclen, rapidly became the leading oral contraceptive sold in the United States. In summary, the more selective progestogens, in combination with ethinyl estradiol, compose the third generation oral contraceptives (Table 2).

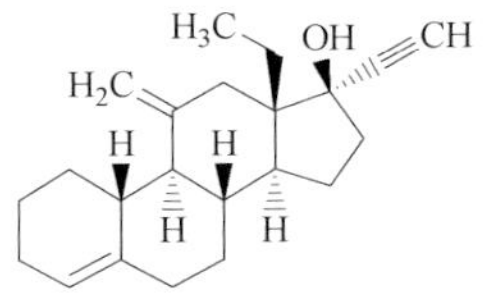

Desogestrel

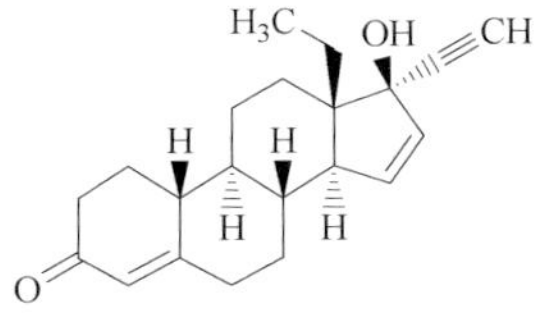

Gestodene

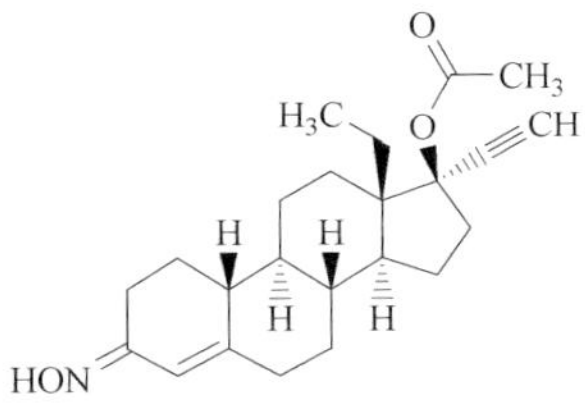

Norgestimate

Synthetic progestogens used in today's oral contraceptives belong to one of three classes. Estranes are 19-nortestosterone derivatives; gonanes are 19-nortestosterone derivatives with a C-13 ethyl group; and pregnanes are 17-α-OH progesterone derivatives similar in structure to progesterone itself.

Only two synthetic estrogens have been used in oral contraceptives to date, ethinyl estradiol [57-53-6], 19-nor-17-pregna-1,3,5(10)-trien-20-yne-3,17-diol, $C_{20}H_{24}O_2$, M_r 269.39, mp (berinhydrate) 141–146 °C, and mestranol [72-33-3], 3-methoxy-19-nor-pregna-1,3,5(10)-trien-20-yn-17-ol, $C_{21}H_{26}O_2$, M_r 310.42, mp 150–151 °C. Ethinyl estradiol differs from naturally occurring estradiol-17β by having an ethinyl group attached to C-17, whereas mestranol differs from ethinyl estradiol by methylation of the hydroxyl group at C-3. Both of these steroids resemble the natural estrogens qualitatively in their actions on the reproductive tract, pituitary and hypothalamus. Mestranol probably exerts most of its pharmacological activity as ethinyl estradiol following demethylation at C-3.

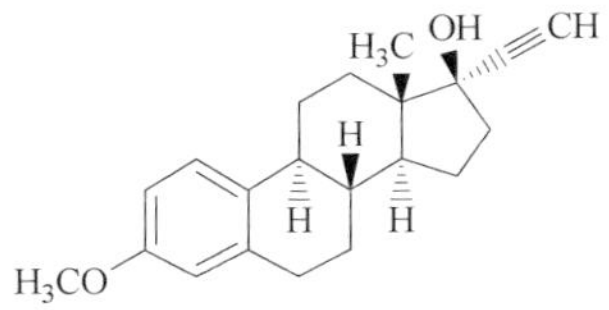

Mestranol

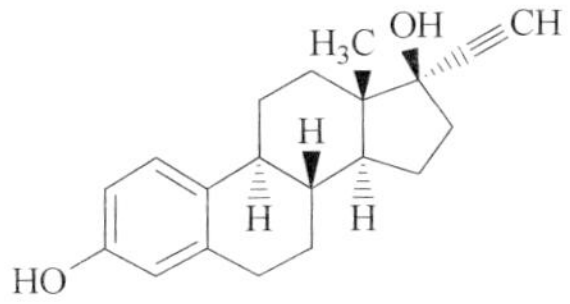

Ethinyl estradiol

Table 2. Representative multiphase combination oral contraceptives

Trade name	Active ingredients		Manufacturer
	Progestogen/dose*, mg/d	Estrogen**, dose, µg/d	
Ortho-Novum 7/7/7	norethindrone/0.5 (7)	35	Ortho McNeil/Janssen Cilag
	norethindrone/0.75 (7)	35	
	norethindrone/1.0 (7)	35	
Ortho Tri-Cyclen/TriCiilest	norgestimate/0.18	35	Ortho McNeil/Janssen Cilag
	norgestimate/0.215	35	
	norgestimate/0.250	35	
Ortho Novum 10/11	norethindrone/0.5 (10)	35	Ortho McNeil/Janssen Cilag
	norethindrone/1.0/(11)	35	
Jenest	norethindrone/0.50 (7)	35	Organon
	norethindrone/1.00 (14)	35	
Tri-Norinyl	norethindrone/0.50 (7)	35	Syntex
	norethindrone/1.0 (9)	35	
	norethindrone/0.5 (5)	35	
Triphasil	levonorgestrel/0.05 (6)	30	Wyeth
	levonorgestrel/0.075	40	
	levonorgestrel/0.125 (10)	30	
Triquilar	levonorgestrel/0.05 (6)	30	Schering
	levonorgestrel/0.075 (5)	40	
	levonorgetrel 0.125 (10)	30	
Tristep	levonorgestrel/0.05 (6)	30	Asche
	levonorgestrel/0.05 (5)	50	
	levonorgestrel/0.125 (10)	40	

* Number of medication days in each phase in parenthesis. ** Ethinyl estradiol in all cases.

Today we know that estrogen receptors in various target tissues differ and that a synthetic non-steroidal estrogen analog can have tissue specific agonist activity, antagonist activity or mixed pharmacological activity. Like that which happened research on progestogens, this discovery opens the possibility of finding a new generation of more selective estrogens for use in contraceptives. As a result, there is an increasing effort within the pharmaceutical industry and in universities directed toward the search for tissue selective estrogens which retain the desired pharmacological activity but minimize any potential side-effects. These efforts are enhanced by our increased knowledge of the pharmacological mechanisms that underlie the selective action of estrogens [57], the recently cloned new estrogen receptor ERβ [58], and tissue distribution of estrogen receptors α and β [59]. Scientists look forward to the discovery of selective estrogens which can be used in contraceptives.

Since oral contraceptives were introduced in the 1960s, knowledge on the benefits and risks of oral contraceptives has increased. First, numerous studies have shown that oral contraception is the most effective and reversible means of preventing pregnancy [60]. Secondly, however, it is generally agreed that the use of oral contraceptives can slightly increase the risk of venous thromboembolism, ischemic heart disease, cerebrovascular disease and hypertension [61]. The thrombolic risk appears to be concentrated among older women who smoke. However, several recently published articles have reported that users of certain oral contraceptives, those containing gestodene and desogestrel, may be at increased risk of venous thromboembolism (VTE) [39]–[44]. This increased risk was not observed for third-generation pills containing the progestin norgestimate [62], [63]. Regulatory reaction has varied from advising patients to restrict use of desogestrel or gestodene containing products to making changes in the products' labeling [63]. Thirdly, many studies have shown that there are other, beneficial effects that women appear to obtain from oral contraceptive use in addition to protection against pregnancy. They include protection against pelvic inflammatory disease, ectopic pregnancy, endometrial cancer, cancer of the ovaries, benign breast disease and benign ovarian cysts [61]. Relief from a wide range of menstrual disorders is also a known beneficial effect of oral contraceptive use [61].

4.2. Progestogen-Only Oral Contraceptives

Sometimes termed the mini-pill, progestogen-only oral contraceptives are available widely throughout the world but are not as extensively used as combination oral contraceptives. These preparations contain progestational 19-norsteroids such as norethindrone, lynoestrenol, ethynodiol diacetate, or norgestrel; or 17-β-acetoxy progestogens such as chlormadinone acetate and megestrol acetate. Progestogen-only methods, including these mini-pills, exhibit variable rates of gonadotrophin and ovulation suppression, and rely instead more upon progestational changes to cervical mucus and endometrial histology for effectiveness. In general, the contraceptive effectiveness of these products is not as high as that of combination oral contraceptives, and intermenstrual or breakthrough bleeding and spotting occur more frequently with these products. However, these products are valuable for women who are intolerant of estrogens, and there are fewer effects on breast feeding.

4.3. Long-Acting Contraceptives

The establishment of fertility regulation with hormonally active oral contraceptives led to the development of long-acting agents which could be delivered parenterally. Long-acting contraceptives avoid compliance issues, and are useful in countries with fewer health professionals. Their popularity is based on simplicity of administration and a relatively high degree of effectiveness. Efforts to develop long-acting contraceptives have included both chemical modifications of known progestogens for injectable administration and the use of new drug-delivery systems [64].

Two well-known injectable long-acting products are medroxyprogesterone acetate (Depo Provera) and norethindrone enanthate [3836-23-5] (NET EN). Both of these products are progestational in nature. Depo Provera (Upjohn Pharmacia Company) is a once-every-three-months injectable administered as a microcrystalline aqueous suspension of the steroid. NET EN (Schering AG) is a two-month injectable administered as solution of steroid in caster oil. Depo Provera is approved for use as a contraceptive in more than ninety countries. NET EN is used in more than forty countries. Other products developed by chemically modifying known progestogens include 17α-hydroxyprogesterone caproate [630-56-8], $C_{27}H_{40}O_4$, and dihydroxyprogesterone acetophenide [24356-94-3], $C_{29}H_{36}O_4$. These monthly injectables are widely used in Latin America.

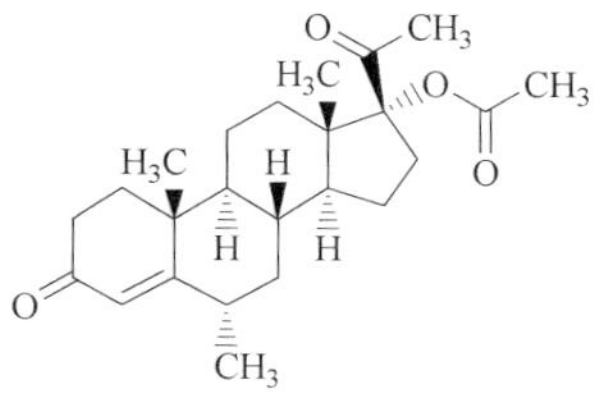

Medroxyprogesterone acetate

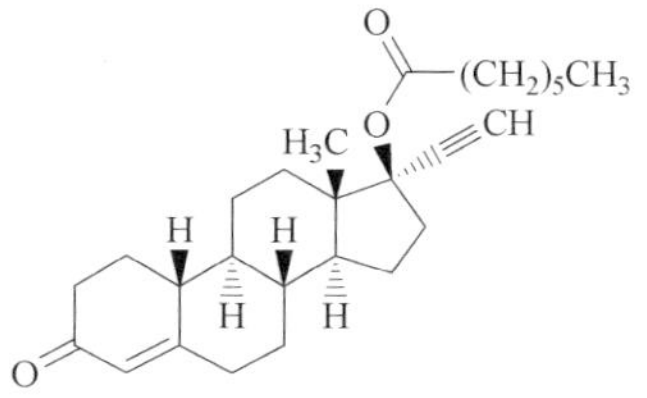

Norethindrone enanthate

Like progestogen-only oral contraceptives, parenteral long-acting progestins act primarily as ovulation inhibitors. An important second mechanism of action is their effect on the cervical mucus and endometrium, achieved at circulating blood drug levels below those required for ovulation inhibition [65].

The major disadvantage of these long-acting injectable contraceptives is the side effect of irregular menses or amenorrhea. In an attempt to reduce this, long-acting injectable contraceptives containing both an estrogen and a progestogen are being studied, including dihydroxprogesterone acetophenide, norethisterone enanthate, medroxyprogesterone acetate and norethisterone in combination with estradiol esters such as their valerate, enanthate, or cypionate esters.

Drug-Delivery Systems. *Drug Releasing Rods.* Recent research efforts have been directed at the use of new drug-delivery systems that release drug slowly over a prolonged period of time, such as silastic capsules, rods, or vaginal rings, and biodegradable implantable systems [66]. The best known contraceptive implant is Norplant, developed by The Population Council in the United States. Norplant is composed of six matchlike silastic cylinders with the progestogen levonorgestrel incorporate inside each cylinder. After implantation under the skin, the product can provide effective contraception for a period of five years [67], [68]. Since silastic is not biodegradable, the implant can be removed. The main reason patients request the removal of Norplant is unpredictable vaginal bleeding episodes followed by amenorrhea. The bleeding problem is an unavoidable sequel of progestogen-only contraception.

Biogradable Implants. Utilization of implantable biodegradable polymers obviates the need for implant removal. Technically, however, it is not as easy to develop such a product as it may appear. For example, biodegradation should not take place before the drug release is essentially finished; before that, structural integrity permitting surgical removal of the implant must be maintained. In addition, release rates for the drug must be sufficient to maintain blood levels of drug which will insure contraceptive effectiveness.

An example of a biodegradable contraceptive implant is Capronor. This device has walls composed of σ-caprolactone; it releases levonorgestrel and is a single implant with projected life span of 12–18 months [66]. The timecourse for biodegradation and the disappearance of the implant are still being defined.

Transdermal Contraceptives. Transdermal administration of contraceptive steroids offers a number of potential advantages over oral administration. It can provide continuous absorption, thus reducing the large swings between "peak" and "trough" concentrations (C_{max} and C_{min}). Like other non-oral delivery modes, it also avoids hepatic first-pass metabolism, which may reduce exposure to unwanted metabolites and can also permit lower delivered doses. Finally, a transdermal contraceptive can be worn for several days, providing convenience for many women and increasing compliance — and yet be easily removed should that become medically necessary.

Transdermal drug delivery systems have been used for administration of a variety of drugs in a number of conditions, but until now, have not been used for delivery of contraceptive steroids. Johnson & Johnson is currently developing a transdermal (patch) system containing two contraceptive steroids, ethinyl estradiol and 17-deacetylnorgestimate — the primary active metabolite of norgestimate. It is expected that the product will be available soon.

Fused pellets. Another form of an implant is a fused pellet. These pellets may be composed of either the drug alone or the drug fused with cholesterol [53], [69], and are formed as small cylinders by melting the drug and then solidifying it under pressure. Clinical studies with norethindrone pellets have been in progress for a number of years. Polymeric implants and pellets require minor surgery for both their insertion and removal.

Microencapsulated Steroids. Microencapsulation of steroids to date have utilized the homopolymers polylactic acid (PLA), polyglycolic acid (PGA), and their copolymers (PLGA). The ratio of PLA to PGA determines the rate of steroid release. These thermoplastic biodegradable forms have been constructed as films, micropellets, microspheres, rods, and fibers [70].

Vaginal Rings. Vaginal epithelium is readily permeable to contraceptive steroids. Since the vascular drainage of the vagina bypasses the liver, this route of administration potentially permits utilization of drugs that have low oral activity due to degradation in the liver.

Contraceptive vaginal rings consist of silastic shells or core rings of various sizes and membrane thicknesses. They have been developed for delivery of progestins alone or progestins combined with estrogens. The biology behind progestin-alone rings is conceptually similar to that of implants, but the use of vaginal rings is under direct control of the patient since they can be inserted or removed at any time without a trained professional being present. After the ring is removed, a predictable withdrawal bleeding takes place. Vaginal rings have been under development for nearly 30 years by the World Health Organization and the Population Council [71], [72]. Progress is being made. Materials have been selected which allow release of the active ingredient. Large clinical studies must still be done, but we know today that higher doses and/or release rates are required for heavier women than for lower weight women. Acceptability of this route of contraceptive drug administration must still be determined.

5. Intrauterine Devices (IUDs)

Intrauterine devices are medical products that prevent conception when placed in the uterus. In spite of their ancient origins, modern intrauterine devices (IUDs) have been widely used only in the last 30 years. The two generic subclasses of IUDs are nonmedicated (inert) devices and medicated IUDs, i.e., progestin-releasing or copper IUDs.

IUDs are used throughout the world, with an estimated 79 million users in 1989 [73]. They are a highly effective contraceptive method with protection rates for some devices reported to be 94–99 per 100 women during one year of exposure to pregnancy. Excellent reviews have been written on the efficacy and safety of this contraceptive method [74].

Complications associated with IUDs include uterine perforation and pelvic inflammatory disease [74]. Uterine bleeding and cramping are the most common causes for discontinuation of this method. The IUD's relationship to pelvic infection, fueled by the high rate of septic abortion and pelvic inflammatory disease (PID) among users of the Dalkon Shield IUD, led to a decline in the popularity of IUDs. For example, all IUDs, except the progesterone-releasing Progestasert, were withdrawn from the U.S. market in the mid-1980s. However, with time, interest in IUDs in the USA grew again and, in 1988, the copper T-380A (ParaGard) was introduced into the USA.

Inert Devices. Inert IUDs act by creating an environment hostile to sperm or fertilized ova and by blocking implantation [75], [76]. The exact mechanism of action of IUDs is not totally clear, but convincing evidence is mounting to support the idea that IUDs act primarily as contraceptives and not as abortifacients [77]. Compared with noncontraceptors, IUD users have fewer recoverable sperm in the uteri and tubes after intercourse. In addition, there are fewer recoverable ova in the uteri and tubes at mid-cycle. Those ova found are rarely fertilized. In nonhuman primates, the rate of recovery of degenerating embryos is not significantly different from that seen among controls, in contrast to what might be expected if the IUD works by preventing embryo implantation. Transient elevations of human chorionic gonadotropin (hCG), which may indicate early pregnancy, were not more frequent among IUD users than among noncontraceptors [77].

After insertion of an IUD, polymorphonuclear leukocytes and macrophages accumulate in the uterine cavity. These cells appear to phagocytize sperm and liberate a blastotoxic toxin [78], [79]. Intrauterine devices also may create a hostile environment, perhaps because antibodies are produced that interfere with implantation of the fertilized ovum [79].

Medicated Devices. Medicated IUDs consist of an inert base reservoir for a uterus-affecting or spermicidal agent such as progesterone or copper. Copper-bearing IUDs have spermicidal activity and interfere with implantation [80]. Two of these, the TCu-200 and the Multiload-250 (MLCu-250), are widely available except in China and the USA. The inclusion of progestogens in IUDs does not appear to improve their efficacy but may reduce menstrual cramping and bleeding associated with IUD usage [81].

The second generation of copper IUDs have more copper wire, copper sleeves, and/or silver core in the copper wire, denoted by Ag in the IUD name. Significant second-generation IUDs include the TCu-380A, TCu-220C, Nova T, and Multiload-375 (MLCu-375); these are available worldwide except China.

Progestasert is a hormone-releasing IUD containing 38 mg progesterone, released at a rate of 65 µg/d for one year. LNG-20, a long-lasting levonorgestrel-releasing IUD, is still under development in Europe.

6. Sterilization

In the last couple of decades, voluntary surgical sterilization of both men and women increased in popularity. Sterilization appears equally popular with the oral contraceptive, each chosen by about one in four couples.

Worldwide, an estimated 42 million couples rely on vasectomy; nearly 140 million rely on female sterilization [73]. Vasectomy is a principal family planning method in only six developed countries, i.e., the USA, New Zealand, Australia, UK, Canada, and the Netherlands; and in three developing countries, i.e., China, India, and South Korea. The method is hardly used in other countries and few people have heard of vasectomy compared with other methods [82].

Voluntary female sterilization is the world's most widely used family planning method. An estimated 138 million women of reproductive age used the method in 1990, 43 million more than in 1984. Millions more were expected to ask for the method during the 1990s [83].

Because of the increasing worldwide interest and demand for simple, effective and inexpensive female sterilization, a variety of procedures and methods have been developed. These approaches differ whether they are performed postpartum, postabortum, or in interval situations. The choice of methods also largely depends upon the physician's prior training, knowledge, and experience. Excellent reviews have been written on sterilization [84].

7. Contragestational Drugs

Pharmacological substances that either inhibit implantation or interrupt pregnancy after implantation have been investigated during the 1980s and 1990s. A number of different terms have been used to describe these compounds [47]–[63], including antiimplantive agents, postcoital contraceptives, morning-after pills, emergency contraceptives, once-a-week pills, interceptives, abortifacients, and contragestational agents. This medical approach to fertility regulation presents several principal advantages, including potentially fewer long-term side effects as a result of short-term periodic administration and greater convenience.

However, in the decision to develop and market contragestational drugs, social, political, legal, ethical, and religious factors have played a significant role.

Post-Coital Contraception. Post-coital contraception historically has been viewed as an emergency measure where regular contraceptive were not used or where the primary contraceptive methods may have failed. A number of different approaches have been utilized in post-coital regimens. Early regimens utilized high doses of the estrogens [84] diethylstilbestrol, ethinyl estradiol, and the conjugated equine estrogens (Premarin). Characteristically, the drugs are taken for a period of three days. When given later than 72 h after coitus, the effectiveness is reduced; the drugs are ineffective if implantation has been established. Although the estrogens are highly effective, users suffer from a high incidence of nausea and vomiting; cycle regularity also is disturbed. A similar approach has been reported for levonorgestrel-containing contraceptives [86], [87].

The U.S. FDA recently approved a specific regimen of combination oral contraceptives containing levonorgestrel and ethinyl estradiol for post-coital pregnancy prevention for women who have had unprotected sexual intercourse. The regimen consisting of two oral contraceptive tablets taken within 72 h of intercourse and two pills taken 11 h later. When used in this manner, the treatment is reportedly about 75 %

effective in preventing/terminating pregnancy. The treatment does not work if a woman is already pregnant.

It has been demonstrated that Danazol is also highly effective in post-coital regimens, and that it has a very low incidence of side effects. Danazol is marketed in many countries for the treatment of endometriosis and its availability is unrestricted. It does not appear that there is a significant difference in effectiveness when doses of 800 to 1200 mg have been utilized for three days [86], [88]. When utilized in the post-coital mode, the precise mechanism of action of Danazol is not well-understood.

A nonsteroidal weekly pill, centchroman [*31477-60-8*] [16], was developed by the Central Drug Research Institute, Lucknow, India, and is being marketed as Choice-7 and Sahali (Hindustan Latex). Centchroman, (3,4-*trans*-2,2-dimethyl-3-phenyl-4-[*p*-β-pyrrolidino-ethoxy)-phenyl]-7-methoxychromane) inhibits implantation of the fertilized egg, thus avoiding pregnancy [89]. It exerts its antifertility effect via weak estrogenic and potent anti-estrogenic activity [90]. Although the synthesis and pharmacological actions of centchroman have been well documented [91], overall efficacy and side effects are not known or described except in a brief description of the product. It is reported that the Pearl index has been calculated to be 3.05 when weekly doses of centchroman (30 mg) were adminstered to approximately 1600 women for a total of 20 000 months [89].

Centchroman

Abortifacients. It is estimated that between 30 and 40 million legal abortions and the same number of illegal abortions are carried out each year worldwide. Most abortions are induced because the pregnancy is unintended and unwanted. In the USA, about 60 % of all pregnancies are unwanted or mistimed, half of which are aborted [92].

In the mammal, removal of the corpus luteum during early pregnancy results in termination of pregnancy. Pharmacologically, this can be accomplished by various methods, including inhibition of gonadotropin support of the corpus luteum, inhibition of progesterone biosynthesis in the ovary, inhibition of progesterone binding to the uterine progestin receptor, and increase in the metabolism of progesterone. Many compounds that affect these processes have been reported [93]; some of these compounds have been introduced for clinical study.

Chemically induced abortion originally involved the administration of the natural prostaglandin $F_{2\alpha}$ [*551-11-1*] (Dinoprost, $C_{20}H_{34}O_5$) and synthetic analogues such as sulprostone [*60325-46-4*], $C_{23}H_{31}NO_7S$, and prostaglandin ONO 802, [*64318-79-2*]

(Gemeprost, $C_{23}H_{38}O_5$) have also been reported [94]. After administration of relatively small amounts of these prostaglandins, the muscular tone of the uterus increases, followed by contractions, cervical dilation and expulsion of the uterine contents. The usefulness of prostaglandins for termination of pregnancy is limited because of a high frequency of gastrointestinal side effects.

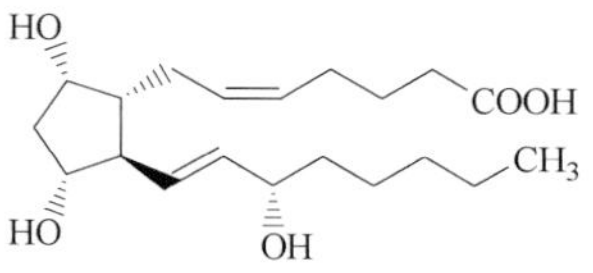

Prostaglandin $F_{2\alpha}$

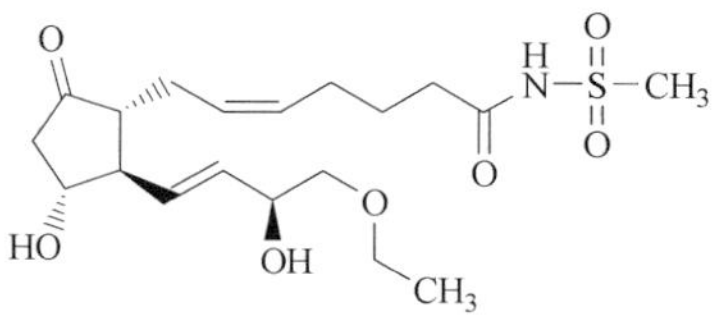

Sulprostone

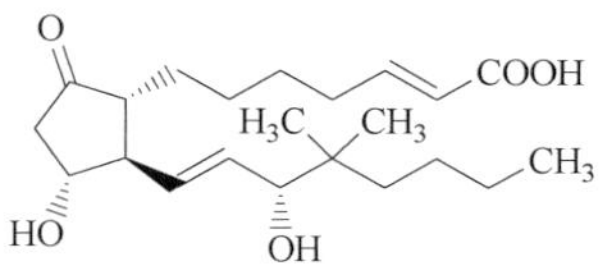

Prostaglandin ONO 802

In 1988, the French government approved the marketing of an abortion pill. RU486 [*84371-65-3*] (Mifepristone), an antiprogestin developed by the French pharmaceutical company Roussel-UCLAF. It was the first clinically useful progesterone antagonist. A review of the chemistry, pharmacology, and clinical applications of this compound [95] is available, as is a review on the use of RU486 alone or in combination with a prostaglandin analogue for termination of early pregnancy [96]. Other studies suggest the use of antiprogestins for contraception and for treatment of gynecological disorders related to hormone production [97]. The discovery of RU486 was followed by laboratory and clinical studies with other antiprogestins, including ZK98 299 and ZK98 734 [98]. Many issues, not only scientific, but also political and religious, surround the clinical application of progesterone antagonists, and it is difficult to project worldwide availability of RU486 and related products.

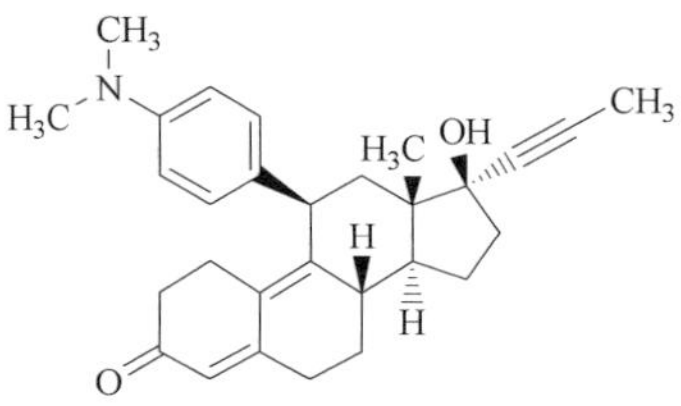

RU468
(Mifepristone)

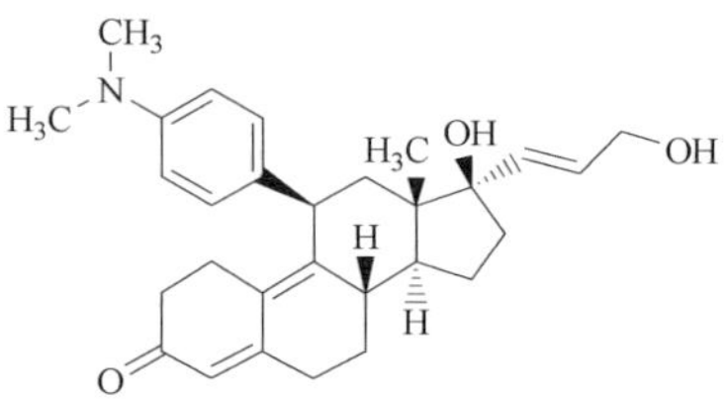

ZK98 734
(Lifopristone)

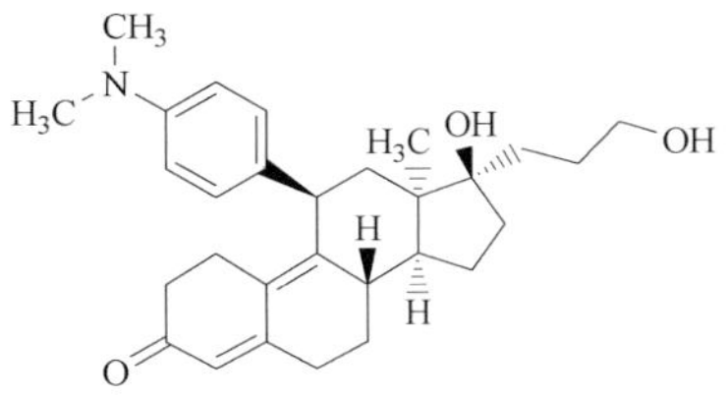

ZK98 299
(Onapristone)

8. New Approaches

Because there are no methods of fertility regulation available today that are totally safe and effective or that appeal to all groups and societies in the world, there continues to be great interest in developing new and improved contraceptives. This has been a difficult task, because to be successful, new contraceptives must be superior to existing products. For example, oral contraceptives, used by millions of women over the last 30 years, are not only safe and effective but even protect women against some cancers. Because oral contraceptives are so effective, they become a very high standard that new products must meet and exceed.

Contraceptive Vaccines. Major research efforts today involve immunological approaches to fertility control; excellent reviews of this area are available [99]–[102]. The development of contraceptive vaccines is directed towards immunoneutralization of reproductive process or the interference of fertilization by inducing antibodies against oocytes and spermatozoa. Attempts have been made to develop vaccines against luteinizing hormone releasing hormone (LHRH) (also known as gonodotropin releasing hormone, or GnRH), LH, follicle stimulating hormone (FSH), human chorionic gonadotropin (hCG), placenta antigen, the zona pellucida of the ovum, and different sperm antigens.

Research on an hCG vaccine has been conducted over the past 20 years. WHO has conducted a Phase I clinical study in Australia, using a vaccine based on a synthetic C-terminal peptide (109-141) of β-hCG conjugated to Diptheria Toxoid (CTP-DT). That study demonstrated that potentially effective contraceptive levels of antibodies were produced in vaccinated women without any adverse side effects. Phase II clinical studies are under consideration to determine if the immune response, raised to its prototype anti-hCH vaccine is capable of preventing pregnancy in fertile women volunteers [103]. While research on the C-terminal peptide from the β-subunit of hCG has been carried out under the auspices of WHO, research supported by the Population Council and the U.S. National Institutes of Health has involved two alternative vaccine candidates [101], [104].

Using recent advanced technologies, unique sperm antigens have been identified and partially characterized. Sperm antigens shown to have high immunocontraceptive potential are human sperm membrane antigen (SP-10) and guinea pig sperm membrane protein (PH-20). SP-10 is a sperm membrane-specific antigen of 24–34 kD, isolated by using a monoclonal antibody (MHS-10) that cross-reacts with the entire acrosomal region. It is associated with the outer aspect of the inner acrosomal membrane and the inner aspect of the outer acrosomal membrane of mature human sperm [104]. It has been produced recombinantly in an *E. coli* expression system. The recombinant SP-10 fusion protein is under study in the baboon.

PH-20, a guinea pig sperm protein of 64 kD, is present on both the plasma membrane and inner acrosomal membrane of sperm. It is essential for adhesion of sperm to the zona pellucida, the initial step in the fertilization process. Active immunization with PH-20 causes infertility in both male and female guinea pigs for a period ranging from six to fifteen months [105].

Another interesting sperm specific antigen is lactic dehydrogenase-x (LDH-x or $LDHC_4$), an isoenzyme of LDH confined to male germ cells. LDH-x is one of the best characterized antigens and its amino acid sequence is known. A synthetic peptide based on a portion of the molecule has been shown to reduce fertility in laboratory animals. The nucleotide sequence coding for human LDH-x has been defined and engineered into an expression vector system [100].

The zona pellucida (ZP) is the complex extracellular glycoprotein matrix that surrounds the oocyte. It plays an important role in sperm penetration and fertilization. It is a composite of several antigenic glycoproteins designated ZPI, ZPII, ZPIII, and, in some species, ZPIV. Immunologically, it does not cross-react with any other body tissues, but interspecies cross-reactivity has been observed among several species, including primates. Numerous studies indicate the immunocontraceptive potential of the zona pellucida [100]. Several other antigens with good immunocontraceptive potential have been identified and investigated in laboratory animals.

In most immunocontraceptive studies to date, however, the rate and duration of immunocontraceptive effects have been less than acceptable. Two problems which are faced in immunological studies to antifertility research are the need for a safe, effective adjuvant and suitable animal models for evaluating the efficacy and safety of the adjuvants [99]. Newer and more effective adjuvants are required for contraceptive vaccines and vaccines in general.

Luteinizing Hormone Releasing Hormone. The isolation and synthesis of luteinizing hormone releasing hormone (LHRH) was an important advance in reproductive research [122]. LHRH is a peptide hormone produced and secreted by the hypothalamus that stimulates the secretion of FSH and LH. A decapeptide with an amino acid sequence of pyroGlu-His-Trp-Ser-Tyr-Gly-Leu-Arg-Pro-Gly-NH_2, it is chemically and functionally similar in both males and females of all mammalian species studied thus far. More than 1000 analogues of LHRH have been synthesized by deletion or substitution of amino acids to introduce agonist or antagonist properties. A large number of reviews have appeared discussing their therapeutic importance [106]–[108].

These agonists and antagonists may provide useful therapy for several clinical conditions such as prostatic carcinoma, precocious puberty, and endometriosis. Scientists are also studying LHRH analogues for contraception [109]. Treatment with LHRH analogues blocks ovulation in women and spermatogenesis in men. However, treatment also results in a loss of estrogen and testosterone and causes other related side effects. Scientists are attempting to assess the use of these agents in combination with replacement estrogen, progesterone, and testosterone to determine if the side effects can be avoided. Several small nonhuman primate and clinical studies have suggested possible utility in this area, but large-scale clinical efficacy studies have not been conducted.

Progesterone Antagonists as Contraceptives. Another area of antifertility research involves progesterone antagonists or inhibitors. The use of one progesterone antagonist, RU-486, for inducing abortion was discussed under the section on abortifacients. There are others as well. Progesterone is required to maintain pregnancy in women, and infertility results from failure of the corpus luteum to produce adequate amounts of progesterone. Inhibitors of progesterone synthesis, such as epostane [110], and inhibitors of progesterone-receptor binding, such as RU486, have been investigated for termination of pregnancy. Studies in the nonhuman primate and nonpregnant women indicate that progesterone antagonists may also have antifertility potential other than as an abortifacient [88], [111].

Inhibin and Activin. Inhibin, a water-soluble, gonadal factor known for over 50 years to inhibit pituitary function, has been isolated and identified [112]–[115]. Inhibin is a glycoprotein hormone that preferentially inhibits the secretion of FSH. It consists of an α-chain subunit, M_r 14 000, linked by disulfide bonds to a β-chain subunit, M_r 18 000. There exist two forms of the β-chain subunit, β-A and β-B. The smaller subunit combines with either the β-A or β-B subunit to form inhibin-A or inhibin-B, respectively.

During the isolation of inhibin from follicular fluid, it was discussed that some chromatographic fractions stimulated FSH release from cultured anterior pituitary cells, suggesting the existence of FSH releasing proteins (FRPs). Two FRPs, given the generic term activins, were subsequently isolated [116], [117]. One is composed of two disulfide-linked β-A subunits (activin A); the other consists of similarly linked β-A and β-B subunits (activin AB).

Studies confirm that inhibin plays a role in regulating FSH secretion. However, the importance of this role in the human has not yet been determined. If inhibin-regulated FSH secretion is pivotal in follicular recruitment and growth, then it may be possible to block ovulation by means of inhibin antagonists.

Activin has the potential to serve the same therapeutic uses as GnRh analogues. This is because of its ability to regulate steroidogenesis directly at the gonadal level. Although the spectrum activities of inhibin and activin are not completely understood at present, this peptide family has already demonstrated, by the nature of its differential subunit association, a powerful mechanism for the generation of dimers with opposing biologic actions. These characteristics of the inhibin peptide family warrant further study and evaluation as alternative approaches to fertility control.

Male Fertility Control. There is interest in male fertility control, both from a scientific as well as a sociological viewpoint. Many compounds have been identified as having male antifertility activity in various species, e.g., gossypol [*303-45-7*], ORF 5513, 5-thio-D-glucose [*20408-97-3*], and 6-chlorodeoxyglucose [118]. A principal program centering around the use of androgens has also been conducted [119].

Organic molecules thus far identified, such as those listed above, appear either to have irreversible antifertility effects, to be inherently toxic, or to affect libido. With respect to androgens, it has been demonstrated that sperm count can be depressed but not eliminated in men injected with large doses of androgens. However, questions about the potential utility of androgens as male antifertility agents are still debated, not the least of which is whether lowering, but not eliminating sperm, will provide effective male contraception.

The ideal male contraceptive would produce azoospermia without compromising libido and sexual potency. While not totally fulfilling the criteria for a perfect male contraceptive, GnRH antagonists appear to hold a greater potential than GnRH agonists. Unlike the agonists, GnRH antagonists inhibit gonodotropin secretion, decrease androgen levels, and induce azoospermia in male primates [120], [121]. Similar effects on hormone secretion have been reported in men [115].

In monkeys, testosterone replacement delays, but does not prevent, GnRH antagonist-induced azoospermia [122]. In men, the combination of testosterone and a GnRH antagonist results in a more complete gonadotropin and gonadal suppression than either agent alone [123]. These results suggest that GnRH antagonists, given in conjunction with androgens to maintain libido and sexual potency, have potential as male contraceptives. More recently the WHO and other research groups are actively pursuing the use of androgen plus progestogens as male contraceptives. Excellent reviews of the area of male contraceptive research are available [124], [125].

9. References

[1] R. O. Greep, M. A. Koblinsky, F. S. Jaffe: *Reproduction and Human Welfare*, MIT Press Cambridge, Mass. 1976 p. 53.

[2] C.-R. Garcia, D. L. Rosenfeld: *Fertility: The Regulation of Reproduction*, FA Davis Company, Philadelphia 1977, p. 59.

[3] L. Mastroiani, Jr., P. J. Donaldson, T. T. Kane: *Developing New Contraceptives*, National Academic Press, Washington, D.C. 1990, p. 11.

[4] P. Kestelman, R. L. Kleinman, *International Planned Parenthood Federation (IPPF) Directory of Contraceptives*. IPPF, London 1981.

[5] R. L. Kleinman: *Periodic Abstinence for Family Planning*, IPPF, London, 1983.

[6] H. Knaus, *Zentralbl. Gynäkol.* **53** (1929) 2193.

[7]   K. Ogino, *Zentralbl. Gynäkol.* **54** (1930) 464.
[8]   T. H. van de Velde: *Die Vollkommene Ehe, eine Studie über ihre Physiologie und Technik,* 21st ed., Benno Konegen, Medizinischer Verlag, Leipzig-Stuttgart 1928.
[9]   P. Miller, M. Soules, *Obstetrics and Gynecology,* **87** (1996) 1 p. 13.
[10] J. Bonnar, in D. R. Mishell, Jr. (ed.): *Advances in Fertility Research,* vol. **1,** Raven Press, New York 1982, p. 1.
[11] K. L. Kennedy, R. Rivera, A. S. McNeilly, *Contraception,* **39** (1989) 477.
[12] J. L. McGuire, F. C. Greenslade, in G. I. Zatuchni, A. J. Sobrero, J. J. Speidel, J. J. Sciarra (eds.): *Vaginal Contraception,* Harper and Row, New York 1980.
[13] A. Kolliker: "Physiologische Studien über die Samenflüssigkeit," *Z. Wiss. Zool.* **7** (1855) 221.
[14] J. R. Baker, *J. Hyg.* **32** (1932) 550.
[15] F. V. Sander, S. D. Cramer, *Human Fertil.* **6** (1941) 134.
[16] U.S. Patent No. 2 752 284, 1956 (V. R. Berliner, W. C. Mende, H. O. Singher).
[17] S. Iwahara, K. Furuse, *Jpn. Public Health* **12** (1965) 123.
[18] "Recommendations for the Development of Vaginal Microbicides, International Working Group on Vaginal Microbicides," *AIDS* **10** (1966) 8, 1–6.
[19] D. W. Hahn, R. E. Homm, B. E. McKenzie, in G. I. Zatuchni, A. J. Sobrero, J. J. Speidel, J. J. Sciarra (eds.): *Vaginal Contraception,* Harper and Row, New York 1980, p. 232.
[20] E. B. Connell: "Vaginal Contraception" in D. R. Mishell, Jr. (ed.): *Advances in Fertility Research,* Raven Press, New York 1982, p. 19.
[21] The Contraceptive Diaphragm, Emory University, School of Medicine, Healthcare Communications Network, New York, 1989.
[22] A. Melanie, Do Gold, *Pediatric Ann.* **24** (1995) 4, 211.
[23] *Contraceptive Technology Update,* **11** (1990) 10, 145.
[24] Condoms, Population Reports, Series H, no. 8, Population Information Program, The Johns Hopkins University, Baltimore, 1990.
[25] R. A. Hatcher, et al.: *Contraceptive Technology,* Irvington Publishers, Inc., New York, 1990–1992, p. 227.
[26] The American College of Obstetricians and Gynecologists (ACOG) Technical Bulletin No. 198, 1994.
[27] L. Haberlandt, *Münch. Med. Wochenschr.* **68** (1921) 1577.
[28] C. Djerassi, L. Miramontes, G. Rosenkranz, F. Sondheimer, *J. Am. Chem. Soc.* **76** (1954) 4092.
[29] C. R. Kay, *J. Royal Col. Gen. Pract.* **30** (1980) 8.
[30] W. H. Inman, M. P. Vessey, B. Westerhom, A. England, *Br. Med. J.* **2** (1970) 203.
[31] Royal College of General Practitioners, Oral Contraceptives and Health: *An Interim Report from the Oral Contraception Study,* Pitman Medical, New York 1974.
[32] M. P. Vessey et al., *J. Biosoc. Sci.* **8** (1976) 373.
[33] A. Rosenfield, *Am. J. Obstet. Gynecol.* **132** (1978) 92.
[34] H. W. Ory, A. Rosenfield, L. C. Landman, *Fam. Plann. Perspect.* **12** (1980) 278.
[35] S. Ramcharan, et al.: *The Walnut Creek Contraceptive Drug Study: A Prospective Study of the Side Effects of Oral Contraceptives,* Vol. **3** U.S. Dept. of Health and Human Services, Government Printing Office, Washington D.C. 1981.
[36] M. Notelovitz, C. S. Kitchens, L. Coone, L. McKenzie, R. Carter, *Am. J. Obstet. Gynecol.,* **141** (1981) 71–75.
[37] D. R. Mishell, Jr., *Am. J. Obstet. Gynecol.* **142** (1982) 809.
[38] "Oral Contraceptives", in *Popul. Rep. [A]* **6** (May-June 1982).

[39] "World Health Organization Collaborative Study of Cardiovascular Disease and Steroid Hormone Contraception," *Lancet* **346** (1995) 1575–1582.
[40] "World Health Organization Collaborative Study of Cardiovascular Disease and Steroid Hormone Contraception," *Lancet* **346** (1995) 1582–1588.
[41] K. W. M. Bloemenkamp et al., *Lancet* **346** (1995) 1593–1596.
[42] H. Jick et al., *Lancet* **346** (1995) 1589–1593.
[43] W. O. Spitzer et al., *BMJ* **312** (1996) 83–87.
[44] M. A. Lewis et al., *BMJ* **312** (1996) 88–90.
[45] S. A. Pasquale, *J. Reprod. Med.* **29** (Suppl.) (1984) 560.
[46] K. Fotherby, *Br. Family Planning* **11** (1985) 86.
[47] G. Silfverstolpe, A. Gustafson, G. Samsoie, A. Svanborg, *Acta Obstet. Gynecol. Scand. (Suppl)* **88** (1979) 89.
[48] B. R. Carr, *Int. J. Fertil.* **42** (1997) suppl. 1, 133.
[49] T. Gordon et al., *Am. J. Med.* **62** (1977) 707.
[50] I. F. Godsland, D. Crook, *Am. J. Obste. Gynec.* **170** (1994) 15–28.
[51] A. Phillips, D. W. Hahn, S. Klimek, J. L. McGuire, *Contraception,* **36** (1987) 181.
[52] A. Phillips, K. Demarest, D. W. Hahn, *Contraception,* **41** (1990) 399.
[53] H. J. Kloosterboer, C. A. Von Knoordegraaf, E. W. Turpijn, *Contraception,* **38** (1988) 325.
[54] D. W. Hahn, A. Phillips, J. L. McGuire, *Aktuelle Aspekte der Hormonalen Kontrazeption,* Karger, New York 1991, p. 46.
[55] S. L. Corson, *Acta Obstet. Gynecol. Scand.* **69** (1990) (suppl. 152) 25–31.
[56] R. S. London, et al., *Acta Obstet. Gynecol. Scand.* **71** (1992) (suppl. 156) 9–14.
[57] J. A. Katzenellenbogen, B. W. O'Malley, B. S. Katzenellenbogen, *Molecular Endocrinology* **10** (1996) no. 2, 119.
[58] G. J. M. Kuiper, E. Enmark, M. Pelto-Huikko, S. Nilsson, J. Gustafsson, *Proc. Nat'l. Acad. Sci. USA* **93** (1996) p. 5925.
[59] G. J. M. Kuiper, et al., *Endocrinology,* **138** (1997) 863.
[60] *"Oral Contraceptives-Update on Usage, Safety and Side Effect,"* Population Reports, Series A, no. 5, Population Information Program, The Johns Hopkins University, Baltimore 1979.
[61] *Oral Contraceptives in the 1980s.* Population Reports, Series A, no. 6, Population Information Program, The Johns Hopkins University, Baltimore 1982.
[62] D. A. Grimes (ed.): *The Contraception Report,* Vol. **7,** No. 1, Emron, Totowa, NJ 1996, p. 1.
[63] J. S. Lippman, G. A. Shangold, *Int. J. Fertil.* **42** (1997) 4, 230.
[64] D. R. Mishell, Jr. (ed.): "Long-Acting Steroid Contraception," *Advances in Human Fertility and Reproductive Endocrinology,* Raven Press New York, 1983.
[65] "Hormonal Contraception: New Long-Acting Methods," Population Reports, Series K no. 3, Population Information Program, The Johns Hopkins University, Baltimore, 1987.
[66] G. I. Zatuchni, A. Goldsmith, J. D. Shelton, J. J. Sciarra (eds.): *Long-Acting Contraceptive Delivery Systems,* Harper and Row, Philadelphia 1984.
[67] Norplant: *A Summary of Scientific Data,* The Population Council, New York 1990.
[68] *Contraceptive Technology Update* **12 (1)** (1991) 1.
[69] F. Michael (ed.): *Safety Requirements for Contraceptive Steriods,* Cambridge University Press, Cambridge, 1980.
[70] L. R. Beck, V. Z. Pope, *Drugs* **27** (1984) 528.
[71] E. Diczfalusy, *Contraception* **33** (1986) 7.
[72] *Contraceptive Technology Update* **12** (1991) 5, 79.
[73] E. Diczfalusy, *Contraception* **43** (1991) 201.

[74] *"Intrauterine Devices," Population Reports,* Series B, no. 3 Population Information Program, The Johns Hopkins University, Baltimore 1979.
[75] R. L. Kleinman: *Intrauterine Contraception,* 4th ed., International Planned Parenthood Federation, London 1977.
[76] F. Alvares et al., *Fertil. Steril.* **49** (1988) 768.
[77] I. Sivin, *Studies in Family Planning* **20** (1989) 355.
[78] D. R. Mishell, Jr., in J. R. Newton (ed.): *Clinics in Obstetrics and Gynaecology,* vol. **11,** W. B. Saunders, London 1984, p. 679.
[79] H. J. Tatum, in D. R. Mishell, Jr. (ed.): *Advances in Fertility Research,* vol. **1,** Raven Press, New York 1982, p. 47.
[80] S. Iwahara, K. Furuse, *Jpn. Public Health* **12** (1965) 123.
[81] D. A. Edelman, L. P. Cole, R. Apelo, P. Lavin, in G. I. Zatuchuni, A. Goldsmith, J. D. Shelton, J. J. Sciarra (eds.): *Long-Acting Contraceptive Delivery Systems,* Harper and Row, Philadelphia 1984, p. 621.
[82] "Vasectomy: New Opportunities," Population Reports, Series D, no. 5, Population Information Program, The Johns Hopkins University, Baltimore, 1992.
[83] "Voluntary Female Sterilization," Population Reports, Series C, no. 10, Population Information Program, The Johns Hopkins University, Baltimore, 1990.
[84] J. R. Newton, in J. R. Newton (ed.): *Clinics in Obstetrics and Gynaecology,* vol. **11,** W. B. Saunders, London 1984, p. 603.
[85] *Outlook* **8** (1990) 3.
[86] G. Zuliani, U. F. Colombo, R. Molla, *European J. OB/GYN and Reprod. Biol.* **37** (1990) 153.
[87] Amer. College Obstet. Gynecol. (ACOG) Practice Patterns, No. 3, Dec. 1996.
[88] M. Fasoli, F. Parazzini, G. Cecchetti, C. LaVacchia, *Contraception* **39** (1990) 459.
[89] S. S. Lyer, *Drug News and Perspectives* **4** (1991) 40, 228.
[90] V. P. Kamboj et al., *Ind. J. Exp. Biol.* **9** (1971) 103.
[91] *Drugs of the Future* **2** (1977) 441.
[92] C. Tietze and S. K. Henshaw: *Induced Abortion: A World Review,* The Alan Guttmacher Institute, New York–Washington, D.C. 1986, p. 29.
[93] G. I. Zatuchni, M. H. Labbok, J. J. Sciarra (eds.): *Research Frontiers in Fertility Regulation,* Harper and Row, Hagerstown, MD 1980, p. 330.
[94] *"Prostaglandins," Population Reports,* Series E, no. 8, Population Information Program, The Johns Hopkins University, Baltimore 1980.
[95] E. E. Baulieu, S. J. Ssegal (eds.): *The Antiprogestin Steroid RU486 and Human Fertility Control,* Plenum Press, New York 1985.
[96] O. M. Aurech et al., *Fert. Ster.* **56** (1991) 385.
[97] G. D. Hodgen, *Fert. Ster.* **56** (1991) 394.
[98] W. Elger et al., *J. Steroid. Biochem.* **25** (1986) 835.
[99] V. C. Stevens, in G. Benagiano, E. Diczfalusy: *Endocrine Mechanisms in Fertility Regulation,* Raven Press, New York 1983, p. 141.
[100] N. J. Alexander, D. Griffin, J. M. Spieler, G. M. H. Waites (eds.): *Gamete Interaction,* Wiley-Liss, New York 1990.
[101] K. N. Sacco, O. Singh, R. Pal, G. P. Talwar: *Human Reproduction Update,* Vol. **1,** no. 1, Oxford Press, 1985, p. 1.
[102] W. R. Jones, *Bailliere's Clin. Obstet. Gynecol.* **10** (1996) 1, 69.
[103] P. D. Griffin, *Human Reprod.* **6** (1991) 166.
[104] J. C. Herr, *Amer. J. Reprod. Immunol.* **35** (1996) 184.

[105] P. Primakoff et al., *Nature* **335** (1988) 543.
[106] A. V. Schally, in F. Labrie, A. Belanger, A. Dupont (eds.): *LHRH and Its Analogues – Basic and Clinical Aspects*, Excerpta Medica, Amsterdam 1984, p. 3.
[107] G. I. Zatuchni, J. D. Shelton, J. J. Sciarra (eds.): *LHRH Peptides as Female and Male Contraceptives*, Harper and Row, Philadelphia 1981.
[108] B. H. Vickery, B. Luneneld: *GnRH Analogs in Cancer and Human Reproduction*, Kluwer Academic, Lancaster 1989.
[109] M. C. Pike, J. R. Daniels, D. V. Spicer, *Endocr. Related Cancer* **4** (1997) 125.
[110] H. P. Schane, J. E. Creange, in G. I. Zatuchni, M. Labbook, J. J. Sciarra (eds.): *Research Frontiers in Fertility Regulation*, Harper and Row, Hagerstown, MD 1980.
[111] A. L. Goodman, G. D. Hodgen, in E. Y. Adashi, J. A. Rock, Z. Rosanulaks (eds.): *Reproductive Endocrinol, Surgery and Technology*, Lippincott Raven Publishers, Philadelphia 1996.
[112] K. Miyamato et al., *Biochem. Biophys. Res. Comm.* **129** (1985) 396.
[113] N. Ling et al., *Proc. Natl. Acad. Sci.* **82** (1985) 7217.
[114] J. Rivier et al., *Biochem. Biophys. res. Comm.* **133** (1985) 120.
[115] D. M. Robertson et al., *Biochem. Biophys. Res. Comm.* **126** (1985) 220.
[116] N. Ling et al., *Nature* **321** (1986) 779.
[117] W. Vale et al., *Nature* **321** (1986) 776.
[118] A. Bartke, D. W. Hahn, R. G. Foldesy, J. L. McGuire in G. I. Zatuchni, A. Goldsmith, J. M. Spieler, J. J. Sciarra (eds.): *Male Contraception*, Harper and Row, Philadelphia 1985, p. 158.
[119] C. A. Paulsen, W. J. Bremner, J. M. Leonard, in D. R. Mishell, Jr. (ed.): *Advances in Fertility Research*, vol. **1**, Raven Press, New York 1982, p. 157.
[120] F. B. Akhtar, G. F. Weinbauer, E. Nieschlag, *J. Endocrinol.* **104** (1984) 345.
[121] G. F. Weinbauer et al., *Fertil. Steril.* **42** (1984) 906.
[122] G. F. Weinbauer, F. J. Surmann, E. Nieschlag, *Acta Endocrinol.* **114** (1987) 138.
[123] C. J. Bagatell et al., *J. Clin. Endocrinol. Metab.* **69** (1989) 43.
[124] F. C. W. Wu, *Bailliere's Clin. Obstet. Gynecol.* **10** (1996) 1, 1.
[125] M. J. Cosentino, S. A. Matlin, *Exp. Opin. Invest. Drugs* **6** (1997) x6, 635.

Thyrotherapeutic Agents

SIMONE M. DAHLMANNS, Forschungszentrum Jülich GmbH, Jülich, Federal Republic of Germany

HANS-WILHELM MÜLLER-GÄRTNER, Forschungszentrum Jülich GmbH, Jülich, Federal Republic of Germany

1.	Introduction	1520	3.2.	Iodization Inhibitors	1525
2.	Thyroid Hormones as Therapeutic Agents	1521	3.3.	Inhibitors of Hormone Incretion	1526
2.1.	Pharmacology of the Thyroid Hormones	1521	3.4.	Inhibitors of the Conversion of T_4 to T_3	1528
2.2.	Chemical and Physical Properties of Thyroid Hormones	1523	3.5.	Radioiodine	1528
3.	Thyroid Depressants as Therapeutic Agents	1524	3.6.	Chemical and Physical Properties of Thyroid Depressants	1529
3.1.	Iodination Inhibitors	1525	4.	References	1531

Abbreviations:
BTU   benzylthiouracil
CBZ   carbimazole
DIT   diiodotyrosine
INN   International nonproprietary name
MIT   monoiodotyrosine
MMI   methimazole, thiamazole
MTU   methylthiouracil
PTU   propylthiouracil
rT_3 reverse L-triiodothyronine
T_3 L-3,5,3′-triiodothyronine, liothyronine
T_4 L-thyroxine, L-3,5,3′,5′-tetraiodothyronine
TBA   thyroxine binding albumin
TBG   thyroxine binding globulin
TBP   thyroxine binding protein
TBPA   thyroxine binding prealbumin
Tg   thyroglobulin
TRH   thyrotropin releasing hormone
TRIAC   triiodothyroacetic acid
TSH   thyroid-stimulating hormone
TSI   thyroid-stimulating immunoglobulins

1. Introduction

Thyrotherapeutic agents are drugs that affect the function and growth of the thyroid gland. The interest in this class of substances is due to the commonness of diseases of the thyroid. For example, 15% of the German population suffer from enlargement of the thyroid gland (goiter) [1].

The thyroid is an endocrine gland which provides the organism with the metabolic hormones L-thyroxine (T_4) and L-triiodothyronine (T_3). In terms of weight, the largest component of the thyroid hormones T_4 and T_3 is the element iodine. Iodine is taken in as iodide in food, resorbed, and actively transported into the thyroid gland, where it is used for the synthesis of the thyroid hormones. The adult human organism requires an iodine supply of 150–300 µg/d for sufficient synthesis of thyroid hormones [1].

The synthesis of thyroid hormones involves several steps: iodide entry into the thyroid gland (iodination) and oxidation to I_2, incorporation of iodine into tyrosine with the formation of MIT and DIT (iodization), and the coupling of iodotyrosines to form iodothyronines. Iodization and coupling occur in thyroglobulin (Tg). This is followed by the storage phase in Tg and the incretion phase (secretion of the thyroid hormones into the bloodstream). Depending on the requirements of the organism, the thyroid gland releases its hormones into the bloodstream, regulated by hypothalamic TRH and hypophysial TSH.

About ten times more T_4 than T_3 is produced and released into the bloodstream. Outside the thyroid gland, T_4 is mostly deiodinated in the peripheral tissue to T_3, which is 100 times more active, and partly to rT_3, which is inactive. Serum levels of the thyroid hormones that are too high result in suppression of TSH secretion in the hypophysis, while serum T_4 and T_3 levels that are too low lead to stimulation of TSH secretion. This feedback mechanism guarantees the maintenance of a constant level of thyroid hormone and, consequently, normal metabolism.

Diseases of the thyroid are classified according to functional and morphological criteria. From a functional standpoint, a distinction is made between hypofunction (hypothyreosis) and hyperfunction (hyperthyreosis). Morphological classification includes diffuse struma, multinodular struma, thyroiditis, Morbus Basedow, thyroid cysts, adenoma, and malignant tumors.

Hyperthyreosis is characterized by an increased peripheral blood level of thyroid hormones. The resulting complaints include insomnia, weight loss, diarrhoea, tachycardia and, in severe cases, central nervous disorders.

Diseases of the thyroid that can be connected to hyperfunction are: Morbus Basedow, functional autonomy, thyroiditis, and secondary hyperthyreosis in the case of pathologically increased hypophysial TSH incretion.

Hypothyreosis is characterized by decreased hormone levels in the blood. The resulting complaints include increase in weight, obstipation, bradycardia, central nervous disorders, depression, sensitivity to cold, myxedema, and cretinism. Hypothyreosis can be congenital due to organ dysgenesis and disturbances of hormone synthesis and can appear in the course of a thyropathy or as a result of medication, radiation therapy, or surgical treatment. The most frequent cause of acquired hypothyreosis is autoimmune atrophy.

Thyroid hyper- and hypofunction can be rectified by the administration of the appropriate preparations.

2. Thyroid Hormones as Therapeutic Agents

2.1. Pharmacology of the Thyroid Hormones (→ Hormones).

Thyroid hormone preparations find therapeutic application for:

1) Substitution in the case of hypothyreosis of any genesis
2) Preventive treatment against recurrence of struma after strumectomy or radioiodine treatment
3) Suppressive treatment of euthyroid struma
4) Substitution therapy after thyroidectomy and radioiodine treatment for thyroid malignancies
5) Treatment accompanying thyroid depressant therapy for hyperthyreosis

Four types of preparations are available for medicamentous therapy [2]:

1) Levothyroxine preparations (T_4)
2) Liothyronine preparations (T_3)
3) Preparations that combine T_3 and T_4
4) Preparations that combine T_4 and iodide

T_4 and T_3 differ fundamentally in their pharmacokinetic properties, as shown in Table 1. Today, monotherapy with T_4 is preferred to treatment with T_3 or combined treatment with T_4 and T_3. In this way a demand-orientated supply of the metabolically active T_3 is guaranteed to a certain extent. The biological half-life of T_4 of 6–7 d is much longer than that of T_3 (ca. 19 h). T_4 is converted extrathyroidally to metabolically active T_3. Thus, even a single dose of T_4 results in essentially constant hormone levels in the serum. Monotherapy with T_3 can easily result in unphysiological T_3 peak values in the serum and overdosage symptoms. Therefore, treatment with T_3 offers no advantages over T_4 therapy. An exception is the short-term application of T_3 if thyroid hormone therapy must be interrupted prior to diagnostic and therapeutic measures.

Combined treatment with both thyroid hormones attempts to imitate physiological conditions. The balanced combination of both thyroid hormones ($T_4:T_3 = 5:1$ or $10:1$) should guarantee physiologically adequate substitution and suppression.

Table 1. Pharmacokinetic data of the hormones T_4 and T_3 (active hormone)

	Levothyroxine T_4	Liothyronine T_3
Incretion rate [25]	94–110 µg/d (120–140 nmol/d)	16–22 µg/d (24.6–33.8 nmol/d)
Utilization rate [25]	80 µg/d (102 nmol/d)	60 µg/d (92 nmol/d)
Blood level [26]	60–80 µg/L (75–105 nmol/L)	1.0–1.5 ng/L (1.5–2.3 nmol/L)
Unbound physiologically active form [26]	0.03–0.05 %	0.3–0.5 %
Plasma protein binding [25]	99.9 % [a]	99.7 % [b]
Dosage	50–300 µg	10–100 µg
Relative activity [c]	1	4
Enteral resorption [d]	75–85 %	78–95 %
Bioavailability	75–85 %	80–95 %
Distribution volume [25]	12 L	46 L
Total metabolic clearance rate [25]	1.17 L/d	26.1 L/d
Plasma half-life	6–7 d	1 d
Onset of action	3–5 d (oral), 6 h (i.v.)	12–48 h
Maximal effect after	ca. 9 d (single dose)	ca. 2–3 h (single dose)
Duration of effect	7–10 d	3–5 d
Plasma transfer	slow	slow
Excretion in breast milk	in small amounts	in small amounts
Biliary excretion	yes	yes
Enterohepatic circulation	yes	yes
Renal unchanged	0 %	0 %

[a] Up to 80 % on TBG, up to 15 % on TBPA, up to 10 % on TBA.
[b] Up to 99.7 % on TBG, up to 0.3 % on TBPA and TBA; binding is ten times weaker than T_4.
[c] For enteral administration.
[d] In fasting state; ingestion of food lowers resorption to 35 %.

Combination preparations are indicated when conversion to T_3 does not occur in the case of monotherapy with T_4. These preparations are used to treat struma caused by lack of conversion. Preparations that combine levothyroxine and iodide are given preference in the treatment of iodine-deficiency struma.

An adequate supply of iodide is required for protection against iodine-deficiency struma. Since in many areas of Germany, the alimentary supply of iodine is less than the iodine requirement and endemic iodine deficiency affects both the function and morphology of the thyroid gland [2] (formation of struma, degeneration, autonomy), adequate prevention of iodine deficiency by administration of iodide preparations is recommended (100–200 µg I^-/d).

As a rule, substitution therapy with thyroid hormones starts very slowly. The dosage is established individually, depending on the age, condition, and indication, the regular determination of the hormone values in the serum being imperative. If the dosage is correct, no undesirable effects occur. If the individual tolerance limit or the proper dose is exceeded, the clinical symptoms typical of hyperthyreosis can occur, especially if the dose is increased too fast at the start of treatment.

During pregnancy, substitution treatment for hypothyreosis of the mother should be rigorously carried out because otherwise the insufficient hormone production in the mother can result in defective development of the child. However, overdosage must be carefully avoided. Even in the case of high-dose therapy, the amount of thyroid hormones secreted into the breast milk during lactation is not sufficient for the development of hyperthyreosis or suppression of TSH secretion in the infant.

2.2. Chemical and Physical Properties of Thyroid Hormones

Levothyroxine (INN), T_4, L-thyroxine, L-3,5,3′,5′-tetraiodothyronine, L-3-[4-(4-hydroxy-3,5-diiodophenoxy)-3,5-diiodophenyl]alanine (IUPAC) [51-48-9], $C_{15}H_{11}I_4NO_4$, M_r 776.93, decomp. 235–236 °C, $[\alpha]_D^{20}$ +4.4° (3% solution in 0.13 N NaOH in 70% ethanol). Colorless, light-sensitive crystals, insoluble in water, ether, ethanol, and chloroform.

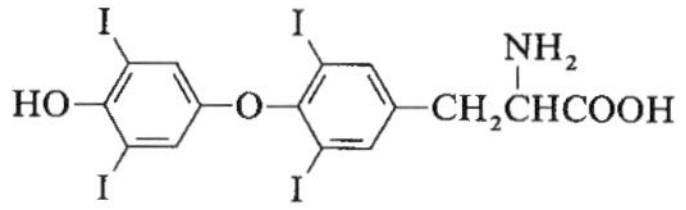

Levothyroxine sodium (INN), L-3-[4-(4-hydroxy-3,5-diiodophenoxy)-3,5-diiodophenyl]alanine sodium (IUPAC) [55-03-8], $C_{15}H_{10}I_4NNaO_4$, M_r 798.85.

Levothyroxine sodium x-water (INN) [25416-65-3], $C_{15}H_{10}I_4NNaO_4 \cdot x\,H_2O$, pH 8.9 (saturated, aqueous solution). White to lightly colored, light-sensitive, crystalline powder. Solubility: very slightly soluble in water (solubility decreases with pH), slightly soluble in ethanol, insoluble in acetone, chloroform, ether, and soluble in alkali metal hydroxide solution. The analytical profile and synthesis of levothyroxine compounds are described in [3].

Liothyronine (INN), T_3, 3,5,3′-triiodothyronine, L-3-[4-(4-hydroxy-3-iodophenoxy)-3,5-diiodophenyl]alanine (IUPAC), [6893-02-3], $C_{15}H_{12}I_3NO_4$, M_r 651.01, mp 233–234 °C, $[\alpha]_D^{29,5}$ +21° (4.75% solution in 1:2 1 M HCl/ethanol, pK_a 8.5, insoluble in water and ethanol, soluble in dilute alkali.

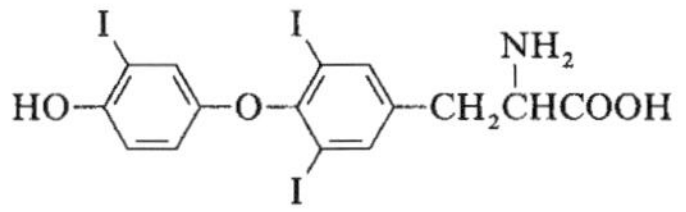

Liothyronine sodium (INN) [55-06-1], $C_{15}H_{11}I_3NNaO_4$, M_r 672.99, UV_{max} 319 nm. White to slightly colored powder, insoluble in water and ether, slightly soluble in ethanol, soluble in dilute alkali metal hydroxide solution. For synthesis, see → Hormones.

Tiratricol (INN), TRIAC, TA3, 3,5,3′-triiodothyroacetic acid, 4-(4-hydroxy-3-iodophenoxy)-3,5-diiodophenylacetic acid (IUPAC) [51-24-1], $C_{14}H_9I_3O_4$, M_r 621.95, *mp* 65 °C from methanol and water; after recrystallization at 110 °C: 180–183 °C. Synthesis is described in [4].

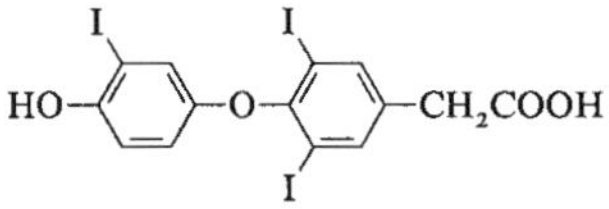

HO—⟨ ⟩—O—⟨ ⟩—CH$_2$COOH (with I substituents)

TRIAC [5], [6] is a natural metabolite of T$_3$. In spite of its high affinity for the T$_3$ receptor, TRIAC evidently has no physiological role. In vivo, it exerts a strong, dose-dependent and reversible inhibitory effect on the hypophysial secretion of TSH. The effects on peripheral tissue are negligibly small. Consequently, TRIAC has been used as an alternative to T$_4$ for hypophysial TSH suppression in patients suffering from thyroid carcinoma. More recent studies show that TRIAC results in increased L-thyroxine activity, even in peripheral tissue.

L-3,5-Diiodothyronine (INN), T$_2$, L-3-[4-(4-hydroxyphenoxy)-3,5-diiodophenyl]alanine (IUPAC) [*1041-01-6*], C$_{15}$H$_{13}$I$_2$NO$_4$, M_r 525.10, *mp* 256–257 °C (decomp.), $[\alpha]_D^{25}$ +25.2° [5% solution in 1 M HCl:95% ethanol (1:2)], colorless crystals. Thyroid hormone with a L-thyroxine-like activity. For synthesis, see [7].

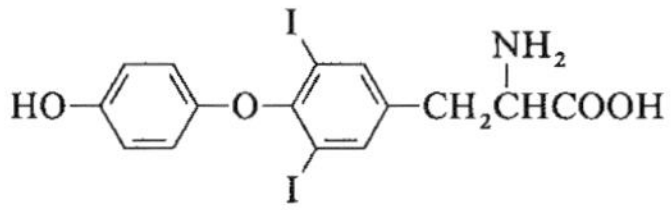

HO—⟨ ⟩—O—⟨ ⟩—CH$_2$CHCOOH (with I substituents and NH$_2$)

3. Thyroid Depressants as Therapeutic Agents

Thyroid depressants are divided into five groups, based on their mechanism of action.

1) Iodination inhibitors, which inhibit iodide transport into the thyroid gland
2) Iodization inhibitors, which directly inhibit the synthesis of thyroid hormones
3) Inhibitors of hormone incretion
4) Inhibitors of T$_4$ and T$_3$ conversion
5) Radioiodine, which can destroy thyroid tissue by means of ionizing radiation

3.1. Iodination Inhibitors

Iodination inhibitors act as thyroid depressants by competitively inhibiting the uptake of iodine by the thyroid cell (iodination). They release accumulated iodide from the thyroid gland provided it is not incorporated into Tg molecules. In this manner, these agents inhibit the synthesis of the thyroid hormones, resulting in iodine depletion.

Iodination inhibitors are administered to block the thyroid gland prior to scintigraphic investigations of other organs with iodine isotopes, during the unavoidable application of iodine-containing contrast medium if an autonomic thyropathy is known to exist, and in the treatment of hyperthyreosis if other thyroid depressants must be discontinued due to unjustifiable side effects. These agents should not be used if treatment with radioiodine or iodine is to be performed. Perchlorate finds therapeutic application in the form of the sodium or potassium salt in a dosage of up to 1500 mg/d. The undesirable effects observed are gastrointestinal disturbances; serious but more seldom side effects are agranulocytosis, aplastic anemia, and nephrotic syndrome.

3.2. Iodization Inhibitors

Iodization inhibitors are derivatives of thiourea. Agents that are therapeutically used are: benzylthiouracil (BTU), methylthiouracil (MTU), propylthiouracil (PTU), carbimazole (CBZ), methimazole (MMI), and thibenzazoline. Today, CBZ and MMI have gained acceptance in the clinical practice [8].

These substances inhibit the iodine oxidation reaction catalyzed by a thyroid peroxidase [9], the incorporation of iodine into tyrosine, and the coupling of tyrosine molecules to form iodothyronines. Thus, they inhibit the synthesis of thyroid hormones. This makes possible the symptomatic treatment of thyroid hyperfunction irrespective of its etiology.

In addition, thiourea derivatives are attributed with a favorable effect on the immunopathogenesis of hyperthyreosis [10]. At present, it cannot be definitely established to what extent the underlying immunopathogenetic process can be suppressed.

Processes that are not influenced are the transport of iodide into the thyroid gland, the incretion of the thyroid hormones, and hyperthyreosis due to hormone liberation after destruction of the thyroid cells (e.g., after radioiodine therapy or thyroiditis). This mechanism of action explains why the clinical effect starts only after the hormone depots in the gland have been emptied and the peripheral hormone concentrations decrease. An additional mechanism of action being discussed for PTU is the blocking of the peripheral deiodination of T_4 to T_3.

Indications for the administration of thiourea thyroid depressants include:

1) Initial and long-term treatment of Basedow hyperthyreosis
2) Initial treatment of thyroid autonomy

3) Treatment of transitory hyperthyreosis caused by inflammation
4) Intravenous treatment of thyrotoxic crisis (preferably PTU)
5) Intermediate therapy after radioiodine treatment until an euthyroid metabolic condition is reached
6) Preoperative and prior to radioiodine therapy in all hyperthyreosis conditions because treatment must always be given in the euthyroid state
7) Not for the treatment of hyperthyreosis caused by inflammation

Due to their goitrogenic action when given over an extended time period, antithyroid substances can be administered in combination with synthetic thyroid hormones.

In principle, all thyroid depressants give rise to the same undesirable effects. The total rate of side effects is about 14.3% [11]. The most frequent are skin reactions (5.6%) and ailments of the joints (1.6%). The most dangerous side effect, which occurs with a frequency of up to 1%, is toxic bone marrow damage (agranulocytosis, leukopenia, thrombopenia). Other rare side effects include gastrointestinal complaints, swelling of the joints, fever, headache, and rash.

The frequency of side effects varies greatly from agent to agent. CBZ, MMI, and PTU are favorably assessed. As a result of the lower dosage and toxicity, thioimidazoles are preferred to PTU, and MMI as an active metabolite to CBZ.

The highest rate of side effects is shown by MTU (14%; agranulocytosis in 1–3% of all cases). Hence, the use of MTU in ready-for-use drugs is no longer recommended [12].

Care must be taken during pregnancy because the drugs pass the placental barrier and can result in the development of struma or cretinism in the child. For this reason, during pregnancy, the lowest possible doses of thyroid depressants must be administered without the addition of thyroid hormones, which pass the placental barrier. In the case of combination therapy with a thyroid depressant and T_4, in spite of the euthyroid metabolic state of the mother, there is a risk of the development of hypothyreosis in the child due to thyroid suppression.

The agent of choice during pregnancy and lactation is PTU in the United States, and MMI in Germany.

A survey of the pharmacokinetic data of antithyroid agents is presented in Table 2.

3.3. Inhibitors of Hormone Incretion

Higher concentrations (10–500 mg/d) of iodide ions influence all the important steps of iodine metabolism and indirectly inhibit the secretion of thyrotropin. The active uptake of iodide by the thyroid gland is inhibited by high concentrations of iodide in the blood. At the same time, the organic binding of iodine is inhibited (Wolff–Chaikoff effect), resulting in the inhibition of hormone synthesis.

The inhibition of hormone release from Tg is clinically most significant and is the cause of the immediate (after ca. 2 h) action of iodide. This inhibition is due to the blocking of the proteases involved. However, the thyroid-depressing effect decreases

Table 2. Pharmacokinetic data of antithyroid substances

	Methylthiouracil	Propylthiouracil	Carbimazole	Methimazole, Thiamazole
Notes	not commercially available in Germany since July 92		prodrug (inactive) enzymatic conversion to methimazole during or directly after resorption	active metabolite
Initial dose	400–600 mg/d	300–600 mg/d	30–60 mg/d	20–40 mg/d
Maintenance dose	100–300 mg/d in single doses	50–200 mg/d in single doses every 8 h [27]	5–15 mg/d in single doses every 8 h [27]	5–10 mg/d in single doses every 8 h [27]
Relative activity		10 mg ≈ 1 mg Methimazole [28]	15 mg ≈ 10 mg Thiamazole [29]	
Enteral resorption	almost complete	fast and almost complete	fast and almost complete	fast and almost complete [7], [8], [30]
Bioavailability (oral)	80–90% [30]	80–95% [30]	80–95% [30]	80–95% [30]
Max. plasma level	1–2 h	1–2 h [30]	3 h	3–6 h [31]
after Plasma level		2–3 µg/mL (dose: 200 mg)		
Distribution volume		20–30 L [28], [30]		40 L [28]
Plasma clearance		120 mL/min [30]		200 mL/min [28]
Plasma half-life	1–1.65 h	75 min [28] 1–2 h [30]		4–6 h [28] 2–6 h [30]
Plasma protein binding		75–80% [28], [30]	high	none [8]
Onset of action		after 1–2 weeks	after 1–3 weeks	after 1–3 weeks
Period of effect		2–3 h	24 h	24–40 h [28]
Placental transfer		low	higher than PTU [28]	higher than PTU [28]
Excretion in breast milk		1/10 of the serum concentration of the mother	higher than PTU [28]	higher than PTU [28]
Distribution		intrathyroidal accumulation	intrathyroidal accumulation	intrathyroidal accumulation
Metabolism				
intrathyroidal		propylthiouracil sulfinic acid PTU–SO$_2$H [8] SO$_4^{2-}$ [7] 6-propyluracil, SO$_4^{2-}$ [7], [16] S-methyl-6-propylthiouracil [16]	3-methylthiohydantoin [8], [31] methimazole, SO$_4^{2-}$ [7]	3-methylthiohydantoin [8], [31]
extrathyroidal	65% glucuronide	65% glucuronide [7], [16] S-methyl-PTU		
Excretion				
biliary		yes	yes	yes
enterohepatic circulation		yes [7], [8]		yes [7], [8]
renal (unchanged)	ca. 30%	<10% [30]		<10% [30]
Rate of side effects	7–14% agranulocytosis in 1–3% of all cases	5%	2–3%	5% agranulocytosis in 0.3–0.6% of all cases

with continuous use. For this reason, therapy with iodides (with the exception of iodine prophylaxis) in preparation for thyroidectomy in a thyrotoxic crisis or hyperthyroidal struma (Plummering) is possible only for a duration of 2–6 weeks. The agents used are KI, Lugol's solution, and organic iodine preparations.

Polonium iodide, which was administered as an i.v. preparation until now, is no longer produced for economic reasons.

Lithium salts, which have attained considerable importance in the treatment of mental diseases, inhibit the proteolytic release of the thyroid hormones [1]. As a result of their side effects and small therapeutic range, lithium salts are reserved for special indications such as thyrotoxic crisis, especially when caused by the i.v. application of contrast media [13].

In preoperative thyroid depressant therapy for hyperthyreosis, the thyroid depressants mentioned in Section 3.2 and symptomatic treatment with β-blockers are preferred [14].

3.4. Inhibitors of the Conversion of T_4 to T_3

Blockers are used in the clinical symptomatic treatment of hyperthyreosis. The agent of choice, propranolol, inhibits the conversion of T_4 to T_3. In the case of PTU (iodization inhibitor), the blockage of peripheral deiodination of T_4 to T_3 is being discussed as an additional mechanism of action. Glucocorticoids, which are intravenously administered in the treatment of thyrotoxic crisis, are also believed to inhibit the peripheral conversion of T_4 to T_3. The principle of action of the conversion inhibitors is preferentially employed in the treatment of thyrotoxic crisis [1].

3.5. Radioiodine

A radioactive form of iodine, ^{131}I [7790-26-3], $T_{1/2}$ 8 d, γ energy 364 keV, is used in the treatment of hyperthyreosis in Morbus Basedow or thyroid autonomy, in the treatment of differentiated thyroid carcinomas after thyroidectomy, in struma therapy, in recurrences after strumectomy, and in metastasizing thyroid carcinomas that store iodine [15], [16]. Like the stable isotope ^{127}I, it enters thyroid metabolism and is incorporated into the thyroid hormones. The radiation comprises 90% of β-particles and 10% of γ-rays. As a result of the low range of β radiation (1 mm), thyroid tissue is largely destroyed selectively. The γ radiation can be used for diagnostic purposes. According to recommendations made by the German Federal Board of Health (BGA), it should be used only in the case of thyroid malignancies (whole body scintigraphy) or to calculate the dosage for radioiodine therapy.

3.6. Chemical and Physical Properties of Thyroid Depressants

Aminothiazole (INN), 2-aminothiazole (IUPAC) [*96-50-4*], $C_3H_4N_2S$, M_r 100.14, *mp* 93 °C from benzene and petroleum ether, *bp* 140 °C (1.47 kPa). Crystalline substance, soluble in hot water, slightly soluble in cold water, ethanol, and ether, readily soluble in dilute hydrochloric acid and in 20% sulfuric acid.

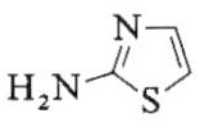

This agent is synthesized according to the King method by the conversion of paraldehyde with thiourea in the presence of sulfuryl chloride [17].

Benzylthiouracil (INN), BTU, 6-benzyl-2,3-dihydro-2-thioxo-4(1*H*)-pyrimidinone (IUPAC) [*33086-27-0*], $C_{11}H_{10}N_2OS$, M_r 218.30, *mp* 223–224 °C. The synthesis is described in [18].

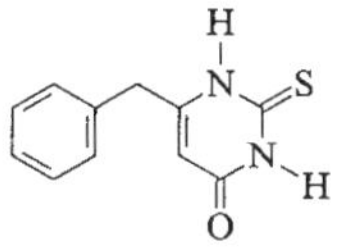

Methylthiouracil (INN), MTU, 2,3-dihydro-6-methyl-2-thioxo-4(1*H*)-pyrimidinone (IUPAC) [*56-04-2*], $C_5H_6N_2OS$, M_r 142.18, *mp* 326–331 °C with decomposition; sublimes. White, finely crystalline, odorless powder with a slightly bitter taste, light-sensitive.

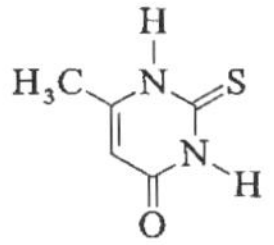

Solubility: 1:2000–2500 in water, 1:130–150 in boiling water, 1:800 in ethanol, and 1:1300 in 85% glycerol. Readily soluble in alkali metal hydroxide and ammonia solutions, and methanol, slightly soluble in acetone, chloroform, and ether. Sparingly soluble in dilute mineral acids, and practically insoluble in benzene. The synthesis is conducted by reacting thiourea with ethyl 3-oxobutyrate [18].

Propylthiouracil (INN), PTU, 2,3-dihydro-6-propyl-2-thioxo-4(1*H*)-pyrimidone (IUPAC) [*51-52-5*], $C_7H_{10}N_2OS$, M_r 170.23, *mp* 219–221 °C. White, finely crystalline, bitter tasting powder.

Solubility: 1:900 in water at 20 °C, 1:100 in boiling water, 1:60 in 90% ethanol, and 1:60 in acetone. Very slightly soluble in ether, chloroform, and benzene. In an aqueous medium, PTU is a monobasic acid with a pK_a of 8.3 (20 °C). It dissolves in aqueous ammonia and aqueous alkali metal hydroxide with formation of salts.

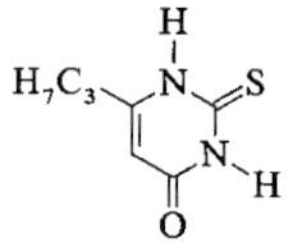

PTU is made by the condensation of thiourea with ethyl 3-oxohexanoate [18]. A summary of the analytical data of PTU is given in [19].

Carbimazole (INN), CBZ, ethyl-3-methyl-2-thioxo-4-imidazoline-1-carboxylate (IUPAC) [*22232-54-8*], $C_7H_{10}N_2O_2S$, M_r 186.23, *mp* 122–125 °C. White or creamy crystalline powder with a characteristic odor.
Solubility: 1:5000 in water at 20 °C, 1:50 in ethanol, 1:330 in ether, 1:3 in chloroform, and 1:17 in acetone.
It is produced by the reaction of methimazole with ethyl chloroformate [20].

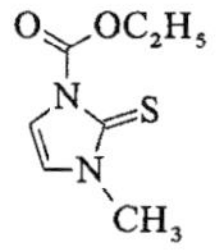

Methimazole, thiamazole (INN), MMI, 1-methyl-2-imidazole thiol (IUPAC) [*60-56-0*], $C_4H_6N_2S$, M_r 114.17, *mp* 146–148 °C from ethanol, *bp* 280 °C with slight decomposition. White, crystalline powder with a weak characteristic odor and a bitter taste, light-sensitive.
Solubility: readily soluble in water (1:5), soluble in ethanol (1:5), and chloroform (1:5), slightly soluble in ether (1:125) and petroleum ether.

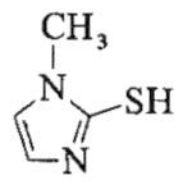

It is produced by the reaction of aminoacetaldehyde diethylacetal with methylisothiocyanate or of thiocyanic acid with N-methylaminoacetals [21]. A collection of analytical data is given in [22].

Thibenzazolin (INN), 1-3-bis(hydroxymethyl)-2-benzimidazolinethione (IUPAC) [*6028-35-9*], $C_9H_{10}N_2O_2S$, M_r 210.26, *mp* 160–162 °C. Colorless crystals having a very bitter taste, soluble in dilute alkali metal hydroxide solutions.

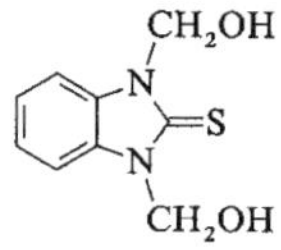

It is synthesized as described in [23], and is a thyroid depressant with a sedative effect. A selection of the common trade names of the substances mentioned above is given in [24].

4. References

[1] P. Pfannenstiel: *Schilddrüsenkrankheiten, Diagnose und Therapie*, Berliner Medizinische Verlagsanstalt GmbH, Berlin 1991.
[2] F. A. Horster: *Schilddrüsenkrankheiten, Diagnose und Therapie in der Praxis*, Deutscher Ärzte Verlag, Köln 1989.
[3] A. Post, R. J. Warren in K. Florey (ed.): *Analytical Profiles of Drug Substances*, vol. **5**, Academic Press, New York 1976, pp. 225–281.
[4] R. I. Meltzer et al., *J. Org. Chem.* **26** (1961) 1418–1428.
[5] C. Mechelany et al., *Clin. Endocrinol. (Oxford)* **35** (1991) 123–128.
[6] H.-W. Müller-Gärtner, C. Schneider, *Clin. Endocrinol.* **28** (1988) 345–351.
[7] J. Elks, G. J. Waller, *J. Chem. Soc.* 1952, 2366–2370.
[8] G. Benker, T. Olbricht, *Krankenhausarzt* **55** (1982) 591–601.
[9] G. Benker, D. Reinwein, *Klin. Wochenschr.* **60** (1982) 531–539.
[10] A. Pinchera et al., *J. Clin. Endocrinol. Metab.* **29** (1969) 231.
[11] M. Meyer-Geßner et al., *DMW Dtsch. Med. Wochenschr.* **114** (1989) 166–171.
[12] R. Thesen, M. Schulz, R. Braun, *Pharm. Ztg.* **25** (1989) 1590.
[13] H. P. Wolff, T. R. Weihrauch: *Internistische Therapie*, 7th ed., Urban & Schwarzenberg, München 1988, pp. 868–890.
[14] F. Raue, *Dtsch. Ärzteblatt* **87** (1990) no. 42, 31.
[15] H. J. Hermann: *Nuklearmedizin*, 2nd ed., Urban & Schwarzenberg, München 1989.
[16] B. Marchant et al., *Pharmacol. Ther.* Part B 3 (1978) 305.
[17] H. Erlenmeyer, L. Herzfeld, B. Prijs, *Helv. Chim. Acta* **38** (1955) 1291–1294.
[18] G. W. Anderson et al., *J. Am. Chem. Soc.* **67** (1945) 2197–2200.
[19] H. Y. Aboul-Enein in K. Florey (ed.): *Analytical Profiles of Drug Substances*, vol. **6**, Academic Press, New York 1977, pp. 457–486.
[20] J. A. Baker, *J. Chem. Soc.* 1958, 2387–2390.
[21] R. G. Jones et al., *J. Am. Chem. Soc.* **71** (1949) 4000–4002.
[22] H. Y. Aboul-Enein, A. A. Al-Badr in K. Florey (ed.): *Analytical Profiles in Drug Substances*, vol. **8**, Academic Press, New York 1979, pp. 351–370.
[23] L. Monti, G. Venturi, *Gazz. Chim. Ital.* **76** (1946) 365.
[24] Arzneibüro der Bundesvereinigung deutscher Apothekerverbände (ABDA): *Pharmazeutische Stoffliste*, Werbe- und Vertriebsgesellschaft deutscher Apotheker mbH, Frankfurt 1993.
[25] K. Oberdisse, E. Klein, D. Reinwein: *Die Krankheiten der Schilddrüse*, 2nd ed., Thieme Verlag, Stuttgart–New York 1980.
[26] W. Forth, D. Henschler, W. Rummel: *Allgemeine und spezielle Pharmakologie und Toxikologie*, Bibliographisches Institut, Mannheim 1988.
[27] J. E. F. Reynolds: *Martindale, The Extra Pharmacopoeia*, Pharmaceutical Press, London 1989, pp. 682–689, 1487–1491.
[28] D. S. Cooper, *N. Engl. J. Med.* **311** (1984) 1353–1362.

[29] R. Jansson et al, *J. Clin. Endocrinol. Metab.* **57** (1983) 129–132.
[30] J. P. Kampmann, J. M. Hansen, *Clin. Pharmacokinet.* **6** (1981) 401–428.
[31] G. Skellern, *Br. J. Clin. Pharmacol.* **9** (1980) 137–143.

Hormones

Separate keyword: → *Peptides and Protein Hormones*

JÜRGEN SANDOW, Hoechst AG, Frankfurt, Federal Republic of Germany (Chap. 1, Section 2.2)
EKKEHARD SCHEIFFELE, Henning Berlin GmbH, Berlin, Federal Republic of Germany (Section 2.1)
MICHAEL HARING, Henning Berlin GmbH, Berlin, Federal Republic of Germany (Section 2.1)
GÜNTER NEEF, Schering AG, Berlin, Federal Republic of Germany (Chaps. 3–5)
KLAUS PREZEWOWSKY, Schering AG, Berlin, Federal Republic of Germany (Chaps. 3–5)
ULRICH STACHE, Hoechst AG, Frankfurt, Federal Republic of Germany (Chaps. 3–5)

1.	Introduction	1535
1.1.	Basic Principles	1535
1.1.1.	Definitions and Classification	1535
1.1.2.	Hormone-Producing Systems	1537
1.1.3.	Structural Analysis and Synthesis	1540
1.1.4.	Receptors and Mechanisms of Action	1541
1.1.5.	Therapeutic Use of Hormones	1542
1.2.	**Peptide and Protein Hormones**	1542
1.2.1.	Hypothalamic Hormones	1542
1.2.2.	Proopiomelanocortin Hormones	1543
1.2.3.	Glycoprotein Hormones	1544
1.2.4.	Growth Hormone, Prolactin, and Placental Lactogen	1545
1.2.5.	Placental Hormones	1546
1.2.6.	Calcium-Regulating Hormones	1546
1.2.7.	Cardiovascular Hormones	1547
1.2.8.	Glucose-Regulating Hormones	1548
1.2.9.	Gastrointestinal Hormones	1549
1.2.10.	Growth Factors and Related Peptides	1549
1.2.11.	New Hormonal Peptides and Proteins	1550
1.3.	**Amino Acid Derivatives**	1550
1.3.1.	Catecholamines and Indolamines	1550
1.3.2.	Thyroid Hormones	1551
1.4.	**Steroid Hormones**	1551
1.4.1.	Gonadal Steroids	1551
1.4.2.	Adrenal Steroids	1552
1.4.3.	Calciferols	1553
1.4.4.	New Synthetic Steroid Hormones	1553
1.5.	**Prostaglandins**	1554
1.6.	**Hormone Antagonists and Analogues**	1554
2.	**Amino Acid Hormones**	1555
2.1.	**Thyroid Hormones**	1555
2.1.1.	Physiology and Therapeutic Uses	1556
2.1.2.	Physical Properties	1558
2.1.3.	Production	1559
2.2.	**Hormones of the Adrenal Medulla**	1560
3.	**Steroid Sex Hormones**	1564
3.1.	Introduction	1564
3.2.	**Partial Chemical Synthesis of Steroid Intermediates**	1566
3.2.1.	Raw Materials	1566
3.2.2.	Partial Synthesis	1570
3.3.	**Microbiological Methods for the Synthesis of Steroid Intermediates**	1577
3.4.	**Total Synthesis of Sex Hormones**	1579
3.5.	**Estrogens**	1582
3.5.1.	General Aspects	1582
3.5.2.	Natural Estrogens and Estradiol Derivatives	1585
3.5.3.	Nonsteroidal Estrogens	1585
3.5.4.	Antiestrogens	1586

3.6.	Androgens and Anabolic Agents	1588
3.6.1.	General Aspects	1588
3.6.2.	Androgens	1589
3.6.3.	Antiandrogens	1591
3.6.4.	Anabolic Agents	1591
3.7.	**Gestagens**	1593
3.7.1.	General Aspects	1593
3.7.2.	Gestagens of the Pregnane and 19-Norpregnane Series	1595
3.7.3.	Gestagens of the Androstane Series	1597
3.7.4.	Gestagens of the 19-Norandrostane Series	1597
3.7.5.	Antigestagens	1600
4.	**Adrenal Steroid Hormones**	1601
4.1.	**Introduction**	1601
4.2.	**Historical Aspects**	1602
4.3.	**Physiology and Mode of Action**	1603
4.4.	**Therapeutic Uses of Glucocorticoids**	1604
4.5.	**Therapy and Side Effects**	1604
4.6.	**Biosynthesis and Metabolism**	1605
4.7.	**Detection Methods**	1606
4.8.	**Naturally Occurring Glucocorticoids**	1607
4.8.1.	Properties and Uses	1607
4.8.2.	Partial Synthesis of Cortisone	1608
4.8.3.	Partial Synthesis of Cortisol	1612
4.9.	**Naturally Occurring Mineralocorticoids and Antagonists**	1613
4.9.1.	Cortexone	1614
4.9.2.	Aldosterone	1615
4.9.3.	Spironolactone	1616
4.10.	**Chemically Modified Adrenal Steroid Hormones**	1617
4.10.1.	1,2-Dehydrocorticosteroids	1617
4.10.2.	9α-Halocorticosteroids	1619
4.10.3.	16-Hydroxycorticosteroids	1620
4.10.4.	6α-Methylcorticosteroids	1622
4.10.5.	16-Methylcorticosteroids	1624
4.10.6.	16-Methylenecorticoids	1627
4.10.7.	6α-Fluoro-, 6α-Chloro-, and 6α,9α-Difluorocorticoids	1629
4.10.8.	17-Deoxycorticosteroids	1632
4.10.9.	9α,11β-Dihalocorticosteroids	1635
4.11.	**Recent Developments**	1635
5.	**Cholecalciferol**	1636
5.1.	**Introduction**	1636
5.2.	**Biosynthesis and Metabolism**	1637
5.3.	**Biological Activity**	1638
5.4.	**Chemical Synthesis**	1639
5.5.	**Synthetic Analogues**	1641
6.	**References**	1641

Abbreviations used in this article:

ACTH	adrenocorticotropic hormone, corticotropin
ACE	angiotensin converting enzyme
ADH, AVP	antidiuretic hormone, vasopressin
ANP	atrionatriuretic peptide
CCK	cholecystokinin
CRH	corticotropin releasing hormon, corticocoliberin
CT	calcitonin
FSH	follicle stimulating hormone, follitropin
GH	growth hormone (also designated STH, somatotropic hormone)
GnRH	gonadotropin releasing hormone, gonadorelin (also designated LHRH, luteinizing hormone releasing hormone)
GHRH, GRH	growth hormone releasing hormone
HCG	human chorionic gonadotropin = choriotropin
HCS	human chorionic somatotropin (also designated HPL, human placental lactogen)

HMG	human postmenopausal gonadotropin = menotropin
HPL	human placental lactogen (also designated HCS, human chorionic somatotropin)
LH	luteinizing hormone, lutropin (also designated ICSH, interstitial cell stimulating hormone)
LHRH	luteinizing hormone releasing hormone
MSH	melanophore stimulating hormone, melanotropin
PIF	prolactin inhibiting factor
PMS, PMSG	pregnant mare serum gonadotropin, equitropin
PRL	prolactin
PTH	parathormone, parathyrin
STH	somatotropic hormone, somatotropin
T_3	triiodothyronine
T_4	thyroxine
TRF	thyrotropin releasing factor (also designated TRH)
TRH	thyrotropin releasing hormone, thyroliberin
TSH	thyroid stimulating hormone, thyrotropin

1. Introduction

1.1. Basic Principles

1.1.1. Definitions and Classification

Definitions. Hormones are discrete substances which regulate biological function, either directly by acting on target organs or indirectly by controlling the secretion and synthesis of secondary or tertiary hormone systems [1], [8], [21]. They appear early in evolution. Hormonal substances have been identified in bacteria and plants which later in phylogenesis acquired functional roles in animal species. Plant hormones are treated elsewhere. The highly selective action of hormones is due to their specific interaction with cell receptors. Adaptation of hormones and receptors is a long evolutionary process, during which the hormone receptors may change and the chemical nature of the hormone remains the same. For example, insulin controls gonadal function in certain invertebrates but in mammals it is a glucose-regulating hormone that also controls the metabolism of fatty acids and amino acids. Precise definition of new hormonal substances is difficult due to rapid progress in hormone biochemistry, physiology, and pharmacology. In classical endocrinology, hormones are defined as products that are synthesized by a group of cells (a hormone-producing organ), secreted into the bloodstream, and transported to their site of action (a hormone-responsive organ or tissue) where they exert specific effects. The hormone-producing organ is identified by the deficiencies observed after surgical removal of the source of hormone synthesis (endocrine gland). The hormone is characterized by isolation, structural analysis, and chemical synthesis; the deficiencies observed after removal of the endocrine gland are corrected by substitution. In this experimental approach, the biological effects of the hormone are used for its initial identification (bioassay) during purification from biological materials. Another source of information about hormones is clinical endocrinology [17]. The disease symptoms resulting from hormone deficiencies or hypersecretion provide a concept for the biological function of a particular hormone, and suggest methods for its bioassay. If the hormone is produced in cells disseminated throughout a nonremovable organ or at multiple sites in the body, a biochemical approach must be employed for identification of the hormone's physicochemical characteristics that can be exploited for its extraction and purification from normal tissue or endocrine tumors. Some tumors produce hormones in much larger quantities [23] than the physiological concentrations found in normal, nonneoplastic tissue.

The definition of hormones and regulatory factors will certainly change, classical hormones will no longer be defined as products associated only with specific endocrine glands, but will turn up in new unexpected locations and functions. Hormonal regulatory factors will acquire a more comprehensive role as endogenous substances secreted by many endocrine and nonendocrine cells. Their definition may then depend on regulation of cell activity rather than on their site of origin. The future classification of hormones may be based on assignment to chemical families, on interaction with receptors and intracellular effector systems, physiological and pharmacological effects, and localization in different tissues.

Classification of Hormones. Hormones can be classified according to their structural and biochemical similarities. This procedure is particularly useful because the wide distribution of hormonal substances in nonendocrine tissues makes it difficult to assign them to specific endocrine organs. Progress in structural analysis and isolation of natural compounds is rapid, especially as regards peptide and protein biosynthesis from genes expressed in different tissues and posttranslational processing to mature hormones [24]. Peptide and protein hormones and their precursors may be secreted directly into the bloodstream (*endocrine secretion*), released from nerve endings into specialized local vascular systems (*neuroendocrine secretion*), or secreted into the surrounding tissue (*paracrine secretion*).

A contemporary definition of hormones must include the peptides and amino acid derivatives (neurotransmitters) produced and secreted by specialized neuroendocrine and endocrine cells in the endocrine hypothalamus [14], the gastrointestinal tract [25], the right atrium of the heart [25], the endocrine pancreas [26], the thymus [27], and other organs not exclusively devoted to hormone secretion and synthesis. Neuroendocrine cells with a dual role of hormone secretion and neurotransmitter release are found in the adrenal medulla and sympathetic ganglia. Surprisingly, many hormone products that resemble the peptides of the central nervous system and the gastrointestinal hormones are found in the skin of amphibia.

Advances in structural elucidation, chemical synthesis, and specific detection methods (radioimmunoassay, immunohistochemistry, high-performance liquid chromatography) have disclosed a surprisingly wide dissemination of hormone-producing cells in tissues that were previously inaccessible to investigation by surgical removal and chemical substitution. The structural analysis of peptides, proteins, and their precursors by gene sequencing is progressing much more rapidly than the biological evaluation of new hormone candidates. It is therefore helpful to discuss hormones according to their chemical and functional families rather than according to hormone-producing glands.

The most important chemical groups are the *steroid hormones* (Chaps. 3, 4, 5), [20] closely related to their common precursor, cholesterol, and the *peptide and protein hormones* [7] composed of amino acids which form chains of different length and conformation, frequently containing discrete subunits (→ Peptides and Protein Hormones). The hormone group of *amino acid derivatives* (Chap. 2) comprises the *thyroid hormones*, and the hormonally active *catecholamines* and *indolamines*. The biogenic amines have multiple functions as hormones and neurotransmitters in the central and autonomous peripheral nervous systems. Hormone candidates are the *kinins* [12], [28] and *prostaglandins* [29], [30], which are a group of widely distributed endogenous substances with marked hormone-like activities.

1.1.2. Hormone-Producing Systems

Many hormones are integrated in complex regulatory systems. Their secretion may be activated or suppressed by hypothalamic and/or pituitary control, and the central regulation is in turn under the influence of feedback signals from peripheral hormone-responsive organs and tissues. Hormones with a homeostatic function maintain a constant internal environment by adaptation to changing external conditions (thyroid hormones and adrenal steroids). Hormones may be secreted throughout life if they are indispensable for metabolic activities (e.g., glucose- and calcium-regulating hormones), or at certain times of ontogenesis such as sexual maturation (gonadotropins and gonadal steroids).

Classical hormone-producing organs are the anterior and posterior pituitary gland, the thyroid gland, the parathyroid glands (epithelial bodies), the endocrine pancreas, the adrenal glands (adrenal cortex and adrenal medulla with chromaffin cells), the gonads, and the pineal gland. Destruction of these glands is associated with characteristic deficiency syndromes. Transient hormone-secreting organs also exist, e.g., the corpus luteum of the ovary and the placenta.

Site of Origin and Biosynthesis. To identify a hormone's site of origin, an endocrine gland is removed to study the resulting functional deficiencies and the effect of hormone substitution. More recent techniques to identify hormone-producing cells in non endocrine organs are immunohistochemistry and active immunization of animals against a hormone to neutralize its effects and define its function. Such studies sometimes disclose characteristic deficiency symptoms, but often fail due to compensatory mechanisms which replace the defective link [15].

Cells of common embryological ancestry may produce similar functional compounds even after their integration into nonendocrine tissues during ontogenesis. One hypothesis assigns a specific role to neuroendocrine Amine Precursor Uptake and/or Decarboxylation cells (APUD cells) which contain peptides and neurotransmitter-like amino acid derivatives. Such cells are widely found in the endocrine hypothalamus, anterior pituitary, parathyroid, and the gastrointestinal tract [31]. Biosynthesis of steroids is limited to the adrenal cortex. testes, and ovaries; the liver and kidneys metabolize steroid precursors to active calciferol derivatives. The thyroid hormones (thyroxine, triiodothyronine) are synthesized from monoiodotyrosine and diiodotyrosine. In the pineal gland, the hormone melatonin is synthesized from the precursor tryptophan. Remarkably, the placenta as a transient hormone-producing structure can synthesize peptides (e.g., LHRH, also called gonadorelin), proteohormones (e.g., HCG, human chorionic gonadotrophin), and steroid hormones (e.g., estrogens and progestins). Similarly, the corpus luteum of the ovary (a transient structure formed after ovulation) can secrete steroid hormones (estrogens and progestins), peptides (oxytocin), proteohormones (relaxin), and tissue hormones derived from arachidonic acid (prostaglandins).

Mode of Secretion. Many hormones are released directly into the bloodstream and reach their target organs over considerable distances. However, special modes of secretion also occur. Tissue hormones with direct (paracrine) action on neighbouring cells are secreted directly into the surrounding tissue and regulate cell-to-cell interaction. Some neurosecretory hormones (oxytocin, vasopressin) are transported to their site of storage in vesicles by axonal transport and released upon appropriate stimuli. The hypothalamic regulatory hormones reach the anterior pituitary gland (their target organ) by a special vascular pathway, the portal vessel system [14]. The pituitary gland is a composite, complex organ that consists of the anterior lobe (adenohypophysis), and a posterior lobe (neurohypophysis). The adenohypophysis contains several distinct cell types that secrete TSH, LH, FSH, growth hormone, prolactin, and corticotropin as well as proteohormones with presently undefined functions (endorphins, lipotropin, melanotropin). The posterior pituitary gland is a storage organ for two oligopeptide hormones, vasopressin and oxytocin that are synthesized in hypothalamic centers and transported to the posterior pituitary gland by axonal transport in neurosecretory cells, together with their binding proteins, the neurophysis.

The hierarchy of hormonal systems is summarized in Figure 1. Hierarchic hormone systems with two or three levels of control are regulated by hypothalamic hormones in the hypophysiotropic zone of the hypothalamus. These regulatory peptides activate or inhibit secretion of the anterior pituitary hormones. The *pituitary glandotropic hormones* (TSH, ACTH, FSH, LH) regulate their target tissues indirectly by hormone secretion of peripheral endocrine glands (thyroid and adrenal glands, gonads). The *pituitary effector*

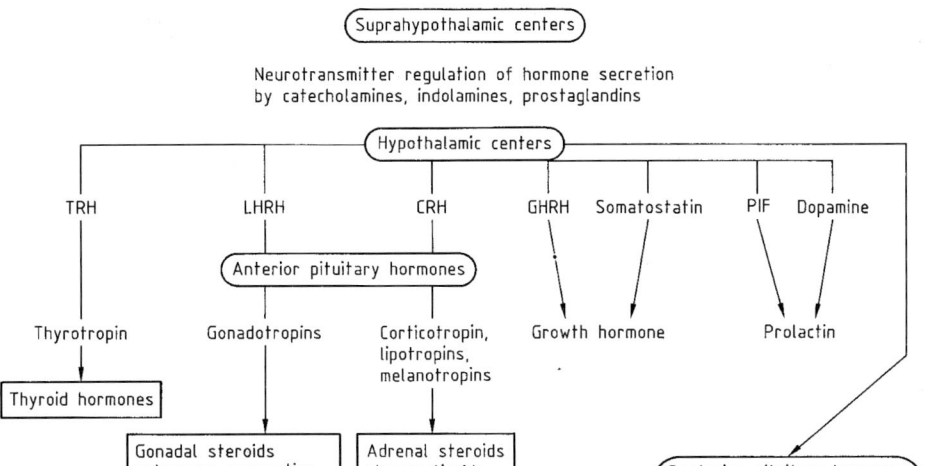

Figure 1. Hierarchy of hormone systems

hormones (growth hormone, prolactin) act directly on target organs (Fig. 2), regulating proliferation and metabolism of peripheral tissues; in this respect they are similar to the calcium-regulating, glucose-regulating, and gastrointestinal hormones. Steroid hormones secreted by the gonads and adrenals act directly on target cells to induce specific proteins which regulate gene expression.

Control of Hormone Secretion. An important principle in the homeostatic regulation of body functions by hormones is their organization in hierarchic *feedback systems*, which may operate independently or in interdependence. Secretion of effector hormones is regulated by the feedback of their respective substrate (e.g., insulin by glucose, or parathormone by calcium). Secretion of glandotropic hormones is controlled by the feedback of their products (e.g., steroids) synthesized by the target organ. The serum concentration is monitored by hypothalamic and/or pituitary receptors. In the endocrine hypothalamus (hypophysiotropic zone), the secretion of regulatory factors (peptides and neurotransmitters) drives the endogenous rhythms of pituitary hormones [32]. The corticotropin releasing hormone (CRH) induces the secretion of ACTH (corticotropin), which has a marked diurnal rhythm with an early morning peak [33]. The gonadotropin releasing hormone GnRH, also known as gonadorelin or luteinizing hormone, LHRH) has a pulsatile circhoral rhythm which is reflected by pulsatile gonadotropin secretion, the amplitude and frequency of gonadotropin pulses being regulated by gonadal steroids [34]. The secretion of gonadotropin in a multiloop feedback system is summarized in Figure 3. The growth hormone releasing hormone (GHRH) activates the secretion of growth hormone (somatotropin), which has a

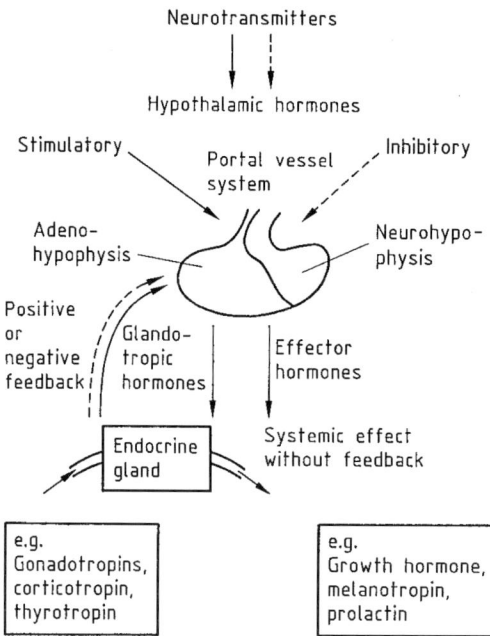

Figure 2. Secretion of pituitary hormones

marked endogenous sleep-associated rhythm closely coupled to rapid eye movement activity of the central nervous system [35]. Hormone secretion can also be controlled by neural regulation (e.g., prolactin by the suckling reflex of nursing), and is often modulated by the autonomous nervous system.

1.1.3. Structural Analysis and Synthesis

In hormone chemistry and biochemistry, the constraints of classical organ-related endocrinology are replaced by new concepts based on chemical, biochemical, and functional similarities, which take full advantage of modern methodology for identification of minute amounts of hormones, their precursors, and their metabolites in various tissues. New natural compounds (hormone candidates) can be identified by classical extraction, purification, and isolation followed by structural analysis and chemical synthesis, or by indirect structural analysis via gene sequencing, and biosynthesis by recombinant DNA technology. Their physiological and pharmacological functions must be established in biological studies.

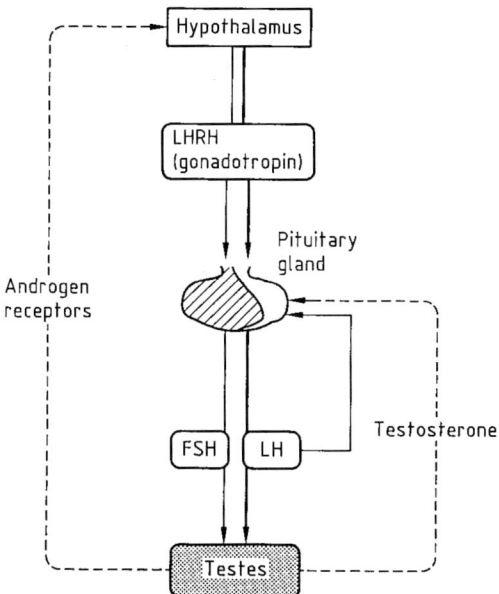

Figure 3. Feedback regulation of gonadotropin secretion

The pituitary secretion of LH can act directly (short feedback) on the secretion rate. The testicular steroid secretion can also modulate gonadotropin secretion (long feedback). In the hypothalamic centers the LHRH pulsatility is controlled by the binding of gonadal steroids to receptive cell areas.

1.1.4. Receptors and Mechanisms of Action

Peptide and protein hormones act on specific receptors located in the cell membrane, whereas steroid hormones act on receptors located in the cytoplasm and cell nucleus. The formation of a hormone–receptor complex activates specific intracellular processes [13].

Membrane receptors for peptides and protein hormones are coupled to adenyl cyclase [36] or phospholipase as the intracellular effector system; they activate protein kinases, which regulate protein synthesis and the calcium permeability of the cell membrane and intracellular structures [37].

The binding capacity of receptors for peptides and protein hormones depends on the concentration of their ligand in the endocrine cells. The receptors are expressed upon physiological stimulation, their number increases with the ligand concentration (up-regulation), and declines when the physiological level of stimulation is exceeded. During supraphysiological stimulation, the receptor response is desensitized, and binding capacity is down-regulated (i.e., the number of receptors is reduced). The binding of hormone analogues with enhanced resistance to degradative enzymes often reduces the concentration of available receptors. Receptors are removed by enhanced coupling between receptors (microaggregation), followed by internalization of hormone–receptor complexes, and intracellular receptor degradation.

Steroid hormones bind to intracellular (cytoplasmic or nuclear) receptors. The activated steroid-receptor complexes are translocated to the cell nucleus, where they regulate the synthesis of enzymes or functional proteins by means of gene expression [18], [38]. The binding capacity of the steroid hormone receptors is regulated in a dose-dependent manner. Steroid receptors are generally induced in the presence of their ligands and decrease with declining secretion of the relevant steroid.

1.1.5. Therapeutic Use of Hormones

Natural hormones provide valuable leads for structure – activity studies. Numerous steroid hormone analogues have been introduced into therapy, and synthetic analogues of peptides have also become valuable diagnostic and therapeutic aids. The larger proteins previously obtained by extracting biological material are now produced more economically by recombinant DNA technology [2]. The enhancement of specific biological activities or modification of the activity spectrum toward more desirable effects are the general aim of structure – activity studies.

1.2. Peptide and Protein Hormones (→ Peptides and Protein Hormones)

1.2.1. Hypothalamic Hormones

The hypothalamic hormones comprise peptides which regulate anterior pituitary function (*neurohumoral mechanism*), and others (vasopressin and oxytocin) which are stored in the posterior pituitary and released upon demand (*neurosecretory mechanism*). Paracrine secretion of hormones into the surrounding tissue is the most ancient phylogenetic transmission principle. The neuroendocrine secretion of vasopressin and oxytocin is analogous to the more specific and effective transmission between nerve cells that involves the synaptic release of amine or peptide neurotransmitters. Vasopressin and oxytocin are synthesized in the hypothalamic neurosecretory cells and are transported along the cell axons to their storage vesicles in the posterior pituitary gland. The hormones are structurally similar nonapeptides that are stabilized by intramolecular sulfhydryl bridges. *Vasopressin* controls water reabsorption by the kidney and osmolality in the intravascular space. At pharmacological concentrations, vasopressin has marked cardiovascular effects (vasoconstriction and hypertension). Numerous vasopressin analogues have been synthesized to enhance antidiuresis, and to reduce hypertensive activity [39]. A second group of vasopressin agonists is used for the local control of bleeding by vasoconstriction. A central role for hormonally inactive vasopressin fragments has been postulated in the control of memory and learning. *Oxytocin* acts primarily on the contractility of smooth muscle of the uterus during parturition, and on the myoepithelium of the lactating mammary gland. Surprisingly, oxytocin has recently been identified in the ovary, but its physiological role in this organ is unknown. Detection of other established hormones in tissues where the occurrence was previously unknown is increasing due to improved microanalytical techniques.

1.2.2. Proopiomelanocortin Hormones

Peptides and proteins are generally synthesized as large prehormone molecules and processed to smaller active substances. A complex hormone precursor found both in hypothalamic and anterior pituitary tissue gives rise to a family of opioid, corticotropic, and melanotropic peptides and proteohormones with diverse actions and functions [40]–[42]. The precursor, preproopiomelanocortin is processed to corticotropin (ACTH), melanotropin (MSH), lipotropins (LPH), enkephalins and endorphins.

Corticotropin (ACTH) is a linear polypeptide of 39 amino acids. This classical pituitary hormone is an interesting example of the posttranslational processing of a precursor to form a highly active product. Structure–activity relations of ACTH analogues have been studied extensively [43]. The full corticotropic activity is expressed in the N-terminal amino acid sequence 1–24; the C-terminal sequence 25–39 does not contribute to biological activity. Corticotropin analogues with protected N- and C-termini significantly increase and prolong cortisol and aldosterone secretion in the adrenal cortex. They consist of 17 or 18 amino acids with enhanced enzyme resistance and a prolonged half-life for plasma elimination. ACTH (1–39) is secreted after hypothalamic stimulation of the corticotropin-producing cells by the corticotropin releasing hormone (CRH), a hypothalamic polypeptide of 41 amino acids [33]. This type of regulation of a small pituitary hormone by a slightly larger regulatory hormone is a surprising exception to the rule that hypothalamic hormones are small molecules which release several multiples of their own weight from the anterior pituitary gland (amplifier mechanism). Secretion of ACTH in the human has a characteristic peak in the early morning hours and can be reproduced by pulsatile injection of human CRH. There are clear indications for a dual control of ACTH by CRH and arginine–vasopressin. A similar redundant dual control system is found for mineralocorticoid secretion, aldosterone being secreted by the adrenal cortex after stimulation by both ACTH and angiotensin II. The latter is formed from angiotensinogen by the successive action of the enzymes renin and angiotensin converting enzyme (ACE). ACTH is a potent stimulator of the adrenal cortex, its hypersecretion by corticotropic pituitary adenomas can override the adrenal feedback completely. ACTH regulates biosynthesis of adrenal steroids from cholesterol via the intermediate Δ-5-pregnenolone [7]. It also has extra-adrenal effects on pigmentary cells (melanotropic activity), and on mobilization of fatty acids from fat cells (lipolytic activity). When ACTH secretion is abolished by hypothalamic or pituitary lesions, the adrenal cortex rapidly involutes resulting in a characteristic hormone deficiency syndrome (Addison's disease) that is also observed after destruction of the adrenal cortex by tumors or infectious diseases. Excessive ACTH secretion results in marked melanotropin activity, inducing intense pigmentation of the skin.

Other polypeptides with significant sequence homologies to ACTH are also expressed on the proopiomelanocortin gene, namely α-, β-, and γ-MSH. These hormones have intense pigmentary activity in animals and humans. α-MSH is an N-terminal derivative that is acetylated by posttranslational processing. β-MSH is found as a circulating hormone in the human, but no physiological role has been established. The *lipotropins* [44] are of intermediate molecular mass (58–92 amino acids) and are also derived from the common ACTH–MSH precursor. Their physiological role is not fully established, they release free fatty acids from adipose tissue in vitro and in vivo. The *enkephalins* and *endorphins* (neuropeptides or peptide neurotransmitters) are regulators of central nervous activities, and of peripheral sensory functions such as perception of pain.

1.2.3. Glycoprotein Hormones

A group of pituitary hormones closely related by common subunits (an α-subunit and a hormone-specific β-subunit) comprises thyroid stimulating hormone (thyrotropin, TSH) and the gonadotropins follicle stimulating hormone (follitropin, FSH) and luteinizing hormone (lutropin, LH). The biological effects of these polypeptides are different despite extensive sequence homologies of their subunits. They undergo posttranslational glycosylation after assembly of the α and β-subunits [45]–[48].

Thyrotropin. The secretion of thyrotropin (TSH) is under hypothalamic control of thyrotropin releasing hormone (TRH, protirelin), a tripeptide with protected N- and C-terminus. To a limited extent, protirelin also stimulates the release of prolactin (PRL) and is involved in other hypothalamic control systems for the release and inhibition of prolactin secretion [14]. Numerous analogues of TRH have been synthesized with enhanced TSH releasing activity, or with significant neuropharmacological and cardiovascular effects but without endocrine activity [11]. TRH is useful in the diagnosis of thyroid function, but has not acquired a therapeutic role. Excess stimulation of TSH secretion by TRH can induce thyroid hyperplasia in animals, whereas human hyperthyroidism is generally caused by receptor-stimulating antibodies. Goiter formation in the human may result from disorders of thyroid hormone synthesis, which cause hypersecretion of TSH and proliferation of thyroid tissue. Human TSH has not found a therapeutic role, because thyroid hormone secretion is more easily substituted by synthetic thyroxin preparations.

Gonadotropins. The secretion of the gonadotropins FSH and LH is regulated by the hypothalamic luteinizing hormone releasing hormone (LHRH), also designated as the gonadotropin releasing hormone (GnRH). The decapeptide LHRH is used for diagnosis and therapy of infertility. With an intact pituitary gland, pulsatile infusion of LHRH stimulates gonadotropin secretion. This provides a method for treating human hypogonadotropic hypogonadism and is an alternative to gonadotropin preparations. Numerous agonists and antagonists of LHRH have been synthesized [9], [49] to control gonadal steroid secretion in diseases such as endometriosis, uterine leiomyoma, and steroid-dependent tumors (prostate and mammary carcinoma). The LHRH agonists induce persistent receptor desensitization and selective inhibition of gonadotropin secretion during daily high dose injections or continuous release from implants. The LHRH antagonists have similar inhibitory pharmacological effects on gonadal steroid secretion although their mechanisms of action are clearly different.

The gonadotropins are glycoproteins with about 180 amino acids, and multiple glycosylation sites [46], [50]. Follitropin (FSH) stimulates estrogen secretion from ovarian follicles and prepares them for subsequent induction of ovulation by lutropin (LH). LH has ovulatory activity and enhances ovarian steroidogenesis in synergy with FSH, in particular the secretion of progesterone by the corpus luteum after ovulation. In men, spermatogenesis is dependent on FSH, which acts on the germinal epithelium in synergy with LH and testosterone. LH stimulates the interstitial cells of the testes to secrete testosterone and small quantities of estrogens.

In animals and humans, FSH and LH are excreted in the urine; similar gonadotropins have been extracted from the serum of pregnant animals (pregnant mare's serum gonadotropin, PMSG). These urinary gonadotropins have been extensively studied, and are used for therapy. Urinary gonadotropins with high FSH- and LH-activities can be extracted from the urine of postmenopausal women for the stimulation of follicular maturation in infertile women (human menopausal gonadotropin, hMG).

1.2.4. Growth Hormone, Prolactin, and Placental Lactogen

Extensive structural homology is found between pituitary growth hormone (GH, somatotropin), pituitary prolactin (PRL), and human placental lactogen (hPL, see under Section 1.2.5). GH and prolactin contain 191 and 198 amino acids, respectively [51]. Both hormones have been extracted from human and animal pituitary glands, but only growth hormone has become a valuable therapeutic agent. It is now produced by recombinant DNA biosynthesis to avoid contamination with the neurotropic virus found in human pituitary tissue [52].

The amino acid sequence of *growth hormone* has marked species specificity [53], [54], therefore substitution therapy in children should only be performed with human growth hormone (hGH) to avoid antibody formation and loss of efficacy. The plasma elimination half-life of injected growth hormone is about 20 min. Growth hormone has many effects on metabolism (e.g., increased glucose utilization and lipid mobilization) and a marked anabolic effect on amino acid transport into cells and subsequent protein biosynthesis. Treatment with hGH can correct growth deficiency and short stature in children provided that epiphyseal closure of the bones has not occurred. Appropriate species-specific GH can also induce growth in farm animals. Many of the metabolic effects of GH are mediated by inducing the synthesis of growth factors [4], particularly the somatomedins. The GH-dependent somatomedins stimulate the formation of new cartilage; increased incorporation of sulfate into newly formed cartilage (sulfation factor) is a sensitive bioassay. During lactation, growth hormone acts in synergy with sexual steroids and prolactin on milk production. Hypersecretion of GH in acromegaly is manifested by pathological growth of the hands and feet, facial skeleton, enlargement of the skull, and heart, frequently followed by cardiac insufficiency.

Despite marked sequence homology with growth hormone, the growth-inducing activities of prolactin (PRL) and human placental lactogen (HPL) are negligible. *Prolactin* contributes to the maintenance of corpus luteum function in animals, and stimulates milk production during lactation, in synergy with GH and sexual steroids. Prolactin secretion appears to be under dual hypothalamic inhibitory control effected by release of dopamine [55] and specific peptide factors (prolactin inhibiting factor, PIF) into the pituitary portal vessels. Hypersecretion of PRL is found in pituitary micro- and macroadenoma. The insufficient endogenous inhibition of PRL secretion by endogenous dopamine can often be remedied by using potent dopamine agonists [56].

1.2.5. Placental Hormones

The placenta secretes structural homologues of TSH, FSH, LH, GH, and prolactin. A placental gonadotropin, human chorionic gonadotropin (choriotropin, hCG), is extracted from urine of pregnant women in large quantities for therapy of infertility; it is a glycoprotein that is similar to FSH and LH [57]. Other placental proteins (human placental lactogen and a TSH-like glycoprotein) have not found therapeutic applications [51]. Numerous oligopeptides occur in placental tissue, but are apparently not secreted during pregnancy.

1.2.6. Calcium-Regulating Hormones

There are two groups of calcium regulating hormones, the rapidly acting polypeptides calcitonin and parathormone (PTH), and the slowly acting steroids (calciferols, see Chap. 5). The polypeptides allow rapid adjustment of the serum calcium concentration to physiological requirements. *Calcitonin* is a species-specific polypeptide of 32 amino acids (M_r 3700); salmon and chicken calcitonin are among the most potent natural calcitonins [58]. The hormone is synthesized in the parafollicular C cells of the human thyroid gland, and in the ultimobranchial gland of several species of fish. Calcitonin lowers elevated serum calcium and phosphate levels, inhibits the activity of osteoclasts (cells inducing the resorption of bone), decreases renal reabsorption of calcium and phosphate ions, and decreases absorption of calcium from the gastrointestinal tract. Calcitonin acts on specific peptide receptors in the bone, repeated administration causes progressive receptor desensitization. In the brain, a calcitonin-gene related peptide is found with unidentified central regulatory effects [24]. Calcitonin has analgesic effects, which may be mediated by brain and spinal cord receptors.

The physiological antagonist of calcitonin, *parathormone* (PTH) is a species-specific polypeptide of 84 amino acids; the N-terminal 34 amino acids are required for full biological activity [59]. Parathormone is synthesized in the parathyroid cells (epithelial bodies), which in the human are found close to the thyroid gland. The effects of PTH include rapid mobilization of calcium from the bone matrix by activation of osteoclasts, activation of phosphate excretion by the kidney via inhibition of reabsorption, concomitantly increased renal excretion of calcium, and enhanced absorption of calcium from the gut. Endocrine-active tumors may secrete parathormone and induce hypercalcemia, which responds rapidly to calcitonin injections.

1.2.7. Cardiovascular Hormones

Three groups of peptides contribute to the regulation of blood pressure and intravasal volume: the angiotensins, the atrionatriuretic peptide (ANP), and the kinins [12]. Marked vasoconstriction and blood pressure increase is found after the stimulation of angiotensin I production. Angiotensin I is a nonapeptide that is produced from the precursor angiotensinogen by the enzyme renin. Renin is secreted by the juxtaglomerular kidney cells upon reduction of the glomerular filtration rate and hypoxia. Angiotensin I is converted to the octapeptide angiotensin II by the angiotensin converting enzyme (ACE). Numerous angiotensin antagonists have been synthesized as antihypertensive (blood pressure-lowering) peptides, but all have intrinsic vasoconstrictor activity. Angiotensin antagonists can be used to diagnose angiotensin-induced hypertension because they competitively inhibit the increased angiotensin concentrations found after renal artery occlusion. The di- and tripeptide ACE inhibitors captopril and enalapril, are used for the long-term therapy of hypertension; they have found great importance due to their high efficacy after oral absorption.

The *atrionatriuretic peptide* (ANP) is a functional antagonist of angiotensin II. It occurs in the right atrium of the heart and is released upon acute dilatation of the right atrium [60], [61]. ANP is synthesized as a preprohormone complex of 152 amino acids, and post-translationally cleaved to an active peptide containing 28 amino acids. ANP agonists consisting of peptides containing 22 or 23 amino acids are available; they have markedly enhanced vasodilatory and blood pressure-lowering activity, as well as acute and sustained stimulatory effects on sodium excretion and urinary flow rate.

The *kinins* are also physiological antagonists of angiotensin and are found in many tissues. They are nona-, deca- or undecapeptides that are released by the enzyme trypsin from the α-globulin fraction of serum [12]. All kinins are potent vasodilators; they lower blood pressure acutely, increase capillary permeability, and rapidly induce local tissue reactions characterized by edema, pain, and acute inflammation. The kinin peptides cause contraction of smooth muscle in the intestinal tract and the uterus. They can be secreted in significant quantities by endocrine-active tumors of the gastrointestinal tract. Many bradykinin-derived peptides have been studied without confirmation of therapeutic usefulness.

1.2.8. Glucose-Regulating Hormones

The protein hormones insulin and glucagon have vital effects on energy metabolism, cell permeability, and the homeostasis of cell function. Insulin and glucagon are the most important physiological factors in the control of cellular glucose concentrations; however, other hormones (e.g., growth hormone, glucocorticoids and somatostatin) also have significant effects. Insulin is used for substitution therapy of diabetes mellitus; glucagon is employed very infrequently.

Insulin is a polypeptide of 51 amino acids. It contains an α- and a β-chain that are synthesized as part of a larger precursor molecule of 84 amino acids (proinsulin). After post-translational removal of the connecting peptide from proinsulin, the two chains are joined by disulfide bonds [3]. Insulin is stored in the β-cells of the pancreas as a hexameric zinc-containing complex. The main action is to regulate cell permeability to glucose and potassium. The primary sequence of insulin is species-specific, but the biological action of human and animal insulins are similar. Porcine insulin is most closely related to human insulin. Human insulin can be obtained from porcine insulin by removing the eight N-terminal amino acids and replacing them with the corresponding human sequence. Human insulin has a plasma elimination half-life of about 20 min after intravenous injection. Numerous derivatives and pharmaceutical formulations have been developed to prolong the action of insulin by delayed absorption and sustained release from the subcutaneous injection site. Insulin pumps can deliver insulin according to a preset daily program; nasal administration of insulin is also being studied. The chemical synthesis of insulin has been described [6], but human insulin produced by recombinant DNA technology is now gradually replacing the previous porcine and bovine insulin preparations. Numerous structure activity studies have been performed with insulin analogues but they are not used in clinical medicine [6].

The physiological antagonist of insulin is *glucagon* [22], a linear polypeptide of 29 amino acids (M_r 3500). The biosynthesis of glucagon proceeds via a large preprohormone. The pancreatic β-cells contain biologically active glucagon and two glucagon-related peptides of unknown physiological significance. The gut contains an inactive precursor form of glucagon (glicentin). Glucagon is released from the pancreas upon a decrease in the blood-glucose concentration and stimulates gluconeogenesis. Its plasma elimination half-life is short (ca. 5 min). Glucagon has a strong inotropic action (i.e., enhances myocardial contractility). In contrast to insulin, glucagon has not acquired an important therapeutic role, although some of its pharmacological effects at higher doses have been widely studied.

1.2.9. Gastrointestinal Hormones

Numerous peptides and proteins with established or possible hormonal function are found in the salivary glands, stomach, duodenum, ileum, and colon. These gastrointestinal hormones influence acid and bile secretion, digestive enzyme production, and gastrointestinal motility [5], [10], [25], [62]. *Tissue hormones* such as kinins and neurotensin have been isolated from the salivary glands, but their physiological role is doubtful [12]. In the stomach, *gastrin* is found in three different molecular forms with 13, 17, and 34 amino acids [16]. It is an N-terminally processed peptide whose biological activity (stimulation of acid secretion in the stomach) is associated with its C-terminal 14–34 sequence. Gastrin has extensive sequence homology with the C-terminal active sequence of *cholecystokinin* (CCK), which is synthesized in the cells of the duodenal mucosa. Cholecystokinin stimulates contraction of the gall bladder and release of bile into the duodenum; it also enhances the secretion of pancreatic enzymes. The biological activity of CCK is attributed to a linear sequence of 33 amino acids (M_r 3883). Several smaller C-terminal forms of CCK are found in the duodenum and in central nervous tissue [63]. The central role of CCK fragments may be in the control of appetite as a satiety signal. Another duodenal hormone is *secretin*, a linear polypeptide with 28 amino acids (M_r 3000). Secretin stimulates the release of bicarbonate-rich pancreatic fluid which neutralizes the acid gastric juice when it reaches the duodenum. Secretin has significant sequence homologies with glucagon and other gastrointestinal peptides such as the vasoactive intestinal polypeptide and gastrointestinal polypeptide. *Motilin* is a further gastrointestinal polypeptide (22 amino acids) which stimulates gastric motility. Tumors of the gastrointestinal tract may contain and secrete a large variety of hormonally active peptides, e.g., growth hormone releasing hormone (GHRH) and its fragments, somatostatin, vasoactive intestinal peptide, and gastrin [5].

1.2.10. Growth Factors and Related Peptides

The growth requirements of cultured cells have been studied extensively. Specific growth factors appear to be involved in cell replication, physiological tissue repair, and pathological tumor growth [4]. Their secretion may be activated by human oncogenes and viruses [64].

The growth factors are produced by the tissues in response to growth hormone and other hormones. For example, the *somatomedins* are produced in response to the action of growth hormone; and *the estromedins* in response to the action of estrogens [65]. Numerous closely related groups of factors have been described such as the nerve growth factors [4], the epidermal growth factor and transforming growth factor alpha, the basic and acidic fibroblast growth factor [66], the endothelial cell growth factor [67], the platelet-derived growth factor, and the transforming growth factor beta [68]. The hormonal roles of these regulatory polypeptides are not clearly established because they have many pharmacological effects in vitro, and definition of their physiological functions is difficult.

1.2.11. New Hormonal Peptides and Proteins

The rapid progress of indirect structural analysis by gene sequencing and recombinant DNA technology has resulted in characterization and biosynthesis of many new hormonal factors. An example is *erythropoietin*, a glycoprotein found in the kidney which stimulates erythrocyte production [69] and hemoglobin synthesis. Local ovarian hormones have been identified such as relaxin [70], [71], and the Muellerian-inhibiting substance (MIS). The MIS is a glycoprotein of 300–350 amino acids that is synthesized as a preprohormone of 560 amino acids but its hormonal role is difficult to assess [72]. It controls the development of the male phenotype by suppressing development of the (female) Muellerian duct during embryonal differentiation. The MIS may have a pharmacological use to suppress proliferation of ovarian and endometrial tumors. Another ovarian protein regulates FSH secretion in vitro and in vivo. It consists of two subunits, which may associate to form either an FSH-inhibiting heterodimeric protein (*inhibin*) or an FSH-releasing homodimeric protein (*activin*) [73], [74].

1.3. Amino Acid Derivatives

The catecholamines, indolamines, and the more complex thyroid hormones are amino acid derivatives.

1.3.1. Catecholamines and Indolamines

Catecholamines have a dual role as cardiovascular and stress hormones. They are found in the chromaffin cells of the adrenal medulla and in sympathetic nerve endings; they also act as neurotransmitters in the central nervous system. *Noradrenaline* (norepinephrine) and *adrenaline* (epinephrine) are neurotransmitters that control hormone secretion by modulating the endocrine activity of the hypothalamus and pituitary gland. They are also involved in the central and peripheral regulation of blood pressure, and in the modulation of metabolic functions by the autonomous nervous system [75]. The most important indolamine is *melatonin*, a substance found in the pineal gland of animals and humans [76]–[78]. Melatonin inhibits sexual development in young animals before puberty (antigonadotropic hormone), its function in adults is not known.

1.3.2. Thyroid Hormones

The thyroid hormones thyroxine (3,5,3′,5′-tetraiodothyronine, T_4, and 3,5,3′-triiodothyronine, T_3) are iodinated derivatives of thyronine and are formed by oxidative coupling of the precursors 3-monoiodotyrosine and 3,5-diiodotyrosine. Thyroid-specific enzymes are responsible for the iodination of tyrosine. Numerous thyroid hormone analogues have been synthesized but do not have practical benefits for therapy [79]. Thyroxine (T_4) is stored in the thyroid follicles in the form of thyroglobulin (a glycoprotein with iodinated tyrosyl residues). The thyroid stimulating hormone (TSH) from the anterior pituitary stimulates release of thyroxine into the blood stream. Thyroxine serves both as a direct stimulator of metabolic activity, and as a prohormone for its more active metabolite, thyronine (T_3). The thyroid hormones regulate the adaptation of body temperature and metabolic rate to exogenous conditions. The pituitary–thyroid system is largely independent of hypothalamic activation and can function in animals with brain lesions that eliminate endogenous TRH secretion. Several synthetic compounds that inhibit the biosynthesis of thyroid hormones have been found, they act primarily by inhibiting thyroid hormone iodination, or by restricting the uptake of iodine into the thyroid gland [75].

1.4. Steroid Hormones

Steroid hormones share cholesterol as their common precursor [18], [20] and can be divided into three main physiological groups: the *gonadal steroids* produced in the ovary (estrogens and progestins) and testis (androgens), the *adrenal steroids* produced in the adrenal cortex (glucocorticoids and mineralocorticoids), and the *calcium-regulating sterols* (calciferols). The calcium-regulating calciferols [80], derivatives of 7-dehydrocholesterol and ergosterol, are activated by hydroxylation in the liver (25-hydroxycalciferol) and in the kidneys (1,25-dihydroxycalciferol and 24,25-dihydroxycalciferol).

There are three structural groups, the C-21, C-19, and C-18 steroids.

1.4.1. Gonadal Steroids

The most important C-21 gonadal steroid is *progesterone*, which is secreted in large amounts by the corpus luteum and the placenta. The most important C-19 gonadal steroid is *testosterone*, which is responsible for the secondary sex characteristics in the male and supports spermatogenesis in synergy with follicle stimulating hormone (FSH). *Estrogens* are C-18 gonadal steroids that are secreted in large quantities by the developing ovarian follicles, and in small quantities by the Leydig cells of the testis. Natural estrogens, progestins, and androgens are of little use for therapy because of poor gastrointestinal absorption and insufficient duration of action. Since 1930 many groups of workers have synthesized steroid hormone analogues to obtain synthetic hormones with improved therapeutic properties for the control of pituitary function (steroid inhibition of estrogen or androgen secretion by gonadotropin suppression), for substitution therapy in deficiency syndromes such as hypogonadism or after the menopause, and particularly for the control of ovulation (contraceptive steroids). In general, the pharmacological activity of synthetic gonadal steroids is never limited to interaction with a particular type of receptor (i.e., estrogenic, androgenic, progestagenic, glucocorticoid, or mineralocorticoid); instead, the spectrum of biological and pharmacological activities is dose-dependent. For example, some synthetic C-19 steroids have a high progestagenic potency, but frequently retain androgenic effects. The synthetic estrogen 17-ethinyl-oestradiol displays high oral absorption due to its low metabolic inactivation and has become a major component of oral contraceptives. Stilbene compounds are highly active nonsteroidal estrogenic substances with high oral absorption rates. Steroidal and nonsteroidal receptor antagonists have also been developed which competitively inhibit the receptor binding and the biological effects of natural androgens, estrogens, and progestins [9]. Of particular importance are the steroidal antiandrogen cyproterone acetate and the nonsteroidal antiandrogen flutamide. In treatment of postmenopausal mammary carcinoma, the partial antiestrogen tamoxifen is most widely used [9]; nonsteroidal antiestrogens are in development. The pharmacological activities of synthetic gonadal steroids have been reviewed extensively [19]. Current work is aimed at developing new specific receptor antagonists (antiprogestins), and steroids with novel therapeutic activities (e.g., inhibitors of aromatase and 5-α-reductase).

1.4.2. Adrenal Steroids

The adrenal cortex contains glucocorticoids (cortisol and cortisone) and mineralocorticoids (aldosterone and desoxicorticosterone). These C-21 steroids [7] are of particular importance because of their life-supporting activity in substitution therapy of adrenal insufficiency.

The *glucocorticoids* regulate gluconeogenesis from amino acids, and serve as stress-adaption hormones. Their medical importance is due to the strong antiinflammatory and immunosuppressive potential. The glucocorticoids are synthesized in the zona fasciculata of the adrenal cortex and their diurnal pattern of secretion is closely linked to that of ACTH (corticotropin). Feedback between the adrenal steroids and the pituitary secretion of ACTH is very effective. During long-term therapy with synthetic glucocorticoids, endogenous ACTH secretion is often suppressed, and may result in reversible adrenal involution or irreversible adrenal atrophy. At higher doses, the glucocorticoids suppress the tissue reaction to inflammation, reduce cell division in the bone marrow and lymphopoetic tissues (thymus gland and lymph nodes), and acutely inhibit exudation by reducing capillary permeability. This pharmacological activity is of importance in the treatment of acute kidney disease because it suppresses pathological permeability of the glomerular filtration system for proteins, and prevents progressive loss of albumin. Glucocorticoids have catabolic and antianabolic activities, they enhance protein catabolism and gluconeogenesis, and mobilize calcium by lowering the anabolic stimulation of bone formation anc bone mineralization. Under glucocorticoid therapy, calcium is lost from bones resulting in steroid-induced osteoporosis.

The *mineralocorticoids* regulate water and electrolyte metabolism by enhancing the renal reabsorption of sodium and water, and the excretion of potassium and hydrogen ions. The most important mineralocorticoid in the human is aldosterone (sodium-retaining hormone), which increases sodium reabsorption in the distal tubule of the kidney. Together with vasopressin it also regulates the osmolarity and ionic composition of the intravascular space. Enhanced secretion of aldosterone by adrenal tumors may lead to hypertension. In cardiovascular disease and hepatic failure, circulating aldosterone may be elevated (secondary aldosteronism) stimulating the formation of edema due to pathological sodium retention. Specific receptor antagonists of aldosterone (spironolactones) are used in therapy to counteract the sodium-retaining effect [9].

1.4.3. Calciferols

The calciferols (calcium-regulating sterols) are secosteroids with an opened ring B (between C 9 and C 10) derived from 7-dehydrocholesterol or ergosterol. They have a significantly delayed onset of action because they induce intestinal calcium-binding proteins [80] by interaction with the genome, they also metabolize calcium and phosphate from the bone to provide a higher concentration in the extracellular fluid for osteoid mineralization.

1.4.4. New Synthetic Steroid Hormones

Recent progress in the development of synthetic steroid hormone analogues has resulted in new possibilities for the specific control of reproductive function. Receptor antagonists of progesterone have been developed for clinical studies on postcoital contraception by reducing or blocking corpus luteum formation [9]. These compounds also inhibit the action of cortisol and may stimulate the immune system by neutralizing the immunosuppressive effects of glucocorticoids.

1.5. Prostaglandins

The prostaglandins are endogenous compounds with hormone-like activities that all have the prostanoic acid skeleton and share arachidonic acid as their common precursor [29], [30]. They influence the regulation of blood flow, kidney function, and hormone secretion. For a detailed description, see → Prostaglandins.

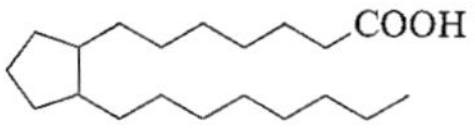

Prostanoic acid

Prostaglandin analogues of the E- and F-series have marked biological activities on reproductive functions, they lower progesterone secretion from the corpus luteum, and increase uterine contractility during pregnancy, and at parturition. Their multiple effects and sites of action make the clear assignment of classical hormonal functions very difficult, they should therefore be viewed as hormone candidates. In contrast to many peptides, proteins, and steroids which can be administered systemically without major tolerance problems, the prostaglandins are more difficult to handle because of their wide spectrum of activities that often includes severe side effects.

1.6. Hormone Antagonists and Analogues

Functional or competitive hormone antagonists (antihormones) can be divided into several groups with different modes of action [9]. The competitive receptor antagonists displace hormones from their receptors, and prevent physiological receptor activation by endogenous hormones. Such antagonists may be structurally related to the natural ligand (e.g., LHRH antagonists), or structurally unrelated, but of similar three-dimensional conformation (e.g., nonsteroidal antiandrogens). The noncompetitive inhibitors of hormone secretion often interfere with biosynthesis. They can act at a multitude of sites, as shown by the antithyroid drugs [75], the inhibitors of adrenal steroid secretion, aminogluthetimid and mitotan [9], and the inhibitor of gonadal and adrenal steroid biosynthesis, ketoconazole [9]. Future developments in drug design will be based on computer modeling of the active sites of hormone receptors, and of the three-dimensional conformation of synthetic hormone analogues capable of interacting with these receptors. Examples for nonsteroidal compounds interacting with steroid receptors are numerous, whereas similar developments are rare in the peptide and protein field, with the exception of structural similarities between opiates and endogenous opioid peptides [81], the encephalins and endorphins. There are until now no examples of nonpeptide drugs capable of blocking peptide receptor binding.

The concept of using natural hormones as lead compounds for synthetic modification by structure–activity studies has been particularly successful. Gonadal and adrenal steroids have been modified by chemical synthesis to obtain more potent compounds with enhanced metabolic resistance, specific receptor antagonists with selective inhibitory potential, and nonsteroidal compounds with simplified structures. The development of new steroidal and nonsteroidal molecules with improved biological and therapeutic characteristics is still in progress; there are indications for selective antiprogestins and antiglucocorticoids [9]. In the field of peptides and proteins, significant progress is being made by the widespread application of molecular biology and genetic engineering. Structural analysis of polypeptides and proteins by gene sequencing has afforded many new lead compounds which can be studied by recombinant DNA synthesis in bacteria, yeast, or mammalian cell lines [2]. The availability of these biosynthetic molecular probes may initiate structure–activity studies on highly selective and specific synthetic peptide hormones. The receptor binding of new synthetic hormone derivatives, and an understanding of the mechanisms of action of natural hormones has been crucial for a more effective approach to drug design. Homogeneous receptor proteins for peptides and steroids are sequenced and synthesized in genetic engineering. The understanding of the molecular basis of cellular activation is increasing, and the mechanisms of action of peptide, protein, and steroid hormones are becoming more transparent.

2. Amino Acid Hormones

2.1. Thyroid Hormones

The thyroid hormones are produced in the thyroid gland, an endocrine gland located in front of the trachea and below the larynx, which consists of two lobes, connected by an isthmus. The thyroid gland stores ingested iodine in the follicular cells and secretes the thyroid hormones thyroxine (L-tetraiodothyronine) and triiodothyronine into the blood, as required. These two iodinated amino acid derivatives affect metabolism in a variety of ways. They not only regulate energy turnover, water balance, and the metabolism of carbohydrates, proteins, fats, and minerals, but also affect growth and maturation.

In addition, the parafollicular C cells of the thyroid produce another type of hormone, calcitonin, a peptide hormone, which participates in the regulation of calcium metabolism. However, the exact physiological function of this hormone is not yet known. Another peptide hormone, parathormone, which also influences calcium metabolism, is secreted by the parathyroid glands.

2.1.1. Physiology and Therapeutic Uses

Biosynthesis of Thyroid Hormones. The healthy thyroid gland takes up approximately 75 µg of iodine per day, provided enough dietary iodine is available (150–300 µg/d). Iodine is used to iodinate the tyrosine residues of a thyroid-specific protein *thyroglobulin*. The resulting 3,5-diiodotyrosine groups in the peptide chain of thyroglobulin react with each other or 3-monoiodotyrosine to form thyroxine (3,5,3′,5′-tetraiodothyronine) or triiodothyronine, respectively. This thyroxine- and triiodothyroninecontaining thyroglobulin is stored in the colloid within the thyroid follicle. The normal human thyroid contains approximately 10 mg of iodine in this form. During secretion, thyroglobulin leaves the colloid and is hydrolyzed by the surrounding follicular cells. The liberated hormones, thyroxine (T_4) and triiodothyronine (T_3), are released into the blood at rates of ca. 100 and 8 µg/d, respectively.

The thyrotropic hormone of the pituitary hypophysis (thyrotropin or thyroid-stimulating hormone, TSH) stimulates the secretory function of the thyroid gland. The hypophyseal secretion of TSH, in turn, is stimulated by the hypothalamic hormone TRH (thyrotropin-releasing hormone; see also → Peptides and Protein Hormones) and inhibited by elevated levels of blood T_4 or T_3 in a feedback mechanism. In the blood, T_4 and T_3 are bound to serum proteins, principally thyroxine-binding globulin. Only 0.03% of T_4 and 0.3% of T_3 are present in the free (physiologically active) form. The biological half-life of T_3 is only ca. 19 h, whereas that of T_4 is ca. 190 h. Further formation of T_3 (approximately 30 µg/d) occurs in the peripheral tissues as a result of the deiodination of T_4. Because T_3 has approximately ten times the biological activity of T_4, thyroxine is considered to be the prohormone of the biologically more active triiodothyronine [87]. In addition to T_3, a second isomeric iodoamino acid, reverse T_3, (L-3,3′,5′-triiodothyronine, rT_3) is formed from T_4 in the periphery but has no apparent physiological activity [88].

Effect of Thyroid Hormones. Free thyroxine and especially free triiodothyronine probably bind to the nuclear and mitochondrial receptors, thereby activating protein synthesis or adenosine triphosphate production, respectively. The primary effects of the thyroid hormones are

1) stimulation of carbohydrate turnover, growth, and maturation in the central nervous system, skeletal and muscular systems, and genital organs; stimulation of heat production, oxygen consumption, cholesterol metabolism, turnover of free fatty acids, muscle contraction, heart rate, and cardiac output; and
2) inhibition of glycogen and protein synthesis (only at nonphysiological concentrations).

The amount of thyroid hormones required by the human organism varies between 50 and 300 µg/d, depending partly on the individual metabolism.

Thyroid Diseases. Thyroid diseases can affect the morphology (goiter, nodules, carcinoma) or the function (hyperthyroidism, hypothyroidism) of the thyroid. In many cases, thyroidal autoantibodies or iodine deficiency is the cause of the thyroid disorder. Hyperthyroidism is characterized by tachycardia, sweating, weight loss, diarrhea, etc. Hypothyroid children show abnormal physical and mental development. Hypothyroid adults tend to react slowly or to be depressed; additional symptoms are dry skin, obstipation, and sensitivity to cold.

Diagnosis. Diagnosis of thyroid disease is based not only on clinical symptoms but also on the measurement of serum concentrations of T_4, T_3, and TSH (basal or stimulated by the addition of TRH, i.e., the TRH test) and, if necessary, on thyroid uptake, iodine turnover, and the functioning of the hypophysis–thyroid feedback control system.

Therapy with Thyroid Hormones. (See also → Thyrotherapeutic Agents). Diseases of the thyroid gland that involve a thyroid hormone deficiency can be treated easily and successfully by oral administration of thyroid hormones. Treatment with levothyroxine is preferred because of the long half-life of this hormone. The sodium salt of levothyroxine in tablet form is absorbed to the extent of 40–60% [89], [90], or 70–85%, depending on the galenic preparation [91]. The daily dose of levothyroxine varies between 50 and 300 µg, depending on the severity of thyroid deficiency. The daily dose of liothyronine is correspondingly lower (10–100 µg). Injectable thyroid hormone solutions are used in cases of severe hypothyroidism (myxedema coma). Preparations containing both thyroid hormones are also available. Treatment with thyroid gland powder is no longer recommended because the standardization and stability of these preparations are problematic.

Therapy with thyroid hormones is sometimes recommended in certain types of obesity. Both dextrothyroxine (4–10 mg/d) and dextrotriiodothyronine (0.5–2 mg/d) are used in the treatment of disorders of fat metabolism, particularly hypercholesterolemia. L-Diiodotyrosine is also administered clinically; however, this preparation functions primarily as an iodine carrier because it is rapidly metabolized.

A number of analogues and derivatives of thyroid hormones have been synthesized and studied. Although substances with many times the thyromimetic activity of levothyroxine have been produced [92], none has found practical use. Triiodothyroacetic acid [*51-24-1*] (TRIAC), a degradation product of the thyroid hormones, occurs only in small amounts in the human body and exhibits a TSH-suppressing effect without the peripheral effects of levothyroxine or liothyronine. This compound has been successfully used in the treatment of hyperthyroidism caused by hypophyseal TSH-producing tumors [93].

2.1.2. Physical Properties

Levothyroxine [51-48-9] (L-3,5,3′,5′-tetraiodothyronine, T_4), $C_{15}H_{11}O_4I_4N$, M_r 776.87, mp 235–236 °C (decomp.), $[\alpha]_D$ +20.4° (2% solution in a 1:4 mixture of HCl and ethanol), $[\alpha]_D$ −5.8° (3% solution in a 1:2 mixture of 1 mol/L NaOH and ethanol), forms colorless, light-sensitive crystals. It is sparingly soluble in water, slightly soluble in methanol, and readily soluble in alkali or dilute mineral acids.

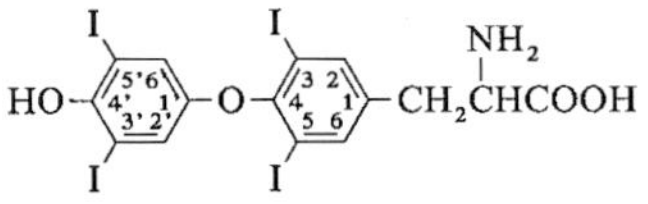

$$HO-\underset{I}{\overset{I}{\underset{3',2'}{\overset{5',6'}{\bigcirc}}}}-O-\underset{I}{\overset{I}{\underset{5\ 6}{\overset{3\ 2}{\bigcirc}}}}-CH_2\overset{NH_2}{\underset{|}{C}}HCOOH$$

The sodium salt of levothyroxine [25416-65-3] crystallizes from a hot solution of the free acid in sodium carbonate: $C_{15}H_{10}O_4I_4NNa \cdot 5\ H_2O$, M_r 888.94 [94, p. 476], [95, p. 642].

Liothyronine [6893-02-3] (L-3,5,3′-triiodothyronine, T_3), $C_{15}H_{12}O_4I_3N$, M_r 650.98, mp 233–234 °C, $[\alpha]_D^{24}$ +23.6° (5% solution in a 1:2 mixture of 1 mol/L HCl and ethanol), forms colorless crystals. It is slightly more water-soluble than thyroxine and readily soluble in dilute alkali [94, p. 268], [95, p. 643].

Reverse T_3 [5817-39-0] (L-3,3′,5′-triiodothyronine, rT_3) $C_{15}H_{12}O_4I_3N$, mp 206 °C (decomp.), $[\alpha]_D^{21}$ +16.7° (1% solution in a 1:1 mixture of 1 mol/L HCl and ethanol), forms colorless needles [96].

L-3,5-Diiodothyronine [1041-01-6] (T_2), $C_{15}H_{13}O_4I_2N$, M_r 525.08, has the following properties: mp 254 °C (decomp.), $[\alpha]_D^{25}$ +25.2° (5% solution in a 1:2 mixture of 1 mol/L HCl and 95% ethanol).

L-3,5-Diiodotyrosine [66-02-4] (DIT, iodogorgoic acid), $C_9H_9O_3I_2N \cdot 2\ H_2O$, M_r 469.02, mp 204° (decomp.), $[\alpha]_D$ +2.75° (4.8% solution in 1 mol/L HCl [97]), has the following structure:

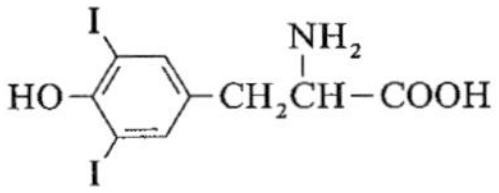

$$HO-\underset{I}{\overset{I}{\bigcirc}}-CH_2\overset{NH_2}{\underset{|}{C}}H-COOH$$

L-3-Iodotyrosine [70-78-0] (monoiodotyrosine, MIT), $C_9H_{10}O_3IN$, M_r 307.09, has the following properties: mp 204–206 °C, $[\alpha]_D^{22}$ −8.8° (4% solution in 1 mol/L HCl).

The corresponding D-isomers of the above iodoamino acids are also known. Dextrothyroxine is used therapeutically to lower cholesterol levels.

2.1.3. Production

In the thyroxin synthesis, first performed in 1927, 4-(4-methoxyphenoxy)-3,5-diiodonitrobenzene was converted into the corresponding aldehyde and then into D,L-diiodothyronine, which was iodinated to yield thyroxine [98]. This method was used commercially for a long time and produced a racemate. Resolution of the racemates of thyroxine has not yet been achieved.

The natural hormones levothyroxine and liothyronine could be produced on a large scale from protected L-3,5-dinitrotyrosine by inserting the iodine substituents after formation of the ether bond [99]. This synthesis also led to the production of D-T_4 and D-T_3. However, optically pure preparations could not be obtained easily.

A method adapted from the biological synthesis was used for some time to produce levothyroxine. Two molecules of L-diiodotyrosine are coupled at pH 9.5 in the presence of manganese dioxide at 60 °C. Levothyroxine is produced with a yield of 2.8%. The reaction of diiodotyrosine with 4-hydroxy-3,5-diiodo-phenylpyruvic acid proceeds similarly except that yields up to 50% can be obtained [100], [101].

A total synthesis of T_4 and T_3 was developed by Hoechst in 1963 and involves racemate resolution at the level of diiodothyronine [102], [103]. The reaction steps are shown below:

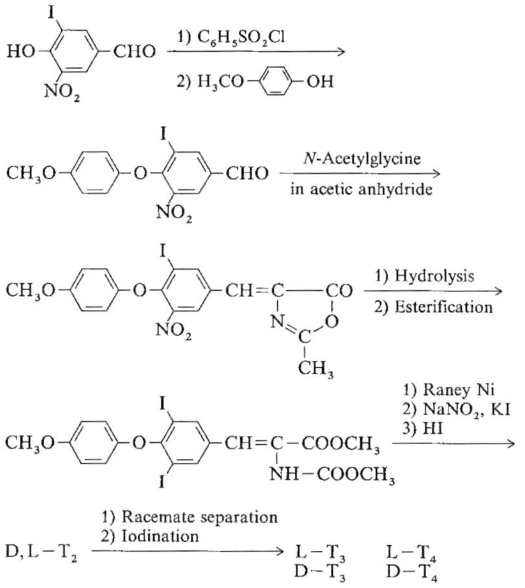

Most other methods used today in the large-scale production of thyroid hormones are modifications of the procedure described by HILLMANN in 1956 [104], [105]. The ether bond in thyronine is synthesized by using diaryliodonium salts and the only slightly reactive diiodotyrosine [106], [107].

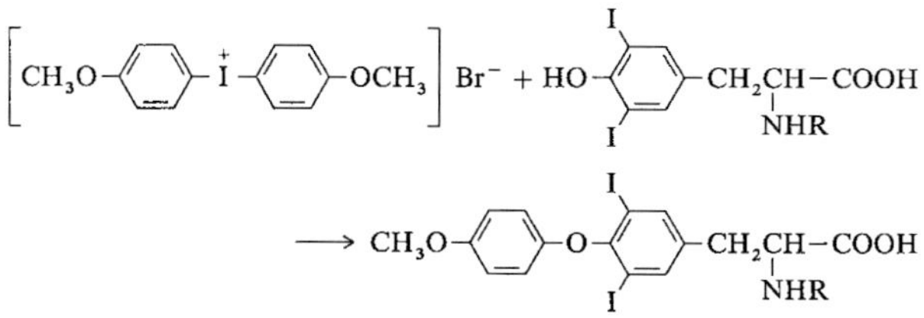

$$\left[CH_3O-\underset{}{\underset{}{\bigcirc}}-\overset{+}{I}-\underset{}{\underset{}{\bigcirc}}-OCH_3\right] Br^- + HO-\underset{I}{\overset{I}{\underset{}{\bigcirc}}}-CH_2\underset{NHR}{CH}-COOH$$

$$\longrightarrow CH_3O-\underset{}{\underset{}{\bigcirc}}-O-\underset{I}{\overset{I}{\underset{}{\bigcirc}}}-CH_2\underset{NHR}{CH}-COOH$$

L-Thyronine is made from L-tyrosine, obtained by protein hydrolysis (e.g., of wool). A number of methods are available for obtaining T_3 and T_4 by subsequent iodination [102], [104], [108]. The arylation of diiodotyrosine with iodonium salts permits the recovery of optically pure levo- and dextrothyroxine preparations. Especially in the case of dextrothyroxine, which is used therapeutically to lower cholesterol levels, the amount of L-isomer should not exceed 0.2% [109]. Chemically pure T_4 and T_3 preparations are obtained by preparative column chromatography [110]. Analogues of the thyroid hormones and ^{14}C-labeled preparations are also produced by the above methods [111]–[113].

Analysis and Testing. The following methods are used to test the identity and purity of iodoamino acids: iodine determination; UV absorption [114]; paper [115], thin-layer [116], and gas chromatography [117], polarography [118]; enzymatic analysis [109]; and radioimmunoassays.

Producers of thyroid hormones in Germany and Austria are Henning (Berlin), Hoechst (Frankfurt), and Sanabo (Vienna).
Trade names are listed in Table 1.

2.2. Hormones of the Adrenal Medulla

The hormones of the adrenal medulla are the catecholamines *adrenaline* (epinephrine) and *noradrenaline* (norepinephrine). The catecholamines act on α and β-adrenoreceptors in the central nervous system, in the autonomous nervous system, and in other tissues [119].

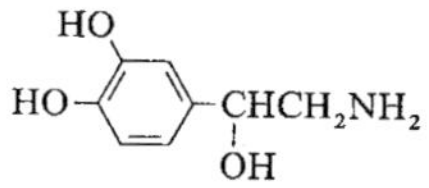

$$HO-\underset{HO}{\underset{}{\bigcirc}}-\underset{OH}{CH}CH_2NH_2$$
Noradrenaline

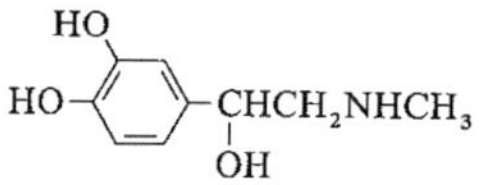

$$HO-\underset{HO}{\underset{}{\bigcirc}}-\underset{OH}{CH}CH_2NHCH_3$$
Adrenaline

They also modify the metabolism of many cells [120]. Both hormones are synthesized in the adrenal medulla. After transsection of the human adrenal gland, the central

Table 1. Selected trade names of thyroid hormone preparations

Trade name	Producer	Country
Sodium levothyroxine		
Eferox	Efeka	FRG
Eltroxin	Glaxo	Denmark, Great Britain, Switzerland
Euthyrox	Merck	Austria, Belgium, FRG
L-Thyroxin Henning	Henning Berlin	FRG
L-Thyroxine Roche	Roche	France
Levoroxine	Bariatric	USA
Levothroid	USV	USA
Levothyrox	Merck-Clevenot	France
Levothyroxine Sodium	Armour-Yamanouchi	Japan
Synthroid	Flint	USA
Thevier	Glaxo	FRG
Thyrex	Sanabo	Austria
Sodium liothyronine		
Cynomel	Uhlmann-Eyrand	Switzerland
Cynomel	Merrell	France
Tertroxin	Glaxo	Denmark, Great Britain
Thybon	Hoechst	FRG
Combination preparations of sodium levothyroxine and sodium liothyronine		
Novothyral	Merck	FRG
Prothyrid	Henning Berlin	FRG
Thyroxin-T$_3$	Henning Berlin	FRG
Combination preparation of sodium levothyroxine and iodine		
Jodthyrox	Merck	FRG
Injection solutions (sodium levothyroxine)		
L-Thyroxin inject	Henning Berlin	FRG
Synthroid	Flint	USA
Injection solutions (sodium liothyronine)		
Tertroxin	Glaxo	Denmark
Thyrotardin inject	Henning Berlin	FRG

adrenal medulla is clearly visible because it is darker than the surrounding adrenal cortex. The adrenal medulla and the adrenal cortex are regulated by different mechanisms, the adrenal cortex being under hormonal control, whereas the adrenal medulla is under nervous control. The blood-pressure increasing effect of an adrenal extract was described in 1985 by OLIVER and SCHÄFER [121]. The first chemical synthesis of adrenaline was performed in 1904 by STOLZ, the definitive elucidation of its structure by FRIEDMANN was in 1906.

Normally only noradrenaline is secreted; the medulla secretion of adrenaline is restricted to acute stressful situations which require brief, maximal activation of cadiovascular function, and enhanced central nervous tone. Noradrenaline administered intravenously or subcutaneously has a predominantly peripheral action on blood pressure and cardiovascular function. After subcutaneous injection, absorption is much delayed due to potent vasoconstriction. The adrenal medulla also produces *dopamine*, a precursor of noradrenaline.

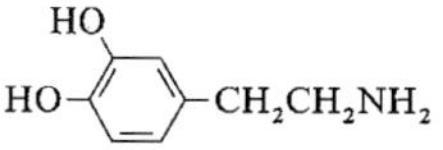

HO-C₆H₃(OH)-CH₂CH₂NH₂

Dopamine

Low concentrations of dopamine have therapeutic effects on specific dopamine receptors in the kidney, increasing the glomerular filtration rate in shock syndromes [122]. At increasing concentrations, dopamine stimulates α-adrenoreceptors, it is therefore administered by infusion at a controlled rate to avoid undesirable vasoconstriction. A lack of dopamine is also responsible for Parkinson's disease (→ Parkinsonism Treatment).

The catecholamines have major endogenous functions as synaptic transmitters in the central regulation of blood pressure [122]. Noradrenaline stored in synaptic vesicles is released from the presynaptic nerve endings and acts on receptors in the postsynaptic membrane. Its local concentration in the synapse is controlled by reuptake mechanisms. Noradrenaline also acts as a neurotransmitter in the autonomous nervous system.

Noradrenaline [*51-41-2*], levarterenol, norepinephrine, R(–)- or D(–)-2-amino-1-(3,4-dihydroxyphenyl)ethanol; $C_8H_{11}NO_3$, M_r 169.2, *mp* 217–218 °C (decomp.), sparingly soluble in water and ethanol;

Hydrochloride: $C_8H_{11}NO_3 \cdot HCl$, M_r 205.6, *mp* 145–146 °C, readily soluble in water, α_D^{25} –40° (6% aqueous solution);

Hydrogen tartrate: $C_8H_{11}NO_3 \cdot C_4H_6O_6$, M_r 319.3, *mp* 158–159 °C, α_D^{25} –10.7° (1.6% aqueous solution).

Adrenaline [*329-65-7*], suprarenin, epinephrine, R(–)- or D(–)-1-(3,4-dihydroxyphenyl)-2-methylaminoethanol $C_9H_{13}NO_3$, M_r 183.2, *mp* 211–212 °C, α_D^{25} –50–53° (5% in 0.5 mol/L HCl);

Hydrogen tartrate: $C_9H_{13}NO_3 \cdot C_4H_6O_6$, M_r 333.3, *mp* 149–152 °C, α_D^{25} –14.5°

Hydrochloride: $C_9H_{13}NO_3 \cdot HCl$, M_r 219.7, *mp* 206 °C.

The levorotatory isomers of both noradrenaline and adrenaline are 15–40 times more potent in several test systems. This is also a general characteristic of many other catecholamines.

Synthesis. Both compounds are obtained from 4-chloroacetylcatechol which is prepared from catechol by reaction with chloroacetic acid and $POCl_3$.

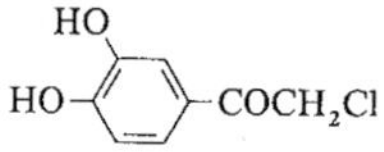

HO-C₆H₃(OH)-COCH₂Cl

For the preparation of adrenaline, 4-chloroacetylcatechol is reacted with methylamine, the keto group is converted to the hydroxy group by catalytic hydration and the racemic adrenaline base thus obtained is cleaved with tartaric acid. For noradrenaline, the 4-chloroacetylcatechol is reacted with ammonia.

Therapeutic Uses. Adrenaline and noradrenaline have limited therapeutic importance, because their duration of action is short and therapy of hypotension would require administration by infusion. Their rapid rate of disappearance after presynaptic release is a physiological prerequisite for rapid synaptic transmission. However, it has promoted the development of numerous synthetic catecholamine derivatives and indirect sympathomimetics [123] that act on the stored neurotransmitters in synaptic vesicles (→ Blood Pressure Increasing Agents).

Many catecholamine derivatives and sympathomimetic compounds have been developed in an attempt to obtain specific stimulation of α-adrenoreceptors and β(2)-adrenoreceptors involved in smooth muscle relaxation, e.g. as bronchodilatory compounds [123]. Research on nonselective β-adrenoreceptor blockers and selective β-blocking agents has been very successful [124] (see also → Beta Blockers). The indirect sympathomimetics displace catecholamines from their storage vesicles and thus temporarily enhance postsynaptic stimulation [125]. This effect is abolished if the storage vesicles are first depleted of catecholamines by administration of reserpine [126].

Clinical Uses. Adrenaline has important clinical applications in internal medicine, surgery, and dental medicine. It provides rapid relief in hypersensitivity reactions caused by anaphylactic and allergic reactions (by injection or infusion), in cardiogenic shock caused by myocardial infarction, and relieves respiratory distress due to bronchospasm when administered by oral inhalation. In surgery and dental medicine, adrenaline is used as a hemostatic agent; it is administered by infiltration or locally application on bleeding surfaces. In combination with local anesthetics it prolongs an anesthesia by preventing diffusion of the anesthetic from the site of injection. Adrenaline acts as a bronchodilatory and antiallergic compound, because it prevents the release of histamine and other mediators of inflammation [126]. Noradrenaline and related synthetic sympathomimetic amines that act predominantly on α-adrenoreceptors have similar clinical uses to adrenaline.

3. Steroid Sex Hormones (→ Steroids)

3.1. Introduction

The sex hormones, like the hormones of the adrenal cortex, are steroids with the following general formula:

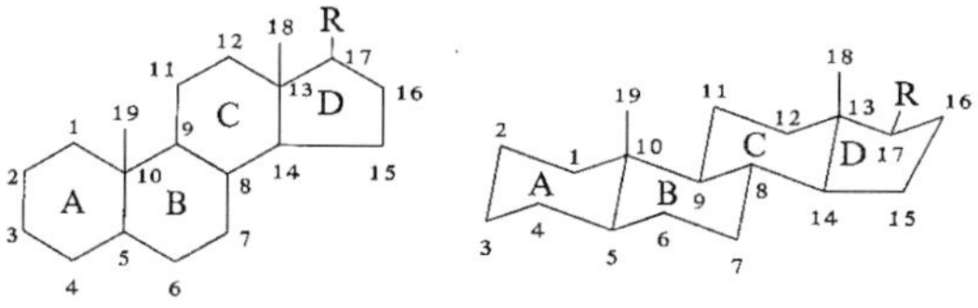

All substituents below the plane of the ring system are designated as α and those above the plane of the molecule as β. These orientations can be indicated with a dashed or solid line, respectively. The substituent R may be an alkyl residue or an oxygen function (for further details on steroid structures, see [127], [128]). These compounds are called "steroids" because they bear a great structural resemblance to the naturally occurring sterols [127]. The sex hormones are classified as

1) *androgens* (male sex hormones, e.g., testosterone);
2) *estrogens* (female sex hormones, e.g., estrone, estradiol, estriol); and
3) *gestagens* (female sex hormones, e.g., progesterone).

Naturally occurring steroids have several asymmetric carbon atoms (centers of chirality) and are optically active. They are defined as belonging to the "normal series" [127], [128]. Synthetic steroids are considered to belong to the normal series when C-10 or, if this position is not asymmetric (as in the estrogens), C-13 has the same configuration as in cholesterol. If all the centers of asymmetry in the steroid molecule have the opposite configuration to normal (i.e., mirror image of the normal form), the compound carries the prefix "ent" (enantiomer). A racemate is an equimolar mixture of the normal and the enantiomeric forms of a steroid molecule and is indicated by the prefix "rac" [128].

Of the large number of possible stereoisomers of sex hormones, only the isomer with the normal (natural) configuration is active (in this respect, estrogen is the simplest sex hormone; it has four centers of asymmetry, 16 stereoisomers, and 8 racemates). All other stereoisomers either are less active, are inactive, or exert other effects.

Cholesterol is the principal mammalian sterol and is biosynthesized from 18 molecules of acetyl coenzyme A (acetyl-CoA) by a complex route involving consecutive formation of mevalonic acid, farnesyl pyrophosphate, squalene, lanosterol, and cholesterol. For further details, see [133, pp. 20–70].

Cholesterol is the biochemical precursor of both the sexual and the adrenal steroid hormones. Figure 4 briefly summarizes the most important steps of cholesterol transformation, as well as the principal enzymes involved.

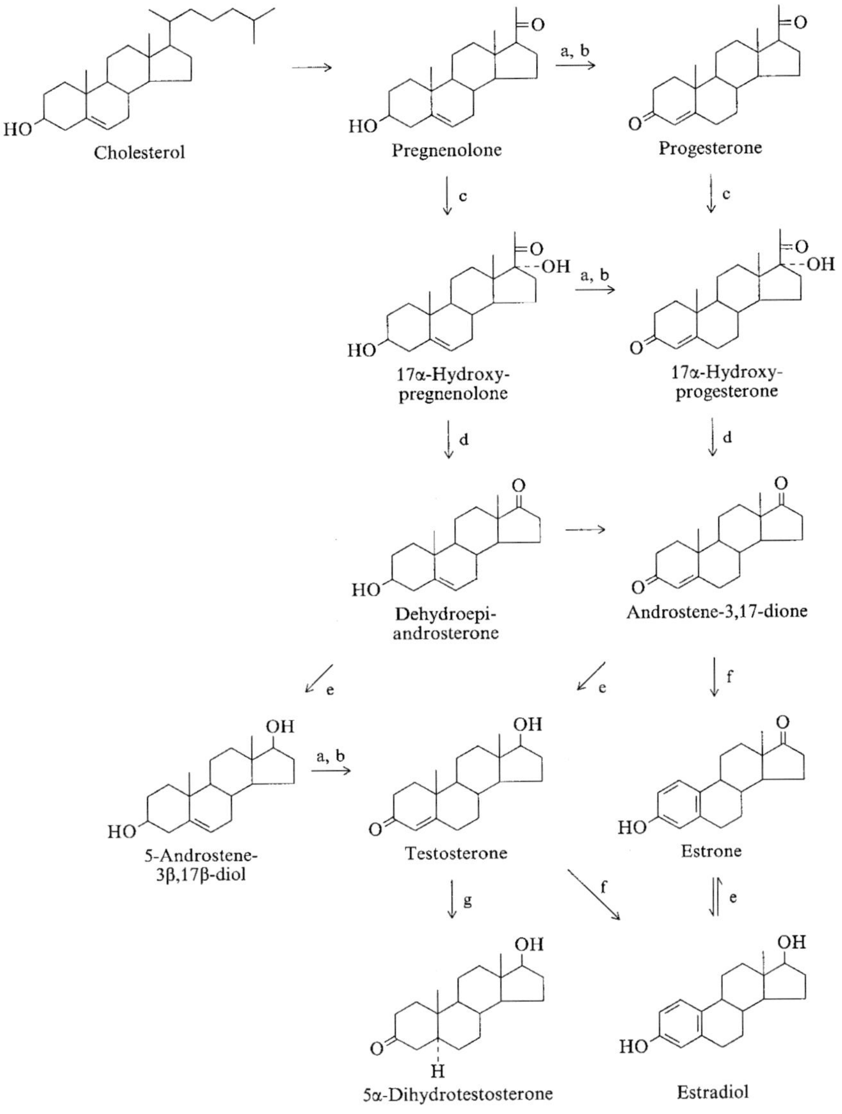

Figure 4. Biosynthesis of steroid sex hormones [130], [133]
Enzymes: a) 3β-Hydroxy-Δ⁵-steroid dehydrogenase; b) Steroid-Δ-isomerase; c) Steroid-17α-monooxygenase; d) C-17,20-Lyase; e) 17β-Hydroxysteroid dehydrogenase; f) Aromatase; g) 5α-Reductase

Commercial methods used in the production of sex hormones or their precursors can be divided into four groups:

1) *Partial chemical synthesis* of steroid hormones from natural materials, that have the steroid nucleus (e.g., sterols, saposterols), was the first method used for commercial production of sex hormones; it still has considerable economic importance.

2) *Microbiological degradation* of sterols, such as cholesterol, stigmasterol, ergosterol, and β-sitosterol, which cannot otherwise be efficiently converted into steroid intermediates, has rapidly become commercially important [141]–[144].
3) The *total synthesis* of steroid hormones involves synthesizing the sterically complex steroid skeleton from inexpensive starting substances. This method is only important for steroids that do not occur in nature (see Section 3.4) [134]–[136].
4) The *extraction* of sex hormones or their conjugates from organic tissue or urine is the oldest method for isolating therapeutically important sex hormones. However, it is also the most unproductive because the hormones occur in very low concentration (10 mg of estrone can be isolated from 1 L of mare urine and 1 mg of progesterone from 35 g of pig corpus luteum). Today, this method is only used to recover a mixture of estrogen conjugates from the urine of pregnant mares, which is required for the menopause preparation Presomen. However, hormone extraction is still a very important analytical tool, e.g., in the identification and isolation of metabolites of sex hormones [132].

3.2. Partial Chemical Synthesis of Steroid Intermediates

3.2.1. Raw Materials

Raw materials for the partial chemical synthesis of steroid intermediates must fulfill the following requirements:

1) they must contain the steroid nucleus,
2) they must belong to the normal series,
3) they must occur in nature in sufficient quantities, and
4) they must be easily convertible into sex hormones by simple chemical reactions.

The most important raw materials are listed in Table 2 [146]. All have a hydroxyl group at position 3 and a double bond between C-5 and C-6. These common structural elements readily permit the introduction of an unsaturated keto group in ring A (e.g., by Oppenauer oxidation). These four compounds differ from each other with respect to the side chain at C-17; cholesterol and β-sitosterol possess a saturated side chain that is not easily accessible to chemical attack and subsequent modification. Stigmasterol has a double bond in its side chain and is thus more susceptible to chemical modification. Diosgenin contains a spiroketal side chain which not only makes it easy to modify, but also permits insertion of a double bond between C-16 and C-17.

The steroid starting materials and their derivatives obtained by chemical or microbiological modification are shown in Figures 5 and 6, respectively. The derivatives, in

Table 2. Starting substances for the partial synthesis of steroid intermediates

Name and natural source	Formula
Diosgenin [512-04-9] (**1**): Dioscoreaceae	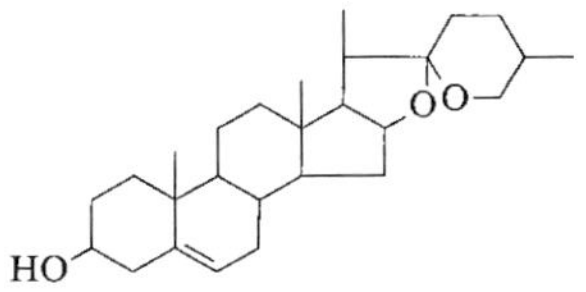 **1**
Stigmasterol [83-48-7] (**2**): soybean oil	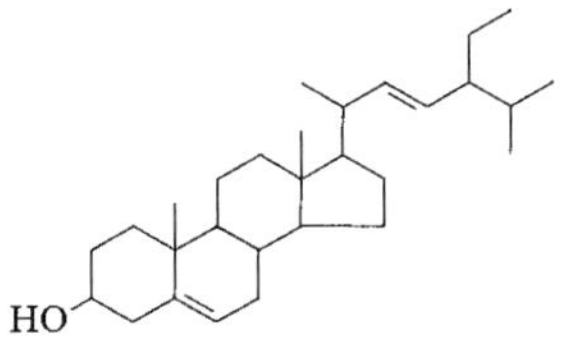 **2**
β-Sitosterol [83-46-5] (**3**): wheat germ oil, soybean oil, sugarcane wax	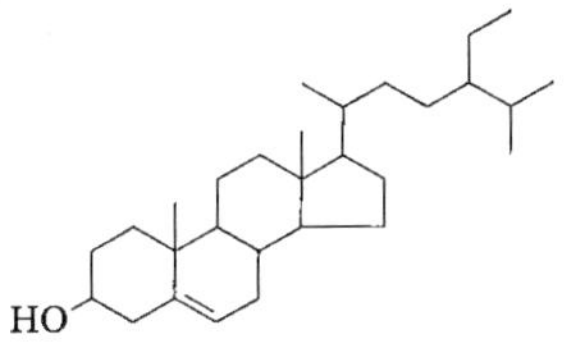 **3**
Cholesterol [57-88-5] (**4**): wool grease, spinal cord, brain	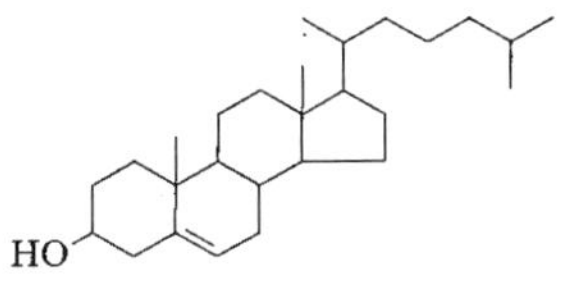 **4**

turn, are starting materials for the synthesis of different series of sex hormones (see Table 3). Almost all steroid hormones belonging to the pregnane, (21 carbon atoms in the steroid skeleton), androstane (19 carbon atoms), and estrane (18 carbon atoms) series can be obtained by chemical modification of diosgenin (**1**) and stigmasterol (**2**). Microbiological modification of β-sitosterol (**3**), cholesterol (**4**), and stigmasterol (**2**) and subsequent chemical conversions have only yielded androstanes and estranes [141]–[144].

Cholesterol [57-88-5], $C_{27}H_{46}O$, M_r 386.6, *mp* 148 °C (for structure see Table 2), was an important starting material in the early industrial production of sex hormones. Oxidation of cholesterol or its 5,6-dibromide derivative with chromic acid gave androstenolone acetate [127]; maximum yields were 7–8 %. Today, both cholesterol and β-

Table 3. Steroid derivatives for the synthesis of steroid hormones

Steroid derivative	Hormone product
Dehydropregnenolone (**5**), progesterone (**7**)	1) Gestagens of the 17α-hydroxyprogesterone series 2) Adrenal cortex hormones (corticoids)
Adrostenolone (**8**)	1) Androgens 2) Gestagens of the 17α-ethynyltestosterone series 3) Anabolic androstane steroids
Estrone (**10**)	1) Estrogens 2) Gestagens of the 17α-ethynyl-19-nortestosterone series 3) Anabolic 19-nortestosterone steroids

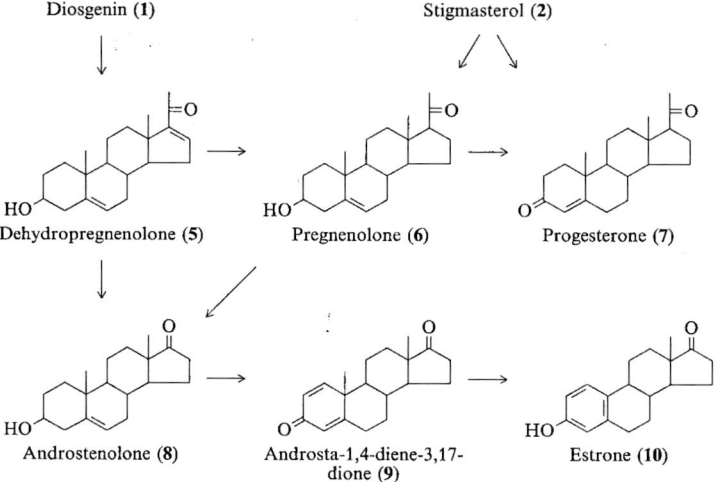

Figure 5. Steroid starting materials and their chemically synthesized derivatives

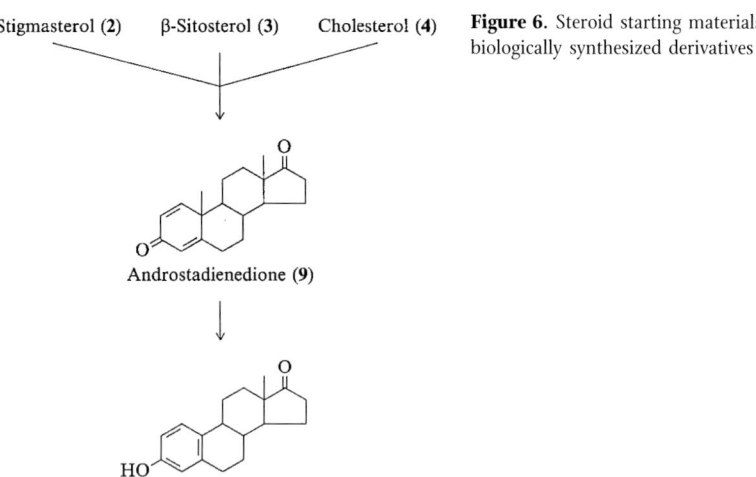

Figure 6. Steroid starting materials and their microbiologically synthesized derivatives

sitosterol have little importance as starting materials for chemical syntheses. However, they are all more important as raw materials for syntheses involving microbiological degradation of the side chain (see Section 3.3) [141]–[144].

Cholesterol occurs in animal tissue at the following concentrations, in weight percent:

Skeletal muscle	0.09 – 0.12
Liver	0.32 – 0.61
Brain (calf)	2.11 – 2.36
Kidney	0.28
Egg yolk	2 – 3.9
Spinal cord (calf)	4
Wool fat	15

It can be recovered from these sources by extraction with chlorinated hydrocarbons, if necessary, after saponification of cholesterol esters. Purification is difficult because cholesterol is often isolated with lanosterol and other triterpenes [145].

Stigmasterol [*83-48-7*], $C_{29}H_{48}O$, M_r 412.7, *mp* 170 °C, and **β-sitosterol** [*83-46-5*], $C_{29}H_{50}O$, M_r 414.7, *mp* 140 °C (for structures see Table 2), are two additional steroid starting materials. After diosgenin, stigmasterol is the most important raw material for steroid hormone synthesis. As a result of the double bond between C-22 and C-23 in the side chain, chemical modification of stigmasterol yields either pregnenolone [145], via bisnorcholic acid, or progesterone [127], via bisnorcholic aldehyde. Both compounds are key substances in the production of pregnane, androstane, and estrane hormones. Microbiological degradation of stigmasterol and β-sitosterol yields androstadienedione (see Fig. 6).

Stigmasterol occurs with β-sitosterol at a concentration of 12 – 25 % in the nonsaponifiable fraction of soybean oil. The sterols are extracted as a mixture, but their similarity makes the purification and isolation of stigmasterol extremely difficult. Thus, stigmasterol is not a starting material for chemical syntheses. However, microbiological degradation can be carried out conveniently with this sterol mixture.

Diosgenin [*512-04-9*], $C_{27}H_{42}O_3$, M_r 414.6, *mp* 208 °C (for structure see Table 2), is a steroid sapogenin. When compared with the sterols mentioned above, diosgenin is the most suitable candidate as a starting material for the production of steroids and sex hormones [127]. Not until the discovery of this compound and the elaboration of a commercially feasible method for degradation of the spiroketal side chain, to give dehydropregnenolone (**5**) (see Fig. 5), was the large-scale industrial production of sex hormones possible. Almost all steroid hormones, including those of the adrenal cortex, can be derived from dehydropregnenolone (see Fig. 5). Since the early 1950s diosgenin has become the most important steroid raw material. Its discovery gave a tremendous technological boost to the production of sex hormones. The price of progesterone dropped within three years from $ 80 to $ 3 per gram, and sufficient amounts of sex hormones became available for medical use.

Diosgenin occurs as a 3-glycoside (dioscin) [*19057-60-4*] in numerous *Liliaceae* and *Dioscoreaceae* species and is extracted from the air-dried rhizomes of *D. mexicana, D. floribunda,* and *D. composita* (also called barbasco) with ethanol [127], [145]. The extracts are concentrated by evaporation, glycoside cleavage is achieved by heating with dilute hydrochloric or sulfuric acid, and crude diosgenin is isolated by filtration. This raw material can be purified by crystallization from alcohol; however, it is usually employed without previous purification in the production of dehydropregnenolone (see Section 3.2.2).

3.2.2. Partial Synthesis

This section describes only the synthesis of intermediates having special importance for the commercial production of sex hormones used as active ingredients in drugs.

Batch synthesis of the active substances is still prevalent; however, both automation by programmed monitoring and semicontinuous or continuous methods are becoming increasingly important.

Synthesis of 3β-Acetoxydehydropregnenolone. 3β-Acetoxydehydropregnenolone [*979-02-2*], also called dehydropregnenolone acetate (**5b**), is synthesized from diosgenin [127], [145] by a method first developed by MARKER and coworkers and later improved upon by several teams.

Diosgenin acetate is heated to 190 – 200 °C with acetic anhydride or a homologous compound and is converted to pseudodiosgenin diacetate which, in turn, on mild oxidation with chromic acid in glacial acetic acid yields the ketone diosone. Hydrolysis with dilute alkali gives dehydropregnenolone or its acetate (**5b**), depending on reaction conditions, with an overall yield of ca. 45 %.

Diosgenin acetate

Diosone

3β-Acetoxy-
dehydropregnenolone (**5b**)

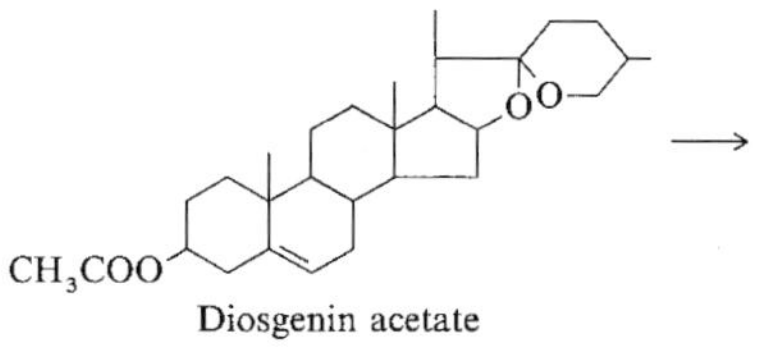

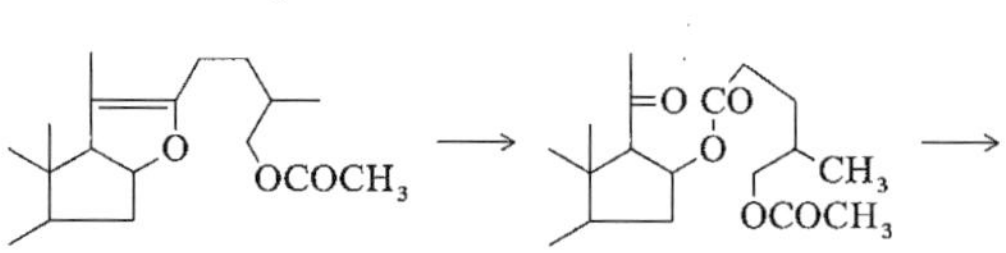

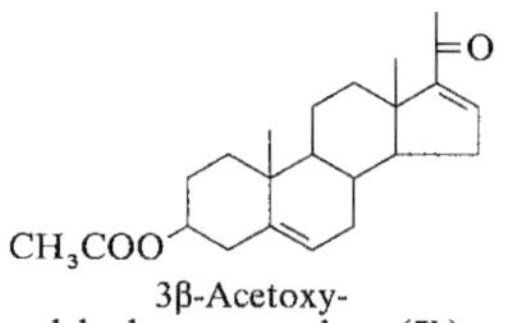

Dehydropregnenolone is the key substance in the synthesis of all the steroid and sex hormones derived from diosgenin (see Fig. 5).

Synthesis of Progesterone. The natural gestagen progesterone [*57-83-0*] (**7**) is used as an active ingredient in gestagen preparations (see Section 3.7.2), and is also an important intermediate in the production of adrenal cortical hormones. Two methods are available for the preparation of progesterone:

Selective hydrogenation of dehydropregnenolone acetate (**5b**) with palladium charcoal results in the retention of the Δ^5 double bond and the formation of pregnenolone acetate (**6b**). This compound is saponified and then oxidized (Oppenauer) in benzene or toluene to progesterone [127], [145] by refluxing with cyclohexanone and aluminum isopropylate. Yields of ca. 85% are obtained.

The *conversion of stigmasterol* (**2**) to progesterone can be carried out in several ways. The Ciba method uses 5,6-dibromostigmasterol (**11**) as the starting compound [148]:

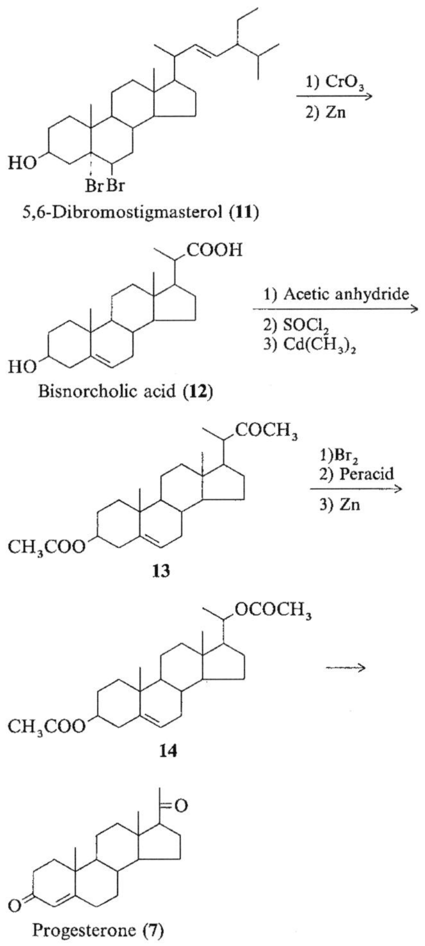

Oxidation of the dibromide with chromic acid followed by zinc reduction gives bisnorcholic acid (**12**), which is then converted to the methyl ketone (**13**). The 5,6-double bond of compound **13** is protected by bromination before the compound is oxidized with peracid to give the 20-acetate derivative (**14**). In the final step, saponification, Oppenauer oxidation, and oxidation with chromic acid give progesterone with a total yield of 40% (based on **12**).

In a method developed at Upjohn [127], the hydroxyl group of stigmasterol (**2**) is first oxidized (Oppenauer) to the ketone. If this ketone is ozonized under specific conditions, only the double bond in the side chain is attacked, yielding the aldehyde (**15**). Compound (**15**) is converted first to the enamine (**16**) by azeotropic distillation with piperidine and, finally, to progesterone (**7**) by oxidation. The overall yield is 60%.

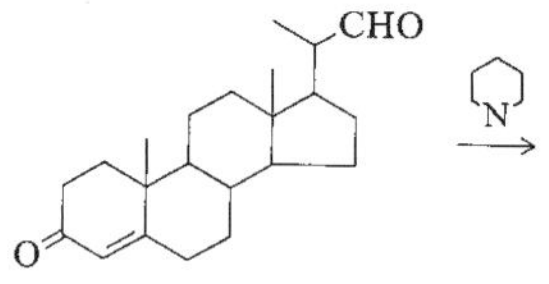

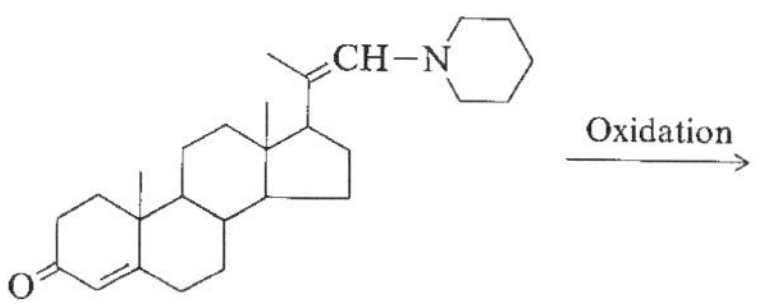

Synthesis of 3β-Acetoxy-5-androsten-17-one. 3β-Acetoxy-5-androsten-17-one [53-43-0] (**8**) (formerly known as androstenolone) and its acetate (**8b**) serve as starting materials for the synthesis of testosterone, estrone, and other androgens and estrogens including 19-norandrostane steroids derived from nortestosterone. Two of the most important production methods are described below.

1) 3β-Acetoxy-5,16-pregnadien-20-one (**5**) (16-dehydropregnenolone acetate), derived from diosgenin (**1**), is heated with hydroxylamine hydrochloride and sodium acetate or pyridine in ethanol to give the oxime (**17**) [149]. Beckmann rearrangement of **17** yields the enamide (**18**), which is then subjected to mild acid hydrolysis to give androstenolone acetate (**8b**).

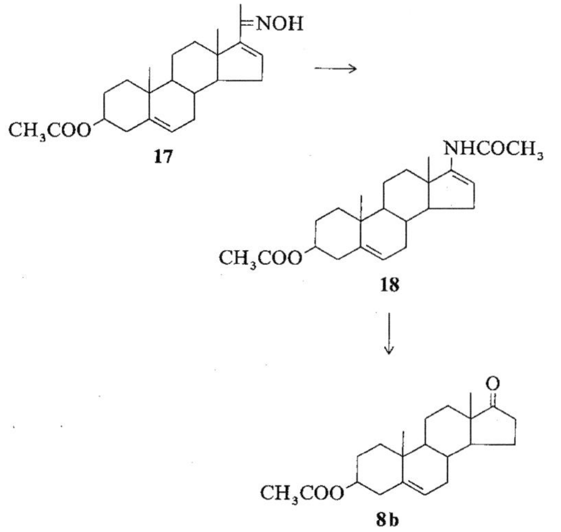

2) The second method uses pregnenolone acetate (**6b**), synthesized as described in Section p. 1571, as the starting material. Compound **6b** is first converted to the oxime and then subjected to a Beckmann rearrangement to give the acetamide (**19**), which in turn undergoes alkaline hydrolysis to yield the corresponding amine [150]. Reaction of the amine with hypochloric acid gives the chloramine (**20**), which is then treated with alkoxide (ketimine), followed by acetylation and acid hydrolysis to produce androstenolone acetate [151].

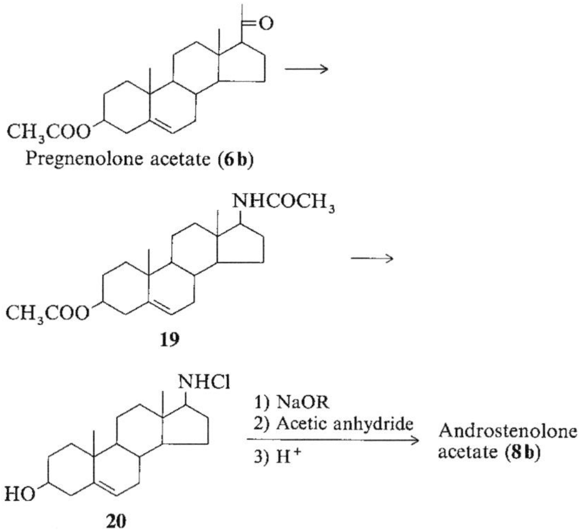

Synthesis of Estrone. Estrone [53-16-7] (**10**) is an active ingredient of commercial hormone preparations; it is also an important starting substance in the production of estradiol, ethinyl estradiol, other estrogenic steroid hormones, and 19-norandrostane compounds.

Estrone is synthesized from androstenolone (**8**), which is first hydrogenated with palladium charcoal to give epiandrosterone (**21**). Subsequent oxidation with chromic acid yields the 3,17-diketone, which is then treated with bromine in acetic acid to give the 2,4-dibromide. The dibromide, in turn, is converted to androstadienedione (**9**) by elimination of two molecules of hydrogen bromide. Dissolution of **9** in tetralin or mineral oil and pyrolysis (i.e., brief exposure to high temperature, ca. 600 °C) in the presence of quartz beads yields estrone (**10**) [152].

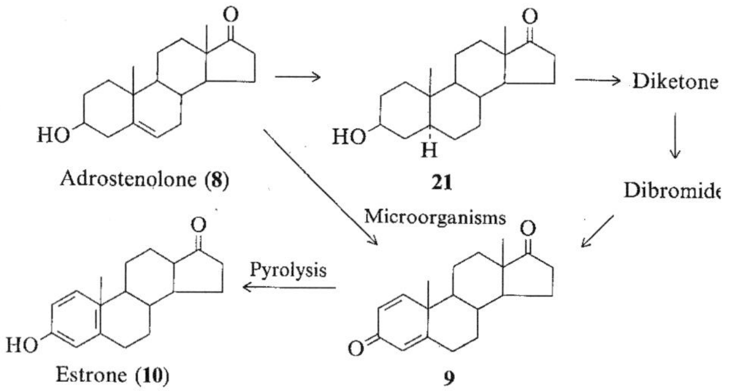

The "aromatization" of androstadienedione to give estrone has long been the subject of intensive research [127], [145]. Chemical synthesis of compound **9** from compound **8** requires four steps, whereas microbiological synthesis proceeds directly [153].

Synthesis of Testosterone. Testosterone [*58-22-0*] (**23**) is used therapeutically as the free steroid or as an ester. Testosterone is synthesized from androstenolone acetate (**8b**) which is reduced to the 17β-alcohol with Raney nickel and then esterified with benzoyl chloride in pyridine. This protecting ester group permits partial saponification of the 3-acetate with methanolic sodium hydroxide solution to yield the 3-hydroxy compound (**22**). Subsequent Oppenauer oxidation produces the testosterone benzoate, which is then subjected to alkaline hydrolysis to give testosterone (**23**) [145]. Testosterone 17-esters can be obtained by normal esterification procedures.

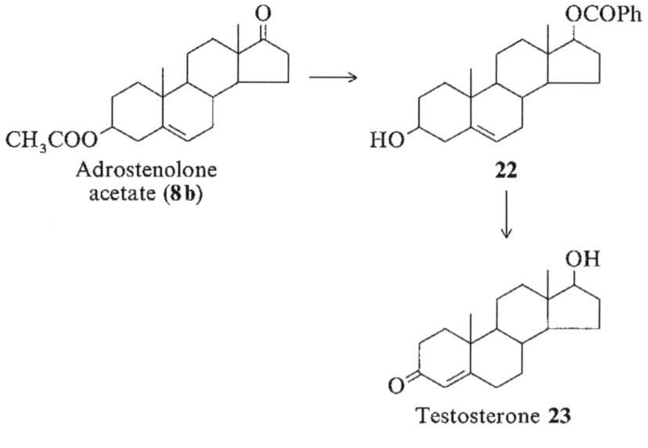

Synthesis of 19-Nortestosterone. 19-Nortestosterone [*434-22-0*] (**26**), which differs from testosterone (**23**) in that it does not have a methyl group at position 19, is the parent compound of the anabolic 19-norandrostane steroids and of the highly gestagenic hormones belonging to the norethisterone series.

Birch reduction of the methyl ether of estrone (**24**) is one of the most important methods used for the partial synthesis of 19-norandrostane compounds [154], [155].

The methyl ether is obtained by methylation of estrone with dimethyl sulfate and methanolic potassium hydroxide solution and then reduced to the alcohol with a complex hydride (e.g., $LiAlH_4$ or $NaBH_4$). The aromatic ring is subsequently reduced with lithium in liquid ammonia and diethyl ether in the presence of a small amount of ethanol to give the enolether (**25**). Treatment with methanolic hydrochloric acid yields 19-nortestosterone (**26**).

24 **25**

19-Nortestosterone (**26**)

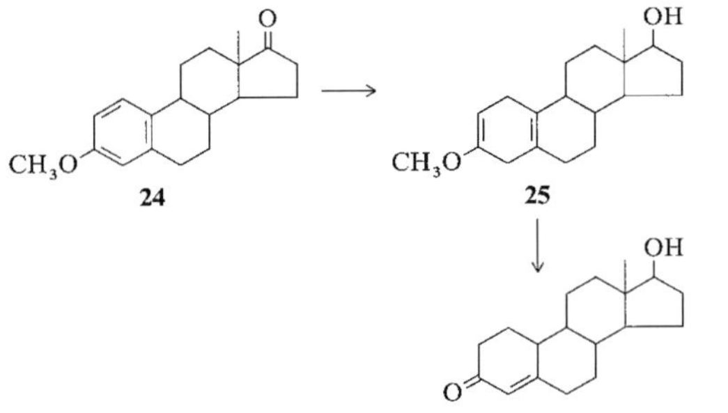

Synthesis of Norethisterone [127], [145], [156]. 17α-Ethynyl compounds may be obtained by ethynylation of 17-keto steroids. Most of these ethynyl compounds not only have interesting biological properties but also have a high oral activity. Norethisterone [*68-22-4*] (**29**) is a highly potent, orally active gestagen; many of its derivatives have characteristic activities.

An exceptionally short synthesis starts with the enolether (**25**) prepared by Birch reduction of estrone methyl ether (**24**). Oppenauer oxidation [157] of **25** yields the 17-keto compound, which can be ethynylated to compound **28** by one of the usual methods [156]. Subsequent acid hydrolysis gives norethisterone.

28

Norethisterone (**29**)

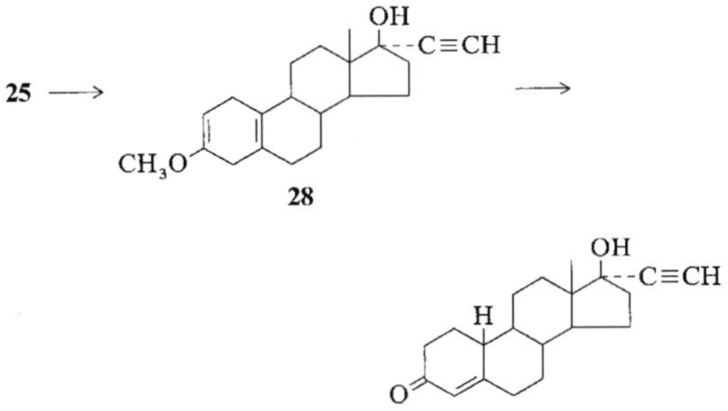

Another method used in the production of 19-norsteroids entails the functionalization and subsequent elimination of the 19-methyl group of 19-methyl steroids to produce the corresponding 19-nor compounds [158]. Both the aromatization of ring A and the subsequent Birch reduction are circumvented. Thus, androstenolone acetate (**8b**) can be converted to norethisterone (**29**), and pregnenolone acetate (**6b**) to gestonorone caproate (**66**) [178].

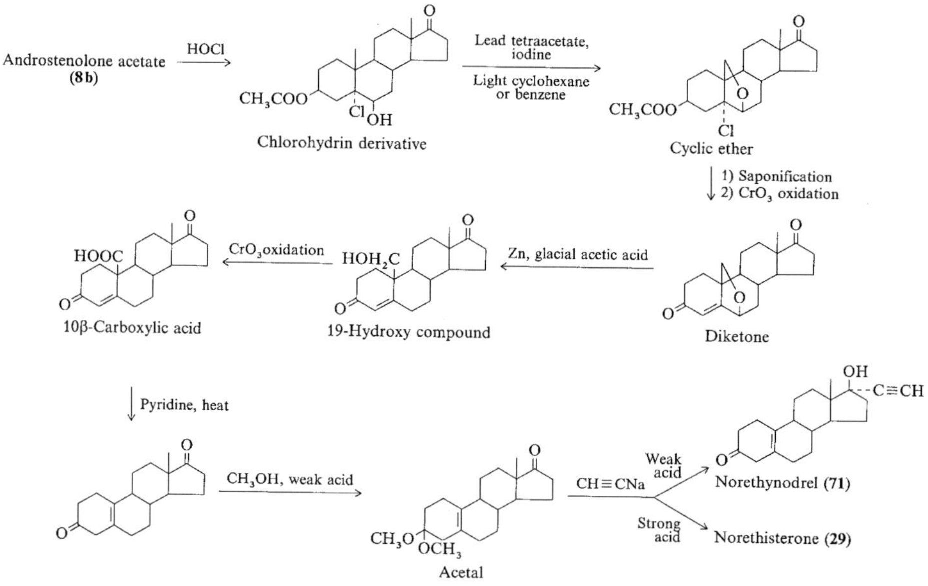

The synthesis of norethisterone (**29**) and norethynodrel (**71**) proceeds as follows. Addition of hypochloric acid to androstenolone acetate (**8b**) gives the corresponding chlorohydrin. On reaction with lead tetraacetate and iodine, the chlorohydrin is converted to a cyclic ether, which in turn is saponified and oxidized to give a diketone. Reductive opening of the cyclic ether ring with zinc in glacial acetic acid produces a 19-hydroxy derivative, and subsequent oxidation with chromic acid gives the 10β-carboxylic acid, which is decarboxylated to the diketone by heating with pyridine. Partial ketalization yields the acetal, and ethynylation with sodium acetylide followed by treatment with either a weak acid (malonic acid) or a mineral acid finally yields norethynodrel (**71**) or norethisterone (**29**), respectively.

Synthesis of 17-Acetoxyprogesterone. Introduction of an acetoxy group at position 17 of progesterone (**7**) considerably increases the oral activity of this gestagen. 17-Acetoxyprogesterone [*302-23-8*] (**32**) is the active ingredient of a large number of commercial gestagen preparations and the parent compound of many gestagens with excellent oral activity [127], [145].

One of the syntheses available for the preparation of 17-acetoxyprogesterone starts with dehydropregnenolone acetate (**5b**), obtained from diosgenin. Compound **5b** is first converted to the 16α,17-

epoxide and then to the bromohydrin (**30**). Subsequent debromination gives the 3-formate (**31**), which is subjected to acid-catalyzed acetylation and Oppenauer oxidation to give compound **32**.

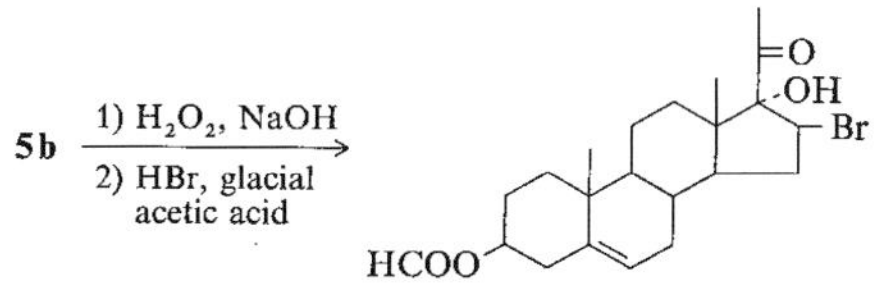

5b
1) H_2O_2, NaOH
2) HBr, glacial acetic acid
→ **30**

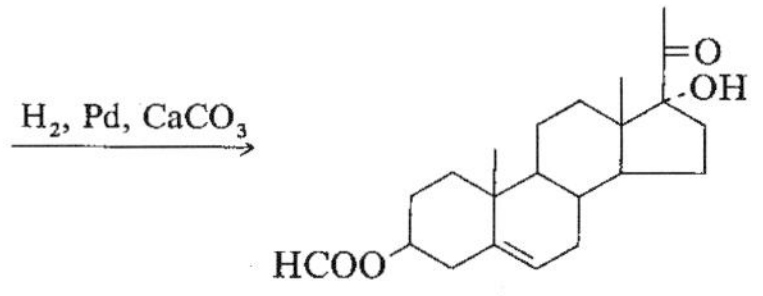

H_2, Pd, $CaCO_3$ → **31**

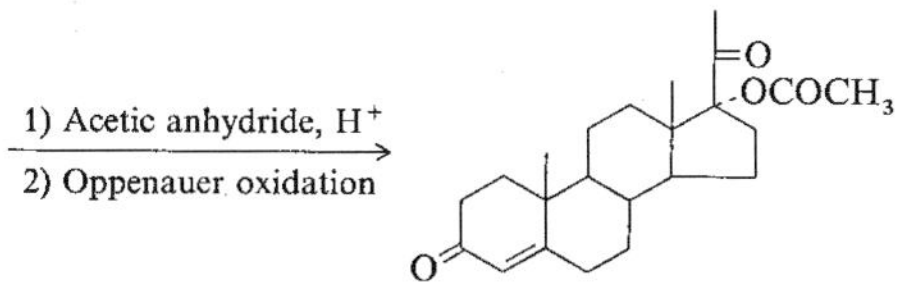

1) Acetic anhydride, H^+
2) Oppenauer oxidation
→ **32**

3.3. Microbiological Methods for the Synthesis of Steroid Intermediates

Since the beginning of the 1950s, microbiological conversion has become increasingly important in the commercial production of steroid hormones. Advantages of this technique include stereospecific reactions at positions that are not accessible to chemical reaction, mild reaction conditions, and reaction coupling (e.g., dehydrogenation and isomerization). These advantages more than compensate for the disadvantages (low substrate concentrations and, therefore, large reaction volumes; higher personnel and energy costs; more sophisticated equipment; and sterility of all media and materials involved in the reaction).

Microbiological reduction has become the key reaction in the total synthesis of sex hormones. This is the only economic way to produce pure enantiomers of the natural, optically active form of these hormones and to circumvent the tedious and inefficient resolution of racemates (see Section 3.4).

Microbiological degradation can be used to convert naturally occurring sterols, such as cholesterol, β-sitosterol, stigmasterol, and ergosterol, to androstadienedione (**9**). Chemical degradation of these abundantly occurring compounds is inefficient and unprofitable. The technology used in the microbiological conversion of steroids is similar to that used in the production of antibiotics (→ Antibiotics). The required quantity of the desired microorganisms is obtained by growing the cells in one or two prefermentation steps. Then, reaction with the relevant substrate occurs in the main fermentor. All steps must be carried out under sterile conditions to avoid contamina-

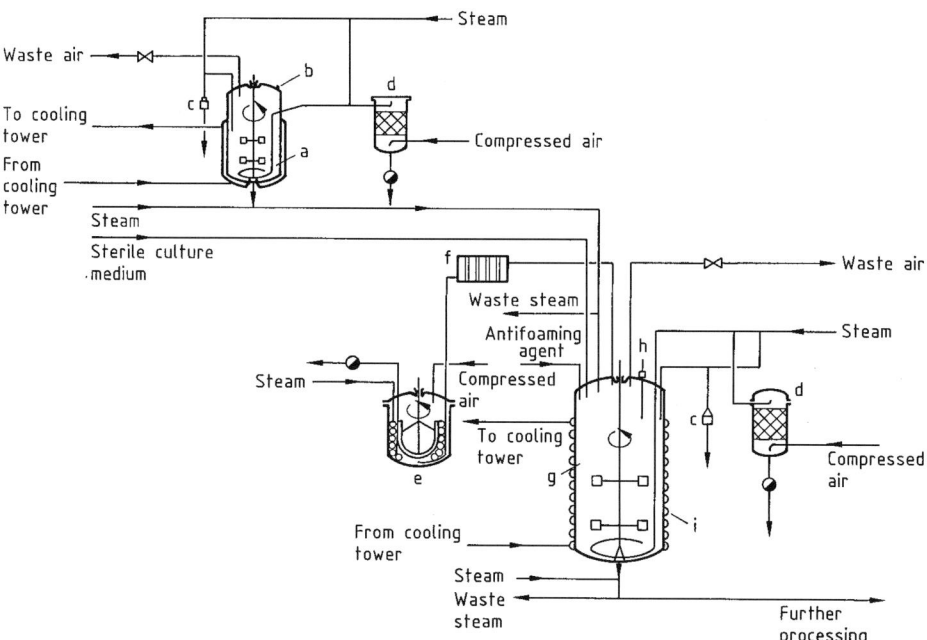

Figure 7. Schematic of a fermentation plant [159]
a) Prefermentor; b) Inoculation port; c) Sampling device; d) Filter; e) Substrate tank; f) Sterile filter; g) Main fermentor; h) Electrode for foam control; i) Cooling coil

tion. The flow diagram of a fermentation plant is shown in Figure 7. The main fermentor can have a volume up to 50 000 L [159].

Degradation to Androstadienedione. The complete fermentative degradation of sterols to carbon dioxide and water by microbial enzymes has been known for a long time. Methods have been developed to prevent complete breakdown of the sterol nucleus after degradation of the side chain, e.g., at the stage of androstadienedione (**9**) [141], [142], [144]. In all these methods, inexpensive sterols (cholesterol, sitosterol, stigmasterol, and ergosterol) serve as starting materials, a variety of microorganisms are used, and the aerobic fermentation must be carried out in the presence of inhibitors of the enzymes that continue degradation to carbon dioxide and water (e.g., inorganic heavy-metal salts, such as nickel sulfate, and organic chelating agents). An overview of the microbiological degradation methods available and their yields of androstadienedione is presented in Table 4.

Table 4. Microbiological degradation used for the synthesis of androstadienedione

Sterol used as starting material	Microorganism	Inhibitor	Yield of androstadienedione, %	Reference
Cholesterol (3)	Mycobacterium phlei	8-hydroxyquinoline	38	[144]
		nickel(II) sulfate	50	[141]
	Nocardia	cobalt(II) chloride	13	[142]
	Arthrobacter	nickel(II) sulfate	25	
β-Sitosterol (4)	Mycobacterium phlei	8-hydroxyquinoline	18	[144]
	Arthrobacter	nickel(II) sulfate	13	[142]
Stigmasterol (2)	Mycobacterium phlei		27	[144]

3.4. Total Synthesis of Sex Hormones

The consumption of sex hormones as oral contraceptive agents as well as the demand for steroids to produce adrenal cortical hormones has increased dramatically since 1960 so that demands could not be met by available raw material sources [147]. For this reason, the total synthesis of steroid substances is very important. In addition, total synthesis of steroids permits the production of structural elements that are impossible or extremely difficult to obtain with partial syntheses (e.g., production of homologous 13-alkyl steroids).

Synthesis of D-Norgestrel. D-Norgestrel [6533-00-2] (38), the gestagenic component of the contraceptives Neogynon and Stediril-d, is produced by total synthesis [159]–[161].

6-Methoxy-1-tetralone is converted to the vinyl alcohol (33) with vinylmagnesium chloride. Condensation of 33 with 2-ethylcyclopentane-1,3-dione gives the secodione (34). Stereospecific microbiological reduction of 34 yields the β-hydroxy ketone (35), which has two asymmetric carbon atoms (C-13 and C-17). Compound 35 is optically active and belongs to the normal series; i.e., C-13 and C-17 have the same configuration as in natural steroid compounds. Because subsequent reactions are stereospecific, the final product (D-norgestrel) retains the optical activity and normal configuration introduced by microbiological reduction [161].

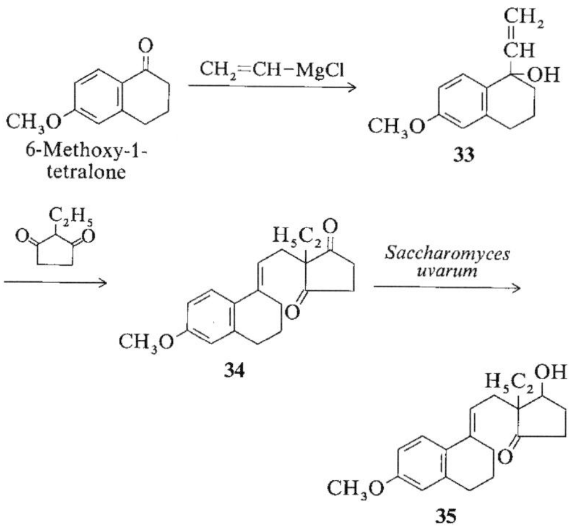

Acidic cyclization of **35** produces **36**, which has a double bond between C-14 and C-15. In the subsequent Birch reduction, the $\Delta^{8,9}$ double bond and the aromatic ring are reduced. Oppenauer oxidation of the 17β-hydroxyl group of the resulting dienolether leads to the corresponding ketone. Subsequent ethynylation of the ketone and cleavage of the enolether yield D-norgestrel (**38**).

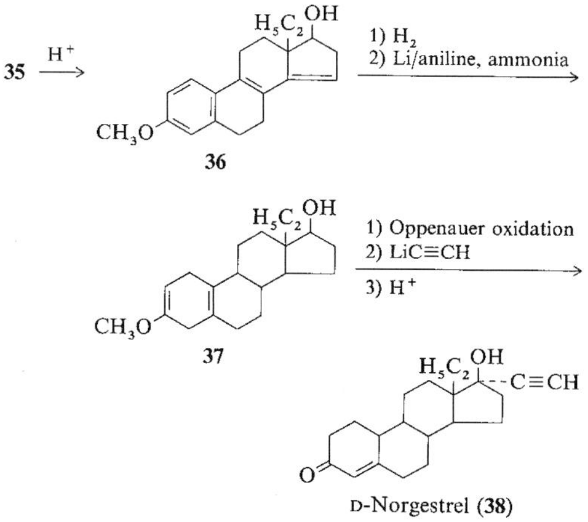

Synthesis of Norgestrienone. Norgestrienone [*848-21-5*] (**46**) is the gestagenic component of the oral contraceptive Planor, registered only in France [162]. Michael addition of methyl 5-oxo-6-heptenoate to 2-methylcyclopentane-1,3-dione produces the methyl ester of the triketocarboxylic acid (**39**). As in the synthesis of D-norgestrel, **39** is then subjected to partial, stereospecific reduction with the help of microorganisms. The resulting optically active β-hydroxyketone (**40**) belongs to the normal series [163]. Again, subsequent reactions proceed stereospecifically, and the final product retains the correct steric configuration attained in one of the early steps.

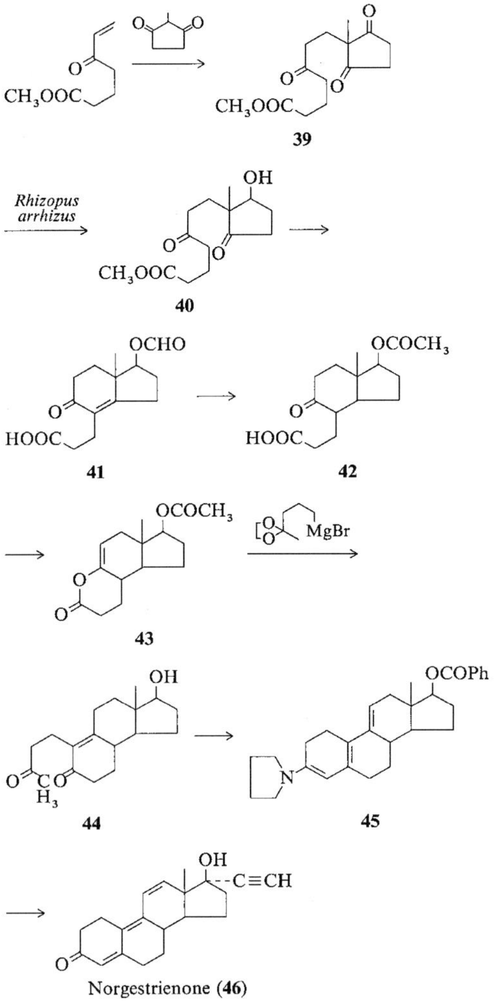

Compound **40** is converted to the formate (**41**) by aldol condensation, acid-catalyzed hydrolysis, and formylation. Compound **41** is converted to the saturated acetate (**42**) by stereoselective hydrogenation, saponification, and acetylation. Treatment with acetic anhydride gives the enol lactone (**43**). Subsequent Grignard reaction yields **44**, which already possesses rings B, C, and D of the steroid skeleton. Compound **44** is benzoylated at position 17 and then converted to compound **45** by aldol condensation and enamine formation with pyrrolidine. Subsequent dehydrogenation with 2,3-dichloro-5,6-dicyano-1,4-benzoquinone, protection of the resulting 3-keto group by oxime formation, oxidation to the 17-ketone, ethynylation, and oxime cleavage finally yield norgestrienone (**46**) [163].

Both syntheses use inexpensive chemicals as starting materials and apply microbiological reduction (production of an asymmetric carbon atom) to skillfully circumvent the formation of racemates and their inefficient resolution. Older versions of both methods produced racemates. Consequently, the contraceptives Eugynon (Schering) and Stediril (Wyeth) contain racemic norgestrel as gestagen.

3.5. Estrogens

3.5.1. General Aspects

Physiology and Clinical Applications. The first sex hormone to be isolated was estrone (**10**). It was recovered from the urine of pregnant humans by DOISY and BUTENANDT in 1929; later, the urine of pregnant mares was used as a richer source of estrogens. The structure was elucidated in 1932 by BUTENANDT, MARRIAN, and HASLEWOOD.

The naturally occurring estrogens estradiol (**47**), estrone (**10**), and estriol (**48**), together with the gestagen, progesterone (**7**), regulate the menstrual cycle in humans and several animal species [164]. In addition, estrogenic hormones influence the growth and blood circulation of the uterus, vagina, and mammary glands. They are used in substitution therapy to control estrogen deficiency caused by ovarian hypofunction, castration, or menopause. They are also used in suppression therapy to suppress gonadotropin-induced ovulation and the endogenous production of estrogens [164]. Because of the latter, estrogens are used together with gestagens as combination contraceptive agents (→ Chemical Contraception). A defined dose ratio of estrogen to gestagen together with a set mode of administration prevents ovulation and, consequently, fertilization with minimal side effects. The hormones are also administered in osteoporosis senilis, prostate carcinoma, mastocarcinoma, and for suppression of lactation [164].

Lists of the most important commercially available estrogenic sex hormones are given in Tables 5 (natural estrogens) and 6 (estradiol derivatives). They are available as single substances or as combination preparations.

Bioassays. Four types of bioassay are used for the quantitative determination of estrogenic hormones:

1) *Vaginal Smear Test.* The vaginal smear test, developed by ALLEN and DOISY [130], [131], is specific for active estrogens and permits the isolation of estrogenic sex hormones. Subcutaneous or oral administration of these hormones leads to a marked proliferation of the vaginal epithelium and keratinization of the surface cell layers (Schollen stage) in small rodents. These characteristic changes in cell structure are visible on microscopic examination of vaginal smears. When equal hormone doses are used, the degree of change depends on the duration of the effect and the potency of the estrogenic substances. The reference substances used are estradiol (**47**) for subcutaneous application, ethinyl estradiol (**51**) for oral applica-

Table 5. Natural estrogens

Formula	Names	CAS registry no.	Trade name (manufacturer)
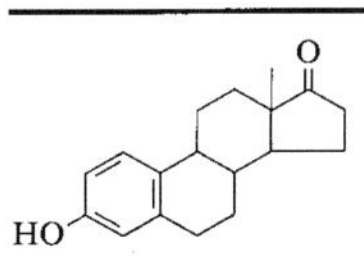 **10**	estrone, 3-hydroxy-1,3,5(10)-estratrien-17-one [a], [b]	[53-16-7]	Estrone (Abbott, Lilly, Wyeth); Glandubolin (Gedeon Richter); Kolpon (Organon); Ovex (Leo Danmark); Theelin (Parke Davis)
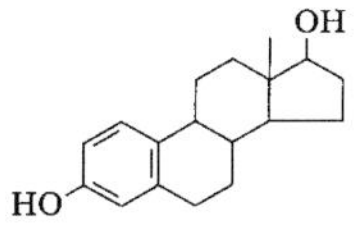 **47**	estradiol, 1,3,5(10)-estratriene-3,17β-diol [a], [b]	[50-28-2]	Dimenformon (Organon); Diogyn (Pfizer); Gynoestryl (Roussel); Ovocyclin (Ciba); Profoliol (Schering Corp.); Progynon-Salbe (Schering AG)
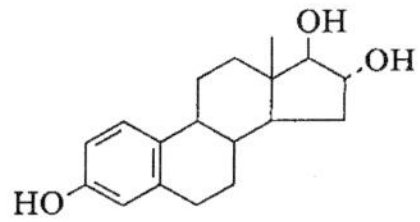 **48**	estriol, 1,3,5(10)-estratriene-3,16α,17β-triol [a], [b]	[50-27-1]	Gynäsan (Bastian); Hormomed (Merckle); Ovestin (Organon); Theelol (Parke Davis); Triovex (Leo)
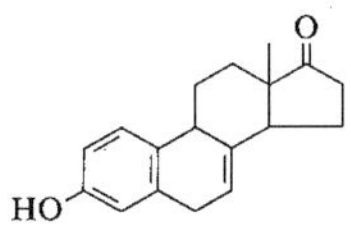 **49**	equilin	[474-86-2]	
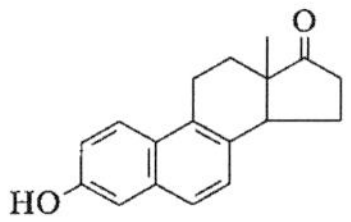 **50**	equilenin	[517-09-9]	
	mixture of estrone, equilin, equilenin, and the sodium 3-sulfate salts of the corresponding 17α-dihydro compounds [c]		Amniotin (Squibb); Conestron (Wyeth); Konogen (Lilly); Premarin (Ayerst); Presomen (Kali-Chemie)

[a] [130], [b] [131], [c] [127].

tion, and estradiol heptanoate or undecenoate for long-acting substances with depot activity. Castrated animals are used to avoid the effects of the natural sexual cycle.

2) *Sialic Acid Content of Mouse Vagina.* Administration of estrogen to castrated mice leads to a dose-dependent reduction of the sialic acid content of the vagina [165]. After subcutaneous or oral treatment, the animals are sacrificed, the vaginas are removed and processed, and the sialic acid content is determined colorimetrically. This test is specific for estrogenic substances; the sialic acid content is a measure of the potency of the estrogen. Reference substances are estradiol for subcutaneous application and ethinyl estradiol for oral administration.

3) *Vaginal Opening Test.* The vagina of rodents remains closed until sexual maturity. Active estrogens cause premature opening of the vagina. This test, however, is not estrogen-specific because androgens and anabolic substances produce a similar effect.

Table 6. Estradiol derivatives

Formula	Names	CAS registry no.	Trade name (manufacturer)
Estradiol esters			
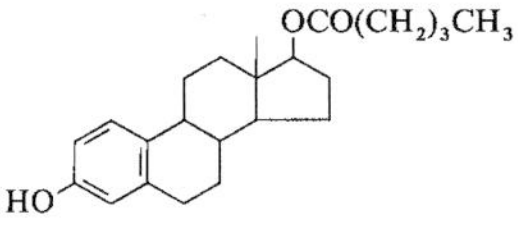 [structure with OCO(CH₂)₃CH₃, HO]	estradiol valeriate, 17β-valeryloxy-1,3,5(10)-estratrien-3-ol	[979-32-8]	Cyclo-Progynova (Schering AG); Delestrogen (Squibb); Östrogynal sine (Asche); Progynova, Primogyn-Depot (Schering AG)
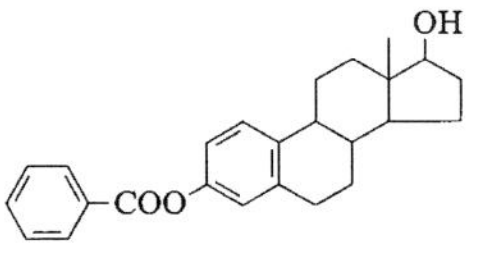 [structure with OH, benzoyl-O]	estradiol benzoate, 3-benzoyloxy-1,3,5(10)-estratrien-17β-ol	[50-50-0]	Benzo-Gynoestryl (Roussel); Diogyn (Pfizer); Follicyclin (Ciba); Östradiol "Vitis" (Vitis); Ostroform (British Drug Houses); Ovocyclin M (Ciba)
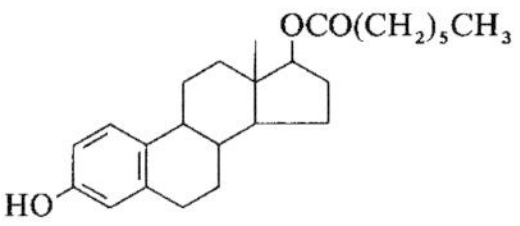 [structure with OCO(CH₂)₅CH₃, HO]	estradiol enanthate, 17β-heptanoyloxy-1,3,5(10)-estratrien-3-ol	[4956-37-0]	Used only in the depot contraceptive Deladroxat (Squibb)
17-Alkylated estradiol derivatives			
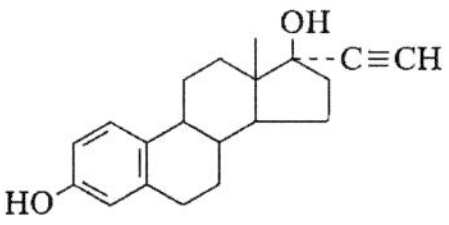 [structure 51, OH, C≡CH, HO]	ethinyl estradiol, 17α-ethynyl-1,3,5(10)-estratriene-3,17β-diol	[57-63-6]	Diogyn-E (Pfizer); Eticyclin (Ciba); Feminone (Upjohn); Gynolett-Tabletten (Labopharma); Kolpolyn (Organon); Progynon C, Progynon M (Schering AG); Ylestrol (Ferndale), used in oral contraceptives
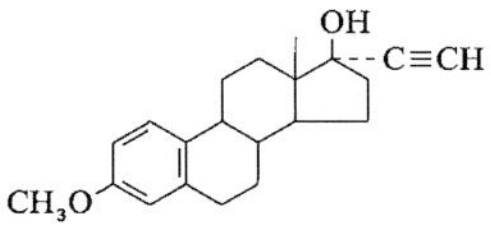 [structure 52, OH, C≡CH, CH₃O]	mestranol, 17α-ethynyl-3-methoxy-1,3,5(10)-estratrien-17β-ol	[72-33-3]	Ovastol (Rendell), used in oral contraceptives
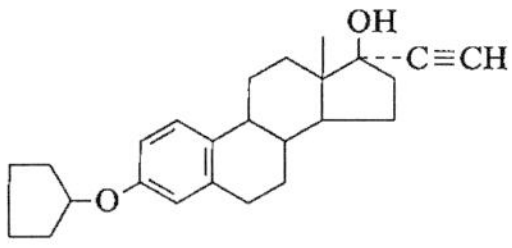 [structure with OH, C≡CH, cyclopentyl-O]	quinestrol, 17α-ethynyl-3-cyclopentyloxy-1,3,5(10)-estratrien-17β-ol *	[152-43-2]	Estrovis, Estrovis 4000 (Gödecke), used in depot contraceptives
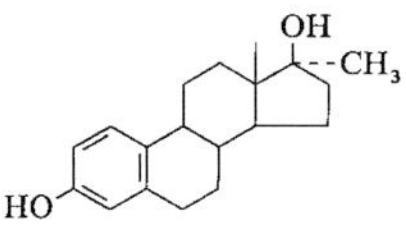 [structure with OH, CH₃, HO]	methylestradiol, 17α-methyl-1,3,5(10)-estratriene-3,17β-diol	[302-76-1]	Klimanosid "R" (Boehringer Mannheim)

* [166].

4) *Uterine Growth Test.* Administration of estrogens to rodents causes increased growth of the endometrium and myometrium of the uterus. The increase in uterine weight is proportional to the dose and potency of the estrogen administered.

3.5.2. Natural Estrogens and Estradiol Derivatives

Natural Estrogens. The naturally occurring estrogens, estrone (**10**), estradiol (**47**), and estriol (**48**), as well as the estrogens from mare urine, equilin (**49**), equilenin (**50**), and their 17-dihydro compounds, are of considerable importance in the treatment of menopausal symptoms (for structures, see Table 5) [161]. However, very high oral doses are required to elicit responses because these compounds are rapidly metabolized and inactivated in the liver. Injection preparations for parenteral administration can only be produced in the form of crystal suspensions because of the low solubility.

Estradiol Esters. Esterification of naturally occurring estrogens with longer chain carboxylic acids is an important method for increasing their duration of action. In contrast with free estrogens, oily solutions of esterified estrogens can be produced for depot injection preparations.

17-Alkylated Estradiol Derivatives. Alkylation of estradiol derivatives at position 17 increases their oral activity many times. The 17-alkyl group imparts stability to biodegradation which, in natural estrogens, begins with oxidation of the secondary 17-hydroxyl group to a ketone [131], [133]. As a result of the high oral activity of these estrogenic compounds, [e.g., ethinyl estradiol (**51**) and its methyl ether, mestranol (**52**)], they are used widely in combination preparations for oral contraception (→ Chemical Contraception).

3.5.3. Nonsteroidal Estrogens

Estrogenic activity is not restricted to the estrogenic sex hormones. Certain plant constituents, bituminous components of the Seefeld oil shale, easily synthesized stilbene derivatives, and their dihydro compounds have estrogenic properties [130]. The therapeutic application of stilbestrol derivatives in clinical medicine is limited. They are of greater importance in veterinary medicine (e.g., in the artificial caponizing of chickens).

The parent compounds of the nonsteroidal estrogens are diethylstilbestrol [*56-53-1*] (**53**), hexestrol [*84-16-2*] (**54**), and dienestrol [*84-17-3*] (**55**).

Table 7. Stilbestrol derivatives

R*	Generic name	CAS registry no.	Trade name (manufacturer)
H	diethylstilbestrol, trans-3,4-bis(4-hydroxyphenyl)-3-hexene	[56-53-1]	Cyren A (Bayer); Klimax Taeschner (Taeschner); Oekolb (Dr. Kade); Östrogen-Holzinger (Holzinger)
H_2PO_3	diethylstilbestrol phosphate	[522-40-7]	Cytonal (VEB Berlin-Chemie); Honvan (Asta)
CH_3CH_2CO	diethylstilbestrol dipropionate	[130-80-3]	Cyren B (Bayer); Syntoestron (Eros, Kuessnacht)
CH_3	dimethoxydiethylstilbene, trans-3,4-bis(4-methoxyphenyl)-3-hexene	[130-79-0]	Depot Östromon (Merck AG); Dimethylöstrogen-Holzinger (Holzinger)

* All derivatives have the following structure:

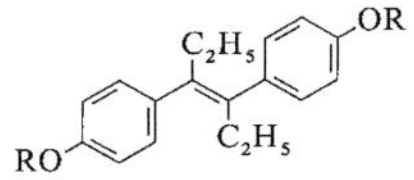

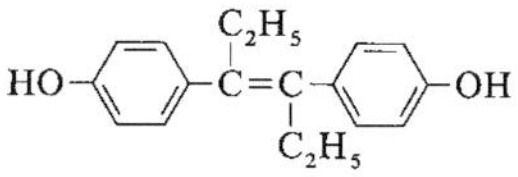

Diethylstilbestrol (**53**)

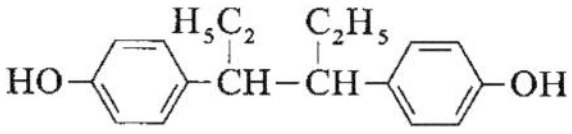

Hexestrol (**54**)

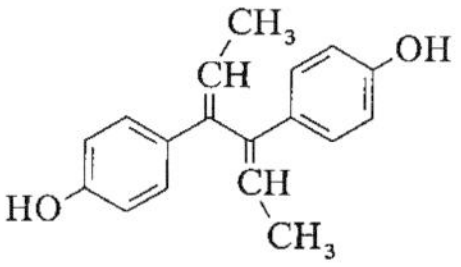

Dienestrol (**55**)

Compounds **53–55** are used in the free form, as their diethers or diesters, or with substituted aromatic rings or double bonds. The most important commercial preparations are listed in Table 7.

3.5.4. Antiestrogens [130], [131]

The term antiestrogen designates a very heterogeneous group of substances that counteract the effects of estrogen. The use of antiestrogens is indicated primarily in the palliative treatment of metastasizing mastocarcinoma. In addition, they can be used to treat benign prostate hyperplasia and endometriosis.

Antiestrogens can be divided into three groups on the basis of their mode of action:

1) *Competitive Antagonists.* These active antiestrogenic substances act by binding competitively to the estradiol receptor. The nonsteroidal stilbene derivatives tamoxifen

[*911-45-5*], cyclofenil [*2624-43-3*], and clomiphene [*911-45-5*] have practical importance. Chemical modification of estradiol has also produced derivatives with antiestrogenic properties. However, these compounds usually exert a mixture of dosage-dependent, antagonistic and agonistic effects.

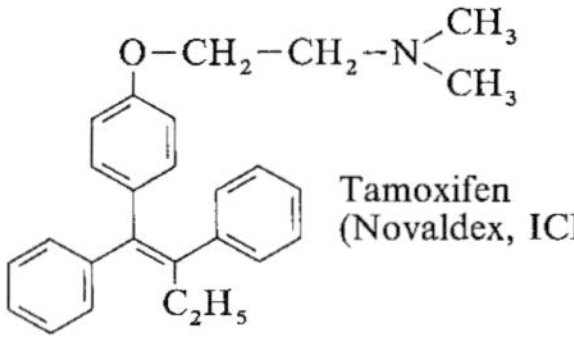

Tamoxifen
(Novaldex, ICI)

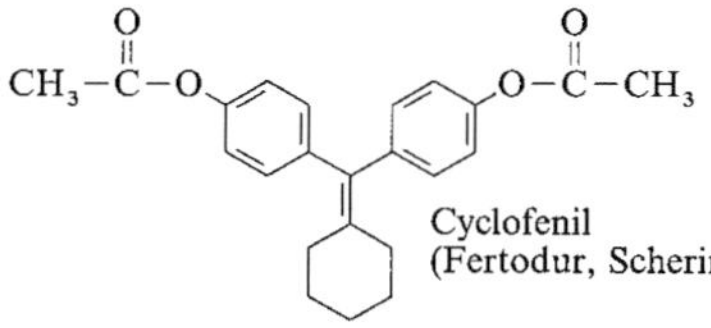

Cyclofenil
(Fertodur, Schering)

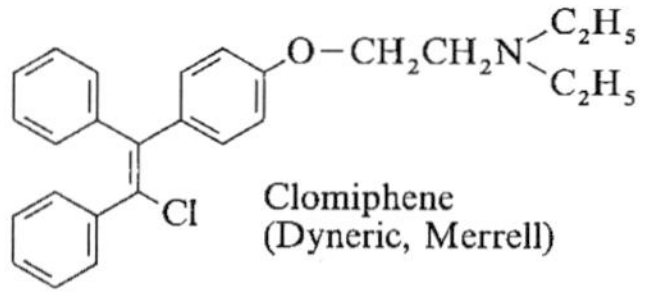

Clomiphene
(Dyneric, Merrell)

2) *Aromatase Inhibitors.* Transformation of androstenedione to estrone is catalyzed by the enzyme aromatase (see Fig. 4). A series of steroidal and nonsteroidal substances reversibly or irreversibly inhibit this enzyme system and are thus suitable agents for lowering the biological level of estrogen. In addition to androstane derivatives, imidazole derivatives also have remarkable inhibitory properties. Examples of aromatase inhibitors follow:

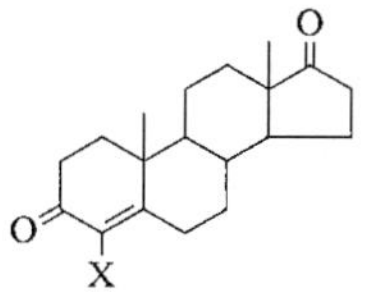

X = OH [*566-48-3*]
NH$_2$ [*105051-81-8*]
F [*98102-30-8*]
CH$_3$COO [*61630-32-8*]

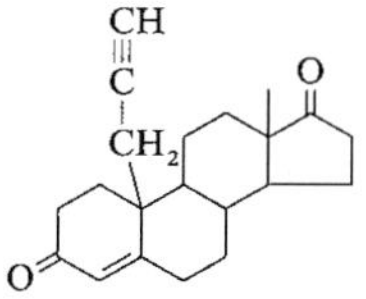

PED [*77016-85-4*]
(Merrel-Dow)

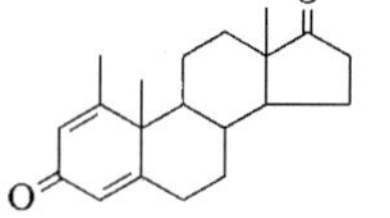

Atamestan [*96301-34-7*]
(Schering)

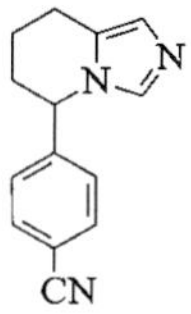

CN

CGS 16949 A
[*102676-47-1*]
(Ciba-Geigy)

3) *Antigonadotropic Substances.* Another way of reducing ovarian estrogen production is by inhibiting hypophyseal gonadotropin secretion. Although Danazol [*17230-88-5*] has a weak androgenic, anabolic component, it primarily exhibits a central gonadotropin-inhibiting effect. Danazol is especially useful for the treatment of endometriosis.

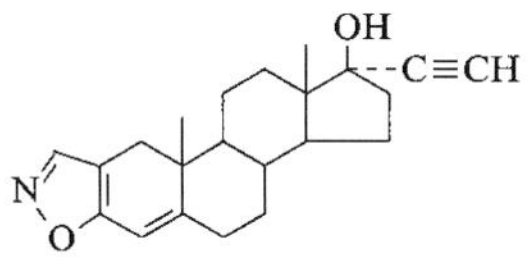

Danazol [*17230-88-5*]
(Sterling Winthrop)

3.6. Androgens and Anabolic Agents [140]

3.6.1. General Aspects

Physiology and Clinical Applications. Testosterone (**23**), the principal male sex hormone (androgen), was isolated in 1935 from bull testes (100 kg of bull testis gave approximately 10 mg of testosterone), and its structure was subsequently elucidated. A weaker androgenic substance androsterone (**56**) was isolated in 1931 from human urine and shown to be a breakdown product of testosterone.

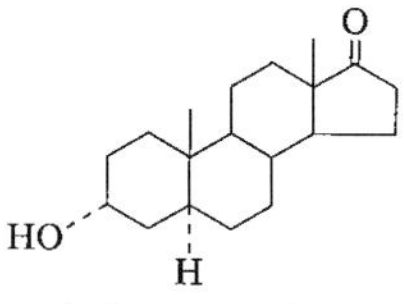

Androsterone (**56**)

Testosterone synthesis occurs in the testis and is regulated by the hypophyseal luteinizing hormone (LH). However, small amounts of androgens are also produced in the ovaries, the adrenal glands, and the placenta. The chief functions of testosterone are to ensure normal development of the seminal vesicles, prostate, epididymis, and Cowper glands and to maintain secondary male characteristics (facial hair, deep voice). Spermatogenesis, on the other hand, is regulated by the follicle-stimulating hormone (FSH).

Clinical use of androgens is indicated in cases of androgen deficiency, such as climacterium virile, hypogenitalism, prostate hypertrophy, and other potency disorders (substitution therapy). The administration of testosterone also has a suppressive effect on the hypophysis; it inhibits the release of LH and FSH and, consequently, spermatogenesis. A "rebound effect" is observed after treatment stops; i.e., the production of LH and spermatogenesis are intensified. This effect may be desirable in cases of oligozoospermia. Androgens are prescribed for carcinoma of the breast, carcinoma of the uterus, or endometriosis, as well as to inhibit lactation.

Testosterone and androsterone, like all androgens, exert an anabolic influence and stimulate protein synthesis in muscle and bone. They are used therapeutically in the treatment of consumptive disease, anorexia, old-age infirmity, osteoporosis, etc. Chemical modification of androgens has produced substances with a high anabolic-to-androgenic potency, which can also be administered to women. However, all anabolic agents do have a certain residual androgenic effect [167].

Bioassays. Two types of bioassay are used for the quantitative determination of androgens and anabolic agents:

Assay of Androgenic and Anabolic Activity [168]. The function and size of the accessory sex glands (seminal vesicles and prostate) and of the musculus levator ani (a muscle of the perineal complex) in castrated male rats depend on the presence of androgens. The fresh weights of the seminal vesicles (androgenic activity) and levator ani (anabolic effect) are measured. The anabolic and androgenic effects are then determined from a dose–response curve. The organ weights are compared with values given by reference substances (testosterone propionate for subcutaneous application and 17α-methyltestosterone for oral application). The relative potencies are expressed as fractions or multiples of the activity of the reference substance.

Chicken Comb Test. The comb of a young cockerel grows on administration of androgens. After standard treatment, the comb and body weights are measured, and the ratio thus obtained is compared with that produced by reference substances. Again, testosterone propionate is used as reference for subcutaneous application and 17α-methyltestosterone for oral administration.

3.6.2. Androgens

Testosterone and Testosterone Esters. A large number of preparations contain testosterone (**23**) in the free form or as one of its esters; examples are listed in Table 8. The rapid breakdown of testosterone (oxidation at position 17, reduction at position 3, and reduction of the double bond) and its subsequent inactivation can be slowed down by esterification with longer chain fatty acids; consequently, the duration of its action can be effectively prolonged (depot effect).

Other Androgens. Commercial preparations available for androgen therapy are listed in Table 9. They are androgens with good oral activity, sufficient duration of action, and a low inhibitory effect on spermatogenesis [140]. Introduction of a methyl group at positions 17 and 1 of the steroid nucleus, a hydroxymethyl group at C-2, and the production of 9α-fluoro-11β-hydroxy compounds have proved particularly interesting. However, 17-methyl derivatives tax the liver to a greater extent and are contraindicated in cases of liver malfunction.

Table 8. Testosterone and testosterone esters

R*	Names	CAS registry no.	Trade name (manufacturer)
H	testosterone; 17β-hydroxy-4-androsten-3-one	[58-22-0]	Androlin (Lincoln); Malestrone (Kirk); Perandren (Ciba); Synandrol F (Pfizer); Testoviron T (Schering AG); Testryl (Squibb)
CH_3CH_2CO	testosterone propionate, 17β-propionyloxy-4-androsten-3-one	[57-85-2]	Anertan (Boehringer Mannheim); Masenate (Schiefelin); Sterandryl (Roussel); Synandrol (Pfizer); Testosid (Boehringer Mannheim); Testoviron (Schering AG)
⬡-CH_2CH_2-CO	testosterone cyclopentylpropionate, 17β-cyclopentylpropionyloxy-4-androsten-3-one	[58-20-8]	Depo-Testadiol (Upjohn); Depovirin (Hoechst); Telipex-Retard (Sanabo)
$CH_3(CH_2)_5CO$	testosterone enanthate, 17β-heptanoyeoxy-4-androsten-3-one	[315-37-7]	Delatestryl (Squibb); Sterandryl Retard (Roussel); Testoviron Depot (Schering)

* Compounds have the following structure:

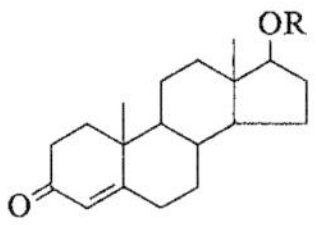

Table 9. Commercially available androgens for therapeutic use

Formula	Name	CAS registry no.	Trade name (manufacturer)
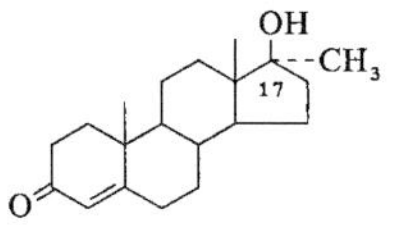	methyltestosterone, 17β-hydroxy-17α-methyl-4-androsten-3-one	[58-18-4]	Andrometh (Central Pharm.); Metandren (Ciba); Neo Hombreol M (Organon); Testosid (Boehringer Mannheim), Testosteron "Berco" (Berco)
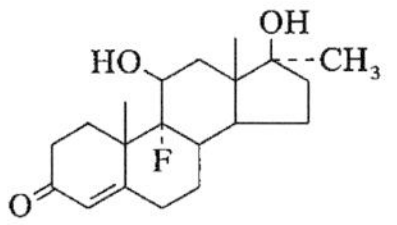	fluoxymestrone, 9-fluoro-11β,17β-dihydroxy-17α-methyl-4-androsten-3-one [a]	[76-43-7]	Afluteston (Arcana); Halotestin (Upjohn); Ora Testryl (Squibb); Ultranden (Ciba)
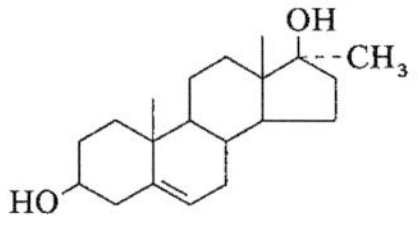	methandriol, 17α-methyl-5-androstene-3β,17β-diol [b]	[521-10-8]	Metandiol (Roussel); Methostan (Schering Corp.); Neosteron (Organon)
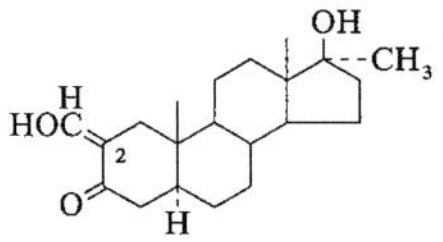	oxymetholone, 17β-hydroxy-17α-methyl-2-hydroxymethylen-5α-androstan-3-one [c]	[434-07-1]	Adroyd (Parke Davis); Anadrol (Syntex); Nastenon (Cassenne)
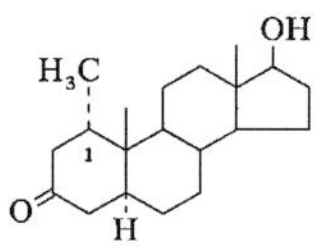	Mesterolone, 17β-hydroxy-1α-methyl-5α-androstan-3-one [d]	[1424-00-6]	Mestoran (Schering Denmark); Proviron (Schering AG)

[a] [169], [b] [170], [c] [171], [d] [172].

3.6.3. Antiandrogens [139], [140]

The term antiandrogen designates any substance that inhibits the action of the male sex hormone testosterone. Examples are cyproterone acetate [*427-51-0*] (for synthesis, see [181]) and flutamide [*13311-84-7*], which represent different classes of substances.

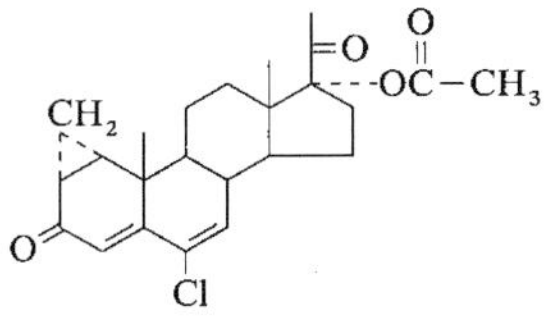

Cyproterone acetate

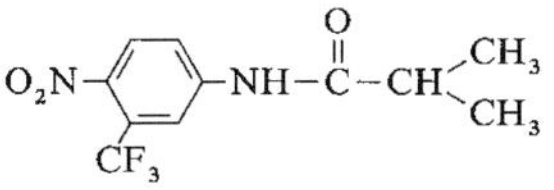

Flutamide

Both antiandrogens function by competitively displacing testosterone from the androgen receptor. Whereas flutamide has no other endocrinal side effects, cyproterone acetate possesses a strong gestagenic activity and, consequently, can also be used in combination preparations as an oral contraceptive. Antiandrogens are applied primarily in the treatment of prostate tumors, hypersexuality in men, pubertas praecox in young men, androgen-dependent acne, seborrhea, and alopecia in women.

Biological Testing of Antiandrogenic Activity. Administration of an antiandrogen to pregnant rodents at the fetal sexual differentiation stage produces an androgen deficit; consequently, normal development of the exterior sex characteristics of the male fetus does not occur. A mixture of male and female characteristics is obtained, depending on the potency of the administered antiandrogen.

3.6.4. Anabolic Agents [167]

In many cases, chemical modification of the steroid nucleus of androgens of the 19-norandrostane, androstane, and androstanolone series increases their anabolic effects at the expense of their androgenic activity. Generally, the production of long-chain esters prolongs the action of subcutaneous preparations (depot effect) and methylation at C-17 increases oral potency.

19-Norandrostane Compounds. 19-Nortestosterone (**26**) is the parent compound of the series of therapeutically important anabolic agents listed in Table 10. Apart from esterification with long-chain acids, the introduction of an 18-methyl group or a 4-hydroxyl group, the reduction or removal of the 3-keto group, and alkylation at C-17 give rise to potent anabolic agents.

Table 10. Anabolic norandrostane derivatives

Formula	Name	CAS registry no.	Trade name (manufacturer)
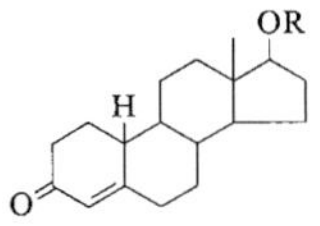 R = H (**26**)	nandrolone, nortestosterone, 17β-hydroxy-4-estren-3-one	[434-22-0]	Nortestonat (Upjohn)
R = ⟨cyclopentyl⟩-CH$_2$-CH$_2$-CO-	nortestosterone cyclopentylpropionate, 17β-cyclopentylpropionyloxy-4-estren-3-one	[601-63-8]	Depo-Nortestonat (Upjohn)
R = CH$_3$-(CH$_2$)$_8$-CO-	nortestosterone decanoate, 17β-decanoyloxy-4-estren-3-one	[360-70-3]	Deca-Durabolin (Organon)
R 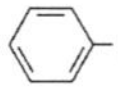 = ⟨phenyl⟩-(CH$_2$)$_2$-CO-	nortestosterone phenylpropionate, 17β-phenylpropionyloxy-4-estren-3-one	[62-90-8]	Durabolin (Organon); Norandrol (Osta); Norstenol (Ravizza)
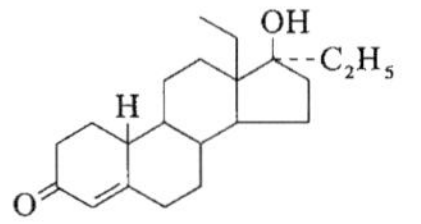 (structure with OH, C$_2$H$_5$)	norbolethone, 17α-methyl-18-methyl-17β-hydroxy-4-estren-3-one [a], [b]	[1235-15-0]	Genabol (Wyeth)
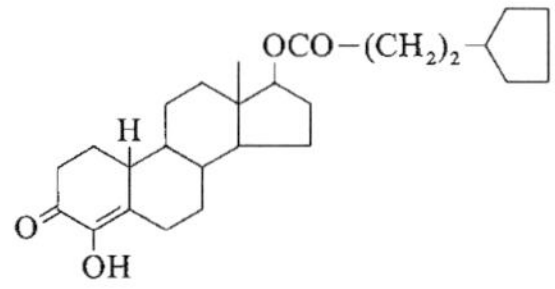 (structure with OCO-(CH$_2$)$_2$-cyclopentyl, OH)	oxabolone cypionate, 17β-cyclopentylpropionyloxy-4-hydroxy-4-estren-3-one [c]	[1254-35-9]	Steranabol ritardo (Farmitalia)
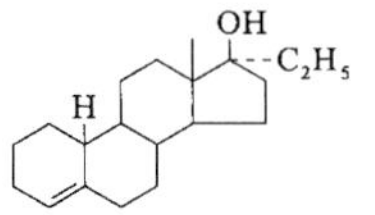 (structure with OH, C$_2$H$_5$)	ethylestrenol, 17α-ethyl-4-estren-17β-ol	[965-90-2]	Duraboral, Durabolin O, Maxibolin, Orabolin (Organon)

[a] [160], [b] [173], [c] [174].

Testosterone Compounds. Some anabolic testosterone derivatives are listed in Table 11. Introduction of a hydroxyl group or a chlorine atom at position 4, methylation at C-7, introduction of a thioacetyl group or an additional double bond between C-1 and C-2 favorably influence activity, and 17-methylation produces high oral potency.

Androstanolone Compounds. Androstanolone (**57**) is the parent compound of another series of anabolic sex hormones [175]; see Table 12. The remaining androgenic activity of androstanolone is further reduced by methylation at C-1 or C-2, or by replacing C-2 with oxygen or the A ring with a heterocyclic ring. Again, methylation at C-17 increases oral activity, but also increases stress on the liver.

Table 11. Anabolic testosterone derivatives

Formula	Name	CAS registry no.	Trade name (manufacturer)
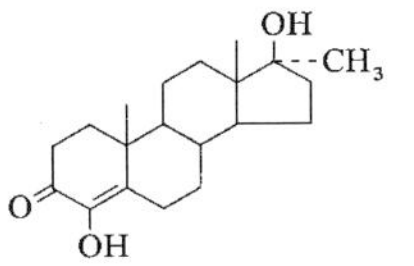	oxymesterone, 4,17β-dihydroxy-17α-methyl-4-androsten-3-one	[145-12-0]	Oranabol (Farmitalia) Theranabol (Theraplix)
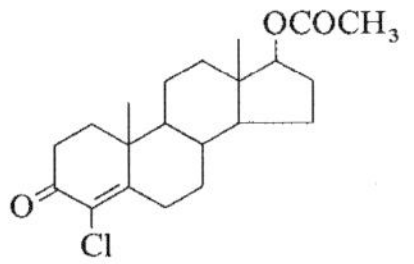	chlorotestosterone acetate, 17β-acetoxy-4-chloro-4-androsten-3-one	[855-19-6]	Steranabol (Farmitalia); Turinabol (VEB Jenapharm)
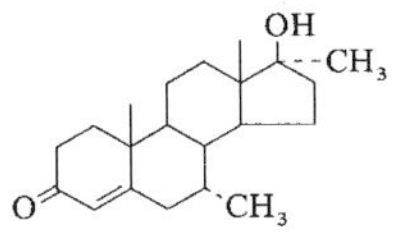	bolasterone, 7α,17α-dimethyl-17β-hydroxy-4-androsten-3-one	[1605-89-6]	Myagen (Upjohn)
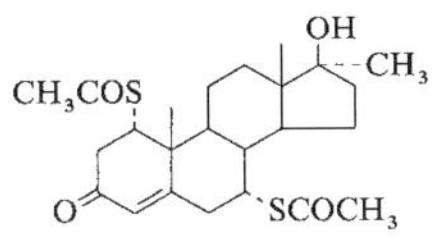	thiomesterone, 17β-hydroxy-17α-methyl-1α,7α-bis(thioacetyl)-4-androsten-3-one	[2205-73-4]	Emdabol (Merck)
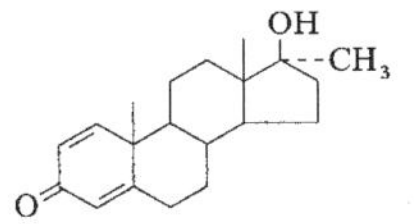	metandienone, 17β-hydroxy-17α-methyl-1,4-androstadien-3-one	[72-63-9]	Dianabol (Ciba)

3.7. Gestagens

3.7.1. General Aspects

Physiology and Clinical Applications. Progesterone (**7**) is the most important representative of a group of female sex hormones known as gestagens. LUDWIG FRENKEL discovered progesterone in 1903, but not until 1934 was progesterone isolated and its structure analyzed.

Progesterone acts with another group of female sex hormones, the estrogens (see Section 3.5), to regulate the sexual cycle. After ovulation, progesterone is continuously formed by the corpus luteum and prepares the endometrium of the uterus for implantation of the fertilized egg (transition from proliferation to secretion) [130], [131]. It prevents renewed ovulation and fertilization during pregnancy, decreases the tonus of the uterus, and thus maintains pregnancy. The inhibition of ovulation is the basis for the use of gestagens as contraceptive agents. The administration of gestagens "imitates" pregnancy—the secretion of FSH by the hypophysis diminishes, thus preventing ma-

Table 12. Anabolic androstanolone derivatives

Formula	Name	CAS registry no.	Trade name (manufacturer)
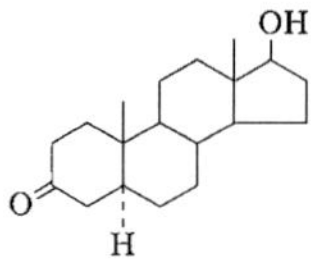 **57**	androstanolone, 17β-hydroxy-5α-androstan-3-one	[521-18-6]	Anaboleen (Uni-Chemie); Anabol-Tablinen (Sanorania); Stanaprol (Pfizer)
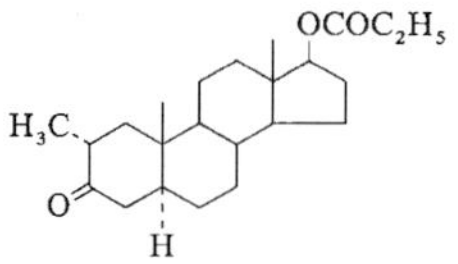	androstanolone propionate, 2α-methyl-17β-propionyloxy-5α-androstan-3-one	[521-12-0]	Drolban (Lilly); Masterid (Grünenthal)
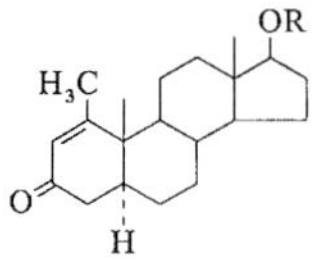	R = CH_3CO: methenolone acetate, 17β-acetoxy-1-methyl-5α-androst-1-en-3-one	[434-05-9]	Primobolan (Schering AG)
	R = $CH_3(CH_2)_5CO$: methenolone enanthate	[303-42-4]	Nibal (Squibb); Primobolan Depot (Schering AG)
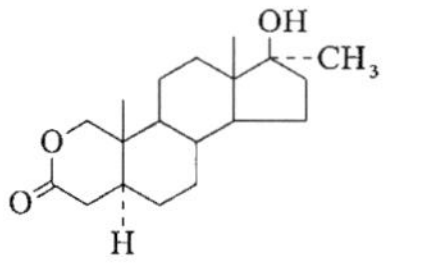	oxandrolone, 17β-hydroxy-17α-methyl-2-oxa-5α-androstan-3-one	[53-39-4]	Anavar (Searle); Protivar (Byla)
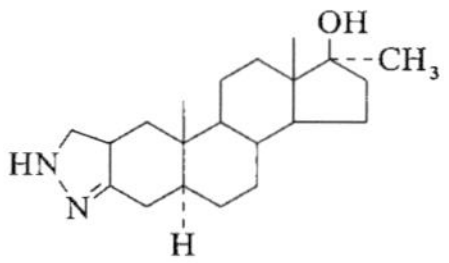	stanazolol, 17α-methyl-5α-androstano-[3,2-c]pyrazol-17β-ol	[10418-03-8]	Stromba, Winstrol (Winthrop)

turation and release of the ovum [176]. The most important application of gestagens, alone or in combination with estrogens, is in the inhibition of ovulation or contraception (suppression therapy). In addition, gestagens are important in the treatment of deficiency symptoms such as imminent abortion, sterility, and irregularities of the menstrual cycle (substitution therapy). The strong inhibition of gonadotropin secretion exerted by some gestagens is also exploited in the treatment of prostate adenoma.

Hormonal Contraception [130], [131]. Hormonal contraception is based on the suppression of ovulation in the middle of the menstrual cycle. The secretion of LH, which is reponsible for ovulation, does not occur under the influence of a gestagen – estrogen combination preparation. An additional contraceptive effect is achieved by the fact that inhibitors of ovulation also influence the development of the endometrium and the consistency of the cervical secretion. The hormonal contraceptives presently available contain different types of gestagenic components and varying doses of individual components. Although older combination preparations had a high proportional estrogen, the current tendency is to reduce estrogen content because this is primarily responsible for side effects. The successful search for new gestagens has produced a generation of low-dosage ovulation inhibitors. New administration regimes (sequential preparations, two- and three-phase preparations) have resulted in further dose reduction; the doses of estrogen and gestagen are adapted to the course of the menstrual cycle (see also → Chemical Contraception) [130], [131].

Bioassays. Two types of bioassay are used for the quantitative determinations of gestagens:

Clauberg Test [130]. The synergistic action of estrogens and gestagens induces characteristic changes of the endometrium. The endometrium proliferates under the action of estrogens and undergoes characteristic morphological changes under the subsequent influence of active gestagens. These changes are most apparent in rabbits; they are evaluated histologically and compared with those produced by reference substances. This test is used routinely in the search for novel gestagenic sex hormones and also permits determination of relative potencies.

Maintenance of Pregnancy. In pregnant rodents, removal of the corpus luteum or ovaries results in the termination of pregnancy. Active gestagens can maintain pregnancy in castrated, pregnant animals. The test is performed on pregnant rats and is used for qualitative characterization of the activity spectrum of a compound. However, it is not suitable for the determination of relative potencies.

3.7.2. Gestagens of the Pregnane and 19-Norpregnane Series

Progesterone (**7**), the naturally occurring gestagen, is administered subcutaneously as an injection solution; very high oral doses are required to elicit responses because of rapid degradation in the liver. For this reason, progesterone is not used as an oral contraceptive.

The extensive search for gestagens with high oral activity and low side effects started with three compounds that were more orally active than progesterone [127]: hydroxyprogesterone acetate (**58**), ethisterone (**68**), and 19-norethisterone (**29**), with approximately 30, 20, and 100 times the potency of progesterone, respectively. Modification of hydroxyprogesterone acetate (**58**), in particular substitution at C-6 and C-17, produced a large number of highly active oral gestagens. Some of these preparations and their trade names are listed in Table 13. Characteristic modifications include introduction of the Δ^6 double bond (**61–65**), substitution at C-6 with chlorine or a methyl group (**60–62, 64**), changing the ring A – B coupling (**65**), and transition to the 19-nor series (**66**). The gestagens listed in Table 13 differ in their effectiveness; those that exert a strong antiovulatory effect (central inhibition) are of considerable importance as contraceptives (**60–62, 67**). Other gestagens exert a stronger proliferative effect on the endometrium [176]. Gestonorone caproate (**66**) is also used in the treatment of prostate adenoma.

Table 13. Pregnane and 19-norpregnane gestagens

Formula	Name	CAS registry no.	Trade name (manufacturer)
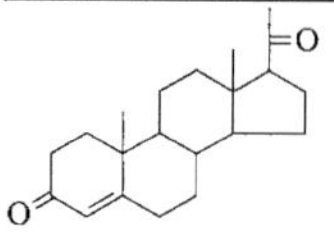 **7**	progesterone, 4-pregnene-3,20-dione [a],[b]	[57-83-0]	Colprosteron (Ayerst); Geston (Paines and Byrne); Luteogan (Henning); Luteosid (Boehringer Mannheim); Progestin (Organon); Proluton (Schering AG)
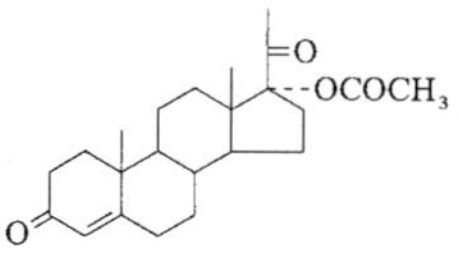 **58**	hydroxyprogesterone acetate, 17-acetoxy-4-pregnene-3,20-dione [a],[b]	[302-23-8]	Prodox (Upjohn)
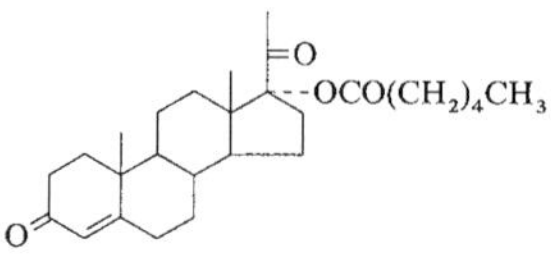 **59**	17α-hydroxyprogesterone caproate, 17-hexanoyloxy-4-pregnene-3,20-dione [a]	[630-56-8]	Delalutin (Squibb); Idrogesten (Farmila); Neolutin forte (Spofa); Proluton-Depot (Schering AG)
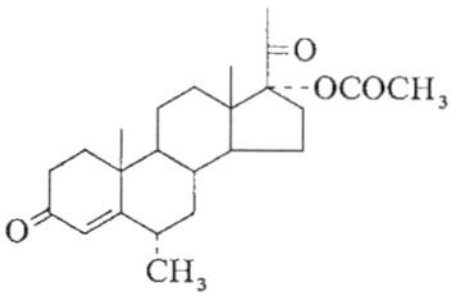 **60**	medroxyprogesterone acetate, 17-acetoxy-6α-methyl-4-pregnene-3,20-dione [a]	[71-58-9]	Clinovir (Upjohn); Gesinal (Hoechst); Provera (Upjohn); combination preparation with ethinyl estradiol: Provest (Upjohn)
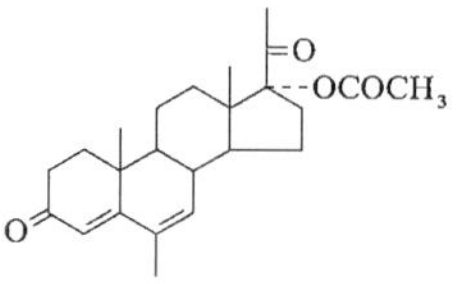 **61**	megestrol acetate, 17-acetoxy-6-methyl-4,6-pregnadiene-3,20-dione [a],[b]	[595-33-5]	Megage (Mead Johnson); Ovarid (Glaxo); combination preparation with ethinyl estradiol or mestranol: Delpregnin (Novo); Volidan (British Drug Houses)
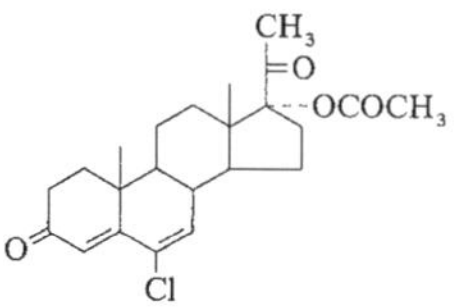 **62**	chlormadione acetate, 17-acetoxy-6-chloro-4,6-pregnadiene-3,20-dione [a],[b]	[302-22-7]	Gestafortin (Merck); Lormin (Lilly); Lutoral (Syntex); combination preparation with mestranol: Ovosiston (VEB Jenapharm)
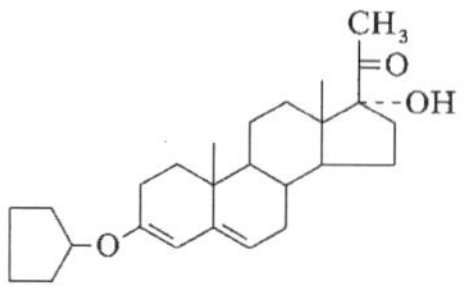 **63**	pentagestrone, 17α-hydroxy-3-cyclopentyloxy-3,5-pregnadien-20-one	[7001-56-1]	Gestovis, Gestovister (Vister)

Table 13. (continued)

Formula	Name	CAS registry no.	Trade name (manufacturer)
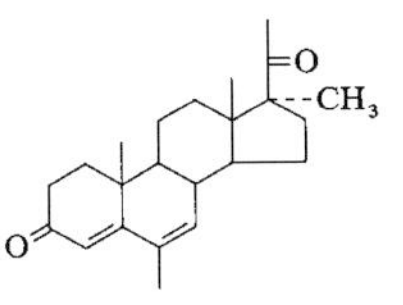 **64**	medrogestone, 6,17-dimethyl-4,6-pregnadiene-3,20-dione [c]	[977-79-7]	Colpro, Colprone (Ayerst); Prothil (Kali-Chemie)
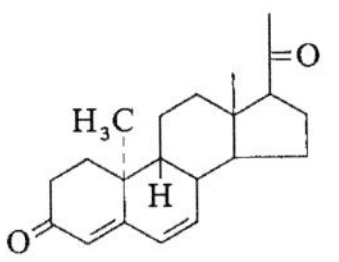 **65**	dydrogesterone, 9β,10α-pregna-4,6-diene-3,20-dione	[152-62-5]	Duphaston (Philips Duphar); Gestatron (Leo); Gynorest (Mead Johnson), Terolut (Ferrosan)
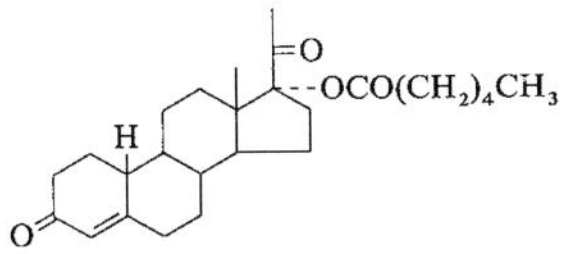 **66**	gestonorone caproate, 17-hexanoyloxy-19-nor-4-pregnene-3,20-dione [d]	[1253-28-7]	Depostat (Schering AG)
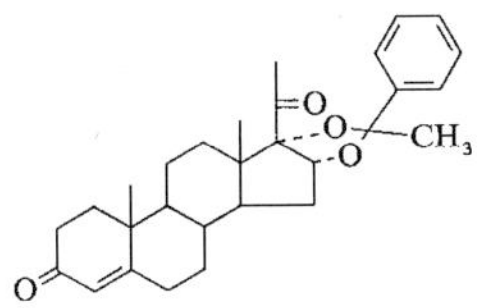 **67**	dihydroxyprogesterone acetophenonide, 16α,17-(1'-methyl-1'-phenylmethylenedioxy)-4-pregnene-3,20-dione [e]	[1179-87-9]	Deladroxone, Droxone (Squibb); combination preparation with estradiol enanthate: Deladroxat (Squibb)

[a] [130], [b] [131], [c] [177], [d] [178], [e] [179].

3.7.3. Gestagens of the Androstane Series

Insertion of a 17α-ethynyl group into testosterone yields ethisterone (**68**) which, surprisingly, is only slightly androgenic but is a highly potent oral gestagen [127]. The synthesis of ethisterone was the first successful step in the search for orally active gestagens. Dimethisterone (**69**) is even more potent than ethisterone; these are the only two products of this series on the market (Table 14).

3.7.4. Gestagens of the 19-Norandrostane Series

Compounds of the 19-norandrostane (or estrone) series can be prepared easily by Birch reduction of estrone methyl ether. Examples commercially available as contraceptive agents are listed in Table 15 [176]. Transition from the androstane to the 19-norandrostane series further increases the biological potency of a large number of these

Table 14. Androstane gestagens

Formula	Name	CAS registry no.	Trade name (manufacturer)	Combination preparations with ethinyl estradiol (manufacturer)
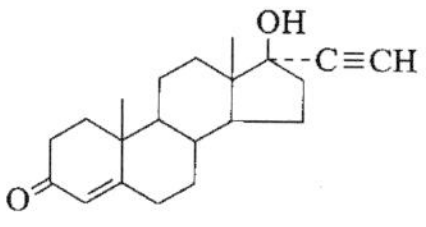 **68**	ethisterone, 17α-ethynyl-17β-hydroxy-4-androsten-3-one	[434-03-7]	Lutocyclin (Ciba); Lutoral (Schieffelin); Pranone (Schering Corp.)	Amenorone (Roussel); Menstrogen (Organon); Norma-sterin (Norma)
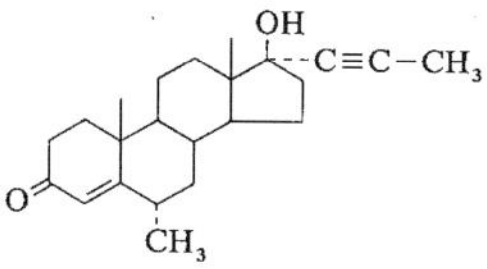 **69**	dimethisterone, 17β-hydroxy-6α-methyl-17α-(1-propinyl)-4-androsten-3-one	[79-64-1]	Secrosteron (British Drug Houses)	Oracon, Ovin (Mead Johnson); Secrodyl (British Drug Houses)

Table 15. 19-Norandrostane gestagens

Formula	Name	CAS registry no.	Trade name (manufacturer)
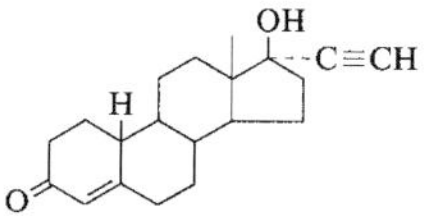 **29**	norethisterone, 17α-ethynyl-17β-hydroxy-4-estren-3-one [a],[b]	[68-22-4]	Micronett (Ortho); Micronor (Cilag); Norlutin (Parke Davis); Proluteasi (Tiber); combination preparation with mestranol: Norinyl (Syntex); Ortho-Novum (Cilag)
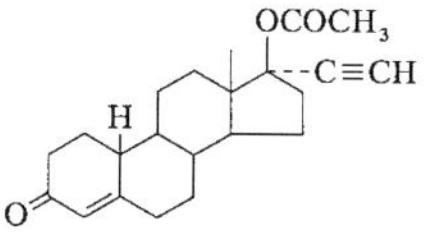 **70**	norethisterone acetate, 17α-ethynyl-17β-acetoxy-4-estren-3-one [a],[b]	[51-98-9]	Norlutate, Norlutin A (Parke Davis); Primolut Nor (Schering AG); combination preparation with ethinyl estradiol: Anovlar (Schering AG); Etalontin, Orlest (Parke Davis)
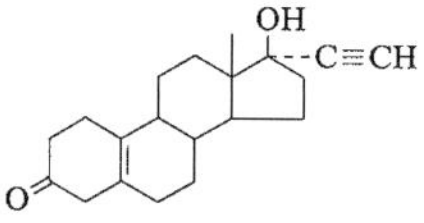 **71**	norethynodrel, 17α-ethynyl-17β-hydroxy-5(10)-estren-3-one [a],[b]	[68-23-5]	combination preparation with mestranol: Enavid, Enovid (Searle)
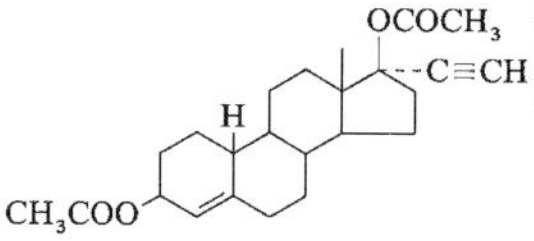 **72**	ethynodiol diacetate, 17α-ethynyl-3β,17β-diacetoxy-4-estrene	[297-76-7]	Demulen (Searle); combination preparation with mestranol: Ovulen (Searle)
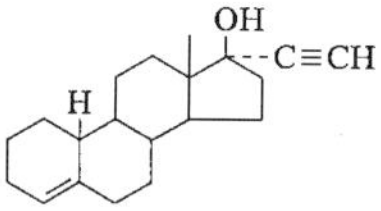 **73**	lynestrenol, 17α-ethynyl-4-estren-17β-ol	[52-6-6]	Orgametril (Organon); combination preparation with mestranol: Lyndiol (Organon); Noracyclin (Ciba)

Table 15. (continued)

Formula	Name	CAS registry no.	Trade name (manufacturer)
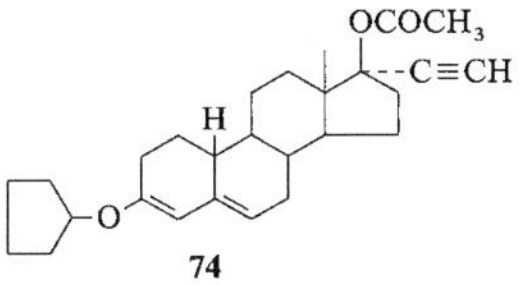 **74** (OCOCH₃, C≡CH, cyclopentyloxy)	quingestanol acetate, 17α-ethynyl-17β-acetoxy-3-cyclopentyloxy-3,5-estradiene	[3000-39-3]	combination preparation with ethinyl estradiol: Reglovis (Vister); combination preparation with ethinyl estradiol-3-cyclopentyl-ether: Unovis (Warner-Chilcott)
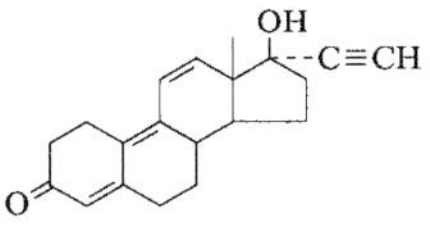 **46** (OH, C≡CH)	norgestrienone, 17α-ethynyl-17β-hydroxy-4,9,11-estratrien-3-one ^{c,d}	[848-21-5]	combination preparation with mestranol: Planor (Roussel-Uclaf)
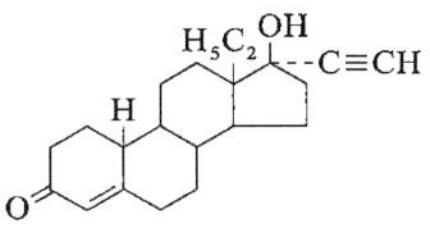 **38** (H₅C₂, OH, C≡CH)	norgestrel (racemate), rac. 17α-ethynyl-13-ethyl-17β-hydroxy-4-gonen-3-one ^{e}		combination preparation with ethinyl estradiol: Duoluton, Eugynon (Schering AG); Stediril (Wyeth); combination preparation with estradiol valerate: Cyclo-Progynova (Schering AG)
	D-norgestrel, 17α-ethynyl-13-ethyl-17β-hydroxy-4-gonen-3-one ^{f}	[6533-00-2]	Microlut (Schering AG); combination preparation with ethinyl estradiol: Microgynon, Neogynon (Schering AG); Stediril-d (Wyeth)
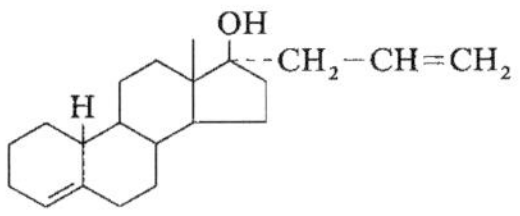 (OH, CH₂–CH=CH₂)	allylestrenol, 17α-allyl-17β-hydroxy-4-estrene ^{g}	[432-60-0]	Gestanin, Gestanon (Organon); Gestanyn (Pharmacia)
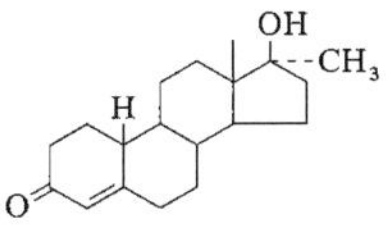 (OH, CH₃)	methylestrenolone, 17β-hydroxy-17α-methyl-4-estren-3-one ^{h}	[514-61-4]	Methalutin (Syntex); Orgasteron (Organon); combination preparation with methylestradiol: Gynaekosid (Boehringer Mannheim)
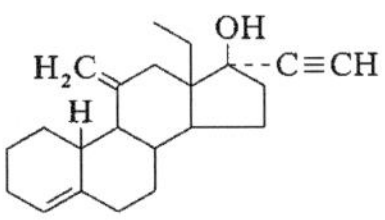 **75** (H₂C, OH, C≡CH)	desogestrel ^{i}	[54024-22-5]	combination preparation with ethinyl estradiol: Marvelon (Organon)
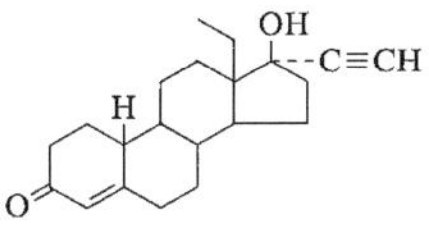 **76** (OH, C≡CH)	gestoden ^{j}	[60282-87-3]	combination preparation with ethinyl estradiol: Femovan (Schering AG)
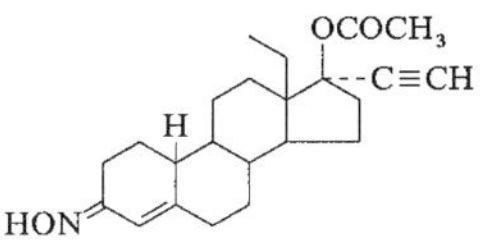 **77** (OCOCH₃, C≡CH, HON)	norgestimat ^{k}	[35189-28-7]	combination preparation with ethinyl estradiol: Cilest (Cilag)

sex hormones. 19-Norethisterone (**29**) was synthesized as an analogue of ethisterone (**68**); like ethisterone, it has only slight androgenic activity, but it is a fivefold stronger oral gestagen. A number of very potent oral gestagens with strong antiovulatory activity has been derived from 19-norethisterone by methods such as acetylation (**70, 72, 74**), isomerization of the double bond (**71**), reduction (**72**), elimination of the 3-keto group (**73**), or enolether formation (**74**) [130], [131]. The totally synthetic gestagens—norgestrienone (**46**), norgestrel (**38**, racemate) and D-norgestrel (**38**), desogestrel (**75**), gestoden (**76**), and norgestimat (**77**)—all belong to this series.

3.7.5. Antigestagens [185]

Antigestagens may be either progesterone antagonists or inhibitors of progesterone synthesis.

Competitive progesterone antagonists are substances that inhibit gestagen function by blocking the progesterone receptor. The first compound found to have this property was mifepristone [*84371-65-3*] Roussel-Uclaf [RU 486, 11β-(4-dimethylaminophenyl)-17β-hydroxy-17α-(1-propynyl)estra-4,9-dien-3-one]. This substance can be administered in fertility control (nidation inhibition, menstruation commencement, abortion induction), obstetrics, and the treatment of hormone-dependent tumors.

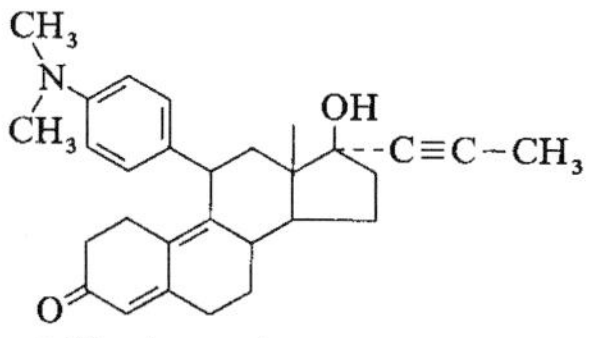

Mifepristone (Roussel-Uclaf)

Inhibitors of progesterone synthesis block the action of progesterone by preventing its formation [139]. At present, attempts are being made to inhibit the conversion of pregnenolone to progesterone by reversibly or irreversibly blocking the enzyme Δ^5-3β-hydroxysteroid dehydrogenase. Epostane [*80471-63-2*] (4α,5α-epoxy-17β-hydroxy-4β,17α-dimethyl-3-oxo-androstane-2α-carbonitrile) is being tested clinically [186]. Some nonsteroidal compounds (e.g., aminogluthetimide and antimycotics of the ketoconazole family) can also suppress steroid biosynthesis by enzyme inhibition.

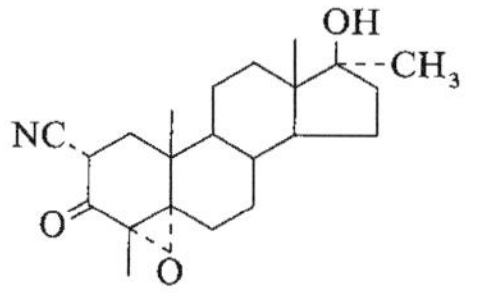

Epostane (Sterling-Winthrop)

4. Adrenal Steroid Hormones
(→ Steroids)

4.1. Introduction

Very few classes of substances exert so many physiologically and therapeutically useful effects as the steroid hormones of the adrenal cortex. These hormones, also called corticosteroids or corticoids, are divided into two groups depending on their physiological mode of action: the *glucocorticoids* promote glycogen formation from carbohydrates and gluconeogenesis from proteins, whereas the *mineralocorticoids* act primarily on electrolyte and water metabolism. The corticoids are thus responsible for the maintenance of important metabolic functions and are essential for life. The life of adrenalectomized animals is maintained by continuous corticoid administration (survival test) [191]. The "liver glycogen test" was developed especially for glucocorticoids [187].

Administration of glucocorticoids results in glycogen accumulation in the liver of normal or adrenalectomized, fed or fasting animals. The liver glycogen test measures the glycogen content of liver homogenate after the treated animals have been sacrificed.

The simultaneous increase of blood glucose and certain gluconeogenetic enzymes in the liver is a further useful parameter for determining the glucocorticoid potency of a test compound.

Both groups of substances are C_{21} steroids and are structurally related to the pregnane series.

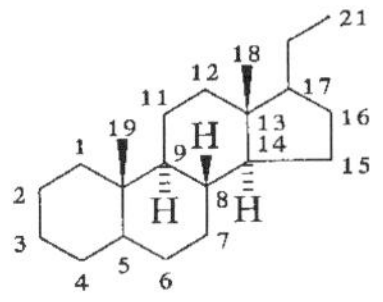

Pregnane skeleton

The important structural characteristics of these compounds are the α,β-unsaturated 3-keto-$\Delta^{4(5)}$ system and an α-hydroxyketone group in the 17β side chain. An 11β-hydroxyl group or 11-keto group is important for a predominantly glucocorticoid effect. In the mineralocorticoids, this group is absent (cortexone) or physiologically inactivated (e.g., lactol formation as in aldosterone). An additional hydroxyl group at the 17α-position increases glucocorticoid activity significantly.

Approximately 50 steroids of the pregnane (steroids with 21 carbon atoms), androstane (19 carbon atoms), and estrane series (18 carbon atoms) have been obtained from the adrenal cortex since the 1930s. These steroids have all been isolated and their structures analyzed, but only seven compounds are considered to be typical corticosteroids:

Glucocorticoids

1) Cortisol (**78**), also called hydrocortisone, 17α-hydroxycorticosterone, 11β,17α,21-trihydroxypregn-4-ene-3,20-dione
2) Cortisone, 17α,21-dihydroxypregn-4-ene-3,11,20-trione
3) Corticosterone, 11β,21-dihydroxypregn-4-ene-3,20-dione
4) 11-Dehydrocorticosterone, 21-hydroxypregn-4-ene-3,11,20-trione

Mineralocorticoids

1) Cortexone, deoxycorticosterone, 21-hydroxypregn-4-ene-3,20-dione
2) Cortexolone, 17α-hydroxydeoxycorticosterone, 17α,21-dihydroxypregn-4-ene-3,20-dione
3) Aldosterone (**79**), 18,21-dihydroxy-11β,18-oxidopregn-4-ene-3,20-dione

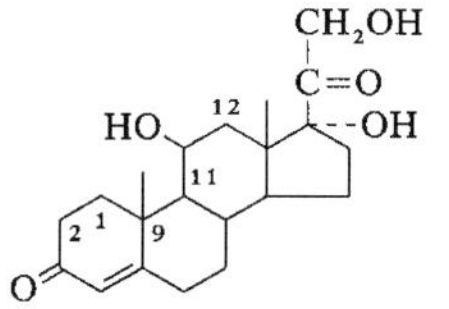

Cortisol (**78**) 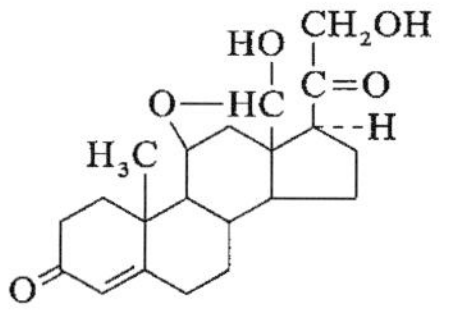   Aldosterone (**79**)

Of the steroids listed above, only cortisol, corticosterone, and aldosterone exhibit hormonal characteristics. The other corticoids, despite their often high physiological activity, are considered precursors or metabolites of the true adrenal cortical hormones.

4.2. Historical Aspects

The isolation and structural analyses of naturally occurring corticoids were carried out by WINTERSTEINER, PFIFFNER, KENDALL, and REICHSTEIN. Crystalline aldosterone was first obtained in 1953 from adrenal cortical tissue by REICHSTEIN and coworkers, who then proceeded to elucidate the structure and configuration of this steroid. The mineralocorticoid cortexone (commercially available since 1938) was the first corticoid to be used in the treatment of Addison's disease, a corticosteroid deficiency caused by damage or destruction of the adrenal glands. Cortexone can be synthesized easily. At first, cortisone and cortisol were employed primarily in the treatment of rheumatic disease, but after the mid-1950s, cortisone, cortisol, and especially the newly developed, chemically modified glucocorticoids achieved tremendous therapeutic and industrial importance.

The former industrial extraction of naturally occurring corticoids, especially cortisone and cortisol, from animal adrenal glands [196] is no longer important. It has been completely replaced by partially synthetic production methods.

4.3. Physiology and Mode of Action

The adrenal cortex is a peripheral endocrine gland whose hormone production (with the exception of aldosterone) is regulated by the anterior pituitary hormone, adrenocorticotropic hormone (ACTH). The formation of aldosterone is independent of ACTH; it is controlled primarily by the angiotensin–renin system and by the functional coupling of adrenal cortex, hypophysis, and hypothalamus.

The adrenal corticosteroid hormones play an important part in the metabolism of carbohydrates, proteins, and minerals. Their importance is exemplified by the fact that even slight stress to the organism causes a rapid increase in the activity of the adrenal cortex.

The adrenal steroid hormones cannot be strictly divided into mineralocorticoids and glucocorticoids because they all produce the effects of both groups. Thus, aldosterone exerts its prime action as a mineralocorticoid but also has some glucocorticoid activity; for example, in the liver glycogen test [191], it is only slightly less potent than cortisone.

The *mineralocorticoids* are responsible for maintaining the correct distribution of sodium and potassium between the cell contents and the extracellular fluid and, consequently, for the maintenance of osmotic equilibrium. Malfunction of the adrenal cortex results in increased or diminished excretion of sodium and chlorine, with correspondingly diminished or increased excretion of potassium. Severe disturbances in water metabolism occur. Aldosterone exerts a much stronger effect on mineral metabolism than the other hormones. It has approximately 25–30 times the sodium-retaining activity and 5 times the potassium excretory activity of deoxycorticosterone. In humans, ca. 100–200 µg of aldosterone is synthesized daily.

In the case of natural or chemically modified *glucocorticoids*, a distinction must be made between their physiological action and their subsequently discovered therapeutically important pharmacodynamic effects. The primary physiological functions of glucocorticoids are to increase glycogen formation in the liver, raise the blood sugar level, and augment gluconeogenesis from amino acids by protein breakdown (catabolic effect). Moreover, glucocorticoids enable the organism to withstand stress (e.g., physical or psychological demands) and to maintain a normal internal state (homeostasis). In humans, 15–40 mg of cortisol is produced daily, but ten times this amount can be secreted under stress conditions. The amount of corticosterone synthesized daily is 1.3–4 mg.

Androgenic hormones, such as 11-hydroxyandrostenedione and androsterone, and to a small extent estrogenic hormones are also produced in the adrenal cortex [194].

4.4. Therapeutic Uses of Glucocorticoids

The therapeutic effects of glucocorticoids, particularly of chemically modified corticoids, are based primarily on the inhibition of proliferative reactions of mesenchymal tissue, especially allergic and inflammatory tissue reactions (e.g., polyarthritis, pleuritis). However, glucocorticoids are used to treat the symptoms but not the cause of these diseases. Glucocorticoids reduce the number of lymphocytes and eosinophilic granulocytes in the blood. Because their main effects are anti-inflammatory, antiexudative, cytotoxic, immunosuppressive, antitoxic, and antipruretic, they may be applied in the treatment of the following syndromes:

Substitution Therapy. Glucocorticoids can be used in acute and chronic adrenal cortical insufficiency, adrenogenital syndrome, thyroidal crises, Addison's disease (in combination with mineralocorticoids), and inhibition therapy of hyperadrenal cortical states.

Pharmacodynamic Therapy. The most important indications for glucocorticoid treatment are acute and chronic polyarthritis; rheumatic fever; allergic reactions such as asthma and obstructive bronchitis (→ Antiasthmatics); cardiogenic, anaphylactic, and traumatic shock; stress situations; collagen disease; autoimmune disease; shock caused by burns and acid burns; poisoning; acute attack of gout; malignant tumors (especially leukemia and lymphogranulomatosis); and local application in a number of skin afflictions and dermatoses (especially eczema and exanthema, → Drugs Used in Dermatology). In bacterial diseases, glucocorticoids can be administered with antibiotics or chemotherapeutic agents.

4.5. Therapy and Side Effects

Types of Therapy. Three types of corticoid therapy are possible: systemic, topical, and local. *Systemic therapy* can in turn be divided into oral application (administration of corticoids in the free 21-alcohol form or in the ester form as tablets, capsules, dragées, etc.); intravenous application (administration as a water-soluble 21-ester, e.g., 21-sodium hemisuccinate, 21-disodium phosphate, or 21-dialkylamino carboxylate); and intramuscular application (administration of corticoids as a free 21-alcohol or a short- or long-chain 21-ester). In *topical corticoid therapy,* free 21-alcohols or their esters are injected as crystal suspensions into a joint or tissue cavity.

Local corticoid therapy is used commonly to treat skin disease. Corticoids are again applied as the free 21-alcohols or as specially structured 21- or 17-esters in ointments, creams, lotions, or sprays. A broad-spectrum antibiotic is usually added to local corticoid preparations to prevent bacterial infection (e.g., 0.5% neomycin sulfate or 0.1% gentamicin).

Table 16. Cortisol equivalence, average maintenance dose, and Cushing threshold dose of the most important glucocorticoids used in systemic human therapy listed in order of increasing effectiveness [193], [197]

Glucocorticoid	Cortisol equivalence (cortisol = 1)	Average maintenance dose, mg/d	Cushing threshold dose, mg/d
Cortisone	0.8	10 – 50	50
Cortisol	1.0	10 – 50	40
Prednylidene	2	12 – 15	18 – 24
Prednison, prednisolone	4	5 – 15	10
Fluocortolone	4	2.5 – 12.5	10 – 15
6α-Methylprednisolone	5	6 – 10	8
Triamcinolone	5 – 10	6 – 8	8
Paramethasone	10	2 – 6	4 – 6
Fludrocortisone	15		
Dexamethasone	20 – 25	1 – 2	2
Betamethasone	25 – 50	1 – 1.5	2

Side Effects. Medium- or long-term systemic application of glucocorticoids can cause the following side effects: hypercorticism or Cushing's syndrome (adiposity of the trunk, moon-shaped face, steroid diabetes, hypertension); tumor formation in the gastrointestinal tract; retardation of granulation and scar formation; osteoporosis; hypophyseal inhibition; atrophy of the adrenal cortex; psychological effects (e.g., euphoria); increased risk of infection; endocrine disturbance; morphological changes of the skeletal muscle and muscular atrophy; or adynamia and fatigue.

The central mineralocorticoid side effects of cortisone and cortisol have been eliminated to a large extent by the development of chemically modified glucocorticoids. To reduce the risk of side effects in systemic anti-inflammatory therapy, glucocorticoids are often combined with nonsteroidal anti-inflammatory agents (→ Analgesics and Antipyretics).

Dosage. The dose equivalence is used to compare the effects of systemically administered glucocorticoids; it is the glucocorticoid or anti-inflammatory activity relative to that produced by cortisol (cortisol equivalence = 1). The average maintenance dose in milligrams per day is also important in establishing the dosage. Both values are listed in Table 16 for the most important glucocorticoids used systemically in medicine today; the Cushing threshold dosage is also given.

4.6. Biosynthesis and Metabolism

Biosynthesis of the adrenal corticosteroid hormones starts with the conversion of cholesterol to pregnenolone, which is subsequently oxidized to progesterone (see Fig. 4). This compound is then converted to the glucocorticoids by primary hydroxylation at C-17 and subsequent hydroxylation at C-21 (cortisone) and C-11 (cortisol). In contrast, biosynthesis of the mineralocorticoids is initiated by primary hydroxylation at C-21. The C-17 hydroxylase is unable to attack the C-21 hydroxylated product. Glucocorticoids are produced in the central zona fasciculata and the mineralocorticoids in the outer zona glomerulosa of the adrenal cortex.

Corticoid metabolism occurs primarily in the liver, the major enzymatic degradation reactions being [190]

1) hydrogenation of the double bonds at the 4,5-positions and, in the case of $\Delta^{1,4}$ corticoids, the 1,2-positions to give products with a saturated ring A;
2) reduction of the 3-keto, 20-keto and, in the case of 11-oxocorticoids, 11-keto groups to the corresponding hydroxyl groups; and
3) oxidative cleavage of the side chains at the 17β-position to give 17-keto steroids.

The metabolites of adrenal cortical steroids are excreted in the urine as 3-glucuronides or 3-sulfates, e.g., as glucuronides of type **80** (tetrahydrocortisol derivatives) or type **81** (tetrahydroandrostane derivatives).

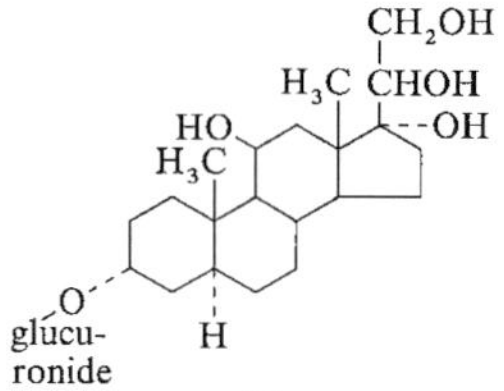

4.7. Detection Methods

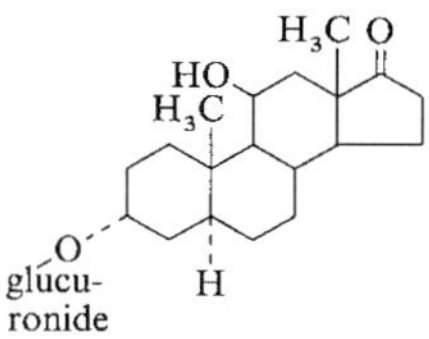

The special detection methods used in the analysis of corticoids are almost all based on chemical reactions specific for the side chain at position 17. Important reactions are degradation with periodic acid and chromic acid [187], and oxidation of the C-17 side chain with sodium bismuthate to give the 17-keto derivative (e.g., androsterone) or 17-ethio acid (17β-carboxylic acid) [198]. Another method for identification of the dihydroxyacetone side chain is the Porter–Silber reaction [199], which is based on the formation of a yellow dye on heating with phenylhydrazine and sulfuric acid. This substance has a characteristic UV absorption band at 410 nm, and its molar extinction can be measured easily. For further details on chemical, spectral, and radioactive detection methods for corticoids, see [194].

4.8. Naturally Occurring Glucocorticoids

4.8.1. Properties and Uses

Cortisone [53-06-5] (**82**), $C_{21}H_{28}O_5$, M_r 360.46, mp 220–224 °C, $[\alpha]_D^{20}$ +209° (95% ethanol),

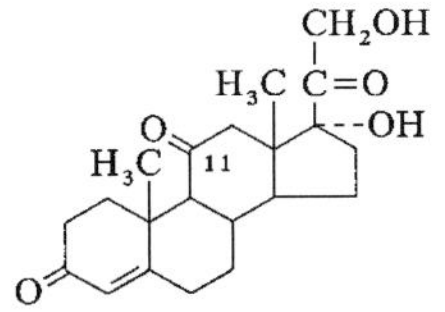

Cortisone (**82**)

and **cortisol** [50-23-7](**78**), $C_{21}H_{30}O_5$, M_r 362.47, mp 217–220 °C, $[\alpha]_D^{20}$ +167° (ethanol), are glucocorticoid hormones whose importance has diminished considerably in recent years because of their relatively low potency and, above all, their pronounced side effects (especially the strong mineralocorticoid action). They have been replaced by chemically modified corticoids and are used only in exceptional cases, namely, in substitution therapy and in special preparations for local application (e.g., in ear, nose, and throat conditions and in ophthalmology). However, cortisone and cortisol are even more important today as intermediates in the synthesis of chemically modified corticoids. Because totally synthetic methods are not economically feasible, the industrial production of cortisone and cortisol is still based on a series of efficient, partially synthetic methods. These methods use inexpensive, simple plant steroids as raw materials.

Total synthesis of cortisone was achieved in 1951 by WOODWARD and coworkers [200] and a year later by SARETT and coworkers [201]. A total synthesis of cortisone had already been published in 1950 by WENDLER and coworkers [202]. These methods are solely of scientific interest and have no industrial importance. Although the large-scale corticoid total syntheses developed by Roussel–Uclaf have been improved considerably, they are still not as economical as the partially synthetic processes [203]–[205].

Trade Names: Despite their minor therapeutic importance, a large number of cortisone and cortisol preparations are still available on the market. Only a few typical representatives can be listed here:
Cortisone: Cortisone "Ciba" (Ciba, Wehr, FRG); Scheroson (Schering).
Cortisol: Scheroson F (Schering); Ficortril (Pfizer, FRG); Hydrocort (Ferring, FRG); Hydrocortisone "Hoechst" (Hoechst).

Corticosterone [50-22-6], $C_{21}H_{30}O_4$, M_r 346.45, mp 180–182 °C, $[\alpha]_D^{25}$ +223° (alcohol), and **11-dehydrocorticosterone** [72-23-1], $C_{21}H_{28}O_4$, M_r 344.43, mp 178–180 °C, $[\alpha]_D^{25}$ +258° (alcohol), are two other glucocorticoids.

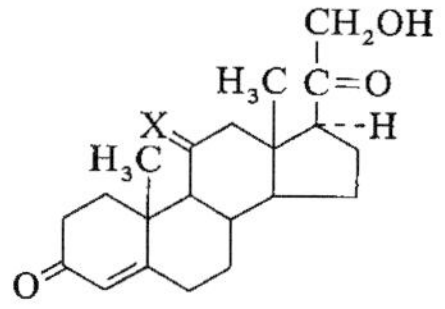

Corticosterone, X = HO⟶ / H⟶
11-Dehydrocorticosterone, X = O

They do not have a hydroxyl group at position 17; therefore, they can be considered as 17-deoxycortisol and 17-deoxycortisone, respectively. In comparison with cortisol and cortisone, these compounds have a lower glucocorticoid activity but exert a stronger mineralocorticoid effect. For this reason, these corticoids have achieved neither industrial nor therapeutic importance. For partial synthesis, see [189].

4.8.2. Partial Synthesis of Cortisone [191]

The most important methods for the industrial production of cortisone (**82**) from steroid raw materials are

1) production from progesterone,
2) production from 16-dehydropregnenolone via cortexolone, and
3) production from bile acid.

The first two methods are more important because they use an elegant microbiological process to insert the oxygen function at C-11.

Production from Progesterone. Production of cortisone from progesterone was developed in 1952 and is important because progesterone can be produced cheaply from diosgenin (see Section 3.2.2). Progesterone is converted to 11α-hydroxyprogesterone (**83**) by a high-yield (90%) microbiological oxidation with the mold *Rhizopus arrhizus* or *R. nigricans* [206], [207]. The method used by Upjohn proceeds as shown in Figure 8 [208]. 11α-hydroxyprogesterone (**83**) is oxi-dized with chromium trioxide to 11-oxoprogesterone, which in turn is converted to the sodium salt of 21-(ethoxalyl)-11-oxoprogesterone (**84**) with diethyl oxalate and sodium ethoxide [209]. The next step involves treatment with 2 molar equivalents of bromine. The resulting 21-dibromo derivative is not further isolated but is subjected to a Favorskii rearrangement in the presence of sodium methoxide, to yield compound **85**. The 3-oxo-Δ^4 group of this compound is protected by formation of the enamine (**86**). Reduction of compound **86** with lithium aluminum hydride leads to the formation of compound **87**, in which both the 21-carbomethoxy and the 11-keto groups are reduced to alcohol groups. After hydrolytic cleavage of the enamine group and esterification of the 21-hydroxyl group with acetic anhydride – pyridine, compound **88** is obtained (the 11β-hydroxyl group is not esterified for steric reasons). Compound **88** is subjected to chromic acid oxidation

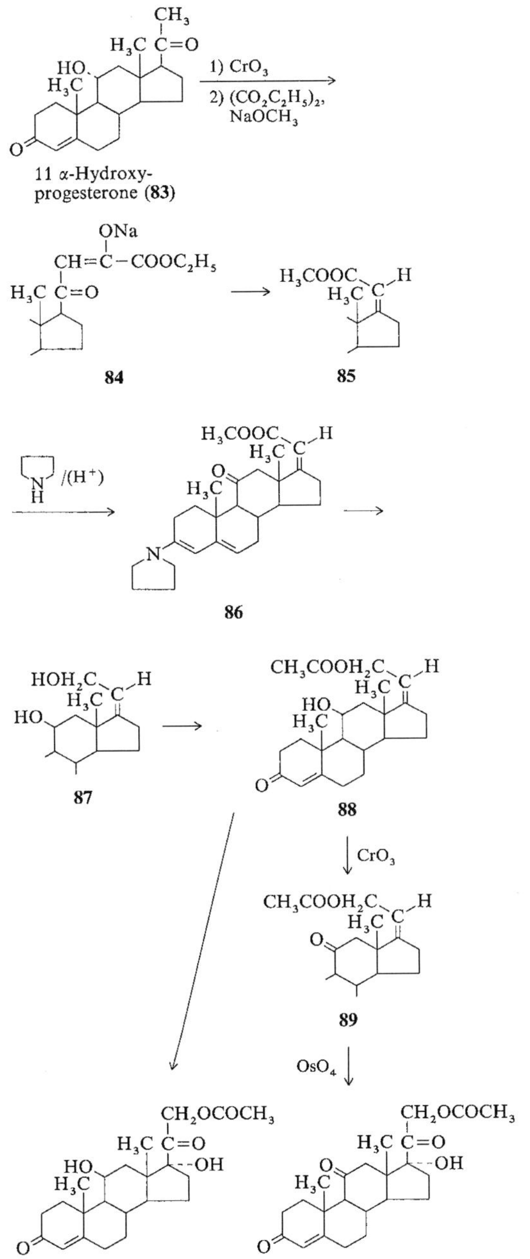

Figure 8. Synthesis of cortisone from 11α-hydroxyprogesterone

and then effectively converted to cortisone acetate (**90**), by using a method developed at Upjohn, with *N*-methylmorpholine *N*-oxide and osmium tetroxide [210]. The total yield is 20–25%, based on 11α-hydroxyprogesterone (**83**).

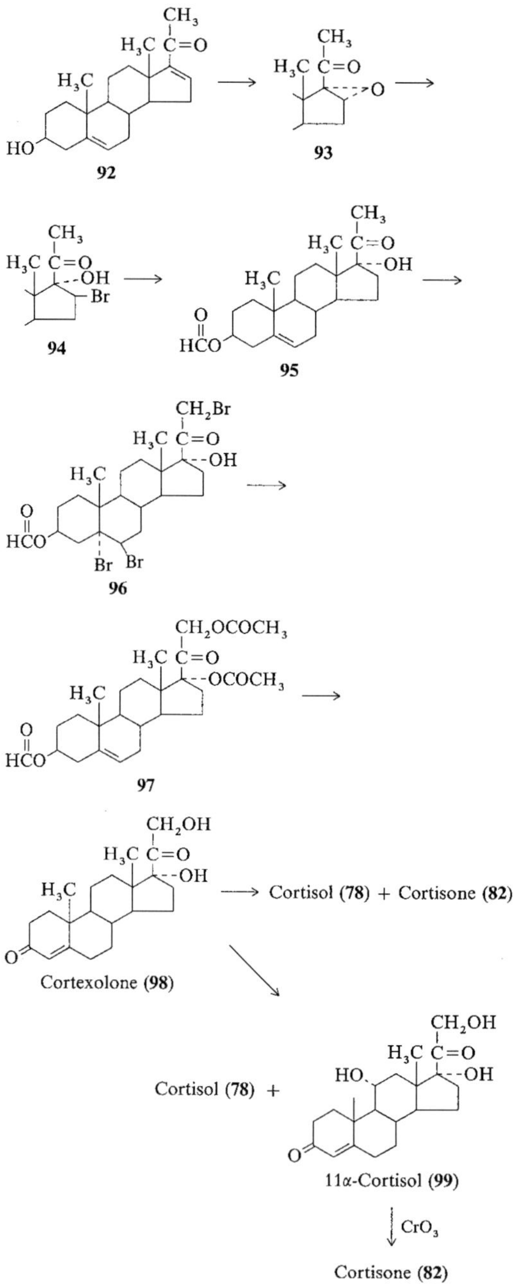

Figure 9. Synthesis of cortisone from 16-dehydropregnenolone

Production from 16-Dehydropregnenolone. 16-Dehydropregnenolone (**92**) is prepared easily on a large scale from diosgenin (see Section 3.2.2). A method developed by JULIAN [211] and modified by Syntex [212] requires several steps to convert this compound to cortexolone (Fig. 9; **98**).

16-Dehydropregnenolone (**92**), obtained by hydrolysis of 16-dehydropregnenolone acetate, is converted with alkaline hydrogen peroxide to 16α,17α-oxidopregnenolone (**93**), which in turn reacts with hydrobromic acid to give the bromohydrin (**94**). After removal of the 16β-bromine atom by hydrogenation with palladium–hydrogen and partial esterification of the 3-hydroxyl group with formic acid, 3β-formoxy-17α-hydroxypregnene (**95**) is formed. This compound can also be prepared by using the method described in [213]. Compound **95** is converted to the tribromo derivative (**96**) with 2 molar equivalents of bromine; then the Δ^5 double bond is regenerated with sodium iodide, and the 21-bromine group is replaced by the 21-acetoxy group with potassium acetate. Esterification of the 17α-hydroxyl group with acetic anhydride and toluene-*p*-sulfonic acid gives the diester (**97**). Direct Oppenauer oxidation (during which the 3-formoxy group near the less vulnerable acetate groups at C-17 and C-21 is partially saponified) and subsequent alkaline saponification of the 17- and 21-acetate groups yield cortexolone (**98**). The total yield of **98** based on **95** is 45–50%. Again, the hydroxyl group at C-11 is introduced microbiologically. Compound **98** is reacted with the fungus *Cunninghamella blakesleeana* or *Curvularia lunata* to yield up to 77% cortisol and 22% cortisone [214], [215]. Microbiological hydroxylation with *Absisia glauca* converts compound **98** to a mixture of equal parts cortisol (**78**) and 11α-cortisol (**99**), the 11α-epimer of cortisol, with a total yield of 90% [216], [217]. Compound **99** is separated from cortisol by means of its acetate and then oxidized to cortisone (**82**). Another Upjohn method permits the specific hydroxylation of compound **98** to 11α-cortisol **99** and gives a yield of ca. 85%. 11α-Cortisol can subsequently be oxidized to cortisone with chromium trioxide [218], [219].

Production from Bile Acid. Cortisone was first produced from bile acid by SARETT [220] and by KENDALL of the Merck Corporation [221]. Bile acid (i.e., cholic acid [*81-25-4*], **100**) is isolated from the bile of slaughtered animals and has a hydroxyl group at position 12. Although the synthesis of cortisone from cholic acid originally entailed 36 reaction steps, compared with the considerably shorter syntheses from progesterone and 16-dehydropregnenolone, it is still used today.

Cholic acid (**100**)

A large number of improvements and changes have been made (reviewed in [188]), and recommendation of any one variation is difficult (cf. [187]–[189], [191], [195]). However, all variations share the following basic concepts:

1) Elimination of the 7α-hydroxyl group of cholic acid to give 7-deoxycholic acid [209]
2) Transfer of the hydroxyl function at position 12 of deoxycholic acid to position 11

3) Partial degradation of the side chain at C-17 of the bile acid and conversion to the 17-dihydroxyacetone side chain characteristic of corticoids
4) Introduction of the 3-keto-4,5-dehydro structure into the corticoid molecule

Miscellaneous Methods. Other commercial methods of cortisone production start with ergosterol (**101**) and stigmasterol [222], phytosterols of yeast and soybean oil, or with the sapogenin hecogenin (**102**), obtained as a waste product in the recovery of sisal [223], [224]. However, these methods no longer have any importance.

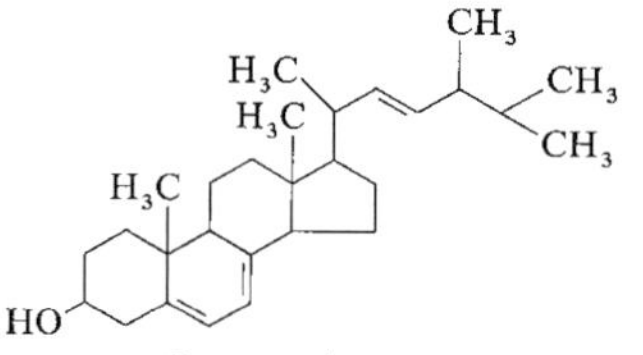

Ergosterol (**101**)

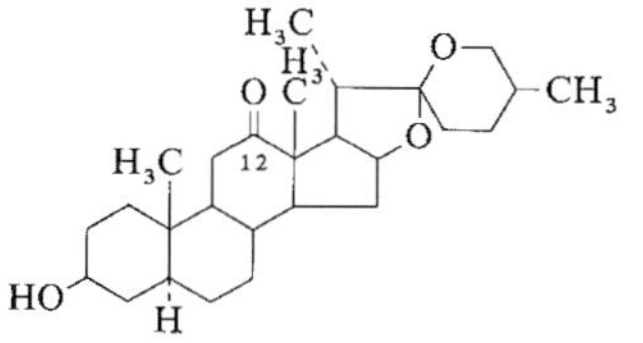

Hecogenin (**102**)

4.8.3. Partial Synthesis of Cortisol

Pregnane derivatives are generally used as starting materials in the partially synthetic commercial production of cortisol. These derivatives already have an 11-keto or 11α-hydroxyl group and are obtained in the production of cortisone from progesterone and 16-dehydropregnenolone described in Section 4.8.2. The Upjohn process (Fig. 8) allows conversion of 11α-hydroxyprogesterone (**83**), obtained microbiologically from progesterone, to cortisol [196], [225].

In the American Cyanamide method, cortisol (**78**) is produced from 11α-cortisol (**99**) or cortisone (**82**) with a yield of 72% [196], [226]. Reaction with ethylene glycol and subsequent oxidation with chromium trioxide yield the 3,20-bisketal (**103**).

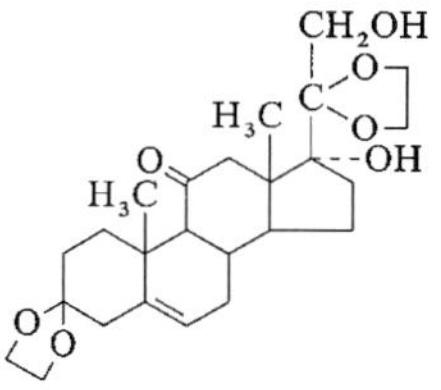

103

Treatment with sodium borohydride stereospecifically reduces the 11-keto group to an 11β-hydroxyl group, which is finally hydrolyzed to cortisol (**78**).

Another cortisol production process, developed by Upjohn [227], starts with pregnane-3α,17α-diol-11,20-dione (**104**) obtained from progesterone [228]. Here, the 3α-hydroxyl group is oxidized to a 3-keto group with *tert*-butyl hypochlorite with simultaneous chlorine substitution at position 4. The bisketal (**105**) is subsequently obtained by reaction with ethylene glycol. Conversion of the bisketal to cortisol requires six additional steps and proceeds with a yield of 15% [196].

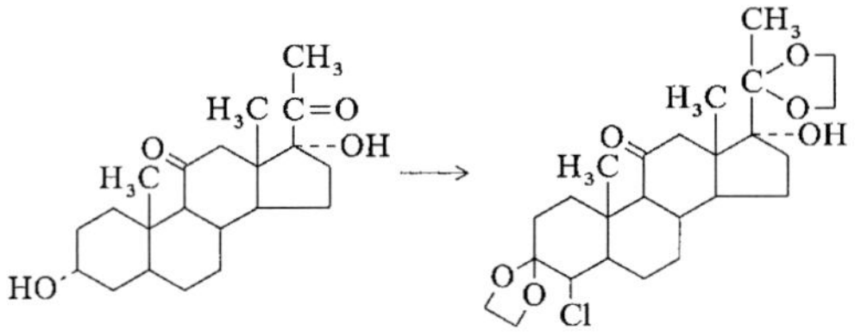

Other commercially important biochemical methods are available in which cortexolone (**98**) can be stereoselectively hydroxylated at the 11β-position to give cortisol (**78**) with yields of approximately 40%. This is achieved microbiologically by using *Cunninghamella blakesleeana* [229] or *Curvularia lunata* [230] (see also Fig. 9 and [231]).

The production of cortisol (**78**) from cortisone acetate (**90**) [232] involves formation of the 3,20-bissemicarbazone. The 11-keto function does not react with semicarbazide for steric reasons. Subsequently, the 11-keto group is reduced with potassium borohydride, whereby the 11-hydroxyl group formed acquires the desired β-configuration and the 21-acetate group is simultaneously hydrolyzed. Cortisol is finally obtained by hydrolytic cleavage of the semicarbazone group with nitric acid.

4.9. Naturally Occurring Mineralocorticoids and Antagonists

The most important mineralocorticoids are deoxycorticosterone (**112**), also called *cortexone*, and the more powerful, physiologically more important *aldosterone* [52-39-1] (**79**). *Spironolactone* (**115**) is an important synthetic aldosterone antagonist. In comparison with the glucocorticoids, the mineralocorticoids are seldom used as therapeutic agents because specific mineralocorticoid activity is required in only a few conditions, e.g., in the treatment of Addison's disease, a rare condition that requires substitution therapy, and (with glucocorticoids) in stress situations. Cortexone is commercially important as an intermediate or starting material in the synthesis of other steroids.

Although a number of interesting studies have been carried out on the synthesis of aldosterone (**79**), it has little importance as a therapeutic agent and intermediate. Aldosterone has 20–30 times the mineralocorticoid potency of cortexone 21-acetate and one-third the glucocorticoid activity of cortisone, whereas cortexone acetate has no significant glucocorticoid activity.

Both mineralocorticoids are administered systemically. Cortexone is generally given parenterally as an oily solution of a 21-ester (e.g., acetate, cyclopentylpropionate, trimethylacetate, or heptanoate). Substitution therapy is indicated in chronic adrenal cortical insufficiency (Addison's disease) and in adrenogenital salt-loss syndrome. Hypertension and advanced cerebral sclerosis are contraindications to the use of mineralocorticoids.

4.9.1. Cortexone

Cortexone [64-85-7], deoxycorticosterone, $C_{21}H_{30}O_3$, M_r 330.45, mp 141–142 °C, $[\alpha]_D^{22}$ +178° (ethanol), is one of the most important mineralocorticoids.

Production. Three methods are used for the production of cortexone:

1) *Production from Pregn-5-en-3β-ol-20-one.* Pregnenolone (**106**) is subjected to Ruschig condensation with diethyl oxalate to give the sodium salt of the corresponding enol (**107**) [209], which is then converted with iodine to the iodine ester (**108**). Subsequent ketonic cleavage with sodium methoxide gives compound **109**, which yields the precursor (**110**) of cortexone on treatment with potassium acetate in aqueous acetone. A 71 % yield is obtained, based on compound **106** as starting material. Oppenauer oxidation of the 3β-hydroxyl group yields the 3-keto-Δ^4 group, and saponification of the 21-acetate group of **111** with 1 mol equivalent of sodium methoxide produces almost quantitative yields of cortexone (**112**).

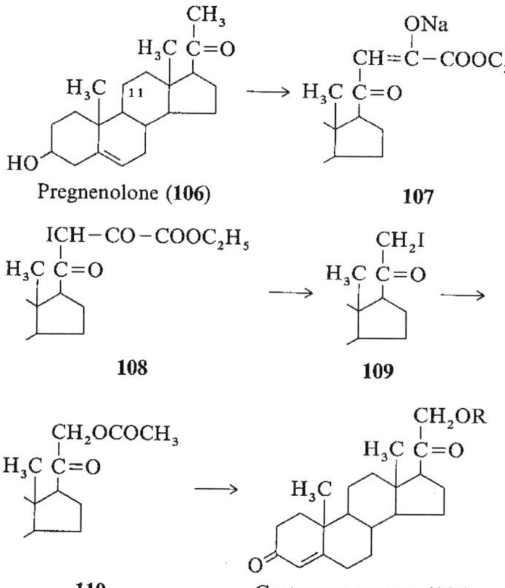

Pregnenolone (**106**)

107

108

109

110

Cortexone acetate (**111**), R = CH_3CO
Cortexone (**112**), R = H

This synthesis can be carried out easily on a large scale because only two of the intermediates, compounds **108** and **110**, must be isolated.

2) *Microbiological Production from Progesterone.* Progesterone is specifically hydroxylated at position 21 by a microbiological method developed by Ciba, which uses fungal strains of scoleco spores, e.g., *Ophiobolus herbotrichus.* Cortexone (**112**) is formed with a yield of 60 % [233].

3) *Chemical Production from Progesterone.* In the commercially important process developed by Syntex, iodine reacts directly with progesterone in tetrahydrofuran – methanol in the presence of calcium carbonate to give 21-iodoprogesterone. The iodoprogesterone is then converted to cortexone 21-acetate (**111**) with potassium acetate in a mixture of acetic acid and acetone [234].

Trade Names: Cortexone is available under the following names: Cortenil depot (cyclopentylpropionate, Hoechst); Cortiron (acetate, Schering); Cortiron depot (heptanoate, Schering); Percoten (Ciba, Wehr).

4.9.2. Aldosterone

Aldosterone [*52-39-1*] (**79**), $C_{21}H_{28}O_5$, M_r 360.44, has the following properties: *mp* 108 –112 °C (contains water of crystallization), *mp* 164 °C (anhydrous), $[\alpha]_D^{23}$ + 152.2 ° (anhydrous acetone).

Production. After the structure of aldosterone had been established, its total synthesis was achieved by Wettstein, Szpilfogel, Prelog, Reichstein, and their coworkers. For a review of these syntheses, see [191].

The partial syntheses of aldosterone are of greater preparative importance [191]; however, only the most significant methods will be mentioned here. One method converts corticosterone acetate (corticosterone, see p. 1607) in four steps to aldosterone acetate [235]. Another method developed by Ciba starts with 11α-acetoxypregnenolone (cf. **106**), which is converted to compound **113** with lead tetraacetate and iodine and, after another 16 reaction steps, to aldosterone (**79**) [236].

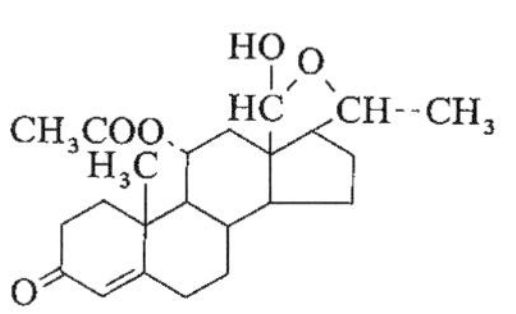

113

The synthesis of aldosterone developed by Velluz and coworkers at Roussel-Uclaf is described in [237].

Trade Name: Aldosterone is marketed unter the trade name Aldocorten (Ciba, Wehr).

4.9.3. Spironolactone

Spironolactone [*52-01-7*] (**115**), $C_{24}H_{32}O_4S$, M_r 416.69, *mp* (double melting point) 134–135 °C and 201–202 °C, $[\alpha]_D^{2}$ −33.5 ° (chloroform), the synthetic aldosterone antagonist, is of considerable therapeutic importance. It increases excretion of sodium and water, and is used as an oral saluretic. Treatment with aldosterone antagonists is indicated in cirrhosis of the liver with ascites and edema, water retention of cardiac and renal origin, hypertension resistant to antihypertonic agents containing saluretics, and in myasthenic and hypokalemic crises. Therapy is contraindicated in kidney insufficiency with anuria and hyperkalemia.

Production. The production of spironolactone from 17α-ethynylandrost-5-ene-3β,17β-diol (**114**) [238], [239] involves the following steps (Searle) [240]:

Spironolactone (**115**)

Potassium aldadiene (**116**)

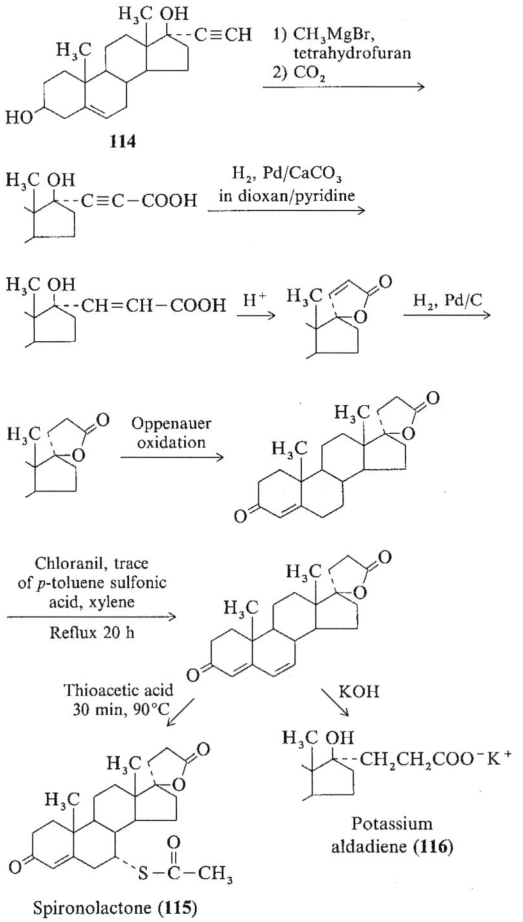

Trade Names: Spironolactone for oral administration is sold under the names Aldactone (Searle, Boehringer Mannheim) and Osyrol (Hoechst).

Potassium aldadiene [*2181-04-6*] (**116**) is used for intravenous administration: Aldactone for injection (Searle, Boehringer Mannheim) and Osyrol for injection (Hoechst).

4.10. Chemically Modified Adrenal Steroid Hormones

A large number of chemically modified, highly potent corticosteroids have been prepared for therapeutic purposes by substitution or dehydrogenation reactions at positions 1, 2, 6, 9, 12, 16, and 21 of the corticosteroid skeleton. These compounds have greatly diminished the importance of naturally occurring cortisone derivatives with their severe mineralocorticoid side effects. However, despite intensive efforts, adequate reduction of the severe side effects that accompany systemic glucocorticoid therapy (e.g., Cushing's syndrome or osteoporosis) has not yet been achieved. The emphasis in corticoid therapy has changed from systemic administration to local or topical application because the associated side effects are then minor. A large number of corticoids have recently been developed for local or topical administration.

4.10.1. 1,2-Dehydrocorticosteroids

Introduction of a second double bond at the 1,2-positions of the cortisol molecule by HERSHBERG and coworkers of the Schering Corporation in 1955 represented a breakthrough in the synthesis of modified corticoids [241]. 1,2-Dehydrogenation yields derivatives with glucocorticoid, anti-inflammatory, and antiallergenic effects that are four to five times higher than those of cortisone. At the same time, these substances have one-third the mineralocorticoid activity of cortisone [242]. The 1,2-dehydro derivatives of cortisone and cortisol, *prednisone* (**117**) and *prednisolone* (**118**), respectively, do not have significant mineralocorticoid activity and are, therefore, widely used in general, systemic pharmacodynamic therapy, particularly as anti-inflammatory and antirheumatic agents.

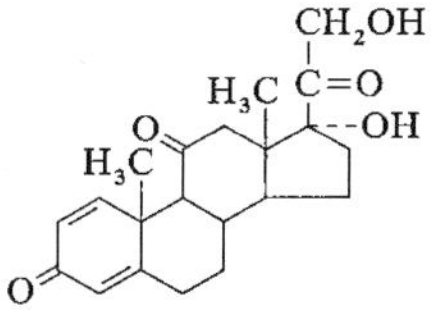

Prednisone (**117**)

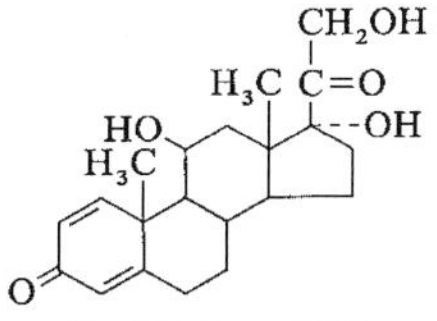

Prednisolone (**118**)

Prednisone [*53-03-2*] (**117**), 1,2-dehydrocortisone, $C_{21}H_{26}O_5$, M_r 368.44, *mp* 233–235 °C, $[\alpha]_D^{25}$ +172° (dioxane), and **prednisolone** [*50-24-8*] (**118**), 1,2-dehydro-cortisol, $C_{21}H_{28}O_5$, M_r 360.44, *mp* 240–241 °C, $[\alpha]_D^{25} + 102°$ (dioxane), are the two most important 1,2-dehydrocorticosteroids.

Production. Insertion of the 1,2-double bond is achieved by chemical or microbiological methods, the latter being more economical. Prednisone can be produced chemically by a process developed by the Schering Corporation from pregnane-17α,21-diol-3,11,20-trione 21-acetate (**119**) involving a 1,4-dibromide and subsequent hydrogen bromide elimination with pyridine [242]. Other methods use the 1,2-dehydrogenation of cortisone with selenium dioxide [243], [244].

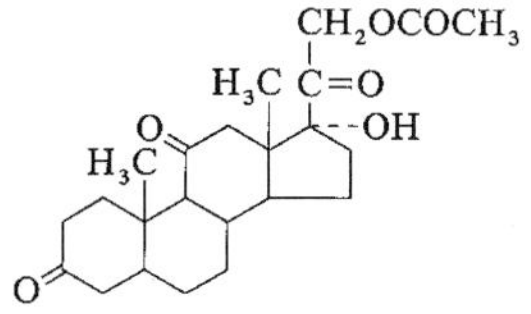

119

The industrially important biological methods permit high-yield production of prednisone and prednisolone by microbiological dehydrogenation at the 1,2-positions of cortisone (**82**) and cortisol (**78**), respectively [245], with *Corynebacterium simplex* [246], *Rhizopus nigricans*, or *Bacillus subtilis* [247].

The chemical conversion of prednisone to prednisolone has also attained some commercial significance. This reaction proceeds via the 3,20-semicarbazone of prednisone and is very similar to the analogous conversion of cortisone to cortisol (see Section 4.8.3) [242].

Effects and Uses. Prednisone and prednisolone have four to five times the anti-inflammatory activity of cortisone when administered orally or parenterally, but only two to three times its activity when applied locally. In contrast, their mineralocorticoid potency is only one-third that of cortisone. The indications listed in Section 4.4 apply to both compounds, but prednisolone has become the "standard" corticoid in general practice. In comparison with cortisone and cortisol, the mineralocorticoid side effects are very slight.

Trade Names: A large number of commercial preparations are derived from prednisolone and prednisone; only a few examples are given here:
Prednisolone: Decortin-H (Merck); Deltacortril (Pfizer); Hostacortin-H (Hoechst); Scherisolon (Schering); and Ultracorten-H (Ciba, Wehr).
Prednisone: Decortin (Merck); Hostacortin (Hoechst); Keteocort (Desitin, Hamburg); and Ultracorten (Ciba, Wehr).

4.10.2. 9α-Halocorticosteroids

In 1953 scientists at Syntex succeeded in inserting halogen atoms at the 9α-position of the cortisone molecule. The resulting modified corticoids had many times the glucocorticoid or anti-inflammatory activity of cortisone [215]. The glucocorticoid activity of hydrocortisones and cortisones with a halogen substituent at position 9α increases dramatically in the order: iodine, bromine, chlorine, fluorine. In comparison with cortisone acetate (activity = 1), the following activities are obtained for the 9α-halocorticoids in the liver glycogen test: I = 0.1, Br = 0.3, Cl = 5, F = 11. The value for 9α-fluorocortisone is 9. The mineralocorticoid effect also increases with the glucocorticoid activity; for example, 9α-fluorocortisol (**123**) has two to four times the mineralocorticoid activity of cortexone. Consequently, the halogen preparations have a very limited therapeutic applicability, despite their high glucocorticoid activities.

9α-Fluorocortisol [*127-31-1*] (**45**) is also known as fluorocortisone, $C_{21}H_{29}FO_5$, M_r 380.46, *mp* 260–262 °C, $[\alpha]_D^{23}$ +139° (95% ethanol).

Production. A method developed by Syntex uses 11α-cortisol 21-acetate (**120**) as the starting material [248]. This compound is converted to the 11α-tosylate, and subsequent elimination of the tosylate group with sodium acetate yields compound **121**. Stereospecific addition of hypochloric or hypobromic acid at the 9,11-double bond yields 9α-chloro- or 9α-bromocortisol 21-acetate, respectively. The 9,11β-epoxide (**122**) is obtained by treating the chloro- or bromohydrin with alkali. Introduction of fluorine or iodine at the 9α-position is achieved by cleavage of **122** with hydrofluoric or hydriodic acid.

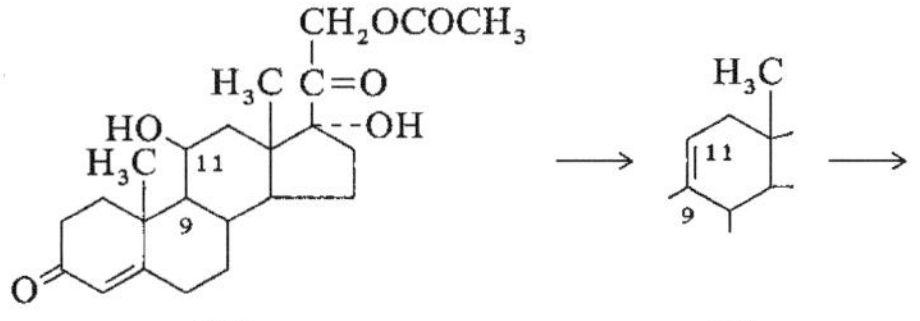

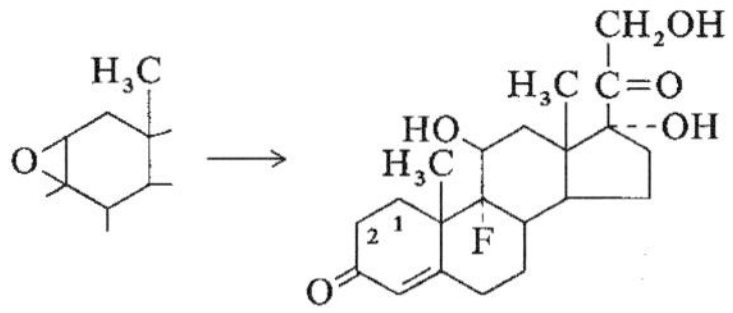

Preparation of the corresponding 1,2-dehydrocorticoid, 9α-fluoroprednisolone (**124**), is described in [196]. The industrial and therapeutic significance of this compound has diminished in recent years.

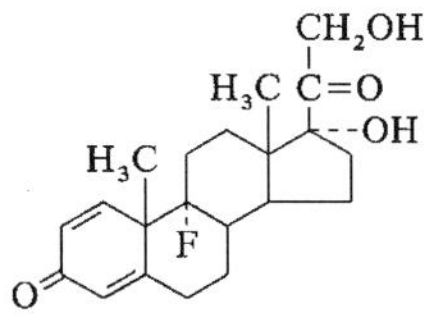

9α-Fluoroprednisolone (**124**)

Effects and Uses. Animal experiments have shown that 9α-fluorocortisol has almost 10 times the glucocorticoid or anti-inflammatory activity of cortisol, but its mineralocorticoid potency is approximately 15–20 times that of cortisone. 9α-Fluoroprednisolone has approximately 20 times the glucocorticoid potency of cortisol; its mineralocorticoid activity is the same as that of 9α-fluorocortisol.

9α-Fluorocortisol and its derivatives are primarily used locally in dermatology and ophthalmology in the form of ointments, cream, or drops. They are also used topically in rheumatology in the form of crystal suspensions for intraarticular injections. Oral substitution therapy in tablet form (0.1 mg per tablet) is indicated in cases of adrenogenital salt-loss syndrome and for raising blood pressure in constitutional hypotension (→ Blood Pressure Increasing Agents).

Trade Names: The following trade names are used:
9α-Fluorocortisol: Astonin (Merck); Scherofluron (21-acetate, Schering); Florinef (Squibb).
9α-Fluoroprednisolone: Predef for veterinary medicine (21-acetate, Upjohn).

4.10.3. 16-Hydroxycorticosteroids

Scientists at Lederle succeeded in almost completely eliminating the mineralocorticoid side effects of 9α-fluoroprednisolone by introducing a 16α-hydroxyl group into the steroid nucleus [249]. Although the glucocorticoid and anti-inflammatory activities of 16α-hydroxycorticoids are approximately one-fifth those of 9α-fluoroprednisolone, they are still of tremendous therapeutic significance because they have practically no mineralocorticoid effects. A large number of derivatives have been obtained from the best known 16α-hydroxycorticoid, triamcinolone (**129**). The 16α,17α-acetonides have a much higher anti-inflammatory effect than the parent compounds. Surprisingly, they even exhibit a diuretic effect when pharmacologically tested for mineralocorticoid activity [250], [251]. For this reason, the 16α,17α-acetonides are now indispensable in local corticoid therapy.

Triamcinolone [*124-94-7*] (**129**), $C_{22}H_{27}FO_6$, M_r 394.45, mp 269–271 °C, $[\alpha]_D^{25} + 75°$ (acetone) and *triamcinolone acetonide* [*76-26-5*] (**130**), also called triamcinolone 21, $C_{24}H_{31}FO_6$, M_r 434.49, mp 292–294 °C, $[\alpha]_D^{23} + 109°$ (chloroform), are the most important hydroxycorticosteroids.

Production. Triamcinolone (**129**) is prepared commercially via a method developed by BERNSTEIN and coworkers [252]. By starting with cortisol 21-acetate (**91**), reaction with ethylene glycol gives the 3,20-bisethylene ketal (**125**). The 11β- and 17α-hydroxyl groups of this compound are then dehydrated with thionyl chloride in pyridine, followed by alkaline saponification of the 21-acetate group, acid deketalization of the cyclic acetal groups at positions 3 and 20, and reacetylation of the 21-hydroxyl group to yield compound **126**. Selective hydroxylation of the double bond at position 16 with osmium tetroxide in pyridine–benzene gives compound **127**. The 9,11-double bond of this compound is converted to a 9α-fluoro-11β-hydroxyl group (**128**) by the method described for 9α-fluorocortisol (**123**) in Section 4.10.2. Triamcinolone (**129**) is obtained after alkaline saponification of the 21-acetate group and microbiological introduction of the Δ^1 double bond with *Corynebacterium simplex*. The total yield is not given in the literature.

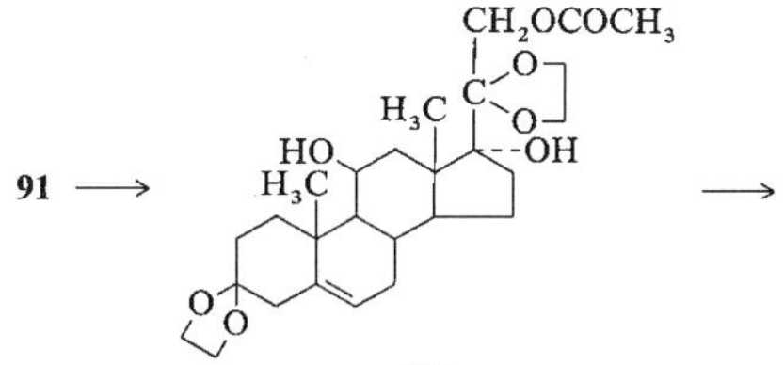

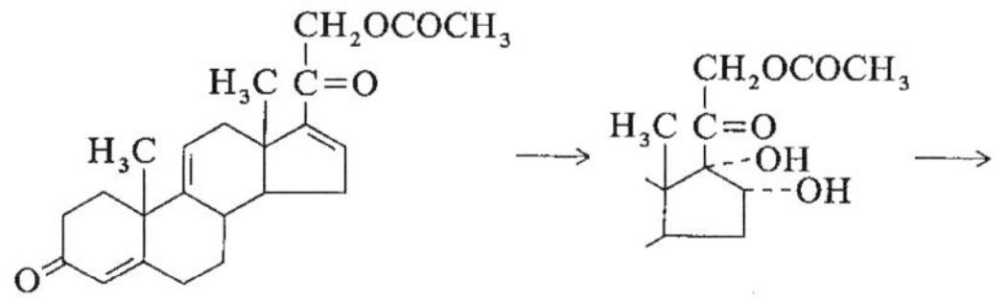

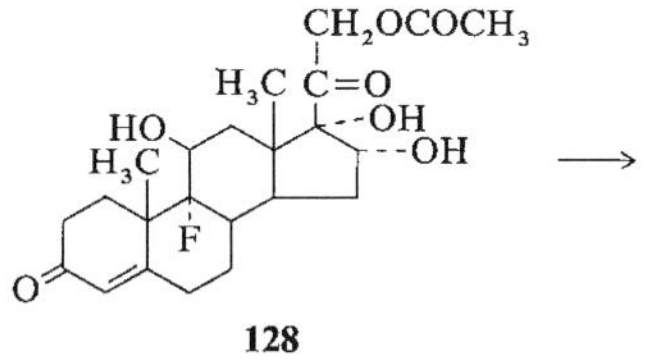

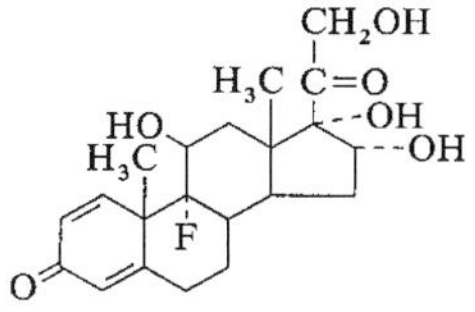

The 16α-hydroxyl group can also be introduced directly into 9α-fluoroprednisolone (**124**) by using a microbiological method originally developed by Squibb [253], [254]. Triamcinolone 16α,17α-acetonide (**130**) is prepared by stirring a suspension of triamcinolone (**129**) in acetone in the presence of catalytic amounts of perchloric acid at approximately 20 °C until the steroid is completely dissolved. Yields of 95% are obtained [255].

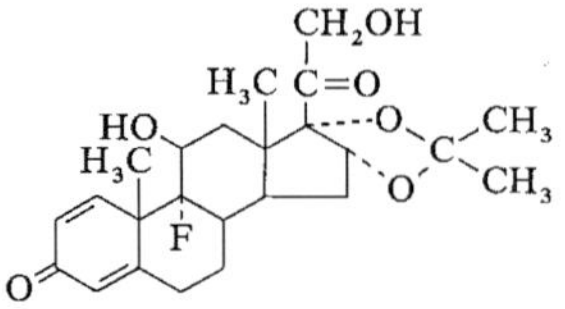

130

Effects and Uses. Systemic administration of 4 mg of triamcinolone produces the same effect as 5 mg of prednisolone or 25 mg of cortisol. Triamcinolone acetonide has 40 times the local anti-inflammatory activity of cortisol. Triamcinolone and its derivatives cause very low sodium and water retention, exert only a slight appetite-stimulating effect, but occasionally produce myopathy. Administration of these preparations is indicated particularly in patients who have a tendency to edema and who are overweight.

Triamcinolone and its derivatives are applied systemically (oral, intravenous, or intramuscular), topically (intraarticular as a crystal suspension), and locally (ointment, cream, or lotion).

Trade Names: The following trade names are used:
Triamcinolone (chiefly for systemic application): Delphicort, Aristocort (Lederle); Volon (Heyden); Kenacort (Squibb).
Triamcinolone 21-(3,3-dimethylbutyrate), also called triamcinolone hexacetonide (for topical long-term therapy): Aristophan (Lederle).
Triamcinolone 16,17-diacetate (for topical therapy): Delphicort (Lederle).
Triamcinolone acetonide (for topical therapy): Volon-A (Heyden); Delphicort (Lederle). Triamcinolone acetonide is used as its 21-disodium phosphate in Aristosol and as its 21-hemisuccinate in Solutedarol for systemic intravenous administration.

4.10.4. 6α-Methylcorticosteroids

Insertion of a 6α-methyl group into the corticosteroid nucleus by scientists at Upjohn [256] and British Drug Houses [257] gave rise to derivatives with enhanced glucocorticoid activity and low mineralocorticoid activity. The most potent orally, parenterally, and topically active compounds are 6α-methylprednisolone (**131**), which has approximately 10 times the anti-inflammatory activity of cortisol in animal experiments, and 6α-methyl-9α-fluoroprednisolone (**132**), which has relatively low mineralocorticoid activity and almost 100 times the glucocorticoid potency of cortisol in animal experiments. 21-Deoxy-6α-methyl-9α-fluoroprednisolone (**133**) is important in the local treatment of inflammatory skin diseases.

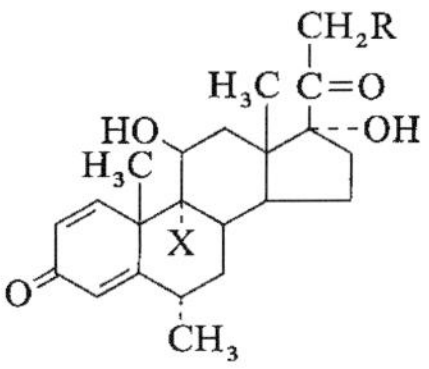

131: X = H, R = OH
132: X = F, R = OH
133: X = F, R = H

6α-Methylprednisolone [*83-43-2*] (**131**), $C_{22}H_{30}O_5$, M_r 374.46, mp 228–237 °C, $[\alpha]_D^{20}$ +83° (dioxane), and *21-deoxy-9α-fluoro-6α-methylprednisolone* [*426-13-1*] (**133**), fluorometholone, $C_{22}H_{29}FO_4$, M_r 376.47, mp 292–303 °C, are the most important 6α-methylcorticosteroids.

Production. 6α-Methylcorticoids are prepared commercially from 11α-hydroxyprogesterone (**83**), as shown below [256]:

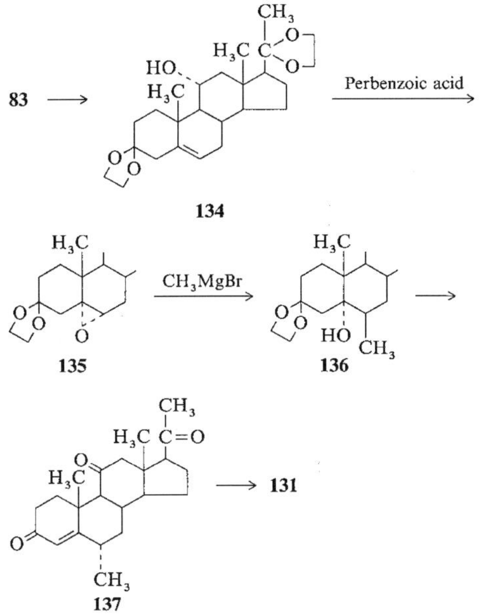

Compound **83** is treated with ethylene glycol to give the 3,20-bis(ethylene ketal). The 6-methyl group is then introduced into the epoxide (**135**) by the Grignard reaction. The 11α-hydroxyl group of compound **136** is subsequently oxidized with chromic acid to give the 11-keto function, and compound **137** is finally obtained after deketalization, dehydration, and isomerization. The 17-acetyl group of compound **137** is then converted to the 17-dihydroxyacetone side chain by using the progesterone method (see Section 4.8.2). The 11-keto group is reduced to an 11β-hydroxyl group, and introduction of the Δ^1 double bond finally leads to 6α-methylprednisolone (**131**).

Another method developed by Merck, Sharp, and Dohme uses cortisone as the starting material. Cortisone is first converted to the 6-keto derivative via suitable intermediates. Subsequent Grignard reaction with methylmagnesium bromide and further reaction steps then yield 6α-methylprednisolone [258].

In the production of 6α-methyl-21-deoxycorticosteroids, the corresponding 11-deoxy-21-hydroxylated compounds are used as starting materials because they can easily be hydroxylated microbiologically at position 11. After 11-hydroxylation, the 21-hydroxyl group is first replaced with an iodine moiety, via the tosylate or mesylate, and then reduced with hydrogen [259]. Preparation of 21-deoxy-6α-methyl-9α-fluoroprednisolone (**133**) is described in [260].

Effects and Uses. 6α-Methylprednisolone (4 mg of which is equivalent to 5 mg of prednisolone) is well tolerated by the stomach, has no mineralocorticoid side effects, and is the corticoid of choice for systemic long-term therapy. It increases appetite slightly and is mildly stimulating psychologically.

6α-Methylprednisolone is administered orally, intravenously (as a water-soluble solution of the 21-sodium hemisuccinate), or topically (as a crystal suspension). Its 21-cyclopentylpropionate derivative is most suitable for intramuscular application and has an extremely long duration of action. Fluorometholone (**133**) is a strong anti-inflammatory substance and is used exclusively in the form of ointments, creams, and lotions in the local treatment of dermatosis.

Trade Names: The following trade names are employed:
6α-Methylprednisolone (for oral application): Urbason (Hoechst); Medrate (Upjohn); Medrol (Upjohn).
6α-Methylprednisolone 21-acetate (for topical application): Urbason crystal suspension (Hoechst); Depo-Medrol (Upjohn).
6α-Methylprednisolone 21-sodium hemisuccinate (for intravenous application): Urbason soluble (Hoechst); Medrate soluble (Upjohn); Solu-Medrol (Upjohn).
6α-Methylprednisolone 21-cyclopentylpropionate (for intramuscular application with long-term effects): Urbason Depot (Hoechst).
21-Deoxy-6α-methyl-9α-fluoroprednisolone (for local application): Delmeson (Hoechst); Oxylone (Upjohn).

4.10.5. 16-Methylcorticosteroids

Introduction of a methyl group at the 16α-position of the corticoid molecule greatly increases glucocorticoid and anti-inflammatory activity, and decreases mineralocorticoid activity [261]. 16α-Methylcorticosteroids include 16α-methylhydrocortisone, 16α-methylprednisolone and 16α-methyl-9α-fluoroprednisolone. The last mentioned, also known as *dexamethasone* (**141**), has very low mineralocorticoid activity and approximately 30 – 40 times the anti-inflammatory potency of hydrocortison. It is, therefore, highly effective when given orally, parenterally, or locally in the treatment of inflammatory, rheumatic, and allergic diseases.

The corresponding 16β-*methylcorticosteroids* also possess high glucocorticoid activity and exert only a slight effect on mineral metabolism. Their anti-inflammatory activity is equal to that of the 16α-methylcorticoids. *Betamethasone* (**68**) [262]–[264], 16β-methyl-9α-fluoroprednisolone, was developed by the Schering Corporation and, like dexamethasone, is of great importance in oral, parenteral, and local therapy.

Dexamethasone [*50-02-2*] (**141**), 9α-fluoro-16α-methylprednisolone, $C_{22}H_{29}FO_5$, M_r 392.45, mp 262–264 °C, $[\alpha]_D^2 + 77.5°$, and **betamethasone** [*378-44-9*] (**146**), 9α-fluoro-16β-methylprednisolone, $C_{22}H_{29}FO_5$, M_r 392.45, mp 231–234 °C, $[\alpha]_D + 108°$ (acetone), are representative of the 16-methylcorticosteroids.

Production. A number of multistage procedures for the synthesis of *dexamethasone* have been developed, mainly by Merck [261], Schering [265], Roussel-Uclaf [266], and Upjohn [269]. The Schering method uses 16-methylpregna-4,16-dien-3β-ol-20-one (**138**) as the starting material (Fig. 10). This compound is obtained by copper(II)-catalyzed 1,4-addition of methylmagnesium bromide to the Δ^{16}-20-keto system of 16-dehydropregnenolone (**92**) [267].

Stereospecific catalytic hydrogenation of the double bonds at the 3,4- and 16,17-positions of compound **138** yields 16α-methyl-3β-hydroxy-5α-pregnan-20-one. Compound **139** is obtained by the stereospecific introduction of a 17α-hydroxyl group into the $\Delta^{17,20}$-20-enolacetate with peracetic acid. The 21-acetate group is then inserted via the 21-bromide. The next steps involve oxidation of the 3β-hydroxyl group with chromium trioxide to the 3-keto group, dibromination in the α,α'-position to give the α,α'-dibromo derivative, elimination of the bromine groups to yield the $\Delta^{1,4}$-dien-3-one system, and alkaline saponification of the 21-acetate group. Compound **140** is then obtained microbiologically by introduction of the required oxygen function at the 11α-position.

The 11α-hydroxyl compound is converted to the 9α-fluoro-11β-hydroxyl derivative as described in Section 4.10.2. Alkaline saponification of the 21-acetate group finally produces dexamethasone (**141**).

The Merck [261] method involves the preparation of 16-pregnen-3α-ol-11,20-dione 3-acetate (**142**), as described in [268]. This compound already has the essential oxygen function at position 11. Conversion to dexamethasone (**141**) requires 14 further reaction steps.

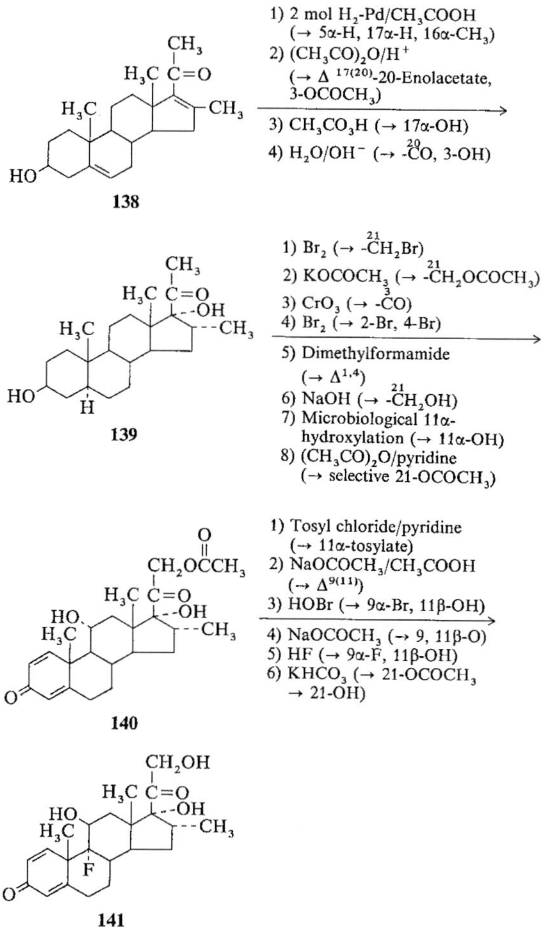

Figure 10. Production of dexamethasone (**141**) from 16-methylpregna-4,16-dien-3β-ol-20-one (**138**)

1) 2 mol H$_2$-Pd/CH$_3$COOH
  (→ 5α-H, 17α-H, 16α-CH$_3$)
2) (CH$_3$CO)$_2$O/H$^+$
  (→ Δ$^{17(20)}$-20-Enolacetate, 3-OCOCH$_3$)
3) CH$_3$CO$_3$H (→ 17α-OH)
4) H$_2$O/OH$^-$ (→ -$\overset{20}{C}$O, 3-OH)

1) Br$_2$ (→ -$\overset{21}{C}$H$_2$Br)
2) KOCOCH$_3$ (→ -$\overset{21}{C}$H$_2$OCOCH$_3$)
3) CrO$_3$ (→ -$\overset{3}{C}$O)
4) Br$_2$ (→ 2-Br, 4-Br)
5) Dimethylformamide
  (→ Δ^{1,4})
6) NaOH (→ -$\overset{21}{C}$H$_2$OH)
7) Microbiological 11α-hydroxylation (→ 11α-OH)
8) (CH$_3$CO)$_2$O/pyridine
  (→ selective 21-OCOCH$_3$)

1) Tosyl chloride/pyridine
  (→ 11α-tosylate)
2) NaOCOCH$_3$/CH$_3$COOH
  (→ Δ$^{9(11)}$)
3) HOBr (→ 9α-Br, 11β-OH)
4) NaOCOCH$_3$ (→ 9, 11β-O)
5) HF (→ 9α-F, 11β-OH)
6) KHCO$_3$ (→ 21-OCOCH$_3$ → 21-OH)

Similar concepts have been developed for the industrial production of *betamethasone* by Merck [270], Schering [271], and Roussel-Uclaf [272].

The 16β-methyl group is inserted into the appropriate precursor. The Merck method uses 3α-acetoxy-16-pregnene-11,20-dione (**142**) as the starting material [273]. A 1,3-dipolar addition reaction with diazomethane at the double bond between positions 16 and 17 gives compound **143**. Alkaline cleavage or pyrolysis at ca. 200 °C then leads to the homologous 16-methyl-16-dehydro-20-keto compound (**144**). Stereospecific, catalytic hydrogenation of the Δ^{16} double bond with palladium and calcium carbonate yields compound **145**, which has a β-methyl group at position 16. The other essential groups (the 17α- and 21-hydroxyl groups, and the 9α-fluoro group) and the dienone structure in ring A are then introduced in a series of steps to give betamethasone (**146**).

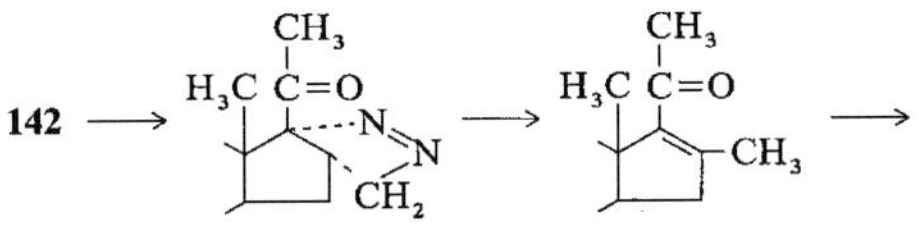

Effects and Uses. A dose of 1 mg of dexamethasone or betamethasone is equivalent to 5 mg of prednisolone or 20 mg of cortisol.

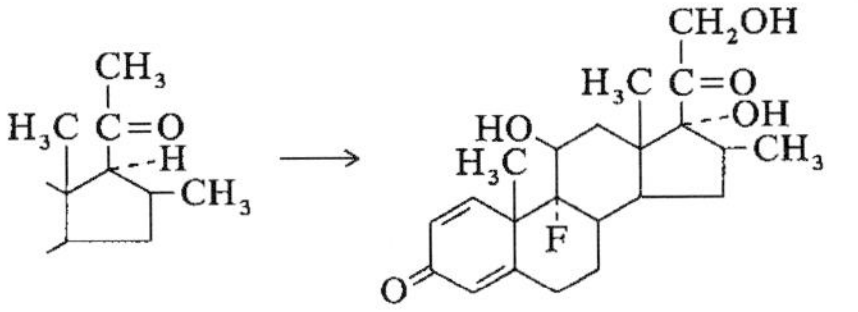

Of all the corticoids used in medicine, dexamethasone and betamethasone exert the strongest glucocorticoid and anti-inflammatory effects. However, these compounds not only have a high glucocorticoid potency, they also have a number of unpleasant side effects that must be considered, especially in systemic therapy. Both compounds are very psychologically stimulating (euphoria) and increase appetite. Other unfavorable side effects include high catabolic activity, strong hypophyseal inhibition, and considerable calcium and phosphate excretion leading to osteoporosis. For these reasons, both of these highly potent corticoids are more suitable for limited, pharmacodynamic therapy than for long-term treatment. Betamethasone is slightly better tolerated by the stomach than dexamethasone.

Trade Names: The following are some trade names for these two compounds:
Dexamethasone and derivatives (21-acetate, 21-trimethylacetate, 21-disodium phosphate, 21-sodium hemisuccinate, 21-isonicotinate): Decadron (Sharp and Dohme); Fortecortin (Merck); Millicorten (Ciba); Auxiloson (Thomae); Dexa-Scheroson (Schering); Oradexon (Organon); Dexa-Cortidelt (Roussel-Pharma);Dexamethason "Ferring" (Ferring); Dexonil (Schering).
Betamethasone and derivatives (as above) for systemic, topical, and local therapy: Betnesol (Glaxo); Celstone (Schering); Celstan (Byk-Essex)
Derivatives for local therapy include betamethasone 17-valerate [Bentnesol V (Glaxo); Celstan V (Byk-Essex) and betamethasone 17,21-dipropionae [Dipcosone (Schering)].
9α-Chloro-16β-methylprednisolone 17,21-dipropionate [*4419-39-0*], also known as beclomethasone, is especially suitable for local therapy: Propaderm, production is described in [274], [275].

4.10.6. 16-Methylenecorticoids

Animal experiments have shown that insertion of a 16-methylene group into the prednisolone nucleus enhances glucocorticoid activity. 16-Methyleneprednisolone, prednylidene (**147**), has an almost total absence of mineralocorticoid activity. Substitution at the 9α-position with fluorine leads to the local corticoid fluprednylidene [*2193-87-5*] (**148**).

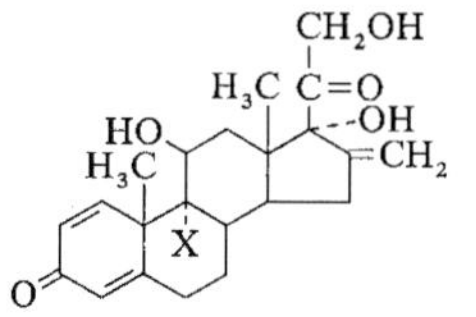

147: X = H
148: X = F

16-Methyleneprednisolone [*599-33-7*] (**147**), prednylidene, $C_{22}H_{28}O_5$, M_r 372.44, mp 233–235 °C, $[\alpha]_D^{23} +31°$ (dioxane), is produced as described below.

Production. The 16-methylene-17-dihydroxyacetone group of the 16-methylenecorticoids is achieved by using the following reaction sequence [276]–[278]:

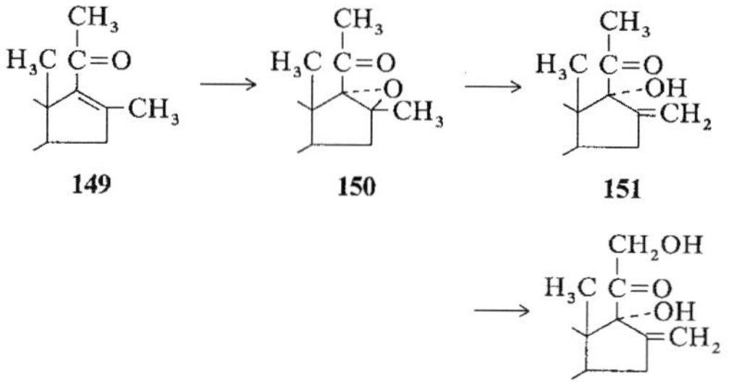

149 → **150** → **151**

→ **152**

The addition of alkaline hydroperoxide to the double bond at the 16,17-positions of the 16-methyl-16-dehydro-20-keto system (**149**) of a suitable pregnane derivative stereospecifically produces the 16α,17α-oxide (**150**). Treatment with toluene-*p*-sulfonic acid in acetone then leads to opening and isomerization of compound **150**, giving the 17α-hydroxyl and 16-methylene groups (**151**). Insertion of a 21-hydroxyl group by conventional means gives the complete 16-methylene-17-dihydroxyacetone structure (**152**). Both prednylidene (**147**) and fluprednylidene (**148**) are synthesized in this way.

Effects and Uses. A dose of 9 mg of 16-methyleneprednisolone is equivalent to 7.5 mg of prednisolone or 30 mg of cortisol. Prednylidene, which is slightly less potent than prednisolone in humans, is similar to 6α-methylprednisolone in that it is very mildly, psychologically stimulating, only slightly stimulating to the appetite, well tolerated by the stomach, and suitable for long-term therapy. Fluprednylidene is more active when applied locally and is used exclusively in local corticoid therapy because of its strong mineralocorticoid activity.

Trade Names: Prednylidene is used primarily in systemic therapy as such or as the water-soluble 21-diethylaminoacetate hydrochloride: Decortilen (Merck).
Fluprednylidene is used exclusively in local corticoid therapy as the 21-acetate: Decoderm (Merck).

4.10.7. 6α-Fluoro-, 6α-Chloro-, and 6α,9α-Difluorocorticoids

Insertion of fluorine or chlorine into the 6α-position of the steroid skeleton increases glucocorticoid potency and eliminates mineralocorticoid activity [279]. Here again, the 6α-fluoro derivatives have a stronger anti-inflammatory effect than the homologous chlorine compounds.

6α-Fluoroprednisolone (**153**), also called fluprednisolone, has approximately the same glucocorticoid activity as triamcinolone. Insertion of other activating groups into the 6α-fluoroprednisolone molecule (primarily a methyl or a hydroxyl group at position 16 or another fluoro group at position 9α) strikingly enhances glucocorticoid activity. However, the undesirable mineralocorticoid side effects also increase. Insertion of a 16α-methyl group into the fluprednisolone molecule yields systemically active paramethasone (**154**). 6α,9α-difluoro-16α-methylprednisolone (**155**), flumethasone, has not only the 6α-fluoro group, but also the strongly activating 9α-fluoro and 16α-methyl groups; it is currently the most active anti-inflammatory local corticoid on the market. The 16,17-acetonides (**156**) are also widely used. The following are some therapeutically important compounds:

6α-Fluoroprednisolone [53-34-9] (**153**), fluprednisolone, $C_{21}H_{27}FO_5$, M_r 378,45, mp 208–213 °C

Paramethasone [53-33-8] (**154**), $C_{22}H_{29}FO_5$, M_r 392.45, 21-acetate: mp 228–241 °C

Flumethasone [2135-17-3] (**155**), $C_{22}H_{28}F_2O_5$, M_r 410.46

Fluocinolone acetonide [67-73-2] (**156**), $C_{24}H_{30}F_2O_6$, M_r 452.50, mp 265–266 °C, $[\alpha]_D$ +95 ° (chloroform)

6α-Fluoro-16α-hydroxycortisol (**158**), fluandrenolone, $C_{21}H_{29}FO_6$, M_r 396.46, mp 234–236 °C, $[\alpha]_D$ +95 ° (dioxane)

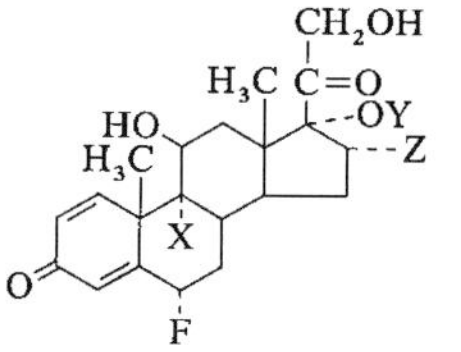

153: X = Y = Z = H
154: X = Y = H; Z = CH$_3$
155: X = F; Y = H; Z = CH$_3$
156: X = F; Y and Z = $-O-C(CH_3)_2-O-$

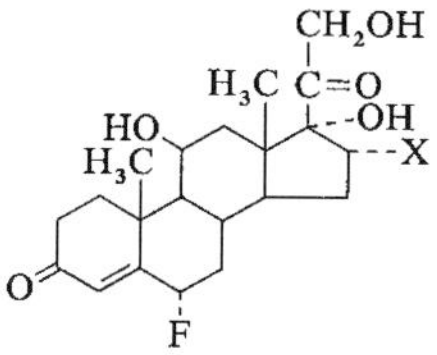

157: X = H
158: X = OH

Production. 6α-*Fluorocortisol* (**157**) is prepared from 3β,17α,21-trihydroxypregn-5-en-20-one 17,21-diacetate [280] with a method developed by Syntex [281]. This compound has no therapeutic importance but typifies insertion of the 6α-fluoro group.

A method developed by Syntex [282] is used to synthesize the therapeutically important compound *paramethasone* (**154**). The double bond at the 5,6-positions of the starting compound, 16α-methylpregnenolone 3-acetate (**159**), is converted stereospecifically with perbenzoic acid to the 5,6α-oxide (**160**). Compound **160** is then cleaved with boron trifluoride to give the 6β-fluoro-5α-hydrin (**161**). After insertion of the 17α- and 21-hydroxyl groups, oxidation with chromium trioxide, and dehydration to the 3-keto-4,5-dehydro system (**162**), the 6β-fluoro group is isomerized to the 6α-fluoro group by treatment with acid. The 11β-hydroxyl group is then inserted enzymatically by incubation with bovine adrenal gland [283], to give a yield of 50–60%. Product **163** is dehydrogenated with selenium dioxide in *tert*-butanol and pyridine to yield paramethasone (**154**).

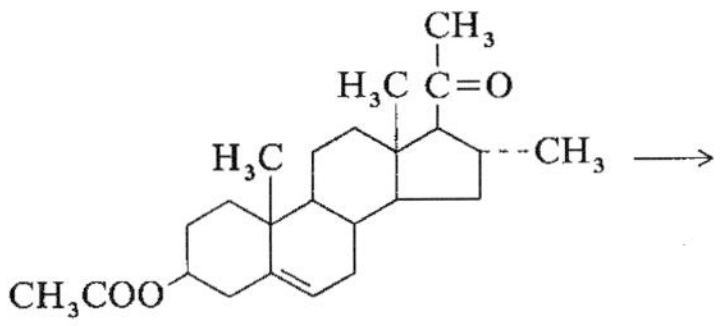

159

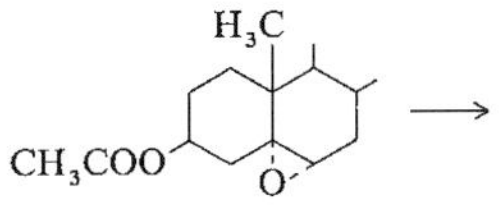

160

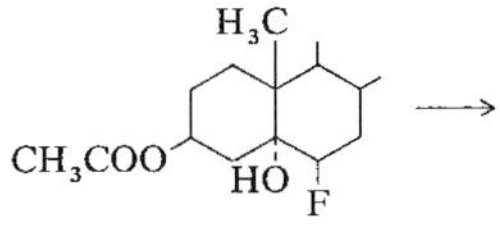

161

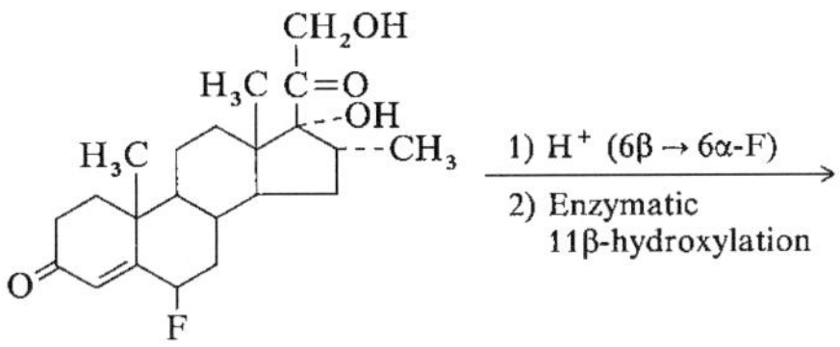

162

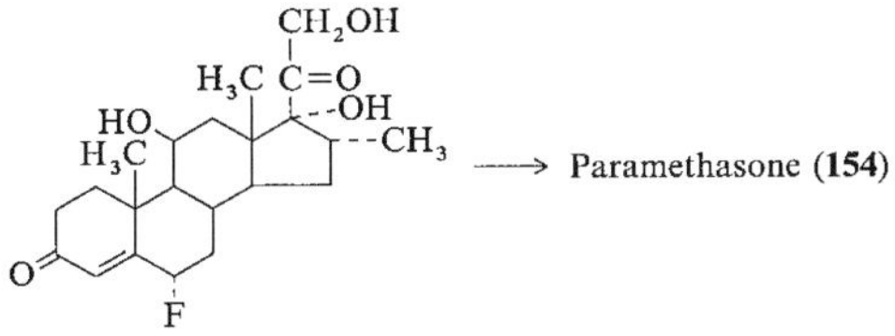

1) H⁺ (6β → 6α-F)
2) Enzymatic 11β-hydroxylation

163 → Paramethasone (**154**)

Compound **163** also serves as an intermediate in the commercial synthesis of the local corticoid *flumethasone* (**155**) as follows:

1) dehydration of **163** with SOCl$_2$ and pyridine to give the $\Delta^{9,11}$ derivative;
2) formation of the 9,11β-oxide, as described in Section 4.10.2;
3) cleavage with hydrogen fluoride to give the 9α-fluoro,11β-hydroxy derivative; and
4) dehydrogenation with selenium dioxide to $\Delta^{1,2}$.

Fluprednisolone (**153**) is synthesized with methods developed by Upjohn [284], Syntex [285], and Bayer [286]. *Fluocinolone acetonide* (**156**) is important in local therapy and can be prepared commercially from 16α,17α-oxidopregn-5-ene-3β,21-diol-20-one 21-acetate (made according to [287]) via fluandrenolone (**158**) in a multistage process (more than 20 reaction steps) [288].

In addition to the cleavage of 5,6α-oxy groups with hydrofluoric acid, other methods have been elaborated to insert the 6α-fluoro group into the corticoid nucleus. In a method developed by Syntex, FBr (from hydrofluoric acid and N-bromosuccinimide) is added to the Δ^5 double bond of compound **164**, forming the corresponding 5α-bromo-6β-fluoro compound (**165**). Subsequent oxidation of the 3-hydroxyl group with chromium trioxide gives the 3-oxo group, and elimination of hydrogen bromide produces 6β-fluoro-Δ^4-3-keto steroids (**166**). Acid isomerization of the 6β-fluoro group yields 6α-fluoro-Δ^4-3-keto steroids [279], [289].

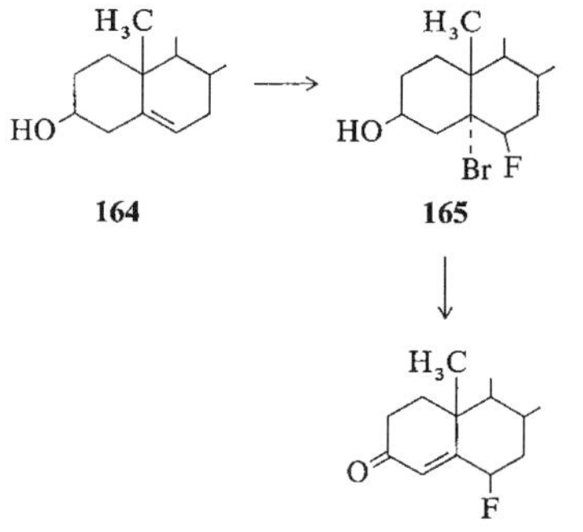

[Structures 164, 165, 166]

The 6-fluoro group can also be inserted by converting the $\Delta^{3,5}$-3-enolacetates of suitable pregnane precursors with perchloryl fluoride (FClO$_3$) in dioxane as solvent [290].

Similarly, the 6α-chlorine atom can be inserted into the steroid skeleton by reacting the $\Delta^{3,5}$-3-enolethers or Δ^5-3-ethylene dioxysteroids with N-chloroacetamide [291].

Effects and Uses. A systemic dose of 2.5 mg of fluoroprednisolone or 2.0 mg of paramethasone is equivalent to 5.0 mg of prednisolone. Both corticoids are used in systemic therapy; they are well tolerated by the stomach and suitable for short- or long-term treatment. However, the relatively strong hypophyseal inhibition exerted by both corticoids must be taken into consideration.

The highly active local corticoids flumethasone and fluocinolone acetonide are widely applied as the free 21-alcohols or their esters in local corticoid therapy. They are not administered systemically.

Trade Names: The following trade names are used:
Fluprednisolone (systemic administration): Alphadrol (Upjohn).
Paramethasone (systemic administration as the 21-acetate or 21-disodium phosphate): Monocortin (Grünenthal); Stemex (Syntex); Haldrone (Lilly).
Flumethasone (applied locally as the 21-pivalate): Locacorten (Ciba and Syntex).
Fluocinolone acetonide (for local application): Jellin (Grünenthal); Synalar (Syntex); Lidex (21-acetate, Syntex).

4.10.8. 17-Deoxycorticosteroids

In 1964 RASPÈ and coworkers found that 16α-methylcorticosteroids lacking the 17α-hydroxyl group are still relatively strong anti-inflammatory substances with very low mineralocorticoid activity [292]. For instance, 16α-methyl-1-dehydrocorticosterone has approximately the same anti-inflammatory activity as prednisolone when applied locally, subcutaneously, or orally. 6α-Fluoro-16α-methyl-1-dehydrocorticosterone, fluocortolone (**169**) is used in systemic corticoid therapy [293]. The similarly structured 9α-fluoro-16α-methyl-1-dehydrocorticosterone, desoximetasone (**173**), is commercially available as a highly active local corticoid [294].

Fluocortolone [152-97-6] (**169**), $C_{22}H_{29}FO_4$, M_r 376.47, mp 188–190.5 °C, $[\alpha]_D^{20}$ +100° (dioxane), and *17-desoximetasone* (**173**), $C_{22}H_{29}FO_4$, M_r 376.47, mp 217 °C, $[\alpha]_D^{20}$ +109° (chloroform), are important 17-deoxycorticosteroids.

Production. *Fluocortolone* (**169**) is prepared from 16α-methylpregn-5-ene-3β,21-diol-20-one 21-acetate (**167**) (obtained as described in [295]) in a method developed by Schering [293], [296]. 6α-Fluoro-16α-methyldeoxycorticosterone (**168**) is formed in a reaction sequence analogous to **164** → **166**, with subsequent isomerization. The 11β-hydroxyl group is introduced microbiologically by using *Curvularia lunata* [297], followed by stereospecific dehydrogenation at the 1,2-positions with *Corynebacterium simplex* [298] or *Bacillus lentus* [299] and saponification of the 21-acetate group to yield fluocortolone (**169**).

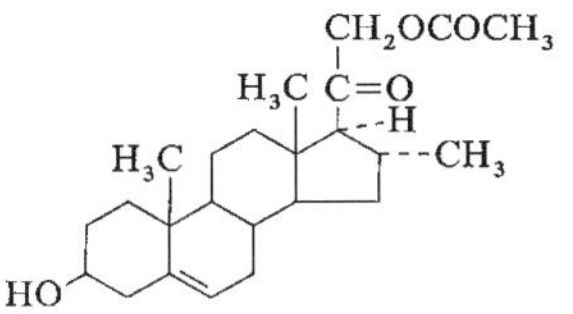

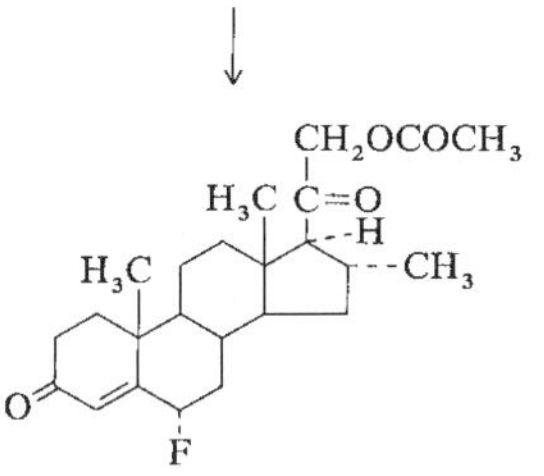

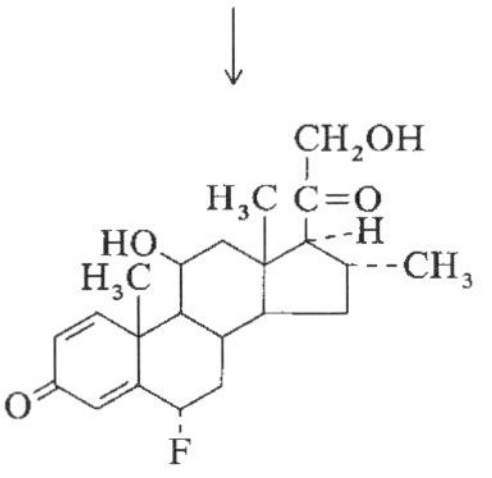

Desoximetasone (**173**) is prepared from compounds **170** or **171** according to two different methods developed by Roussel–Uclaf [294], [300]. These processes require six and eight reaction steps, respectively, to produce the important intermediate **172**, which is then converted to desoximetasone (**173**) in four further steps:

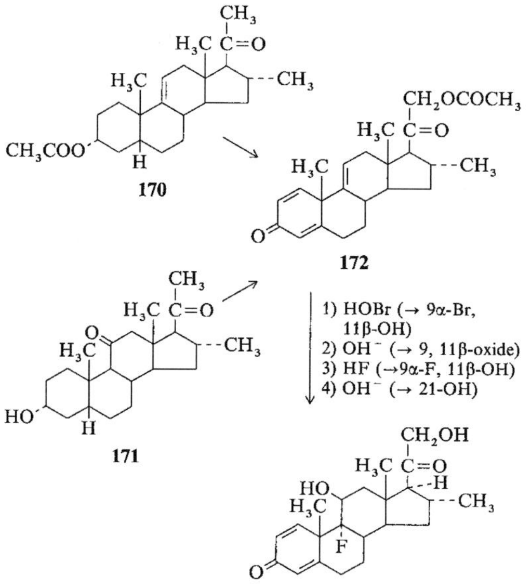

[Structures 170, 171, 172, 173 with reaction scheme:
171 → 1) HOBr (→ 9α-Br, 11β-OH)
2) OH⁻ (→ 9, 11β-oxide)
3) HF (→9α-F, 11β-OH)
4) OH⁻ (→ 21-OH)
→ 173]

Effects and Uses. A systemic dose of 7.5 mg of fluocortolone is equivalent to 7.5 mg of prednisolone. Although fluocortolone is a corticosterone derivative, it exerts the same systemic glucocorticoid and anti-inflammatory effects in humans as prednisolone. It is well tolerated by the stomach, exhibits only slight catabolic activity, and is thus highly suitable for long-term treatment. In addition, fluocortolone is a powerful local anti-inflammatory substance and, consequently, is used in local corticoid therapy. Because 17-desoximetasone has an unusually high local anti-inflammatory activity, it is an excellent local corticoid.

Trade Names: The following trade names are used:
Fluocortolone (systemic, local, and topical): Ultralan (Schering). It is used locally as such, as the 21-trimethylacetate, as the 21-capronate, or as a mixture of these compounds.
17-Desoximetasone (local): Topisolon (Hoechst); Topicort (Roussel).

Chlorocortolone [4828-27-7] (**174**) is another 17-deoxycorticoid that has a chlorine atom at position 9α and a fluorine atom at position 6α. This compound is prepared by a method patented by Schering and is used exclusively in local corticoid therapy [301].

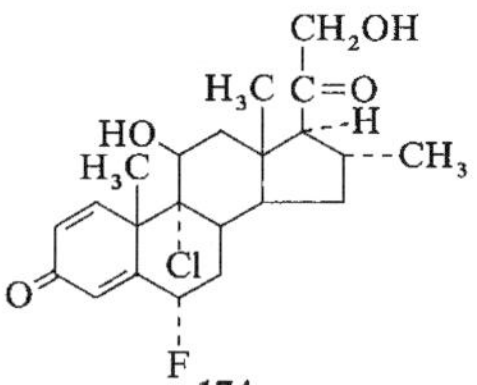

[Structure 174]

Trade Name: Chlorocortolone is applied locally as the 21-trimethylacetate or 21-capronate: Kaban (Asche, Hamburg).

4.10.9. 9α,11β-Dihalocorticosteroids

Scientists at Schering found that the oxygen at position 11 in corticosteroids can be replaced by a halogen without loss of glucocorticoid or anti-inflammatory activity. Thus 9α,11β-dichloroprednisolone (**175**), also called dichlorisone, not only has a very low mineralocorticoid activity but also has ten times the anti-inflammatory activity of prednisolone [302]. It was launched on the market as the 21-acetate.

Dichlorisone [*7008-26-6*] (**175**), $C_{21}H_{26}Cl_2O_4$, M_r 413.35, has the following properties: *mp* 238–241 °C, $[\alpha]_D^{20} +134°$ (pyridine).

Production. The synthesis of dichlorisone involves selective addition of equimolar amounts of the desired halogens to steroids with a double bond at position $\Delta^{9,11}$ [302], [303].

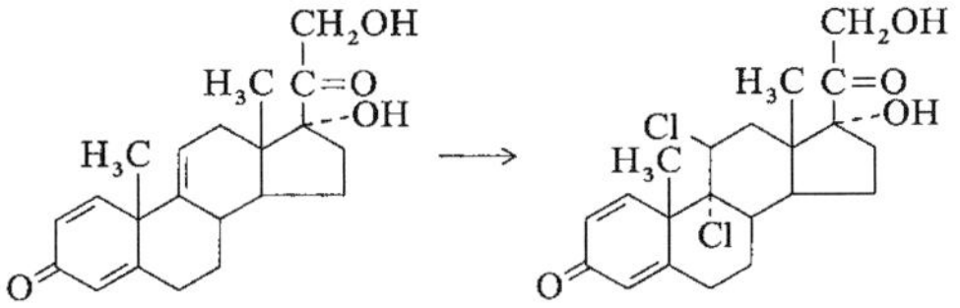

Dichlorisone (**175**)

Effects and Uses. In animal experiments, dichlorisone has ten times the anti-inflammatory potency of prednisolone. The 9α,11β-dichlorocorticoids have not been used in systemic therapy because they exert exceptionally strong (specific) glucocorticoid effects (Cushing's syndrome). However, they also have a strong local and topical anti-inflammatory effect. Dichlorisone is especially useful in local antipruritic therapy.

Trade Names: Dichlorisone is used for topical and local therapy only: Diloderm, Disoderm (Schering).

4.11. Recent Developments

Extensive efforts to obtain better separation of systemic and topical effects have produced a series of new preparations with little or no systemic activity. Therefore, these compounds are especially suitable for treating dermatosis in children and large cutaneous areas. These preparations are derivatives of known, active corticoids.

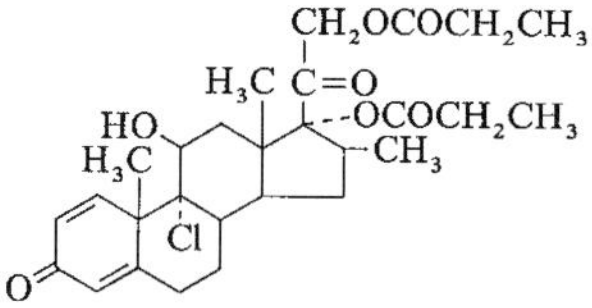

Beclomethasone dipropionate [5534-09-8] (Sanasthmyl, Glaxo) [4.118]

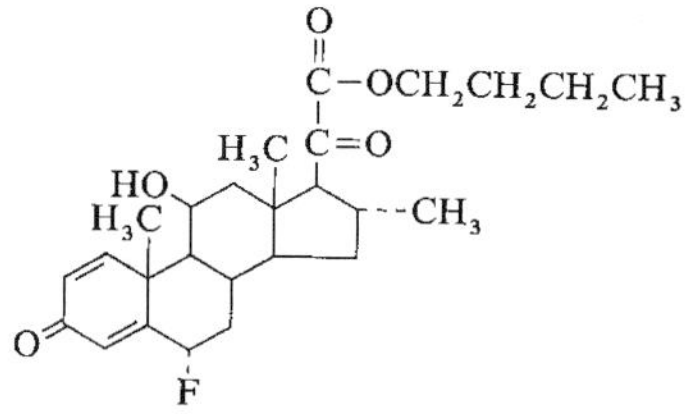

Fluocortin butyl [41767-29-7] (Vaspit, Schering AG) [4.119]

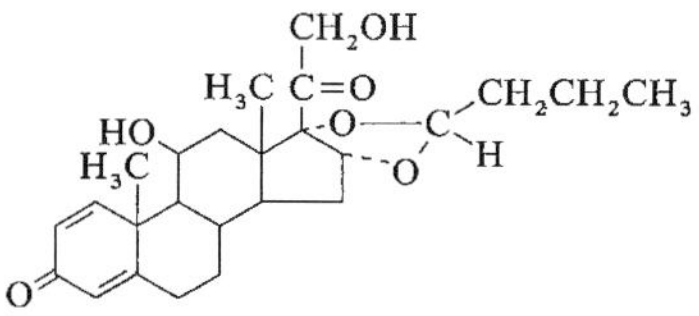

Budesonide [51333-22-3] (Pulmicort, Astra) [4.120]

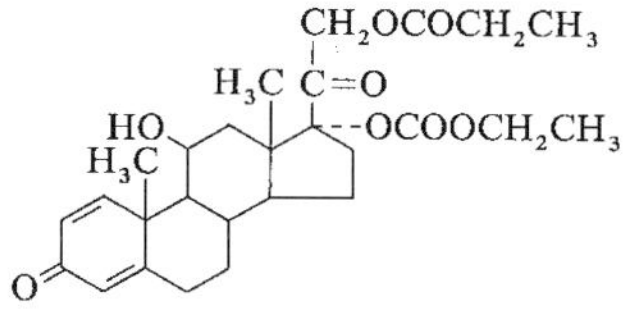

Prednicarbate [73771-04-7] (Hoe 777, Hoechst) [4.121]

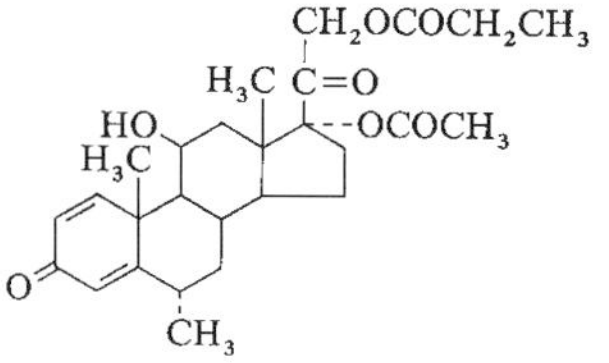

6α-Methylprednisolone aceponate [83-43-2] (Schering AG) [4.122]

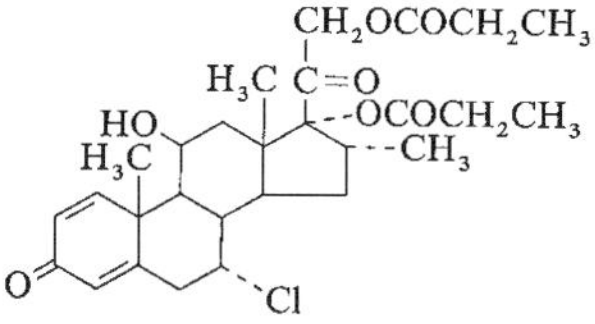

Alclometasone dipropionate [66734-13-2] (Delanol, Modrasone, Schering Corp.) [4.123]

Fundamentally new preparations or replacement of corticoids with another class of substances is not in sight. Improvements are to be found in application protocols (the circadian rhythm is taken into account) and in the wide range of galenical preparations.

5. Cholecalciferol

5.1. Introduction

Cholecalciferol [67-97-0] (vitamin D_3) is formed by UV irradiation of 7-dehydrocholesterol in the skin [310]. In 1971, DeLuca and coworkers succeeded in identifying 1α,25-dihydroxyvitamin D_3 [32222-06-3] (calcitriol) as the basic active principle formed by consecutive hydroxylation of vitamin D_3 in the liver and kidney [311]; it functions as a regulator of calcium and phosphate transport.

Cholecalciferol and its biologically active metabolite calcitriol must be classified as hormones rather than as vitamins (i.e., substances in food required for normal development). However, certain vitamin D deficiencies can be treated by diets containing vitamin D_3 or D_2.

Historical Aspects. The first description of rickets, a vitamin D deficiency disease, dates to antiquity. During the Industrial Revolution the incidence of rickets increased to epidemic proportions, especially in Europe and North America. The systematic search for an antirachitic factor in food began around 1920, stimulated by the observation that the disease could be cured by either UV irradiation or the intake of fish-liver oil [312]. A major breakthrough was achieved when vitamin D_2 [50-14-6] (ergocalciferol) was identified as a potent antirachitic principle that could be obtained by irradiating natural ergosterol [313], [314]:

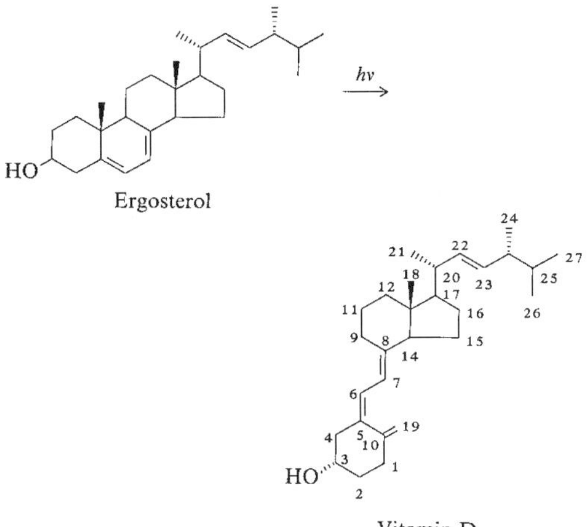

Ergosterol $\xrightarrow{h\nu}$ Vitamin D_2

Although vitamin D_2 occurs naturally (in fish-liver oil), it proved nonidentical with the factor produced by photochemical conversion in the skin of mammals. WINDAUS and coworkers succeeded in identifying and synthesizing vitamin D_3 as the natural principle formed by photolysis of 7-dehydrocholesterol [315], [316].

5.2. Biosynthesis and Metabolism

The metabolism of vitamin D_3 is summarized in Figure 11. In a reaction catalyzed by UV light (wavelength 250 – 300 nm), 7-dehydrocholesterol accumulates in the skin and is transformed into previtamin D_3 which undergoes isomerization to vitamin D_3 by a thermal 1,7-hydrogen shift [317].

Alternatively, dietary vitamin D_3 is taken into the bloodstream by the intestinal route. A specific vitamin D_3 binding protein transports vitamin D_3 to the liver where a cytochrome P 450 enzyme hydroxylates the compound at position C-25 to yield 25-hydroxyvitamin D_3. In the kidney, another cytochrome P 450 steroid hydroxylase (1α-hydroxylase) introduces a hydroxyl group at position C-1 to form the active metabolite 1α,25-dihydroxyvitamin D_3 (calcitriol). This step appears to be modulated by parathyroid hormone [310], [318]. Calcitriol is bound with high affinity to a specific protein receptor (1α,25-dihydroxyvitamin D_3 receptor, M_r 50 000 – 60 000), which binds to

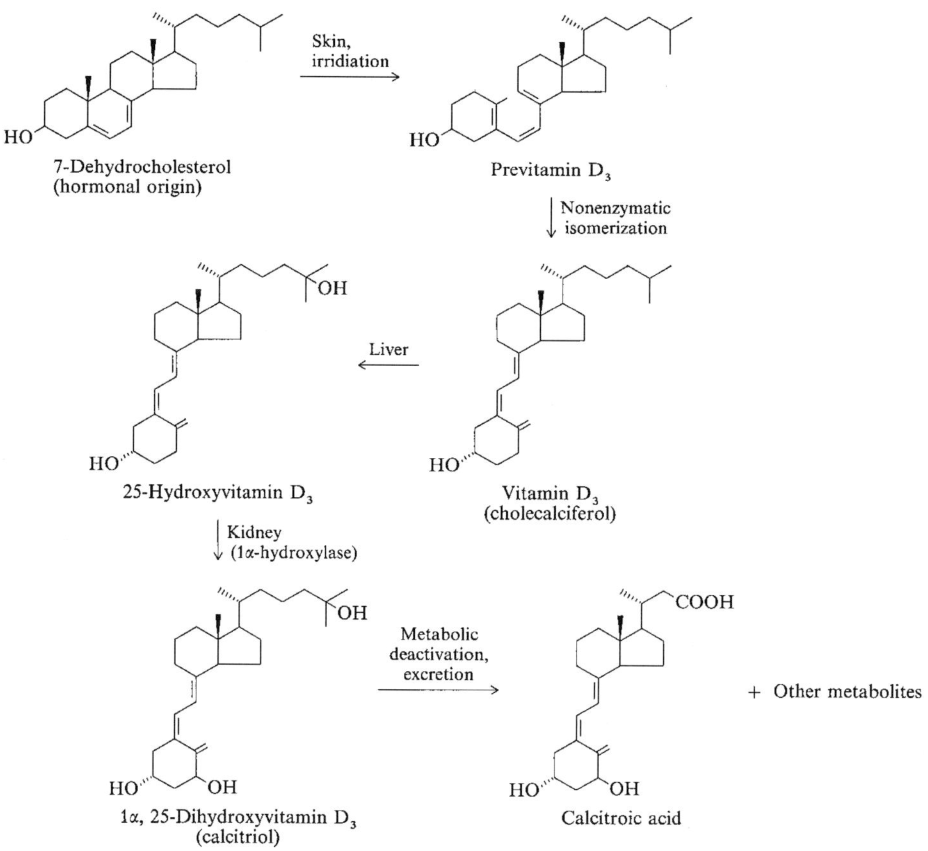

Figure 11. Metabolism of vitamin D_3

DNA. $1\alpha,25$-Dihydroxyvitamin D_3 receptors have been detected in many tissues (intestine, bone, pancreas, thyroid, muscle, gonads).

Calcitriol is inactivated by further enzymatic hydroxylation and oxidation. Approximately 20 metabolic derivatives of calcitriol have been found in human and animal tissue, all of them considerably less active than $1\alpha,25$-dihydroxyvitamin D_3.

5.3. Biological Activity

DeLuca and coworkers demonstrated that the biological activity previously ascribed to vitamine D_3 was actually produced by $1\alpha,25$-dihydroxyvitamin D_3 (calcitriol).

Calcitriol stimulates bone growth and mineralization by increasing the intestinal absorption of calcium and phosphate, thereby maintaining the necessary serum levels. The discovery of $1\alpha,25$-dihydroxyvitamin D_3 receptors in bone indicates a direct action of calcitriol on calcium transport in bone tissue. Bone undergoes constant remodeling; both synthesis and resorption are influenced by calcitriol in a dose-dependent manner.

Calcitriol can be used to treat rickets and osteomalacia (softening of the bones), and is indispensable in the treatment of chronic renal failure. Controversy still exists as to whether osteoporosis in postmenopausal women is amenable to calcitriol therapy.

In addition to its effects on mineral homeostasis, 1α,25-dihydroxyvitamin D_3 influences cell proliferation and differentiation in tissues not related to the classical sites of action (intestine, bone, kidney). The irregular growth and maturation of epidermal cells (a symptom of psoriasis) can be inhibited by topical administration of calcitriol. Inhibitory effects have been demonstrated for various cancer cell lines, so that calcitriol and its derivatives may find new therapeutic uses [310], [318].

5.4. Chemical Synthesis

Following the pioneering work of WINDAUS and coworkers [314], [315], numerous research groups have been engaged in the synthesis of vitamin D_3, its hydroxylated metabolites, and analogues [319].

Classical photolysis of a steroidal 5,7-diene precursor remains the basis for several recent strategies, although other nonphotochemical concepts are being developed. The principle of convergence (i.e., separate construction of major molecular portions that are combined in a final step to form the structural entity) appears to be a promising approach to an economical, large-scale production of vitamin D_3 and derivatives.

Examples are given for two calcitriol syntheses based on the classical concept [320] and the convergent strategy [321].

Classical Concept. BARTON et al. started their synthesis of calcitriol with 25-hydroxycholesterol (**176**) [320]. Dehydrogenation with dichlorodicyano-p-benzoquinone gave the trienone **177**, which was selectively epoxidized to form the 1α,2α-epoxide **178**. Lithium – ammonia reduction of **178** followed by acetylation yielded the triacetate **179**, which was transformed to the key intermediate **180** by allylic bromination – dehydrobromination. Electrocyclic ring opening by irradiation via the intermediate **181**, thermal isomerization, and saponification resulted in the formation of 1α,25-dihydroxyvitamin D_3 (calcitriol).

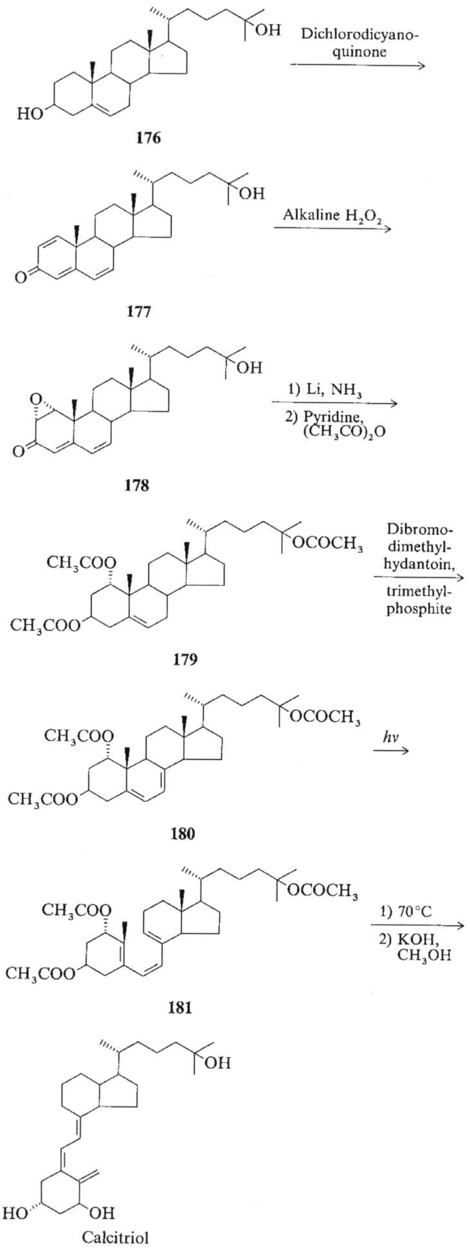

Convergent Strategy. Uskokovic and coworkers achieved a highly stereoselective, total synthesis of calcitriol by combining synthons **182** and **183**, prepared from easily accessible chiral starting materials, in a Wittig–Horner reaction as the final step [321]:

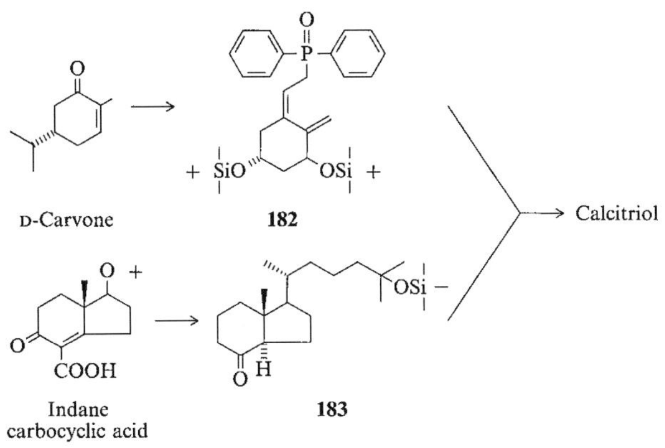

5.5. Synthetic Analogues

Numerous attempts have been made to find derivatives of vitamin D_3 and $1\alpha,25$-dihydroxyvitamin D_3 with better separation of the systemic effect on calcium metabolism and the antiproliferative, cell-differentiating activities [322]. Fluorination at position C-24, C-26, or C-27 substantially increases potency, compared with fluorine-deficient compounds. Other chemical modifications have led to derivatives with much lower activity, so that synthetic efforts in this field continue [322].

6. References

Chapter 1.

General References

[1] "Endocrinology," *Handbook of Physiology*, vol. **1–7**, Sect. 7, American Physiological Society, Washington 1974.
[2] *Cold Spring Harbor Symposia on Quantitative Biology*, vol. **LI/1, LI/2,** Cold Spring Harbor Laboratory, Cold Spring Harbor 1986.
[3] S. Baba, T. Kaneko, N. Yanaihara: *Proinsulin, Insulin, C-Peptide*. Excerpta Medica, Amsterdam 1979.
[4] R. Baserga: *Handbook of Experimental Pharmacology*, "Tissue Growth Factors," vol. **57,** Springer Verlag, Berlin 1981.
[5] S. R. Bloom: *Gut Hormones*, Churchill Livingstone, Edinburgh 1978.
[6] D. Brandenburg, A. Wollmer: *Insulin. Chemistry, Structure and Function of Insulin and Related Hormones*, Walter de Gruyter, Berlin 1980.
[7] I. Chester Jones, I. W. Henderson: *General, Comparative and Clinical Endocrinology of the Adrenal Cortex*, vol. **2,** Academic Press, London 1978.
[8] L. DeGroot et al.: *Endocrinology*, vol. **1–3,** Grune & Statton, New York 1979.

[9] B. J. A. Furr, A. E. Wakeling: *Pharmacology and Clinical Uses of Inhibitors of Hormone Secretion and Action*, Baillière Tindall, Eastbourne 1987.

[10] G. B. J. Glass: *Gastrointestinal Hormones*, Raven Press, New York 1980.

[11] E. C. Griffiths, G. W. Bennett: *Thyrotropin-Releasing Hormone*, Raven Press, New York 1983.

[12] F. Gross, H. G. Vogel: *Enzymatic Release of Vasoactive Peptides*, Raven Press, New York 1980.

[13] G. Litwack: *Biochemical Actions of Hormones*, vol. **13**, Academic Press, London 1986.

[14] J. B. Martin, S. Reichlin, G. M. Brown: *Clinical Neuroendocrinology*, F. A. Davis Co., Philadelphia 1977.

[15] E. Nieschlag: *Immunization with Hormones in Reproduction Research*, North-Holland/American Elsevier, Amsterdam 1975.

[16] J. A. Parsons: *Peptide Hormones*, MacMillan Press, London 1976.

[17] D. Rabin, T. J. McKenna: *Clinical Endocrinology, Principles and Practice*, Grune & Stratton, New York 1982.

[18] D. Schulster, S. Burnstein, B. A. Cooke: *Molecular Endocrinology of the Steroid Hormones*, John Wiley & Sons, London 1976.

[19] M. Tausk: *Pharmacology of the Endocrine System and Related Drugs*, "Progesterone, Progestational Drugs and Antifertility Agents," vol. **1–2**, Pergamon Press, Oxford 1972.

[20] L. Träger: "Biosynthese, Stoffwechsel, Wirkung," *Steroidhormone*, Springer Verlag, Berlin 1977.

[21] J. D. Wilson, D. W. Foster: *Textbook of Endocrinology*, 7th Ed., W. B. Saunders Company, Philadelphia 1985.

[22] R. H. Unger, L. Orci: "Physiology, Pathophysiology, and Morphology of the Pancreatic A-Cells," *Glucagon*, Elsevier, New York 1981.

Specific References

[23] S. B. Baylin, G. Mendelsohn: "Ectopic (inappropriate) hormone production by tumors: Mechanisms involved and the biological and clinical implications," *Endocr. Rev.* **1** (1980) 45–77.

[24] J. F. Habener: "Genetic control of hormone formation," in [21], pp. 9–32.

[25] G. F. Grossman: "Gastrointestinal hormones," in [16] pp. 105–117.

[26] D. H. Coy: "Somatostatins: Pharmacology and potential clinical uses," in [9] pp. 449–460.

[27] A. L. Goldstein et al.: "Current status of thymosin and other hormones of the thymus gland," *Recent Prog. Horm. Res.* **37** (1981) 369–416.

[28] S. E. Leeman, N. Aronin, C. Ferris: "Substance P and neurotensin," *Recent Prog. Horm. Res.* **38** (1982) 93–132.

[29] M. P. L. Caton, in G. P. Ellis, G. B. West (eds.): *Progress in Medicinal Chemistry*, "The prostaglandins," vol. **8**, Butterworths, London 1971, pp. 317–376.

[30] M. P. L. Caton, K. Crowshaw in G. P. Ellis, G. B. West (eds.): *Progress in Medicinal Chemistry*, "Prostaglandins and thromboxanes," vol. **15**, Elsevier/North Holland Biomedical Press, Amsterdam 1978, pp. 356–423.

[31] A. G. E. Pearse: "Evolutionary and developmental relationships among the cells producing peptide hormones," in [16], pp. 33–47.

[32] D. W. Lincol et al: "Hypothalamic pulse generators," *Recent Prog. Horm. Res.* **41** (1985) 369–420.

[33] F. A. Antoni: "Hypothalamic control of adrenocorticotropin secretion: Advances since the discovery of 41-residue corticotropin-releasing factor," *Endocr. Rev.* **7** (1986) 351–378.

[34] E. Knobil: "The neuroendocrine control of the menstrual cycle," *Recent Prog. Horm. Res.* **36** (1980) 53–88.

[35] L. A. Frohman, J.-O. Jansson: "Growth-hormone releasing hormone," *Endocr. Rev.* **7** (1986) 223–253.

[36] J. Roth, C. Grunfeld: "Mechanism of action of peptide hormones and catecholamines," in [21], pp. 76–122.

[37] R. V. Farese: "Phosphoinositide metabolism and hormone action," *Endocr. Rev.* **4** (1983) 78–96.

[38] J. H. Clark, W. T. Schrader, B. W. O'Malley: "Mechanisms of steroid action," in [21], pp 33–75.

[39] V. J. Hruby: "Structure-activity of the neurohypophyseal hormones and analogs and implications for hormone-receptor interactions," in [16] pp. 192–241.

[40] B. A. Eipper, R. E. Mains: "Structure and biosynthesis of pro-ACTH/endorphin and related peptides," *Endocr. Rev.* **1** (1980) 1–27.

[41] D. T. Krieger, A. S. Liotta, M. J. Brownstein, E. A. Zimmermann: "ACTH, β-lipotropin, and related peptides in brain, pituitary, and blood," *Recent Prog. Horm. Res.* **36** (1980) 277–335.

[42] J. L. Roberts et al.: "Glucocorticoid regulation of proopiomelanocortin gene expression in rodent pituitary," *Recent Prog. Horm. Res.* **38** (1982) 227–256.

[43] K. Hofmann: "Chemistry of adenohypophysial hormones," vol. **4**, The Pituitary Gland and Its Neuroendocrine Control, Part 2 (1974), in [1] pp. 28–131.

[44] M. Chrétien in S. A. Berson, R. S. Yalow (eds.): "Lipotropins (LPH)," *Methods in Investigative and Diagnostic Endocrinology*, vol. 2 A: Peptide Hormones, North Holland Publ. Co., Amsterdam 1973, pp. 617–632.

[45] W. H. Bishop, A. Nureddin, R. J. Ryan in J. A. Parsons (ed.): *Peptide Hormones*, "Pituitary luteinising and follicle-stimulating hormones," The Macmillan Press Ltd., London 1976, pp. 273–298.

[46] S. C. Chappel, A. Ulloa-Aguirre, C. Coutifaris: "Biosynthesis and secretion of follicle-stimulating hormone," *Endocr. Rev.* **4** (1983) 179–212.

[47] J. C. Fiddes, K. Talmadge: "Structure, expression, and evolution of the genes for the human glycoprotein hormones," *Recent Prog. Horm. Res.* **40** (1984) 43–78.

[48] B. D. Weintraub et al.: "Glycosilation and posttranslational processing of thyroid-stimulating hormone: Clinical implications," *Recent Prog. Horm. Res.* **41** (1985) 577–606.

[49] M. J. Karten, J. E. Rivier: "Gonadotropin-releasing hormone analog design. Structure-function studies toward the development of agonists and antagonists: Rationale and perspectives," *Endocr. Rev.* **7** (1986) 44–66.

[50] P. Licht et al.: "Evolution of gonadotropin structure and function," *Recent Prog. Horm. Res.* **33** (1977) 169–248.

[51] A. G. Frantz: "Prolactin, growth hormone and human placental lactogen," in [16] pp. 199–231.

[52] J. T. Potts, G. D. Auerbach: "Chemistry of the calcitonins," Parathyroid Gland, vol. **7** (Endocrinology, Sect. 7) (1976) in [1], pp. 423–430.

[53] W. L. Miller, N. L. Eberhardt: "Structure and evolution of the growth hormone gene family," *Endocr. Rev.* **4** (1983) 97–130.

[54] C. C. Nicoll, G. L. Mayer, S. M. Russell: "Structural features of prolactins and growth hormones that can be related to their biological properties," *Endocr. Rev.* **7** (1986) 169–203.

[55] N. Ben-Jonathan: "Dopamine: A prolactin-inhibiting hormone," *Endocr. Rev.* **6** (1985) 564–589.

[56] R. Markstein: "Pharmacology of dopamine receptor agonists," in [9], pp. 461–498.

[57] E. G. Armstrong, S. Birken, W. R. Moyle, R. E. Canfield (1986): "Immunochemistry of human chorionic gonadotropin," in [13], pp. 91–128.

[58] P. H. Seeburg: "Human growth hormone: From clone to clinic," in [2], pp. 669–678.

[59] D. V. Cohn, R. R. MacGregor: "The biosynthesis, intracellular processing, and secretion of parathormone," *Endocr. Rev.* **2** (1981) 1–26.

[60] S. A. Atlas: "Atrial natriuretic factor: A new hormone of cardiac origin," *Recent Prog. Horm. Res.* **42** (1986) 207–242.

[61] M. Cantin, J. Genest: "The heart and the atrial natriuretic factor, *Endocr. Rev.* **6** (1985) 107–127.

[62] J. D. Gardner, R. T. Jensen: "Gastrointestinal peptides: The basis of action at the cellular level," *Recent Prog. Horm. Res.* **39** (1983) 211–244.

[63] E. Straus, S. W. Ryder, J. Eng, R. Yalow: "Immunochemical studies relating to cholecystokinin in brain and gut," *Recent Prog. Horm. Res.* **37** (1981) 447–476.

[64] M. Ellis, K. Sikora: "Genes and cancer: A clinical perspective," *J. R. Coll. Physicians London* **21** (1987) 122–128.

[65] R. B. Dickson, M. E. Lippman: "Estrogenic regulation of growth and polypeptide growth factor secretion in human breast carcinoma," *Endocr. Rev.* **8** (1987) 29–43.

[66] D. Gospodarowicz, G. Neufeld, L. Schweigerer: "Fibroblast growth factor," *Mol. Cell. Endocrinol.* **46** (1986) 187–204.

[67] M. Jaye et al.: "Human endothelial cell growth factor: Cloning, nucleotide sequence, and chromosome localization," *Science* **233** (1986) 541–545.

[68] M. B. Sporn, A. B. Roberts, L. M. Wakefield, R. K. Assolan: "Transforming growth factor-β: Biological function and chemical structure," *Science* **233** (1986) 532–534.

[69] J. K. Browne et al. (1986): "Erythropoietin: Gene cloning, protein structure, and biological properties," in [2], vol. **LI/1,** pp. 693–702.

[70] G. D. Bryant-Greenwood: "Relaxin as a new hormone," *Endocr. Rev.* **3** (1982) 62–90.

[71] C. Schwabe et al.: "Relaxin," *Recent Prog. Horm. Res.* **34** (1978) 123–212.

[72] R. L. Cate et al. (1986): "Development of Mullerian inhibiting substance as an anti-cancer drug," in [2], vol. **LI/1,** pp. 641–648.

[73] A. J. Mason et al., "Complementary DNA sequences of ovarian follicular fluid inhibin show precursor structure and homology with transforming growth factor-β," *Nature (London)* **318** (1985) 659–663.

[74] W. Vale et al.: "Purification and characterization of an FSH releasing protein from porcine ovarian follicular fluid," *Nature (London)* **321** (1986) 776–779.

[75] F. Murad, R. C. Haynes in A. Goodman Gilman, L. S. Goodman, A. Gilman (eds.): "Hormones and hormone antagonists," *The Pharmacological Basis of Therapeutics,* 6th ed., Sect. **15,** Macmillan Publ. Co., New York 1980, pp. 1367–1550.

[76] D. P. Cardinali: "Melatonin. A mammalian pineal hormone," *Endocr. Rev.* **2** (1981) 327–346.

[77] J. P. Preslock: The pineal gland: Basic implications and clinical correlations," *Endocr. Rev.* **5** (1984) 282–307.

[78] R. J. Reiter: "The pineal and its hormones in the control of reproduction in mammals," *Endocr. Rev.* **1** (1980) 109–131.

[79] C. S. Pittman, J. A. Pittman: "Relation of chemical structure to the action and metabolism of thyroactive substances," Thyroid, vol. **3** (Endocrinology, Sect. 7) 1974, in [1] pp. 233–253.

[80] A. W. Norman, J. Roth, L. Orci: "The vitamin D endocrine system: Steroid metabolism, hormone receptors, and biological response (calcium binding proteins)," *Endocr. Rev.* **3** (1982) 331–366.

[81]   T. J. Cicero: "Basic endocrine pharmacology of opioid agonists and antagonists," in [9], pp. 518–537.

Chapter 2.

General References

[82]   K. Oberdisse, E. Klein, D. Reinwein: *Die Krankheiten der Schilddrüse*, 2nd ed., Thieme Verlag, Stuttgart 1980.
[83]   S. H. Ingbar, L. E. Braverman: *Werner's The Thyroid*, 5th ed., Lippincott, Philadelphia 1986.
[84]   P. Pfannenstiel: *Schilddrüsenkrankheiten – Diagnose und Therapie*, Henning GmbH, Berlin 1985.
[85]   A. G. Gilman, L. S. Goodman, T. W. Rall, F. Murad (eds.): *The Pharmacological Basis of Therapeutics*, 7th ed., Macmillan, New York 1985, p. 1397.
[86]   R. O. Greep, E. B. Astwood, M. A. Greer, D. H. Solomon (eds.): Thyroid; *Handbook of Physiology*, vol. **3**, Section 7, Endocrinology,Am. Physiol. Soc., Washington 1974.

Specific References

[87]   I. J. Chopra, D. H. Solomon, G. N. Chua Teco, *J. Clin. Endocrinol. Metab.* **36** (1973) 1050.
[88]   I. J. Chopra, *J. Clin. Invest.* **54** (1974) 583–592.
[89]   M. T. Hays, *J. Clin. Endocrinol. Metab.* **28** (1968) 749.
[90]   M. I. Surks et al., *J. Clin. Invest.* **52** (1973) 805.
[91]   K.-W. Wenzel, H. Kirschsieper, *Metabolism.* **26** (1977) 1.
[92]   C. M. Greenberg et al., *Am. J. Physiol.* **205** (1963) 821.
[93]   P. Lind, O. Eber, *Acta Med. Austriaca* **13** (1985) no. 1, 13.
[94]   British Pharmacopeia 1973.
[95]   US Pharmacopeia 18 (1970).
[96]   J. S. Varcoe, W. K. Warburton, *J. Chem. Soc.* 1960, 2711.
[97]   B. Block Jr., G. Powell, *J. Am. Chem. Soc.* **65** (1943) 1430.
[98]   *Biochem. J.* **21** (1949) 3424.
[99]   *J. Chem. Soc.*, 1949, 3424.
[100] R. I. Meltzer, R. J. Stanaback, *J. Org. Chem.* **26** (1961) 1977.
[101] S. Lissitzky, C. Cheftel, *C. R. Hebd. Séances Acad. Sci.* **256** (1963) 3898.
[102] H. Nahm, W. Siedel, *Chem. Ber.* **96** (1963) 1.
[103] Hoechst, DE 1 077 673, 1958 (W. Siedel, H. Nahm, J. König).
[104] G. Hillmann, *Z. Naturforsch.* **11 b** (1956) 319.
[105] G. Hillmann, DE 1 065 855, 1956. G. Hillmann DE 1 064 529, 1956.
[106] Hoffmann-La Roche, US 2 895 927, 1957.
[107] Sandoz, DE 1 221 646, 1963.
[108] R. Wegner, K. Rudnick, DL 65 933, 1968.
[109] H. Neudecker, E. Scheiffele, *Arzneim. Forsch.* **21** (1971) 432.
[110] A. D. Williams, D. E. Freeman, W. H. Florsheim, *J. Chromatogr.* **45** (1969) 371.
[111] A. Dibbo et al., *J. Chem. Soc.* 1961, 2645.
[112] Chemie Grünenthal, DE-OS 1 493 533, 1964; 1 793 776, 1965.
[113] T. Shiba, H. J. Cahnmann, *J. Org. Chem.* **27** (1962) 1773. P. Block Jr., *J. Med. Chem.* **19** (1976) no. 8, 1067.
[114] E. Wachholz, S. Pfeifer, *Pharmazie* **24** (1969) 459.
[115] C. A. Johnson et al., *Analyst London* **92** (1967) 328.
[116] R. H. Osborn, T. H. Simpson, *J. Chromatogr.* **40** (1969) 219.

[117] B. M. R. Heinl, H. M. Ortner, H. Spitzy, *J. Chromatogr.* **60** (1971) 51.

[118] E. Wachholz, S. Pfeifer, *Pharmazie* **27** (1972) 43, 97.

[119] R. F. Furchgott: "The Classification of Adrenoceptors (Adrenergic Receptors). An Evaluation from the Standpoint of Receptor Theory," in H. Blaschko, E. Muscholl (eds.): *Catecholamines. Handbuch der Experimentellen Pharmakologie*, vol. **33**, Springer Verlag, Berlin 1972, pp. 283–335.

[120] J. Himms–Hagen: "Sympathic Regulation of Metabolism," *Pharmacol. Rev.* **19** (1967) 367–461.

[121] A. Burger: "Adrenergic Hormones and Drugs," in *Medicinal Chemistry, Part II*, Chap. 46, 3rd. ed., J. Wiley & Sons, New York 1970.

[122] D. Palm, D. Hellenbrecht, K. Quiring: "Pharmakologie des noradrenergen und adrenergen Systems Katecholamine, Andrenozeptor-Agonisten und -Antagonisten Antisympathotonika und andere Antihypertensiva," in W. Forth, D. Henschler, W. Rummel (eds.): *Allgemeine und spezielle Pharmakologie und Toxikologie*, 5th ed., BI Wissenschaftsverlag, Mannheim 1987, pp. 124–168.

[123] A. Weiner: "Norepinephrine, Epinephrine, and the Sympathomimetic Drugs," in A. Goodman Gilman, L. S. Goodman, A. Gilman (eds.): *The Pharmacological Basis of Therapeutics*, 6th ed., Macmillan Publ. Co., New York 1980, pp. 138–175.

[124] A. Weiner: "Drugs that Inhibit Adrenergic Nerves and Block Adrenergic Receptors," in [121], p. 176–210.

[125] J. H. Burn, M. J. Rand: "The Action of Sympathomimetic Amines in Animals Treated with Reserpine," *J. Physiol. (London)* **144** (1958) 314–336.

[126] L. I. Goldberg: "Cardiovascular and Renal Actions of Dopamine: Potential Clinical Applications," *Pharmacol. Rev.* **24** (1972) 1–29.

Chapter 3.

General References

[127] L. Fieser, M. Fieser: *Steroids*, Reinhold Publishing Corp. New York 1959; *Steroide*, Verlag Chemie, Weinheim 1961.

[128] IUPAC-IUB 1967: "Tentative Rules for Steroid Nomenclature," *Biochim. Biophys. Acta* **164** (1968) 453–486.

[129] J. Fried, J. A. Edwards (eds.): *Organic Reactions in Steroid Chemistry*, vol. **1** and **2**, Van Nostrand Reinhold Comp., New York, Cincinatti, Toronto, Melbourne 1972.

[130] W. Forth, D. Hentschler, W. Rummel (eds.): *Allgemeine und Spezielle Pharmakologie und Toxikologie*, 5th ed., Wissenschaftsverlag, Mannheim, Wien, Zürich 1987, pp. 397–435.

[131] Goodman and Gilman's: *The Pharmacological Basis of Therapeutics*, 7th ed., Macmillan Publ. Co., New York 1985, pp. 1360–1459.

[132] B. Green, R. E. Leake: *Steroid Hormones, a practical approach*, IRL Press, Oxford, Washington DC 1987.

[133] H. L. J. Makin (ed.): *Biochemistry of Steroid Hormones*, 2nd ed., Blackwell Scientific Publications, Oxford, London, Edinburgh, Boston, Palo Alto, Melbourne 1984.

[134] R. T. Blickenstaff, A. C. Ghosh, G. C. Wolf: *Total Synthesis of Steroids*, Academic Press, New York, London 1974.

[135] A. A. Akhrem, Y. A. Titov: *Total Steroid Synthesis*, Plenum Press, New York, London 1970.

[136] M. B. Groen, F. J. Zeelen: "Steroid Total Synthesis," *Recl. J. R. Neth. Chem. Soc.* **105** (1986) 465–487.

[137] Bundesverband der Pharmazeutischen Industrie e. V.: Rote Liste 1987, Editio Cantor, Aulendorf (Württ.).

[138] E. R. Barnhart (publisher): *Physicians' Desk Reference,* 4th ed., Medical Economics Company Inc., Oradell (New Jersey) 1987.

[139] M. K. Agarwal (ed.): *Adrenal Steroid Antagonism,* Walter de Gruyter, Berlin, New York 1984.

[140] L. Martini, M. Motto (eds.): *Androgens and Antiandrogens,* Raven Press, New York 1977.

Specific References

[141] Kon. Nederland, Gist-en Spiritusfabr., Delft, DE-AS 1 568 932, 1966.

[142] Kon. Nederland. Gist-en Spiritusfabr., DE-OS 1 768 215, 1968.

[143] Noda Inst. for Scient. Res., DE-AS 1 543 269, 1965 (K. Arima, G. Tamura, M. Bae, M. Nagasawa).

[144] Richter Gedeon Vegyeszeti Gyar Rt., DE-AS 1 593 327, 1966 (G. Wix, K. Büki, E. Tömörkeny, G. Ambrus).

[145] N. Applezweig: *Steroid Drugs,* McGraw-Hill, New York 1962.

[146] K. S. Markley: *Soybean and Soybean Products,* vol. **2**, Interscience Publ., New York 1951, p. 837.

[147] N. Applezweig, *Chem. Week* **115** (1974) no. 2, 31.

[148] C. Meystre, K. Miescher, *Helv. Chim. Acta* **32** (1949) 1758, 1764. P. Wieland, K. Miescher, ibid, 1922.

[149] Parke Davies, US 2 335 616, 1941 (F. H. Tendick, E. J. Lawson); G. Rosenkranz, O. Mancera, F. Sondheimer, C. Djerassi, *J. Org. Chem.* **21** (1956) 520.

[150] J. Schmidt-Thomé, *Chem. Ber.* **88** (1955) 895.

[151] G. Ehrhard, H. Ruschig, W. Aumüller, *Angew. Chem.* **52** (1939) 363.

[152] H. H. Inhoffen, *Naturwissenschaften* **25** (1937) 125. E. B. Hersberg, M. Rubin, E. Schwenk, *J. Org. Chem.* **15** (1950) 292. Schering Corp., US 2 361 847, 1941 (H. H. Inhoffen).

[153] Schering AG, DE-AS 1 135 899, 1960 (G. Raspé, K. Kieslich, E. Olivar, R. Müller, B. Wagner). Eli Lilly, US 3 128 238, 1962 (G. E. Mallet).
Schering Corp., US 3 134 718, 1963 (A. Nobile).

[154] A. J. Birch, H. Smith, *Q. Rev. Chem. Soc.* **12** (1958) 17.

[155] A. L. Wilds, N. A. Nelson, *J. Am. Chem. Soc.* **75** (1953) 5366. C. Djerassi, L. Miramontes, G. Rosenkranz, F. Sondheimer, *ibid* **76** (1954) 4092.

[156] D. Onken, D. Heublein, *Pharmazie* **25** (1970) 3.

[157] F. B. Colton, L. N. Nysted, B. Riegel, A. L. Raymond, *J. Am. Chem. Soc.* **79** (1957) 1123. Searle, US 2 655 518, 1952 (F. B. Coltan).

[158] T. B. Windholz et al., *Angew. Chem.* **76** (1964) 249. H. Ueberwasser et al., *Helv. Chim. Acta* **46** (1963) 344. A. Popper et al., *Arzneim. Forsch.* **19** (1969) 352.

[159] R. Müller, K. Kieslich: "Technologie der Darstellung organischer Substanzen mit Mikroorganismen," *Chem. Ing. Tech.* **38** (1966) 813. K. Kieslich, *Synthesis* 1969, 120, 147.

[160] H. Smith et al., *J. Chem. Soc.* 1964, 4472.

[161] C. Rufer, H. Kosmol, E. Schröder, K. Kieslich, H. Gibian, *Justus Liebigs Ann. Chem.* **702** (1967) 141.

[162] L. Velluz et al., *C. R. Hebd. Seances Acad. Sci.* **257** (1963) 3086.

[163] P. Bellet, G. Nominé, J. Matthieu, *C. R. Hebd. Seances Acad. Sci. Ser. C* **263** (1966) 88.

[164] J. Ufer: *Hormontherapie in der Frauenheilkunde,* 4th ed., De Gruyter, Berlin 1972.

[165] L. Svennerholm, *Biochem. Biophys. Acta* **24** (1957) 604.

[166] A. Ercoli, E. Gardi, *Chem. Ind. (London)* 1961, 1037. Warner-Lambert Pharmac. Corp., DE-AS 1 157 610, 1961 (A. Ercoli).

[167] H. L. Krüskemper: *Anabole Steroide*, 2nd ed., Thieme Verlag, Stuttgart 1965.

[168] L. G. Hershberger, E. G. Shipley, R. K. Meyer, *Proc. Soc. Exp. Biol. Med.* **83** (1963) 175.

[169] Upjohn, US 3 029 263, 1959 (A. Campbell, J. C. Babcock). Ciba, De-AS 1 037 447, 1955 (A. Wettstein, G. Anner, Ch. Mestre).

[170] K. R. Bharucha, *Experientia* **14** (1958) 5. L. Ruzicka, M. W. Goldberg, H. R. Rosenberg, *Helv. Chim. Acta* **18** (1935) 1487.

[171] Syntex, DE-AS 1 159 943, 1957 (H. W. Ringold, G. Rosenkranz). R. O. Clinton et al., *J. Am. Chem. Soc.* **83** (1961) 1478.

[172] F. Neumann, R. Wiechert, *Arzneim. Forsch.* **15** (1965) 1168.

[173] H. Smith, *Experientia* **19** (1963) 394.

[174] Soc. Farmaceutici Italia, DE-AS 1 146 491, 1960 (B. Camerino, B, Patelli).

[175] A. Butenandt, K. Tscherning, G. Hanisch, *Ber. Dtsch. Chem. Ges.* **68** (1935) 2097.

[176] J. Haller: *Ovulationshemmung durch Hormone*, 2nd ed., Thieme Verlag, Stuttgart 1968.

[177] R. Deghenghi et al., *Tetrahedron* **19** (1963) 289; *J. Med. Chem.* **6** (1963) 301.

[178] A. Popper et al., *Arzneim. Forsch.* **19** (1969) 352.

[179] Olin Mathieson, DE-AS 1 125 423, 1959 (J. Fried).

[180] Syntex, GB 748 824, 1952.

[181] R. Wiechert, H. Steinbeck, W. Elger, F. Neumann, *Arzneim. Forsch.* **17** (1967) 1103.

[182] A. J. van den Broek, C. van Bokhoven, P. M. J. Hobbelein, J. Lemhuis, *Recl. Trav. Chim. Pays Bas* **94** (1975) 35.

[183] H. Hofmeister et al., *Arzneim. Forsch.* **36** (1986) 781.

[184] Ortho Pharmaceut, DE-AS 1 620 102, 1976 (A. P. Schroff).

[185] E.-E. Baulieu, S. J. Segal: *The Antiprogestin Steroid RU 486 and Human Fertility Control*, Plenum Press, New York, London 1985.

[186] K. Hiller, *Drugs of the Future* **7** (1982) 661; ibid **9** (1984) 714; ibid **10** (1985) 793.

Chapter 4.

General References

[187] G. Erhart, H. Ruschig (eds.): *Arzneimittel*, 2nd ed., Verlag Chemie, Weiheim 1972.
W. Forth, D. Hentschler, W. Rummel (eds.): *Allgemeine und Spezielle Pharmakologie und Toxikologie*, 5th ed., Wissenschaftsverlag, Mannheim, Wien, Zürich 1987.
Goodman and Gilman's: *The Pharmacological Basis of Therapeutics*, 7th ed., Macmillan Publishing Company, New York 1985.
L. Fieser, M. Fieser: *Steroids*, Reinhold Publishing Corp., New York 1959; *Steroide*, Verlag Chemie, Weinheim 1961.
W. Stalmans, M. Laloux in J. D. Baxter, G. G. Rousseau (eds.): *Glucocorticoid Hormone Action*, Springer Verlag, Berlin 1979, pp. 517–533.

[188] N. Applezweig, *Chem. Week* **115** (1974) no. 2, 31.

[189] Z. Buděšsinský, M. Protiva: "Corticoide,"*Synthetische Arzneimittel*, Chapt. 19, Akademie-Verlag, Berlin, GDR, 1961.

[190] A. Burger: "Steroids, I. The Adrenal Cortex Hormones," *Medical Chemistry*, 3rd. ed., Chapt. 34, J. Wiley, New York, London, Sydney, Toronto 1970.

[191] L. F. Fieser, M. Fieser: *Steroids*, Reinhold Publishing Corp., New York 1959; *Steroide*, Verlag Chemie, Weinheim 1961.

[192] B. Helwig: *Moderne Arzneimittel,* "Nebennierenrindenhormone," Chapt. 5 A, 4th ed., Wissenschaftl. Verlags GmbH, Stuttgart 1972.

[193] H. Kaiser: *Cortisonderivate in Klinik und Praxis,* 5th ed., Thieme Verlag, Stuttgart 1968.

[194] R. T. Curnow, J. Larner in J. Litwack (ed.): *Biochemical Actions of Hormones,* Academic Press, New York 1979, pp. 77–119.

N. R. Slaunwhite, A. A. Sandberg in E. Diezfalusy (ed.): *Steroid Assay by Protein Binding,* Karolinska Symposium on Research Methods in Reproductive Endocrinology, Stockholm 1970, pp. 144–154 (Acta Endocrinologica, Suppl. 147).

[195] H. Lettré, H. H. Inhoffen, R. Tschesche: *Über Sterine, Gallensäuren und verwandte Naturstoffe,* 2nd ed., Enke Verlag, Stuttgart 1959.

Specific References

[196] *Ullmann,* 3rd ed., **8,** 613 ff.
[197] R. Schüppel, *Pharm. Unserer Zeit* **2** (1973) 161.
[198] C. J. Brooks, J. N. Norymberski, *Biochem. J.* **55** (1953) 371.
[199] C. C. Porter, R. M. Silber, *J. Biol. Chem.* **185** (1952) 201.
[200] R. B. Woodward, F. Sondheimer, D. Taub, *J. Am. Chem. Soc.* **73** (1951) 3547.
[201] L. H. Sarett et al., *J. Am. Chem. Soc.* **74** (1952) 4974.
[202] N. L. Wendler, R. Richard, P. Traber, R. E. Jones et al., *J. Am. Chem. Soc.* **72** (1950) 5793.
[203] L. Velluz et al., *Angew. Chem.* **72** (1960) 725; **77** (1965) 185.
[204] R. Bucourt, G. Nominé, *Bull. Soc. Chim. Fr.* 1966, 1537.
[205] J. C. Gase, L. Nédélec, *Tetrahedron Lett.* 1971, 2005.
[206] D. H. Peterson et al., *J. Am. Chem. Soc.* **74** (1952) 5933.
[207] Upjohn, US 2 602 769, 1952. (H. C. Murray, D. H. Peterson).
[208] J. A. Hogg et al., *J. Am. Chem. Soc.* **77** (1955) 4436.
[209] H. Ruschig, *Angew. Chem.* **60** (1948) 247; *Chem. Ber.* **88** (1955) 878.
[210] Upjohn, US 2 769 823, 1954 (W. P. Schneider, A. R. Hanze).
[211] P. L. Julian, E. W. Meyer, W. J. Karpel, J. R. Waller, *J. Am. Chem. Soc.* **72** (1950) 5145.
[212] H. J. Ringold, B. Löken, G. Rosenkranz, F. Sondheimer, *J. Am. Chem. Soc.* **78** (1956) 816.
[213] J. Romo, G. Rosenkranz, F. Sondheimer, *J. Am. Chem. Soc.* **79** (1957) 5034.
[214] F. R. Hanson et al., *J. Am. Chem. Soc.* **75** (1953) 5369.
[215] J. Fried et al., *Recent Prog. Horm. Res.* **11** (1955) 149.
[216] J. Schmidt-Thomé, *Angew. Chem.* **69** (1957) 238.
[217] Hoechst, DE 1 009 627, 1955 (F. Lindner et al.).
[218] Upjohn, US 2 602 769, 1952 (H. C. Murray, D. H. Peterson).
[219] D. H. Peterson, *Research (London)* **6** (1953) 309/319.
[220] L. H. Sarett, *J. Biol. Chem.* **162** (1946) 601.
[221] B. F. McKenzie, V. R. Mattox, L. L. Engel, E. C. Kendall, *J. Biol. Chem.* **173** (1948) 271.
[222] G. Rosenkranz, F. Sondheimer, *Fortschr. Chem. Org. Naturst.* **10** (1953) 274.
[223] C. Djerassi, H. J. Ringock, G. Rosenkranz, *J. Am. Chem. Soc.* **73** (1951) 5513.
[224] J. H. Chapman, J. Elks, L. J. Wyman, *Chem. Ind. (London)* 1955, 603.
[225] J. A. Hogg et al., *J. Am. Chem. Soc.* **77** (1955) 4436.
[226] W. S. Allen, S. Bernstein, R. Littell, *J. Am. Chem. Soc.* **76** (1954) 6116. Amer. Cyanamid, US 2 622 081, 1951, 2 666 069, 1951 (S. Bernstein, R. M. Antonucci, M. D. Heller), 2 700 666, 1953 (S. Bernstein, R. Littel).
[227] R. H. Levin et al., *J. Am. Chem. Soc.* **75** (1953) 502.
[228] L. H. Sarett, *J. Am. Chem. Soc.* **70** (1948) 1454.

[229] F. R. Hanson et al., *J. Am. Chem. Soc.* **75** (1953) 5369.

[230] G. M. Shull, D. A. Kita, *J. Am. Chem. Soc.* **77** (1955) 763.

[231] A. Wettstein, *Experientia* **11** (1955) 465.

[232] E. Oliveto et al., *J. Am. Chem. Soc.* **78** (1956) 1736.

[233] C. Meystre, E. Visher, A. Wettstein, *Helv. Chim. Acta* **37** (1954) 1548.

[234] H. J. Ringold, G. Storck, *J. Am. Chem. Soc.* **80** (1958) 250.

[235] D. H. R. Barton et al., *J. Am. Chem. Soc.* **82** (1960) 2640, 2641.

[236] K. Heusler et al., *Helv. Chim. Acta* **44** (1961) 502; **45** (1962) 1317.

[237] L. Velluz, *C. R. Hebd. Seances Acad. Sci.* **250** (1960) 725.

[238] L. Ruzicka, K. Hofmann, *Helv. Chim. Acta* **20** (1937) 1280.

[239] H. H. Inhoffen, W. Logemann, W. Hohlweg, A. Serini, *Ber. Dtsch. Chem. Ges.* **71** (1938) 1024.

[240] J. A. Cella et al., *J. Org. Chem.* **24** (1959) 743, 1109. Searle, US 3 013 012, 1960 (J. A. Cella, R. C. Tweit).

[241] H. L. Herzog et al., *Science (Washington D.C.)* **121** (1955) 176.

[242] H. L. Herzog et al., *J. Am. Chem. Soc.* **77** (1955) 4781.

[243] S. A. Szpilfogel et al., *Rec. Trav. Chim. Pays Bas* **75** (1956) 475.

[244] C. Meystre, H. Frey, W. Voser, A. Wettstein, *Helv. Chim. Acta* **39** (1956) 734.

[245] E. Vischer, C. Meystre, A. Wettstein, *Helv. Chim. Acta* **38** (1955) 835, 1502.

[246] A. Nobile et al., *J. Am. Chem. Soc.* **77** (1955) 4184.

[247] F. Lindner et al., *Naturwissenschaften* **43** (1956) 39.

[248] J. H. Fried, E. F. Sabo, *J. Am. Chem. Soc.* **75** (1953) 2273; **76** (1954) 1455.

[249] S. Bernstein et al., *J. Am. Chem. Soc.* **78** (1956) 5693.

[250] J. Fried, W. B. Kessler, P. Grabowich, E. F. Sabo, *J. Am. Chem. Soc.* **80** (1958) 2338.

[251] S. Bernstein et al., *J. Am. Chem. Soc.* **81** (1959) 1689.

[252] S. Bernstein et al., *J. Am. Chem. Soc.* **77** (1955) 1028; **78** (1956) 5693; **81** (1959) 1689.

[253] R. W. Thoma, J. Fried, S. Bonanno, P. Grabowich, *J. Am. Chem. Soc.* **79** (1957) 4818.

[254] L. L. Smith, H. Mendelsohn, T. Foell, J. J. Goodman, *J. Org. Chem.* **26** (1961) 2859.

[255] J. Fried et al., *J. Am. Chem. Soc.* **80** (1958) 2338.

[256] G. B. Spero et al., *J. Am. Chem. Soc.* **78** (1956) 6213; **79** (1957) 1515.

[257] G. Cooley, B. Ellis, D. N. Kirk, V. Petrow, *J. Chem. Soc.* 1957, 4112.

[258] J. H. Fried et al., *J. Am. Chem. Soc.* **81** (1959) 1235.

[259] Upjohn, DE 1 056 605, 1957 (F. H. Lincoln, W. P. Schneider, G. P. Spero); 1 123 321, 1958 (B. J. Magerlein, W. P. Schneider, G. B. Spero, J. A. Hogg).

[260] Upjohn, US 2 867 638, 1957 (F. H. Lincoln, W. P. Schneider, G. B. Spero); DE-AS 1 082 261, 1959 (F. H. Lincoln, G. B. Spero, W. P. Schneider).

[261] G. E. Arth et al., *J. Am. Chem. Soc.* **80** (1958) 3161.

[262] E. P. Oliveto et al. *J. Am. Chem. Soc.* **80** (1958) 6687.

[263] D. Taub et al., *J. Am. Chem. Soc.* **82** (1960) 4012.

[264] Schering Corp., US 3 164 618, 1958 (R. Rausser, E. P. Oliveto).

[265] E. P. Oliveto, *J. Am. Chem. Soc.* **80** (1958) 4431.

[266] Lab. Français de Chimiothérapie, US 3 007 923, 1961 (G. Müller, R. Bardoneschi, J. Jolli).

[267] R. E. Marker, H. M. Crooks, *J. Am. Chem. Soc.* **64** (1942) 1280.

[268] W. R. Nes, H. L. Mason, *J. Am. Chem. Soc.* **73** (1951) 4765.

[269] Upjohn, GB 869 511, 1959.

[270] D. Taub et al., *J. Am. Chem. Soc.* **80** (1958) 4435; **82** (1960) 4012.

[271] E. P. Oliveto et al., *J. Am. Chem. Soc.* **80** (1958) 4428, 6678.

[272] Roussel-Uclaf, US 3 104 246, 1963 (G. Amiard, V. Torelli, J. Cérède).

[273] H. L. Slates, N. L. Wendler, *J. Org. Chem.* **22** (1957) 498.
[274] Merck Corp., GB 912 378, 1959.
[275] Schevico, GB 901 093, 1958.
[276] Merck AG, DE 1 134 074, 1959 (H.-J. Mannhardt, K.-H. Bork, K. Brückner, H. Metz).
[277] H. J. Mannhardt et al., *Tetrahedron Lett.* 1960, no. 16, 21.
[278] D. Taub et al., *J. Org. Chem.* **25** (1960) 2258.
[279] J. A. Hogg et al., *Chem. Ind. (London)* 1958, 1002.
[280] H. J. Ringold, G. Rosenkranz, F. Sondheimer, *J. Am. Chem. Soc.* **78** (1956) 820.
[281] A. Bowers, H. J. Ringold, *J. Am. Chem. Soc.* **80** (1958) 4423.
[282] J. A. Edwards, H. J. Ringold, C. Djerassi, *J. Am. Chem. Soc.* **82** (1960) 2318.
[283] A. Zaffaroni et al., *J. Am. Chem. Soc.* **80** (1958) 6110. Syntex, US 2 671 752, 1951 (A. Zaffaroni).
[284] Upjohn, US 2 841 600, 1957 (J. A. Hogg, G. B. Spero).
[285] Syntex, DE 1 079 042, 1958 (E. Batres et al.).
[286] Bayer, DE 1 088 953, 1958 (H. Lettré, D. Hotz).
[287] Sterling Drug Inc., US 2 678 932, 1951 (J. S. Buck, R. O. Clinton).
[288] J. S. Mills, A. Bowers, C. Djerassi, H. J. Ringold, *J. Am. Chem. Soc.* **82** (1960) 3399. Syntex, US 3 014 938, 1959 (J. S. Mills, A. Bowers).
[289] A. Bowers et al., *J. Am. Chem. Soc.* **81** (1959) 4107, 5991.
[290] B. M. Bloom, V. V. Bogert, R. Pinson Jr., *Chem. Ind. (London)* 1959, 1317.
[291] H. L. Ringold et al., *J. Am. Chem. Soc.* **80** (1958) 6464.
[292] G. Raspé, K. Kieslich, U. Kerb, *Arzneim. Forsch.* **14** (1964) 450.
[293] A. Doménico et al., *Arzneim. Forsch.* **15** (1956) 46.
[294] R. Joly et al., *Arzneim. Forsch.* **24** (1974) 1.
[295] V. Petrow, D. M. Williamson, *J. Chem. Soc.* 1959, 3595.
[296] Schering AG, DE-AS 1 169 444, 1961 (K. Kieslich, U. Kerb, G. Raspé).
[297] G. N. Shull, A. Kita, *J. Am. Chem. Soc.* **77** (1955) 763.
[298] A. Nobile et al., *J. Am. Chem. Soc.* **77** (1955) 4184.
[299] Schering AG, DE 1 135 899, 1960 (G. Raspé et al.).
[300] Roussel-Uclaf, DE 1 159 441, 1961 (R. Joly, J. Warnant, B. Goffinet). DE 1 205 096, 1961 (R. Joly, J. Warnant, J. Jully, B. Goffinet).
[301] Schering AG, DE-AS 1 249 270, 1963 (K. Kieslich, U. Kerb, G. Raspé).
[302] C. H. Robinson, L. Finckenor, E. P. Oliveto, D. Gould, *J. Am. Chem. Soc.* **81** (1959) 2191.
[303] Schering Corp., US 2 894 963, 1959 (D. H. Gould, H. Reimann, L. E. Finckenor).
[304] Glaxo, US 3 312 590, 1967 (J. Elks, P. J. May, N. G. Weir).
[305] Schering AG, US 3 824 260, 1974 (H. Laurent et al.).
[306] A. Thalen, R. Brattsand, E. Gruvstad, *Acta Pharm. Suec.* **21** (1984) 109.
[307] Schering Corp., US 3 444 217, 1969 (E. L. Shapiro, L. E. Finckenor).
[308] S. Sugai et al., *Chem. Pharm. Bull* **33** (1985) 1889.
[309] H.-J. Shue, M. J. Green, *J. Med. Chem.* **23** (1980) 430.
[310] H. F. DeLuca in H. L. J. Makin (ed.): *The Metabolism and Function of Vitamin D in Biochemistry of Steroid Hormones*, 2nd ed., Blackwell Scientific Publ., Oxford 1984, pp. 71–116.
[311] M. F. Holick, H. F. DeLuca, J. Suda, R. J. Cousins, *Biochemistry* **10** (1971) 2799.
[312] K. Huldshinsky, *Dtsch. Med. Wochenschr.* **45** (1919) 712. H. Chick, E. J. Palzell, E. M. Hume, Studies of rickets in Vienna 1919–1922. *Med. Res. Counc. (G.B.) Spec. Rep. Ser.* **77** (1923).
[313] F. A. Askew et al., *Proc. R. Soc. London, B* **107** (1931) 76.
[314] A. Windaus, O. Linsert, A. Lüttringhaus, G. Weidlich, *Justus Liebigs Ann. Chem.* **492** (1932) 226.

[315] A. Windaus, F. Schenck, F. von Werder, *Hoppe-Seylers Z. Physiol. Chem.* **241** (1936) 100.

[316] L. F. Fieser, M. Fieser: *Steroids,* Reinhold Publ. Co., New York 1959.

[317] R. B. Woodward, R. Hoffmann: *The Conservation of Orbital Symmetry,* Verlag Chemie, Weinheim and Academic Press, New York 1970.

[318] H. L. Henry, A. W. Norman, Vitamin D: Metabolism and Biological Actions, *Annu. Rev. Nutr.* **4** (1984) 493. H. F. DeLuca, H. K. Schnoes, Vitamin D: Recent Advances, *Annu. Rev. Biochem.* **52** (1983) 411.

[319] T. Kametani, H. Furuyama: *Synthesis of Vitamin D_3 and Related Compounds, Medicinal Research Reviews,* vol. **7**, no. 2, J. Wiley & Sons, New York 1987, pp. 147–171.

[320] D. H. R. Barton, R. H. Hesse, M. M. Pechet, E. Rizzardo, *J. Chem. Soc. Chem. Commun.* 1974, 203.

[321] E. G. Baggiolini, J. A. Iacobelli, B. M. Hennessy, M. R. Uskokovic, *J. Am. Chem. Soc.* **104** (1982) 2945.

[322] S.-J. Shiuey, J. J. Partridge, M. R. Uskokovic, *J. Org. Chem.* **53** (1988) 1040. A. Kutner et al., *J. Org. Chem.* **53** (1988) 3540. H. S. Gill et al., *J. Steroid Biochem.* **31** (1988) 147. M. J. Calverley, *Tetrahedron* **43** (1987) 4609.

Oral Antidiabetic Drugs

Insulin → *Peptide and Protein Hormones; GLP-1* → *Peptide and Protein Hormones*

HILMAR BISCHOFF, Bayer AG, Wuppertal, Federal Republic of Germany

1.	Introduction............	1653	4.	**Inhibitors of Intestinal**
2.	**Stimulators of Insulin**			**Carbohydrate Digestion....** 1661
	Secretion.............	1654	4.1.	α-Glucosidase Inhibitors ... 1661
2.1.	**Sulfonylureas and Related**		4.1.1.	Mode of Action 1662
	Compounds...........	1654	4.1.2.	Pharmacology 1662
2.2.	**Benzoic Acid Derivatives ...**	1654	4.1.3.	Individual Compounds 1662
2.3.	**Amino Acid Derivatives....**	1659	4.2.	**Plant Fibers............** 1664
2.4.	**Mode of Action.........**	1659	5.	**Enhancers of Insulin Action .** 1665
3.	**Biguanides.............**	1660	6.	**References.............** 1667

1. Introduction

Noninsulin dependent diabetes mellitus (NIDDM) or type 2 diabetes mellitus is a chronic disease of impaired carbohydrate, lipid, and protein metabolism, resulting from inadequate insulin action. Patients may have diminished insulin secretion, i.e., insulin deficiency, or insulin is secreted but does not act properly because of impaired sensitivity of muscle and liver to insulin (insulin resistance) [1], [2]. The treatment with antidiabetic drugs is aimed at achieving glycemic control to prevent the development of diabetic complications: neuropathy, nephropathy, and retinopathy. Oral antidiabetic drugs reduce or normalize elevated concentrations of glucose in blood and urine. Antidiabetic drugs available today aim at various target tissues and can be characterized by their different mode of actions. Basically two major pharmacological principles can be distinguished: (1) stimulation of insulin secretion (A) and (2) mechanisms where no elevated plasma insulin levels are required for glycemic control (B – D):

A) Insulin secretagogues act on the B-cells of the pancreas and stimulate insulin secretion.

B) Biguanides affect the glucose metabolism in the liver and peripheral tissues (e.g., skeletal muscle).

C) Inhibitors of carbohydrate digestion are active in the small intestine and delay carbohydrate digestion and glucose absorption.

D) Insulin sensitizers primarily act on peripheral tissues and enhance insulin action.

2. Stimulators of Insulin Secretion

2.1. Sulfonylureas and Related Compounds

When testing sulfonamide derivatives for antibacterial activity LOUBATIERES recognized their blood glucose lowering potenital already in 1944 [3]. However, it took more than 10 years before the first nonantibacterial compounds carbutamide (Boehringer Mannheim) [4] and tolbutamide (Hoechst) [5] (Table 1) were introduced into the market for the treatment of type 2 diabetic patients in 1956.

Sulfonylurea (SU) derivatives of the first generation were active only in high dosages of up to 2000 mg/d. Sulfonylureas of the "second generation" showed much higher pharmacological potency and were active in significantly lower doses [6]. The first one, glibenclamide (glyburide) (Table 1), was introduced in the market by Boehringer Mannheim and Hoechst in 1969 requiring daily doses of only 2.5 – 15 mg.

Although the efforts in pharmaceutical sulfonylurea research decreased in the 1970s, investigations in the last two decades have been concentrated on increasing, prolonging, or shortening the action of insulin secretagogues or accelerating their onset of action. Glimepiride (Table 1), a structure closely related to glibenclamide in which the phenyl substituent of glibenclamide is replaced by pyrroline, was the latest SU derivative which was introduced into the market in 1996. In contrast to glibenclamide, in several animal species glimepiride is characterized by a faster onset of blood glucose lowering activity, and a longer-lasting glucose lowering effect [7].

Also in clinical trials, glimepiride showed longer-lasting hypoglycemic effects than glibenclamide [8]. Due to these properties, glimepiride was introduced as a drug for once-daily treatment. Although all 14 compounds of Table 1 are available at least in parts of the world, today more than 90% of the SU world market is dominated by only five SU derivatives — glibenclamide/glyburide, gliclazide, glipizide, chlorpropamide, and glimepiride (Table 1).

2.2. Benzoic Acid Derivatives

Research on the structure-activity relationship of potent "second generation" SUs like gliblenclamide has revealed that the nonsulfonylurea moiety of glibenclamide also stimulated insulin secretion. The benzoic acid derivative meglitinide (HB 699) stimulated insulin release in vitro and in vivo, but the potency was only comparable to tolbutamide [8]. Repaglinide, the active (S)-enantiomer of AG-EE 388 ZW (Boehringer Ingelheim) (Table 1) is a benzoic acid derivative with remarkably higher potency, comparable with glibenclamide [23]. Due to its different pharmacokinetic behavior

Table 1. Examples of insulin secretion simulators (daily dose for humans and threshold dose for rabbits) and elimination half life $\tau_{1/2}$

Generic name M_r, mp, Ref.	Examples of commercial preparations formula	Daily dose, mg	Threshold dose, mg	$\tau_{1/2}$ h
Sulfonylureas				
Acetohexamide [968-81-0] $C_{15}H_{20}N_2O_4S$ 324.2, 188–190 °C, [9]	Dimelor (Lilly) 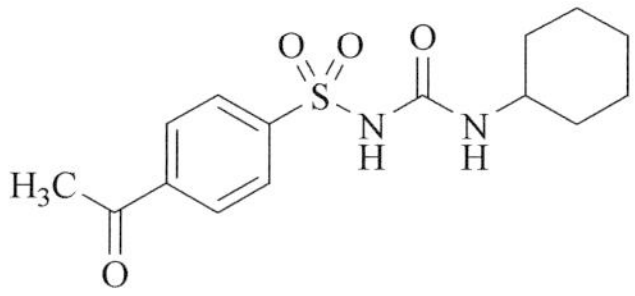	250–500	10	4–7
N-(4-Acetylbenzenesulfonyl)-*N'*-cyclohexylurea				
Carbutamide [339-43-5] $C_{11}H_{17}N_3O_3S$ 271.3, 140–143 °C, [10], [11]	Nadisan (Boehringer Mannheim) Invenol (Hoechst) 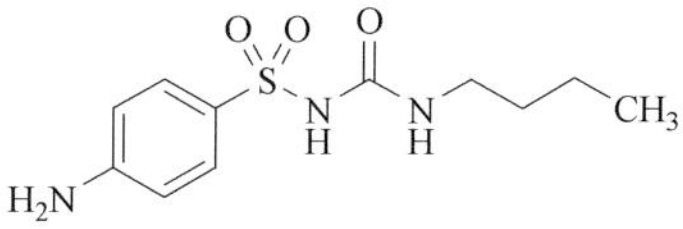	500–2000	200	40
N-Sulfanilyl-*N'*-*n*-butylurea				
Chlorpropamide [94-20-2] $C_{10}H_{13}ClN_2O_3S$ 276.6, 126–128 °C, [5]	Chloronase (Hoechst) Diabetoral (Boehringer Mannheim) Diabenese (Pfizer) 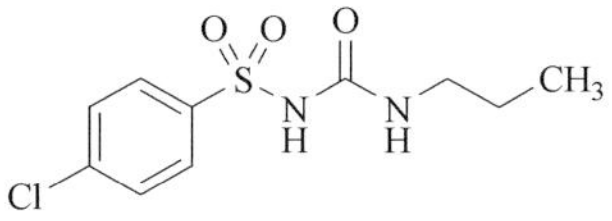	125–500	10	35
N-(4-Chlorobenzenesulfonyl)-*N'*-*n*-propylurea				
Glibornuride [26944-48-9] $C_{18}H_{26}N_2O_4S$ 366.4, 192–195 °C, [12]	Gluborid (Grünenthal) Glutril (Hoffmann-La Roche) 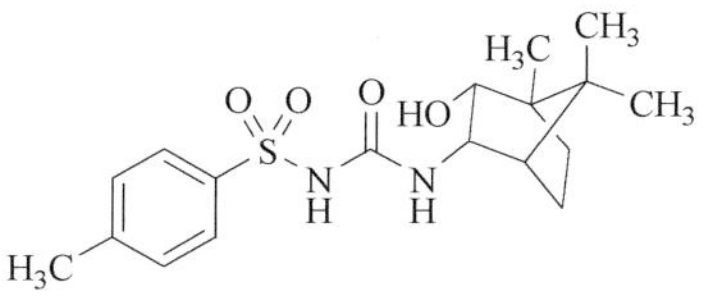	12.5–75	1	5–7
N-(Tosyl)-*N'*-(2-endo-hydroxy-3-endo-DL-bornyl)-urea				

Table 1. (continued)

Generic name M_r, mp, Ref.	Examples of commercial preparations formula	Daily dose, mg	Threshold dose, mg	$\tau_{1/2}$ h
Glyburide (Glibenclamide) [10238-21-8] $C_{23}H_{28}ClN_3O_5S$ 495.0, 172–174 °C, [6]	Euglucon (Boehringer Mannheim, Hoechst) Daonil (Hoechst) 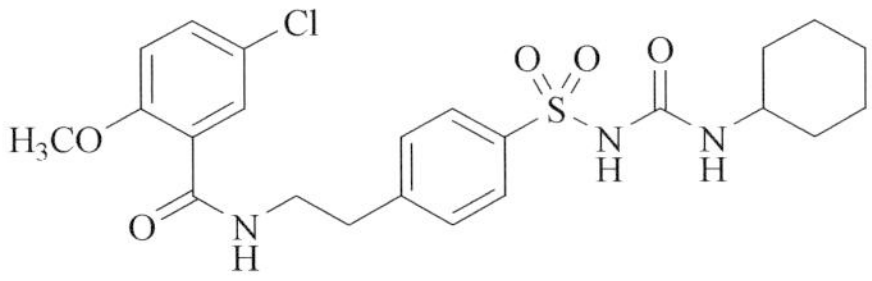	2.5–15	0.01	5–8

N-4-[2-(Chlor-2-methoxybenzamido)-ethyl]-benzenesulfonyl)-N'-cyclohexylurea

| **Gliclazide** [21187-98-4] $C_{15}H_{21}N_3O_3S$ 323.2, 180–182 °C, [13] | Diamicron (Science Union) 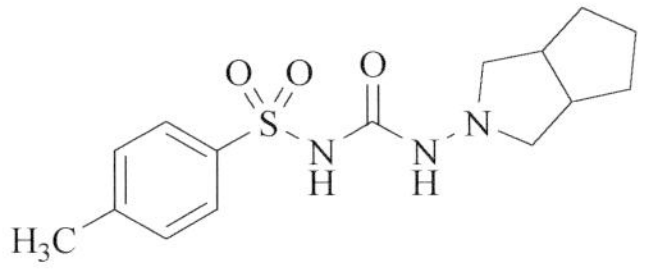 | 40–160 | 1.0 | 7 |

N-(Tosyl)-N'-(3-aza-bicyclo-[3.3.0.]-3-octyl)-urea

| **Glimepiride** [93479-97-1] $C_{24}H_{34}N_4O_5S$ 490.62, 207 °C, [14] | Amaryl (Hoechst/Marion Roussel) 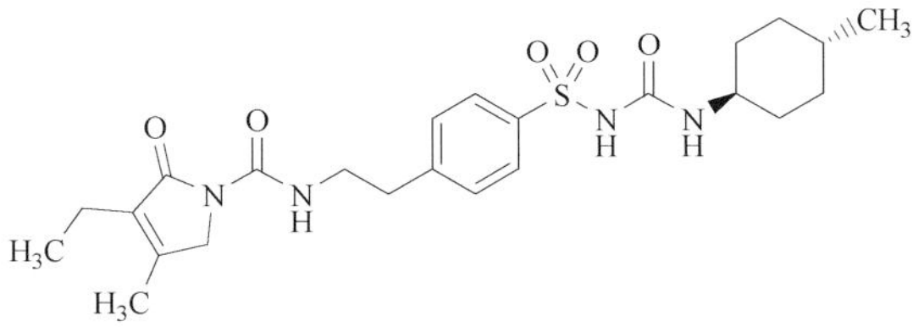 | 1–6 | | 1.5–3.5 |

Trans-3-ethyl-2,5-dihydro-4-methyl-N-[2-[4-[[[[(4-methylcyclohexyl) amino]carbonyl]amino]sulfonyl]phenyl]ethyl]-2-oxo-1H-pyrrole-1-carboxamide

| **Glipizide** [29094-61-9] $C_{21}H_{27}N_5O_4S$ 445.5, 208–209 °C, [15] | Glibenese (Pfizer) Mini Diab (Carlo Erba) 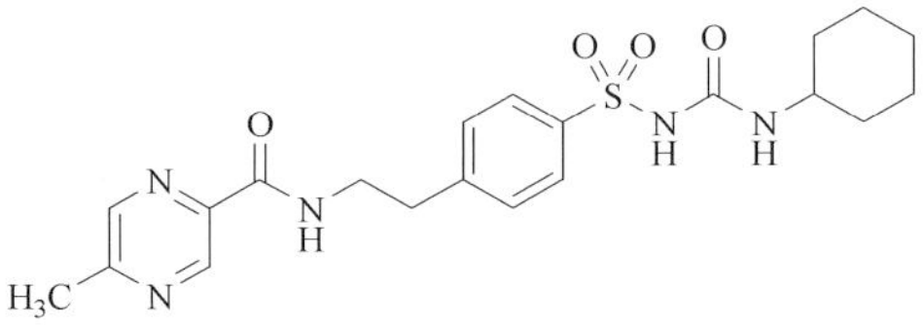 | 2.5–20 | 0.02 | 2–4 |

N-{4-[5-methylpyrazine-2-carboxamido)-ethyl]-benzenesulfonyl}-N'-cyclohexylurea

Table 1. (continued)

Generic name M_r, mp, Ref.	Examples of commercial preparations formula	Daily dose, mg	Threshold dose, mg	$\tau_{1/2}$ h
Gliquidone [*33342-05-1*] $C_{27}H_{33}N_3O_6S$ 527.3, 180–182 °C, [16]	Glurenorm (Thomae)	15.0–120	2.5	2–4

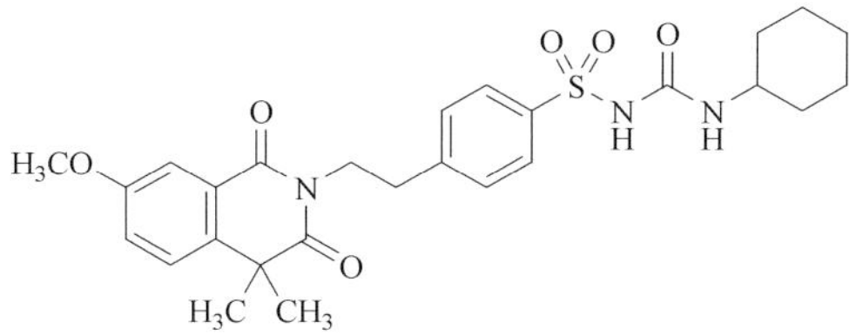

N-{4-[2-(3,4-Dihydro-7-methoxy-4,4-dimethyl-1,3-dioxa-2(1H)-isoquinolyl)-ethyl]-benzenesulfonyl}-N-cyclohexylurea

| **Glisoxepide** [*25046-79-1*]
$C_{20}H_{27}N_5O_5S$
449.5, 194–198 °C,
[17], [18] | Pro-Diaban (Bayer, Schering) | 2.0–16 | 0.05 | 2–4 |

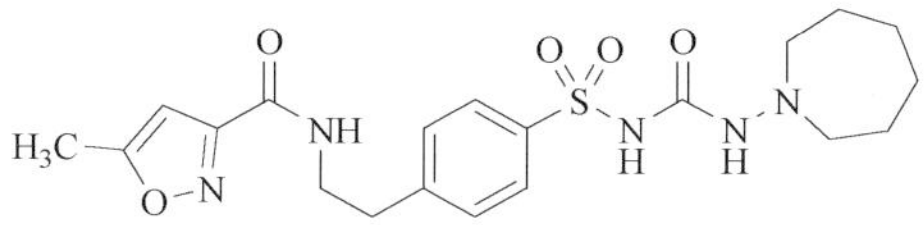

4-{4-[β-(5-Methylisoxazolyl-3-carboxamido)-ethyl]-benzenesulfonyl}-1,1-hexamethylenesemicarbazide

| **Glymidine (Glycodiazine)** [*3459-20-9*]
$C_{13}H_{14}N_3O_4SNa$
331.2, 221–226 °C,
[19] | Redul S (Schering) | 500–1500 | 20 | 4 |

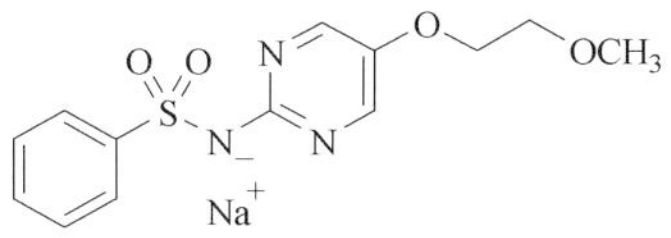

2-Benzenesulfonamido-5-(2-methoxyethoxy)-pyrimidine sodium

| **Tolazamide** [*1156-19-0*]
$C_{14}H_{21}N_3O_3S$
311.4, 165–173 °C,
[20] | Norglycin (Upjohn) Tolinase (Upjohn) | 125–1000 | 5 | 7 |

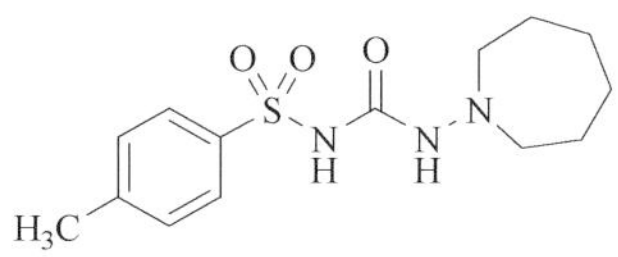

N-(Tosyl)-N′-(hexahydro-1H-azepin-1-yl)-urea

Table 1. (continued)

Generic name M_r, mp, Ref.	Examples of commercial preparations formula	Daily dose, mg	Threshold dose, mg	$\tau_{1/2}$ h
Tolbutamide [*64-77-7*] $C_{12}H_{18}N_2O_3S$ 270.4, 127–129 °C, [5] *N*-(Tosyl)-*N'*-*n*-butylurea	Artosin (Boehringer Mannheim) Rastinon (Hoechst) Orinase (Upjohn) 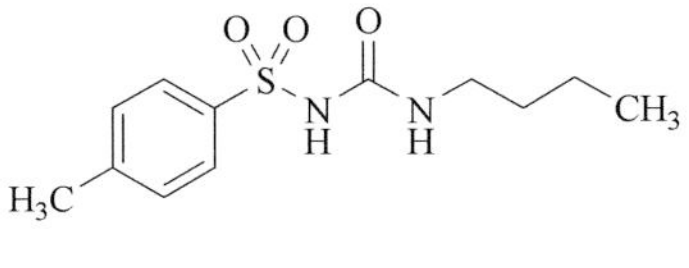	500–2000	10	5–7
Tolcyclamide (Glyciclamide) [*664-95-9*] $C_{14}H_{20}N_2O_3S$ 296.4, 174–175 °C, [5] *N*-(Tosyl)-*N'*-cyclohexylurea	Diaboral (Carlo Erba) 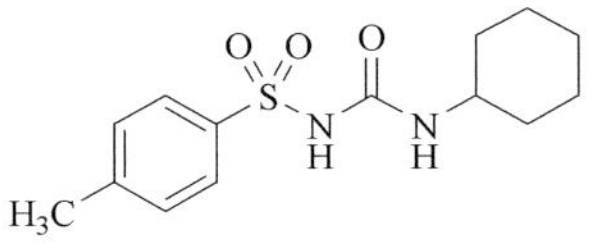	150–1000	10	6–7

Benzoic Acid Derivative

Repaglinide [*135062-02-1*] $C_{27}H_{36}N_2O_4$ 452.59, 126–131 °C, [21]	Novonorm (Novo Nordisk)	1.5–16		0.5–1

(*S*)-2-Ethoxy-4-[2-[[3-methyl-1-[2-(1-piperidinyl)phenyl]butyl]amino]-2-oxoethyl]benzoic acid

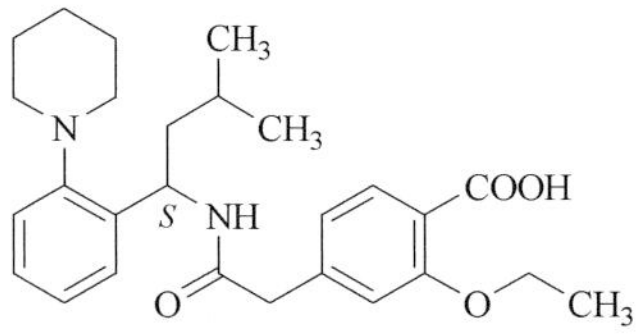

Amino Acid Derivative

Nateglimide [22]	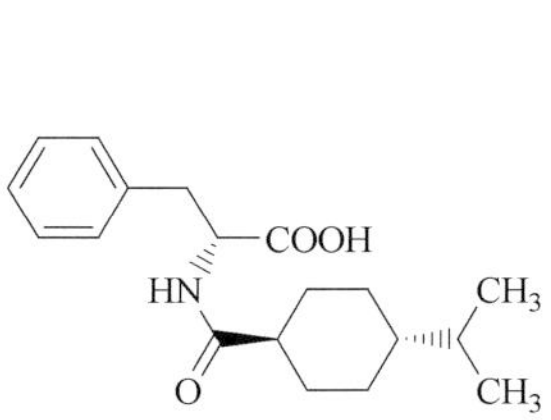			1–2

from glibenclamide, more rapid absorption, and shorter half-life, repaglinide showed interesting pharmacological properties in terms of quick insulin release and short-acting blood glucose lowering effects [24]. Repaglinide was introduced into the market in 1998.

2.3. Amino Acid Derivatives

Amino acid derivatives derived from D-phenylalanine are structurally a new type of insulin secretagogues [23].

Nateglinide (A-4166), Starlix, shows stereospecific activity, and it was found that the R-configuration was essential for hypoglycemic activity.

It is suggested that the phenylmethyl substituent of the amino acid residue is essential for activity. Variation of the amino acid moeity, e.g., glycine, alanine, phenylglycine, tyrosine, and others reduced the hypoglycemic activity [23]. Nateglinide produces a very quick insulin release followed by a rapid onset of blood glucose decrease and shows a short duration of action [25], [26]. Clinical trials have demonstrated mild blood glucose lowering activity after single doses of 60–180 mg. The first market introduction was in 1999 in Japan, therefore experience in daily practice is not available yet.

2.4. Mode of Action

Insulin secretion is physiologically stimulated by the binding of ATP to a cytosolic nucleotide binding site of the membrane bound ATP-sensitive K^+ channel which leads to closure of the K^+ channel. The inhibition of K^+ permeability depolarizes the plasma membrane, subsequently the voltage-dependent Ca^{2+} channel opens to promote the Ca^{2+} influx which finally results in insulin secretion [27].

The insulin secretagogues stimulate insulin secretion by closure of this ATP-sensitive K^+ channel of pancreatic B-cells but the binding of the secretagogues is suggested to occur to a separate regulator protein containing the binding sites for SUs (termed "sulfonylurea receptor") and other compounds [27].

3. Biguanides

The hypoglycemic activity of guanidine derivatives was firstly reported in 1918 [28]. The first blood glucose lowering diguanidine derivative decamethylenediguanidine (Synthalin A) was synthetized in the 1920s. Because of liver toxicity development of these compounds was discontinued. Only in the 1950s three biguanides were developed. The disubstituted dimethylbiguanide metformin [*657-24-9*] was introduced in 1957.

$$(H_3C)_2N-\underset{NH}{C}(=)-\underset{}{N}H-\underset{NH}{C}(=)-NH_3^+ \, Cl^-$$

Monosubstituted compounds with a longer side chain, the phenylethylbiguanide *phenformin* [*114-86-3*] and the *N*-butylbiguanide *buformin* [*692-13-7*], were introduced in the same year and 1958, respectively [29].

Pharmacology and Mode of Action. Biguanides are ineffective in normal animals, unless very high doses are used and there is no risk of hypoglycemia. However, in diabetic animals blood glucose-lowering effects of biguanide can be demonstrated, depending on the model used. Biguanides do not stimulate insulin secretion, but need insulin for pharmacological hypoglycemic activity. In severe insulinopenic animal models, high doses have to be used and effects are modest. In mild diabetic animals or in insulin resistant animals, lower doses are already efficacious in lowering hyperglycemia and hyperinsulinemia [30].

The mechanism of action of biguanides is still not fully understood. Three major tissues have been identified as pharmacological sites of action: (1) the small intestinal wall, (2) the liver, and (3) peripheral tissues, mainly the skeletal muscle:

1) For the small intestine an inhibition of glucose absorption was described, however, this is, at least for metformin, of minor significance and not important for the blood glucose lowering effect. However, the intestinal glucose metabolization to lactate is stimulated and reduces the postprandial uptake of glucose by the liver.
2) Numerous studies have shown that biguanides inhibit hepatic gluconeogenesis and this may contribute to the blood glucose lowering effect, particularly in the fasting state. Again, metformin has probably less impact on gluconeogenesis than phenformin and buformin.
3) In the peripheral tissues, metformin increases the glucose disposal and utilization particularly in the skeletal muscle, which is probably the major contribution to the blood glucose lowering activity. In vitro studies using cell cultures have shown that metformin potentiates insulin action. In vivo studies in animals and diabetic patients have demonstrated that metformin reduces insulin resistance, at least in obese individuals [30], [31].

Phenformin and buformin may cause lactic acidosis, which is the most serious adverse event of biguanides. Due to very serious cases of lactic acidosis both drugs were withdrawn from the markets in most parts of the world during the 1970s. Today they are available only in a few countries of Eastern Europe and South America. Owing to a more favorable pharmacokinetic profile of metformin, metformin associated lactic acidosis is very rare [31]. Therefore, only metformin is available today in all important countries. In the 1990s metformin experienced a renaissance in many countries, particularly in the USA, where metformin was approved for the treatment of type 2 diabetes only in 1994.

4. Inhibitors of Intestinal Carbohydrate Digestion

Delaying absorption as a therapeutic principle is well accepted for dietary therapy to avoid high postprandial blood glucose excursions [32]. Carbohydrates represent quantitatively the major portion of the human diet, in the form of mono-, di- and polysaccharides. However, only the relatively infrequently occurring monosaccharides such as glucose and fructose can be absorbed and are readily taken up from the small intestine. The more important dietary carbohydrate components, the complex starches and also sucrose have to be hydrolyzed enzymatically to their constituting monosaccharides by specific intestinal α-glucosidases (α-amylase, glucoamylase, dextrinase, maltase, isomaltase and sucrase) before they can be absorbed. The high density of intestinal membrane bound a-glucosidase enzymes in the jejunum and the high proportion of refined carbohydrates in the Western diet are the main reasons that the digestive process usually takes place very rapidly in the proximal small intestine. As a result, a high carbohydrate digestion rate leads to a rapid postprandial rise in blood glucose which is followed by a glucose stimulated insulin increase [33].

4.1. α-Glucosidase Inhibitors

α-Glucosidase inhibitors are of various chemical structures (polypeptides and carbohydrates). The most potent inhibitors are from a carbohydrate-derived backbone containing a nitrogen atom in the molecule. Polypeptides are less active.

4.1.1. Mode of Action

Essential for the action of the carbohydrate-type α-glucosidase inhibitors is their intramolecular nitrogen atom, which binds to the carbohydrate binding site of the intestinal α-glucosidases [34]. Compared with the oxygen atom of the α-glycosidic bond of dietary complex carbohydrates, the affinity of the nitrogen atom to the carbohydrate binding site of the enzyme is much higher (factor $10^4 - 10^7$). Despite their high affinity for these enzymes, the binding of α-glucosidase inhibitors is reversible and their inhibition kinetics are competitive [35].

4.1.2. Pharmacology

Pharmacological inhibition of intestinal α-glucosidases effectively retards the digestion of carbohydrates, slows down the absorption of glucose and avoids the often very high postprandial blood glucose peaks particularly in diabetic patients. As a result the postprandial glycemia and glucose stimulated hyperinsulinemia are flattened and the metabolic control will be improved. From the primary pharmacological effects, it can be concluded that α-glucosidase inhibitors act to reduce postprandial hyperglycemia without stimulating insulin secretion, thus avoiding any risk of hypoglycemia. Because of their unique mode of action in the small intestine, α-glucosidase inhibitors can be used in combination with all other antidiabetic treatments and act in an additive manner together with other antidiabetic drugs [33].

4.1.3. Individual Compounds

The first active compounds were inhibitors of the *soluble pancreatic α-amylase*. BAY D 7791 was a protein isolated from wheat flour [36]. This *α-amylase* inhibitor inhibited the postprandial blood glucose rise in oral loading tests with raw starch, but was much less effective when used together with cooked starch.

BAY E 4609, an α-amylase inhibitor of microbial origin, was characterized to be of carbohydrate nature and showed a significantly higher inhibitory potency [36]. However, also this compound demonstrated only modest efficacy in clinical trials. Because α-amylase inhibitors affect only starch digestion, the degradation of important oligo- and disaccharides by the membrane bound enzymes is not inhibited, the overall efficacy is only modest.

Further research aimed at inhibitors of the membrane bound *α-glucosidases*. This approach was very successful and lead to the discovery of acarbose (Table 2) [37], the first α-glucosidase inhibitor, which was introduced into the market for the treatment of type 2 diabetic patients in 1990 (in many countries also for type 1 in addition to insulin).

Table 2. α-Glucosidase inhibitors

Generic name M_r, mp, Ref.	Examples of commercial preparations formula	Daily dose
Acarbose [56180-94-0] $C_{25}H_{43}NO_{18}$ 645.61; amorph [38]	Glucobay, Precose, Glucor (Bayer) 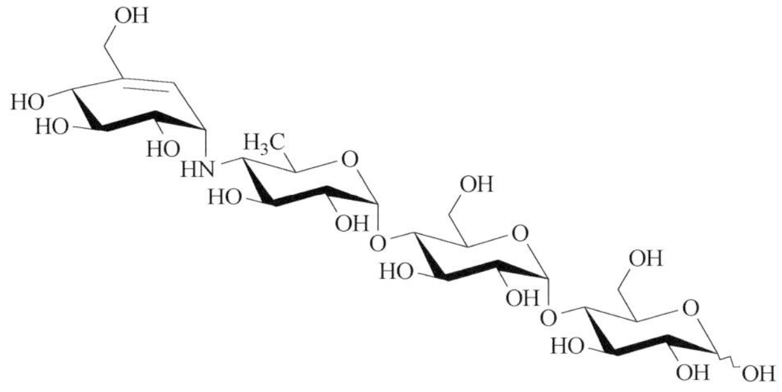	100 – 300 mg

O-{4,6-Dideoxy-4-[1(S)-(1,4,6/5)-4,5,6-trihydroxy-3-hydroxymethyl-2-cyclohexene-1-yl]amino-α-D-glucopyranosyl}-(1 → 4)-O-α-D-glucopyranosyl-(1 → 4)-D-glucose

Miglitol [72432-03-2] $C_8H_{17}NO_5$ 207.23; 114 °C [39]	Diastabol (Sanofi) Glyset (Pharmacia-Upjohn) 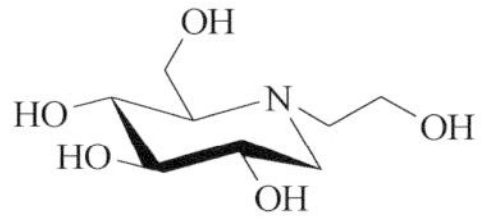	100 – 300 mg

1,5-Dideoxy-1,5-[(2-hydroxyethyl)imino]-D-glucitol

Voglibose [83480-29-9] $C_{10}H_{21}NO_7$ 267.28; 162 – 163 °C [40]	Basen (Takeda) 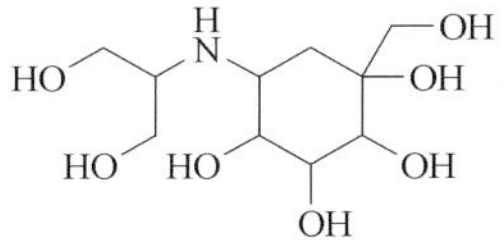	0.3 – 0.9 mg

3,4-Dideoxy-4-[[2-hydroxy-1-(hydroxymethyl)ethyl]amino]-2-C-(hydroxy-methyl)-D-epiinositol

Acarbose is a pseudotetrasaccharide of microbial origin, and was isolated from bacterial culture broths (*Actinoplanes* spec.) by Bayer [34], [35]. Acarbose consists of an acarviosine unit and maltose. The acarviosine nitrogen is responsible for the high affinity to the active binding site of α-glucosidases, however, the C – N linkage cannot be hydrolyzed and the enzymatic reaction stops.

Further research generated smaller molecules with different structures. The synthetic N-substituted valiolamine and the N-substituted 1-deoxynojirimycin derivatives showed even higher affinity to various α-glucosidases [34], [36].

The 1-deoxynojirimycin derivative *miglitol* (BAY M 1099) (Table 2) shows similar activity to acarbose with daily dosages of 100 – 300 mg [35].

Emiglitate (BAY O 1248) [*80879-63-6*], another deoxynojirimycin analogue is characterized by at least five-fold higher inhibitory potency than acarbose and additionally shows a long-lasting inhibitory effect [34], [35].

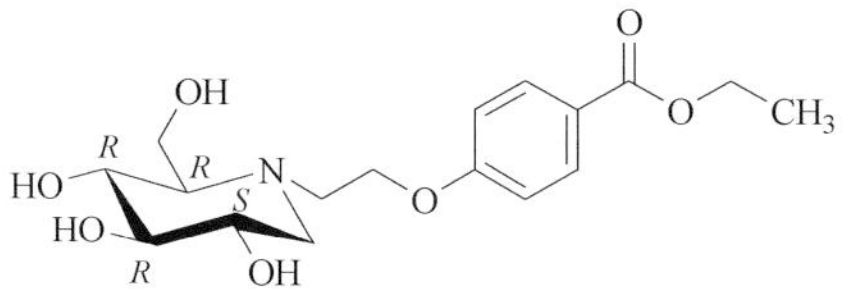

In animal studies emiglitate was still active up to 17 h after a single oral administration due to its pharmacokinetic behavior (formation of an active metabolite) [36], [41]. It is in clinical development for once daily treatment in Japan. Both miglitol and emiglitate differ in their pharmacokinetic behavior from acarbose, which is poorly absorbed (1 – 2 %). The deoxynojirimycin derivatives are structurally related to monosaccharides and will be absorbed in the jejunum [41].

The most potent α-glucosidase inhibitor is the valiolamine derivative *voglibose* (Takeda) (Table 2). Compared with acarbose its inhibitory potency in vitro, depending on the α-glucosidase enzyme, is up to 200 times higher [34], [36]. In Japan, daily dosages are only 0.3 – 0.9 mg, however, therapeutic efficacy on HbA1c reduction could not be clearly demonstrated with those dosages. HbA1c, a glycated hemoglobin species, which occurs in blood, is a diagnostic measure for the non-enzymatic glycation process, and depends on the plasmaglucose concentration. The process of protein glycation is discussed to be responsible for the development of diabetic complications (neuropathy, retinopathy, nephropathy). In Europe voglibose is under evaluation with distinctly higher daily dosages of 1.5 to 6 mg [42]. In contrast to acarbose, the smaller inhibitors of the monosaccharide type (miglitol, emiglitate, voglibose) do not inhibit pancreatic α-amylase [34], [36].

4.2. Plant Fibers

A more dietary approach for delaying absorption of nutrients is the use of undigestable carbohydrate polymers. Guar gum [*9000-30-0*] is a galactomannan obtained from the seeds of the guar plant (*Cyamopsis tetragonoloba*). Guar gum has no specific pharmacological site of action, but delays the absorption of digestible carbohydrates because it increases the viscosity in the small intestine and additionally hinders physically the enzymatic degradation of digestible carbohydrates. This approach is less effective in delaying the carbohydrate absorption than the pharmacological approach by the use of α-glucosidase inhibitors.

5. Enhancers of Insulin Action

Thiazolidine-2,4-Dione Derivatives. The thiazolidinediones or glitazones are insulin sensitizing compounds which improve glucose utilization without stimulating pancreatic insulin secretion. However, the thiazolidinediones are only active in conjunction with insulin and show the best results in hyperinsulinemic, insulin resistant, obese individuals (animals and humans). The first compound *ciglitazone*, synthetized by Takeda and reported in 1982, was a result of screening of clofibrate analogues for hypolipidemic activity [43], [44].

Mode of Action. The mechanism of blood glucose lowering activity by thiazolidinedione derivatives has not been fully elucidated yet. Only minor or no effects are observed in normal animals or insulin sensitive diabetes models. In obese or hyperinsulinemic insulin resistant rodents thiazolidinediones lower insulin and triglyceride levels after treatment for 2 to 8 d. If hyperglycemia occurs, also the blood glucose concentration is reduced [23], [44].

Mechanistic in vitro studies with adipocytes have shown that the thiazolidinediones are ligands for the peroxisome proliferator activated receptor γ [45]. Peroxisome proliferator activated receptors (PPARs) belong to the nuclear receptor superfamily of ligand-activated transcription factors.

So far, three receptor subtypes have been identified: PPARα, PPARδ, and PPARγ. The PPARs are believed to play a physiological role in the regulation of lipid metabolism [46], [47]. Recent studies have demonstrated that the in vivo antihyperglycemic activity of thiazolidinediones strongly correlates with their potency as PPARγ agonists in vitro [48]. The mechanisms responsible for the blood glucose effects are controversally discussed, however, as long as there are no more mechanistic data available it is suggested that PPARγ is a primary site of action of the thiazolidinediones [49].

Pharmacology. Ciglitazone reduced plasma triglycerides, hyperglycemia and hyperinsulinemia in several insulin resistant diabetic and nondiabetic rodent models after repeated oral dosing at doses of 100–300 mg/kg body weight, while having little or no effect in normal rodents or those with insulin deficient diabetes [44]. In long-term animal studies lens cataracts occurred and ciglitazone development was discontinued. Further synthetic research on thiazolidinedione analogues yielded compounds of distinctly higher in vitro and in vivo activity. Depending on the animal model, after repeated dosing troglitazone lowered hyperglycemia as well as levels of free fatty acids, triglycerides, and insulin at daily doses of 50 mg/kg p.o. or slightly lower [23], [50]. The glitazones are not only able to reduce insulin resistance, in vitro and in vivo studies have demonstrated increased synthesis of glucose transporter 1 (GLUT1) in various tissues. Little or no effect was observed on the insulin sensitive glucose transporter 4 (GLUT4), synthesis or insulin stimulated translocation of GLUT4, particularly in adipocytes, were only slightly increased. These effects may contribute to increased cellular glucose uptake [51].

In accordance with the PPARγ mechanism, which primarily affects the adipose tissue, all obese animals gain weight under treatment with glitazones [47].

Table 3. Thiazolidine-2,4-dione derivatives

Generic name M_r, mp, Ref.	Examples of commercial preparations formula	Daily dose
Pioglitazone [112529-15-4] $C_{19}H_{20}N_2O_3S \cdot HCl$ 392.90; 183–184 °C [53]	Actos (Ely Lilly) (Takeda) 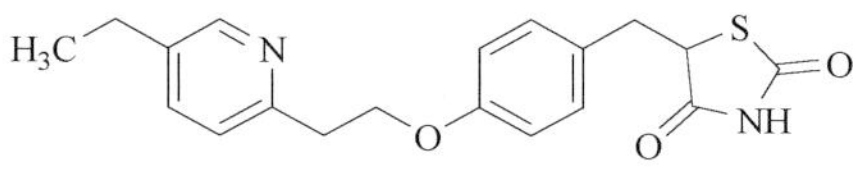	15–45 mg

(±)-5-[[4-[2-(5-Ethyl-2-pyridinyl)ethoxy]phenyl]methyl]-2,4-thiazolidinedione, hydrochloride

Rosiglitazone [122320-73-4] $C_{18}H_{19}N_3O_3S$ 357.43; 122–123 °C [54]	Avandia (SKB) 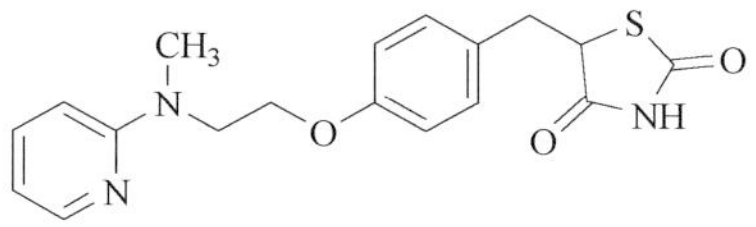	4–8 mg

5-(4-[2-(Methylpyridin-2-ylamino)ethoxy]benzyl)thiazolidine-2,4-dione

Troglitazone [97322-87-7] $C_{24}H_{27}NO_5S$ 441.55; 184–186 °C [55]	Noscal (Sankyo) Rezulin (Warner-Lambert) 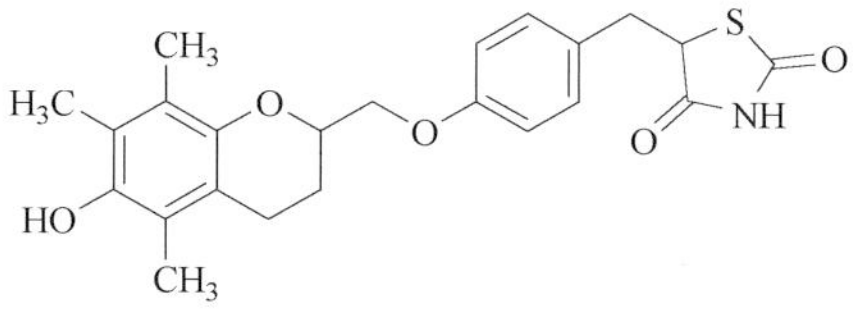	400–600 mg

5-[[4-[(3,4-Dihydro-6-hydroxy-2,5,7,8-tetramethyl-2H-1-benzopyran-2-yl) methoxy]phenyl]methyl]-2,4-thiazolidinedione

Clinical trials have confirmed the preclinical results. Best effects were observed in obese, highly insulin resistant type 2 diabetic patients and particularly in combination with insulin secretagogues (SU). However, the increase in body weight gain was also observed [47].

Troglitazone (Sankyo, see Table 3) was firstly approved 1995 in Japan and 1997 in USA and UK. It was launched in these countries in 1997. The recommended starting dose is 400 mg/d and the highest dose 600 mg. After being marketed, the daily practice revealed adverse events in the liver with cases of fatal liver failure. This lead to withdrawal of troglitazone from the market in the UK already after two months [52]. In the USA and Japan, troglitazone was labeled to closely monitoring liver transaminases in the first year of treatment.

Pioglitazone (Table 3) is approximately ten times more potent in both, animals, and humans than troglitazone. The highest potency shows *rosiglitazone* (Table 3), which is approximately 100-fold more active than troglitazone [52]. Both compounds showed better tolerability in clinical trials than troglitazone. Both drugs were introduced into the US market in 1999, therefore, experience in daily practice is not available yet.

6. References

[1] R. A. DeFronzo, *Diabetes* **37** (1988) 667–681.
[2] A. J. Scheen, P. F. Lefèbvre in J. Kuhlmann, W. Puls (eds.): *Handbook of Experimental Pharmacology*, vol. **119**, Springer, Berlin–Heidelberg–New York 1996, pp. 7–42.
[3] C. R. Loubatières, *Soc. Biol. (Paris)* **138** (1944) 766–767.
[4] H. Franke, J. Fuchs, *Dtsch. Med. Wochenschr.* **80** (1955) 1449–1452.
[5] H. Ruschig et al., *Arzneimittelforsch.* **8** (1958) 448–454.
[6] W. Aumüller et al., *Arzneim.-Forsch./Drug Res.* **16** (1966) 1640–1641.
[7] K. Geisen, *Arzneim.-Forsch./Drug Res.* **38** (1988) 1120–1130.
[8] C. M. Clark, A. W. Helmy, *Drugs of Today* **34** (1998) 401–408.
[9] Hoechst, DE-AS 1135891, 1960; Eli Lilly, DE-AS 1177631, 1961.
[10] E. Haack, *Arzneim.-Forsch.* **8** (1958) 444–448.
[11] Boehringer Mannheim, DE-AS 1117103, 1953.
[12] Hoffmann-La Roche, DE-AS 1695201, 1967.
[13] Sciences Union et Cie., BE 693702, 1966; DD 113223, 1973.
[14] R. Weyer, V. Hitzel, *Arzneim.-Forsch./Drug Res.* **38** (1988) 1079–1080.
[15] V. Ambrogi et al., *Arzneim.-Forsch.* **21** (1971) 200–204.
[16] Thomae, DE-AS 2000339, 1970.
[17] Bayer, DE-AS 1670952, 1967.
[18] H. Plümpe, H. Horstmann, W. Puls, *Arzneim.-Forsch./Drug Res.* **24** (1974) 363–374.
[19] K. Gutsche et al., *Arzneim.-Forsch./Drug Res.* **14** (1964) 373–376.
[20] Upjohn, US 3063903, 1961.
[21] PCT Int.- Pat. Appl. WO 93/00337, 1993 (W. Grell et al.).
[22] H. Shinkai et al., *J. Med. Chem.* **32** (1989) 1436–1441.
[23] H. Bischoff, H. E. Lebovitz in J. Kuhlmann, W. Puls (eds.): *Handbook of Experimental Pharmacology*, vol. **119**, Springer, Berlin–Heidelberg–New York 1996, pp. 650–696.
[24] J. A. Balfour, D. Faulds, *Drugs & Aging* **13** (1998) 173–180.
[25] T. Ikenoue et al., *Br. J. Pharmacol.* **120** (1997) 137–145.
[26] A. H. Karara, B. E. Dunning, *J. Clin. Pharmacol.* **39** (1999) 172–179.
[27] U. Panten, M. Schwanstecher, C. Schwanstecher in J. Kuhlmann, W. Puls (eds.): *Handbook of Experimental Pharmacology*, vol. **119**, Springer, Berlin–Heidelberg–New York 1996, pp. 129–159.
[28] C. K. Watanabe, *J. Biol. Chem.* **33** (1918) 253.
[29] R. Beckmann in H. Maske (ed.): *Handbook of Experimental Pharmacology*, vol. **29**, Springer, Berlin–Heidelberg–New York 1971, pp. 439–596.
[30] N. F. Wiernsperger in J. Kuhlmann, W. Puls (eds.): *Handbook of Experimental Pharmacology*, vol. **119**, Springer, Berlin–Heidelberg–New York 1996, pp. 305–358.
[31] L. S. Hermann in J. Kuhlmann, W. Puls (eds.): *Handbook of Experimental Pharmacology*, vol. **119**, Springer, Berlin–Heidelberg–New York 1996, pp. 373–407.

[32] W. Creutzfeld, W. R. Fölsch (eds.): *Delaying Absorption as a Therapeutic Principle in Metabolic Disease*, Thieme Verlag, Stuttgart – New York 1983.
[33] H. Bischoff, *Clin. Invest. Med.* **18** (1995) 303 – 311.
[34] B. Junge, M. Matzke, J. Stoltefuss in J. Kuhlmann, W. Puls (eds.): *Handbook of Experimental Pharmacology*, vol. **119,** Springer, Berlin – Heidelberg – New York 1996, pp. 411 – 482.
[35] E. Truscheit et al., *Prog. Clin. Biochem. Med.* **7** (1988) 17 – 99.
[36] W. Puls in J. Kuhlmann, W. Puls (eds.): *Handbook of Experimental Pharmacology*, vol. **119,** Springer, Berlin – Heidelberg – New York 1996, pp. 497 – 534.
[37] W. Puls et al., *Naturwissenschaften* **64** (1977) 536.
[38] B. Junge et al., *Carbohydr. Res.* **128** (1984) 235 – 268.
[39] EP 49858, 1982 (prior DE 3038901, 1980) (G. Kinast, M. Schedel, W. Köbernick).
[40] S. Horii et al., *J. Med. Chem.* **29** (1986) 1038 – 1046.
[41] H. P. Krause, H. J. Ahr in J. Kuhlmann, W. Puls (eds.): *Handbook of Experimental Pharmacology*, vol. **119,** Springer, Berlin – Heidelberg – New York 1996, pp. 541 – 555.
[42] B. Göke, H. Fuder, *Digestion* **56** (1995) 493 – 501.
[43] T. Sohda et al., *Chem. Pharm. Bull.* **30** (1982) 3563.
[44] A. Y. Chang, B. M. Wyse, B. J. Gilchrist, T. Peterson, A. R. Diani, *Diabetes* **32** (1983) 830 – 838.
[45] M. J. Reginato, M. A. Lazar, *Trends Endocrinol. Metab.* **10** (1999) 9 – 13.
[46] K. Schoonjans, B. Staels, J. Auwerx, *Biochim. Biophys. Acta* **1302** (1996) 93 – 109.
[47] J. Auwerx, G. Martin, M. Guerre-Millo, B. Staels, *J. Mol. Med.* **74** (1996) 347 – 351.
[48] T. M. Willson et al., *J. Med. Chem.* **39** (1996) 665 – 668.
[49] J. Berger et al., *Endocrinology* **137** (1996) 4189 – 4195.
[50] T. Fujiwara, S. Yoshioka, T. Yoshioka, I. Ushiyama, H. Horikoshi, *Diabetes* **37** (1988) 1549 – 1558.
[51] B. Hulin, P. A. McCarthy, E. M. Gibbs, *Curr. Pharmac. Des.* **2** (1996) 85 – 102.
[52] C. Day, *Diabetic Medicine* **16** (1999) 179 – 192.
[53] T. Sohda et al., *Arzneim.-Forsch./Drug Res.* **40** (1990) 37 – 42.
[54] B. C. Cantello et al., *J. Chem. Soc. Perkin Trans. 1,* **22** (1994) 3319 – 3324.
[55] T. Yoshioka et al., *J. Med. Chem.* **32** (1989) 421 – 428.